HOW EVERYTHING WORKS

How Everything Works

Making Physics Out of the Ordinary

Louis A. Bloomfield

The University of Virginia

 John Wiley & Sons, Inc.

SENIOR ACQUISITIONS EDITOR	Stuart Johnson
EXECUTIVE MARKETING MANAGER/ CHANNEL DEVELOPMENT	Lisa Culhane
VICE PRESIDENT/EXECUTIVE PUBLISHER	Kaye Pace
PRODUCTION MANAGER	Pam Kennedy
SENIOR PRODUCTION EDITOR	Sarah Wolfman-Robichaud
ASSISTANT DIRECTOR, DOMESTIC RIGHTS	Adam Levison
COORDINATOR, DOMESTIC RIGHTS	Ashley Ginter
CREATIVE DIRECTOR	Harry Nolan
SENIOR MEDIA EDITOR	Tom Kulesa
PRODUCTION ASSISTANT	Jenna Belisonzi
COPYEDITORS	Elana Bloomfield, Aaron Bloomfield
EDITORIAL ASSISTANT	Alyson Rentrop
COVER PHOTO	©Mike Powell/Stone/Getty Images

This book was designed and set in 10pt Adobe Warnock Pro and Adobe Trajan Pro by Louis A. Bloomfield and printed and bound by Courier Westford. The cover was printed by Courier Westford.

This book is printed on acid free paper. ∞

To order books or for customer service please, call 1-800-CALL WILEY (225-5945).

ISBN-13 978- 0-471-74817-5
ISBN-10 0-471-74817-X

Printed in the United States of America

10 9 8 7 6 5 4 3 2 1

To Karen for your friendship,
wisdom, and support,

to Elana and Aaron for your enthusiasm,
thoughtfulness, and camaraderie,

and to Sadie for being you.

FOREWORD

In today's world we are surrounded by science and by the technology that has grown out of that science. For most of us, this is making the world increasingly mysterious and somewhat ominous as technology becomes ever more powerful. For instance, we are confronted by many global environmental questions such as the dangers of greenhouse gases and the best choices of energy sources. These are questions that are fundamentally technical in nature and there is a bewildering variety of claims and counterclaims as to what is "the truth" on these and similar important scientific issues. For many people, the reaction is to throw up their hands in hopeless frustration and accept that the modern world is impossible to understand, and one can only huddle in helpless ignorance at the mercy of its mysterious and inexplicable behavior.

In fact, much of the world around us and the technology of our everyday lives is governed by a few basic physics principles, and once these principles are understood, the world and the vast array of technology in our lives become understandable and predictable. How does your microwave oven heat up food? Why is your radio reception bad in some places and not others? And why can birds happily land on a high-voltage electrical wire? The answers to questions like these can be obvious once you know the relevant physics. Unfortunately, you are not likely to learn that from a standard physics course or physics textbook. There is a large body of research showing that instead of providing this improved understanding of everyday life, most introductory physics courses are doing quite the opposite. In spite of the best intentions of the teachers, most students are "learning" that physics is abstract, uninteresting and unrelated to the world around them.

How Everything Works is a dramatic step towards changing that by presenting physics in a new way. Instead of starting out with abstract principles that leave the reader with the idea that physics is about artificial and uninteresting ideas, Lou Bloomfield starts out talking about real objects and devices that we encounter in our everyday lives. He then shows how these seemingly magical devices can be understood in terms of the basic physics principles that govern their behavior. This is much the way that most physics was discovered in the first place; people asked why the world around them behaved as it did and as a result discovered the principles that explained and predicted what they observed.

I have been using *How Things Work*, the textbook from which *How Everything Works* developed, in my classes for several years and I continue to be impressed with how Lou can take seemingly highly complex devices and strip away the complexity to show how at their heart are simple physics ideas. Once these ideas are understood, they can be used to understand the behavior of many devices we encounter in our daily lives, and often even fix things that before had seemed impossibly complex. In the process of teaching from Lou's book, I have increased my own understanding of the physics behind much of the world around me. In fact, after consulting *How Things Work*, I have had the confidence to confront both plumbers and air-conditioner repairmen to tell them (correctly as it turned out) that their diagnosis did not make sense and they needed to do something different to solve my plumbing and AC problems. Now I am regularly amused at the misconceptions some trained physicists have about some of the physics they encounter in their daily lives, such as how a microwave oven works and why it can be made out of metal walls, but putting aluminum foil in it is bad. It has convinced me that we need to take the approach used in this book in far more of our science texts.

Of course, the most important impact is on the students in my classes that use Bloomfield's book. These are typically nonscience students majoring in fields such as film studies, classics, English, business, etc. They often come to physics with considerable trepidation. It is inspiring to see many of them discover to their surprise that physics is very different from what they thought—that physics can actually be interesting and useful and makes the world a much less mysterious and more understandable place. I remember many examples of seeing this in action: the student who, after learning how both speakers and TVs work, was suddenly able to understand that it was not magic that putting his large speaker next to the TV distorted the picture but in fact it was just physics, and now he knew just how to fix it; the young woman scuba diver who, after learning about light and color, suddenly interrupted class to announce that now she understood why it was that you could tell how deep you were by seeing what color lobsters appeared; or the students who announced that suddenly it made sense that the showers on the first floor of the dorm worked better than those on the second floor. In addition, of course everyone is excited to learn how a microwave oven works and why there are these strange rules as to what you can and cannot put in it. These examples are particularly inspiring to a teacher, because they tell you that the students are not just learning the material presented in class, but they are then able to apply that understanding to new situations in a useful way, something that happens far too seldom in science courses.

Whether a curious layperson, a trained physicist, or a beginning physics student, most everyone will find this book an interesting and enlightening read and will go away comforted in that the world is not so strange and inexplicable after all.

Carl Wieman
Nobel Laureate in Physics 2001
CASE/Carnegie US University Professor of the Year 2004

TABLE OF CONTENTS

PREFACE

My purpose in writing this book is to bring together two seemingly separate worlds: the world of our everyday experience and the world of physics. Both are of great importance to me, and I find it sad that they have drifted so far apart. The whole purpose of physics is to explain real observations, after all.

Most of us would agree that you don't need to be an artist to appreciate art, a mason to appreciate brickwork, or a historian to appreciate the significance of events past. But while it's clear to *me* that you don't need to be a physicist to appreciate the physics of things around you, most people would maintain otherwise.

Physicists are at least partly to blame for the situation, being all too eager to take laypersons on longwinded tours of physics' esoteric frontiers and arcane mysteries. One consequence of our behavior is that the remark "Oh, I'm a physicist" almost always ends a casual conversation. The present arrangement is doing none of us any good.

In this book I'll take you on a tour of much more accessible and relevant territory: the physics of everyday life. Some of the physics we'll encounter along the way is fully mature stuff and some is practically brand new. Since my primary goal is to explain how ordinary things work, I am willing to dwell at length on whatever physics is necessary, even if it originated a century or two ago. Physics, unlike milk, doesn't have an expiration date.

By the end of the tour, you should truly understand how myriad ordinary things work. You'll know not merely what is inside them, but why they are fashioned as they are and how the laws of physics govern their behaviors. You'll have picked up a wide range of physics concepts as well and seen references to the quantitative machinery (i.e. formulaic physics) associated with those concepts. Please don't be intimidated by an occasional formula; the insights and understandings in this book are almost entirely in its words.

On a broader scale, I hope that this book will help bring physics back into polite conversation. What's the point of lamenting our children's poor science preparation if none of us ever talk about science or show any interest in it ourselves? With this book, I intend to lead by example, bringing together the real world, real physics, and real people. If I'm lucky, the day will come when the words "Oh, I'm a physicist" will invariably lead to an interesting and lively conversation.

Lou Bloomfield
Charlottesville, Virginia

ACKNOWLEDGEMENTS

Acknowledgments are more than just thanks to individuals who have contributed to a book's development; they're also hurried introductions to fascinating people, ones you'd probably enjoy meeting, if only there were time and opportunity.

First among them is my editor, Stuart Johnson, whose insight, confidence, and hard work made this book possible. Books need their champions to move from pipedream to reality, and Stuart was this book's champion. With his help it earned the support of my intrepid publisher, Kaye Pace, and so came to be.

During the book's writing and composition, I talked almost daily with my production editor, Sarah Wolfman-Robechaud, and frequently with her colleague, Geraldine Osnato. With that cheerful, supportive, and knowledgeable pair only a phone call away, I never felt like I was working alone. Assisting them was Alyson Rentrop, whose kind and thoughtful comments always helped to keep me heading forward.

A book is of no value if its intended readers don't like it. I'm thus grateful to Lisa Culhane and Dana Kasowitz for helping me to understand my readers. Lisa and Harry Nolan together produced the beautiful exterior of this book, a triumph that I greatly appreciate. Thanks also to Harry for his design guidance on the book's interior.

I'd like to express my fond gratitude finally to my wife, Karen, and to our children, Elana and Aaron. Not only did the four of us discuss this book, its contents, its purpose, and its audience regularly, but we also worked together on many of its details. Elana and Aaron copyedited most of the chapters and improved them enormously.

This is a book about the real world and real physics; I merely report on both. I am therefore indebted to a great many colleagues, friends, and fellow scientists who have helped me to understand physics more thoroughly and to explain it more effectively. Among those generous people are Bascom Deaver, Michael Fowler, Tom Gallagher, Bob Jones, Richard Lindgren, Despina Louca, John Malone, Rick Marshall, Mike Timmins, and Rob Watkins at the University of Virginia and Bob Anderson, Nora Berrah, Katy Disney, Ursula Gibson, Laura Green, Robert Hubel, Larry Hunter, Edwin Jones, Julian Krolik, John Krupczak, Laura Lising, Alan Nathan, Mike Noel, David Ollis, Promod Pratap, Chuck Stone, Richard Superfine, Kristin Wedding, Bob Welsh, and Carl Wieman at other institutions. I am particularly grateful to Carl Wieman for writing the foreword to this book and expressing therein his vision for how to make physics accessible and valuable to everyone.

The best way to discover what people want to know about science and examine how they learn it is to teach science. I am ever so grateful to the students of the University of Virginia for being such eager, enthusiastic, and interactive participants in a long educational experiment: a case-study approach to the teaching of physics, one conducted in the context of everyday objects. It has been a delight and a privilege to get to know so many of my students as individuals, and their influence on this enterprise is immeasurable.

CHAPTER 1

THINGS THAT MOVE

As a physicist, I am especially fond of how and why questions. Knowing what, where, and when may be interesting, but it is while trying to answer how and why that people originally discovered physics and the other sciences. In keeping with that tradition, this book rediscovers physics in response to a simple but far-reaching question: how does everything work?

Although your journey through this book will expose you to a great deal of physics, don't let that worry you. The physics will be introduced gently in the context of real objects and ordinary situations as we investigate how they work. Such a case-study approach does more, however, than merely ease the process of learning physics. It also demonstrates the value of understanding physics. Not only does physics explain much of how our world works at present, it also allows us to make useful predictions about the future.

Since physics is the study of the material world and the rules that govern its behavior, I am puzzled that physics principles are routinely taught in the abstract. It's as though any appearance of physics in daily life must be uninteresting and inconsequential, or that physics itself is too delicate for everyday use. My own view is exactly the opposite. And without its myriad real-world contexts, physics is unsupported and shapeless. It's like a milkshake without a cup.

If discussing physics amidst ordinary things makes physics itself seem terribly ordinary, then you know the secret: physics *is* ordinary. It's not some special activity reserved for research laboratories—it's something we encounter all the time. The purpose of this book is to show you the role physics plays in your everyday life and to allow you to use physics to explain and influence the things around you.

To get us started, this chapter and the one following it will do two main things: they'll introduce the language of physics, which we'll be using throughout the book, and they'll present the basic laws of motion on which everything else will rest. In later chapters, we'll explore objects that are more interesting and important, both in their own right and because of the scientific issues they raise. But these first two chapters are special because they must provide an orderly introduction to the discipline of physics itself. Whether you're an old hand at physics or encountering it for the first time, examining THINGS THAT MOVE will help set the stage for everything that follows.

Chapter Itinerary

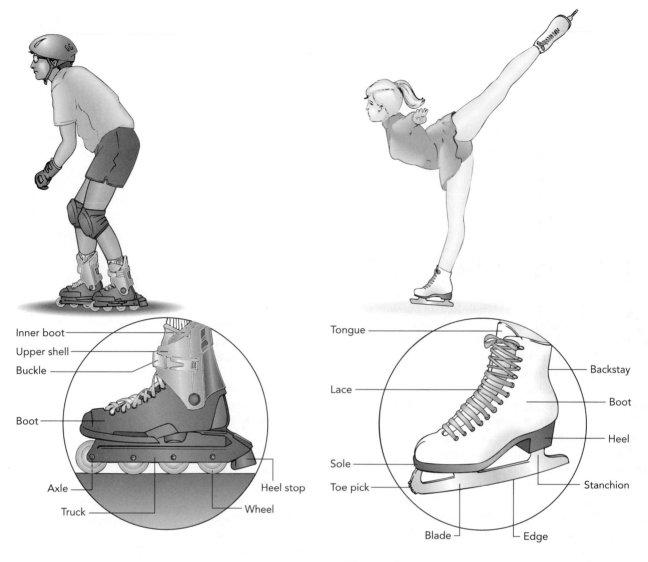

Inner boot
Upper shell
Buckle
Boot
Axle
Truck
Heel stop
Wheel

Tongue
Lace
Sole
Toe pick
Blade
Edge
Backstay
Boot
Heel
Stanchion

SECTION 1.1 **Skating**

Like many sports, skating is trickier than it appears. If you're a first-time skater, you'll likely find yourself getting up repeatedly from the ground or ice, and it will take some practice before you can glide smoothly forward or come gracefully to a stop. But whether you're wearing ice skates or Rollerblades®, the physics of your motion is surprisingly simple. When you're on a level surface with your skates pointing forward, you coast!

Coasting is one of the most basic concepts in physics and it is our starting point in this book. Joining it in this section will be starting, stopping, and turning, which together will help us understand the first few laws of motion. Our exploration of skating will get us well on the way to an understanding of the fundamental principles that govern all movement and thereby prepare us for many of the objects we'll examine in the rest of this book.

Gliding Forward: Inertia and Coasting

While you're putting on your skates, let's take a moment to think about what happens to a person who has nothing pushing on her at all. When she's completely free

❶ Aristotle (Greek philosopher, 384–322 BC) theorized that objects' velocities were proportional to the forces exerted on them. While this theory correctly predicted the behavior of a sliding object, it incorrectly predicted that heavier objects should fall faster than lighter objects. Nonetheless, Aristotle's theory was respected for a long time, in part because finding the simpler and more complete theory was hard and in part because the scientific method of relating theory and observation took time to develop.

of outside influences (Fig. 1.1.1), free of pushes and pulls, does she stand still? Does she move? Does she speed up? Does she slow down? In short, what does she do?

The correct answer to that apparently simple question eluded people for thousands of years; even Aristotle, perhaps the most learned philosopher of the classical world, was mistaken about it ❶. What makes this question so tricky is that objects on earth are never truly free of outside influences; instead, they all push on, rub against, or interact with one another in some way.

As a result, it took the remarkable Italian astronomer, mathematician, and physicist Galileo Galilei many years of careful observation and logical analysis to answer that question ❷. The solution he came up with, like the question itself, is simple: if the person is stationary, she will remain stationary; if she is moving in some particular direction, she will continue moving in that direction at a steady pace, following a straight-line path. This property of steady motion in the absence of any outside influence is called inertia.

Inertia
A body in motion tends to remain in motion; a body at rest tends to remain at rest.

The main reason why Aristotle failed to discover inertia, and why we often overlook inertia ourselves, is friction. When you slide across the floor in your shoes, friction quickly slows you to a stop and masks your inertia. To make inertia more obvious, we must get rid of friction. That's why you're wearing skates.

Skates almost completely eliminate friction, at least in one direction, so that you can glide effortlessly across the ice or roller rink and experience your own inertia. For simplicity, imagine that your skates are perfect and experience no friction at all as you glide. Also, for this and the next couple of sections, let's forget not only about friction but also about air resistance. Since the air is calm and you're not moving too fast, air resistance isn't all that important to skating anyway.

Now that you're ready to skate, we'll begin to examine five important physical quantities relating to motion and look at their relationships to one another. These quantities are position, velocity, mass, acceleration, and force.

Let's start by describing where you are. At any particular moment, you're located at a position—that is, at a specific point in space. Whenever we report your position, it's always as a distance and direction from some reference point: how many meters north of the refreshment stand or how many kilometers west of Cleveland.

Position is an example of a vector quantity. A vector quantity consists of both a magnitude and a direction; the magnitude tells you how much of the quantity there is, while the direction tells you which way the quantity is pointing. Vector quantities are common in nature. When you encounter one, pay attention to the direction part; if you're looking for buried treasure 30 paces from the old tree but forget that it's due east of that tree, you'll have a lot of digging ahead of you.

You're on your feet and beginning to skate. If you're moving, then your position is changing. In other words, you have a velocity. Velocity measures how quickly your position changes; it's our second vector quantity and consists of the speed at which you're moving and the direction in which you're heading. Your speed is the distance you travel in a certain amount of time,

$$\text{speed} = \frac{\text{distance}}{\text{time}},$$

and the direction you're heading might be east, north, or down—if you're taking a spill.

But when you're gliding freely, with nothing pushing you horizontally, your velocity is particularly easy to describe. Since you travel at a steady pace along a straight-line path, your velocity never changes—it is constant. For example, if you're

Fig. 1.1.1 This skater glides without any horizontal influences. If she's stationary, she'll tend to remain stationary; if she's moving, she'll tend to continue moving.

❷ While a professor in Pisa, Galileo Galilei (Italian scientist, 1564–1642) was obliged to teach the natural philosophy of Aristotle. Troubled with the conflict between Aristotle's theory and observations of the world around him, Galileo devised experiments that measured the speeds at which objects fall and determined that all objects fall at the same rate.

❸ In 1664, while Sir Isaac Newton (English scientist and mathematician, 1642–1727) was a student at Cambridge University, the university was forced to close for 18 months because of the plague. Newton retreated to the country, where he discovered the laws of motion and gravitation and invented the mathematical basis of calculus. These discoveries, along with his observation that celestial objects such as the moon obey the same simple physical laws as terrestrial objects such as an apple (a new idea for the time), are recorded in his *Philosophiæ Naturalis Principia Mathematica*, first published in 1687. This book is perhaps the most important and influential scientific and mathematical work of all time.

heading west at a speed of 10 meters-per-second (33 feet-per-second), you will have that same velocity indefinitely. A speed of 10 meters-per-second means that, if you travel for 1 second at your present speed, you'll cover a distance of 10 meters. Since your velocity is constant, you'll travel 100 meters in 10 seconds, 1000 meters in 100 seconds, and so on. Furthermore, the path you'll take is a straight line. In a word, you coast.

Thanks to your skates, we can now restate the previous description of inertia in terms of velocity: an object that is not subject to any outside influences moves at a constant velocity, covering equal distances in equal times along a straight-line path. This statement is frequently identified as Newton's first law of motion, after its discoverer, the English mathematician and physicist Sir Isaac Newton ❸. The outside influences referred to in this law are called forces, a technical term for pushes and pulls. (For several common examples of inertia, see ❹.)

Newton's First Law of Motion

An object that is not subject to any outside forces moves at a constant velocity, covering equal distances in equal times along a straight-line path.

Intuition Alert: Coasting

Intuition says that when nothing pushes on an object, that object slows to a stop; you must push it to keep it going.

Physics says that when nothing pushes on an object, that object coasts at constant velocity.

Resolution: objects usually experience hidden forces, such as friction or air resistance, that tend to slow them down. Eliminating those hidden forces is difficult, so that you rarely see the full coasting behavior of force-free objects.

The Alternative to Coasting: Acceleration

As you glide forward with nothing pushing you horizontally, what prevents your speed and direction from changing? The answer is your mass. Mass is the measure of your inertia, your resistance to changes in velocity. Almost everything in the universe has mass. Because you have mass, your velocity will change only if something pushes on you—that is, only if you experience a force. You'll keep moving steadily in a straight path until something exerts a force on you to stop you or send you in another direction. Force is our third vector quantity, having both a magnitude and a direction. After all, a push to the right is different from a push to the left.

❹ Many common activities depend on inertia to keep stationary things in place. The sharp blade of a rotary lawnmower cuts through stationary grass because inertia prevents the grass from moving out of its way. Similarly, blenders, food processors, and some coffee grinders can dice or puree foods that are held in place only by their own inertia. Even the act of snatching sheets off a roll of paper towels depends on inertia to keep the roll in place.

When something pushes on you, your velocity changes; in other words, you accelerate. Acceleration, our fourth vector quantity, measures how quickly your velocity changes. *Any* change in your velocity is acceleration, whether you're speeding up, slowing down, or even turning. If either your speed or direction of travel is changing, you're accelerating!

Like any vector quantity, acceleration has a magnitude and a direction. To see how these two parts of acceleration work, imagine that you're at the starting line of a speed-skating race, waiting for it to begin. The starting buzzer sounds and you're off! You dig your skates into the surface beneath you and begin to accelerate—your speed increases and you cover ground more and more quickly. The magnitude of your acceleration depends on how hard the skating surface pushes you forward. If it's a long race and you're not in a hurry, you take your time getting up to full speed. The surface pushes you forward gently and the magnitude of your acceleration is small. Your velocity changes slowly. But if the race is a sprint and you need to reach top speed as quickly as possible, you spring forward hard and the surface exerts an enormous forward force on you. The magnitude of your acceleration is large and your velocity changes rapidly. In this case, you can actually feel your inertia opposing your efforts to pick up speed.

But acceleration has more than just a *magnitude*. When you start the race, you also select a *direction* for your acceleration—the direction toward which your velocity is shifting with time. This acceleration is in the same direction as the force causing it. If you obtain a forward force from the surface, you'll accelerate forward—your velocity will shift more and more forward. If you obtain a sideways force from the surface, the other racers will have to jump out of your way as you career into the wall. They'll laugh all the way to the finish line at your failure to recognize the importance of direction in the definitions of force and acceleration.

Once you're going fast enough, you can stop fighting inertia and begin to glide. You coast forward at a constant velocity. Now inertia is helping you; it keeps you moving steadily along even though nothing is pushing you forward. (Recall that we're neglecting friction and air resistance. In reality, those effects push you backward and gradually slow you down as you glide. However, since we're ignoring them in this section, your motion is smooth and steady.)

But even when you're not trying to speed up or slow down, you can still accelerate. As you steer your skates or go over a bump, you experience sideways or up–down forces that change your *direction of travel* and thus cause you to accelerate.

Finally the race is over and you skid to a stop. You're accelerating again. This time you're accelerating backward, in the direction opposite your forward velocity. While we often call this process *deceleration*, it's just a special type of acceleration. Your forward velocity gradually diminishes until you come to rest.

To help you recognize acceleration, here are some accelerating objects:

1. A runner who's leaping forward at the start of a race—the runner's velocity is changing from zero to forward so the runner is accelerating *forward*.
2. A bicycle that's stopping at a crosswalk—its velocity is changing from forward to zero so it's accelerating *backward* (that is, it's decelerating).
3. An elevator that's just starting upward from the first floor to the fifth floor—its velocity is changing from zero to upward so it's accelerating *upward*.
4. An elevator that's stopping at the fifth floor after coming from the first floor—its velocity is changing from upward to zero so it's accelerating *downward*.
5. A car that's beginning to shift left to pass another car—its velocity is changing from forward to left-forward so it's accelerating mostly *leftward*.
6. An airplane that's just beginning its descent—its velocity is changing from level-forward to descending-forward so it's accelerating almost directly *downward*.
7. Children riding a carousel around in a circle—while their speeds are constant, their directions of travel are always changing. We'll discuss the directions in which they're accelerating in Section 3.3.

Here are some objects that are *not* accelerating:

1. A parked car—its velocity is always zero.
2. A car traveling straight forward on a level road at a steady speed—no change in its speed or direction of travel.
3. A bicycle that's climbing up a smooth, straight hill at a steady speed—no change in its speed or direction of travel.
4. An elevator that's moving straight upward at a steady pace, halfway between the first floor and the fifth floor—no change in its speed or direction of travel.

❺ The easiest way to determine an object's mass is to shake it—to push it rhythmically back and forth and observe its acceleration. The greater its mass, the less it will accelerate in response to your pushes. Even when a closed container is too heavy to lift or weigh, you can often figure out how full it is by shaking it.

Seeing acceleration isn't as easy as seeing velocity. You must watch skaters closely for some time to see whether or not they're accelerating. If their paths aren't straight or if their speeds aren't steady, then they're accelerating.

How Forces Affect Skaters

Now that we've learned what acceleration is, let's see how you accelerate in response to a particular force. First, your acceleration depends on the strength of that force: the stronger the force, the more you accelerate. But your acceleration also depends on your mass: the more massive you are, the less you accelerate (Fig. 1.1.2). For example, it's easier to change your velocity before you eat Thanksgiving dinner than afterward.

There is a simple relationship between the force exerted on you, your mass, and your acceleration. Your acceleration is equal to the force exerted on you divided by your mass or

$$\text{acceleration} = \frac{\text{force}}{\text{mass}}. \qquad (1.1.1)$$

Your acceleration, as we've seen, is in the same direction as the force on you.

This relationship was deduced by Newton from his observations of motion and is referred to as Newton's second law of motion. Structuring the relationship this way sensibly distinguishes the causes (force and mass) from their effect (acceleration). However, it has become customary to rearrange this equation to eliminate the division. The relationship then takes its traditional form:

$$\text{force} = \text{mass} \cdot \text{acceleration}. \qquad (1.1.2)$$

Remember that in Eq. 1.1.2 the direction of the acceleration is the same as the direction of the force. (For a useful way to measure mass using this relationship, see ❺.)

Newton's Second Law of Motion

The force exerted on an object is equal to the product of that object's mass times its acceleration. The acceleration is in the same direction as the force.

Because it's an equation, the two sides of Eq. 1.1.1 are equal. Your acceleration equals the force on you divided by your mass. Since your mass is constant unless you visit the snack bar, Eq. 1.1.1 indicates that an increase in the force on you is accompanied by a similar increase in your acceleration. That way, as the right side of the equation increases, the left side increases to keep the two sides equal. Thus the harder something pushes on you, the more rapidly your velocity changes in the direction of that push.

We can also compare the effects of equal forces on two different masses, for example, you and the former sumo wrestler to your left. I'll assume, for the sake of argument, that you're the least massive of the two. Equation 1.1.1 indicates that an increase in mass must be accompanied by a corresponding decrease in acceleration. Sure enough, your velocity changes more rapidly than the velocity of the sumo wrestler when the two of you are subjected to identical forces.

So far we've explored five principles:

1. Your position indicates exactly where you're located.
2. Your velocity measures how quickly your position changes.
3. Your acceleration measures how quickly your velocity changes.
4. In order for you to accelerate, something must exert a force on you.
5. The greater your mass, the less acceleration you experience for a given force.

Fig. 1.1.2 When this player kicks the soccer ball, it will accelerate easily because its mass is relatively small. But imagine what would happen if she kicked a much more massive bowling ball. Why would it matter?

We've also encountered five important physical quantities—mass, force, acceleration, velocity, and position—as well as some of the rules that relate them to one another. Much of the groundwork of physics rests on these five quantities and on their interrelationships.

Skating certainly depends on these quantities. We can now see that, in the absence of any horizontal forces, you either remain stationary or coast along at a constant velocity. To start, stop, or turn, something must push you horizontally and that something is the ice or pavement. We haven't talked about how you obtain horizontal forces from the ice or pavement and we'll leave that problem for later sections. But as you skate, you should be aware of these forces and notice how they change your speed, direction of travel, or both. Learn to watch yourself accelerate.

Several Skaters: Frames of Reference

While skating alone is peaceful, it's usually more fun with other skaters around. That way, you have people to talk to and an audience for your athleticism and artistry.

However, with several skaters coasting on the ice at once, there's a question of perspective. As you glide steadily past a friend, the two of you see the world somewhat differently. From your perspective, you are motionless and your friend is moving. But from your friend's perspective, your friend is motionless and you are moving. Who is right?

It turns out that you're both right and that physics has a way of accommodating this apparent paradox. Each of you is observing the world from a different inertial frame of reference, the viewpoint of an inertial object—an object that is not accelerating and that moves according to Newton's first law. One of the remarkable discoveries of Galileo and Newton is that the laws of physics work perfectly in any inertial frame of reference. From an inertial frame, everything you see in the world around you obeys the laws of motion that we're in the process of exploring. Though you may find it odd to think of scenery as moving, your inertial frame of reference is as good as any and in your frame you are at rest amidst the moving landscape.

Since both you and your friend are coasting, each of you views the world from an inertial frame of reference and sees the surrounding objects moving in perfect accordance with the laws of motion. Some objects travel at constant velocity while others accelerate in response to forces. But because the two of you are observing those objects from different inertial frames, you will disagree on the particular values of some of the physical quantities you might measure.

In the present case, you see yourself as motionless because you view the world from your own inertial frame. In that frame, your friend is coasting westward at 2 meters-per-second (6.6 feet-per-second). However, your friend sees things differently. In your friend's inertial frame, your friend is motionless and you yourself are coasting eastward at 2 meters-per-second. As long as the two of you don't try to compare the positions or velocities of objects you observe, or certain physical quantities derived from those values, there will be no disagreements and no inconsistencies. But if you forget to watch where you're going and crash into a wall, don't expect your friend to sympathize when you claim that you were motionless and that the moving wall ran into you. That's not how your friend saw it.

Each time we examine an object in this book, we'll pick a specific inertial frame of reference from which to view that object. We'll normally select an inertial frame that makes the object and its motions appear as simple as possible and then stick with that frame consistently. The best choice of inertial frame will usually be so obvious that we'll adopt it without even a moment's thought. But on occasion, we'll have to pick the frame carefully and deliberately. Finally, while there are formal methods for working with two or more inertial frames at once, I'll leave that for another book.

Measure for Measure: The Importance of Units

If you went to the grocery store and asked for "6 of sugar," the clerk wouldn't know how much sugar to give you. The number 6 wouldn't be enough information; you need to specify which units—cups, pounds, cubes, or tons—you have in mind. This need to specify units applies to almost all physical quantities—velocity, force, mass, and so on—and has led our society to develop units that everyone agrees on, also known as standard units.

For example, when you say that a skater's speed is 20 miles-per-hour, you have chosen "miles-per-hour" as the standard unit of speed and you're asserting that the skater is moving 20 times that fast. You can report the skater's speed as a multiple of any standard unit of speed—feet-per-second, yards-per-day, or inches-per-century, to name only a few—and you can always find a simple relationship to convert from one unit of speed to another. For example, to convert from miles-per-hour to kilometers-per-hour, you multiply by 1.609 kilometers/mile. Using that technique, you'll find that the skater's speed is 32.2 kilometers-per-hour.

Many of the common units in the United States come from the old English system of units, which most of the world has abandoned in favor of the SI units (Systéme Internationale d'Unités). The continued use of English units in the United States often makes life difficult. If you have to triple a cake recipe that calls for ¾ cup of milk, you must work hard to calculate that you need 2¼ cups. Then you go to buy 2¼ cups of milk, which is slightly more than half a quart, but end up buying two pints instead. You now have 14 ounces of milk more than you need. But is that 14 fluid ounces or 14 ounces of weight? And so it goes.

The SI system has two important characteristics that distinguish it from the English system and make it easier to use. In the SI system:

1. Different units for the same physical quantity are related by factors of 10.
2. Most of the units are constructed out of a few basic units: the meter, the kilogram, and the second.

Let's start with the first characteristic: different units for the same physical quantity are related by factors of 10. When measuring volume, 1000 milliliters is exactly 1 liter and 1000 liters is exactly 1 cubic meter (1 meter^3). When measuring mass, 1000 grams is exactly 1 kilogram and 1000 kilograms is exactly 1 metric ton. Because of this consistent relationship, enlarging a recipe that's based on the SI system is as simple as multiplying a few numbers. You never have to think about converting pints into quarts, teaspoons into tablespoons, or ounces into pounds. Instead, if you want to triple a recipe that calls for 500 milliliters of sugar, you just multiply the recipe by 3 to obtain 1500 milliliters of sugar. Since 1000 milliliters is 1 liter, you'll need 1.5 liters of sugar. Converting milliliters to liters is as simple as multiplying by 0.001 liter/milliliter. (See Appendix B for more conversion factors.)

SI units remain somewhat mysterious to many U.S. residents, even though some of the basic units are slowly appearing on our grocery shelves and highways. As a result, while the SI system really is more sensible than the old English system, developing a feel for some SI units is still difficult. How many of us know our heights in meters (the SI unit of length) or our masses in kilograms (the SI unit of mass)? If your car is traveling 200 kilometers-per-hour and you pass a police car, are you in trouble? Yes, because 200 kilometers-per-hour is about 125 miles-per-hour. Actually, the hour is not an SI unit—the SI unit of time is the second—but the hour remains customary for describing long periods of time. Thus the kilometer-per-hour is a unit that is half SI (the kilometer part) and half customary (the hour part).

The second characteristic of the SI system is its relatively small number of basic units. So far, we've noted the SI units of mass (the kilogram, abbreviated kg), length (the meter, abbreviated m), and time (the second, abbreviated s). One kilogram is about the mass of a liter of water; one meter is about the length of a long stride; one second is about the time it takes to say "one banana." From these three basic units, we can create several others, such as the SI units of velocity (the meter-per-second, abbreviated m/s) and acceleration (the meter-per-second^2, abbreviated m/s^2). One meter-per-second is a healthy walking speed; one meter-per-second^2 is about the acceleration of an elevator after the door closes and it begins to head upward. This conviction that many units are best constructed out of other, more basic units dramatically simplifies the SI system; the English system doesn't usually suffer from such sensibility.

The SI unit of force is also constructed out of the basic units of mass, length, and time. If we choose a 1-kilogram object and ask just how much force is needed to make that object accelerate at 1 meter-per-second^2, we define a specific amount of force. Since 1 kilogram is the SI unit of mass and 1 meter-per-second^2 is the SI unit of acceleration, it's only reasonable to let the force that causes this acceleration be the SI unit of force: the kilogram-meter-per-second^2. Since this composite unit sounds unwieldy but is very important, it has been given its own name: the newton (abbreviated N)—after, of course, Sir Isaac, whose second law defines the relationship among mass, length, and time that the unit expresses. One newton is about the weight of 18 U.S. quarter dollars; if you hold 18 quarters steady in your hand, you'll feel a downward force of about 1 newton.

Because a complete transition to the SI system will take generations, this book uses both unit systems whenever possible. Although it will emphasize the SI system, English and customary units may give you a better intuitive feel for a particular physical quantity. A bullet train traveling "67 meters-per-second" doesn't mean much to most of us, while one moving "150 miles-per-hour" (150 mph) or "240 kilometers-per-hour" (240 km/h) should elicit our well-deserved respect.

Quantity	SI Unit	English Unit	SI → English	English → SI
Position	meter (m)	foot (ft)	1 m = 3.2808 ft	1 ft = 0.30480 m
Velocity	meter-per-second (m/s)	foot-per-second (ft/s)	1 m/s = 3.2808 ft/s	1 ft/s = 0.30480 m/s
Acceleration	meter-per-second^2 (m/s^2)	foot-per-second^2 (ft/s^2)	1 m/s^2 = 3.2808 ft/s^2	1 ft/s^2 = 0.30480 m/s^2
Force	newton (N)	pound-force (lbf)*	1 N = 0.22481 lbf	1 lbf = 4.4482 N
Mass	kilogram (kg)	pound-mass (lbm)*	1 kg = 2.2046 lbm	1 lbm = 0.45359 kg
*The English units of force and mass are both called the pound. To distinguish these two units, it has become standard practice to identify them explicitly as pound-mass and pound-force.				

Section 1.2 **Falling Balls**

We've all dropped balls from our hands or seen them arc gracefully through the air after being thrown. These motions are simplicity itself and, not surprisingly, they're governed by only a few universal rules. We encountered several of those rules in the previous section, but we're about to examine our first important type of force: gravity. Like Newton, who reportedly began his investigations after seeing an apple fall from a tree, we'll start simply by exploring gravity and its effects on motion in the context of falling objects.

Weight and Gravity

Like everything else around us, a ball has a weight. For example, a golf ball weighs about 0.45 N (0.10 lbf). But what is weight? Evidently it's a force, since both the newton (N) and the pound-force (lbf) are units of force. But to understand what weight is—and, in particular, where it comes from—we need to look at gravity.

Gravity is a physical phenomenon that produces attractive forces between every pair of objects in the universe. In our daily lives, however, the only object massive enough and near enough to have obvious gravitational effects on us is our planet, the earth. Gravity weakens with distance and the moon and sun are so far away that we notice their gravities only through such subtle effects as the ocean tides.

The earth's gravity exerts a downward force on any object near its surface. That object is attracted directly toward the center of the earth with a force we call the object's weight (Fig. 1.2.1). Remarkably enough, this weight is exactly proportional to the object's mass—if one ball has twice the mass of another ball, it also has twice the weight. Such a relationship between weight and mass is astonishing because weight and mass are very different attributes: weight is how hard gravity pulls on a ball, and mass is how difficult that ball is to accelerate. Because of this proportionality, a ball that's heavy is also hard to shake!

An object's weight is also proportional to the local strength of gravity, which is measured by a downward vector called the acceleration due to gravity—an odd name that I'll explain shortly. At the surface of the earth the acceleration due to gravity is about 9.8 N/kg (1.0 lbf/lbm). That value means that a mass of 1 kilogram has a weight of 9.8 newtons, and that a mass of 1 pound-mass has a weight of 1 pound-force.

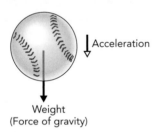

Fig. 1.2.1 The earth's gravity exerts a downward force on this ball, a force we call the ball's weight. Its weight causes the ball to accelerate downward.

More generally, an object's weight is equal to the product of its mass times the acceleration due to gravity or

$$\text{weight} = \text{mass} \cdot \text{acceleration due to gravity}. \qquad (1.2.1)$$

But why *acceleration* due to gravity? What acceleration do we mean? To answer that question, let's consider what happens to a ball when you drop it.

If the only force on the ball is its weight, the ball accelerates downward; in other words, it falls. While a ball moving through the earth's atmosphere encounters additional forces due to air resistance, let's ignore those forces for the time being. Doing so costs us only a little in terms of accuracy—the effects of air resistance are negligible as long as the ball is dense and its speed relatively small—and allows us to focus exclusively on the effects of gravity.

How much does the falling ball accelerate? According to Eq. 1.1.1, the ball's acceleration is equal to the force exerted on it divided by its mass. But because the ball is *falling*, the only force on it is its own weight. That weight, according to Eq. 1.2.1, is equal to the ball's mass times the acceleration due to gravity. Using a little algebra, we get

$$
\begin{aligned}
\text{falling ball's acceleration} &= \frac{\text{ball's weight}}{\text{ball's mass}} \\
&= \frac{\text{ball's mass} \cdot \text{acceleration due to gravity}}{\text{ball's mass}} \\
&= \text{acceleration due to gravity}.
\end{aligned}
$$

As you can see, the falling ball's acceleration is equal to the acceleration due to gravity. So acceleration due to gravity really is an acceleration after all: it's the acceleration of a freely falling object. Moreover, the units of acceleration due to gravity can be transformed easily from those relating weight to mass, 9.8 N/kg (1.0 lbf/lbm), into those describing the acceleration of free fall, 9.8 m/s² (32 ft/s²).

Thus a ball falling near the earth's surface experiences a downward acceleration of 9.8 m/s² (32 ft/s²), regardless of its mass. This downward acceleration is substantially more than that of an elevator starting its descent. When you drop a ball, it picks up speed very quickly in the downward direction.

Because all falling objects at the earth's surface accelerate downward at exactly the same rate, a billiard ball and a bowling ball dropped simultaneously from the same height will reach the ground together ❶. (Remember that we're not considering air resistance yet.) Although the bowling ball weighs more than the billiard ball, it also has more mass; so while the bowling ball experiences a larger downward force, its larger mass ensures that its downward acceleration is equal to that of the lighter and less massive billiard ball.

The Velocity of a Falling Ball

We're now ready to examine the motion of a falling ball near the earth's surface. A falling ball is one that has only the force of gravity acting on it and gravity, as we've seen, causes any falling object to accelerate downward at a constant rate. But we're usually less interested in the falling object's acceleration than we are in its position and velocity. Where will the object be in 3 seconds, and what will its velocity be then? When you're trying to summon up the courage to jump off the high dive, you want to know how long it'll take you to reach the water and how fast you'll be going when you hit.

The first step in answering these questions is to look at how a ball's velocity is related to the time you've been watching it fall. To do that, you'll need to know

❶ The observation that all things accelerate downward at the same rate is more familiar than it sounds. When you throw a handful of coins or pebbles, you're not surprised when they travel together and hit the ground as a group. As long as air resistance is insignificant, objects that you throw as a group fall as a group.

the ball's *initial velocity*, that is, its speed and direction at the moment you start watching it. If you drop the ball from rest, its initial velocity is zero.

You can then describe the ball's present velocity in terms of its initial velocity, its acceleration, and the time that has passed since you started watching it. Because a constant acceleration causes the ball's velocity to change by the same amount each second, the ball's present velocity differs from its initial velocity by the product of the acceleration times the time over which you've been watching it:

$$\text{present velocity} = \text{initial velocity} + \text{acceleration} \cdot \text{time.} \qquad (1.2.2)$$

For a ball falling from rest, the initial velocity is zero, the acceleration is downward at 9.8 m/s² (32 ft/s²), and the time you've been watching it is simply the time since it started to drop (Fig. 1.2.2). After one second, the ball has a downward velocity of 9.8 m/s (32 ft/s). After two seconds, the ball has a downward velocity of 19.6 m/s (64 ft/s). After three seconds, its downward velocity is 29.4 m/s (96 ft/s), and so on. Since the ball's motion is strictly vertical, we often put a negative sign in front of the acceleration to indicate the direction. By convention, we say that a negative sign means "down."

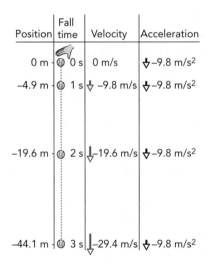

Position	Fall time	Velocity	Acceleration
0 m	0 s	0 m/s	−9.8 m/s²
−4.9 m	1 s	−9.8 m/s	−9.8 m/s²
−19.6 m	2 s	−19.6 m/s	−9.8 m/s²
−44.1 m	3 s	−29.4 m/s	−9.8 m/s²

Fig. 1.2.2 The moment you let go of a ball that was resting in your hand, it begins to fall. Its weight causes it to accelerate downward. After 1 second, it has fallen 4.9 m and has a velocity of 9.8 m/s downward. After 2 seconds, it has fallen 19.6 m and has a velocity of 19.6 m/s downward, and so on. As the ball continues to accelerate downward, its velocity continues to increase downward. Negative values for the position and velocity are meant to indicate downward movement, caused by a negative or downward acceleration.

The Position of a Falling Ball

The ball's velocity continues to increase as it falls, but where exactly is the ball located? To answer that question, you need to know the ball's *initial position*, that is, where it was when you started to watch it fall. If you dropped the ball from rest, the initial position was your hand and you can define that spot as 0.

You can then describe the ball's present position in terms of its initial position, its initial velocity, its acceleration, and the time that has passed since you started watching it. However, because the ball's velocity is changing, you can't simply multiply its present velocity by the time that it's been falling to determine how much the ball's present position differs from its initial position. Instead, you must use the ball's average velocity during the whole period you've been watching it. Since the ball's velocity has been changing uniformly from its initial velocity to its present velocity, the ball's average velocity is exactly halfway in between the two individual velocities:

$$\text{average velocity} = \text{initial velocity} + \tfrac{1}{2} \cdot \text{acceleration} \cdot \text{time.}$$

The ball's present position differs from its initial position by the product of this average velocity times the time over which you've been watching it:

$$\text{present position} = \text{initial position} +$$

$$\text{initial velocity} \cdot \text{time} + \tfrac{1}{2} \cdot \text{acceleration} \cdot \text{time}^2. \quad (1.2.3)$$

For a ball falling from rest, the initial velocity is zero, the acceleration is downward at 9.8 m/s² (32 ft/s²), and the time you've been watching it is simply the time since it started to drop (Fig. 1.2.2). After one second, the ball has fallen 4.9 m (16 ft). After two seconds, the ball has fallen downward a total of 19.6 m (64 ft). After three seconds, the ball has fallen a total of 44.1 m (145 ft), and so on.

Equations 1.2.2 and 1.2.3 depend on the definition of acceleration as the measure of how quickly *velocity* changes and the definition of velocity as the measure of how quickly *position* changes. Because the acceleration of a falling ball doesn't change with time, the two equations can be derived using algebra. But in more complicated situations, where an object's acceleration changes with time, predicting position and velocity usually requires the use of calculus. Calculus is the mathematics of change, invented by Newton to address just these sorts of problems.

We've been discussing what happens to a falling ball, but we could have chosen another object instead. Everything falls the same way; heavy or light, large or small, all objects take the same amount of time to fall some distance near the earth's surface, as long as they're dense enough to overcome air resistance. If there were no air, this statement would be exactly true for any object; a feather and a lead brick would plummet downward together if you dropped them simultaneously.

Now that we've explored acceleration due to gravity, we can see why a ball dropped from a tall ladder is more dangerous than the same ball dropped from a short stool. The farther the ball has to fall, the longer it takes to reach the ground and the more time it has to accelerate. During its long fall from the tall ladder, the ball acquires a large downward velocity and becomes very hard to stop. If you try to catch it, you'll have to exert a very large upward force on it to accelerate it upward and bring it to rest quickly. Exerting that large upward force may hurt your hand. (For a discussion of how air resistance affects falling, see ❷.)

The same notion holds if you're the falling object. If you leap off a tall ladder, a substantial amount of time will pass before you reach the ground. By the time you arrive, you'll have acquired considerable downward velocity. The ground will then accelerate you upward and bring you to rest with a very large and unpleasant upward force. (For an interesting and less painful application of long falls, see ❸.)

How a Thrown Ball Moves: Projectile Motion

If the only force acting on an object is its weight, then the object is falling. So far, we've explored this principle only as it pertains to balls dropped from rest. However, a thrown ball is falling, too; once it leaves your hand, it's subject only to the downward force of gravity and it falls. It may seem odd but even though it's initially traveling upward, a ball tossed upward is accelerating downward at 9.8 m/s² (32 ft/s²). As a result, the tossed ball's upward velocity diminishes, it stops rising, its velocity becomes downward, and it eventually returns to the ground.

Equation 1.2.2 still describes how the ball's velocity depends on the fall time, but now the initial velocity isn't zero; it points in the upward direction! If you toss a ball straight up in the air, it leaves your hand with a large upward velocity (Fig. 1.2.3). As soon as you let go of it, it begins to accelerate downward. If the ball's initial upward velocity is 29.4 m/s (96 ft/s), then after one second its upward velocity is 19.6 m/s (64 ft/s). After another second, its upward velocity is only 9.8 m/s (32 ft/s). After a third second, the ball momentarily comes to a complete stop with

❷ As a dropped object descends, an upward force of air resistance opposes its downward weight and reduces its downward acceleration. Since air resistance increases speed, the object eventually stops accelerating altogether and coasts downward at terminal velocity. While a penny dropped from the Empire State Building would reach 340 km/h (210 mph) in the absence of air resistance, it actually tumbles downward at a terminal velocity of less than 80 km/h (50 mph) and you can catch it. But watch out for a ball-point pen!

❸ In 1782, William Watts, a plumber from Bristol, England, patented a technique for forming perfectly spherical, seamless lead shot for use in guns. His idea was to pour molten lead through a sieve suspended high above a pool of water. The lead droplets cool in the air as they fall, solidifying into perfect spheres before reaching the water. Shot towers based on this idea soon appeared throughout Europe and eventually in the United States. Nowadays, iron shot has all but replaced environmentally dangerous lead shot. Iron shot is cast, rather than dropped, because the longer cooling time needed to solidify molten iron would require impractically tall shot towers.

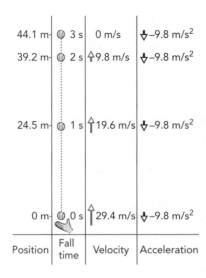

Position	Fall time	Velocity	Acceleration
44.1 m	⊙ 3 s	0 m/s	↓ −9.8 m/s²
39.2 m	⊙ 2 s	↑ 9.8 m/s	↓ −9.8 m/s²
24.5 m	⊙ 1 s	↑ 19.6 m/s	↓ −9.8 m/s²
0 m	⊙ 0 s	↑ 29.4 m/s	↓ −9.8 m/s²

Fig. 1.2.3 The moment you let go of a ball thrown straight upward, it begins to accelerate downward at 9.8 m/s². The ball rises but its upward velocity diminishes steadily until it momentarily comes to a stop. It then descends with its downward velocity increasing steadily. In this example, the ball rises for 3 s and comes to rest. It then descends for 3 s before returning to your hand in a very symmetrical flight.

a velocity of zero. It then descends from this peak height, falling just as it did when you dropped it from rest.

The ball's flight before and after its peak is symmetrical. It travels upward quickly at first, since it has a large upward velocity. As its upward velocity diminishes, it travels more and more slowly until it comes to a stop. It then begins to descend, slowly at first and then faster and faster as it continues its constant downward acceleration. The time the ball takes to rise from its initial position in your hand to its peak height is exactly equal to the time it takes to descend back down from that peak to your hand. Equation 1.2.3 indicates how the position of the ball depends on the fall time, with the initial velocity being the upward velocity of the ball as it leaves your hand.

The larger the initial upward velocity of the ball, the longer it rises and the higher it goes before its velocity is reduced to zero. It then descends for the same amount of time it spent rising. The higher the ball goes before it begins to descend, the longer it takes to return to the ground and the faster it's traveling when it arrives. That's why catching a high fly ball with your bare hands stings so much: the ball is traveling very, very fast when it hits your hands, and a large force is required to bring the ball to rest quickly.

What happens if you don't toss the ball exactly straight up? Suppose you throw the ball upward at some angle. The ball still rises to a peak height and then begins to descend; but as it rises and descends, it also travels away from you horizontally so that it strikes the ground at some distance from your feet. How much does this horizontal travel complicate the motion of a falling ball?

The answer is not very much. One of the beautiful simplifications of physics is that we can often treat an object's vertical motion independently of its horizontal motion. This technique involves separating the vector quantities—acceleration, velocity, and position—into components, those portions of the quantities that lie along specific directions (Fig. 1.2.4). For example, the vertical component of an object's position is that object's altitude.

If you know a ball's altitude, you know only part of its position; you still need to know where it is to your left or right and to your front or back. In fact, you can completely specify its position (or any other vector quantity) in terms of three components along three directions that are perpendicular (or at right angles) to one another. This means that you can specify the ball's position by its vertical altitude, its horizontal distance to your left or right, and its horizontal distance in front or in back of you. For example, the ball might be 10 m above you, 3 m to your left, and 2 m behind you. These three distances indicate just where the ball is located.

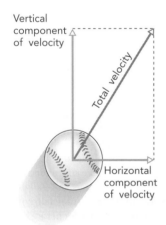

Fig. 1.2.4 Even if the ball has a velocity that is neither purely vertical nor purely horizontal, its velocity may nonetheless be viewed as having a vertical component and a horizontal component. Part of its total velocity acts to move this ball upward, and part of its total velocity acts to move this ball in a horizontal direction.

Up until now, we've actually been examining only the vertical components of position, velocity, and acceleration. But now we're going to let the ball move horizontally so that we also have to consider what happens to the two horizontal components of each vector quantity. Keeping track of all three components of each quantity is difficult. Since we're interested in the motion of a tossed ball, we can eliminate the left–right component by throwing the ball forward; that way, the ball will move only in a vertical plane that extends directly in front of us. We can then specify the ball's position as its altitude and its horizontal distance in front of us.

Because the falling ball's acceleration is constant and independent of where the ball is near the earth's surface, the ball's horizontal motion is independent of its vertical motion. We already know about the falling ball's vertical motion, but what is its horizontal motion? It appears we need some new relationships to describe how the horizontal components of position, velocity, and acceleration change with time. Fortunately, we can reuse Eqs. 1.2.2 and 1.2.3.

Although Eqs. 1.2.2 and 1.2.3 were introduced to describe a ball's vertical motion, they're actually more general. They relate three vector quantities—position, velocity, and constant acceleration—to one another. They describe how an object's position and velocity change with time when the object undergoes constant acceleration, regardless of the direction of that constant acceleration.

These equations also apply to the components of position, velocity, and constant acceleration. If you add the words "vertical component of" in front of each vector quantity in Eqs. 1.2.2 and 1.2.3, the equations correctly describe the object's vertical motion. The same can be done with the object's horizontal motion. Our previous look at vertical motion implicitly inserted the words "vertical component of" into the equations. Now we'll try inserting the words "horizontal component of" to understand the tossed ball's horizontal motion.

As soon as the ball leaves your hand, its motion can be broken into two parts: a vertical motion and a horizontal motion (Fig. 1.2.4). Part of the ball's initial velocity is in the upward direction, and that vertical velocity component determines the object's ascent and descent. Part of the ball's initial velocity is in the horizontal or downfield direction, and that horizontal velocity component determines the ball's drift downfield.

Because gravity is the only force on the ball and it acts only in the downward vertical direction, there is no horizontal component of acceleration. The horizontal velocity component therefore remains constant, and the ball travels downfield at a steady rate throughout its flight (Fig. 1.2.5). Overall, the upward vertical component of the ball's initial velocity determines how high the ball goes and how long it stays aloft before striking the ground, while the horizontal component of the initial velocity determines how quickly the ball travels downfield during its time aloft (Fig. 1.2.6).

Just before the ball hits the ground it still has its original horizontal velocity component, but its vertical velocity component is now in the downward direction. The total velocity of the ball is composed of these two components. The ball starts with its velocity up and forward and ends with its velocity down and forward.

If you want a ball or shot put to hit the ground as far from your feet as possible, you should keep it aloft for a long time *and* give it a sizable horizontal component of velocity; in other words, you must achieve a good balance between the vertical and horizontal velocity components (Fig. 1.2.7). These components of velocity together determine the ball's flight path, its trajectory. If you throw the ball straight up, it will stay aloft for a long time but will not travel downfield at all (and you will need to wear a helmet). If you throw the ball directly downfield, it will have a large initial horizontal velocity component but will hit the ground almost immediately.

Neglecting air resistance and the altitude difference between your throwing arm and the ground that the ball will eventually hit, your best choice is to throw the ball

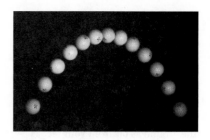

Fig. 1.2.5 This golf ball drifts steadily to the right after being thrown because gravity affects only the ball's vertical component of velocity.

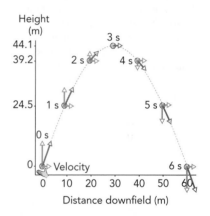

Fig. 1.2.6 If you throw a ball upward, at an angle, part of the initial velocity will be in the vertical direction and part will be in the horizontal direction. The vertical and horizontal motions will take place independently of one another. The ball will rise and fall just as it did in Figs. 1.2.2 and 1.2.3; at the same time, however, it will move downfield. Because there is no horizontal acceleration (gravity acts only in the vertical direction), the horizontal velocity remains constant, as indicated by the velocity arrows. In this example, the ball travels 10 m downfield each second and strikes the ground 60 m from your feet after a 6-s flight through the air.

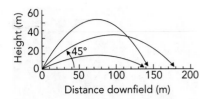

Fig. 1.2.7 If you want the ball to hit the ground as far from your feet as possible, given a certain initial speed, throw the ball at 45°. Halfway between horizontal and vertical, such a throw gives the ball equal initial vertical and horizontal components of velocity. The ball then stays aloft for a relatively long time and makes good use of that flight time to travel downfield.

❹ Archers and marksmen must aim high because arrows and bullets fall en route to their targets. This effect becomes more significant as the target distance and travel time increase. At least in principle, an arrow or bullet shot horizontally would take the same time to fall to the ground as would an arrow or bullet dropped from rest. The projectile's horizontal coasting motion has no effect on its vertical falling motion.

at an angle of 45° above horizontal. At that angle, the initial upward velocity component will be the same as the initial downfield velocity component. The ball will stay aloft for a reasonably long time and will make good use of that time to move downfield. Other angles won't make such good use of the initial speed to move the ball downfield. (A discussion of how to determine the horizontal and vertical components of a velocity appears in Appendix A.)

These same ideas apply to two baseballs, one dropped from a cliff and the other thrown horizontally from that same cliff. If both leave your hands at the same time, they will both hit the ground below at the same time. The fact that the second ball has an initial horizontal velocity doesn't affect the time it takes to descend to the ground, because the horizontal and vertical motions are independent. Of course, the ball thrown horizontally will strike the ground far from the base of the cliff, while the dropped ball will land directly below your hand ❹.

Mass Versus Weight

Since many grocery store items are sold in units of mass (e.g., grams or kilograms) after being weighed on a scale, it's easy to think that weight and mass are the same thing. Nothing could be further from the truth. Weight is the measure of how hard gravity pulls on an object while mass is the measure of how difficult that object is to accelerate. Weight depends on something outside the object, typically the earth, so it isn't entirely a property of the object itself. In contrast, mass is an intrinsic property of the object and has nothing at all to do with what's around that object. While a 1-kilogram box of chocolates will be just as difficult to accelerate on the moon as it is to accelerate in your living room, it won't weigh nearly as much.

One illustration of the difference between mass and weight involves a trip to the bicycle shop. Suppose you are in that shop and you want to determine which bicycle has the least mass. Mass is what matters to you if you're going to be riding only on level pavement and must start and stop often. A massive bicycle will be difficult both to start and stop so keeping the mass small is a good idea. To assess the masses of the bicycles, don't pick them up. Just roll them quickly back and forth along the floor. The more massive the bicycle, the harder you'll have to push it to get a certain acceleration to occur. You'll easily pick out the low-mass bicycles because they'll change speed and direction particularly easily.

On the other hand, if you bicycle into the mountains regularly, you'll want a light bicycle because you shouldn't be lifting any more weight than necessary as you climb the steep slopes. To measure the weights of the bicycles, you simply lift them off the ground. The heavier bicycles will require larger upward forces before they'll break contact with the ground.

Drawing this careful distinction between mass and weight isn't always important because of one remarkable fact: at a given location in space, an object's mass and its weight are exactly proportional to one another. The proportionality constant is the local acceleration due to gravity. Because of this proportionality, you can quantify a bicycle either by measuring its mass (rolling it back and forth) or its weight (picking it up). If you find one bicycle that's twice as massive as another, you can be sure that it's also twice as heavy as the other. This result allows us to become lazy and to confuse mass and weight as though they were the same. It's important to remember that they're different and that they're associated with different physical concepts.

Hand
truck

Ramp

Section 1.3 **Ramps**

In the previous section, we looked at what happens to an object experiencing only a single force: the downward force of gravity. But what happens to objects that experience two or more forces at the same time? Imagine, for example, an object resting on the floor. That object experiences both the downward force of gravity and an upward force from the floor. If the floor is level, the object doesn't accelerate; but if the floor is tilted, so that it forms a ramp, the object accelerates downhill. In this section, we'll examine the motion of objects traveling along ramps. In addition to making sports such as skateboarding and skiing more fun, ramps are common tools that help us lift and move heavy objects.

A Piano on the Sidewalk

Imagine that you have a friend who's a talented but undiscovered pianist. She's renting a new apartment, and because she can't afford professional movers (Fig. 1.3.1), she's asked you to help her move her baby grand. Fortunately, her new apartment is only on the second floor. But the two of you still face a difficult challenge: how do you get that heavy piano up there? More importantly, how do you keep it from falling on you during the move?

The problem is that you can't push upward hard enough to lift the entire piano at once. One solution to this problem, of course, would be to break the piano into pieces and carry them up one by one. But this method has obvious drawbacks: your friend isn't expecting a firewood delivery. A better solution would be to find something else to help you push upward, and one of your best choices would be the simple machine known as a ramp.

Fig. 1.3.1 A ramp would make this move much easier.

Throughout the ages, ramps, also known as inclined planes, have made tasks like piano-moving possible. Because ramps can exert the enormous upward forces needed to lift stone and steel, they've been essential building equipment since the days of the pyramids. To see how ramps provide these lifting forces, we'll continue to explore the example of the piano, looking first at the force that the piano experiences when it touches a surface. For the time being, we'll continue to ignore friction and air resistance; they would needlessly complicate our discussion. Besides, as long as the piano is on wheels, friction is negligible.

The Sidewalk Pushes Back: Newton's Third Law

With the piano resting on the sidewalk outside the apartment you make a startling discovery: the piano is *not* falling. Has gravity disappeared? The answer to that question would be painfully obvious if your foot were underneath one of the piano's wheels. No, the piano's weight is still all there. But something is happening at the surface of the sidewalk to keep the piano from falling. Let's take a careful look at the situation.

To begin with, the piano is clearly pushing down hard on the sidewalk. That's why you're keeping your toes out of the way. But the presence of a new downward force *on the sidewalk* doesn't explain why *the piano* isn't falling. Instead, we must look at the sidewalk's response to the piano's downward push: the sidewalk pushes upward on the piano! You can feel this response by leaning over and pushing down on the sidewalk with your hand—the sidewalk will push back. Those two forces, your downward push on the sidewalk and its upward push on your hand, are exactly equal in magnitude but opposite in direction.

This observation, that two things exert equal but opposite forces on one another, isn't unique to sidewalks, pianos, or hands; in fact, it's always true. If you push on any object, that object will push back on you with an equal amount of force in exactly the opposite direction. This rule—often expressed as "for every action, there is an equal but opposite reaction"—is known as Newton's third law of motion, the last of his three laws.

Newton's Third Law of Motion
For every force that one object exerts on a second object, there is an equal but oppositely directed force that the second object exerts on the first object.

The universality of this law is astounding. Whether an object is large or small, hard or soft, stationary or faster than a rocket, if you can push on it, it *will* push back on you with an equal but oppositely directed force.

In the present case, the sidewalk and piano push on one another with equal but oppositely directed forces. Of this pair of equal-but-opposite forces, only one force acts *on the piano*: the sidewalk's upward push. This upward push on the piano is what keeps the piano from falling. We've solved the mystery.

Intuition Alert: Action and Reaction
Intuition says that when you push an object that's moving away from you, it pushes back more gently on you than you push on it and when you push an object that's moving toward you, it pushes back harder on you than you push on it.
Physics says that when you push on an object, it always pushes back on you exactly as hard as you push on it.
Resolution: It's difficult to push on an object that's moving away from you, so you naturally push on it more gently than you expect. The gentle force it exerts on you is simply an equal but oppositely directed response to your gentle force on it. In contrast, it's difficult not to push strongly on an object that's moving

toward you, so you naturally push on it harder than you expect. The strong force it exerts on you is again an equal but oppositely direct response to your strong force on it.

Looking for Support and Adding Up the Forces

Although we've figured out why the piano isn't falling, we still don't know what type of force the sidewalk is using to hold it up or why that upward force so perfectly balances the piano's downward weight.

Let's begin with the type of force: since two objects can't occupy the same space at the same time, their surfaces push apart whenever they're in contact. They exert support forces on one another, each pushing the other directly away from its surface—the direction normal (or perpendicular) to its surface ❶. Since the sidewalk is horizontal, the support force it exerts on the piano is vertical—straight up (Fig. 1.3.2).

But how large is that support force? To answer this question, suppose the sidewalk's support force were strong enough to make the piano accelerate upward. The piano would soon lift off the sidewalk and, as their surfaces stopped touching, the sidewalk's support force on the piano would grow weaker. Alternatively, suppose the sidewalk's support force was weak enough to let the piano accelerate downward. The piano would soon drop into the sidewalk and, as their surfaces overlapped more, the sidewalk's support force on the piano would grow stronger.

Because of these behaviors, the sidewalk's upward support force on the piano adjusts automatically until it exactly balances the piano's downward weight and the piano accelerates neither up nor down. And when you sit on the piano during a break, the sidewalk's upward support force quickly readjusts to balance your weight as well.

Another way to state that the upward support force on the piano exactly balances the piano's downward weight is to say that the net force on the piano is zero, meaning that the sum of all the forces on the piano is zero. Objects often experience more than one force at a time, and it's the net force, together with the object's mass, that determines how it accelerates. When you and your friend push the piano in the same direction, your forces add up, assisting one another so that the piano accelerates in that direction (Fig. 1.3.3a). When the two of you push the piano in opposite directions, your forces oppose and at least partially cancel one another (Fig. 1.3.3b).

❶ Forces that are directed exactly away from surfaces are called normal forces, since the term normal is used by mathematicians to describe something that points exactly away from a surface—at right angles or perpendicular to that surface.

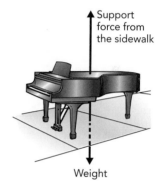

Fig. 1.3.2 A piano resting on the sidewalk. The sidewalk exerts an upward support force that exactly balances the piano's downward weight. The net force on the piano is zero, so the piano doesn't accelerate.

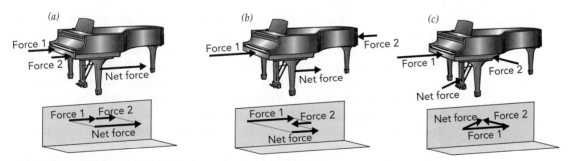

Fig. 1.3.3 When several forces act simultaneously on an object, the object responds to the sum of the forces. This sum is called the net force, and it has both a magnitude and a direction. Here, as elsewhere in this book, the length of each force arrow indicates its magnitude. You can sum the forces graphically by arranging the force arrows in sequence, head to tail. The net force arrow then points from the tail of the first arrow to the head of the last arrow. Some of the force arrows are shown displaced slightly for clarity, a shift that doesn't affect the summing process.

And when the two of you push the piano at an angle with respect to one another, the net force points somewhere in between. For example, if you push the piano eastward while your friend pushes it northward, the net force will point to the northeast and the piano will accelerate in that direction (Fig. 1.3.3c). The precise angle of the net force and the piano's subsequent acceleration depends on exactly how hard each person pushes. For most of the following discussion, we'll need only a rough estimate of the net force's magnitude and direction, and we'll obtain that estimate using common sense.

Apart from direction, there's one crucial difference between the force gravity exerts on the piano and the support force the sidewalk exerts on it. While the piano's weight is dispersed throughout the piano, the sidewalk's upward support force acts only on the piano's wheels. Even when the net force on the piano is zero, having individual forces act on it at different locations can lead to considerable stress within the piano. If it weren't built so well, the piano might lose a leg or two during the move.

Common Misconceptions: Newton's Third Law and Balanced Forces

Misconception: When you push on an object and it pushes back on you, those two equal-but-opposite forces somehow cancel one another perfectly and therefore have no effect on either you or the object.

Resolution: The two forces described by Newton's third law always act on *two different things*. Your push acts on the object while the object's push acts on you. Since the object accelerates in response to the net force it experiences—the sum of all the individual forces acting *on it*—it is affected only by your force *on it*, not by its force *on you*. If you are the only thing pushing on it, it will accelerate. And if the object is the only thing pushing on you, you'll accelerate, too!

What the Piano Needs: Energy

As you approach the task of lifting your friend's piano into her apartment, you might begin to worry about safety. There is clearly a difference between the piano resting on the sidewalk and the piano suspended on a board just outside the second floor apartment. After all, which one would you rather be sitting beneath? The elevated piano has something that the piano on the sidewalk doesn't have: the ability to produce motion and even to break itself and the things beneath it. This capacity to make things happen is called energy, and the process of making them happen is called work.

Energy and work are both important physical *quantities*, meaning that both are measurable. For example, you can measure the amount of energy in the suspended piano, and the amount of work the piano does when the board breaks and it falls to the sidewalk. As you may suspect, the physical definitions of energy and work are somewhat different from those of common English. Physical energy isn't the exuberance of a child at the amusement park or the contents of a large cup of coffee; instead, it's defined as the capacity to do work. Similarly, physical work doesn't refer to activities at the office or in the yard; instead, it refers to the process of transferring energy.

Energy is what's transferred, and work does the transferring. The most important characteristic of energy is that it's conserved. In physics, a conserved quantity is one that can't be created or destroyed but can be transferred between objects or, in the case of energy, converted from one form to another. Conserved quantities are very special in physics; there are only a few of them. An object that has energy can't simply make that energy disappear; it can only get rid of energy by giving it to another object, and it makes this transfer by doing work on that object.

The relationship between energy and work is analogous to the relationship between money and spending: money is what is transferred, and spending does the transferring. Sensible, law-abiding citizens don't create or destroy money; instead they transfer it among themselves through spending. Just as the most interesting aspect of money is spending it, so the most interesting aspect of energy is doing work with it. We can define money as the capacity to spend, just as we define energy as the capacity to do work.

So far we've been using a circular definition: work is the transfer of energy and energy is the capacity to do work. But what is involved in doing work on an object? You do work on an object by exerting a force on it as it moves in the direction of that force. As you throw a ball, exerting a forward force on the ball as the ball moves forward, you do work on the ball; as you lift a rock, pushing the rock upward as it moves upward, you do work on the rock. In both cases, you transfer energy from yourself to an object by doing work on it.

This transferred energy is often apparent in the object. When you throw a ball, it picks up speed and undergoes an increase in kinetic energy—energy of motion that allows the ball to do work on whatever it hits. And when you lift a rock, it shifts farther from the earth and undergoes an increase in gravitational potential energy—energy stored in the gravitational forces between the rock and the earth that allows the rock to do work on whatever it falls on. In general, potential energy is energy stored in the forces between or within objects.

Returning to the task at hand, it's now apparent that raising the piano to the second floor apartment is going to increase the piano's gravitational potential energy by a substantial amount. Since energy is a conserved quantity, this additional energy must come from something else. Unfortunately, that something is you! To deliver the piano, you are going to have to provide it with the gravitational potential energy it needs by doing exactly that amount of work on it. And as we'll see, you can do that work the hard way by carrying it up a ladder or the easy way by pushing it up a ramp.

Lifting the Piano: Doing Work

To do work on an object, you must push on it while it moves in the direction of your push. The work you do on it is the product of the force you exert on it times the distance it travels along the direction of your force:

$$\text{work} = \text{force} \cdot \text{distance}. \qquad (1.3.1)$$

Calculating the work you do is easy if the object moves exactly in the direction of your push: you simply multiply your force times the distance it travels. But if the object doesn't move in the direction of your push, you must multiply your force times the component (i.e., the portion) of the object's motion along the direction of your force.

As long as the angle between your force and the object's motion is small, you can often ignore this complication. But as the angle becomes larger, the work you do on the object decreases. When the object moves at right angles to your force, the work you do on it drops to zero—it's not moving along the direction of your force at all (see ❷). And for angles larger than 90°, the object moves *opposite* your force and the work you do on it actually becomes negative!

Recalling that forces always come in equal but oppositely directed pairs, we can now explain why energy is conserved: whenever you do work on an object, that object simultaneously does an equal amount of negative work on you! After all, if you push an object and it moves along the direction of your force, then it pushes back on you and you move along the direction opposite its force. You do positive work on it and it does negative work on you.

❷ When calculating how much work you do while pushing on a moving piano bench, you must carefully consider the direction of your push and the direction of the bench's motion. Only the component of the bench's motion in the direction of your push contributes to the work you do on the bench and therefore to the change in the bench's energy. If the bench moves exactly in the direction of your push, then its entire distance of travel contributes to the work you do on it. If the bench moves opposite to your push, then you do negative work on it. And if the bench moves perpendicular to your push, then none of its distance of travel contributes and you do no work on it at all. An example of the latter is carrying the bench horizontally at constant velocity. Since you are pushing the bench directly upward to support its weight, its motion and your push are perpendicular to one another. You are therefore doing no work on the bench and its energy isn't changing. You may be getting hot and sweaty, but whatever energy you're expending is staying within your body and is not being transferred to the bench via work.

For example, when you lift the piano to judge its weight, you push it up as it moves up and thus do work on it. At the same time, the piano pushes your hand down but your hand moves up, so it does negative work on your hand. Overall, the piano's energy increases by exactly the same amount that your energy decreases—a perfect transfer! The energy that you're losing is mostly food energy, a form of chemical potential energy, and the energy the piano is gaining is mostly gravitational potential energy.

When you lower the piano after realizing that it's too heavy to carry up a ladder, the process is reversed and the piano transfers energy back to you. Now the piano does work on you and you do an equal amount of negative work on the piano. The piano is losing mostly gravitational potential energy and you are gaining mostly thermal energy—a disordered form of energy that we'll examine in Section 2.2. Unlike a rubber band, your body just isn't good at storing work done on it, so it simply gets hotter. Nonetheless, it's usually easier to have work done on you than to do work on something else. That's why it's easier to lower objects than to lift them.

Finally, when you hold the piano motionless above the pavement, while waiting for your friend to reinstall the wheel that fell off, you and the piano do no work on one another. You are simply converting chemical potential energy from your last meal into thermal energy in your muscles and get overheated, probably in more ways than one.

***Conserved Quantity:* Energy**   ***Transferred By:* Work**

Energy: The capacity to do work. Energy has no direction. It can be hidden as potential energy.
Kinetic Energy: The form of energy contained in an object's motion.
Potential Energy: The form of energy stored in the forces between or within objects.
Work: The mechanical means for transferring energy; work = force·distance .

Gravitational Potential Energy

How much work would you do on the piano while lifting it straight up a ladder to the apartment? Apart from a little extra shove to get the piano moving upward, lifting it would entail supporting the piano's weight while it coasted upward at constant velocity from the sidewalk to the second floor. Since you would be pushing upward on the piano with a force equal in amount to its weight, the work you would do on it would be the product of its weight times the distance you lifted it.

As the piano rises in this scenario, its gravitational potential energy increases by an amount equal to the work you do on it. If we agree that the piano has zero gravitational potential energy when it rests on the sidewalk, then the suspended piano's gravitational potential energy is simply its weight times its height above the sidewalk. Since the piano's weight is equal to its mass times the acceleration due to gravity, its gravitational potential energy is its mass times the acceleration due to gravity times its height above the sidewalk.

These ideas aren't limited to pianos. You can determine the gravitational potential energy of any object by multiplying its mass times the acceleration due to gravity times its height above the level at which its gravitational potential energy is zero:

$$\text{gravitational potential energy} =$$

$$\text{mass} \cdot \text{acceleration due to gravity} \cdot \text{height.} \qquad (1.3.2)$$

Of course, if you know the object's weight, you can use it in place of the object's mass times the acceleration due to gravity.

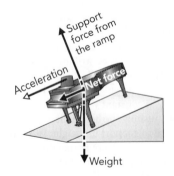

Fig. 1.3.4 A piano sliding on a frictionless ramp while experiencing a force due to gravity. It accelerates down the ramp more slowly than it would if it were falling freely.

So what is the piano's gravitational potential energy when it reaches the second floor? If it weighs 2000 N (450 lbf) and the second floor is 5 m (16 ft) above the sidewalk, you will have done 10,000 N·m (about 7200 ft·lbf) of work in lifting it up there, and the piano's gravitational potential energy will thus be 10,000 N·m. The newton-meter is the SI unit of energy and work; it's so important that it has its own name, the joule (abbreviated J). At the second floor, the piano's gravitational potential energy is 10,000 J.

A few everyday examples should give you a feeling for how much energy a joule is. Lifting a liter bottle of water 10 centimeters (4 inches) upward requires about 1 J of work. A 100-watt lightbulb needs 100 J every second to operate. Your body is able to extract about 2,000,000 J from a slice of cherry pie. When you're bicycling or rowing hard, your body can do about 1000 J of work each second. A typical flashlight battery has about 10,000 J of stored energy.

Quantity	SI unit	English unit	SI → English	English → SI
Energy	joule (J) = newton-meter (N·m)	foot-pound (ft·lbf)	1 J = 0.73757 ft·lbf	1 ft-lbf = 1.3558 J

Lifting the Piano with a Ramp

Unfortunately, you probably can't carry a grand piano up a ladder by yourself. You need a ramp, both to help you support the piano and to make it easier for you to raise the piano to the second floor.

Like the sidewalk, a ramp exerts a support force on the piano to prevent the piano from passing through its surface. However, since the ramp isn't exactly horizontal, that support force isn't exactly vertical (Fig. 1.3.4). The piano's weight still points straight down, but since the ramp's support force doesn't point straight up, the two forces can't balance one another. There is a nonzero net force on the piano.

This net force can't point into or out of the ramp. If it did, the piano would accelerate into or out of the ramp and the two objects would soon either lose contact or travel through one another. Instead, the net force points exactly along the surface of the ramp—a direction tangent or parallel to the surface. More specifically, it points directly downhill so the piano accelerates down the ramp!

But because this net force is much smaller than the piano's weight, the piano's acceleration down the ramp is slower than if it were falling freely. This effect is familiar to anyone who has bicycled downhill or watched a cup slip slowly off a tilted table. While gravity is still responsible, these objects accelerate more slowly than falling and in the direction of the downward slope.

Therein lies the beauty of the ramp. By putting the piano on a ramp, you let the ramp support most of the piano's weight. The piano only experiences a small residual net force pushing it downhill along the ramp. If you now push uphill on the piano with a force that exactly balances that downhill force, the net force on the piano drops to zero and the piano stops accelerating. If you push uphill a little harder, the piano will accelerate up the ramp!

How does the ramp change the job of moving the piano? Suppose that you build a ramp 50 m (164 ft) long that extends from the sidewalk to the apartment's balcony, 5 m (16.4 ft) above the pavement (Fig. 1.3.5). This ramp is sloped so that traveling 50 m uphill along its surface lifts the piano only 5 m upward. Because of the ramp's 10 to 1 ratio between distance travelled along its surface and altitude change, you can push the 2000-N (450 lbf) piano up it at constant velocity with a force of only 200 N (45 lbf). Most people can push that hard, so the moving job is now realistic. To reach the apartment, you must push the piano 50 m along this ramp with a force of 200 N so that you will do a total of 10,000 J of work.

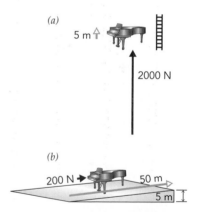

Fig. 1.3.5 To lift a piano weighing 2000 N, you can either (a) push it straight up or (b) push it along a ramp. To keep the piano moving at a constant velocity, you must make sure it experiences a net force of zero. If you lift it straight up the ladder in (a), you must exert an upward lifting force of 2000 N to balance the piano's downward weight. If you push it up the ramp shown in (b), you will only have to push the piano uphill with a force of 200 N in order to give the piano a net force of zero.

By pushing the piano up the ramp, you've used physical principles to help you perform a task that would otherwise have been nearly impossible. But you didn't get something for nothing. The ramp is much longer than the ladder, and you have had to push the piano for a longer distance in order to raise it to the second floor. Of course, you have had to push with less force.

Remarkably, the amount of work you do in either case is 10,000 J. In carrying the piano up the ladder, the force you exert is large but the distance the piano travels in the direction of that force is small. In pushing the piano up the ramp, the force is small but the distance is large. Either way, the final result is the same: the piano ends up on the second floor with an additional 10,000 J of gravitational potential energy and you have done 10,000 J of work. Expressed graphically in an equation, this relationship would appear as follows:

$$\text{work} = \text{large force} \cdot \text{\small small distance} = \text{\small small force} \cdot \text{large distance}.$$

In the absence of friction, the amount of work you do on the piano to get it to the second floor doesn't depend on how you raise it. No matter how you move that piano up to the second floor, its gravitational potential energy will increase by 10,000 J so you'll have to do 10,000 J of work on it. Even if you disassemble the piano into parts, carry them individually up the stairs, and reconstruct the piano in your friend's living room, you will have done 10,000 J of work lifting the piano.

Unless you're an experienced piano tuner, you'll probably be better off sticking with the ramp. It offers an easy method for one person to lift a baby grand piano. The ramp provides mechanical advantage—the process whereby a mechanical device redistributes the amounts of force and distance that go into performing a specific amount of mechanical work. In moving the piano with the help of the ramp, you've performed a task that would normally require a large force over a small distance by supplying a small force over a large distance. You might wonder whether the ramp itself does any work on the piano; it doesn't. Although the ramp exerts a support force on the piano and the piano moves along the ramp's surface, this force and the distance traveled are at right angles to one another. The ramp does no work on the piano.

Mechanical advantage occurs in many situations involving ramps. For example, it appears when you ride a bicycle up a hill. Climbing a gradual hill takes far less uphill force than climbing a steep hill of the same height. Since your pedaling ultimately provides the uphill force, it's much easier to climb the gradual hill than the steep one. Of course, you must travel a longer distance along the road as you climb the gradual hill than you do on the steep hill, so the work you do is the same in either case.

Ramps and inclined planes show up in many devices, where they reduce the forces needed to perform otherwise difficult tasks. They also change the character of certain activities. Skiing wouldn't be very much fun if the only slopes available were horizontal or vertical. By choosing ski slopes of various grades, you can select the net forces that set you in motion. Gentle slopes leave only small net forces and small accelerations; steep slopes produce large net forces and large accelerations.

Finally, our observation about mechanical advantage is this: mechanical advantage allows you to do the same work, but you must make a trade-off—you must choose whether you want a large force or a large distance. The product of the two parts, force times distance, remains the same.

⚏ Fitness Equipment

When you talk about "working out" at a fitness center, you probably don't have physics in mind. Nonetheless, much of what you do during a workout involves

Fig. 1.3.6 This man is raising and lowering a heavy bar. Since he always pushes the bar upward, he does (positive) work on the bar while raising it and negative work on the bar while lowering it. At the same time, the bar always pushes his hands downward, so the bar does negative work on his hands as they rise and (positive) work on his hands as they descend. Overall, the man transfers energy to the bar while raising it and the bar transfers energy to him while he lowers it. Because receiving energy from the bar is difficult and even dangerous, a friend is assisting this man to ensure that he doesn't get injured.

physics work (i.e., mechanical transfers of energy) and understanding that involvement may help you make better use of the fitness equipment.

As you lift and lower free weights (Fig. 1.3.6), you exchange energy with them. You do work on a free weight as you lift it steadily upward—you push it upward as it moves upward—and it does work on you as you lower it steadily downward—it pushes your hands downward and your hands move downward. Because you use chemical potential energy from your food to do the work of lifting the weight and convert the work it does on you while lowering it into thermal energy, the overall effect of a 10-lift set is to convert your food energy into your thermal energy. No wonder you become hot and hungry!

Unfortunately, it's relatively easy to injure yourself while lowering a free weight because your muscles and joints aren't good at handling the influx of energy. That's why many fitness machines are designed not to return energy to you. A common example of this sort of energy absorbing machine is one that offers you a bar to lift and lower. Just as with a free weight, you must push this bar upward as it moves upward and thereby do work on it. But unlike with a free weight, you must pull this bar downward as it moves downward and thereby do work on it again. Instead of returning energy to you on its way down, as a free weight would, this bar extracts still more energy from you.

With every movement of the bar, you transfer energy to the machine. The machine turns this energy into thermal energy, which it uses to warm the room air. Although you become hot while working on this machine, that's merely because your body is inefficient at turning food energy into work. A substantial fraction of your food energy is wasted in your muscles as thermal energy and you're literally "working up a sweat."

If you explore the fitness center, you'll find that many of its machines make you do far more work on them than they do on you. They extract energy from you over the long run by making you push hardest on parts that move away from you and pull hardest on parts that move toward you. For example, when you pedal a stationary bicycle (Fig. 1.3.7), you push downward hardest on each pedal as it moves downward. And if you can pull on the pedals with the help of toe clips or straps, you pull upward hardest on each pedal as it moves upward. The bicycle thus extracts

Fig. 1.3.7 Each time this man pushes a pedal downward, he does work on that pedal and thereby transfers energy to the stationary bicycle. While the bicycle turns most of that energy into thermal energy, which it releases into the room, it also generates electricity to operate its electronics.

Fig. 1.3.8 As this woman pushes the pedals of the stair machine downward, she does work on them and transfers energy to the machine. With her hands off the handrails, she would exert a downward force equal to her weight on each descending pedal and thereby maximize the work she did on it. She is tired, however, after a long workout and is supporting part of her weight with her hands. As a result, she pushes less hard on each descending pedal and does less work on it. She is actually getting less exercise.

Fig. 1.3.9 As this woman draws the handle of the rowing machine toward her, she does work on that handle and transfers energy through the cable to the machine itself. The machine uses that energy to warm the air and blow it around the room.

energy from you and warms the room with it. It may also use some of your energy to power its own electronics, with the aid of an electrical generator.

Another popular type of fitness machine simulates stair climbing. In a typical stair climber or step machine (Fig. 1.3.8), a pedal moves downward when you step on it and then rises gently back to its starting position when you lift your foot. Each time a pedal descends, you do a large amount of work on it; the downward force you exert on the pedal is approximately equal to your weight and the pedal moves a step's height in the direction that force. As the pedal returns upward, however, the two of you barely push on one another and therefore do almost zero work. In effect, you have climbed one step of a strange staircase. Step after step, you do work on the machine and it warms the room.

After a few minutes on a stair climbing machine, it's tempting to start leaning on the handrails for support. Support your weight in this manner feels good because doing so reduces the forces you exert on the pedals and therefore the work you do on them as they descend. The more you lean the less exercise you get, which is exactly why you do it. In fact, if you look around the fitness center, you'll see fatigued people clinging to the non-moving parts of their machines in order to do as little work as they can on the moving parts. Most are blissfully unaware that they are unconsciously avoiding or minimizing their exercise.

On a rowing machine (Fig. 1.3.9), you do work on a handle as you pull it toward you and it moves toward you. The harder you pull on the handle and farther you pull it, the more work you do on it with each stroke. Much of that work comes from your legs as you straighten them out; they push your body backward as your body moves backward in the seat. You similarly do work on the pedals and handles of an elliptical machine as you push each in the direction it's moving.

Finally, you do work on the belt of a treadmill (Fig. 1.3.10) each time your foot pushes it backward and it moves backward. The harder you push and the longer your stride, the more work you do with each step. You can increase your exercise by making yourself push back harder on the belt. For example, by choosing a steeper uphill setting you make it necessary to push back harder with your feet to remain in place on the treadmill. But you can easily avoid some exercise by pulling yourself forward with the stationary handgrips. The harder you pull yourself forward with your hands, the more gently you push backward on the belt with your feet and the less work you do with each stride.

In general, the stationary handgrips on exercise machines are meant for balance and safety, not for serious forces. As a rule of thumb, if part of an exercise machine doesn't move in the direction you push it or pull it, it's probably not part of the exercise and clinging to it may well reduce your overall workout.

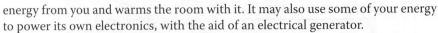 Downhill Sports

If you're in the moving business or building a pyramid in your backyard, you probably use ramps to lift heavy objects. But ramps play a more exciting role in sports such as skiing, sledding, and snowboarding. The physics of these downhill sports is now well within our reach, so let's take a look at them.

Each of these sports uses a slope's downhill force to accelerate you downhill. As we've seen, the downhill force you experience while on a slope is the net result of two individual forces: your downward weight and a support force that the slope exerts on you perpendicular to its surface. Since the slope isn't horizontal, the support force it exerts on you isn't vertical and so the two individual forces don't cancel. Instead, you experience a moderate net force that points directly downhill.

Fig. 1.3.10 These women do work on the belts of the treadmills whenever they push backward on a belt as it moves backward. Since the treadmill on the right has the steeper slope, the woman on the right pushes her belt backward harder and does more work with every stride than the woman on the left.

If you were standing on a slope wearing rubber-soled shoes, the slope's surface would push you uphill with a force of static friction. This third force would oppose the downhill force and prevent you from accelerating. But when friction is eliminated, as it is in skiing, sledding, snowboarding, tobogganing, tubing, mountain biking, luging, and bobsledding, the downhill force is unopposed and you accelerate downhill (Fig. 1.3.11).

How quickly you accelerate depends on the steepness of slope. At the two extremes of steepness, a horizontal plateau and a vertical cliff, your acceleration is easy to predict: you would have zero acceleration on the plateau and the full acceleration due to gravity on the cliff. But at more reasonable steepness, predicting your acceleration requires some thought.

Fortunately, our observations about work, distance, and force along a ramp provide an answer. As we saw with ramps, the downhill force you experience on a slope depends on the ratio of your altitude change to the distance you travel along its surface. Specifically, the downhill force acting on you is equal in amount to your weight times the altitude change divided by the distance along the surface. For example, on a slope that descends 1 meter for every 10 meters travelled along its surface, you would experience a downhill force equal to $1/10^{th}$ of your weight and you would consequently accelerate downhill at $1/10^{th}$ the acceleration due to gravity.

OK, so we've rediscovered the obvious: the steeper the slope, the quicker you pick up speed. But now we know exactly how quickly you pick up speed and we can also start to understand the strategies you use while descending these slopes. When you head straight down a slope, you are experiencing its maximum steepness and you accelerate as quickly as the slope will allow. In effect, you are losing altitude as efficiently as possible and are maximizing the downhill force you experience. On a steep hill, heading directly down the slope is either exhilarating or terrifying, depending on your personality. You accelerate like crazy.

If rapid accelerations don't appeal to you, you do better to cut across a steep slope instead of heading straight down it. That technique stretches out the descent and effectively makes the slope less steep. Just because a hill descends 1 m for every 3 m you travel along its surface, doesn't mean that you must accelerate at a heart stopping $1/3^{rd}$ the acceleration due to gravity. If you cut sharply across the face

Fig. 1.3.11 Although you coast while skating on level pavement, that's not the case when you're on a hill. This boy's downward weight and the support force that the road exerts on him perpendicular to its surface don't balance one another. Instead, he experiences an overall downhill force and accelerates downhill as a result.

Fig. 1.3.12 When the snow is slippery enough, you can sled down a slope on almost anything. Although this woman has reached terminal velocity, with the downhill force now balanced by uphill resistance forces, she continues to coast down the hill in a laundry basket.

of that hill, you can easily reduce its effective slope so that it descends 1 m for every 30 m you travel along its surface. You then accelerate at a mild 1/30th the acceleration due to gravity. Thus while a racer normally takes the steepest path that a course will allow, maximizing his downhill accelerations in hopes of reaching the finish line first, people with more time than temerity usually take meandering, leisurely paths that involve less dramatic accelerations.

But controlling acceleration isn't the same as controlling speed. With only the downhill force acting on you as you descend a hill, you'll arrive at the bottom with the same final speed regardless of path. That's because the steeper your path, the quicker you'll accelerate but the shorter the time over which you'll experience that acceleration. Your fixed final speed is also a consequence of the conservation of energy, a topic that we'll explore more fully in the next chapter.

The downhill force, however, is not the only force that acts on you as you descend. You also experience unavoidable resistance forces such as air resistance and friction that push you opposite your velocity. Since some of these resistance forces grow stronger as your speed increases, you eventually reach a maximum downhill speed or terminal velocity. At that terminal velocity, the uphill resistance forces balance the downhill force and you no longer accelerate (Fig. 1.3.12). You can increase this terminal velocity by choosing a steeper path or by reducing the resistance forces. If maximizing your speed is important, you should stay in contact with the slope whenever possible. Downhill ski racers know that time in the air after a jump is time when the only horizontal force they experience is the slowing force of air resistance.

Although we'll examine friction in Chapter 2 and air resistance in Chapter 6, you probably know already how to reduce them. Reducing air resistance involves making yourself smaller and more streamlined. That's why skiers and snowboarders crouch into tuck positions, bicyclists bend low over their cycles, tobogganers and bobsledders duck behind the fronts of their sleds, and lugers lie flat on their backs. Reducing friction usually requires carefully shaped surfaces or lubrication. That's why skiers carefully wax their skis, bobsledders use sharp runners on ice tracks, and bicyclists use carefully greased bearings in their wheels. And there is an uphill resistance force produced by crushing the surface on which you ride, so avoiding soft snow and spreading out your weight on long skis, a large toboggan, or a big, properly inflated tube let you barrel down the hill at a furious pace.

I have one last downhill sport to discuss: surfing. Like sloping ground, the sloping water on the face of a wave pushes you perpendicular its surface and you thus experience a downhill force. As usual, you accelerate downhill on the wave at a rate that increases with its steepness but can reduce your acceleration by cutting across the wave at an angle. And because you experience resistance forces that grow stronger as you move faster through the water, you soon reach a terminal velocity.

What most distinguishes a wave from a hill is that the wave rises upward beneath you as you surf on its front face. As long as you don't outrun the wave or it doesn't crumble near the shore, you can slide down its front face without actually losing any altitude! If you aim directly down the wave, however, you'll accelerate quickly and your high terminal velocity will soon take you ahead of the wave. To stay on the wave's face, you must cut across it so that your terminal velocity keeps you moving shoreward at the same speed as the wave itself. You can then stay on the wave's face until it crumbles. Then it'll be time to head out to find another ride.

CHAPTER 2
MORE THINGS THAT MOVE

One consequence of spending so many years in the classroom and laboratory is that I no longer remember what it was like to encounter physics for the first time. Though I try to relive that experience each time I start out with a new crop of students, it's just not the same. The feelings of mystery and disorientation that I had in my youth have long since given way to familiarity and comfort.

The good news is that I've learned a few things over the years and mostly know what I don't know. That means that I'm less likely to tell you nonsense or to make things up as I go along. The bad news is that, like a veteran tour guide or actor, I'm a little set in my ways. Oh, it's not as if I'll ever abandon you in an alley or "phone in" a performance. But it does mean that I'll occasionally indulge myself with a visit to a favorite shop or ask you to listen to my interpretation of Hamlet.

It also means that I'll sometimes present things in the wrong order or skip them altogether. It would be better if we were having a conversation, but since I haven't heard a peep out of you since the start of Chapter 1, I'll just have to forge ahead and try to anticipate your questions and concerns as we go along. And while I may not be surprised by what we find around every corner, I usually learn something new each time I show someone a familiar place. I'll try to pass along those insights and hope that you'll get as much out this book as I get from writing it.

In the previous chapter, we saw how things move from place to place and encountered energy, an important conserved quantity. But motion doesn't always involve a change of position, and energy isn't nature's only conserved quantity. In this chapter, we'll take a look at a second type of motion—rotation—and at two other conserved quantities—momentum and angular momentum. Spinning objects are quite common, and we'll do well to explore their laws of motion before proceeding much further. With those additional concepts under our belts, we'll be ready to explore the physics behind a broad assortment of mechanical objects.

Chapter Itinerary

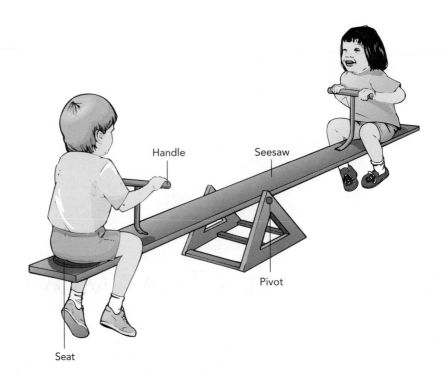

Handle   Seesaw

Pivot

Seat

Section 2.1 **Seesaws**

The ramp that we examined in Section 1.3 isn't the only tool that provides mechanical advantage. In this section, we'll look at another such device: the type of lever known as a seesaw. As we discuss seesaws, we'll revisit many of the laws of motion that we encountered in the previous chapter. However, we'll see these laws in a new context: rotational motion.

The Seesaw

Any child who has played on a seesaw with friends of different sizes knows that the toy works best for two children of roughly the same weight (Fig. 2.1.1*a*). Evenly matched riders balance each other, and this balance allows them to rock back and forth easily. In contrast, when a light child tries to play seesaw with a heavy child, the heavy child's side of the seesaw drops rapidly and hits the ground with a thud (Fig. 2.1.1*b*). The light child is at risk of becoming an astronaut.

There are several solutions to the heavy child/light child problem. Of course, two light children could try to balance one heavy child. But most children eventually figure out that if the heavy child sits closer to the seesaw's pivot, the seesaw will balance (Fig. 2.1.1*c*). The children can then make the seesaw tip back and forth easily, just as it does when two evenly matched children ride at its ends. This is a pretty useful trick, and we'll explore it later in this section. First, though, we'll need to look carefully at the nature of rotational motion.

For simplicity, let's ignore the mass and weight of the seesaw itself. There are then only three forces acting on the seesaw shown in Fig. 2.1.1: two downward forces (the weights of the two children) and one upward force (the support force of the central pivot). Seeing those three forces, we may immediately think about net forces and begin to look for some overall acceleration of this toy and its riders. We know, however, that the seesaw remains where it is in the playground and isn't likely to head off for Kalamazoo or the center of the earth anytime soon. Because the seesaw's fixed pivot always provides just enough upward or sideways force to keep the seesaw from accelerating as a whole, the seesaw always experiences zero net force and never leaves the playground. Overall movement of an object from one place to another is called translational motion. While the seesaw never experiences translational motion, it can turn around the pivot and thus experiences a different kind of motion. Motion around a fixed point is called rotational motion. The hands of a clock experience rotational motion as they go around in a circle.

Rotational motion is what makes a seesaw interesting. The whole point of a seesaw is that it can rotate so that one child rises and the other descends. (You may not think of going up and down as rotating, but if the ground weren't there to get in its way, the seesaw would be able to rotate in a big circle about its central pivot.) What causes the seesaw to rotate, and what observations can we make about the process of rotation?

To answer those questions, we'll need to examine several new physical quantities associated with rotation and explore the laws of rotational motion that relate them to one another. We'll do these things both by studying the workings of seesaws and other rotating objects and by looking for analogies between translational motion and rotational motion.

Imagine holding onto the seesaw in Fig. 2.1.1*a* to keep it level for a moment while the child on the left climbs off the seesaw. Now imagine letting go of the seesaw. As soon as you let go, the seesaw will begin to rotate, and the child on the right will descend toward the ground. The seesaw's motion will be fairly slow at first, but it will move more and more quickly until that child strikes the ground with a teeth-rattling thump.

If we focus only on the rotation itself, we might describe the motion of the seesaw in the following way:

> "The seesaw starts out not rotating at all. When we release the seesaw, it begins to rotate clockwise. The seesaw's rate of rotation increases steadily in the clockwise direction until the moment the seesaw strikes the ground."

This description sounds a lot like the description of a falling ball released from rest:

> "The ball starts out not moving at all. When we release the ball, it begins to move downward. The ball's rate of translation increases steadily in the downward direction until the moment the ball strikes the ground."

The statement about the seesaw involves rotational motion, while the statement about the ball involves translational motion. Their similarity isn't a coincidence; the concepts and laws of rotational motion have many analogies in the concepts and laws of translational motion. The familiarity that we've acquired with translational motion will help us examine rotational motion.

The Motion of a Dangling Seesaw

In the previous chapter we looked at the concept of translational inertia, which holds that a body in motion tends to stay in motion and a body at rest tends to stay at rest. This concept led us to Newton's first law of translational motion. Inserting the word "translational" here is a useful revision because we're about to encounter the corresponding concepts associated with rotational motion.

We'll begin that encounter by observing a seesaw that's free of outside rotational influences. We'll then examine how the seesaw responds to outside influences such as its pivot or a handful of young riders. Because of the similarities between rotational and translational motions, this section will closely parallel our earlier examination of skating and falling balls.

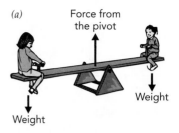

(*a*) Force from the pivot / Weight / Weight

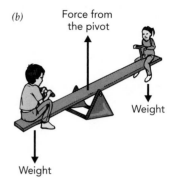

(*b*) Force from the pivot / Weight / Weight

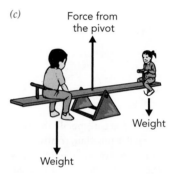

(*c*) Force from the pivot / Weight / Weight

Fig. 2.1.1 (*a*) When two children of equal weight sit at opposite ends of a seesaw, it balances. (*b*) When their weights are not equal, the heavy child descends. (*c*) If the heavy child moves closer to the pivot, the seesaw can balance.

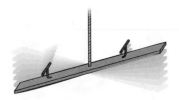

Fig. 2.1.2 A seesaw that's dangling from a rope at its middle. Since nothing twists it, the seesaw rotates steadily about a fixed line in space.

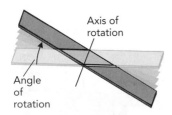

Fig. 2.1.3 You can specify this seesaw's angular position, relative to its horizontal reference orientation, as the axis about which it was rotated to reach its new orientation and the angle through which it was rotated.

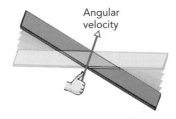

Fig. 2.1.4 This seesaw is spinning about the rotation axis shown. The direction of the seesaw's angular velocity is defined by the right-hand rule.

Let's suppose that your local playground is installing a new seesaw and that this seesaw is presently dangling from a rope. The rope is attached to the middle of the seesaw in such a way that it supports the seesaw's weight but exerts no other influences on the seesaw. Most importantly, let's suppose that the dangling seesaw can spin and pivot with complete freedom—nothing pushes on it or twists it—and that the rope doesn't get in its way as it moves (Fig. 2.1.2). This dangling seesaw is free to turn in any direction, even completely upside down. You, the observer, are standing motionless near the seesaw. When you look over at the seesaw, what does it do?

If the seesaw is stationary, then it will remain stationary. However, if it's rotating, it will continue rotating at a steady pace, about a fixed line in space. What keeps the seesaw rotating? Its rotational inertia. A body that's rotating tends to remain rotating; a body that's not rotating tends to remain not rotating. That's how our universe works.

To describe the seesaw's rotational inertia and rotational motion more accurately, we'll need to identify several physical quantities associated with rotational motion. The first is the seesaw's orientation. At any particular moment, the seesaw is oriented in a certain way—that is, it has an angular position. Angular position describes the seesaw's orientation relative to some reference orientation; it can be specified by determining how far the seesaw has rotated away from its reference orientation and the axis or line about which that rotation has occurred. The seesaw's angular position is a vector quantity of relatively minor importance, pointing along the rotation axis with a magnitude equal to the rotation angle (Fig. 2.1.3).

The SI unit of angular position is the radian, the natural unit for angles. It's a natural unit because it follows directly from geometry, not from an arbitrary human choice or convention the way most units do. Geometry tells us that a circle of radius 1 has a circumference of 2π. By letting arc lengths around that circle's circumference specify angles, we are using radians. For example, there are 2π radians (or 360°) in a full circle and $\pi/2$ radians (or 90°) in a right angle. Since the radian is a natural unit, it is often omitted from calculations and derived units.

If the seesaw is rotating, then its angular position is changing; in other words, it has an angular velocity. Angular velocity is our first important vector quantity of rotational motion and measures how quickly the seesaw's angular position changes; it consists of the angular speed at which the seesaw is rotating and the axis about which that rotation proceeds. The seesaw's angular speed is its change in angle divided by the time required for that change:

$$\text{angular speed} = \frac{\text{change in angle}}{\text{time}}.$$

The SI unit of angular velocity is the radian-per-second (abbreviated 1/s).

The seesaw's axis of rotation is the line in space about which the seesaw is rotating. But just knowing that line isn't quite enough: is the seesaw rotating clockwise or counterclockwise?

To resolve this ambiguity, we take advantage of the fact that any line has two directions to it. Once we have identified the line about which the seesaw is rotating, we can look down that line at the seesaw from either direction. From one direction, the seesaw appears to be rotating clockwise; from the other direction, counterclockwise. By convention, we choose to view the seesaw from the direction in which it appears to be rotating clockwise and say that the seesaw's rotation axis points away from our eye toward the seesaw. This convention is called the right-hand rule because if the fingers of your right hand are curling around the axis in the way the seesaw is rotating, then your thumb is pointing along the seesaw's rotation axis (Fig. 2.1.4). (While experimenting with the *right* hand rule, make sure to hold this book in your *left* hand ❶.)

Remembering this convention isn't as important as understanding why you must specify the direction about which rotation occurs when describing a rotating object's angular velocity. Just as translational velocity consists of a translational speed and a direction in which the translational motion occurs, so angular velocity consists of a rotational speed and a direction about which the rotational motion occurs.

We're now prepared to describe the rotational motion of the dangling seesaw. Because of its freedom from outside influences and its rotational inertia, its angular velocity is constant. The dangling seesaw just keeps on turning and turning, always at the same angular speed, always about the same axis of rotation.

As you might suspect, this observation isn't unique to seesaws. It is Newton's first law of rotational motion, which states that a rigid object that is not wobbling and is not subject to any outside influences rotates at a constant angular velocity, turning equal amounts in equal times about a fixed axis of rotation. The outside influences referred to in this law are called torques—a technical term for twists and spins. When you twist off the lid of a jar or spin a top with your fingers, you're exerting a torque.

This law excludes objects that wobble or can change shape as they rotate because those objects have more complicated motions. They are covered instead by a more general principle—the conservation of angular momentum—that we'll meet up with in Section 2.3.

Newton's First Law of Rotational Motion

A rigid object that is not wobbling and is not subject to any outside torques rotates at a constant angular velocity, turning equal amounts in equal times about a fixed axis of rotation.

The Seesaw's Center of Mass

Even without visiting the playground, you can find many objects that are nearly free from torques: a baton thrown overhead by a baton twirler, for example, or a juggler's club whirling through the air between two clowns. These motions, however, are complicated because those freely moving objects rotate and translate at the same time. The spinning baton travels up and down, the turning club arcs through the air, and, if the rope breaks, our seesaw will fall as it spins. How can we distinguish their translational motions from their rotational motions?

Once again, we can make use of a wonderful simplification of physics. There's a special point in or near a free object about which all its mass is evenly balanced and about which it naturally spins—its center of mass. The axis of rotation passes through this point so that, as the free object rotates, the center of mass doesn't move unless the object has an overall translational velocity. The center of mass of a typical ball is at its geometrical center, while the center of mass of a less symmetrical object depends on how the mass of that object is distributed. You can begin to find a small object's center of mass by spinning it on a smooth table and looking for the fixed point about which it spins (Fig. 2.1.5).

Center of mass allows us to separate an object's translational motion from its rotational motion. As a juggler's club arcs through space, its center of mass follows the simple path we discussed in Section 1.2 on falling balls (Fig. 2.1.6). At the same time, the club's rotational motion about its center of mass is that of an object that's free of outside torques: if it's not wobbling, it rotates with a constant angular velocity.

Many of the objects we'll examine in this book translate and rotate simultaneously, and it's worth remembering that we can often separate these two motions by paying attention to an object's center of mass. For example, the workers install-

❶ Like most physics instructors, I always have a few students who get the right-hand rule backwards. The cause for their confusion escaped me until someone at a workshop pointed out that most people take notes with their right hands. To avoid putting down their pens, some students use their *left* hands to visualize the *right*-hand rule and never get over it. If their shoes feel uncomfortable half the time, perhaps it's for a similar reason.

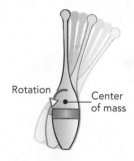

Fig. 2.1.5 This club spins about its center of mass, which remains stationary.

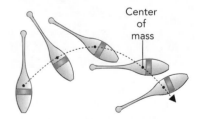

Fig. 2.1.6 A juggler's club that is traveling through space rotates about its center of mass as its center of mass travels in the simple arc associated with a falling object.

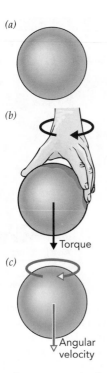

(a)

(b)

Torque

(c)

Angular
velocity

Fig. 2.1.7 If you start with a ball that's not spinning (*a*) and twist it with a torque (*b*), the ball will acquire an angular velocity (*c*) that points in the same direction as that torque.

Fig. 2.1.8 Before capturing and repairing INTELSAT on May 16, 1992, three astronauts used their hands to stop it from spinning. Because the satellite had a large rotational mass, its angular velocity decreased slowly even though they exerted substantial torques on it.

ing our seesaw will locate its pivot strategically at or very near the seesaw's center of mass. As a result, the pivot will prevent any translational motion of the seesaw while permitting nearly free rotational motion of the seesaw about its center of mass, at least about one axis.

How the Seesaw Responds to Torques

The workers are eating lunch, so the seesaw is still hanging from the rope. Why can't this dangling seesaw change its rotational speed or axis of rotation? Because it has rotational mass ❷. Rotational mass is the measure of an object's *rotational* inertia, its resistance to changes in its *angular* velocity. An object's rotational mass depends both on its ordinary mass and on how that mass is distributed within the object. The SI unit of rotational mass is the kilogram-meter^2 (abbreviated kg·m^2). Because the seesaw has rotational mass, its angular velocity will change only if something twists it or spins it. In other words, it must experience a torque.

Torque—our second important vector quantity of rotational motion—has both a magnitude and a direction. The more torque you exert on the seesaw, the more rapidly its angular velocity changes. Depending on the direction of the torque, you can make the seesaw turn more rapidly or less rapidly or even rotate about a different axis. But how do you determine the direction of a particular torque? One way is to imagine exerting this torque on a stationary ball floating in water (Fig. 2.1.7*a,b*). The ball will begin to rotate, acquiring a nonzero angular velocity (Fig. 2.1.7*c*). The direction of this angular velocity is that of the torque. The SI unit of torque is the newton-meter (abbreviated N·m).

The larger an object's rotational mass, the more slowly its angular velocity changes in response to a specific torque (Fig. 2.1.8). You can easily spin a basketball with the tips of your fingers, but it's much harder to spin a bowling ball. The bowling ball's larger rotational mass comes about primarily because it has a greater ordinary mass than the basketball.

But rotational mass also depends on an object's shape, particularly on how far each portion of its ordinary mass is from the axis of rotation. The farther a portion of mass is from that axis, the more rapidly it must accelerate as the entire object undergoes angular acceleration and the more leverage it has with which to oppose that acceleration. We'll examine levers shortly, but the consequence of these two effects of distance from the rotation axis is that each portion of mass contributes to the object's rotational mass in proportion to the square of its distance from that axis. That's why an object that has most of its mass located near the axis of rotation will have a much smaller rotational mass than an object of the same mass that has most of its mass located far from that axis. Thus a ball of pizza dough has a smaller rotational mass than the finished pizza. And the bigger the pizza gets, the harder it is to start or stop spinning.

Because an object's rotational mass depends on how far its mass is from the axis of rotation, changes in the axis of rotation are likely to change its rotational mass. For example, less torque is required to spin a tennis racket about its handle (Fig. 2.1.9*a*) than to flip the racket head-over-handle (Fig. 2.1.9*b*). When you spin the tennis racket about its handle, the axis of rotation runs right through the handle so that most of the racket's mass is fairly close to the axis and the rotational mass is small. When you flip the tennis racket head-over-handle, the axis of rotation runs across the handle so that both the head and the handle are far away from the axis and the rotational mass is large. The tennis racket's rotational mass becomes even larger when you hold it in your hand and make it rotate about your shoulder rather than its center of mass (Fig. 2.1.9*c*).

When something exerts a torque on the dangling seesaw, its angular velocity changes; in other words, it undergoes angular acceleration, our third important vector quantity of rotational motion. Angular acceleration measures how quickly the seesaw's *angular* velocity changes. It's analogous to acceleration, which measures how quickly an object's *translational* velocity changes. Just as with acceleration, angular acceleration involves both a magnitude and a direction. An object undergoes angular acceleration when its angular speed increases or decreases or when its angular velocity changes directions. The SI unit of angular acceleration is the radian-per-second^2 (abbreviated $1/s^2$).

There is a simple relationship between the torque exerted on the seesaw, its rotational mass, and its angular acceleration. The seesaw's angular acceleration is equal to the torque exerted on it divided by its rotational mass or

$$\text{angular acceleration} = \frac{\text{torque}}{\text{rotational mass}}. \qquad (2.1.1)$$

The seesaw's angular acceleration, as we've seen, is in the same direction as the torque exerted on it.

This relationship is Newton's second law of rotational motion. Structuring the relationship this way distinguishes the causes (torque and rotational mass) from their effect (angular acceleration). Nonetheless, it has become customary to rearrange the relationship to eliminate the division. In its traditional form, the relationship becomes

$$\text{torque} = \text{rotational mass} \cdot \text{angular acceleration}. \qquad (2.1.2)$$

It's like Newton's second law of translational motion ($\text{force} = \text{mass} \cdot \text{acceleration}$), except that torque has replaced force, rotational mass has replaced mass, and angular acceleration has replaced acceleration. This new law doesn't apply to wobbling objects, however, because they're being affected by more than one rotational mass simultaneously (see the discussion of tennis rackets above) and follow a much more complicated law.

Newton's Second Law of Rotational Motion

The torque exerted on an object that is not wobbling is equal to the product of that object's rotational mass times its angular acceleration. The angular acceleration points in the same direction as the torque.

Because it's an equation, the two sides of Eq. 2.1.1 are equal. Any change in the torque you exert on the seesaw must be accompanied by a proportional change in its angular acceleration. As a result, the harder you twist or spin the seesaw, the more rapidly its angular velocity changes.

We can also compare the effects of a specific torque on two different rotational masses. Equation 2.1.1 indicates that a decrease in rotational mass must be accompanied by a corresponding increase in angular acceleration. If we replace the playground seesaw with one from a dollhouse, the rotational mass will decrease and the angular acceleration will increase. The angular velocity of a doll's seesaw thus changes more rapidly than the angular velocity of a playground seesaw when the two experience identical torques.

In summary:

1. Your angular position indicates exactly how you're oriented.
2. Your angular velocity measures how quickly your angular position changes.
3. Your angular acceleration measures how quickly your angular velocity changes.
4. In order for you to undergo angular acceleration, something must exert a torque on you.
5. The more rotational mass you have, the less angular acceleration you experience for a given torque.

❷ For clarity and simplicity, this book refers to the measure of an object's rotational inertia as "rotational mass." However, this quantity is known more formally as "moment of inertia."

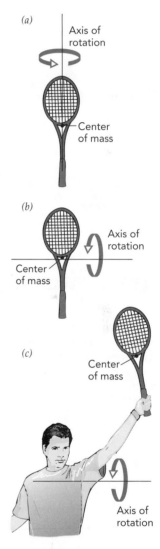

Fig. 2.1.9 A tennis racket's rotational mass depends on the axis about which it rotates. Its rotational mass is small (*a*) when it rotates about its handle and large (*b*) when it rotates head-over-handle. (*c*) If you make it rotate about your shoulder, its rotational mass becomes even larger.

Quantity	SI unit	English unit	SI → English	English → SI
Angular Position	radian (1)	radian (1)		
Angular Velocity	radian-per-second (1/s)	radian-per-second (1/s)		
Angular Acceleration	radian-per-second^2 (1/s^2)	radian-per-second^2 (1/s^2)		
Torque	newton-meter (N·m)	foot-pound (ft·lbf)	1 N·m = 0.73757 ft·lbf	1 ft·lbf = 1.3558 N·m
Rotational Mass	kilogram-meter^2 (kg·m^2)	pound-foot^2 (lbm·ft^2)	1 kg·m^2 = 23.730 lbm·ft^2	1 lbm·ft^2 = 0.042140 kg·m^2

This summary of the physical quantities of rotational motion is analogous to the one for translational motion on p. 6. In fact, I have structured this entire introduction to rotation so as to highlight the similarities between translational and rotational motions. Emerson may have felt that "a foolish consistency is the hobgoblin of little minds," but I'll risk his disapproval to clarify my point.

Forces, Torques, and Seesaws

The workers have finally installed the seesaw. They have mounted it on a pivot that passes directly through its center of mass, so that the pivot coincides with a natural rotation axis of the seesaw. The pivot thus supports the seesaw's weight while leaving it free to obey Newton's first law of rotational motion. That is, the unoccupied seesaw rotates with constant angular velocity about its pivot.

The unoccupied seesaw is balanced, meaning that it has zero torque on it. As a result, it experiences no angular acceleration. You might think that a balanced seesaw always remains horizontal, but that isn't necessarily so. What is certain is that its angular velocity is constant. If it's rotating, then it continues to rotate steadily about the pivot; if it's stationary, then it remains stationary at its current tilt, whether horizontal or not.

To change the seesaw's angular velocity, you must exert a torque on it. But how do you actually exert a torque? You put your hand on one end of the seesaw and push that end down (Fig. 2.1.10a). The seesaw begins to rotate, and your end soon hits the ground. You have exerted a torque on the seesaw.

But you started by exerting a *force* on the seesaw—you pushed on it—so forces and torques must be related somehow. Sure enough, a force can produce a torque and a torque can produce a force. To help us explore that relationship, let's think of all the ways *not* to exert a torque on a seesaw.

What happens if you push on the seesaw right where the pivot passes through it (Fig. 2.1.10b)? Nothing—no angular acceleration. If you move a little away from the pivot, you can get the seesaw rotating but you have to push hard. You do much better to push on the end of the seesaw, where even a small force can start the seesaw rotating. The distance from the pivot to the place where you push on the seesaw is called the lever arm; in general, the longer the lever arm, the less force it takes to cause a particular angular acceleration. Our first observation about producing a torque with a force is this: you obtain more torque by exerting that force farther from the pivot or axis of rotation. In other words, the torque is proportional to the lever arm.

Another ineffective way to start the seesaw rotating is to push its end directly toward or away from the pivot (Fig. 2.1.10c). A force exerted toward or away from the axis of rotation doesn't produce any torque about that axis. At least a component of the force you exert must be perpendicular to the lever arm, which is actually a vector pointing along the seesaw's surface from the pivot to the place where you push on the seesaw. Our second observation about producing a torque with a force is that you must exert at least a component of that force perpendicular to the lever arm. Only that component of force contributes to the torque. To produce the most torque, push perpendicular to the lever arm.

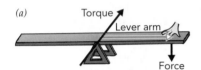
(a)

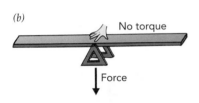
(b)

(c)

Fig. 2.1.10 (*a*) When you push down on the seesaw, far from its pivot, you exert a torque on it. But when you (*b*) exert your force at the pivot or (*c*) exert your force toward the pivot, you exert no torque.

We can summarize these two observations as follows: the torque produced by a force is equal to the product of the lever arm times that force, where we include only the component of the force that is perpendicular to the lever arm:

$$\text{torque} = \text{lever arm} \cdot \text{force perpendicular to lever arm.} \qquad (2.1.3)$$

The directions of the force and lever arm also determine the direction of the torque. The three directions follow another right-hand rule (Fig. 2.1.11). If you point your right index finger in the direction of the lever arm and your bent middle figure in the direction of the force, then your thumb will point in the direction of the torque. Thus in Fig. 2.1.11*a*, the lever arm points to the right, the force points downward, and the resulting torque points into the page so that the seesaw undergoes angular acceleration in the clockwise direction. In Fig. 2.1.11*b*, the lever arm has reversed directions and so has the torque.

What happens if you and a friend push down simultaneously on both seats at once? Then you produce two torques on the seesaw about its pivot, and these torques have opposite directions. The seesaw responds to the net torque it experiences, the sum of all the individual torques on the seesaw. Since your two torques oppose one another, they at least partially cancel. If you carefully exert identical downward forces at identical distances from the pivot, the magnitudes of the two torques will be exactly equal and the torques will sum to zero. The seesaw will experience zero net torque, and it will be balanced.

This observation explains the need for careful seating of the children on the seesaw. Each child's weight exerts a downward force on the seesaw and by properly distributing those weights on both sides of the pivot, the torques that they produce can be made to sum to zero. With zero net torque about its pivot, the seesaw balances.

In fact, the weight of the seesaw itself is balanced in this manner. While each end's weight exerts a torque on the seesaw, those two torques sum to zero and have no overall effect on the seesaw's rotation.

Net Torque and Mechanical Advantage

The amount of torque that a child's weight produces on a seesaw depends on that child's distance from the pivot. If the child sits on the pivot, the lever arm is zero and she produces no torque; but if she sits at the extreme end of the seesaw, the lever arm is long and she produces a large torque. She can adjust her torque by moving along the seesaw because the seesaw provides her with mechanical advantage. As we saw in Section 1.3, mechanical advantage appears when a device redistributes the amounts of force and distance used to produce a particular amount of work. The seesaw allows a small force exerted at its end to do the same work as a large force exerted near its pivot.

To see how mechanical advantage appears in a seesaw, think of what happens when two children sit on its ends. If two 5 year olds, each weighing 200 N (45 lbf), sit at opposite ends of the seesaw, 2 m (6.6 ft) from the pivot (Fig. 2.1.12*a*), each one exerts a torque of 400 N·m (300 ft·lbf) on the seesaw about its pivot ($200\,\text{N} \cdot 2\,\text{m} = 400\,\text{N} \cdot \text{m}$). But because these torques are oppositely directed, they add to zero. The net torque on the seesaw is zero and the seesaw balances.

If you replace one of the 5 year olds with a 400-N (90-lbf) teenager, the teenager must sit at half the distance from the pivot (Fig. 2.1.12*b*). Doubling the force while halving the lever arm leaves the torque unchanged at 400 N·m. The two children again produce equal but oppositely directed torques about the pivot, so that the net torque on the seesaw is zero and the seesaw balances. This effect explains how a small child at the end of the seesaw can balance a large child nearer the pivot.

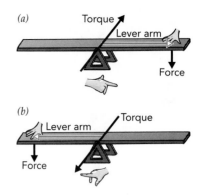

Fig. 2.1.11 The torque on a seesaw obeys a right-hand rule: if your index finger points along the lever arm and your middle finger points along the force, your thumb points along the torque.

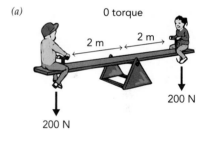

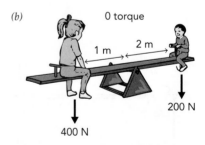

Fig. 2.1.12 (*a*) When two children of equal weight sit at equal distances from the pivot, they produce equal but oppositely directed torques about the pivot. These torques sum to zero so that the seesaw experiences zero net torque. (*b*) When one child weighs twice as much as the other, the seesaw balances when the heavy child sits at half the distance from the pivot.

With the seesaw balanced, nothing is accelerating. Each child experiences zero net force; the seesaw pushes up on the small child with a force of 200 N (45 lbf) and on the large child with a force of 400 N (90 lbf). Ultimately, it's the small child's 200-N (45-lbf) weight that leads to the 400-N (90-lbf) supporting force experienced by the large child. The seesaw's mechanical advantage allows the small child to support and lift the much heavier child. But the small child moves twice as far as the large child, as required to ensure that the work the smaller child does on the seesaw is the same as the work the seesaw does on the larger child. This effect, where a small force exerted over a long distance on one part of a rotating system produces a large force exerted over a small distance elsewhere in that system, is an example of the mechanical advantage associated with levers.

Riding a Seesaw

Each seesaw in Fig. 2.1.12 is balanced, meaning that the net torque on it is zero. Although each child's weight exerts a torque on the seesaw, the two torques sum to zero. Since the seesaw experiences zero net torque and no angular acceleration, it continues rotating at constant angular velocity.

However, as it presently stands, a balanced seesaw should either remain motionless forever or else rotate endlessly in the same direction. Children are unlikely to wait motionless forever, and endless rotation implies that the children will be upside-down periodically. We've obviously neglected a few details.

What do the children do when the seesaw is motionless? To start the seesaw moving, they have to unbalance the seesaw. One of the children must change the torque she exerts on it. She can either change the downward force she exerts on the seesaw or change the distance between that force and the pivot. Actually, children change both the force and the lever arm frequently without even thinking about it. If a child leans inward, toward the pivot, the lever arm decreases and the child exerts less torque on the seesaw; as a result, the seesaw begins to rotate and the child rises. If the child pushes on the ground with his feet, the ground exerts an upward force on him, reducing the force and torque he exerts on the seesaw; again, the seesaw begins to rotate and the child rises.

So either by leaning or by pushing on the ground, the children can start an initially motionless, balanced seesaw rotating. Similarly, when one end of the seesaw hits the ground, the ground exerts a strong, upward support force on it. Located far from the pivot and almost perpendicular to that lever arm, this force produces a huge torque on the seesaw and abruptly stops it from rotating. The angular acceleration is so uncomfortably large that most children push on the ground with their feet to cushion the impact. The child on the ground can continue to push down with her feet until the seesaw rotates in the opposite direction. That child begins to rise and the other child descends. When the other end of the seesaw reaches the ground, this cycle begins again.

As they play on a seesaw, the two children frequently change the torques they exert on it so that it tips back and forth. During the moments when a child is pushing on the ground or leaning inward or outward to get a stationary seesaw moving, the seesaw is no longer balanced. A balanced seesaw has zero angular acceleration; it's only by unbalancing the seesaw that the children can change the angular velocity of the seesaw.

⚐ Levers, Tools, and Sports

A light child can use a seesaw to lift a much heavier child, albeit a reduced distance. In Fig. 2.1.12b, the light child is half the weight of the heavy child, but we could imagine more extreme differences. If the heavy child were very close to the

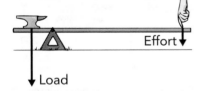

Fig. 2.1.13 This lever pivots about its middle so that it can balance when you push downward on the right end (the effort) and the anvil pushes downward on the left end (the load). Because the load is exerted closer to the pivot than is the effort, the lever balances when the effort is much smaller than the load.

pivot, you would be able to lift him by exerting a very small force on the opposite end of the seesaw.

That's how a pry-bar or crowbar works. A pry-bar is a lever that's designed so that the pivot is very close to one end of the bar (Fig. 12.1.13). When you exert a modest downward force (the *effort*) on the long end of the pry-bar and the object you're lifting exerts an enormous downward force (the *load*) on the short end of the pry-bar, the pry-bar balances and rotates at constant angular velocity about its pivot.

The torques these two forces produce are equal but oppositely directed, however, because the *load* is exerted so much closer to the pivot, it must be much larger than the *effort*. While you and the long end of the balanced pry-bar are pushing gently on one another, the object and the short end of the pry-bar are pushing hard on one another. That's why the pry-bar makes it so easy to pull a nail out of a board or to lift the corner of heavy furniture.

The pry-bar provides you with mechanical advantage and with its associated tradeoff between force and distance. You must still do all the work associated with lifting the heavy object, but you can do it with a small force exerted over a *large* distance.

Many ordinary tools use the pry-bar's pivot-in-the-middle lever concept, including scissors (Fig. 12.1.14). Each scissor blade rotates about a central pivot under the influence of opposing torques: one exerted by you and the other exerted by the paper you're cutting. The paper is quite close to the pivot so as you squeeze gently on the scissors' handles, the paper must push outward hard on the blades to balance your torque. The paper eventually gives up and the blades cut through it. Similar tools include shears, wire and bolt cutters, paper punches, and pliers.

A wheelbarrow uses another type of lever (Fig. 12.1.15) to lift heavy objects. In this case, a wheel supports the pivot at one end of the lever while you lift upward (the *effort*) on the other end of the lever. The objects resting in the wheelbarrow's basket push downward (the *load*) near the middle, but because the load is exerted so much nearer the pivot than the effort, the torques cancel and the wheelbarrow balances when the load is much larger than the effort. Using a wheelbarrow, you can lift and transport heavy things without having to exert unreasonable forces yourself.

Many tools use the wheelbarrow's pivot-at-the-end lever concept, including a garlic press (Fig. 12.1.16). As you squeeze the press's handles, the garlic clove opposes your actions. Because the clove is so close to the pivot, it must exert large outward forces on the press to balance the torques you produce with gentle inward forces. The clove can only push so hard before it smashes flat and begins to leak its juice. Similar tools include simple nutcrackers, jar openers, and juicers.

In fact, even wrenches and screwdrivers use levers that are essentially pivot-at-the-end in character. The center of the target bolt or screw is the pivot about which everything turns and as you and the tool push gently on one another far from that pivot, the tool and its target push fiercely on one another much nearer that pivot. Since the torque you can exert on the bolt or screw increases with the lever arm available, the most effective wrenches are those with long handles and the most successful screwdrivers are those with wide grips.

Some wrenches incorporate devices to measure the torques they're exerting on bolts and screws. Known as "torque wrenches," these tools help you to tighten bolts appropriately so that they don't come loose because of under-tightening or break because of over-tightening.

Having looked at both pivot-in-the-middle and pivot-at-the-end levers, we seem to have covered all the possibilities. But what about effort-in-the-middle levers? Shown in Fig. 2.1.17, this kind of lever may seem pointless because you push on it harder than it pushes on its object. But think about how far you push the lever as

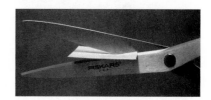

Fig. 2.1.14 When you squeeze the handles of these scissors together, you produce torques that act to close the scissors and cut the paper. The paper pushes outward, producing torques that act to open the scissors. If you squeeze hard, the closing torques overwhelm the opening torques. The scissor blades then undergo angular acceleration and rotate closed, cutting the paper.

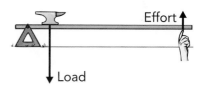

Fig. 2.1.15 This lever pivots about its left end so that it can balance when you push upward on the right end (the effort) and the anvil pushes downward on the middle (the load). Because the load is exerted closer to the pivot than is the effort, the lever balances when the effort is much smaller than the load.

Fig. 2.1.16 Squeezing the handles of this garlic press causes it to rotate closed about the pivot on the left. The garlic clove inside the press is unable to withstanding the crushing forces and extrudes through small holes in the press as juice and pulp.

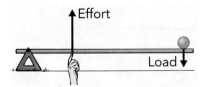

Fig. 2.1.17 This lever pivots about its left end so that it can balance when you push upward on the middle (the effort) and the ball pushes downward on the right end (the load). Since the effort is exerted much closer to the pivot than is the load, the lever balances when the effort is much larger than the load. If the lever is rotating counterclockwise, however, a small rise in the point at which you exert the effort is accompanied by a much larger rise in the point at which the ball exerts the load.

compared to how far the lever pushes its object. This lever is useful not for pushing an object *harder* but for pushing it *farther*.

When you're sweeping. mopping, or raking, you're using this type of lever. It's not that you aren't strong enough to push the broom, mop, or rake head across the ground, it's that you don't really want to move your hands that far over and over again. You'd also have to get down on your hands and knees, which would make the task even more unpleasant. So you use a lever that allows small, forceful motions of your hands, which act as both pivot and effort, to produce large, less-forceful motions of the broom, mop, or rake head.

Along with increased distance, an effort-in-the-middle lever can also provide increased speed. The tip of the lever travels faster than the part on which you push, an effect that lets you propel small things faster than your hands themselves can move. No matter how hard you try, you simply can't throw an object faster than about 160 km/h (100 mph) because that's as fast as you can move your hands. But you can easily get the tip of a lever moving at twice that speed and use it to throw things. Lacrosse sticks, ice and field hockey sticks, jai alai baskets, and fishing rods all take advantage of this effect to hurl their projectiles faster than you can throw them by hand. As a child, I spent many a happy afternoon at my grandparents' house, flinging fallen apples into the woods with a long, springy stick. Using only my arm, I was nobody. But with that stick to double or triple the apple's launch speed, I was suddenly Sandy Koufax at the top of his game.

Axle —
Hub —
Spoke —
Iron rim —
Brake

Section 2.2 **Wheels**

Like ramps and levers, wheels are simple tools that make our lives easier. But a wheel's main purpose isn't mechanical advantage, it's overcoming friction. Up until now, we've ignored friction, looking at the laws of motion as they apply only in idealized situations. But our real world does have friction, and most objects in motion tend to slow down and stop because of it. One of our first tasks in this section will therefore be to understand friction—though, for the time being, we'll continue to neglect air resistance.

Moving a File Cabinet: Friction

When we imagined moving your friend's piano into a new apartment back in Section 1.3, we neglected a familiar force—friction. Luckily for us, your friend's piano had wheels on its legs, and wheels facilitate motion by reducing the effects of friction. We'll focus on wheels in this section. But first, to help us understand the relationship between wheels and friction, we'll look at another item that needs to be moved—your friend's file cabinet.

The file cabinet is resting on a smooth and level hardwood floor; it's full of sheet music and weighs about 1000 N (225 lbf). Despite its large mass, you know that it should accelerate in response to a horizontal force, so you give it a gentle push toward the door. Nothing happens. Of course, the file cabinet accelerates in response to the net force it experiences, not to each individual force acting on it. Something else must be pushing on the file cabinet in just the right way to cancel your force and keep it from accelerating. Undaunted, you push harder and harder until finally, with a tremendous shove, you manage to get the file cabinet sliding across the floor. But the cabinet moves slowly even though you continue to push on it. Something else is pushing on the file cabinet, trying to stop it from moving.

That something else is friction, a phenomenon that opposes the relative motion of two surfaces in contact with one another. Two surfaces that are in relative motion are traveling with different velocities so that a person standing still on one surface would observe the other surface as moving. In opposing relative motion, friction exerts forces on both surfaces in directions that tend to bring them to a single velocity.

For example, when the file cabinet slides by itself toward the left, the floor exerts a rightward frictional force on it (Fig. 2.2.1). The frictional force exerted on the file cabinet, *toward the right*, is in the direction opposite the file cabinet's velocity,

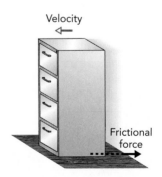

Velocity

Frictional force

Fig. 2.2.1 A file cabinet sliding to the left across the floor. The file cabinet experiences a frictional force toward the right that gradually brings it to a stop.

toward the left. Since the file cabinet's acceleration is in the direction opposite its velocity, the file cabinet slows down and eventually comes to a stop.

According to Newton's third law of motion, an equal but oppositely directed force must be exerted by the file cabinet on the floor. Sure enough, the file cabinet does exert a leftward frictional force on the floor. However, the floor is rigidly attached to the earth, so it accelerates very little. The file cabinet does almost all the accelerating, and soon the two objects are traveling at the same velocity.

Frictional forces always oppose relative motion, but they vary in strength according to (1) how tightly the two surfaces are pressed against one another, (2) how slippery the surfaces are, and (3) whether or not the surfaces are actually moving relative to one another. First, the harder you press two surfaces together, the larger the frictional forces they experience. For example, an empty file cabinet slides more easily than a full one. Second, roughening the surfaces generally increases friction, while smoothing or lubricating them generally reduces it. Riding a toboggan down the driveway is much more interesting when the driveway is covered with snow or ice than when the driveway is bare asphalt. We'll examine the third issue later on.

A Microscopic View of Friction

As the file cabinet slides by itself across the floor, it experiences a horizontal frictional force that gradually brings it to a stop. But from where does this frictional force come? The obvious forces on the file cabinet are both vertical, not horizontal: the cabinet's weight is downward and the support force from the floor is upward. How can the floor exert a horizontal force on the file cabinet?

The answer lies in the fact that neither the bottom of the file cabinet nor the top of the floor is perfectly smooth. They both have microscopic hills and valleys of various sizes. The file cabinet is actually supported by thousands of tiny contact points, where the file cabinet directly touches the floor (Fig. 2.2.2). As the file cabinet slides, the microscopic projections on the bottom of the file cabinet pass through similar projections on the top of the floor. Each time two projections collide, they experience horizontal forces. These tiny forces oppose the relative motion and give rise to the overall frictional forces experienced by the file cabinet and floor. Because even an apparently smooth surface still has some microscopic surface structure, all surfaces experience friction as they rub across one another.

Increasing the size or number of these microscopic projections by roughening the surfaces generally leads to more friction. If you put sandpaper on the bottom of the file cabinet, it will experience larger frictional forces as it slides across the floor. On the other hand, a microscopically smoother "nonstick" surface, like that used in modern cookware, would let the file cabinet slide easily.

Increasing the number of contact points by squeezing the two surfaces more tightly together also leads to more friction. The microscopic projections simply collide more often. That's why adding more sheet music to the file cabinet would make it harder to slide. Doubling the file cabinet's weight would roughly double the number of contact points and make it about twice as hard to move across the floor. A useful rule of thumb is that the frictional forces between two surfaces are proportional to the forces pressing those two surfaces together.

Friction also causes wear when the colliding contact points break one another off. With time, this wear can remove large amounts of material so that even seemingly indestructible stone steps are gradually worn away by foot traffic. The best way to reduce wear between two surfaces (other than to insert a lubricant between them) is to polish them so that they are extremely smooth. The smooth surfaces will still touch at contact points and experience friction as they slide across one another, but their contact points will be broad and round and will rarely break one another off during a collision.

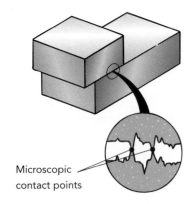

Microscopic contact points

Fig. 2.2.2 Two surfaces that are pressed against one another actually touch only at specific contact points. When the surfaces slide across one another, these contact points collide, producing sliding friction and wear.

Static Friction, Sliding Friction, and Traction

There are really two kinds of friction—sliding and static. When two surfaces are moving across one another, sliding friction acts to stop them from sliding. But even when those surfaces have the same velocity, static friction may act to keep them from starting to slide across one another in the first place.

You find it particularly hard to start the file cabinet sliding across the floor. Contact points between the cabinet and floor have settled into one another, so a small push does nothing. Static friction always exerts a frictional force that exactly balances your push. Since the net force on the file cabinet is zero, it doesn't accelerate.

However, the force that static friction can exert is limited. To get the file cabinet moving, you need to give it a mighty shove and thereby exert more horizontal force on it than static friction can exert in the other direction. The net force on the file cabinet is then no longer zero and it accelerates.

Once the file cabinet is moving, static friction is replaced by sliding friction. Because sliding friction acts to bring the file cabinet back to rest, you must push on the cabinet to keep it moving. With the file cabinet sliding across the floor, however, the contact points between the surfaces no longer have time to settle into one another, and they consequently experience weaker horizontal forces. That's why the force of sliding friction is generally weaker than that of static friction and why it's easier to keep the file cabinet moving than it is to get it started.

Both forms of friction are incorporated in the concept of traction—the largest amount of frictional force that the file cabinet can obtain from the floor at any given moment. When the cabinet is stationary, its traction is equal to the maximum amount of force that static friction can exert on it. But once it begins to slide across the floor, its traction reduces to the amount of force that sliding friction exerts.

While the file cabinet's traction is a nuisance that you must overcome, the traction of your shoes on the floor is crucial. Unless you can push against the wall, your shoes are going to need enough traction to provide the horizontal force required to move the file cabinet. Let's hope you're wearing your Dr. Martens®!

Work, Energy, and Power

There is another difference between static and sliding friction: sliding friction wastes energy. It can't make that energy disappear altogether because energy, as we've seen, is a conserved quantity: it can't be created or destroyed. But energy can be transferred between objects or converted from one form to another. What sliding friction does is convert useful, ordered energy—energy that can easily be used to do work—into relatively useless, disordered energy. This disordered energy is called thermal energy and is the energy we associate with temperature. It's sometimes called internal energy or heat. Sliding friction makes things hotter by turning work into thermal energy.

As we saw in Section 1.3, energy is the capacity to do work and is transferred between objects by doing that work. Energy can also change forms, appearing as either kinetic energy in the motions of objects or as potential energy in the forces between or within those objects. With practice, you can "watch" energy flow through a system just as an accountant watches money flow through a company.

The most obvious form of energy is kinetic energy, the energy of motion. It's easy to see when kinetic energy is transferred into or out of an object. As kinetic energy leaves an object, the object slows down; thus moving water slows down as it turns a gristmill, and a bowling ball slows down as it knocks over bowling pins. Conversely, as kinetic energy enters an object, the object speeds up. A baseball moves faster as you do work on it during a pitch; you're transferring energy from your body into the baseball, where the energy becomes kinetic energy in the baseball's motion.

Potential energy is stored in the forces between or within objects, and usually isn't as visible as kinetic energy. It can take many different forms, some of which appear in Table 2.2.1. In each case nothing is moving; but because the objects still have a great potential to do work, they contain potential energy.

Table 2.2.1 Several Forms and Examples of Potential Energy

Form of Potential Energy	Example
Gravitational potential energy	A bowling ball at the top of a hill
Elastic potential energy	A wound clock spring
Electrostatic potential energy	A cloud in a thunderstorm
Chemical potential energy	A firecracker
Nuclear potential energy	Uranium

We measure energy in many common units: joules (J), calories, food Calories (also called kilocalories), and kilowatt-hours, to name only a few. All of these units measure the same thing, and they differ from one another only by numerical conversion factors, some of which can be found in Appendix B. For example, 1 food Calorie is equal to 1000 calories or 4187 J. Thus a jelly donut with about 250 food Calories contains about 1,000,000 J of energy. Since a joule is the same as a newton-meter, 1,000,000 J is the energy you'd use to lift your friend's file cabinet into the second-floor apartment 200 times (1000 N times 5 m upward is 5000 J of work per trip). No wonder eating donuts is hard on your physique!

Of course, you can eventually use up the energy in a jelly donut; it just takes time. You can only do so much work each second. The measure of how quickly you do work is power—the amount of work you do in a certain amount of time, or

$$\text{power} = \frac{\text{work}}{\text{time}}.$$

The SI unit of power is the joule-per-second, also called the watt (abbreviated W). Other units of power include Calories-per-hour and horsepower; like the units for energy, these units differ only by numerical factors, which are again listed in Appendix B. For example, 1 horsepower is equal to 745.7 W. Since a 1-horsepower motor does 745.7 J of work each second, and since it takes 5000 J of work to move the file cabinet to the second floor, that motor has enough power to do the job in about 6.7 s.

Friction and Thermal Energy

But what about the *thermal* energy produced by sliding friction? Is thermal energy a new kind of potential energy or an alternative to kinetic energy?

In truth, it's neither. Thermal energy is actually a mixture of ordinary kinetic and potential energies. But unlike the kinetic energy in a moving ball or the potential energy in an elevated piano, the kinetic and potential energies in thermal energy are disordered at the atomic and molecular level. Thermal energy makes every microscopic particle in an object jiggle independently; at any moment, each particle has its own tiny supply of potential and kinetic energies, and this dispersed energy is collectively referred to as thermal energy.

As you push the file cabinet across the floor, you do work on it, but it doesn't pick up speed. Instead, sliding friction converts your work into thermal energy, so that the cabinet becomes hotter as the energy you transfer to it disperses among its particles. But while sliding friction easily turns work into thermal energy, there's no easy way to turn thermal energy back into work. Disorder not only makes things harder to use, but it is also difficult to undo. When you drop your favorite coffee mug on the floor and it shatters into a thousand pieces, the cup is still all there, but it's disordered and thus much less useful. Just as dropping the pieces on the floor a second time isn't likely to reassemble your cup, energy converted into thermal energy can't easily be reassembled into useful, ordered energy.

Sliding friction always converts at least some work into thermal energy. Since two surfaces sliding across one another experience frictional forces that oppose their relative motion, sliding friction does negative work on them; it extracts energy from a sliding object and converts that energy into thermal energy. Thus while you do work on the file cabinet by pushing it across the floor, sliding friction does negative work on it. The file cabinet's kinetic energy doesn't change very much, but its thermal energy continues to increase.

In contrast, static friction doesn't convert work into thermal energy. Since two surfaces experiencing static friction don't move relative to one another, there is no distance traveled and thus no work done. You can push against the stationary file cabinet all day without doing any work on it. Even if you lift the file cabinet upward with your hands (no easy task), static friction between your hands and the file cabinet's sides merely assists you in doing work on the file cabinet itself. As you lift the file cabinet upward, all of your work goes into increasing the file cabinet's gravitational potential energy.

Wheels

You've wrestled your friend's file cabinet out the door of the old apartment and are now dragging it along the sidewalk. You're doing work against sliding friction the whole way, producing large amounts of thermal energy in both the bottom of the cabinet and the surface of the sidewalk. You're also damaging both objects, since sliding friction is wearing out their surfaces. The four-drawer file cabinet may be down to three drawers by the time you arrive at the new apartment.

Fortunately, there are mechanical systems that can help you move one object across another without sliding or sliding friction. The classic example is a roller (Fig. 2.2.3). If you place the file cabinet on rollers, those rollers will rotate as the file cabinet moves so that their surfaces never slide across the bottom of the cabinet or the top of the sidewalk. To see how the rollers work, make a fist with one hand and roll it across the palm of your other hand. The skin of one hand doesn't slide across the skin of the other hand; since this silent motion doesn't convert work into thermal energy, your skin remains cool. Now slide your two open palms across one another; this time, sliding friction warms your skin.

Although the rollers don't experience sliding friction, they do experience static friction. The top of each roller is touching the bottom of the cabinet, and the two surfaces move along together because of static friction; they grip one another tightly until the roller's rotation pulls them apart. A similar process takes place between the rollers and the top of the sidewalk; static friction exerts torques on the rollers and hence is what makes them rotate in the first place. Again, you can illustrate this behavior with your hands. Try to drag your fist across your open palm. Just before your fist begins to slide, you'll feel a torque on it. Static friction between the skins of your two hands, acting to prevent sliding, causes your fist to begin rotating just like a roller.

Once you get the file cabinet moving on rollers, you can keep it rolling along the level sidewalk indefinitely. Without any sliding friction, the cabinet doesn't lose kinetic energy, so it continues at constant velocity without your having to push it. However, the rollers move out from under the file cabinet as it travels, and you frequently have to move a roller from the back of the cabinet to the front. In fact, you need at least three rollers to ensure that the file cabinet never falls to the ground when a roller pops out the back. Although the rollers have eliminated sliding fric-

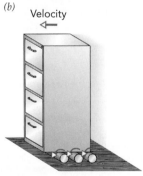

Fig. 2.2.3 (*a*) A file cabinet that's supported on turning rollers experiences only static friction. (*b*) Since the top surface of a roller moves forward with the file cabinet, while its bottom surface stays behind with the sidewalk, the roller's center of mass moves only half as fast as the file cabinet. As a result, the rollers are soon left behind.

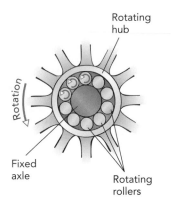

Fig. 2.2.5 In a roller bearing, the hub of the wheel doesn't touch the axle directly. Instead, the two are separated by a set of rollers that turn with the hub. The bottom few rollers bear most of the load since the hub pushes up on them and they push up on the axle. As the wheel turns, the rollers recirculate, traveling up to the right and over the top of the axle before returning down to the left to bear the load once again. The rollers, wheel, and axle experience only static friction, not sliding friction. In a ball bearing, the cylindrical rollers are replaced by spherical balls.

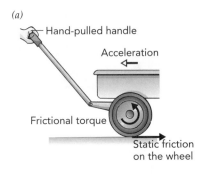

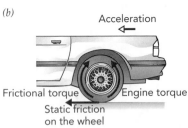

Fig. 2.2.6 (*a*) When a wagon accelerates forward, static friction from the ground exerts the torques that spin its wheels. (*b*) When a car's engine spins its wheels, static friction from the ground pushes the car forward and produces torques that oppose the wheels' rotations.

tion, they've created another headache—one that makes the prospect of a long trip unappealing. Is there another device that can reduce sliding friction without requiring constant attention?

One alternative would be a four-wheeled cart. The simplest cart rests on fixed poles or axles that pass through central holes or hubs in the four wheels (Fig. 2.2.4). The ground exerts upward support forces on the wheels, the wheels exert upward support forces on the axles, and the axles support the cart and its contents. As the cart moves forward, its wheels turn so that their bottom surfaces don't slide or skid across the ground; instead, each wheel lowers a portion of its surface onto the sidewalk, leaves it there briefly to experience static friction, and then raises it back off the sidewalk, with a new portion of wheel surface taking its place. Because of that touch-and-release behavior, there is only static friction between the cart's wheels and the ground.

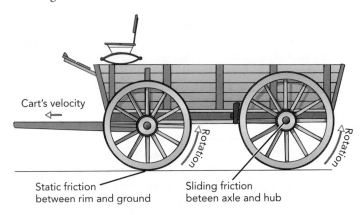

Fig. 2.2.4 As this cart moves toward the left, its wheels rotate counterclockwise. Although the wheel rims experience only static friction with the ground, the wheel hubs slide around the axles and convert the cart's kinetic energy into thermal energy. To reduce this wasted energy, the cart has narrow axles that are lubricated with axle grease.

Unfortunately, as each wheel rotates, its hub slides across the stationary axle at its center. This sliding friction wastes energy and causes wear to both hub and axle. However, the narrow hub moves relatively slowly across the axle so that the work and wear done each second are small. Still, this sliding friction is undesirable and can be reduced significantly by lubricating the hub and axle with "axle grease."

A better solution is to insert rollers between the hub and axle (Fig. 2.2.5). The result is a roller bearing—a mechanical device that minimizes sliding friction between a hub and an axle. A complete bearing consists of two rings separated by rollers that keep those rings from rubbing against one another. In this case, the bearing's inner ring is attached to the stationary axle while its outer ring is attached to the spinning wheel hub. The nondriven wheels of an automobile are supported by such bearings on essentially stationary axles. When the vehicle starts forward, static friction from the ground exerts torques on its free wheels and they begin to turn (Fig. 2.2.6a). A nondriven bicycle wheel is similarly supported on a stationary axle, but its bearings use balls instead of rollers—ball bearings (Fig. 2.2.7).

A car's driven wheels are also supported by roller bearings, but these bearings act somewhat differently. Because the engine must be able to exert a torque on each driven wheel, those wheels are rigidly connected to their axles. As the engine spins these axles, the axles spin their wheels (Fig. 2.2.7b). A bearing prevents each spinning axle from rubbing against the car's frame. This bearing's outer ring is attached to the stationary car frame while its inner ring is attached to the spinning axle.

As a driven wheel begins to spin, it experiences static friction with the ground and the ground pushes horizontally on the wheel's bottom to keep it from skidding. Since that is the only horizontal force on the automobile, the automobile accelerates forward.

Recognizing a good idea when you think of it, you load the file cabinet into the passenger seat of your red convertible sports car and start the engine. The car isn't quite as responsive as usual because of the added mass, but it's still able to accelerate respectably. In a few seconds, you're cruising down the road toward the new apartment and a very grateful friend.

Kinetic Energy

As you near your destination, you begin thinking about the car's brakes. They're designed to stop the car by turning its kinetic energy into thermal energy. They'll perform their task by rubbing stationary brake pads against spinning metal discs, so that sliding friction can transform the energy. Although you're confident that those brakes are up to the task, just how much kinetic energy are they going to have to convert into thermal energy?

One way is to determine the car's kinetic energy is to calculate the work its engine did on it while bringing it from rest to its current speed. The result of that calculation is that the moving car's kinetic energy is equal to one-half of its mass times the square of its speed or

$$\text{kinetic energy} = \tfrac{1}{2} \cdot \text{mass} \cdot \text{speed}^2. \tag{2.2.1}$$

With you and the file cabinet on board, the sports car has a mass of about 1500 kg (3300 lbm). At a speed of 100 km/h (62 mph), it has over 575,000 J of kinetic energy. That enormous energy is four times what it would be at 50 km/h (31 mph), so put down your cell phone and drive carefully. The dramatic increase in kinetic energy that results from a modest increase in speed explains why high-speed crashes are far deadlier than those at lower speeds and why that police officer is checking out your car with a radar gun. Red cars get all the attention.

You're traveling safely within the speed limit and exchange a polite wave with the officer. However, you soon pass another car that has been stopped for a ticket. The light on the nearby police car spins round and round, and rotating objects have kinetic energy, too. Like the kinetic energy of translational motion, the kinetic energy of rotational motion depends on the light's inertia and speed. But for a spinning light, it's the rotational inertia and rotational speed that matter. The light's kinetic energy is equal to one-half of its rotational mass times the square of its angular speed or

$$\text{kinetic energy} = \tfrac{1}{2} \cdot \text{rotational mass} \cdot \text{angular speed}^2. \tag{2.2.2}$$

With the ticket complete, the police car pulls out into traffic with its light still spinning. The light's total kinetic energy is now the sum of two parts: translational kinetic energy and rotational kinetic energy. Its translational kinetic energy depends on the speed of the light's center of mass, which is equal to the police car's speed through traffic. And its rotational kinetic energy depends on the angular speed at which the light turns about its center of mass.

As the police car disappears in the distance, it occurs to you that the spinning wheels of your car also have rotational kinetic energy that adds to the car's substantial translational kinetic energy. Still, you trust your brakes. In a few minutes, you arrive at your destination and brake to a stop. Although you're aware of the added mass as the car decelerates less quickly than usual, the brakes successfully transform the car's kinetic energy into thermal energy. You've reached your goal safely and are now a hero.

Fig. 2.2.7 The cylindrical rollers use in *roller* bearings are often replaced with balls to form *ball* bearings. While these balls serve exactly the same purpose as rollers, they're lighter and easier to work with; as a result, they're perfect for applications that don't require heavy loads, and they appear in bicycles and many other household machines with rotating parts. Still, ball bearings have some drawbacks. If they're overloaded, for example, the balls inside them may be crushed, since only a small region of each ball's surface exerts the forces that support the axle inside the hub. Once its balls have been damaged in this way, a bearing will begin to grind itself up. To reduce the chances of damage, the balls used in ball bearings are manufactured with truly remarkable precision into nearly perfect spheres of uniform size.

Traction, Skidding, and Four-Wheel Drive

Fortunately, you helped your friend move on a warm, sunny day and traction was never a problem. But what if the weather had been bad and the roadway icy? In that case, traction would have been an issue and you'd have had to think carefully about where in the car to put the file cabinet.

Your sports car is a rear-wheel drive car, meaning the engine exerts torque on and provides power to the rear wheels only. As the engine's torque twists those wheels forward, static frictional forces from the ground produce opposing torques on the wheels and it is those frictional forces that push the car forward. The front wheels, in contrast, respond passively to the car's forward motion and are spun when static frictional forces from the ground produce torques on them.

In good weather, your car has plenty of traction. As long as you don't press the accelerator to the floor, slam on the brakes, or turn the steering wheel violently to the side, the ground is able to exert enough static frictional force on each tire to keep it from sliding. But with ice covering the pavement, you need a delicate touch on the pedals and steering wheel. If you try to accelerate too rapidly, whether forward, backward, or sideways, you'll begin to skid. The tires simply won't have enough traction to produce the acceleration you want.

In general, the traction between two surfaces is proportional to the forces squeezing them together. In this case, each tire's static traction—the maximum static frictional force it can obtain—is proportional to how much weight that tire is supporting. The constant relating static traction to weight is often called the coefficient of static friction and depends on both the tire and road surface. For a treaded tire on dry pavement, the measured coefficient of static friction is typically 0.9. On icy pavement, the coefficient of static friction may drop as low as 0.5.

And once skidding has started, the tire's sliding traction is governed by the coefficient of sliding friction—the constant relating sliding traction to weight. While that constant is typically 0.8 for treaded tires on dry pavement, it drops almost to 0 on ice because sliding friction melts a thin layer of ice and the resulting liquid water acts as a lubricant. In short, once the tires on your car begin to slide on ice, you're in trouble.

Since only the rear tires are powered by the engine, you can't accelerate forward quickly on a slick surface unless they're pressed hard against the pavement. With the file cabinet in the passenger seat, the front wheels are supporting some of its weight and it isn't helping the rear-wheel traction as much as it can. If you put the file cabinet in the trunk, so that its weight rests entirely on the rear wheels, the rear-wheel traction will increase. This increased traction explains why people with rear-wheel drive cars and trucks sometimes put extra weight over their rear wheels in icy weather.

Fig. 2.2.8 The engine of this four-wheel drive sport utility vehicle (SUV) can exert torques on all four wheels, so that it can use their static tractions to propel itself forward. Like most modern vehicles, it also has an anti-lock braking system (ABS) that acts to prevent the wheels from skidding while braking. With the help of ABS, it can use static traction throughout the stopping process, rather than the weaker sliding traction that takes over when the vehicle is skidding across the pavement.

Your car's steering, however, is controlled almost entirely by the front wheels, so they need good traction, too. You're in a bind: shifting weight toward the car's rear improves its ability to accelerate forward and shifting weight toward the car's front improves its ability to steer. That's why you'd be better off with a front-wheel drive car on an icy day. With its front-wheels responsible for both forward acceleration and steering, and with its heavy engine pressing those front wheels into the pavement, a front-wheel drive car is much more manageable on a slick road than a rear-wheel drive one. Four-wheel drive vehicles go one step further by allowing the engine to exert torques on all four wheels (Fig. 2.2.8). With no passive wheels to squander traction during forward acceleration, a four-wheel drive vehicle can pick up speed even on relatively slick surfaces.

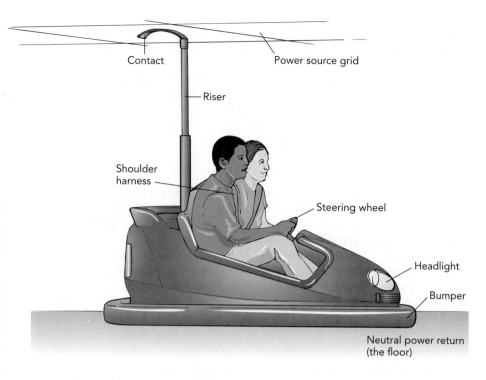

Contact

Power source grid

Riser

Shoulder
harness

Steering wheel

Headlight

Bumper

Neutral power return
(the floor)

SECTION 2.3 **Bumper Cars**

While car crashes normally aren't much fun outside of movies or television, there is one delightful exception: bumper cars. For a few minutes, drivers in this amusement park ride race madly about an oval track, deliberately crashing their vehicles into one another and laughing hysterically at the violent impacts. Jolts, jerks, and spins are half the fun, and it's a wonder that no one gets whiplash. But hidden in the excitement are several important physics concepts that influence everything from tennis to billiards.

Coasting Forward: Linear Momentum

Bumper cars are small, electrically powered vehicles that can turn on a dime and are protected on all sides by rubber bumpers. Each car has only two controls: a pedal that activates its motor and a steering wheel that controls the direction in which the motor pushes the car. Since the car itself is so small, its occupants account for much of the car's total mass and rotational mass.

Imagine that you have just sat down in one of these cars and put on your safety strap. The other people also climb into their cars, usually one person per car, and the ride begins.

With your car free to move or turn, you quickly become aware of its translational and rotational inertias. The car's translational inertia makes it hard to start or stop, and its rotational inertia makes it difficult to spin or stop from spinning. While we've seen these two types of inertia before, let's take another look at them and at how they affect your bumper car. This time, we'll see that they're associated with two new conserved quantities—linear momentum and angular momentum. As promised, energy isn't the only conserved quantity in nature!

When fast-moving bumper cars crash into one another, they exchange more than just energy. Energy is directionless—it's not a vector quantity—yet these cars seem to be exchanging some quantity of motion that has a direction associated with it. For example, if your car is hit squarely by a rightward moving car, then your car's motion shifts somewhat rightward in response. Your car is receiving a vector quantity of motion from the other car, a conserved vector quantity known as linear momentum.

Linear momentum, usually just called momentum, is the measure of an object's translational motion—its tendency to continue moving in a particular direction. Roughly speaking, your car's momentum indicates which way it's heading and just how difficult it was to get the car moving with its current velocity.

The car's momentum is its mass times its velocity or

$$\text{momentum} = \text{mass} \cdot \text{velocity}. \qquad (2.3.1)$$

Note that momentum is a vector quantity and that it has the same direction as the velocity. As we might expect, the faster your car is moving or the more mass it has, the more momentum it has in the direction of its motion. The SI unit of momentum is the kilogram-meter-per-second (abbreviated kg·m/s).

To physicists, conserved quantities are rare treasures that make it easier to understand otherwise complicated motions. Like all conserved quantities, momentum can't be created or destroyed. It can only be transferred between objects. Momentum plays a very basic role in bumper cars: the whole point of crashing them into one another is to enjoy the momentum transfers. During each collision, momentum shifts from one car to the other so that they abruptly change their speeds or directions or both. As long as these momentum transfers aren't too jarring, everyone has a good time.

You've stopped your car, so it has zero velocity and zero momentum. To begin moving again, something must transfer momentum to your car. While you could press the pedal and let the motor gradually transfer momentum from the ground to your car, that's not much fun. Instead, you let two grinning couch potatoes in an overloaded white car slam into you at breakneck speed (Fig. 2.3.1).

The white car was heading westward, and in a few moments your car is moving westward, too, while the white car has slowed significantly. Before you recover from the jolt, your car pounds a child's car westward and your car slows down abruptly. Finally, its impact with a wall stops the child's car. Despite disapproving looks from the child's parents, there's no harm done. Overall, westward momentum has flowed from the spudmobile to your car, to the child's car, and into the wall. No momentum has been created or destroyed; you've all simply enjoyed passing it along from car to car.

Fig. 2.3.1 Your car is hit by a fast-moving, massive white car with westward momentum. Much of that westward momentum is transferred to your car. You crash into a child's car, transferring the westward momentum to it. It then crashes into the wall, transferring the westward momentum to the wall.

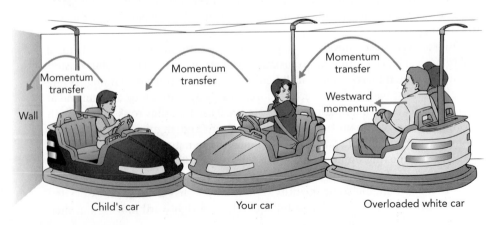

Exchanging Momentum in a Collision: Impulses

Momentum is transferred to a car by giving it an impulse, that is, a force exerted on it for a certain amount of time. When the motor and floor push your bumper car forward for a few seconds, they give your car an impulse and transfer momentum to it. This impulse is the change in your car's momentum and is equal to the product of the force exerted on the car times the duration of that force or

$$\text{impulse} = \text{force} \cdot \text{time}. \qquad (2.3.2)$$

The more force or the longer that force is exerted, the larger the impulse and the more your car's momentum changes. Remember that an impulse, like momentum itself, is a vector quantity and has a direction. If your aim is off and the misdirected impulse you obtain from the floor sends you crashing into the wall, don't say you hadn't been warned!

Different forces exerted for different amounts of time can transfer the same momentum to a car:

$$\text{impulse} = \text{large force} \cdot \text{short time} = \text{small force} \cdot \text{long time}. \qquad (2.3.3)$$

Thus you can get your car moving with a certain forward momentum either by letting the motor and floor push it with a small forward force of long duration or by letting the colliding white car push it with a large forward force of short duration.

We can now explain why bumper cars have soft rubber bumpers. If the bumpers were hard steel, the collision between the white car and your car would last only an instant and would involve an enormous forward force. You'd be in need of a neck brace and the services of a personal injury lawyer. However, amusement parks don't like lawsuits and sensibly limit the car impact forces. To do this, they use rubber bumpers and rather slow-moving cars.

Nonetheless, you can get a pretty good jolt when you collide head-on with another car. Your two cars then start with oppositely directed momenta, and the collision roughly exchanges those momenta between cars. In almost no time, you go from heading forward to heading backward. The impulse that causes this reversal of motion is especially large because it not only stops your forward motion, it also causes you to begin heading backward.

Why should momentum be a conserved quantity? It's conserved because of Newton's third law of motion. When one car exerts a force on a second car for a certain amount of time, the second car exerts an equal but oppositely directed force on the first car for exactly the same time. Because of the equal but oppositely directed nature of the two forces, cars that push on one another receive impulses that are equal in amount but opposite in direction. Since the momentum gained by one car is exactly equal to the momentum lost by the other car, we say that momentum is transferred from one car to the other.

The more mass a car has, the less its velocity changes as a consequence of a momentum transfer. That's why the white car doesn't stop completely when it crashes into your car, while your car speeds up dramatically. Just a fraction of the white car's forward momentum causes a large change in your car's velocity. Like a bug being hit by an automobile windshield, your bumper car does most of the accelerating.

Conserved Quantity: **Momentum**　　　　*Transferred By:* **Impulse**

Momentum: The measure of an object's translational motion—its tendency to continue moving in a particular direction. Momentum is a vector quantity, meaning that it has a direction. It has no potential form and therefore cannot be hidden; momentum = mass · velocity.

Impulse: The mechanical means for transferring momentum; impulse = force · time.

Spinning in Circles: Angular Momentum

When bumper cars are set spinning during crashes, they are exchanging yet another conserved quantity. Like momentum, it's a conserved *vector* quantity, but now it's associated with the angular speed and direction of rotational motion around a specific pivot. For example, when your car receives a glancing blow from a car that is circling you clockwise, your car's rotational motion or spin shifts somewhat clockwise in response. Your car is receiving a vector quantity of motion from the other car, a conserved vector quantity known as angular momentum.

Angular momentum is the measure of an object's rotational motion—its tendency to continue spinning about a particular axis. Simply put, your car's angular momentum indicates the direction of its rotation and just how difficult it was to get it spinning with its current angular velocity. The car's angular momentum is its rotational mass times its angular velocity or

$$\text{angular momentum} = \text{rotational mass} \cdot \text{angular velocity.} \qquad (2.3.4)$$

Note that angular momentum is a vector quantity and that it has the same direction as the angular velocity. The faster your car is spinning or the larger its rotational mass, the more angular momentum it has in the direction of its angular velocity. The SI unit of angular momentum is the kilogram-meter²-per-second (abbreviated $\text{kg} \cdot \text{m}^2/\text{s}$).

Angular momentum is another conserved quantity; it can't be created or destroyed, only transferred between objects. For your car to begin spinning, something must transfer angular momentum to it, and your car will then continue to spin until it transfers this angular momentum elsewhere. But to study angular momentum properly, we must pick the pivot about which all the spinning will occur. In the present situation, a good choice for this pivot is your car's initial center of mass.

Your car is stationary again, so it has zero angular velocity and zero angular momentum. Suddenly, a black car sweeps by and strikes your car a glancing blow (Fig. 2.3.2). Because the black car was circling your car counterclockwise, it had counterclockwise angular momentum about the pivot. Its impact transfers some of this angular momentum to your car, which begins spinning counterclockwise itself. Since it has given up some of its angular momentum, the black car circles your car less rapidly. Your car gradually stops spinning as its wheels and friction transfer the angular momentum to the ground and earth. Overall, no angular momentum was created or destroyed during the collision. Instead, it was transferred from the black car to your car to the earth.

Glancing Blows: Angular Impulses

Angular momentum is transferred to a car by giving it an angular impulse, that is, a torque exerted on it for a certain amount of time. When the black car hits your car and exerts a torque on it briefly, it gives your car an angular impulse and transfers

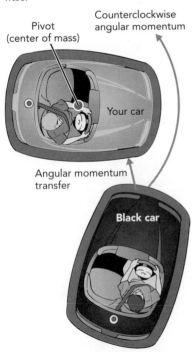

Fig. 2.3.2 Since the black car is circling your car counterclockwise, it has counterclockwise angular momentum. When it hits your car, it transfers some of that angular momentum to your car. Because of this transfer, the black car stops circling so quickly as your car begins to spin counterclockwise.

Counterclockwise angular momentum

Pivot (center of mass)

Your car

Angular momentum transfer

Black car

angular momentum to it. This angular impulse is the change in your car's angular momentum and is equal to the product of the torque exerted on your car times the duration of that torque or

$$\text{angular impulse} = \text{torque} \cdot \text{time.} \tag{2.3.5}$$

The more torque or the longer that torque is exerted, the larger the angular impulse and the more your car's angular momentum changes. Once again, an angular impulse is a vector quantity and has a direction. Had the black car been circling your car clockwise when it struck the glancing blow, its angular impulse would have been in the opposite direction and you'd be spinning the other way.

Different torques exerted for different amounts of time can transfer the same angular momentum to a car:

$$\text{angular impulse} = \text{large torque} \cdot \text{short time} = \text{small torque} \cdot \text{long time.} \tag{2.3.6}$$

Thus you can get your car spinning with a certain angular momentum either by letting the motor and floor twist it with a small torque of long duration or by letting the colliding black car twist it with a large torque of short duration. As with linear momentum, sudden transfers of angular momentum can break things, so the cars are designed to limit their impact torques to reasonable levels. Even so, you may find yourself reaching for the motion sickness bag after a few spinning collisions.

Why should angular momentum be a conserved quantity? Like linear momentum, angular momentum is conserved because of Newton's third law of motion. In this case, we're referring to Newton's third law of rotational motion: if one object exerts a torque on a second object, then the second object will exert an equal but oppositely directed torque on the first object.

Newton's Third Law of Rotational Motion
For every torque that one object exerts on a second object, there is an equal but oppositely directed torque that the second object exerts on the first object.

When one car exerts a torque on a second car for a certain amount of time, the second car exerts an equal but oppositely directed torque on the first car for exactly the same amount of time. Because of the equal but oppositely directed nature of the two torques, cars that exert torques on one another receive angular impulses that are equal in amount but opposite in direction. Since the angular momentum gained by one car is exactly equal to the angular momentum lost by the other car, we say that angular momentum is transferred from one car to the other.

Because a car's angular momentum depends on its rotational mass, two different cars may end up rotating at different angular velocities even though they have identical angular momenta. For example, when the black car hits the overloaded white car and transfers angular momentum to it, the white car's enormous rotational mass makes it spin relatively slowly. The same sort of behavior occurs with linear momentum, where a car's mass affects how fast it travels when it's given a certain amount of linear momentum. But while a bumper car can't change its mass, it can change its rotational mass. If it does so while it's spinning, its angular *momentum* won't change, but its angular *velocity* will!

To see this change in angular velocity, consider the overloaded white car. Its two large occupants are disappointed with the ride because their huge mass and rotational mass prevent them from experiencing the intense jolts and spins that you've been enjoying. Suddenly they get a wonderful idea. As their car slowly spins, one of them climbs into the other's lap and the two sit very close to the car's center of mass. By rearranging the car's mass this way, they have reduced the car's overall rotational mass and the car actually begins to spin faster than before.

Quantity	SI unit	English unit	SI → English	English → SI
Momentum	kilogram-meter-per-second (kg·m/s)	pound-foot-per-second (lbm·ft/s)	1 kg·m/s = 7.2329 lbm·ft/s	1 lbm·ft/s = 0.13826 kg·m/s
Angular momentum	kilogram-meter²-per-second (kg·m²/s)	pound-foot²-per-second (lbm·ft²/s)	1 kg·m²/s = 23.730 lbm·ft²/s	1 lbm·ft²/s = 0.042140 kg·m²/s

As the white car's mass is being redistributed, it's not a freely turning *rigid* object covered by Newton's first law of rotational motion. However, it is freely turning and thus covered by a more general and equally useful rule: an object that is not subject to any outside torques has constant angular momentum. As the car's rotational mass becomes smaller, its angular velocity must increase in order to keep its angular momentum constant. That's just what happens. This effect of changing one's rotational mass explains how an ice skater can achieve an enormous angular velocity by pulling himself into a thin, spinning object on ice.

Conserved Quantity: Angular Momentum

Transferred By: Angular Impulse

Angular Momentum: The measure of an object's rotational motion—its tendency to continue spinning about a particular axis. Angular momentum is a vector quantity, meaning that it has a direction. It has no potential form and therefore cannot be hidden; angular momentum = rotational mass · angular velocity

Angular Impulse: The mechanical means for transferring angular momentum; angular impulse = torque · time .

The Three Conserved Quantities

As you drive your bumper car around the oval track, its motion is governed in large part by three conserved quantities: energy, linear momentum, and angular momentum (Table 2.3.1). While you can exchange those quantities with the earth and the power company by steering your car or switching on its motor, most of the interesting exchanges involve collisions.

Table 2.3.1 The Three Conserved Quantities of Motion and Their Transfer Mechanisms

Conserved Quantity	Transfer Mechanism
Energy	Work
Linear momentum	Impulse
Angular momentum	Angular impulse

Each time your car shoves another car forward, your car does work on that other car and transfers energy to it. Each time your car pushes another car northward briefly, your car gives a northward impulse to that other car and transfers northward momentum to it. And each time your car twists another car clockwise about its center of mass, your car gives a clockwise angular impulse to the other car and transfers clockwise angular momentum to it. These exchanges of energy, momentum, and angular momentum are fast and furious and make for an exciting ride.

Potential Energy and Acceleration

Shortly before the ride stops, you notice that there is a low point in the floor. After years of use, its metal surface has dented into a bowl-shaped depression and you observe that cars naturally tend to roll into this bowl and accelerate toward its bottom. As each car tries to coast through the depression, it accelerates toward the bottom of the bowl and its velocity changes as a result. While some cars pass directly through the bowl's bottom, accelerating forward during their descents and backward during their subsequent rises, most cars encounter the bowl at an angle and have their paths bent by the accelerations they experience. Each deflected car emerges from the bowl traveling in a somewhat altered direction (see ❶).

We've seen this tendency to accelerate downhill before with ramps, but now let's look at it in terms of energy: a car always accelerates in the direction that reduces its total potential energy as quickly as possible. Since a lone car's only potential energy is gravitational potential energy, it accelerates in such a way as to reduce its gravitational potential energy as quickly as possible: down the steepest route to the bottom of the bowl.

This behavior of accelerating in the direction that reduces total potential energy as quickly as possible is universal. Potential energy and forces are intimately related and the direction of quickest total potential energy decrease *is* the direction of the net force—the sum of all forces on an object. An object accelerates in the direction of quickest potential energy decrease because that is the direction of the net force and vice versa. This rule is a useful way to determine how motion will proceed: which way a spring will leap, a chair will tip, or a bumper car will roll. We'll use it frequently in this book.

Remember, however, that acceleration and velocity are different physical quantities and that just because a bumper car is *accelerating* toward the bottom of a bowl doesn't mean that its *velocity* points in that direction or that it will ever reach the bottom of the bowl. In my experience, I can't state those observations too emphatically or too often, although I occasionally have to dodge flying vegetables while repeating them for the hundredth time. I presume, of course, that those projectiles are intended as experimental tests of the distinction between acceleration and velocity.

Potential Energy and Acceleration

An object accelerates in the direction that reduces its total potential energy as quickly as possible.

⚏ Airbags, Seatbelts, and Crumple Zones

While collisions are fun when they involve bumper cars, they're less so when they involve automobiles. The sudden transfers of momentum and releases of energy that occur during car crashes can cause injuries or death, and managing them has become an important part of automobile safety engineering.

Although there are many types of automobile collisions, most involve similar physics issues. During a crash, you want to avoid sudden transfers of momentum and to have all the released energy go somewhere else. To illustrate these ideas, let's focus on a particularly simple type of crash: a head-on collision with a tree. And to show you just how much confidence I have in the safety features of my modern car, I'll be the one who drives into the tree. I just hope it won't affect my insurance.

Prior to hitting the tree, I am driving directly northward at 50 km/h (30 mph). My car and I have considerable momentum in northward direction as well as a large amount of kinetic energy. Recall that kinetic energy, like energy in general, has no direction.

All of a sudden, there's a tree in the middle of the road. OK, so I drifted off the shoulder and the tree only looks like it's in the middle of the road. Before I can respond, my car hits the tree head-on. Watch what happens.

❶ Golf and croquet are played on open grass, which is never perfectly level or flat. As a golf or croquet ball rolls forward across uneven or sloping terrain, it tends to accelerate in whichever direction is locally downhill and its path bends as a result of that acceleration. Skillful golfers or croquet players can "read" the terrain and anticipate the ball's accelerations in order to predict the path the ball will take and use it to their advantage.

Since the tree and car cannot occupy the same space at the same time, they begin to push on one another with equal but oppositely directed forces. The car pushes the tree northward while the tree pushes the car southward. Since the tree is sturdy and rooted firmly in the ground, it barely accelerates at all. However, my car accelerates southward at a tremendous rate. In a fraction of a second, my car has decelerated from 50 km/h to 0 km/h.

Before looking inside to see what's happening to me, let's consider what's happening to the car's momentum and energy. Before the crash, the car had both northward momentum and substantial kinetic energy. After the crash, it has zero momentum and zero kinetic energy. Since momentum and energy are both conserved, where did they go?

The car's momentum was transferred to the tree via an impulse: the car pushed the tree northward for a short period of time and thereby conveyed all of its northward momentum to the tree. The tree, in turn, pushed the ground northward for a short period of time and thus pass that northward momentum along to the ground. That's why the tree is still in place—it didn't retain any of the northward momentum. And because these transfers of momentum took place so quickly, the forces involved in the impulses were enormous. Remember, impulse is the product of force times time, so the shorter the time, the greater the force.

The car's energy, however, wasn't transferred to the tree. Because the tree didn't move as the car pushed on it, the tree had zero work done on it. Instead, the car's kinetic energy was transformed into other forms within the car.

That observation brings us to the first safety feature of my modern car: crumple zones (Fig. 2.3.3). My car crushed itself strategically in a few designated places. Crushing something always involves work: forces push inward on the crumpling object and that object's surfaces move inward. So as the car crumples, the kinetic energy it had before the collision becomes thermal and chemical potential energies in the deforming materials. My car will never drive again, but it's dissipating its kinetic energy in a way that doesn't involve me and I'm sure happy about that.

It's now time to look inside the car. Although everything around me has come to a stop, I'm still coasting forward at 50 km/h. I still have my northward momentum and my kinetic energy and I'm eager to get rid of both as gracefully as possible. Fortunately, I have a seatbelt and an airbag to help me.

If I were unrestrained, I might coast northward through the windshield and beyond. But my seatbelt soon pulls taut and I begin pushing it northward as it pushes me southward. Since these forces are exerted over a substantial period of time, I am gradually transferring my northward momentum to the seatbelt and the forces aren't too severe. So far, so good.

But the steering wheel still looms threateningly in front of me. If I hit it directly, I would transfer my remaining northward momentum to it in a very short period of time. Such a rapid impulse would involve huge forces and I'd lose more than my sense of humor.

Before I coast into the steering wheel, however, the airbag inflates and I hit it instead. I still transfer all of my northward momentum to the airbag, but I do it over a relatively long period of time. While the impulse is the same whether I hit the airbag or the steering wheel, the forces involved are not. Transferring my momentum slowly to the airbag requires small forces while transferring my momentum quickly to the steering wheel requires large forces. Thank goodness for the airbag.

When all the dust has settled, I step from my car shaken but not stirred. My northward momentum has left me through the seatbelt and airbag and much of my kinetic energy went into the airbag through the work I did on its surface.

Before leaving these topics, I should mention that the mechanisms that lock seatbelts and trigger airbags are based on the same physics we've just discussed. They all sense a vehicle's rapid acceleration during a crash by comparing the motion of

Fig. 2.3.3 The US National Highway Transportation Safety Administration studies automobile crashes to determine how well different vehicles perform during collisions. As these cars and trucks undergo head-on collisions, their front ends crumple to absorb energy and slow the transfers of momentum. At the same time, airbags deploy to cushion the occupants. While each occupant eventually loses all of his or her forward momentum, the force involved in that transfer decreases as the duration of the transfer increases. Because it extracts momentum much more slowly than a steering wheel or dashboard, an airbag reducing the impact force and the risk of serious injury.

the car to the motion of an inertial test object. As the car accelerates abruptly on impact and the test object coasts or tries to coast onward, sensors watch the two objects move relative to one another or measure the forces needed to make the two objects accelerate together. Though modern acceleration sensors have been miniaturized to microscopic dimensions, they still depend on Newton's second law and the relationship between forces and accelerations.

Ⓜ Hammers, Wedges, and Related Tools

Pushing a thick nail into wood requires a huge force, which is why you can't do it with your hand alone. But with a few blows from a hammer, you can drive the nail deep into the wood (Fig. 2.3.4). How does that trick work?

At first, you might think that the hammer's weight is somehow involved, but that's not the case. After all, you weigh far more than the hammer yet you can't push the nail into the wood and besides, a hammer can pound a nail upward into the ceiling just as easily as it can pound one downward into the floor. It's not the hammer's weight that matters here but its mass and the key issue is momentum.

When you push the hammer toward the nail, you are investing it with momentum via an impulse: you are pushing the hammer toward the nail over a period of time. When the hammer subsequently encounters the nail, it transfers its momentum to the nail via another impulse. But while your impulse on the hammer involved a modest force exerted over a long time, the impulse the hammer gives the nail involves an enormous force exerted over a split second. Transferring momentum quickly requires large forces and it is just such an impact force that propels the nail into the wood.

Hammers push hardest when they transfer their momentum fastest, which is why most hammers have hardened steel faces. The harder the hammer face, the faster it conveys its momentum to whatever it strikes and the greater the peak im-

Fig. 2.3.4 All of the downward momentum that you invest in this hammer as you swing it is transferred to the nail during their brief collision. Since the transfer time is so short, the hammer must use an enormous force to convey its momentum to the nail. That impact force pushes the nail into the wood.

Fig. 2.3.5 As this chisel moves to the left, it lifts a chip of wood up its ramplike face. Because that ramp provides mechanical advantage, a modest leftward force yields a much larger lifting force on chip and tears it free from the rest of the wood. If the chisel were wider or the wood were harder, a stronger leftward force might be needed to move the chisel and lift the chip. That stronger force could be obtained by hitting the chisel with a hammer. The resulting impact force would then drive the chisel to the left.

pact forces involved. For driving nails, more force is usually better so a hardened hammer with lots of momentum (i.e. plenty of mass and plenty of speed) is your best bet. For really big projects, you might use a sledge hammer—its massive head and long handle allow you to invest it with a tremendous amount of momentum before it strikes its target.

But if you don't want to mar the surface you're hitting, try a somewhat softer tool. Hardened hammers concentrate their momentum transfers in both time and space, producing intense impact forces that can ruin the things they hit. Mallets are designed to spread out their momentum transfers: they extend the time of their impulses to lessen the peak impact forces and they exert those impact forces over more space to avoid denting the surfaces on which they push. Instead of using hardened steel faces, mallets use softer materials, ranging from mild steel, through copper, lead, or plastic, all the way to rubber and soft wood. By picking a mallet head that is significantly softer than whatever you're tapping into place, you can usually avoid damaging your target.

Hammers are particularly useful for driving wedges into things. A wedge is a ramp-shaped tool that you can push into a crack to separate its two halves. A wedge exhibits the same mechanical advantage as a ramp: pushing the wedge a long distance into the crack with a modest force results in the two sides of the crack being pushed outward a short distance with a tremendous force. Moreover, the work you do on the wedge with that small force exerted over a long distance is conveyed to the crack as a huge force exerted over a small distance.

While there are some wedge tools that you push on directly, including chisels (Fig. 2.3.5), planes (Fig. 2.3.6), and saws, or by way of a rotating mechanism, such as drills (Fig. 2.3.7) and screws, the most potent wedges make use of impact forces to do their jobs. To split a log, you can use a sledge hammer to drive a wedge into its end. The wedge's mechanical advantage, combined with the astonishing impact force that the sledge can exert on its end during their collision can tear apart even the strongest wood. An axe combines the sledge hammer and wedge into a single object and a hatchet is just a smaller version of the same idea. Chiseling hard wood and stone becomes much easier when you tap the chisel with a mallet so that it is driven into the wood with brief but intense impact forces.

Fig. 2.3.6 When this wood plane slides to the left across a board, the ramplike blade projecting from its smooth bottom cuts and lifts a wafer-thin slice of wood from the board's surface. The gradual slope of the blade provides mechanical advantage so that a gentle leftward push on the plane is all that's needed to lift the slice. This planing process is so precise that the resulting board surface is almost perfectly smooth.

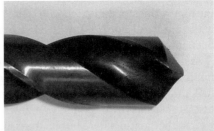

Fig. 2.3.7 When this drill bit rotates against a metal object to its right, its two ramplike cutting edges lift chips of metal and separate them from the surface. The sloping cutting edges provide mechanical advantage, so that a relatively gentle torque on the bit is enough to push the cutting edges around in a circle across the metal surface and thereby lift the chips.

CHAPTER 3

MECHANICAL THINGS

One of my favorite scenes in the movie *Butch Cassidy and the Sundance Kid* is that in which Sundance is being interviewed for a security job in Bolivia and is asked to shoot at a coin lying on the ground. Sundance fires stiffly at the distant coin and misses it several times before asking: "Can I move?" Once permitted to do so, Sundance flies into motion and hits the coin three times in rapid succession. Freedom to move is evidently central to Sundance's skill.

I feel that same need to move when discussing the science in ordinary things. I'm surely not Sundance's equivalent when it comes to explaining everyday physics, but the more flexibility I have and the more conceptual angles from which I can approach each object, the more I feel like I can get it right.

Those first two chapters were tough for me because I couldn't move much yet. I was trying to play fair and introduce concepts methodically and without jumping about wildly or, worse, jumping ahead. If my languorous exposition up to now has made this book useful for treating your insomnia, I hope that will begin to change.

Having surveyed the laws of motion, we can begin using those laws to explain the behaviors of everyday objects. But while we can already address some of the central features at work in a toy wagon, a weight machine, or a ski lift, we're still missing a number of mechanical concepts that are important in the world around us. In this chapter, we'll look at some of those additional concepts.

One of the most important new concepts will be the feeling of acceleration. If we treat acceleration passively, it can be fairly uninteresting: we push on the cart and the cart accelerates. But if we think of it more actively—for example, if we envision ourselves on a roller coaster as it plummets down that first big hill—then the experience of acceleration becomes much more intriguing. In fact, we might even need to hold on to our hats.

Chapter Itinerary

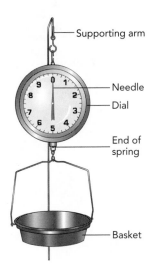

Supporting arm

Needle

Dial

End of spring

Basket

SECTION 3.1 **Spring Scales**

How much of you is there? From day to day, depending on how much you eat, the amount of you stays approximately the same. But how can you tell how much that is? The best measure of quantity is mass: kilograms of gold, grain, or you. Mass is the measure of an object's inertia and, as we saw in Section 1.1, doesn't depend on the object's environment or on gravity. A kilogram box of cookies always has a mass of 1 kilogram, no matter where in the universe you take it.

But mass is difficult to measure directly. Moreover, the very concept of mass is only about 300 years old. Consequently, people began quantifying the material in an object by measuring its weight. Spring scales eventually became one of the simplest and most practical tools for accomplishing this task, and they are still found in bathrooms and grocery stores today. They really do contain springs, although these are normally hidden from view.

Why You Must Stand Still on a Scale

Whenever you stand on a scale in your bathroom or place a melon on a scale at the grocery store, you are measuring weight. An object's weight is the force exerted on it by gravity, usually the earth's gravity. When you stand on a bathroom scale, the scale measures just how much upward force it must exert on you in order to keep you from moving downward toward the earth's center. As in most scales you'll encounter, the bathroom scale uses a spring to provide this upward support. If you're stationary, you're not accelerating, so your downward weight and the upward force from the spring must cancel one another; that is, they must be equal in magnitude but opposite in direction so that they sum to zero net force. Consequently, although the scale actually displays how much upward force it's exerting on you, that amount is also an accurate measure of your weight.

This subtle difference between your actual weight and what the scale is reporting is important. While an object's weight depends only on its gravitational environment, not on its motion, the weighing process itself is extremely sensitive to motion. If anything accelerates during the weighing process, the scale may not report the object's true weight. For example, if you jump up and down while you're standing on a scale, the scale's reading will vary wildly. You're accelerating and, since the net force exerted on you is no longer zero, your downward weight and the upward force from the scale no longer cancel. If you want an accurate measurement of your weight, therefore, you have to stand still.

But even if you stand still, weighing is not a perfect way to quantify the amount of material in your body. That's because your weight depends on your environment. If you always weigh yourself in the same place, the readings will be pretty consistent, as long as you don't routinely eat a dozen jelly doughnuts for lunch. But if you moved to the moon, where gravity is weaker, you'd weigh only about one-sixth as much as on the earth. Even a move elsewhere on the earth will affect your weight: the earth bulges outward slightly at the equator, and gravity there is about 0.5% weaker than at the poles. That change, together with a small acceleration effect due to the earth's rotation, means that a scale will read 1.0% less when you move from the north pole to the equator. Obviously, moving south is not a useful weight-loss plan.

Stretching a Spring

So you know now that when you put a melon in the basket of a scale at the grocery store and read its weight from the scale's dial, the scale is actually reporting just how much upward force its spring is exerting on the melon. While your shopping cart could support the melon equally well, there's no simple way to determine just how much upward force the cart exerts on the melon. Therein lies the beauty, and the utility, of a spring: a simple relationship exists between its length and the forces it's exerting on its ends. The spring scale can therefore determine how much force it's exerting on the melon by measuring the length of its spring.

The springs shown in Fig. 3.1.1 consist of a wire coil that pulls inward on its ends when it's stretched and pushes outward on them when it's compressed. If a coil spring is neither stretched nor compressed, it exerts no forces on its ends.

The top spring (Fig. 3.1.1a) is neither stretched nor compressed, so that when it lies on a table to keep it from falling, its ends remain motionless. Those ends are in equilibrium—experiencing zero net force. As the phrase "zero net force" suggests, equilibrium occurs whenever the forces acting on an object sum perfectly to zero so that the object doesn't accelerate. When you sit still in a chair, for example, you are in equilibrium.

The spring in Fig. 3.1.1a is also at its equilibrium length, its natural length when you leave it alone. No matter how you distort this spring, it tries to return to this equilibrium length. If you stretch it so that it's longer than its equilibrium length, it will pull inward on its end. If you compress it so that it's shorter than its equilibrium length, it will push outward on its ends.

Let's attach the left end of our spring to a post (Fig. 3.1.1b) and look at the behavior of its free right end. With nothing pushing or pulling on the spring, this free end will be in equilibrium at a particular location—its equilibrium position. Since the spring's end naturally returns to this equilibrium position if we stretch or compress it and then let go, the end is in a stable equilibrium.

But what happens if we pull the free end to the right and don't let go? The spring now exerts a steady inward force on that end, trying to return it to its original equilibrium position. The more we stretch the spring, the more inward force it exerts on the end. Remarkably, this inward force is exactly proportional to how far we stretch the end away from its original equilibrium position. Since the end of the spring in Fig. 3.1.1c has been pulled 1 cm to the right of its original equilibrium position, the spring now pulls this end to the left with a force of 1 N; if the end is instead pulled 2 cm to the right, as it has been in Fig. 3.1.1d, the spring pulls it to the left with a force of 2 N. This proportionality continues to work even when we compress the spring: in Fig. 3.1.1e, the end has been pushed 1 cm to the left, and the spring is pushing it to the right with a force of 1 N.

The force exerted by a coil spring thus has two interesting properties. First, this force is always directed so as to return the spring to its equilibrium length. We call this kind of force a restoring force because it acts to restore the spring to equilibrium. Second, the spring's restoring force is proportional to how far it has been distorted (stretched or compressed) from its equilibrium length.

These two observations are expressed in Hooke's law,

$$\text{restoring force} = -\text{spring constant} \cdot \text{distortion}, \qquad (3.1.1)$$

named after the Englishman Robert Hooke, who discovered it in the late seventeenth century. Here the spring constant is a measure of the spring's stiffness. The larger the spring constant—that is, the firmer the spring—the larger the restoring force the spring exerts for a given distortion. The negative sign in this equation indicates that a restoring force always points in the direction opposite the distortion.

Hooke's Law

The restoring force exerted by an elastic object is proportional to how far it has been distorted from its equilibrium shape.

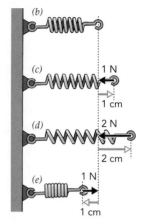

Fig. 3.1.1 Five identical springs. The ends of spring (a) are free so that it can adopt its equilibrium length. The left ends of the other springs are fixed so that only their right ends can move. When the free end of a spring (b) is moved away from its original equilibrium position (c, d, and e), the spring exerts a restoring force on that end that is proportional to the distance between its new position and the original equilibrium position.

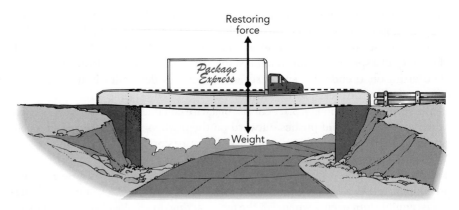

Fig. 3.1.2 A steel bridge sags under the weight of a truck. The bridge bends downward until the upward restoring force it exerts on the truck exactly balances the truck's weight.

Springs are distinguished by their stiffness, as measured by their spring constants. Some springs are soft and have small spring constants—for example, the one that pops the toast out of your toaster, which you can easily compress with your hand. Others, like the large springs that suspend an automobile chassis above the wheels, are firm and have large spring constants. But no matter the stiffness, all springs obey Hooke's law.

Hooke's law is remarkably general and isn't limited to the behavior of coil springs. Almost anything you distort will pull or push back with a force that's proportional to how far you've distorted it away from its equilibrium length—or, in the case of a complicated object, its equilibrium shape. Equilibrium shape is the shape an object adopts when it's not subject to any outside forces. If you bend a tree branch, it will push back with a force proportional to how far it has been bent. If you pull on a rubber band, it will pull back with a force proportional to how far it has been stretched, up to a point. If you squeeze a ball, it will push outward with a force proportional to how far it has been compressed. If a heavy truck bends a bridge downward, the bridge will push upward with a force proportional to how far it has been bent (Fig. 3.1.2).

There is a limit to Hooke's law, however. If you distort an object too far, it will usually begin to exert less force than Hooke's law demands. This is because you will have exceeded the elastic limit of the object and will probably have permanently deformed it in the process. If you pull on a spring too hard, you'll stretch it forever; if you push on a branch too hard, you'll break it. But as long as you stay within the elastic limit, almost everything obeys Hooke's law—a rope, a ruler, an orange, and a trampoline.

Distorting a spring requires work. When you stretch a spring with your hand, pulling its end outward, you transfer some of your energy to the spring. The spring stores this energy as elastic potential energy. If you reverse the motion, the spring returns most of this energy to your hand, while a small amount is converted to thermal energy by frictional effects inside the spring itself. Work is also required to compress, bend, or twist a spring. In short, a spring that is distorted away from its equilibrium shape always contains elastic potential energy.

How a Hanging Grocery Scale Measures Weight

We're now ready to understand how spring scales work. Imagine a hanging spring scale of the kind used to weigh produce. Inside this scale is a coil spring, suspended from the ceiling by its upper end (Fig. 3.1.3). Hanging from its lower end is a basket. For the sake of simplicity, imagine that this basket has little or no weight.

With no force pulling down on it, the scale's spring adopts its equilibrium length, and the basket, experiencing zero net force, is in a position of stable equilibrium. If you shift the basket up or down and then let go of it, the spring will push it back to this position.

When you place a melon in the basket, the melon's weight pushes it downward. The basket starts descending and as it does, the spring stretches and begins to exert an upward force on the basket. The more the spring stretches, the greater this upward force so that eventually the spring is stretched just enough so that its upward force exactly supports the melon's weight. The basket is now in a new stable equilibrium position—again experiencing zero net force.

But how does the scale determine the melon's weight? It uses Hooke's law. Once the basket has adopted its new equilibrium position, where the melon's weight is exactly balanced by the upward force of the spring, the amount the spring has stretched is an accurate measure of the melon's weight.

The scales in Fig. 3.1.3 differ only in the way they measure how far the spring has stretched beyond its equilibrium length. In Fig. 3.1.3a, the scale uses a pointer attached to the end of the spring, while in Fig. 3.1.3b, the scale uses a rack and pinion gear system that converts the small linear motion of the stretching spring into a much more visible rotary motion of the dial needle. The rack is the series of evenly spaced teeth attached to the lower end of the spring; the pinion is the toothed wheel attached to the dial needle. As the spring stretches, the rack moves downward, and its teeth cause the pinion to rotate. The farther the rack moves, the more the pinion turns, and the higher the weight reported by the needle.

Each of these spring scales reports a number for the weight of the melon you put in the basket. In order for that number to mean something, the scale has to be calibrated. Calibration is the process of comparing a local device or reference to a generally accepted standard to ensure accuracy. To calibrate a spring scale, the device or its reference components must be compared against standard weights. Someone must put a standard weight in the basket and measure just how far the spring stretches. Each spring is different, although spring manufacturers try to make all their springs as identical as possible.

Bouncing Bathroom Scales

As we noted earlier, the most common type of bathroom scale is also a spring scale (Fig. 3.1.4). When you step on this kind of scale, you depress its surface and levers inside it pull on a hidden spring. That spring stretches until it exerts, through the levers, an upward force on you that is equal to your weight. At the same time, a rack and pinion mechanism (see Fig. 3.1.3b) inside the scale turns a wheel with numbers printed on it. When the wheel stops moving, you can read one of these numbers through a window in the scale. Because which number you see depends on how far the spring has stretched, this number indicates your weight.

However, the wheel rocks briefly back and forth around your actual weight before it settles down. The wheel moves because you're bouncing up and down as the scale gradually gets rid of excess energy. When you first step on the scale's surface, its spring is not stretched and it isn't pushing up on you at all. You begin to fall. As you descend, the spring stretches and the scale begins to push up on your feet. But by the time you reach the equilibrium height, where the scale is exactly supporting your weight, you are traveling downward quickly and coast right past that equilibrium. The scale begins to read more than your weight.

The scale now accelerates you upward. Your descent slows, and you soon begin to rise back toward equilibrium. Again you coast past the proper height, but now the scale begins to read less than your weight. You are bouncing up and down because you have excess energy that is shifting back and forth between gravitational

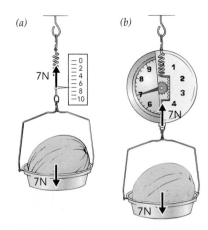

Fig. 3.1.3 Two spring scales weighing melons. Each scale balances the melon's downward weight with the upward force of a spring. The heavier the melon, the more the spring will stretch before it exerts enough upward force to balance the melon's weight. In (a), a pointer indicates how far the spring has stretched and thus how much the melon weighs. In (b), a rack and pinion mechanism rotates a needle on a dial. As the comblike rack moves up and down, it turns the toothed pinion gear.

Fig. 3.1.4 When you step on this bathroom scale, its surface moves downward slightly and compresses a stiff spring. The extent of this compression is proportional to your weight, which is reported by the dial on the left. Levers inside the scale make it insensitive to exactly where you stand.

potential energy, kinetic energy, and elastic potential energy. This bouncing continues until sliding friction in the scale has converted it all into thermal energy. Only then does the bouncing stop and the scale read your correct weight.

The bouncing that you experience about this stable equilibrium is a remarkable motion, one that we'll study in detail when we examine clocks in Chapter 9. You are effectively a mass supported by a spring and your rhythmic rise and fall is that of a harmonic oscillator. Harmonic oscillators are so common and important in nature that Chapter 9 is devoted to them. The details can wait, but there are two features of your present situation that I'll point out now.

First, your total potential energy is at its minimum when you're at the stable equilibrium. Even though both gravitational and elastic potentials are involved, their sum increases as you shift away from the equilibrium. Since an object always accelerates so as to reduce its total potential energy as quickly as possible, you always accelerate toward the stable equilibrium.

Second, your kinetic energy reaches its peak as you pass through the stable equilibrium. Having accelerated toward that equilibrium until the moment of arrival, you're moving fast and coast right through it. But once you leave it, you begin to accelerate toward it again. That acceleration is backward, opposite your velocity, so you are decelerating. Therefore, you reached your peak speed and kinetic energy at the moment you passed through equilibrium. As you bounce up and down, waiting for the scale to waste your excess energy, that excess transforms back and forth rhythmically between kinetic and potential forms.

Common Misconceptions: Equilibrium and Motionlessness
Misconception: an object at equilibrium is always motionless.
Resolution: an object at equilibrium is not accelerating, but its velocity may not be zero. If it was moving when it reached equilibrium it will coast through that equilibrium at constant velocity.

Using Several Scales at Once

One spring scale is enough for you, but how can you weigh an upright piano? It's too heavy and awkward for a single scale, but two scales will do the trick. If you put one scale under each side of the piano, the scales will work together to support its weight. Each scale will report just how much upward force it's exerting on the piano, so the sum of the two measurements will equal the piano's overall weight (Fig. 3.1.5).

The specific readings of the two scales will depend on the position of the piano's center of gravity. Its center of gravity is the effective location of its weight and coincides with its center of mass. Because the piano's longest and heaviest strings are on its left side, the piano's center of gravity is to the left of its middle. As a result, the left scale must support more of the piano's weight and it reads higher than the right scale.

We can explain the different readings by considering rotational motion. Like the seesaw in Section 2.1, the piano can rotate about its center of mass and will undergo angular acceleration in response to a net torque. To avoid angular acceleration so that it can rest motionless on the scales, the piano must be in rotational equilibrium—that is, it must experience zero net torque.

Because the piano's weight effectively acts at the piano's center of gravity, it has no lever arm and exerts no torque on the piano about its center of mass. However, the scales do exert torques on the piano about its center of mass. The left scale pushes up on the piano's left side, thereby producing a clockwise torque on the piano. From Eq. 2.1.3, the amount of this torque is the product of the left horizontal

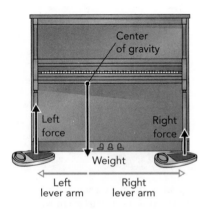

Fig. 3.1.5 You can weigh an upright piano by placing a spring scale under each end. Each scale exerts an upward force on the piano to support it, and the piano's weight is equal to the sum of those two forces, as measured by the scales.

lever arm times the left upward force. The right scale similarly produces a counterclockwise torque on the piano and the amount of that torque is the product of the right horizontal lever arm times the right upward force.

For the piano to be in rotational equilibrium, these two torques must cancel; they must be equal in amount, but opposite in direction. Their amounts will be equal when:

$$\text{left lever arm} \cdot \text{left force} = \text{right lever arm} \cdot \text{right force}\,.$$

Since the left lever arm is shorter than the right lever arm, the left force must be proportionately larger than the right force. That's why the left scale reads higher than the right scale.

This effect, where the scale that is more nearly beneath an object's center of gravity must support more of the object's weight, is familiar to anyone who has moved heavy objects. If you and a friend try to carry the piano in Fig. 3.1.5, the person carrying the piano's left side will bear more of the burden. And if an object is tipped so that its center of gravity is almost directly above one of the movers, as it is in Fig. 1.3.1, that mover will support almost the entire weight of the object.

Shortcomings and Solutions

Spring scales are popular because they're both simple and practical. Yet they have several inevitable shortcomings. As a spring ages, its equilibrium length may shift so that an old spring scale may no longer read zero when its basket is empty. To correct for this drift, most spring scales have an adjustment that raises the upper end of the spring so that its lower end is back at the proper position to read zero. Harder to remedy is a another age-related defect: a spring's tendency to grow less stiff with time, which causes a scale to overestimate the weight in its basket. Because of this second problem, grocery scales have to be checked periodically to ensure that they are still accurate; otherwise you get less than you pay for.

A third problem lies in the mechanical reporting mechanisms, such as the rack and pinion scheme in Fig. 3.1.3b, which inevitably reduce these scales' accuracy. Because of friction, the rack and pinion mechanism exerts small vertical forces on the spring and basket. Unavoidable and unpredictable, these forces move the basket's equilibrium position up or down slightly. As a result, each time you put a melon in the grocery store's hanging scale, it reports a slightly different weight. The melon's weight isn't really changing, just the scale's measurement of it.

Fortunately, there is a single solution to all three problems: a modern version of the spring scale that uses a load cell. Mechanically, a load cell is just another type of spring that exerts a restoring force proportional to how far it has been bent, compressed, or stretched. But the load cell can indicate the extent of its distortion in a unique way. Instead of turning a wheel or moving a needle, it changes its electrical properties. The load cell's distortion is measured completely electronically, so the actual amount of that distortion can be exquisitely small—so small, in fact, that it's usually hard to tell that anything in the scale is bending or compressing.

Load cell scales are gradually replacing conventional spring scales. They're already common in grocery stores, bathrooms, and kitchens (Fig. 3.1.6), not to mention hospitals and research laboratories. Because load cell scales don't use mechanical mechanisms, they don't suffer from the inaccuracies caused by friction; consequently, they're often advertised as containing "no moving parts" and report consistent weights. Most reset their zero positions before every measurement, so they don't need to be manually adjusted to zero. And because load cells don't age nearly as quickly as metal springs, load cell scales maintain their accuracies far longer than ordinary spring scales.

Fig. 3.1.6 This kitchen scale uses a load cell, a spring that reports its distortion electronically. A small computer in the scale measures how much the load cell distorts when you put items on the scale's surface and uses that measurement to determine the weight of those items. A tare button allows you to tell the scale what load-cell shape should correspond to the zero of weight.

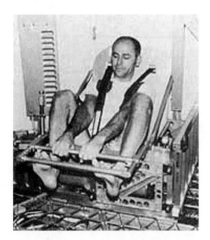

Fig. 3.1.7 Since a spaceship and its occupants are falling together, the astronauts can't measure their weights with a normal scale. Instead, this apparatus measures an astronaut's mass by pushing on him with a known force and recording his acceleration.

♒ Weighing Astronauts

While spring scales work well as long as there is little or no acceleration, they become useless when everything is falling freely. Imagine an astronaut orbiting the earth in a spaceship with its engines turned off. Although gravity still pulls on the astronaut, so that he actually does have weight, both he and his ship are in free fall and his acceleration makes him experience a sense of weightlessness (a feeling that we'll discuss in Section 3.3). If he tries to substantiate this feeling by standing on a spring scale, the scale will report—erroneously—that he is indeed weightless. In reality, the scale reads zero because he and the spring scale don't push on one another. They're both falling at the same rate and don't need any forces to keep them moving as a group. For that reason it's hard to weigh the astronaut in space, even though he has a weight.

It is relatively easy, however, to measure an astronaut's mass. The astronaut is given an impulse and the resulting change in his velocity is used to determine his mass (Fig. 3.1.7). Of course, the astronaut has to be careful not to wiggle around, since measuring his total mass requires that he move as a single object, not as a jumble of independent parts. To help him keep still, the astronaut clings tightly to a rigid frame and the force is exerted on that frame. The acceleration of the frame and astronaut is measured and used to determine the mass of the frame and astronaut. Since the mass of the frame is known, the astronaut's mass is easy to calculate.

♒ Balance Scales

A balance compares the weights of two objects by putting those two objects at opposite ends of a seesaw-like lever. This lever is supported by a central pivot and *balances* when the torques exerted by the two objects cancel one another. To distinguish the device known as a balance from the concept of *balancing*, I will italicize the latter throughout this section. As you'll see, I must occasionally resort to italicizing a concept in this book because people have unimaginatively named an object or material after it. In a typical beam balance (the device), the two objects are equidistant from the pivot (Fig. 3.1.8) so that the lever will *balance* (the concept) if the objects have identical weights.

Because weight and mass are proportional to one another, two objects with equal weights also have equal masses. Thus while a balance uses weight in its operation, it's also accurately comparing the objects' masses. If one object is a set of reference masses, then the balance will indicate the mass of the other object—the balance will only *balance* when the mass of the other object is equal to the total mass of the set of reference masses.

Since the strength of gravity affects both objects equally, a balance can measure mass accurately despite changes in the local strength of gravity. It's also insensitive to overall accelerations of the balance. A balance will continue to make accurate comparisons of mass, even when you take it from the North Pole to the equator. Although the earth's gravity is weaker at the equator and the ground there is accelerating inward due to the earth's rotation, a balance will be unaffected. Its accuracy and insensitivity to location explains why a balance is used in doctors' offices and at weigh-ins for sporting events.

There are two problems, however, with the simple beam balance. First, you have to be sure that the unknown and reference objects are at exactly the same distance from the pivot. If the distances are even slightly different, then equal weights will not produce equal but oppositely directed torques and the balance will be inaccurate. Second, you must have reference masses with exactly the same mass as your unknown object. This requirement is an important issue when measuring large objects such as a person or a moving van or small objects such as a fly.

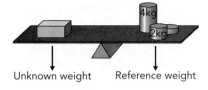

Unknown weight   Reference weight

Fig. 3.1.8 If the objects are at equal distances from the pivot, the beam will *balance* when the unknown object has the same weight as the reference objects. Because weight is proportional to mass, the unknown object and the reference objects also have equal masses. In this case, the unknown object has a mass of 7 kg and weighs 68.6 N.

Let's begin by solving the distance to the pivot problem. The easiest way to make sure that the objects exert their weights at equal distances from the pivot is to use two hanging baskets (Fig. 3.1.9). These hanging baskets are attached to the beam at fixed points that are exactly equidistant from the pivot. Now, no matter where in the basket you put the objects, the basket will shift until the object's weight is centered exactly under the point of attachment to the beam.

This automatic centering dramatically improves the accuracy of the balance. It occurs because placing the basket's center of gravity directly below the attachment point reduces the altitude of that center of gravity to its lowest possible value and thereby minimizes the basket's total potential energy. If you then tip the basket, its center of gravity will rise and its total potential energy will increase. Since an object accelerates in the direction that reduces its total potential energy as quickly as possible, the basket will always accelerate back toward center and, once there, it won't accelerate at all. The centered basket is in stable equilibrium. Although it may swing back and forth briefly after you add objects to it, it will soon settle down with its center of gravity directly below its attachment point. (For applications of this effect, see ❶.)

We can now see why hanging baskets improve the balance. No matter where in the basket you place the objects, the basket will tilt so that its overall center of gravity is directly below its attachment to the beam. The accuracy of the balance depends only on where the baskets attach to the beam and not on the loading of the baskets. The Egyptians used balances of this type more than 7000 years ago, although the addition of a sharp fulcrum as the center support is a more recent Roman improvement.

Another way to make a balance insensitive to where you place the objects is to use a trick devised by the French mathematician Gilles Personne de Roberval (1602–1675) in 1669. The Roberval movement has two separate beams arranged so that the evenly spaced weighing pans remain exactly horizontal as the beams rotate (Fig. 3.1.10). During rotation, the two pans move exactly equal vertical distances, but in opposite directions. If the weights (and masses) in the two pans are equal, then the work done on the balance by the descending pan is exactly the work needed to lift the ascending pan. Since there is no work left over, nothing accelerates and the balance *balances*. If the weights (and masses) in the two pans aren't equal, one pan will accelerate downward. We noticed a similar effect with seesaws in Section 2.1—a seesaw only *balances* if the work needed to lift one child is exactly equal to the work provided by lowering the other child.

Fig. 3.1.9 If this hanging basket beam balance *balances*, then the object in the left basket has the same weight and mass as the reference objects in the right basket. The baskets move until their centers of gravity are located directly below their points of attachment to the beam.

❶ To find the vertical when constructing a building, builders attach a heavy plumb bob to a string and hang it from a support. The plumb bob and string form a single object whose center of gravity is located inside the plumb bob itself. In order to minimize its total potential energy, the plumb bob settles down in stable equilibrium with its center of gravity directly below the support point. The builders can then use the string as a reference for vertical. If you suspend a more complicated object—a chandelier or a wall hanging—from a point other than its center of mass, that object will also orient itself so that its center of gravity is directly below the support point.

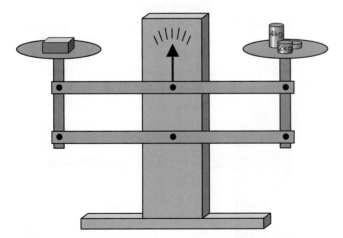

Fig. 3.1.10 This Roberval balance *balances* when the weights of objects on the left and right pans are identical. When that is the case, the work needed to raise one pan is exactly the work done by lowering the other pan.

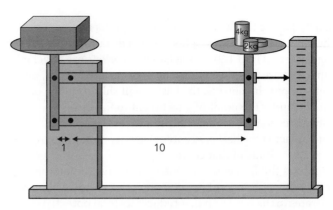

Fig. 3.1.11 This version of the Roberval style balance allows a heavy object to be weighed with light reference objects. Because the right pan is 10 times as far from the pivot as the left pan, the balance *balances* when the object in the left pan is 10 times as heavy as the object in the right pan. The left pan itself must be heavier than the right pan so that the balance *balances* when empty.

The Roberval movement makes it easy to solve the second problem: the awkward need for a reference object with the same mass as the object you're measuring. If you move the pivot away from the middle a Roberval balance, it will *balance* when the products of each object's mass times the length of its side of the balance are equal. For example, if the right side of the balance is 10 times as long as the left side, as it is in Fig. 3.1.11, it will *balance* when the object on the left side has 10 times the mass of the object on the right side. As before, it *balances* when the energy transfer between objects is perfect but now the balance is using mechanical advantage. The object on the right moves vertically 10 times as far as the object on the left but the object on the right weighs only 1/10th as much as the object on the left. The balance *balances* because the work done by the descending object is completely used in raising the other object.

Many other styles of balances are possible. They all use some technique that makes them insensitive to the exact positions of the objects and many of them use mechanical advantage to facilitate measuring very large or very small masses.

Some balances use a fixed set of reference objects but change their mechanical advantages (their distances from the pivot) in order to measure mass (Fig. 3.1.12). The balance in a doctor's office works this way. You slide three different-sized reference masses along a beam until it *balances*—your weight on one side and the reference masses on the other. You can then determine your mass from the positions of the reference masses. There are also some balances that automatically adjust the mechanical advantage of their reference objects and read out your weight on a dial. The weights swing on mechanical levers and don't have to be moved by hand.

Some balances introduce very small restoring torques that act to orient the beam horizontally. Without these additional torques, the beam of a *balanced* balance experiences no angular acceleration—it might be stationary and horizontal, but it might not. The tiny added restoring torques cause the beam to adopt a horizontal orientation when it's approximately balanced. Although these extra torques reduce the balance's sensitivity somewhat, they make it much easier to tell when the beam is approximately balanced.

Fig. 3.1.12 This version of the Roberval balance uses a single sliding reference object to measure the weight and mass of an unknown object. When the weighing pan is empty, the beam *balances* with the sliding reference object at the left mark. Now that there is a weight in the pan, the reference object must be moved farther from the pivot to give it more mechanical advantage. The object's weight and mass are determined by the position of the reference object when the beam *balances*.

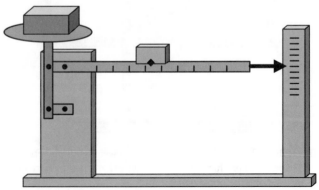

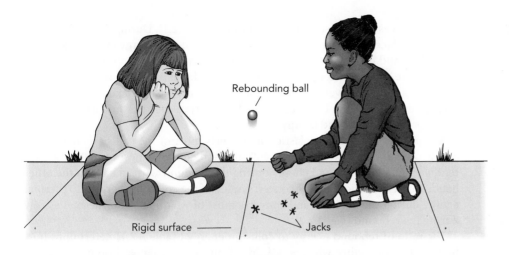

Rebounding ball

Rigid surface ——————— Jacks

Section 3.2 **Bouncing Balls**

If you visit a toy or sporting goods store, you'll find many different balls—almost a unique ball for every sport or ball game. These balls differ in more than just size and weight. Some are very hard, others very soft; some are smooth, others rough or ridged.

In this section we'll focus primarily on another difference: the ability to bounce. A superball, for example, bounces extraordinarily well, while a foam rubber ball hardly bounces at all. Even balls that appear identical can be very different; a new tennis ball bounces much better than an old one. We'll begin this section by exploring these differences.

The Way the Ball Bounces: Balls as Springs

In many ways, balls are perfect objects. What would most sports be like without them? How would industrial machines function without ball bearings to keep them from grinding to a halt? Their simple shapes, uncomplicated motions, and ability to bounce make balls both fascinating and useful. Most balls are spherical, meaning that when no outside forces act on them they adopt spherical equilibrium shapes. But some balls, such as those used in U.S. football and rugby, have equilibrium shapes that are not spherical but oblong.

The term "equilibrium shape," of course, is one we've seen already: the previous section used it to describe springs. This is no coincidence, for a spherical ball behaves like a spherical spring. In fact, everything we associate with springs has some place in the behavior of balls. For example, when you push a ball's surface inward, it exerts an outward restoring force on you. When you do work on the ball as you distort its surface, it stores some of this work as elastic potential energy and when you let the ball return to its equilibrium shape, it releases that stored energy.

This springlike behavior is evident when a ball collides with the floor or with a bat. The ball's surface distorts during the collision, giving it elastic potential energy that is released when the ball rebounds. If the ball is moving, then some of this stored energy comes from the ball's kinetic energy. If the ball hits a moving surface, then some of this stored energy comes from the surface's kinetic energy. Much of the stored energy reappears as kinetic energy in the ball and surface as the ball rebounds.

Some balls bounce better than others. I call a very bouncy ball "lively" and a ball that doesn't bounce well "dead." One way to look at a ball's liveliness is to compare kinetic energies before and after the bounce. We can do that by dividing the bounce

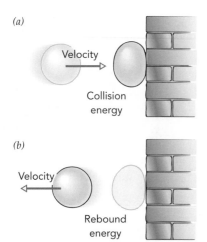

(a)

Velocity

Collision energy

(b)

Velocity

Rebound energy

Fig. 3.2.1 A bounce from a wall has two halves: (*a*) the collision and (*b*) the rebound. During the collision between the ball and the wall, some of their kinetic energy is transformed into other forms—an amount called the collision energy. During the rebound, some stored energy is released as kinetic energy—an amount called the rebound energy. The rebound energy is always less than the collision energy because some energy is lost as thermal energy. However, a lively ball wastes less energy than a dead one.

Fig. 3.2.2 When a tennis ball hits the floor, it dents inward to store energy and then rebounds somewhat more slowly than it arrived. These images show the ball's position at twelve equally spaced times. Is the ball bouncing to the left or the right? How can you tell?

into two halves: the collision and the rebound (Fig. 3.2.1). During the collision, the ball and surface convert a portion of their overall kinetic energy into elastic potential energy and thermal energy. The amount of kinetic energy absorbed at impact is called the collision energy. During the rebound, the ball and surface push away from one another, converting elastic potential energy back into kinetic energy and thermal energy. The total amount of kinetic energy released as the surface and ball push apart is the rebound energy.

A lively ball is extremely elastic—it converts most of the collision energy into elastic potential energy during the collision and converts most of that elastic potential energy into rebound energy during the rebound. A dead ball, on the other hand, is barely elastic at all—it converts much of the collision energy into thermal energy during the collision and converts most of what's left into thermal energy during the rebound (Fig. 3.2.2).

The ratio of rebound energy to collision energy (Table 3.2.1) determines how high a ball will bounce when you drop it from rest onto a hard floor (Fig. 3.2.3). An ideally elastic ball would have a ratio of 1.00 and would rebound to its initial height. But a real ball wastes some of the collision energy and rebounds to a lesser height. The height from which you drop the ball is proportional to its initial gravitational potential energy and therefore to its collision energy. The height to which it rebounds is similarly proportional to its rebound energy. The ratio of these two heights is therefore a good measure of the ratio of the rebound energy to the collision energy. The smaller this ratio (Table 3.2.1), the less kinetic energy the ball receives during the rebound and the weaker the bounce.

Table 3.2.1 Approximate Energy Ratios and Coefficients of Restitution for a Variety of Balls

Type of Ball	Rebound Energy/ Collision Energy	Coefficient of Restitution
Superball	0.81	0.90
Racquet ball	0.72	0.85
Golf ball	0.67	0.82
Tennis ball	0.56	0.75
Steel ball bearing	0.42	0.65
Baseball	0.30	0.55
Foam rubber ball	0.09	0.30
"Unhappy" ball	0.01	0.10
Beanbag	0.002	0.04

While that energy ratio is often useful, a ball is traditionally characterized by its coefficient of restitution—the ratio of its rebound *speed* to its collision *speed* when it bounces off a hard, immobile surface:

$$\text{coefficient of restitution} = \frac{\text{outgoing speed of ball}}{\text{incoming speed of ball}}. \quad (3.2.1)$$

Scientists have found that, for most balls, this speed ratio remains constant over a wide range of collision speeds. A ball that rebounds with the same speed that it had when it collided with the surface has a coefficient of restitution of 1.00. A superball is almost this lively, with a coefficient of restitution of about 0.90. Thus, when a superball traveling at 10 km/h collides with a concrete wall, it rebounds at about 9 km/h. In contrast, a foam rubber ball's coefficient of restitution is about 0.30, while that of a beanbag is almost zero.

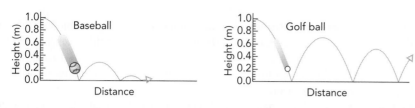

Fig. 3.2.3 A baseball wastes 70% of the collision energy as thermal energy and bounces weakly. In contrast, a golf ball wastes only 30% of the collision energy and bounces well.

Actually, the energy ratio and the coefficient of restitution are directly related. Remember, a ball's kinetic energy equals half its mass times its velocity squared. Even if we don't know the masses of the balls, we know that the energy ratio is equal to the square of the speed ratio. Thus if a foam rubber ball rebounds at only 0.30 times its collision speed, it retains only 0.30^2—0.09 times, or 9%—of its original kinetic energy, and the remaining 91% is converted to thermal energy in the rubber and air that make up the ball. A superball, in contrast, retains about 81% of its original kinetic energy after a bounce.

Balls bounce best when they store energy through compression rather than through surface bending. That's because most ball materials, such as leather or leather-like plastics, experience lots of wasteful internal friction during bending. Since solid balls involve compression, they usually bounce well, whether they're made of rubber, wood, plastic, or metal. But air-filled balls bounce well only when properly inflated. A normal basketball, which stores most of its energy in its compressed air, has a high coefficient of restitution. In contrast, an underinflated basketball, which experiences lots of surface bending during a collision, barely bounces at all. Similarly, a tennis ball bounces best when new; after a while, the compressed air inside leaks out and the ball's coefficient of restitution drops.

How the Surface Affects the Bounce

Since the surface on which a ball bounces isn't perfectly hard, that surface also contributes to the bouncing process. It distorts and stores energy when the ball hits it and returns some of this stored energy to the rebounding ball. Overall, the collision energy is shared between the ball and the surface, both of which behave as springs, and each provides part of the rebound energy.

Just how the collision energy is distributed between the surface and ball depends on how stiff each one is. During the bounce, they push on one another with equal but oppositely directed forces. Since the forces denting them inward are equal, the work done in distorting each object is proportional to how far inward it dents. Whichever object dents the farthest receives the most collision energy.

Since the ball usually distorts more than the surface it hits, most of the collision energy normally goes into the ball. As a result, you might expect the ball to provide most of the rebound energy, too. However, that's not always true. Some lively elastic surfaces store collision energy very efficiently and return almost all of it as rebound energy. Since a relatively dead ball wastes most of the collision energy it receives, a lively surface's contribution to the rebound energy can be very important to the bounce. A lively racket is critical to the game of tennis because the racket's strings provide much of the rebound energy as the ball bounces off the racket (Fig. 3.2.4). Trampolines and springboards are even more extreme examples, with surfaces so lively that they can even make people bounce. People, like beanbags, have coefficients of restitution near zero; when you land on a trampoline, it receives and stores most of the collision energy and then provides most of the rebound energy.

The stiffnesses of the ball and surface also determine how much force each object exerts on the other and thus how quickly the collision proceeds. Hard objects

Fig. 3.2.4 When a tennis ball bounces from a moving racket, both the ball and racket dent inward and share the collision energy. The racket, however, is actually more elastic than the ball, so storing more collision energy in a loosely strung racket leads to a faster outgoing ball. A loosely strung racket's extra speed comes at the expense of angular control; a powerful player who wants accuracy instead will select a tightly strung racket.

❶ How much a particular ball affects its target depends on the ball's liveliness. When a beanbag hits its target and stops dead, it transfers all of its forward momentum to the target. But when a superball hits its target and rebounds elastically, it transfers almost twice its forward momentum to the target. The superball's enhanced momentum transfer makes it the projectile of choice when you're trying to knock down targets at the state fair. Similarly, a rubber-tipped battering ram is much more effective at knocking down doors than one that's tipped with sand.

resist denting much more strongly than soft objects. When both objects are very hard, the forces involved are large and the acceleration is rapid. Thus a steel ball rebounds very quickly from a concrete floor because the two exert enormous forces on one another. If the ball and/or surface are relatively soft, the forces are weaker and the acceleration is slower.

What if the surface that a ball hits isn't very massive? In that case, the surface may do part or all of the "bouncing." During the bounce, the ball and the surface accelerate in opposite directions and share the rebound energy. Massive surfaces, such as floors and walls, accelerate little and receive almost none of the rebound energy. But when the surface a ball hits is not very massive, you may see it accelerate. When a ball hits a lamp on the coffee table, the ball will do most of the accelerating, but the lamp is likely to fall over, too (see ❶).

Similarly, when a baseball strikes a baseball bat, the ball and bat accelerate in opposite directions. The more massive the bat, the less it accelerates. To ensure that most of the rebound energy went to the baseball, the legendary hitters of the early twentieth century used massive bats. Such bats are no longer in vogue because they're too difficult to swing. But in the early days of baseball, when pitchers were less skillful, massive bats drove many long home runs.

How a Moving Surface Affects the Bounce

The last paragraph describes the act of hitting a baseball with a moving baseball bat as though it were a case of "bouncing." That might sound a little strange. When a baseball hits a stationary bat, the ball bounces. But if a moving bat hits a stationary baseball, is it proper to say that the ball bounces?

The answer is yes. In fact, which object is moving and which is stationary depends on your point of view—your inertial frame of reference. A fly resting on the baseball will claim that the baseball is stationary and that it's about to be struck by a moving bat. Another fly resting on the baseball bat will claim that the bat is stationary and that it's about to be struck by a moving baseball. Which fly has the correct inertial frame of reference?

As we noted in Section 1.1, both frames of reference are equally valid. An inertial frame of reference is one that's not accelerating and is thus either stationary or moving at a constant velocity. As long as you view the world around you from an inertial frame of reference, the laws of motion will accurately describe what you see, and energy, momentum, and angular momentum will all be conserved.

But frames of reference aren't the first things you think of during a baseball game. When you swing a bat and drive the pitch toward center field, your main concern is how fast the outgoing ball is traveling toward the bleachers. A speedy ball will be a home run, while a slower ball will be an out. What determines the ball's outgoing speed?

With both bat and ball moving relative to the playing field, there are several useful inertial frames of reference from which to study the collision. However, we'll find it easier to focus on how quickly the bat and ball move toward or away from one another. This relative motion is what matters most in a collision. After all, whether a rock hits a bottle or a bottle hits a rock, it's going to be bad for the bottle.

When a ball bounces off a *moving* surface that's rigid and massive, the ball's coefficient of restitution still applies. But now we must use a more general form of that speed ratio. This version divides the speed at which the ball and surface separate after the bounce by the speed at which they approach before the bounce:

$$\text{coefficient of restitution} = \frac{\text{speed of separation}}{\text{speed of approach}}. \quad (3.2.2)$$

When the surface is stationary, Eq. 3.2.2 is equivalent to Eq. 3.2.1.

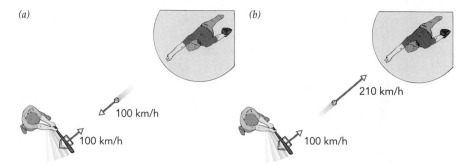

(a)   (b)

100 km/h
100 km/h   210 km/h
100 km/h

Fig. 3.2.5 (*a*) Before they collide, the bat and ball are approaching one another at an overall speed of 200 km/h. (*b*) After the collision, the two are separating from one another at a speed of 110 km/h. However, because the bat is moving toward the pitcher at 100 km/h, the outgoing ball is traveling at 210 km/h in that same direction.

To see how this generalization allows us to explain why the baseball you hit is now traveling over the center fielder's head, let's examine the collision between bat and ball. Let's suppose that, just before the collision (Fig. 3.2.5*a*), the pitched baseball was approaching home plate at 100 km/h (62 mph) and that, as you swung to meet the ball, your bat was moving toward the pitcher at 100 km/h. Since each object is moving toward the other, their speed of approach is the sum of their individual speeds, or 200 km/h (124 mph).

The baseball's coefficient of restitution is 0.55, so after the collision (Fig. 3.2.5*b*) the speed of separation will be only 0.55 times the speed of approach, or 110 km/h. The outgoing ball and the swinging bat separate from one another at 110 km/h. Since the bat is still moving toward the pitcher at 100 km/h, the ball must be traveling toward the pitcher even faster: at 100 km/h plus 110 km/h or a total speed of 210 km/h (130 mph)! That's why the baseball flies past everyone in the outfield and into the stands.

Surfaces Also Bounce... and Twist and Bend

As you can see, a surface's motion has a large effect on a ball bouncing off it. A surface that moves toward an incoming ball will strengthen the rebound, while one that moves away from the ball will weaken it. But we're neglecting the ball's effect on the surface itself. Sometimes that surface bounces, too.

A baseball bat is a case in point. When you swing your bat into a pitched ball, the bat doesn't continue on exactly as before. The ball pushes on the bat during the collision and the bat responds in a number of interesting ways.

First, as we noted before, the bat rebounds from the ball. The bat decelerates slightly during the collision so that its speed after the impact is a little less than before it. Since the ball's final speed depends on the bat's final speed, a slower bat means a slower ball. Thus we've slightly overestimated the ball's speed as it heads toward the bleachers.

Second, the ball's impact sets the bat spinning. When the ball pushes on the bat and makes it accelerate backward, it also exerts a torque on the bat about its center of mass and makes it undergo angular acceleration (Fig. 3.2.6). While these two types of acceleration might seem inconsequential, your hands notice their effects. The bat's acceleration tends to yank its handle toward the catcher, while its angular acceleration tends to twist its handle toward the pitcher. The extent of these two motions depends on just where the ball hits the bat. If the ball hits a point known as the center of percussion, the handle experiences no overall acceleration (Fig. 3.2.6*c*). The smooth feel of such a collision explains why the center of percussion, located about 7 inches from the end of the bat, is known as a "sweet spot."

(a)

(b)

Center of mass

Center of percussion

(c)

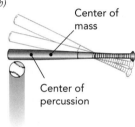

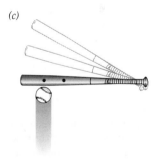

Fig. 3.2.6 When a ball hits a bat, the bat experiences both acceleration and angular acceleration. (*a*) If the ball hits near the bat's middle, the angular acceleration is small and the bat's handle accelerates backward. (*b*) If the ball hits near the bat's end, the angular acceleration is large and the bat's handle accelerates forward. (*c*) But if the ball hits the bat's center of percussion, the angular acceleration is just right to keep the handle from accelerating.

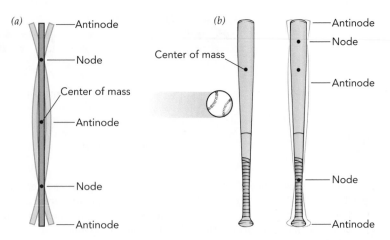

Fig. 3.2.7 (*a*) When struck by a mallet, a xylophone bar vibrates with its middle and ends moving back and forth in opposite directions. The parts that move farthest are antinodes, while the points that don't move at all are nodes. (*b*) When struck by a ball, a baseball bat vibrates in a similar fashion. However, an impact at one of the bat's nodes causes no vibration.

Finally, the collision often causes the bat to vibrate. Like a xylophone bar struck by a mallet (Fig. 3.2.7*a*), the bat bends back and forth rapidly with its ends and center moving in opposite directions (Fig. 3.2.7*b*). These vibrations sting your hands and can even break a wooden bat. However, near each end of the bat, there's a point that doesn't move when the bat vibrates—a vibrational node. When the ball hits the bat at its node, no vibration occurs. Instead, the bat emits a crisp, clear "crack" and the ball travels farther. Fortunately, the bat's vibrational node and its center of percussion almost coincide so you can hit the ball with both sweet spots at once.

As these handle motions and bending vibrations illustrate, the science and engineering of bats is surprisingly complicated. That's why bat manufacturers are forever developing better, more potent ones. Like the makers of golf clubs, tennis rackets, bowling equipment, and billiard tables, they dream of bounces so perfect that all of the collision energy is stored and returned as rebound energy. Known as elastic collisions, such perfect bounces are common among the tiny atoms in a gas, but unattainable for ordinary objects; there are just too many ways for any large item to divert or dissipate energy, including as thermal energy, sound, vibration, and light.

Although equipment manufacturers must resign themselves to inelastic collisions—collisions that fail to return some of the collision energy as rebound energy—they get closer to perfection every year. They also push right up to the limits of regulation, and occasionally beyond, in their quests for maximum performance. (For a discussion of one of the oldest schemes for improving bats, see ❷.)

Multiwalled aluminum, titanium, and composite bats are a case in point. Each of these hollow bats is soft enough to dent considerably during its impact with the ball so that it receives much of the collision energy. Its circular barrel flattens into an oval, storing the collision energy beautifully, and then kicks back to circular over a time that is well matched to the timing of the bounce itself. And because the hi-tech bat's outer wall is thin and light, it retains only a tiny fraction of the rebound energy. Unlike the hard surface of a wooden bat, which barely dents and thus barely participates in the energy storage process, one of these hi-tech bats acts as a trampoline—it stores and returns so much of the collision energy that it substantially increases the outgoing speed of the batted ball. You can find equally hi-tech equipment at the golf shop or tennis store. We're in an era when scientific analysis and design are radically altering sports.

❷ Long before the high-tech revolution in sports, baseball players secretly tampered with their wooden bats in hopes of making them more potent. The most infamous of these forbidden bat modifications was "corking," in which the bat was hollowed out and the missing wood replaced with cork. But despite its mystique, a number of scientists, including American physicists Robert Adair, Alan Nathan, Dan Russell, and Lloyd Smith, have concluded that a corked bat offers a player no measurable advantages that couldn't be obtained legally simply by choosing a slightly light, shorter, or thinner bat.

Billiards, Pool, and Croquet: Balls Hitting Balls

Baseball is played with a single ball, but games such as billiards, pool, and croquet are played with several (Fig. 3.2.8). Moreover, those balls often hit one another with remarkable results. In billiards and pool, when a moving cue ball strikes a stationary object ball squarely, they appear to exchange motions: the cue ball stops and the object ball moves onward in its place (Fig. 3.2.9). In croquet, there is an equivalent maneuver known as "sending" an opponent's ball: a player drives her ball directly into her opponent's stationary ball at very high speed and the two exchange motions: the player's ball stops and the opponent's ball sails across the field and out-of-bounds. How do these tricks work?

During each of these collisions, the two balls push on one another with equal but oppositely directed forces and thereby exchange momentum and energy. By pushing on one another for a period of time, the balls exchange momentum: the incoming ball transfers forward momentum to the outgoing ball. By pushing on one another as both balls move a distance in the forward direction, the balls exchange energy: the incoming ball transfers translational kinetic energy to the outgoing ball. And if the incoming ball gives *all* of its forward momentum and *all* of its translational kinetic energy to the outgoing ball, then the outgoing ball will appear to take over the incoming ball's motion; the incoming ball will stop after the collision and the outgoing ball will continue forward in its place.

Such a perfect transfer, however, can only occur if the two balls have equal masses and are perfectly elastic—they must have coefficients of restitution equal to 1. When their masses are equal, conservation of momentum requires the outgoing ball's forward speed to increase by the same amount the incoming ball's forward speed decreases. When they are perfectly elastic, conservation of energy requires that the two balls have the same total translational kinetic energy after the collision as before it. A collision can satisfy both requirements simultaneously if and only if the incoming ball conveys all of its translational kinetic energy and forward momentum to the outgoing ball. Therefore, if two perfectly elastic balls with equal masses collide head-on, they'll exchange motions perfectly.

Even when the two balls aren't perfectly matched or aren't perfectly elastic, or when they don't quite hit squarely, they still transfer their motions remarkably well. They don't even have to be balls. Try sliding two similar coins into one another on a smooth table and watch them exchange motions! Lawn bowling, shuffleboard, and marbles all depend on this motion-transfer effect to let you knock an opponent's ball or disk out of place with one of your own.

But this nearly perfect transfer of forward momentum and energy isn't quite the whole story, even for two well-matched and elastic balls. All too often the incoming ball stops only momentarily and then slowly follows the outgoing ball. Its renewed motion is caused by something I've neglected so far: rotation. If the incoming ball was spinning before the collision, it will be spinning after the collision and that rotation may cause it to move.

Suppose, for example, that the incoming ball was rolling forward prior to the collision. Although the collision gives away the incoming ball's forward momentum and translational kinetic energy, it has relatively little effect on the ball's angular momentum and rotational kinetic energy. The incoming ball stops translating, but it doesn't stop rotating. And like the wheel of a dragster at the start of a race, the spinning ball then uses sliding friction with the ground to propel itself forward (Fig. 3.2.10*a*). The incoming ball converts some of its rotational kinetic energy into translational kinetic energy and slowly follows the outgoing ball forward. When you've just used the cue ball to knock an object ball into a pocket, the cue ball's slow roll into that same pocket is heartbreaking. A skillful pool or billiards player learns to control not only the cue ball's linear momentum, but also its angular momentum.

Fig. 3.2.8 This pool player uses a cue stick to propel the white cue ball across the felt-covered table so that it collides with the colored object balls and knocks them into pockets. The player transfers momentum and energy through the stick and cue ball to the object balls. Pool and billiard balls demonstrate the laws of motion so directly that physicists frequently draw analogies to "billiard-ball physics."

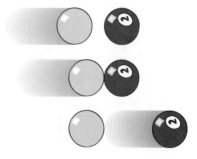

Fig. 3.2.9 When the cue ball strikes a stationary object ball directly, they exchange energy and momentum almost perfectly. The object ball begins to move with the velocity of the cue ball and the cue ball stops. That simple behavior, however, neglects the cue ball's spin.

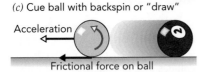

(a) Cue ball with topspin or "follow"

Acceleration

Frictional force on ball

(b) Cue ball with no spin

(c) Cue ball with backspin or "draw"

Acceleration

Frictional force on ball

Fig. 3.2.10 The cue ball's spin prior to a collision affects its motion after the collision. (*a*) If the cue ball had topspin or "follow," friction with the table will accelerate it forward after the collision so that it follows the object ball. (*b*) If the cue ball had zero spin, it will stop after the collision. (*c*) If the cue ball had backspin or "draw," friction with the table will accelerate it backward after the collision so that it draws back away from the object ball.

Just because the cue ball is *moving* forward doesn't mean it has to be *rolling* forward. In fact, if you aim the cue stick directly at the center of the cue ball and strike that ball hard, the stick will exert no torque on the ball and the ball will begin translating without rotating. If the cue ball still isn't rotating when it hits an object ball, it will stop forever after the collision because it will have no way to propel itself forward (Fig. 3.2.10*b*).

A ball that isn't rolling across the table is skidding across the table. The table is therefore pushing the bottom of the ball opposite the bottom's velocity with a horizontal force of sliding friction. This force produces a torque on the ball that will gradually transform its motion from skidding to rolling. However, while the sliding frictional force is getting the ball rolling, it is also making the ball accelerate horizontally. If the ball was initially spinless, as it was in the previous paragraph, its horizontal acceleration will be backward and it will gradually slow down. But if the ball was set spinning by the cue stick, its horizontal acceleration can be in any direction and the ball may follow a curving trajectory as it heads across the table.

Spinning the cue ball with the cue stick requires both an off-center hit from the cue stick and friction. Because the ball is spherical, any support force that the stick exerts on it points directly at its center and produces no torque on it about its center. To spin the ball during the impact, the stick must push the ball's surface sideways with a frictional force. Chalking the cue stick gives it more traction and helps it exert torque on the ball. By carefully controlling the initial spin and velocity of the cue ball, a skillful player can determine both its path across the table and its motion after a collision with an object ball.

If the cue stick hits the cue ball above or below center, it will put topspin or backspin on the ball, respectively. A ball with topspin is similar to a rolling ball and follows any stationary ball it strikes squarely (Fig. 3.2.10*a*). Not surprisingly, topspin is called follow in billiards and pool. In contrast, a ball with backspin is drawn backward after striking a stationary ball squarely (Fig. 3.2.10*c*), so backspin is referred to as draw. Hitting the cue ball left or right of center imparts sidespin on the ball and is known as "putting English on the ball." Pure English makes the ball spin like a toy top and doesn't cause any horizontal acceleration. But combinations of English and either follow or draw make the cue ball accelerate horizontally and curve as it travels along the table.

The post-collision self-propulsion of a spinning ball is important in other sports, most notably in croquet when a player sending an opponent's ball wants to position his ball strategically after the collision. Of much wider applicability, however, is the horizontal acceleration that accompanies a ball's transition from skidding to rolling. This effect is useful in a great many sports, even those with only a single ball.

Nowhere is this horizontal acceleration more important than in bowling. When the bowler releases the ball onto the bowling lane, the ball is spinning rapidly and skidding forward along the surface. The first portion of the lane is oiled so that the ball has little traction and basically coasts straight ahead. But once the oil ends and the ball's traction increases, the skidding to rolling transition begins in earnest. The ball than accelerates horizontally and its path curves. If the ball is properly aimed and spun, it will hit the triangle of bowling pins from the side and at an angle, and it will knock all the pins over.

Still More Bouncing

At this point, it's becoming clear that bouncing is everywhere. Although you might not always recognize it as bouncing, almost any time two things bump into one another, they'll exhibit the phenomena we've been examining. Before turning to a different topic, I'll point out a few more cases where bouncing is important and also a few more concepts that make bouncing so interesting.

For starters, let's return to pool and billiards and consider what happens when two balls *don't* collide; what happens when a moving ball instead hits a bumper or nothing at all? Hitting a bumper sure sounds like a bounce, so let's begin there.

When a ball hits a bumper at right angles (i.e., perpendicular to the bumper's surface), it bounces directly backward. According to what we've already discussed, the ball and bumper dent during the impact, store collision energy in their wonderfully elastic surfaces, and return most of that stored energy to the ball as rebound energy. The ball rebounds from the bumper with its velocity and momentum reversed. And since the bumper is much softer than the ball, it does most of the denting and most of the energy storage. So far, we're simply reviewing old stuff.

But what if the ball doesn't hit the bumper at right angles? If we assume for the moment that the bumper is perfectly elastic, then the ball rebounds at the mirror angle: it "reflects" off the bumper on the opposite side of the perpendicular (Fig. 3.2.11a). That's because the bounce reverses only the components of the ball's velocity and momentum that are perpendicular to the bumper. Moreover, only part of the ball's kinetic energy—that associated with its motion perpendicular to the bumper—is stored and returned. The part of the ball's kinetic energy associated with its motion parallel to the bumper remains in the ball's motion throughout the bounce.

So while the ball's motion perpendicular to the bumper undergoes a bounce and reverses direction, the ball's motion parallel to the bumper continues unaffected. Since the bounce reverses only the perpendicular component of the ball's motion, not the parallel component, the ball rebounds at the mirror angle.

Unfortunately, the bumper isn't perfectly elastic, so the ball loses some of its perpendicular kinetic energy during the bounce. The ball's components of velocity and momentum perpendicular to the bumper are thus less after the bounce than before it and the ball rebounds at an angle farther from the perpendicular than the mirror angle (Fig. 3.2.11b). Billiard and pool players learn to work with this not-quite-mirrorlike reflection; however, they expect it to be uniform around a table and consistent between tables. Table manufacturers go to great lengths to make their bumpers as uniform and elastic as possible.

Once again, I've neglected the ball's rotation. During the time the ball is in contact with the bumper, traction between the two allows a spinning ball to use friction to propel itself along the bumper. Depending on which way the ball is spinning and how fast, the bumper will exert frictional forces on it that can alter the ball's components of velocity and momentum parallel to the bumper. A ball with backspin (Fig. 3.2.12a) bounces at an angle that's less forward than it would be without spin and at a reduced speed. A ball with topspin (Fig. 3.2.12b) bounces at an angle that's more forward and at an increased speed. Skilled players use these effects to their advantage.

(a) Bounce from perfect bumper

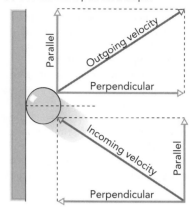

(b) Bounce from real bumper

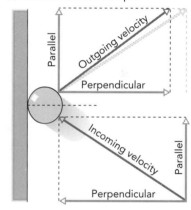

Fig. 3.2.11 (*a*) When a cue ball collides with a perfectly elastic bumper, the component of its velocity perpendicular to the bumper reverses and the ball "reflects" from the bumper at the mirror angle. (*b*) But a real bumper isn't perfectly elastic, so the ball has less perpendicular speed after the bounce. The real bounce angle is farther from the perpendicular than the mirror angle. This figure assumes no spin effects.

Fig. 3.2.12 (*a*) When a cue ball with backspin collides with a perfectly elastic bumper, the component of its velocity parallel to the bumper decreases and the ball bounces weakly and less forward than it would without spin. (*b*) For a cue ball with topspin, the component of its velocity parallel to the bumper increases and the ball bounces strongly and more forward than it would without spin. This figure assumes perfect elasticity.

(a) Bounce of ball with backspin

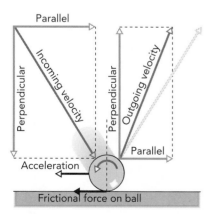

(b) Bounce of ball with topspin

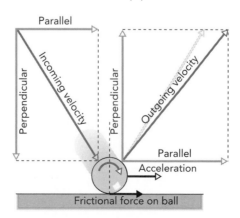

❸ Since an underinflated tire is both softer and less elastic than a properly inflated tire, it also experiences more rolling friction. One of the easiest ways to improve your car's fuel efficiency is to keep those tires inflated to the highest recommended pressure. The process of denting and undenting that underlies rolling friction also rubs small portions of a tire's surface across the pavement and causes wear. Softer, underinflated tires experience more wear than properly inflated tires.

These bounce-angle and spin effects are ubiquitous in ball sports. Basketball players use them when passing the ball to one another off the floor and when taking shots off the backboard or rim. Golfers use them to control how their balls rebound after landing on the green. Tennis, handball, squash, and Ping Pong players use them to outmaneuver their opponents. And players in soccer, baseball, field hockey, and lacrosse learn to predict how a ball will bounce based on the angle at which it approaches the ground and on its spin.

Most of us are so used to these bounce effects that when a ball fails to bounce as we expect, usually because of a defect in the surface, we say it took a "bad bounce." Of course, this predictability is mostly lost when a ball isn't spherical. When an American football or rugby ball hits the ground, its oblong shape couples together its translational and rotational motions and there's almost no telling which way it will bounce. Half the fun (or terror, if it's your team) of watching a fumble is the crazy scramble that ensues as players respond to the ball's unpredictable bounces.

So what about when a billiard or pool ball doesn't hit anything at all, and why does this have anything to do with bouncing? Well, the ball gradually comes to a stop, of course, and it does so because of an extended bounce it has with the felt surface of the table. The ball experiences a phenomenon known as rolling friction, in which the ball and felt continually dent and undent as the ball rolls forward. If the surfaces were perfectly elastic, this extended bounce wouldn't matter and the ball would roll indefinitely. But they're not perfectly elastic and so rolling gradually wastes the ball's energy and brings it to a stop. The felt is most responsible for this rolling friction because it's softer than the ball and it's not very elastic. If the table were bare slate, which is hard and elastic, the ball would roll much longer and farther.

Rolling friction is part of most ball sports and becomes stronger as the ball and surface become less elastic. Tall grass, for example, is a soft and relatively dead surface and a soccer ball doesn't roll very far on it. Underinflated balls and tires are also relatively dead and don't roll very far even on hard surfaces (see ❸). And while running may not seem like rolling, it has its own version of rolling friction. With each stride, your feet encounter the ground and the two colliding surfaces dent and undent in a form of bounce. When those surfaces are highly elastic, as they are with good running shoes and a modern synthetic track, little energy is wasted during each bounce and running is a pleasure. But when you run on a soft, dead surface such as loose sand or dirt, each step costs you considerable energy and running becomes exhausting.

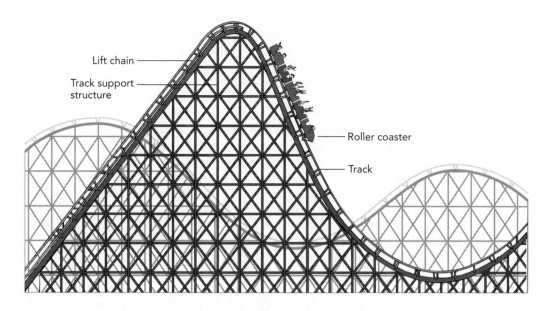

Lift chain

Track support structure

Roller coaster

Track

SECTION 3.3 Carousels and Roller Coasters

As your sports car leaps forward at a green light, you're pressed firmly back against your seat. It's as though gravity were somehow pulling you down and backward at the same time. But it's not gravity that pulls back on you; it's your own inertia preventing you from accelerating forward with the car.

When this happens, you're experiencing the feeling of acceleration. We encounter this feeling many times each day, whether through turning in an automobile or riding up several floors in a fast elevator. But nowhere are the feelings of acceleration more acute than at the amusement park. We accelerate up, down, and around on the carousel, back and forth in the bumper cars, and left and right in the scrambler. The ultimate ride, of course, is the roller coaster, which is one big, wild experience of acceleration. When you close your eyes on a straight stretch of highway, you can hardly tell the automobile is moving. But when you close your eyes on a roller coaster, you have no trouble feeling every last turn in the track. It's not the speed you feel, but the acceleration. What is often called motion sickness should really be called acceleration sickness.

The Experience of Acceleration

Nothing is more central to the laws of motion than the relationship between force and acceleration. Up until now, we've looked at forces and noticed that they can produce accelerations; in this section we'll take the opposite perspective, looking at accelerations and noticing that they require forces. For you to accelerate, something must push or pull on you. Just where and how that force is exerted on you determine what you feel when you accelerate.

The backward sensation you feel as your car accelerates forward is caused by your body's inertia, its tendency not to accelerate (Fig. 3.3.1). The car and your seat are accelerating forward, and since the seat acts to keep you from traveling through its surface, it exerts a forward support force on you that causes you to accelerate forward. But the seat can't exert a force uniformly throughout your body. Instead, it pushes only on your back, and your back then pushes on your bones, tissues, and internal organs to make them accelerate forward. Each piece of tissue or bone is

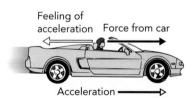

Feeling of acceleration   Force from car

Acceleration

Fig. 3.3.1 As you accelerate forward in a car, you feel a gravity-like feeling of acceleration in the direction opposite to the acceleration. This feeling of acceleration is really the mass of your body resisting acceleration.

❶ The experience of acceleration that this book calls a feeling of acceleration is known elsewhere as a fictitious or apparent force. Since it's not a real force, our term aims to reduce the confusion.

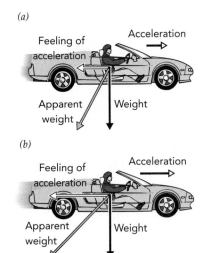

(a)

(b)

Fig. 3.3.2 (*a*) When you accelerate forward gently, the backward feeling of acceleration is small and your apparent weight is mostly downward. (*b*) When you accelerate forward quickly, you experience a strong backward feeling of acceleration and your apparent weight is backward and down.

responsible for the forward force needed to accelerate forward the tissue in front of it. A whole chain of forces, starting from your back and working forward toward your front, causes your entire body to accelerate forward.

Let's compare this situation with what happens when you're standing motionless on the floor. Since gravity exerts a downward force on you that's distributed uniformly throughout your body, each part of your body has its own independent weight; these individual weights, taken together, add up to your total weight. The floor, for its part, is exerting an upward support force on you that keeps you from accelerating downward through its surface. But the floor can't exert a force uniformly throughout your body. Instead, it pushes only on your feet, and your feet then push on your bones, tissues, and internal organs to keep them from accelerating downward. Each piece of tissue or bone is responsible for the upward force needed to keep the tissue above it from accelerating downward. A whole chain of forces, starting from your feet and working upward toward your head, keeps your entire body from accelerating downward.

As you probably noticed, the two previous paragraphs are very similar. But so are the sensations of gravity and acceleration. When the ground is preventing you from falling, you feel "heavy"; your body senses all the internal forces needed to support its pieces so that they don't accelerate, and you interpret these sensations as weight. When the car seat is causing you to accelerate forward, you also feel "heavy"; your body senses all the internal forces needed to accelerate its pieces forward, and you interpret these sensations as weight. This time you experience the weightlike sensation toward the back of the car.

Try as you may, you can't distinguish the weightlike sensation that you experience as you accelerate from the true force of gravity. And you're not the only one fooled by acceleration. Even the most sophisticated laboratory instruments can't determine directly whether they are experiencing gravity or are accelerating. However, despite the convincing sensations, the backward heavy feeling in your gut as you accelerate forward is the result of inertia and is not due to a real backward force. We'll call this experience a feeling of acceleration **❶**. It always points in the direction opposite the acceleration that causes it, and its strength is proportional to that acceleration.

If you accelerate forward quickly, the backward feeling of acceleration you experience can be quite large. However, you don't experience this feeling of acceleration all by itself; you also experience your downward weight, and together these effects feel like an especially strong weight at an angle somewhere between straight down and the back of the car. We'll call the combined experience of weight and feeling of acceleration your apparent weight. The faster you accelerate, the stronger the backward feeling of acceleration, and the more your apparent weight points toward the back of the car (Fig. 3.3.2).

Feelings of acceleration don't have to point backward. They can also point forward or even to the side. When you turn your car to the left, you're accelerating leftward and experience a strong rightward feeling of acceleration. When you're not driving and can safely close your eyes, you should be able to feel acceleration in any direction.

Carousels

When you ride on a carousel, you travel in a circular path around a central pivot. That's not the motion of an object that's simply coasting forward and exhibiting inertia. If you were experiencing zero net force, you would travel in a straight line at a steady pace in accordance with Newton's first law. But since your path is circular instead of straight, your direction of travel is changing; you must be experiencing a nonzero net force and you must be accelerating.

Which way are you accelerating? Remarkably, you are always accelerating toward the center of the circle. To see why that's so, let's look down on a simple carousel that's turning counterclockwise at a steady pace (Fig. 3.3.3). At first, the boy riding the carousel is directly east of its central pivot and is moving northward (Fig. 3.3.3a). If nothing were pulling on the boy, he would continue northward and fly off the carousel. Instead, he follows a circular path by accelerating toward the pivot, that is, toward the west. As a result, his velocity turns toward the northwest and he heads in that direction. To keep from flying off the carousel, he must continue to accelerate toward the pivot, which is now southwest of him (Fig. 3.3.3b). His velocity turns toward the west, and he follows the circle in that direction. And so it goes (Fig. 3.3.3c).

The boy's body is always trying to go in a straight line, but the carousel keeps pulling him inward so that he accelerates toward the central pivot. The boy is experiencing uniform circular motion. "Uniform" means that the boy is always moving at the same speed, although his direction keeps changing. "Circular" describes the path the boy follows as he moves, his trajectory.

Like any object undergoing uniform circular motion, the boy is always accelerating toward the center of the circle. An acceleration of this type, toward the center of a circle, is called a centripetal acceleration and is caused by a centrally directed force, a centripetal force. A centripetal force is not a new, independent type of force like gravity, but the net result of whatever forces act on the object. Centripetal means "center-seeking" and a centripetal force pushes the object toward that center. The carousel uses support forces and friction to exert a centripetal force on the boy, and he experiences a centripetal acceleration. Amusement park rides often involve centripetal acceleration.

The amount of acceleration the boy experiences depends on his speed and the radius of the carousel. The faster the boy is moving and the smaller the radius of his circular trajectory, the more he accelerates. His acceleration is equal to the square of his speed divided by the radius of his path.

But we can also determine the boy's acceleration from the carousel's angular speed and its radius. The faster the carousel turns and the larger the radius of the boy's circular trajectory, the more he accelerates. His acceleration is equal to the square of the carousel's angular speed times the radius of his path. These two relationships can be written:

$$\text{acceleration} = \frac{\text{speed}^2}{\text{radius}} = \text{angular speed}^2 \cdot \text{radius}. \qquad (3.3.1)$$

Since the boy is accelerating inward, toward the center of the circle, he experiences a feeling of acceleration outward, away from the center of the circle. The boy feels as though his weight is pulling him outward as well as down and he clings tightly to the carousel to keep from falling off.

Weight and feeling of acceleration can differ significantly in strength. While your weight is the specific gravitational force you experience near the earth's surface, the feeling of acceleration that you experience can have any strength or direction. If you are moving very rapidly around a small circle, you can easily experience a feeling of acceleration that is stronger than your weight.

Your feeling of acceleration can be measured relative to your weight. When a feeling of acceleration matches the feeling of your weight, it is said to be 1 gravity or 1g, for short. To experience a 1g feeling of acceleration, you must accelerate in the opposite direction at 9.8 m/s² (32 ft/s²), the acceleration due to gravity. If you accelerate 5 times that quickly, on the scrambler or on an airplane maneuvering sharply, you'll experience a 5g feeling of acceleration.

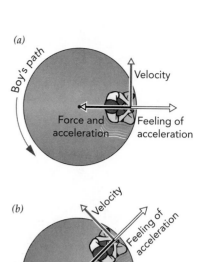

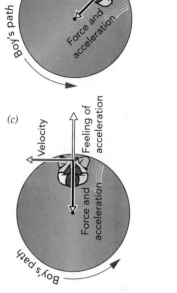

Fig. 3.3.3 A boy riding on a turning carousel is always accelerating toward the central pivot. His velocity vector shows that he is moving in a circle but his acceleration vector points toward the pivot. When he is heading north (a), he is accelerating toward the west. His velocity gradually changes direction until he is heading northwest (b), at which time he is accelerating toward the southwest. He turns further until he is heading west (c) and is then accelerating toward the south. (North is upward.)

Roller Coaster Acceleration

While roller coasters offer interesting visual effects, such as narrowly missing obstacles, and strange orientations, such as upside-down, the real thrill of roller coasters comes from their accelerations. Plenty of other amusement park rides suspend you sideways or upside down, so that you feel ordinary gravity pulling at you from unusual angles. But why pay for these when you can stand on your head for free? For a *real* thrill, you need acceleration to give you the weightless feeling you experience as a roller coaster dives over its first big hill or the several-g sensation you feel as you go around a sharp corner. A change in the amount of "gravity" you feel is much more exciting than a change in its direction. We're now prepared to look at a roller coaster and understand what you feel as you go over hills and loop-the-loops.

Every time the roller coaster accelerates, you experience a feeling of acceleration in the direction opposite your acceleration. That feeling of acceleration gives you an apparent weight that's different from your real weight. As we saw with a car, rapid forward acceleration tips your apparent weight backward toward the rear of the vehicle, while rapid deceleration tips it toward the front. But a roller coaster can do something a car can't: it can accelerate downward rapidly! In that case, the feeling of acceleration you experience is upward and opposes your downward weight so that they partially cancel. As a result, your apparent weight is less than your real weight and points downward or perhaps, if the downward acceleration is fast enough, points upward.

Consider that last possibility: if you accelerate downward at just the right rate, the upward feeling of acceleration will exactly cancel your downward weight. You will feel perfectly weightless, as though gravity didn't exist at all. The rate of downward acceleration that causes this perfect cancellation is that of a freely falling object. Your weight won't have changed, but the roller coaster will no longer be supporting

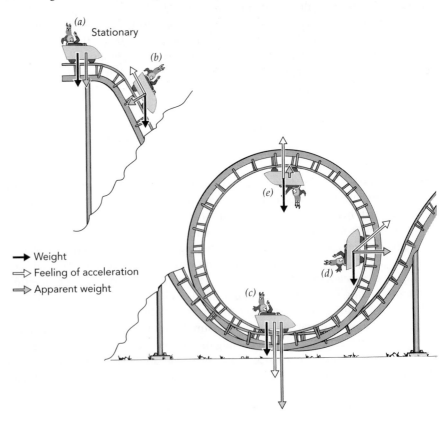

Fig. 3.3.4 A single-car roller coaster going over the first hill and a loop-the-loop. At each point along the track, the car experiences its weight, a feeling of acceleration due to its current acceleration, and an apparent weight that is the sum of those two. The apparent weight always points toward the track and the car doesn't fall off it.

you and you will be accelerating downward at 9.8 m/s² (32 ft/s²). You will experience the same sensations as a diver who has just stepped off the high dive.

Since freely falling objects are subject only to the forces of gravity, they don't have to push on one another to keep their relative positions. As you fall, your hat and sunglasses will fall with you and won't require any support forces from your head. Even if your sunglasses come off, they will hover in front of you as the two of you accelerate downward together. Similarly, your internal organs don't need to support one another, and the absence of internal support forces gives rise to the exhilarating sensation of free fall.

Drop towers offer exactly this sensation—they let you fall for a second or two and experience the sensation of pure weightlessness. Of course, they have to slow you gradually to a stop before you hit the ground so you experience an upward acceleration and a sensation of heaviness as you near the bottom of a drop tower. Bungee jumping offers the same sequence of sensations: weightlessness as you fall freely and heaviness as the bungee slows your descent.

However, because a roller coaster is attached to a track, its rate of downward acceleration can actually exceed that of a freely falling object. In those special situations, the track will be assisting gravity in pushing the roller coaster downward. As a rider, you will feel less than weightless. The upward feeling of acceleration will be so large that your apparent weight will be in the upward direction, as though the world had turned upside down! If the car you're traveling in is upright, you'll feel pulled out of your seat and you better have on a safety harness. But if the car you're in is inverted, so that it's actually above you, you'll feel pushed into your inverted seat.

Roller Coaster Loop-the-Loops

Figure 3.3.4 shows a single-car roller coaster at various points along a simple track with one hill and one loop-the-loop. Weight, the feeling of acceleration, and apparent weight are all vector quantities, as are the car's velocity and acceleration. These quantities are indicated with arrows of varying lengths that show each vector's direction and magnitude. The longer the arrow, the greater the magnitude of the quantity it represents.

At the top of the first hill (Fig. 3.3.4a), the single-car roller coaster is almost stationary. You, the rider, feel only your weight, straight down—nothing exciting yet. But as soon as the car begins its descent, accelerating down the track, a feeling of acceleration appears pointing up the track (Fig. 3.3.4b). The combination of your weight and the feeling of acceleration gives you an apparent weight that is very small and points down and into the track. Most people find this sudden reduction in apparent weight to be terrifying. Our bodies are very sensitive to partial weightlessness, and this falling sensation is half the fun of a roller coaster. An astronaut, falling freely in space, has this disquieting weightless feeling for days on end. No wonder astronauts have such frequent troubles with motion (or rather acceleration) sickness.

On the earth, the weightless feeling can't last. It occurs only during downward acceleration and disappears as your car levels off near the bottom of the hill (see ❷). By the time the car begins its rise into the loop-the-loop, it is traveling at maximum speed and has begun to accelerate upward (Fig. 3.3.4c). This upward acceleration creates a downward feeling of acceleration so that your apparent weight is huge and downward. You feel pressed into your seat as you experience 2 or 3g's.

The trip through the loop-the-loop resembles uniform circular motion. It's roughly like taking a single turn around a vertical carousel. However, as the car rises into the loop-the-loop, some of its kinetic energy becomes gravitational potential energy and the car slows down. As the car descends out of the loop-the-loop, this

❷ As part of its Reduced Gravity Research Program, NASA adapted two KC-135 Stratotanker aircraft to fly in the parabolic trajectories of freely falling objects. For about 25 seconds out of every 65 seconds of flight, each of these "Weightless Wonders" experienced zero net aerodynamic force and thus moved under the influence of its weight alone. The aircraft and everything inside its cavernous fuselage simply fell and the passengers experienced the sensation of pure weightlessness. Not surprisingly, these planes acquired the nickname "vomit comets." Although the KC-135s have been replaced by more modern aircraft, little else has changed.

Fig. 3.3.5 As I swing the wineglass over my head, it's accelerating rapidly toward the center of a circle. The pizza platter is pushing it downward, so the glass is firmly pressed against the platter and the wine is firmly pressed against the glass. If I look nervous, it's not about having wine poured on my head; it's about stopping everything without breaking the glass.

gravitational potential energy returns to kinetic form and the car speeds up. As a result of these speed changes, your acceleration is not exactly inward toward the center of the loop-the-loop and the feeling of acceleration you experience is not exactly outward. Still, an inward acceleration and an outward feeling of acceleration are good approximations for what occurs.

Halfway up the loop-the-loop, your true acceleration is inward and downward, so the feeling of acceleration you experience is outward and upward (Fig. 3.3.4*d*). Your apparent weight is still much more than your weight and is directed outward. You feel pressed into your seat, and the car itself is pressed against the track (Fig. 3.3.5).

Finally, you reach the top of the loop-the-loop (Fig. 3.3.4*e*). The car has slowed somewhat as the result of its climb against the force of gravity. But it's still accelerating toward the center of the circle, and you still experience a feeling of acceleration outward and, in this case, upward. Your weight is downward, but the upward feeling of acceleration exceeds your weight. Your apparent weight is upward!

Not only does the inverted car stay on its track, but you feel a weak weightlike sensation pressing you into your seat. Actually, the car is pushing on you to help gravity accelerate you around the loop. If your hat were to come off at the top of the loop, it would land in your seat, even though that seems to involve some sort of upward movement. In fact, the car is accelerating almost directly downward at a rate that's faster than that of your freely falling hat. Gravity and the track together push the car downward so fast that it overtakes the hat—your hat is really falling but the car is plummeting even faster.

In truth, a typical loop-the-loop isn't perfectly circular; it's more sharply curved on top than on its sides or bottom. This varying radius curve, known as a clothoid, is chosen for safety and comfort. By sharpening the curve only on top, the clothoid track maximizes the downward acceleration there while reducing the acceleration elsewhere on the track. High acceleration is important only when the roller coaster is upside-down. Everywhere else, it simply makes the riders feel heavy and uncomfortable, particularly at the bottom of the loop where the coaster is traveling fastest and accelerating upward rapidly.

Many roller coaster tracks are designed so that acceleration leaves the riders pressed into their seats even when the cars go upside down. In principle, roller coasters that travel on such tracks don't need seat belts to prevent riders from falling out (although seat belts are comforting to the passengers and insurance companies). But some roller coaster tracks have special cars and seat belts that permit them to have apparent weights away from the track. These roller coasters can and do go upside down without enough downward acceleration to keep the riders in their seats. In such roller coasters, the riders feel like they are hanging and if one of them were to lose a hat, it would fall to the ground rather than into the car.

What about a roller coaster with more than one car? For the most part, the same rules apply. However, new forces now act on each car: forces exerted by the other cars in the train. The effects of these cars are most pronounced at the top of the first and biggest hill. As the train disconnects from the lift chain and approaches the descent, it is rolling forward slowly and the first cars are well over the crest of the hill before they pick up much speed (Fig. 3.3.6*a*). They're pulling hard on the cars behind them, and those cars are pulling back, slowing their descent. By the time the train is moving fast, the first car is well down the hill. By then, the track is beginning to turn upward and the first car's riders experience mostly upward acceleration and downward feeling of acceleration. That's why riders in the first few cars of a roller coaster don't feel as much weightlessness.

In contrast, the last car is moving at high speed early in its descent. It undergoes a dramatic downward acceleration as it's yanked over the crest of the first hill by the cars in front of it (Fig. 3.3.6*b*). As a result, its riders feel large upward feelings

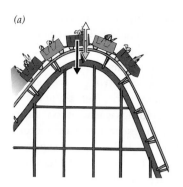 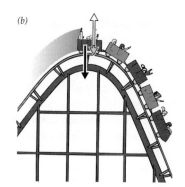

➡ Weight
⇒ Feeling of acceleration
⟹ Apparent weight

Fig. 3.3.6 When a multicar roller coaster descends the first hill, the ride experienced in the first cars is different from that in the last cars. (*a*) The first cars travel over the crest of the hill slowly and reach high speed only well down the hill. The cars behind them slow their descent. (*b*) The last cars are whipped over the top and are traveling very rapidly early on. The last cars accelerate downward dramatically as they go over the first hill and their riders experience a strong feeling of weightlessness.

of acceleration and quite extreme weightlessness. In fact, the designers of the track must be careful not to make the downward acceleration too rapid, or the roller coaster will flick the riders in the last car right out of their seats.

Obviously, it does matter where you sit on a roller coaster. The first seat offers the most exciting view, but it provides less than spectacular weightless feelings. The last car almost always offers the best weightless feelings. Probably the dullest seat in the roller coaster is the second; it offers a relatively tame ride and an unchanging view of the people in the front seat.

⋔ Centrifuges and Spin-Dryers

A carousel is an example of a centrifuge, a machine that spins objects in a circle and exposes them to strong inward accelerations. Another example of a centrifuge is a clothes washer with a spin-dry cycle. When the basket is spinning to extract the water, the clothes undergo uniform circular motion. They move in a circular path, roughly 0.25 m in radius, at a speed of about 20 m/s. According to Eq. 3.3.1, their acceleration is then $(20 \text{ m/s})^2/0.25$ m, or 1,600 m/s^2. Since the acceleration due to gravity is 9.8 m/s^2, the clothes experience about 163g's. If the wet clothes have a mass of 5 kg, they really weigh 49 N (11 pounds) but their apparent weight in the spinning basket is about 163 times that much or almost 8000 N (1800 pounds).

If inertia had its way, the clothes would go straight. The washer's basket, however, exerts a centripetal force of 8000 N on the clothes and they respond by accelerating around in a circle. But water, being a fluid, is able to flow through the fabric and the holes in the basket, and escape from the centrifuge. To the extent it can, given the circumstances, the water is exhibiting inertia and coasting in a straight line instead of accelerating in a circle. The water leaves the centrifuge traveling with the velocity it had when the basket stopped pushing on it: it heads off tangent to the basket's surface (and begins to fall). It doesn't travel outward away from the basket's center because it wasn't heading that way when it left the basket!

Because the water fails to accelerate with the clothes, the clothes are spun dry. This same effect would happen if you placed a shallow pan of water on your lap in

❸ Whenever a car accelerates, objects that aren't properly attached to the car will tend to move about inside it. They are simply exhibiting inertia and failing to accelerate along with the car. During a sudden left turn, for example, loose coins on the dashboard shift toward the right as the car accelerates leftward and leaves them behind. Be careful as you rush off to a potluck dinner party with a juicy casserole in the backseat; there will be messy consequences if you cause the car and container to accelerate out from underneath the food!

❹ An amusement park rotor is a large cylindrical chamber that spins rapidly enough to press its occupants tightly against the inside of its outer wall. When the floor of the rotor chamber is later moved downward, the riders are suspended on the wall by static friction.

❺ Fascinated by the large fictitious forces created by high-speed centrifuges, Jesse W. Beams (U.S. physicist, 1898–1977) developed a number of remarkable rotating devices. These included ultracentrifuges for biological research and gas centrifuges for separating the isotopes of uranium for nuclear power. During his studies, he built a tiny disk that spun over 1 million times per second and experienced fictitious forces of almost 1 billion g's.

a stopped car and then accelerated forward quickly. The car would exert a force on the pan to accelerate it forward, but the water would fail to accelerate with the pan. As the car picked up speed, the water would exhibit inertia (see ❸) and be left behind—on you! You would now be ready to try the spin-dryer.

In addition to merry-go-rounds and spin-dryers, centrifuges are found in water pumps, research labs, doctors' offices, amusement parks (see ❹), and even wineries. They can extract liquids from solids, as in a clothes washer or a winery, or they can separate heavy solids from light solids, as in research labs or doctors' offices. Labs use ultracentrifuges that produce as much as 100,000 g's to separate biological molecules (see ❺). Doctors' offices use less extreme centrifuges to separate components of a patient's blood to look for various abnormalities.

CHAPTER 4

MORE MECHANICAL THINGS

While the natural machinery of our solar system remains visible to anyone who observes the heavens, the machinery of human invention is now often deliberately hidden from view. That wasn't always the case. Prior to the industrial revolution, machinery was celebrated for its intrinsic beauty and the craftsmen who built it took pride in making it attractive as well as functional. But for a century or so, we have taken machinery for granted and done our best to keep it out of sight, like a rich but eccentric great uncle who helps us maintain our lifestyle but embarrasses us too much to invite to a dinner party.

The vanishing of machinery coincided with the transition of science from a quest for pure knowledge and understanding to the engine of industry and commerce. While the humanities and the sciences were once pursued as equals and often by the same individuals—for example, by artist-scientist Leonardo da Vinci, philosopher-scientist René Descartes, and statesman-scientist Benjamin Franklin—the two are now viewed rather differently. The humanities haven't changed much, but the sciences are increasingly seen in terms of making money; we now "invest" in science. What happened to the pursuit of science for its own beauty?

Since you're reading this book, you probably still recognize the fundamental beauty of science and I'm preaching to the choir. I hope that you also share my delight at the recent reappearance of visible machinery. Transparent cases are no longer unusual on elevators, escalators, and watches. To me, it's an encouraging sign that not everyone thinks science and engineering have no value beyond their impact on the next quarterly report.

No matter how sophisticated the machines around us appear, most are based in large part on the simple principles that we have already encountered. In this chapter, we'll take a look at a variety of fascinating machines to see what makes them tick. As we do, we'll find ourselves revisiting familiar issues and exploring a few new ones all the way to the frontiers of science and the cosmos.

Chapter Itinerary

Seat | Handlebars | Extension | Shifter | Brake lever | Stem | Shock absorber | Front brake | Rear brake | Front derailleur | Crank sprocket | Fork | Hub | Rim | Tire | Crank | Pedal | Chain | Idler sprocket | Spoke | Rear derailleur

Section 4.1 Bicycles

A bicycle is a wonderfully energy efficient, human-powered vehicle. Its wheels allow its rider to coast forward easily on level surfaces and accelerate effortlessly down hills. Compare the easy motion of a bicycle to that of walking, which requires effort every step of the way. Bicycles are very simple machines, and most of their moving parts are quite visible: the pedals, sprockets, brakes, and steering mechanisms, to name a few. Their simplicity and visibility make bicycles relatively easy to fix, even for a novice.

Tricycles and Static Stability

Bicycles have a stability problem. With only two wheels to support them, stationary bicycles tip over easily. Why then do we use two-wheeled bicycles for transportation?

We can begin to answer that question by looking at static stability, an object's stability at rest. To have static stability, an object needs a stable equilibrium. Equilibrium always means zero net force or torque, but near a *stable* equilibrium an object experiences restoring influences—forces and/or torques—which push it back toward that equilibrium. Aided by such influences, a statically stable object returns to its stable equilibrium after being disturbed slightly. A stool exhibits such static stability, but a balancing bicycle most definitely does not. If you want a pedal-powered vehicle that's statically stable, try a tricycle (Fig. 4.1.1).

An upright tricycle has static stability because it experiences restoring influences following a tip (Fig. 4.1.2*a*). Actually, the tricycle has only static *rotational* stability—it experiences restoring *torques* that act to return it to upright, a particular angular position. In contrast, a marble in a bowl has only static *translational* stability—it experiences restoring *forces* that act to return it to the bottom of the bowl, a particular (translational) position. The fact that a tricycle doesn't have static translational stability is a good thing; a tricyclist doesn't want to stay near one place, she only wants to stay near one orientation: upright!

But why does the tipped tricycle experience restoring torques? Rather than look for those torques directly, let's take a much more general approach. Part of my training as a physicist was to always look for the big-picture concept hiding behind every simple example and sometimes I just have to do it.

Fig. 4.1.1 A tricycle is very stable when it's standing still. However, it tips over easily during a high-speed turn because the rider can't lean in the direction that the tricycle is turning. The rider must also pedal furiously to move at a reasonable speed.

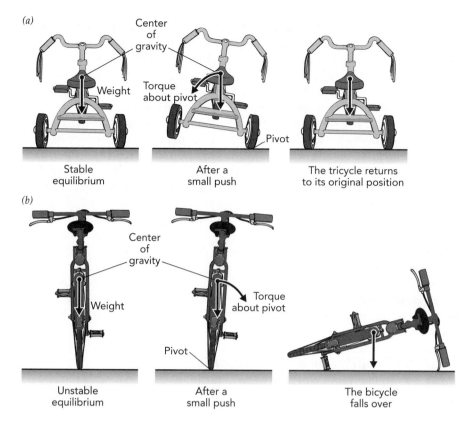

Fig. 4.1.2 (*a*) A tricycle is in a stable equilibrium. When it's disturbed, its center of gravity and gravitational potential energy rise, and it experiences restoring forces that return it to upright. (*b*) A bicycle is in an unstable equilibrium. Any disturbance causes it to fall.

Recall that an object accelerates in whichever direction reduces its total potential energy as quickly as possible. In this case, the tricycle undergoes *angular* acceleration in whichever *rotational* direction reduces its total potential energy as quickly as possible. Since small tips always raise the tricycle's center of gravity and gravitational potential energy, the tipped tricycle can certainly reduce its total potential energy by rotating back to upright. The upright tricycle is thus in a stable rotational equilibrium; if it isn't tipping yet, it doesn't start and if it is tipping, it undergoes angular acceleration so as to return to upright. No wonder children love tricycles!

This relationship between static stability and total potential energy is universal. There is no need to look directly for the restoring influences, which may be complicated anyway. When an object is situated so that its total potential energy rises whenever it shifts, it's in a stable equilibrium and will tend to remain there. That useful rule applies not only to static rotational stability and tricycles, but also to static translational stability and everything from canoes to bridge spans to mobiles. If you want an object to be stable at rest, ensure that any small shift increases its total potential energy.

Stable Equilibrium and Potential Energy

An object is in a stable equilibrium when any small shift increases its total potential energy.

That general observation underlies a simple rule of thumb for the behavior of an independent object like a tricycle resting on a surface: that object will be in a stable rotational equilibrium so long as its center of gravity is positioned over its base of support—the polygon that forms when you draw lines between its contact points with the ground. This rule follows from geometry: if the object's center of gravity is positioned over its base of support, then tipping it raises its center of gravity and increases its gravitational potential energy. Assuming no other potential energies are significant, the tipped object will undergo angular acceleration back toward that rotational equilibrium and the equilibrium will be stable.

Geometry also sets the limits of the object's static rotational stability: if its center of gravity moves outside the original base of support, it can reduce its total potential energy by tipping over and will promptly do so. That's why leaning back too far while on a tricycle is a recipe for disaster! The tricycle's three wheels define a triangular base of support, so it will be statically stable only so long as its rider keeps the overall center of gravity above that triangle.

But with just two wheels touching the ground, a bicycle has no base of support (Fig. 4.1.2b) and therefore no static rotational stability. Although it's at rotational equilibrium when perfectly upright, this equilibrium isn't stable. If the bicycle tips to one side, its center of gravity will descend and its gravitational potential energy will decrease. The bicycle will actually undergo angular acceleration *away* from equilibrium! The upright bicycle is thus in an unstable equilibrium; instead of experiencing restoring torques, the disturbed bicycle will tip farther and faster until it hits the ground.

Unstable Equilibrium and Potential Energy
An object's equilibrium is unstable when a small shift can decrease its total potential energy.

Bicycles and Dynamic Stability

Static rotational stability matters most to people who have difficulty balancing. That's why children learn to ride tricycles first. But when a tricycle is moving, static rotational stability doesn't guarantee safety. If a child rolls down a steep hill and then makes a sudden sharp turn, he or she will probably flip over. What has gone wrong?

A moving tricycle stays upright only if the girl riding it avoids sudden accelerations, such as a left turn at high speed. To make such a turn, she steers the tricycle's wheels so that friction with the pavement pushes the tricycle to the left (Fig. 4.1.3). Frictional forces on the wheel accelerate the tricycle leftward, redirecting its speed so that it turns. Of course, the girl needs to turn, too, so the tricycle pushes her along with it. As long as the turn is slow, a gentle push is all that's required and the tricycle and girl turn together safely.

But if the turn is too abrupt, the girl doesn't complete the turn along with the tricycle. Instead, her body goes straight as the tricycle drives out from under her. Crash. As you can see, a tricycle has good static stability but poor dynamic stability—stability in motion.

The tricycle flips because it can't handle the enormous torque that friction exerts on it during a sharp turn. Because the horizontal frictional force that turns the tricycle is exerted well below the tricycle and rider's combined center of mass (Fig. 4.1.3b), it produces a torque about that center of mass. If this torque is small, the tricycle's static rotational stability will provide a restoring torque in the opposite direction that prevents any angular acceleration. But if the turn is too sharp, the huge frictional torque will overwhelm the limited restoring torque and the tricycle and rider will flip. During high-speed turns, the tricycle is dynamically *unstable*.

Since the goal of a wheeled vehicle is to go somewhere, dynamic rotational stability is ultimately more important than static rotational stability. And while

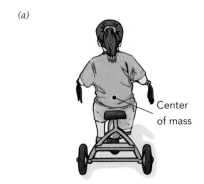

(a)

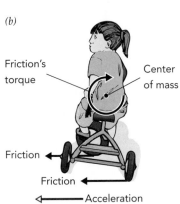

(b)

Fig. 4.1.3 (*a*) A tricycle that is heading straight is stable because any tip causes its center of gravity to move upward. (*b*) During a fast left turn, however, friction exerts large leftward forces on the wheels and the tricycle accelerates leftward. These frictional forces produce a torque about the tricycle's and rider's combined center of mass and can cause the tricycle to tip over.

a bicycle lacks static stability, a moving bicycle is remarkably stable. Its dynamic stability is so good that it's almost hard to tip over and can even be ridden without any hands on the handlebars. This feat is a popular daredevil stunt among children who haven't yet realized how easy it is.

As British physicist David Jones discovered, the bicycle's incredible dynamic stability results from its tendency to steer automatically in whatever direction it's leaning ❶. For example, if the bicycle begins leaning to the left, the front wheel will automatically steer toward the left so as to return the bicycle to an upright position. While a stationary bicycle falls over when it's disturbed from its unstable equilibrium, a forward-moving bicycle naturally drives under the combined center of mass and returns to that unstable equilibrium.

There are two physical mechanisms acting together to produce this automatic steering effect: one involving rotation and one involving potential energy. The first is based on the wheels alone: they behave as gyroscopes. Because each wheel is spinning, it has angular momentum and tends to continue spinning at a constant angular speed about a fixed axis in space. Since a wheel's angular momentum can be changed only by a torque, it naturally tends to keep its upright orientation.

But angular momentum alone doesn't prevent the bicycle from tipping over, any more than it prevents a tricycle from doing so. Instead it prompts the bicycle to steer automatically by way of gyroscopic precession—the pivoting of a gyroscope's rotational axis caused by a torque exerted perpendicular to its angular momentum. When a bicycle is upright, the pavement exerts no perpendicular torque on the front wheel. But when the bicycle leans to the left, the pavement's upward support force no longer points at the wheel's center of mass and it produces a perpendicular torque on the wheel. This torque is what makes that wheel precess; its axis of rotation pivots toward the left and thereby steers the bicycle to safety!

Assisting gyroscopic precession in this automatic steering process is a second effect due to potential energy. Because of the shape and angle of the fork supporting its front wheel, a leaning bicycle can lower its center of gravity and total potential energy by steering its front wheel in the direction that the bicycle is leaning. When the bicycle leans to the left, its front wheel accelerates toward the left—the direction that lowers the bicycle's total potential energy as quickly as possible. Once again the bicycle automatically steers in the direction that it's leaning and avoids falling over. These self-correcting effects explain why a riderless bicycle stays up so long when you roll it forward or down a hill.

There is little a bicycle designer can do to change the gyroscopic effect, but the potential energy effect depends on fork shape and angle. To be stable, the front wheel must touch the ground behind the steering axis (Fig. 4.1.4). If the fork is flawed, so that the wheel touches the ground ahead of the steering axis, the bicycle will steer the wrong way when it leans and be virtually unridable.

The front fork of a typical adult's bicycle arcs forward so that the wheel touches the ground just behind the steering axis. This situation leaves the bicycle dynamically stable enough to ride yet highly maneuverable. In contrast, the front fork of a typical child's bicycle is relatively straight so that the wheel touches the ground far behind the steering axis. The child's bicycle is therefore more dynamically stable than the adult's bicycle but also less easy to turn. That trade-off between dynamic stability and maneuverability is universal, appearing not only in pedal-powered vehicles, but in cars, boats, and aircraft as well. The more stable a vehicle is in motion, the harder it is to turn.

Leaning While Turning

So why does a bicycle rider lean during a turn? The answer is that leaning can balance out the torque that friction exerts on him during the turn, the same frictional

❶ In 1970, British physicist David Jones investigated the origins of a bicycle's dynamic stability. He built a series of "unridable" bicycles, including one that had a front wheel so small that it often skidded and became almost red hot as the bicycle was ridden. Jones found that gyroscopic effects alone did not account for a bicycle's stability. He discovered that this stability also comes from the shape of the front fork.

Wheel touches ground
Steering axis touches ground

Fig. 4.1.4 A bicycle is stable when moving forward in part because its front wheel touches the ground behind the steering axis. As a result, the front wheel naturally steers in the direction that the bicycle is leaning and returns the forward-moving bicycle to an upright position.

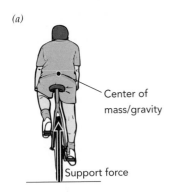

(a)

Center of
mass/gravity

Support force

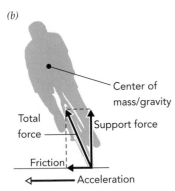

(b)

Center of
mass/gravity

Total
force

Support force

Friction

Acceleration

Fig. 4.1.5 (*a*) A bicycle that is heading straight is in rotational equilibrium when it's perfectly upright. The support force from the road produces no torque about its center of mass. (*b*) A bicycle that is turning left is in rotational equilibrium when it's tilted to the left. Together, the support and frictional forces from the road produce zero net torque about its center of mass.

torque that flipped our unfortunate tricycle rider. With the proper lean, the bicyclist can safely complete even the sharpest turn.

As he rides along, the bicyclist tries to keep himself and the bicycle in rotational equilibrium. Since they experience a frictional torque about their combined center of mass each time they turn, the bicyclist balances that torque by leaning himself and/or the bicycle toward the inside of the turn. The pavement's upward support force on the wheels then produces a torque on them about their combined center of mass and that new torque opposes the frictional torque. When those two torques sum to zero, the bicyclist and bicycle are safely in rotational equilibrium.

Since these opposing torques are both produced by pavement forces on the wheels, we can combine them; we can add the friction and support forces and see how much torque that overall force produces on the bicyclist. What we'll find in doing this addition is that the torque drops to zero when the pavement pushes the wheels directly toward the bicycle and rider's combined center of mass. As we saw in Section 2.1, a force exerted directly toward a pivot produces zero torque about that pivot. To stay in rotational equilibrium as he rides, the bicyclist places the combined center of mass directly in line with the pavement's force on the wheels.

For example, when the bicycle is heading straight, the rider can avoid a net torque by keeping the bicycle upright. The pavement then pushes the wheels straight upward, directly toward the combined center of mass (Fig. 4.1.5*a*). But when the bicycle is turning left, the rider must lean himself and/or the bicycle to the left to stay in rotational equilibrium. That's because each wheel of the turning bicycle is experiencing not only an upward support force, but also the leftward frictional force it needs to accelerate through the turn (Fig. 4.1.5*b*). The road's total force on the turning bicycle therefore points up and to the left. By leaning leftward just the right amount, the rider ensures that this total force points directly toward the combined center of mass and exerts no torque about it.

After a while, leaning the bike while turning becomes so automatic, so habitual, that you don't even think about it. You simply can't ride a bicycle or motorcycle without leaning as you turn. Even if you turn one of these vehicles so sharply that it skids, leaning can still keep it safely in rotational equilibrium.

All this discussion of leaning during turns begs the question: how do you make the bicycle lean prior to a turn? Actually, you cause that lean unconsciously by steering the bicycle briefly in the wrong direction—the direction opposite the turn itself! The bicycle then drives out from under you and the two of you begin to lean in the desired manner. You then steer in the correct direction and remain in rotational equilibrium throughout the turn. When you're ready to stop turning, you steer extra hard briefly in the direction of the turn. The bicycle then drives under you and the two of you return to upright. The turn is over.

Since only a statically unstable vehicle can lean, a statically stable one must rely on restoring torques to keep it near its rotational equilibrium. As we saw with tricycles, restoring torques are limited and a vehicle can tip out of its rotational equilibrium during rapid accelerations. A car, truck, or SUV will flip if it turns too sharply, and some are particularly prone to such catastrophes. The higher a vehicle's center of mass and the narrower its base of support, the more limited its restoring torques and the more easily it flips during turns. SUVs are surprisingly vulnerable to such rollover accidents, and small trucks that have been boosted up to resemble tractor-trailer cabs are truly hazardous. Even some ordinary cars have been found unsafe in this regard.

Pedaling Bicycles

So far, we have only discussed stability. But once we've settled on a bicycle as the most likely configuration for a useful person-powered vehicle, we need to figure out how to person-power it. The rider could push his feet on the ground, but that would be pretty inconvenient and even dangerous at high speeds. Instead, we use foot pedals to produce a torque on one of the wheels. But which wheel, and how do we produce that torque?

The original answer was to power the front wheel, using cranks attached directly to its axle. A crank is simply a lever that projects from the axle and produces a torque on that axle as you push its free end around in a circle. Each bicycle crank has a pedal installed on its free end so that you can use your foot to push on it. With its axle suspended in bearings, the front wheel turns as you pedal it. Friction between the turning wheel and the ground then pushes the bicycle forward.

Of course, you can't get something for nothing: ground friction also pushes back on the wheel, opposing its rotation. That's why you have to keep pedaling: your pedaling twists the wheel forward, while ground friction twists it backward. When those two torques balance, you'll move along at a steady pace.

This pedaling method is still used in children's tricycles, but it has three drawbacks. First, pedaling the front wheel interferes with its other responsibility: steering. Second, you can't take a break from pedaling; if the vehicle is moving, so are the pedals. Third, you can produce more than enough torque on the front wheel, but you often have trouble moving your legs quickly enough to keep up with the pedals. When you ride on level ground, you find yourself pedaling furiously and feeling very little resistance from the pedals.

The frantic pedaling problem goes to the heart of all pedal-powered vehicles, vehicles that draw power from you, the rider, to overcome inertia, air resistance, and perhaps an uphill slope. You provide that power by doing work on the pedals—a certain amount of work each second. Since work is the product of force times distance, you can do the same work on the pedals each second—that is, you can provide the same power to the vehicle—by pushing the pedals down hard as they turn slowly or by pushing them down gently as they turn quickly. However, for reasons having to do with physiology more than physics, your legs are best at providing power when you are pushing the pedals down medium hard as they turn medium fast.

Unfortunately, the pedals of an ordinary tricycle turn too quickly and too easily to make good use of your pedaling power. When the tricycle is moving fast on level ground, you can barely keep up with its pedals, let alone do work on them. The only time a tricycle makes good use of your capacity to supply power is when it's going uphill at moderate speed; only then do the pedals move medium fast and require that you push on them medium hard.

An early solution to the frantic pedaling problem was to use a gigantic front wheel. In such a configuration, one turn of the wheel would take you a considerable distance, so that you no longer had to pedal furiously while traveling at high speed along a level road. At the same time, ground friction exerted a larger backward torque on the huge wheel, so that you had to push harder on the pedals to keep the wheel turning steadily. At last you could reach your peak performance on level ground, pushing medium hard on the pedals as they turned medium fast. The pennyfarthing of the mid-nineteenth century was this sort of bicycle (Fig. 4.1.6). But pedaling still interfered with steering and you couldn't stop pedaling while the bicycle was moving. Furthermore, this bicycle had a new problem: you couldn't push hard enough with your feet to keep its front wheel turning steadily on uphill stretches.

These problems were solved by removing the cranks from the front wheel's axle and using an indirect drive scheme to convey power to the rear wheel. Employing toothed sprockets and a chain loop (Fig. 4.1.7), that indirect drive allowed the pedals and the wheels to turn at different rates. This change lets you to use mechanical advantage to choose how you supply power to the bicycle: whether you exert large

Fig. 4.1.6 The pennyfarthing used a large, directly driven front wheel to permit the rider to travel at a reasonable speed without having to pedal very rapidly. Its name came from its resemblance to two coins, the large English penny and the smaller farthing.

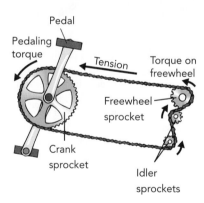

Fig. 4.1.7 The drive system for a modern bicycle. Pedaling produces torque on the crank sprocket, which in turn produces tension in the top segment of the chain. That segment of chain produces torque on the freewheel sprocket, which conveys that torque to the bicycle's rear wheel (not shown). The idler sprockets handle the extra chain.

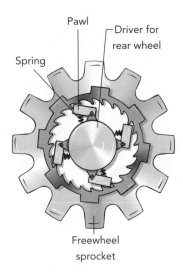

forces on slowly moving pedals or small forces on rapidly moving pedals or, ideally, medium forces on medium-fast moving pedals. Whether you're zooming along a level road or grinding slowly up a steep hill, you can always find a drive setting or "gear" that lets you comfortably supply your maximum power.

Lastly, the nonstop pedaling problem is solved by incorporating a one-way drive or freewheel in the hub of the rear wheel. This freewheel (Fig. 4.1.8) allows the rear wheel to turn freely in one direction so that you can stop pedaling as you coast forward. A modern bicycle with these improvements appears in Fig. 4.1.9.

Bicycle Brakes

Coasting downhill is certainly lots of fun, but sometimes you have to stop. Many bicycles have caliper brakes: rubber pads that rub against the wheel rims to slow their rotation (Fig. 4.1.10). The harder you squeeze the brakes, the more quickly sliding friction slows the wheels. Since the sliding friction force is approximately proportional to the force pressing the rubber pads against the wheel rims, you can control the bicycle's deceleration by how tightly you squeeze the brakes.

Fig. 4.1.8 The ratchet in a bicycle freewheel. If the relative rotation of the inner and outer parts is in the correct direction, the pawls transmit torque from the outer part to the inner part. If the relative rotation direction is reversed, the pawls compress the springs and skip along the teeth on the inside of the outer part. No torque is transmitted.

Fig. 4.1.9 A modern bicycle. The rear wheel is driven by a chain that allows the rider to vary the mechanical advantage between the pedals and the rear wheel. A freewheel in the hub of the rear wheel lets that wheel turn freely in one direction. This free motion allows the bicycle to coast forward, even when the pedals are stationary.

Fig. 4.1.10 When you activate this caliper brake, its rubber pads press against the wheel rim and experience sliding friction. The resulting frictional torque acts to slow the wheel's rotation. The harder you squeeze the brake, the more slowing torque it exerts. The brake pads are pressed against the rim by levers similar to the one shown in Fig. 2.1.15; as the metal break cable near the top pulls each lever end toward the rim a long distance with a small force, that lever pushes a pad toward the rim a small distance with a large force.

High-end bicycles often have disc brakes: sturdy metal rings that turn with the wheels and are gripped by friction plates during braking (Fig. 4.1.11). Disc brakes use the same physics concepts as ordinary caliper brakes, but separating the friction-based braking system from the wheels themselves makes it possible to optimize each to its own task. Bicycles with disk brakes can use wheel rims with any shapes and even heavy mud on those rims won't affect braking.

As the wheels begin to turn more slowly, the pavement exerts a backward static frictional force on the tires. It's this frictional force that ultimately slows the bicycle's forward motion. If you try to stop too quickly, you will exceed the limits of static traction and the wheels will begin to skid. Sliding friction will continue to slow the skidding bicycle, but you will have no control over direction any more and will probably travel in a more or less straight line. This is a good way to have an accident.

Fig. 4.1.11 In this disc brake, friction pads straddle a metal disc that rotates with the wheel. During braking, the pads grip the disc and experience sliding friction. The resulting frictional torque acts to slow the wheel's rotation. The friction pads are housed in the structure on the right and are squeezed against the metal disc by forces produced when the brake cable is pulled upward. The disc's holes help it dissipate heat and shed water during braking.

Most bicycles have independent brakes on their two wheels, in part because those wheels are not equivalent. The rear wheel skids much more easily during braking than the front wheel. The braking forces on the tires produces torques on the bicycle about its center of mass and tend to lift the rear wheel off the ground so that it skids. This same torque, however, presses the front wheel more tightly against the pavement and improves its traction. As long as you don't brake so hard that you throw yourself over the handlebars, the front brake is more effective at slowing the bicycle than the rear brake.

A typical child's bicycle doesn't have exposed brakes. Its coaster brake is located inside the hub of its rear wheel. As the rider pedals backward, the coaster brake activates and sliding friction inside the hub slows the rear wheel's rotation. Since the rear wheel does all of the braking, this type of bicycle tends to skid very easily. Some children enjoy slamming on their brakes at high speeds and skidding to a stop, leaving patches of black rubber on the pavement. To reduce this tendency to skid, some coaster brake bicycles also have a caliper brake for the front wheel.

When you brake, the rotating wheel is doing work against sliding friction in the brake pads. Where is that energy going? The answer is thermal energy. Whenever you do work against sliding friction, you are creating thermal energy. In a bicycle with disc brakes, this thermal energy appears in the discs and friction plates, which then transfer it to the passing air. In the bicycle with caliper brakes, this thermal energy appears in the metal wheel rims, in the rubber brake pads, and in the air inside the tires. They all become hotter. During a long downhill descent, you risk overheating the pads, the wheels, and the tires as your gravitational potential energy is converted into thermal energy by the brakes. Air that is heated in a confined space will experience a rise in pressure so that an overheated tire may explode. Instead of using the brakes to slow your descent, you do better to use air resistance. You will still be creating thermal energy, but in the air around you, not the air in your tires.

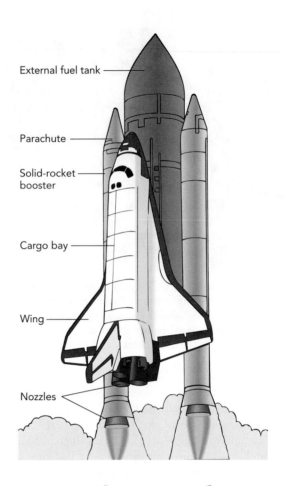

External fuel tank

Parachute

Solid-rocket booster

Cargo bay

Wing

Nozzles

Section 4.2 Rockets and Space Travel

Despite the complexity of modern spacecraft, the rocket is one of the simplest of all machines. It's based on the principle that every action has a reaction. A rocket is propelled forward by pushing material out of its tail. But despite that simplicity, people have been developing better rockets for over 700 years. They're used for such pursuits as space exploration, weaponry, rescue operations, and amusement.

Rocket Propulsion

Among a rocket's most impressive features are its ability to propel itself forward even in the complete isolation of space and its capacity to reach astonishing speeds using that propulsion. It somehow manages to push itself forward without any outside help and to use that forward push to accelerate seemingly without limits.

Of course, a rocket can't really push itself forward, any more than you can lift yourself up by your boots, and it can't accelerate forever. In reality, it obtains a forward force, a thrust force, by pushing against its own limited store of fuel, and when that fuel runs out, it stops accelerating. To understand how a rocket obtains thrust from its fuel supply, let's look at how Newton's third law, the one describing action and reaction, applies to rockets.

Imagine that you're sitting in the middle of a frozen pond with zero velocity and no momentum. It's a warm day and the wet ice is remarkably slippery. Try as you may, you can't seem to get moving at all. How can you get off the ice?

❶ Swedish inventor and engineer Carl Gustaf de Laval's (1845–1913) invention of the converging–diverging nozzle predates the modern development of rockets by several decades. He invented this nozzle as a way to make steam turbines more efficient and is credited with laying the foundation for all future turbine technology. De Laval is also known for his invention of the cream separator for milk.

Because of your inertia, the only way you can start moving is if something pushes on you. Sure, you could order a pizza and then push against the delivery truck when it arrives. But instead you follow the ideas we discussed on page 51. You remove a shoe and throw it as hard as you can toward the east side of the pond (Fig. 4.2.1). As you throw the shoe, you exert a force on it with your hand. The shoe accelerates and heads off across the ice.

What happens to you? You head off toward the west side of the pond! You're moving because when you pushed the shoe toward the east side of the pond, it pushed you equally hard toward the west side of the pond. In the process, you transferred momentum to the shoe and it transferred momentum in the opposite direction to you. Momentum isn't being created or destroyed, it's only being redistributed. Even after you let go of the shoe, your combined momentum remains at zero. The shoe has as much momentum in one direction as you have in the other.

Of course, you are much more massive than the shoe, so you end up traveling slower than it does. Momentum is the product of mass times velocity, so the more massive the object, the less velocity it needs for the same amount of momentum. Still, you've achieved what you set out to do: you're sliding slowly toward the west side of the pond.

Your final speed is limited because you only managed to transfer a small amount of momentum to the shoe and thus received only a small amount of opposite momentum in return. If you'd been able to throw the shoe faster or if you'd thrown a whole box of shoes, you'd have transferred more momentum and would be going faster.

Instead of throwing shoes, you'd have done better to throw very fast-moving gas molecules. Even at room temperature, the molecules in air are traveling about 1800 km/h (1100 mph). When gas molecules are heated to roughly 2800 °C (5000 °F), as they are in a liquid-fuel rocket engine, they move about three times that fast. If you throw something in one direction at that kind of speed, you receive quite a lot of momentum in the other direction.

That's what a conventional rocket engine does (Fig. 4.2.2). It uses a chemical reaction to create very hot exhaust gas from fuels contained entirely within the rocket itself. What started as potential energy in the stored chemical fuels becomes thermal energy in the hot, burned exhaust gas. This thermal energy is mostly kinetic energy, hidden in the random motion of the tiny molecules themselves. The rocket engine's nozzle steers most of this random motion in one direction, and the engine obtains thrust in the opposite direction.

If you've ever watched the launch of a large rocket, you've probably noticed the bell-shaped nozzles through which the exhaust flows (Fig. 4.2.3). Each nozzle allows the rocket to obtain as much forward momentum as possible from its exhaust by directing that exhaust backward and accelerating it to the greatest possible speed. As we'll see in Chapter 6, nozzles allow gases to convert their various internal energies into kinetic energy and are ideally suited for directing and accelerating gases. In the case of rocket exhaust, the most effective nozzle shape is a converging–diverging one, called a de Laval nozzle after its Swedish inventor, Carl Gustaf de Laval ❶.

To understand fully why this complicated nozzle structure works so well for rocket engines, we'd need to examine the physics of flowing gases up to and beyond the speed of sound. We'll encounter some of these issues later in this book, but for now a brief summary will have to suffice.

Inside the rocket and before the de Laval nozzle, the hot exhaust gas is tightly packed and its pressure is enormous. Like the gas in a spray bottle, this exhaust gas accelerates rapidly through the nozzle toward the lower pressure environment outside. The narrowing throat of that nozzle aids its acceleration, up to a point. When the gas reaches the narrowest part of the nozzle, it's traveling at the speed

(a)

Man and shoe
(Stationary)

(b)

Man's
velocity

Shoe's
velocity

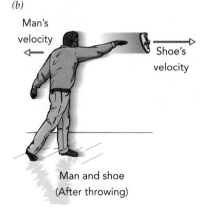

Man and shoe
(After throwing)

Fig. 4.2.1 A man who is holding a shoe while standing still on ice has zero momentum. Once he has thrown the shoe to the right, the shoe has a momentum to the right and the man has a momentum to the left. The total momentum of the man and shoe is still zero. Because the man is much more massive than the shoe, the shoe moves much faster than the man.

Unburned
fuel   Nozzle

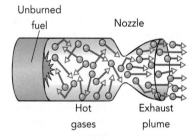

Hot   Exhaust
gases   plume

Fig. 4.2.2 A molecular picture of what happens in a chemical rocket engine. The engine burns its fuel in a confined chamber, and the exhaust gas flows out of a nozzle. The nozzle converts the random, thermal motions of the exhaust gas molecules into directed motion away from the rocket engine.

Fig. 4.2.3 The space shuttle's bell-shaped de Laval nozzles are designed to push its rocket exhaust downward as long glowing plumes. The gas pushes back, lifting the shuttle upward into space.

❷ On January 13, 1920, *The New York Times* ran an editorial attacking Robert Goddard for proposing that rockets could be used for travel in space. With modest financial support from the Smithsonian Institution, Goddard was pioneering the development of liquid fuel rockets. The editorial began: "That Professor Goddard, with his 'chair' in Clark College and the countenancing of the Smithsonian Institution, does not know the relation of action to reaction, and of the need to have something better than a vacuum against which to react—to say that would be absurd. Of course he only seems to lack the knowledge ladled out daily in high schools."

of sound and its characteristics begin to change dramatically. To coax the supersonic exhaust gas to accelerate still further, the nozzle stops narrowing and begins to widen. The tightly packed exhaust gas expands in volume as it flows through that widening bell and thereby prepares to enter the more open environment outside the nozzle.

Just how wide the diverging end of the de Laval nozzle must be to obtain the maximum thrust from its exhaust gas depends the nozzle's surroundings. Near sea level, the exhaust gas flows into ordinary air outside the nozzle and a relatively narrow de Laval nozzle works best. At high altitude or in space, the exhaust gas enters thinner air or nothing at all, so a wider de Laval nozzle is more ideal. Rockets typically make a compromise in their nozzle shapes so as to operate reasonably well in all environments.

By the time the gas reaches the end of the de Laval nozzle, it has converted most of its original energy into kinetic energy, with its velocity directed away from the nozzle. In fact, because the gas actually continues to burn even as it flows through the nozzle, its kinetic energy and speed keep rising until they reach fantastic levels. With the help of the de Laval nozzle, exhaust gas leaves the rocket's engine at an exhaust velocity or backward-directed flow speed of between 10,000 and 16,000 km/h (6,000 and 10,000 mph).

As it creates this plume of exhaust, the rocket pushes the gas backward and gives it backward momentum. The gas completes the momentum transfer by pushing the rocket forward. The very act of ejecting the exhaust is all that's required to obtain forward thrust; the rocket doesn't need anything external to push "against" and will operate perfectly well in empty space (see ❷). When it pushes hard enough on its exhaust, the rocket can not only support its own weight, it can even accelerate upward. The space shuttle weighs about 20,000,000 N (4,500,000 lbf) at launch but its thrust is about 30,000,000 N (6,750,000 lbf). That means that the space shuttle can accelerate upward at about half the acceleration due to gravity! As the shuttle consumes its fuel, so that its weight and mass diminish, it can accelerate upward even more rapidly.

Common Misconceptions: Action and Reaction in Rockets

Misconception: A rocket needs some external object to react against in order to push itself forward.

Resolution: While rocket propulsion does involve a pair of equal but opposite forces, action and reaction, the rocket is pushing its exhaust backward (action) and the exhaust is pushing the rocket forward (reaction). What this exhaust plume hits, if anything, makes no difference to the propulsion effect.

A Rocket's Stability

While it's flying through the atmosphere, a rocket had better travel nose-first. A bird flying tail first would be silly, but a rocket flying tail first would be dangerous. To keep it oriented correctly, a rocket needs dynamic rotational stability. Many a rocket has been destroyed by the launch safety officer because it exhibited dynamic instability and twisted wildly in flight.

To be dynamically stable, the rocket must experience zero net torque about its center of mass whenever it's flying nose first and it must return to that nose-first arrangement whenever it strays. Whatever torques the rocket experiences should return it to nose-first, or else be extremely small.

There are two sources of torque that rocket designers must consider. First, there are torques due to the engine's thrust. A rear-mounted engine pushes the rocket from behind, a situation that tends to cause trouble. After all, it's easier to steer a cart straight by pulling it from in front than by pushing it from behind. To keep the rocket pointed nose-first, the engine must exert its thrust directly at the rocket's center of mass and thereby avoid producing any torque on the rocket. If one of the engines is misaligned, its thrust may exert a torque on the rocket and start it spinning. Torques due to misaligned engines are a common cause of catastrophes in sophisticated rockets. Although their engines are usually steerable, an engine failure or a failure in the engine steering system can make a rocket fly wildly.

Second, a rocket in the atmosphere can experience aerodynamic torques. We'll study aerodynamics in Chapter 6, but for now it's enough to say that if you want the air passing a rocket to help it fly nose-first, you should make sure the rocket's tail experiences more air resistance than the rocket's nose. That way, aerodynamic effects will tend to push the rocket's tail behind its center of mass and ensure that the rocket flies nose-first.

Primitive rockets rely exclusively on aerodynamic force to remain stable. They have tail fins or sticks so that aerodynamic effects keep those tails at the rear. Their exhaust nozzles are also carefully aligned so that their thrusts exert no torques about their centers of mass. These simple rockets fly straight but are difficult to steer.

Sophisticated rockets abandon fins and sticks and use their thrusts to stabilize them in flight. These rockets sense their orientations and swivel their engine nozzles to make corrections. They may also have small rocket engines attached to their sides to exert torques on them and keep them oriented properly. The Space Shuttle and most other modern launch vehicles have essentially no fins. Their stability and maneuverability are governed entirely by carefully controlled rocket engines.

Using rocket exhaust for steering is crucial once a spacecraft leaves the atmosphere. Without air, there are no aerodynamic torques at all and the ship's orientation is handled exclusively with rocket engines. Special attitude control rockets are used to rotate the ship, emitting short bursts of gas to keep it pointed in the desired direction. The Space Shuttle's wings and tail are only effective during re-entry, when it becomes a glider in the atmosphere. In orbit around the earth, these surfaces serve no purpose because there is no air against which to push.

Still, any self-respecting starship commander wants a vessel that's as streamlined as possible, or so movies would have you believe. The uselessness of fins and wings in the vacuum of space doesn't prevent most movie spacecraft from having them. However, next time you see an intergalactic cruiser with wings and a tail, remember that it would work just fine if it were shaped like an oversized school bus.

The Ultimate Speed of a Spacecraft

At rest on the launch pad, a rocket consists principally of a spacecraft and a supply of fuel. Once the rocket's engine begins to fire, exhaust from the burned fuel accelerates backward and the spacecraft accelerates forward. The fuel is gradually consumed until eventually it runs out and the spacecraft coasts along on its own. Although weight and air resistance influence this story, let's neglect both for now to see what determines the spacecraft's eventual speed.

It might seem that the ultimate speed of the spacecraft is limited to the rocket's exhaust speed. Remarkably, there is no such limit. As long as the rocket keeps pushing exhaust backward, it will continue to accelerate forward. However, for the spacecraft to reach extremely high speed, the rocket must push the vast majority of its initial mass backward as exhaust. For example, if the rocket's initial mass is 90% fuel (i.e., it starts with 9 parts fuel and 1 part spacecraft), then we might expect the spacecraft to end up heading forward at about 9 times the speed of the exhaust gas. After all, that arrangement seems to satisfy the conservation of momentum.

Unfortunately, that analysis overestimates the spacecraft's speed. Because the rocket still has fuel on board as it accelerates forward, some of the forward mo-

❸ In 1989, the U.S. government began a program to develop a reusable rocket vehicle that could achieve earth orbit with only a single stage. Nothing but fuel would be jettisoned during launch so that the vehicle could travel to and from orbit repeatedly with only refueling and minimal maintenance between flights. The challenges facing this program are formidable. Even with liquid hydrogen and oxygen as its fuels, almost 90% of this vehicle's launch weight must be fuel. Nonetheless, efforts to produce such Single Stage To Orbit (SSTO) vehicles are proceeding.

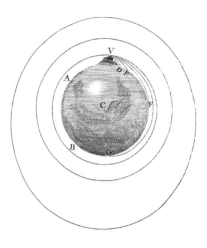

Fig. 4.2.4 Newton's drawing of a cannonball fired horizontally from the top of a tall mountain (V) at various speeds. At the lowest speed, the cannonball hits the ground near the base of the mountain (D). As the cannonball's speed increases, however, it travels farther from the mountain before hitting the earth (P). At still higher speeds, the curved earth drops away beneath the cannonball and it travels part way around the earth before landing (F). And at the highest speed shown, the cannonball never hits the earth at all (B). Instead, it orbits the earth in a circular path (A). The two outermost loops, a circle and an ellipse, show orbits that could be reached from taller mountains.

mentum it obtains from its exhaust goes into its remaining fuel rather than into the spacecraft. Since that fuel will soon be ejected from the rocket as exhaust, giving it precious forward momentum is wasteful and the spacecraft therefore ends up with less forward momentum. While the total momentum of the spacecraft and fuel still always sum to zero, this inopportune momentum transfer from the exhaust to the fuel reduces both the spacecraft's speed forward and the exhaust's average speed backward.

Despite this problem, a spacecraft can still travel faster than the speed of its rocket exhaust; it just needs more fuel. Neglecting air resistance and weight, the spacecraft's final speed is given by the rocket equation:

$$\text{spacecraft speed} = \text{exhaust speed} \cdot \log_e(\frac{\text{mass}_{\text{spacecraft}} + \text{mass}_{\text{fuel}}}{\text{mass}_{\text{spacecraft}}}). \quad (4.2.1)$$

For a rocket that is 90% fuel at launch, its spacecraft can reach 2.3 times the speed of its exhaust gas. If it can eject more than 90% of its initial mass as exhaust, it can go even faster.

But there's a problem with trying to burn up and eject a huge fraction of the rocket's original mass as exhaust. It's difficult to construct a rocket that is 99.99% fuel and 0.01% spacecraft. Instead, space-bound rockets use several separate stages, each stage much smaller than the previous stage. Once the first stage has used up all of its fuel, the whole stage is discarded and a new, smaller rocket begins to operate. In this manner, the rocket behaves as though it's ejecting almost all of its mass as rocket exhaust. With the help of stages and lots of fuel, rockets can travel substantially faster than their exhaust velocities and reach earth orbit or the solar system beyond. (For ongoing developments in single-stage rockets, see ❸.)

Orbiting the Earth

If the spacecraft were heading straight up when it ran out of fuel, it will either fall back to the ground or leave the earth forever (more on that later). But if it were heading primarily horizontally when its engine turned off, it may find itself circling the earth endlessly. With no atmosphere to affect it, the spacecraft follows a path determined only by inertia and gravity, and since the spacecraft's weight causes it to accelerate toward the center of the earth, its trajectory can bend into a huge elliptical loop around the earth.

The spacecraft is following an orbit around the earth. An orbit is the path an object takes as it falls freely around a celestial object. Although the spacecraft accelerates directly toward the earth's center at every moment, its huge horizontal speed prevents it from actually hitting the earth's surface. In effect, the spacecraft perpetually misses the earth as it falls (Fig. 4.2.4). To orbit the earth just above the atmosphere, a spacecraft must travel at the enormous speed of 7.9 km/s (about 17,800 mph) and will circle the earth once every 84 minutes.

However, the farther the spacecraft's orbit is from the earth's surface, the longer its orbital period—the time it takes to complete one orbit. First, since the spacecraft must travel farther to complete the larger orbit, the trip takes longer. Second, the spacecraft must travel slower in order to follow a circular path around the earth because the earth's gravity becomes weaker with distance.

In Chapter 1, I noted that gravity attracts every object in the universe toward every other object in the universe (see ❹). In particular, objects are attracted toward the earth. Near the earth's surface, an object's weight is simply its mass times 9.8 N/kg, the acceleration due to gravity. But as the object's distance from the center of the earth increases, the acceleration due to gravity diminishes. Equation 1.2.1 is only an approximation, valid for objects near the earth's surface and for situations where the altitude doesn't change significantly.

A more general formula relates the gravitational forces between two objects to their masses and the distance separating them. These forces are equal to the gravitational constant times the product of the two masses, divided by the square of the distance separating them. This relationship, discovered by Newton and called the law of universal gravitation, is:

$$\text{force} = \frac{\text{gravitational constant} \cdot \text{mass}_1 \cdot \text{mass}_2}{(\text{distance between masses})^2}. \qquad (4.2.2)$$

Note that the force on mass_1 is directed toward mass_2 and the force on mass_2 is directed toward mass_1. Those two forces are equal in magnitude but oppositely directed. The gravitational constant is a fundamental constant of nature, with a measured value of 6.6720×10^{-11} N·m²/kg².

The Law of Universal Gravitation

Every object in the universe attracts every other object in the universe with a force equal to the gravitational constant times the product of the two masses, divided by the square of the distance separating the two objects.

This relationship describes any gravitational attraction, whether it's between two planets or between the earth and you. The effective location of an object's mass is its center of mass, so the distance used in Eq. 4.2.2 is the distance separating the two centers of mass. For a spacecraft orbiting the earth just above its atmosphere, that distance is roughly the earth's radius of 6378 km (3964 miles). But for a spacecraft far above the atmosphere, the distance is larger and the force of gravity is weaker. That spacecraft experiences a smaller acceleration due to gravity. To give it the additional time it needs for its path to bend around in a circle, the high-altitude spacecraft must travel more slowly than the low-altitude spacecraft. This reduced speed explains the long orbital periods of high-altitude spacecraft.

At 35,900 km (22,300 miles) above the earth's surface, the orbital period reaches 24 hours. A satellite traveling eastward in such an orbit turns with the earth and is said to be geosynchronous. If a geosynchronous satellite orbits the earth around the equator, it's also geostationary—it always remains over the same spot on the earth's surface. Such a fixed orientation is useful for communications and weather satellites.

Not all orbits are circular. The orbits of some spacecraft are elliptical, so that their altitudes vary up and down once per orbit. At apogee, its greatest distance from the earth's center, a spacecraft travels relatively slowly because it has converted some of its kinetic energy into gravitational potential energy. At perigee, its smallest distance from the earth's center, the spacecraft travels relatively rapidly because it has converted some of its gravitational potential energy into kinetic energy. Of course, the perigee should not bring the spacecraft into the earth's atmosphere or it will crash.

The orbit of a spacecraft can also be hyperbolic. If the spacecraft is traveling too fast, the earth will be unable to bend its path into a closed loop and the spacecraft will coast off into interplanetary space. The spacecraft's path near the earth is then a hyperbola. The spacecraft only follows this hyperbolic path once and then drifts away from the earth forever.

A spacecraft usually enters a hyperbolic orbit by firing its rocket engine. It starts in an elliptical orbit around the earth and uses its rocket engine to increase its kinetic energy. The spacecraft then arcs away from the earth and its kinetic energy gradually transforms into gravitational potential energy. But earth's gravity becomes weaker with distance and the spacecraft's gravitational potential energy slowly approaches a maximum value even as its distance from the earth becomes infinite. If

❹ English physicist Henry Cavendish (1731–1810) proved that terrestrial objects do exert gravitational forces on one another. His experiment, performed in 1798, measured the tiny forces that two metal spheres exert on one another, using a very sensitive torsion balance. Comparing the forces between the two spheres with those between the earth and those same spheres (their weights), Cavendish was able to deduce the mass of the earth.

the spacecraft has more than enough kinetic energy to reach this maximum gravitational potential energy, it will be able to escape completely from earth's gravity.

The speed that a spacecraft needs in order to escape from the earth's gravity is called the escape velocity. This escape velocity depends on the spacecraft's altitude and is about 11.2 km/s (25,000 mph) near the earth's surface. A spacecraft traveling at more than the escape velocity follows a hyperbolic orbital path and heads off toward the other planets or beyond.

Common Misconceptions: Astronauts and "Weightlessness"

Misconception: An astronaut orbiting the earth is too far from the earth to experience gravity and is truly weightless.

Resolution: The astronaut is still so near the earth's surface that he experiences almost his full earth weight. He only feels weightless because he is in free fall.

Orbiting the Sun: Kepler's Laws

Once it escapes from the earth's gravity and again turns off its rocket engine, the spacecraft behaves like a tiny planet and orbits the sun. If you watch it patiently as it travels and compare its orbital motion with the motions of the planets themselves, you may begin to notice three universal features of all these solar orbits. First recognized by German astronomer Johannes Kepler (1571–1630) through his careful analysis of the extensive observational data collected by Danish astronomer Tycho Brahe, those three orbital behaviors are known as Kepler's laws.

Kepler's first law is already rather familiar to us from our examination of earth orbits. This law describes the shape of the spacecraft's looping orbit around the sun: it's an ellipse, with the sun at one focus of that ellipse (Fig. 4.2.5). An ellipse isn't an arbitrary oval; it's a planar curve with two foci and a rule stating that each point on the curve has the same sum of distances to the two foci. In this case, the sun occupies one focus and the other focus is empty. If you add the distance from the spacecraft to the sun plus the distance from the spacecraft to the empty focus, that sum will remain constant as the spacecraft orbits the sun. A circular orbit around the sun is a particularly simple elliptical one; its two foci coincide and the sun occupies them both.

Kepler recognized that every object orbiting the sun follows such an elliptical path. The planets move along nearly circular ellipses while the comets travel in highly elongated ones. Our spacecraft's orbit may be circular or elongated, depending on its position and velocity at the time its engine stopped firing. To reach another planet, the spacecraft's solar orbit and that of its destination planet must overlap and the two objects must reach that overlapping point at the same time. Traveling from planet to planet is clearly a tricky business.

Newton later recognized that these elliptical orbits are a direct consequence of the law of universal gravitation (Eq. 4.2.2) and its inverse square relationship between force and distance (force $\propto$ 1/distance²). (Note: the "$\propto$" symbol indicates proportionality.) Any other relationship be-

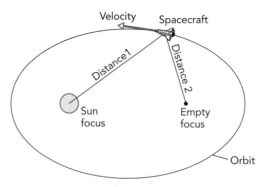

Fig. 4.2.5 A spacecraft's orbit around the sun is an ellipse, with the sun occupying one focus and the other focus empty. The sum of Distance 1 and Distance 2 is the same for all points on this ellipse.

tween force and distance would yield curving paths that don't close on themselves at all, let alone form ellipses. As you follow the spacecraft's elliptical orbit around the sun, you are witnessing an elegant exhibition of the law of universal gravitation.

Kepler's First Law: Orbits
All planets move in elliptical orbits, with the sun at one focus of the ellipse.

Kepler's second law describes the area swept out by a line stretching from the sun to the spacecraft: that line sweeps out equal areas in equal times (Fig. 4.2.6). Regardless of how circular or elongated the spacecraft's orbit is, or where the spacecraft is along that orbit, the area marked off each second by that moving line is always the same.

This observation demonstrates another physical law: conservation of angular momentum. Since the sun's gravity pulls the spacecraft directly toward the sun, it exerts no torque on the spacecraft and

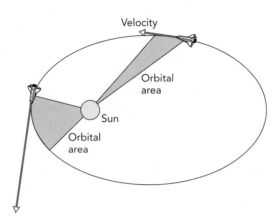

Fig. 4.2.6 The orbiting spacecraft sweeps out the same orbital area each second, despite variations in its distance from the sun. This steady sweep is a result of the spacecraft's constant angular momentum about the sun.

the spacecraft's angular momentum about the sun is constant. Remarkably, the rate at which this line sweeps out area is proportional to the spacecraft's angular momentum, so the steadiness of that sweep demonstrates the constancy of the spacecraft's angular momentum.

Kepler's Second Law: Areas
A line stretching from the sun to a planet sweeps out equal areas in equal times.

Kepler's third law describes the spacecraft's orbital period around the sun: the square of its orbital period is proportional to the cube of its mean distance from the sun, that is, the average of its perihelion (closest distance to the sun, Fig. 4.2.7) and aphelion (farthest distance to the sun). This relationship can be derived from the law of universal gravitation (Eq. 4.2.2), the equations describing centripetal acceleration (e.g. Eq. 3.3.1), and Newton's second law (Eq. 1.1.2).

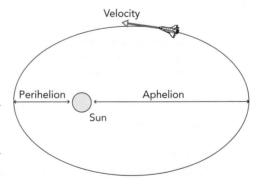

Fig. 4.2.7 The square of the spacecraft's orbital period is proportional to the cube of its mean distance from the sun—the average of its perihelion and its aphelion.

Kepler's Third Law: Periods
The square of a planet's orbital period is proportional to the cube of that planet's mean distance from the sun.

❺ In 1905, while Albert Einstein (German-born Swiss then American physicist, 1879–1955) was working as a patent examiner in Bern, he published four revolutionary papers in three different areas of physics. For the 26-year-old doctoral student from Germany, it was quite a banner year. Though Einstein is often portrayed as an elderly, wild-haired gentleman, his most important contributions to science were made when he was a vibrant young man who had only just married his first wife two years earlier.

◆ Travel to the Stars: Special Relativity

Despite formidable challenges, it may one day be possible for manned spacecraft to venture away from the solar system and travel to the stars. The distances involved are so vast that the only way to cover them in an astronaut's lifetime would be to move at fantastic speeds, speeds comparable to that of light itself.

Should spacecraft one day be able to attain such enormous speeds, they'll find that the basic laws of motion, the Galilean and Newtonian laws that we have been learning up to this point, are incomplete. Although extremely accurate at ordinary speeds, those laws falter near the speed of light (exactly 299,792,458 m/s). They turn out to be low-speed approximations for the more accurate laws of motion developed by Einstein in 1905 (❺). Built on the observation that light always travels at the same speed, regardless of an observer's inertial frame of reference, these relativistic laws of motion are accurate at any speed. They are part of Einstein's special theory of relativity, the conceptual framework that describes space, time, and motion in the absence of gravity.

We saw in Section 1.1 that observers in different inertial frames of reference can disagree on an object's position and velocity. Special relativity recognizes that those observers can also disagree on the distance and time separating two events. More broadly, two inertial observers who are in relative motion perceive space and time somewhat differently. If they're moving at ordinary speeds, that difference in perceptions is negligible and the Newtonian laws of motion are nearly perfect. But if they're moving relative to one another at a substantial fraction of the speed of light, then they perceive space and time quite differently. In that case, the Newtonian approximations fail and the full laws of relativity are required.

Special relativity has many consequences for high-speed space travel, but we'll concentrate on how relativity alters two familiar conserved quantities: momentum and energy. At low speeds, our spacecraft's momentum takes its usual Newtonian value: mass times velocity (Eq. 2.3.1). But with increasing speed, a new relativistic factor enters the picture: $(1 - \text{speed}^2/\text{light speed}^2)^{-\frac{1}{2}}$. Relativistic momentum is the product of the object's mass times its velocity times the relativistic factor or

$$\text{relativistic momentum} = \frac{\text{mass} \cdot \text{velocity}}{\sqrt{1 - \text{speed}^2/\text{light speed}^2}}. \qquad (4.2.3)$$

At ordinary speeds, the relativistic factor is so nearly equal to 1 that this relativistic relationship is beautifully approximated by the Newtonian one. But as the spacecraft's speed begins to approach that of light itself, the relativistic factor spoils the simple proportionality between momentum and velocity. Momentum then increases more rapidly than velocity. One result of this change is that it becomes impossible to reach the speed of light, let alone exceed it. Even if the spacecraft's thrust increases its forward momentum at a steady rate, its speed will increase less and less quickly. It will approach but never reach the speed of light.

A similar change happens to the spacecraft's energy as it approaches the speed of light. At low speeds, our isolated spacecraft's kinetic energy takes its usual Newtonian value: half its mass times the square of its speed (Eq. 2.2.1). But at high speeds, we must begin using relativistic energy. Relativistic energy is the product of the object's mass times the square of the speed of light times the relativistic factor. This relationship can be written:

$$\text{relativistic energy} = \frac{\text{mass} \cdot \text{light speed}^2}{\sqrt{1 - \text{velocity}^2/\text{light speed}^2}}. \qquad (4.2.4)$$

At ordinary speeds, the spacecraft's relativistic energy can be approximated as:

$$\text{relativistic energy} \simeq \text{mass} \cdot \text{light speed}^2 + \tfrac{1}{2} \cdot \text{mass} \cdot \text{speed}^2.$$

The usual Newtonian kinetic energy appears at the right in this approximation, but to its left is a new energy that we've never seen before. Called the rest energy, it's present even when the spacecraft is motionless. Because the rest energy is constant, it doesn't affect low-speed motion and was overlooked in the Newtonian laws. However, this energy associated with mass itself (stated symbolically as $E = mc^2$) does have consequences and is surely the most famous feature of the special theory of relativity.

The relativistic version of energy has two implications for our spacecraft. First, the spacecraft's energy increases so quickly as it nears the speed of light that it can never reach that speed. Second, the spacecraft's initial store of energy before launch is associated with its initial mass. That mass and energy are so closely related is something to which we'll return in Chapter 16.

◆ Visiting the Stars: General Relativity

In its travels near stars and other massive celestial objects, our spacecraft is likely to encounter another surprise: the Newtonian view of gravity is also an approximation! Near extremely dense, massive objects, gravity is no longer accurately described by Newton's law of universal gravitation. Instead, understanding gravity requires a new conceptual framework that Einstein first presented in 1916, the general theory of relativity.

This new framework is based on the observation that you can't distinguish between downward gravity and upward acceleration. As we saw in Section 3.3, they feel exactly the same. For example, if you feel heavy as you stand inside a closed spacecraft, you can't be sure whether you're experiencing a weight due to downward gravity or a feeling of acceleration due to upward acceleration. In fact, the spacecraft's scientific instruments can't help you because they can't distinguish the effects of gravity from those of acceleration, either. No matter how hard you try, you can't tell the difference.

At the heart of this problem is the concept of mass. Up to this point, we have seen mass play two apparently different roles that we could refer to as gravitational mass and inertial mass. When you're experiencing a weight, your gravitational mass is acting together with gravity to make you feel heavy. When you're experiencing a feeling of acceleration, your inertial mass is acting together with acceleration to make you feel heavy. But, in spite of their different roles, these two masses seem to be related.

Without exception, an object that has a large inertial mass and is therefore difficult to shake also has a large gravitational mass and is therefore hard to support against gravity. In fact, the two masses always seem to be exactly the same. That observation led Einstein to propose the principle of equivalence: that these two masses, gravitational and inertial, are truly identical and therefore that no experiment you perform inside your spacecraft can distinguish between free fall and the absence of gravity. The general theory of relativity is based on this principle of equivalence.

As long as your spacecraft stays in regions of weak gravity, Newtonian's law of universal gravitation will adequately describe its motion. But at the extremes of gravity, the general theory of relativity is necessary. That theory describes a universe in which massive objects distort the structure of nearby space and time, and in which extreme masses produce extreme distortions. One of the most startling predictions of this theory is the existence of objects so radical in their gravitational warping of nearby space and time that they are black holes—spherical or nearly spherical surfaces from which not even light can escape. A number of black holes have been discovered, including an enormous one at the center of our galaxy. You might want to avoid them.

History and Types of Rockets

Rockets date from 13th century China, as a follow-up to the invention of gunpowder. Burning gunpowder sent hot exhaust gas out of a nozzle and propelled the rocket forward. To make these rockets stable in flight, a guide stick was attached to the engine. This long stick trailed behind the rocket and allowed aerodynamic forces to provide the rocket with dynamic rotational stability. Guide stick rockets flew straight once they were moving quickly but had to be carefully supported during launch. This type of rocket reached its peak of development in the 18th Century when the British colonel William Congreve designed a 28 kg guide stick rocket that could travel several kilometers. Congreve rockets were common weapons used, for example, in the bombardment of Fort McHenry during the War of 1812 and described by Frances Scott Key in his poem *The Star Spangled Banner*.

In the mid-18th Century, Englishman William Hale replaced the guide stick with vanes attached to the sides of the rocket. The Hale rocket was stable for the same reason an arrow is stable—aerodynamic forces push the vanes to the rear of the moving rocket/arrow so that it flies nose first. Moreover, Hale canted the vanes so that air pressure caused the rocket to spin. This rotation made the rocket more accurate than the guide stick rocket. Any imperfections in the rocket engine would be averaged out as the rocket spun around and it would go quite straight in its path.

The Russian schoolteacher Konstantin Tsiolkovsky first pointed out the value of liquid fuel in a rocket in 1895 (❻). While solid-fuel rockets are reliable and easy to make (Fig. 4.2.8), they don't contain as much chemical potential energy per kilogram as liquid fuels. Liquid-fuel rockets can also vary their thrusts at will, while solid fuel rockets cannot. Tsiolkovsky advocated the use of liquid oxygen and liquid hydrogen as fuels, realizing that they contain more chemical energy per kilogram than any other substances (except for the exotic pairing, fluorine and hydrogen).

But in the early 1900's, few people understood what could be done with rockets and financial support for rocket development was difficult to obtain. It wasn't until 1926 that Robert Goddard launched the first liquid fuel rocket. Despite Goddard's successes, rocket development in the United States proceeded very slowly with little government interest. It was Germany that finally brought about the rapid development of liquid fuel rockets in the late 1930's and early 1940's. The German V2 rocket, developed under the direction of Wernher von Braun (1912–1977), was the first missile to travel faster than the speed of sound, covering approximately 125 miles (200 km) in 5 minutes. The V2 was a potent terror weapon and thousands were launched toward Antwerp and London. After the war, the Allies carted off every bit of V2 technology they could find, including von Braun himself.

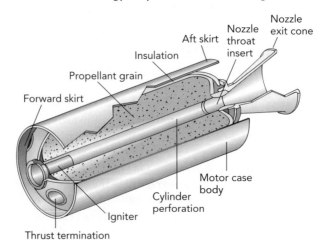

Fig. 4.2.8 A solid fuel rocket engine is a canister filled with a propellant grain. The core of the propellant is hollow, so that the flame burns outward from the middle of the cylinder. Hot gas created in the combustion accelerates out of the nozzle and pushes the rocket engine forward.

Since then, liquid fuel rockets have advanced enormously. Engineers have developed pumps that can deliver hundreds or even thousands of kilograms of liquid fuels to the engines each second (Fig. 4.2.9). Nozzle cooling systems now keep the nozzles from burning up during launch so that they can be used for several flights. Hypergolic liquid fuels, chemicals that ignite spontaneously when mixed, make it possible to construct reliable engines that can be turned on and off thousands of times during a flight. And guidance systems and computers have allowed engineers to control rocket flights with enormous precision. In the past decades, rockets have left the earth far behind, taking objects toward other parts of the solar system or even the stars.

Ion Propulsion

Spraying hot or high-pressure gas out of a nozzle isn't the only way to make a rocket engine. One alternative rocket engine, actually employed in several spacecraft, uses a stream of electrically charged particles to push the engine forward. In this ion-propulsion engine, a material such as mercury is heated to form a gas. The atoms in this gas are then exposed to electrical processes that remove some of their negatively charged electrons. Since they then have an overall positive electric charge, these charged atoms or ions are attracted toward a negatively charged screen that opens into space beyond. The ions accelerate toward the screen. While some hit the screen, most pass through the holes and sail off into space at fantastic speeds.

Instead of leaving at a mere 16,000 km/h (10,000 mph), like the exhaust from a chemical rocket engine, ions leaving the ion engine are traveling perhaps 250,000 km/h (160,000 mph). If you were stuck on a pond of frictionless ice and could throw your shoe eastward that fast, you would leave the ice heading westward at about the speed of sound.

In fact, you could have gotten off the ice by throwing a grain of rice backward at 250,000 km/h! Despite its tiny mass, the rice grain's extraordinarily high backward velocity would mean that it still carried away plenty of backward momentum. You would be left with an equal amount of momentum in the opposite direction and would slide forward off the ice.

Therein lies the attraction of the ion engine: it uses remarkably little material to obtain its thrust. The most precious commodity on many long-duration spacecraft isn't energy, which they typically obtain from solar panels, it's fuel mass. A spacecraft on a long voyage must get as much thrust as possible from every kilogram of its fuel. The ion engine makes efficient use of this fuel by accelerating it to enormous speeds.

Ion engines are used to propel spacecraft that need small but steady propulsion that lasts for years and years. They emit tiny streams of ions and electrons to obtain the required thrust. Since they use so little material to obtain their thrust, these engines can run for years on a small bottle of mercury or another material. The thrust of such an engine is too feeble to lift a rocket off the earth's surface and must be assisted at launch by a more powerful engine. Once in space, however, a modest thrust is all that is required to slowly change a ship's velocity.

Solar Sails

There is one last means of spacecraft propulsion that deserves mention. While not strictly a rocket engine, a solar sail offers the possibility of nearly limitless acceleration to a ship wishing to cross the vast distances between stars. A solar sail is an enormous surface that obtains its thrust by reflecting light from the sun. Light carries momentum and by reflecting that light, the solar sail manages to harness that momentum.

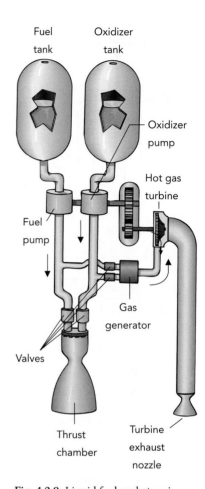

Fig. 4.2.9 Liquid fuel rockets mix and ignite two liquids directly in the thrust chamber. Hot gas produced by combustion accelerates out the nozzle and pushes the rocket forward. The complex turbine-powered pumping scheme shown above is needed to deliver the huge quantities of fuel burned each second.

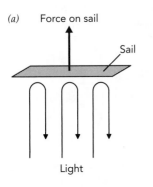

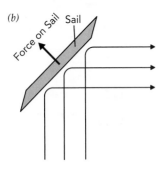

Fig. 4.2.10 A reflective solar sail obtains thrust by redirecting sunlight. *(a)* When the sail reflects sunlight directly backward, it receives twice the sunlight's original momentum and experiences a relatively large thrust force away from the sun. *(b)* When it reflects sunlight at right angles, it obtains a smaller thrust force at a 45° angle.

Perhaps the easiest way to think of this propulsion process is to recognize that light interacts with matter as a stream of particles called photons. Each photon of light carries a tiny amount of momentum that's related to its color. If a photon hits a surface and is absorbed, it conveys all of its momentum to that surface. This effect is analogous to a beanbag hitting a target; as it stops, the beanbag transfers 100% of its forward momentum to its target.

But if a photon hits a surface and reflects straight back, it conveys twice its forward momentum to the surface. This time, it's analogous to an elastic ball hitting a target and rebounding perfectly; as it rebounds, the elastic ball transfers 200% of its forward momentum to its target and ends up with backward momentum.

If the solar sail were black and absorbed sunlight, it would obtain 100% of the sunlight's momentum. The light would simply deposit all of its momentum in the sail. But a highly reflective solar sail is more effective. Instead of absorbing the light, it reverses the light's direction of travel and therefore the direction of the light's momentum. Like a ball bouncing elastically from a wall, the light transfers more forward momentum than it has to the sail and ends up heading backward with backward momentum. The reflective sail thus obtains 200% of the sunlight's momentum (Fig. 4.2.10*a*). Moreover, a reflective sail can tack; just as a sailboat can proceed at various angles to the wind by deflecting that wind to the side, a reflective solar sail can proceed at various angles to the sunlight by deflecting that light to the side (Fig. 4.2.10*b*).

The tiny momentum of light limits the amount of force a solar sail can obtain. A sail many kilometers on a side is needed to obtain a reasonable thrust. Since the sail's mass slows the acceleration of the entire ship, the sail must be extremely light. Current solar sail designs involve ultrathin mirrored plastic, supported in space by a thin frame (Fig. 4.2.11). Other designs have the sail rotating slowly like a giant pinwheel, so that inertial effects hold it open.

Despite the tiny thrust that a solar sail can achieve, that thrust doesn't consume fuel. The solar sail can continue to accelerate for years and its upper speed is limited only by the speed of light itself. When the sail is too far from the sun to obtain much thrust, one could imagine using a laser from earth to bounce additional light from it and aid it in its journey.

Fig. 4.2.11 In this artist's conception of a solar sail, four thin arms extend outward from a small spacecraft and support a huge reflective membrane. When sunlight reflects off that membrane, it transfers momentum to the spacecraft and thus provides a small but steady thrust.

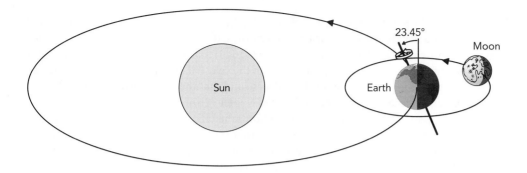

Section 4.3 The Earth, Moon, and Sun

One of Newton's most remarkable discoveries was that celestial objects obey the same laws of motion that govern objects on earth. Until that discovery, the universe seemed to have two separate components: the earth and the heavens. With his discovery, Newton united two formerly independent realms under a single set of rules. The dynamics of the planets and stars became knowable, predictable, and understandable.

The Solar System

The study of the solar system dates back to ancient times when people first discovered heavenly objects that moved relative to the fixed stars. Except for the sun, the true stars in the night sky are so distant that it's virtually impossible to detect their motion. They form a pattern of lights that is essentially fixed, night after night, like a huge picture painted on the sphere of the sky. Against this backdrop of motionless stars, the planets move slowly across the sky.

Early on, people noticed that the fixed stars circle the earth a little more than once a day. We now know that this "motion" is actually due to the rotation of the earth, not to any movement of the stars themselves. If it were not for the planets, people might have assumed the night sky really was nothing more than a picture. But careful observers found that the planets drift slowly past one group of stars after another along a ring that goes all the way around the sky. The five easily visible planets, Mercury, Venus, Mars, Jupiter, and Saturn, as well as the sun and the moon, were seen to travel endlessly around this circular racetrack at different rates, sometimes passing one another and sometimes almost standing still.

The band of stars that makes up this celestial circuit was called the zodiac and was divided into 12 equal sections according to the seasons and the 12 months. Each of these sections was named after the constellation of stars that aligned behind the sun during that particular month. When Greek scientist Ptolomy named the constellations around 150 AD, the sun was in front of the constellation Aries at the vernal equinox—the moment at the start of the northern hemisphere's spring when day and night are equally long—and Libra at the autumnal equinox—the analogous moment at the start of the northern hemisphere's autumn. Since then, however, a slow motion of the earth's rotational axis called the precession of the equinoxes has caused the 12 sections of the zodiac to drift westward across the sky relative to the equinoxes. The sun is now in front of Pisces at the vernal equinox and in front of Virgo at the autumnal equinox.

Originally, people believed that the stars, the planets, the moon, and the sun simply traveled around the earth, which remained motionless. The erratic motions of the planets as they circled the earth were attributed to some complicated machinery that drove these heavenly bodies. It was not until the writings of Polish astronomer Nicolaus Copernicus (1473–1543) that anyone considered the possibility that the earth actually participates in the celestial motion. Copernicus suggested that the sun is the center of the universe and that the earth travels around the sun. Later, German astronomer Johannes Kepler (1571–1630) used the careful measurements of his former master Tycho Brahe (1546–1601) to prove Copernicus's theories correct and to deduce three important laws of planetary motion. These empirical laws—based on observation rather than physical principles—were later proven correct by the work of Sir Isaac Newton.

While I presented Kepler's three laws in Section 4.2, I'll take a moment to remind you of a few highlights. First, Kepler observed that all the planets follow elliptical orbits as they travel around the sun. Moreover, most planets have elliptical orbits that are almost perfectly circular. Second, Kepler found that the larger a planet's orbit, the more time it takes the planet to complete a full circuit around the sun, the orbital period. For a circular orbit, he found that the orbital period is proportional to the orbital radius to the 3/2 power. For example, doubling the radius of the earth's orbit around the sun would increase the time needed to complete that orbit by a factor of $2^{3/2}$ (about 2.8) so that it would require about 2.8 of our present years to complete a full orbit around the sun.

I can now explain why the planets appear to drift across the stars of the Zodiac. Let's follow the motion of two planets, the earth and Mars, over a period of several months (Fig. 4.3.1). At the start of this time, an observer on the earth looks for Mars and finds that it's located in front of a particular group of stars. A month and a half later, the earth has moved 12.5% of the way around its orbit while Mars has moved only 6.6% of the way around its orbit. Mars has a larger orbital radius than the earth and its orbital period is 1.88 years. The observer on the earth finds that Mars has drifted in front of a different group of stars. Another month and a half later, Mars appears in front of yet another group of stars. Thus, each planet's position relative to the fixed stars depends on where that planet and the earth are in their orbits around the sun.

The sun also appears to move against the fixed stars. Since the earth orbits the sun once each year, the sun seems to pass in front of the whole zodiacal ring of fixed stars once per year.

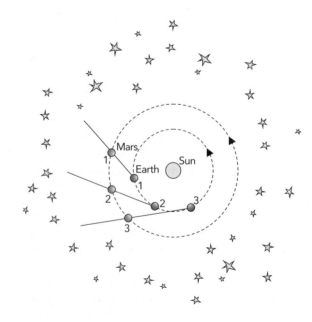

Fig. 4.3.1 The planets of the solar system orbit the sun at different rates. The earth orbits the sun once per year while Mars, with its larger orbital radius, takes 1.88 years to complete the circuit. To an observer on the earth, Mars appears first in front of one group of stars, then another, then another.

The earth's moon is a somewhat different case. As we shall soon see, the moon orbits the nearby earth and follows the latter on its journey around the sun. Since the moon goes completely around the earth once every 27.32 days, it appears to move across the zodiac very quickly.

Another Look at Orbits

Since elliptical orbits establish the basic structure of the solar system, they warrant another look. For the moment, let's imagine that the solar system consists only of the earth and the sun. The earth and the sun exert gravitational forces on one another so that the earth accelerates toward the sun and the sun accelerates toward the earth. But the sun is 333,000 times more massive than the earth and barely accelerates at all. Thus the earth does virtually all of the moving.

We might expect the earth to accelerate quickly, reach an enormous velocity, and smash catastrophically into the sun. Fortunately, the earth and the sun didn't start out at rest or that is exactly what would have happened (Fig. 4.3.2a). Actually, the earth really does accelerate directly toward the sun but it has an initial sideways velocity that prevents it from hitting the sun (Fig. 4.3.2b). This sideways velocity would carry the earth away from the sun were it not for gravity. The earth accelerates toward the sun, so its velocity shifts toward the sun.

The gravitational force the sun exerts on the earth is centripetal, always toward the nearly stationary sun. The earth's acceleration is also centripetal. The earth's velocity continues to change and the planet's path is bent around in a huge elliptical arc through space. This arc eventually closes on itself after a period of one year to form an almost circular ellipse. This ellipse is the earth's orbital path through space.

Because the earth's orbital path is almost circular, the earth undergoes uniform circular motion about the sun. But the earth is so large that the sun's gravity varies slightly from the earth's near side to its far side. The earth's near side, the side closest to the sun, experiences the strongest solar gravity and would travel in a smaller

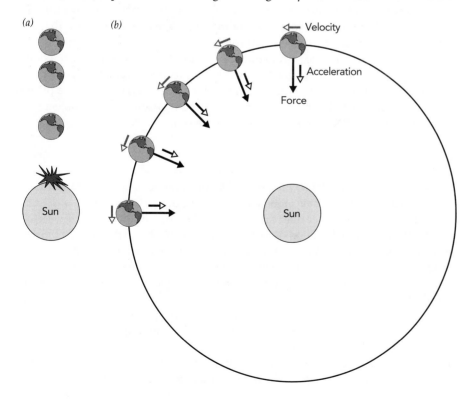

(a)   (b)

Velocity

Acceleration

Force

Sun

Sun

Fig. 4.3.2 The gravitational attraction between the earth and the sun causes both objects to accelerate directly toward one another. The less massive earth does most of the accelerating. *(a)* If the earth had started out stationary, then it would have accelerated directly toward the sun, picked up speed, and smashed into the sun's surface. *(b)* However, the earth has a large velocity relative to the sun, at right angles to the line separating them. The earth accelerates directly toward the sun, but it misses the sun as it heads toward the lower left corner of this figure and circles the sun in an elliptical orbit. The earth always accelerates directly toward the sun but its sideways velocity prevents it from hitting the sun.

orbit if it could. The earth's far side, the side farthest from the sun, experiences the weakest solar gravity and would travel in a larger orbit if it could. Because the earth is essentially rigid, it all travels in the same orbit. The earth's oceans, however, are less rigid and exhibit the effects of trying to follow these different orbits. The oceans bulge out toward the sun on the earth's near side and away from the sun on the earth's far side. As we'll discuss in Section 9.3, the nonuniform solar and lunar gravities produce tidal forces on the earth and the resulting ocean bulges are associated with the tides.

It's as though the tiny earth swings around the gigantic sun on a cord, following a nearly circular path. The cord is actually the force of gravity. As long as the earth continues to travel sideways, gravity can't pull the earth any closer to the sun. Because the force of gravity grows weaker as a planet moves farther from the sun, a larger orbit must take longer than a smaller orbit. Planets close to the sun can orbit it rapidly, as though held by a strong rope. Planets far from the sun must travel very slowly around it, as though held by a finer, more delicate thread.

That the earth's orbit closes on itself perfectly after one trip around the sun is a remarkable result. It occurs because the force of gravity is exactly proportional to one divided by the square of the distance between the sun and the earth, an observation we discussed in Section 4.2. Other types of centripetal forces can create orbits, but these orbits don't necessarily close on themselves. For example, if you attach a ball to a rubber band and swing it around with your hand, the ball will follow a complicated orbital path that never quite closes on itself. That's because the rubber band's centripetal force isn't proportional to 1 divided by the square of the distance between your hand and the ball.

The sun also accelerates toward the earth, but its enormous mass prevents it from picking up much speed. Instead, the sun only wobbles ever so slightly as the tiny earth circles around it. To be precise, the two objects orbit together about their combined center of mass. Because the mass of the sun dominates the combined system, their center of mass is almost exactly at the center of the sun. The other planets orbit the sun in the same manner as the earth. The giant planets, Jupiter, Saturn, Uranus, and Neptune, are more massive than the earth and make the sun wobble more than the earth does. Still, for most purposes we can imagine that the objects in the solar system merely circle about a stationary sun.

The planets are too far apart from one another for their gravitational attractions to be very noticeable. The earth does attract Mars (and vice versa), but so weakly that this gravitational force is nearly impossible to detect. However, the planet Pluto was found in 1930 because people observed that some unseen object was distorting the planet Neptune's orbit around the sun.

The Moon

While the planets seem largely unaffected by one another's presence in the solar system, that isn't true of their moons. For example, the earth's moon is so close to the earth—400 times closer than the sun—that it's strongly affected by the earth's gravity. The earth and the moon move together as a pair. The earth is about 80 times as massive as the moon, so the moon does most of the moving. The moon accelerates toward earth and orbits it, just as the earth orbits the sun. Because of the small size of the earth-moon system, the moon completes its near-circular orbit in only 27.32 days.

This 27.32 day cycle approximately coincides with the phases of the moon. Sometimes the moon appears to be a complete white disk and other times it is only a thin crescent. What causes these phases?

The phases occur because we can only see the portion of the moon that is illuminated by the sun (Fig. 4.3.3). As the moon orbits the earth, the lighted side of the

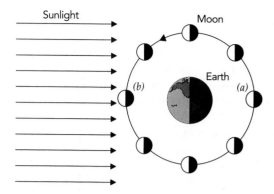

Sunlight

Moon

Earth

(b)   (a)

Fig. 4.3.3 The phases of the moon arise because we only see the lighted portion of the moon. When the moon is farther from the sun than is the earth, most of the half that we can see is illuminated by the sun and we see a full (*a*) or gibbous moon. When the moon is closer to the sun than is the earth, we see a large amount of unilluminated moon and it appears as a new (*b*) or a crescent moon.

moon is sometimes facing toward the earth and sometimes facing away from the earth. When the moon is as far away from the sun as its orbit allows, the moon's lighted side is facing toward us. We then see the complete lighted disk of a full moon. When the moon is as close to the sun as its orbit allows, the moon's lighted side is facing away from us. We can't see the moon at all and call this the new moon. In between, we see portions of the lighted half of the moon. These portions appear as crescent moons, half moons, or gibbous moons.

Occasionally, the earth gets in the way of light illuminating the moon. During those times, the moon is in the earth's shadow and a lunar eclipse occurs. A person on earth, looking at the moon, sees a huge round shadow drift slowly across the moon's surface. For a time, this shadow may completely cover the moon's surface. Because a lunar eclipse can occur only when the moon is as far as possible from the sun (Fig. 4.3.3*a*), the moon appears full just before and just after the eclipse.

Similarly, there are rare occasions when the moon blocks the sunlight on its way to the earth's surface. A person standing on the earth will see the dark lunar sphere pass in front of the sun and block its light, and a solar eclipse occurs. The moon is small and is just barely able to obscure the entire sun from view for people at the right location on earth. That's why solar eclipses are rarer than lunar eclipses. Because a solar eclipse can occur only when the moon is as close as possible to the sun (Fig. 4.3.3*b*), the moon appears new just before and just after the eclipse.

The Day

To a person standing on the surface of the earth, the sun appears to rise in the east every morning and set in the west every evening. Of course, people long ago realized that this daily schedule doesn't reflect any movement of the sun, but rather the rotation of the earth. The earth rotates on its axis slightly more than once a day so that every object in the sky appears to circle the earth in approximately that same amount of time.

The earth keeps on spinning because it has an incredible amount of angular momentum. It obtained this angular momentum during its formation and has retained this angular momentum because there are few torques exerted on it. The sun and the moon exert small torques on the earth as they drive the ocean tides, but for the most part, the earth just keeps on rotating at a very steady rate determined only by its total angular momentum and its rotational mass. For about half a day any point on the earth's surface is turned toward the sun, and for about half a day that same point is turned away from the sun.

It's interesting that the solar day—the time between successive moments when the sun reaches its highest point overhead—is a little longer than the sidereal day—the time it takes the earth to turn exactly once on its axis relative to the fixed stars. That's because the earth travels about 1/365th of the way around the sun over the course of a day and therefore has to complete roughly an extra 1/365th of a rotation before the sun returns to its highest point in the sky. The mean solar day is 86,400 s, while the sidereal day is only 86,164 s.

But how long is a day on the moon? Before answering that question, I'll note that the moon orbits the earth with the same half of its surface always pointing toward the earth. This locking of its rotation to its orbit occurred long ago as the result of tidal effects (see Section 9.3) that deformed the moon so that it isn't perfectly spherical.

The moon takes 27.32 earth days to complete one full orbit of the earth relative to the fixed stars—the lunar sidereal month. Because the moon's rotation is locked to its orbit, it takes that same amount of time to complete one full rotation on its axis—the lunar sidereal day. But because the earth and moon travel about 1/12th of the way around the sun during a lunar month, the moon has to complete about an extra 1/12th of an orbit around the earth and an extra 1/12th of a rotation before the sun returns to its highest point in the lunar sky. As a result, a person on the moon will notice that the sun appears to rise and set once every 29.53 earth days.

We can't see the other half of the moon from the earth, since it always faces away from the earth. The far side of the moon is only visible from spacecraft that travel to that side. People mistakenly interchange the expressions "far side of the moon" and "dark side of the moon." The far side of the moon is the one not visible from earth and is always the same region of the moon's surface. The region of the moon that is dark changes over the course of the 709-hour lunar solar day.

The Seasons

The earth rotates on its axis about once a day and orbits the sun once a year. Apart from changes in the sun's position against the fixed stars, it seems that every day should be identical to the day before. Why are there seasons? Why is a winter's day so much shorter than a summer's day?

The answers to these questions lie in the orientation of the earth's rotational axis relative to its orbit around the sun. The earth's orbit around the sun describes a huge, nearly circular ellipse that you could imagine as defining the edges of a gigantic glass disk. We call the geometrical surface represented by that glass disk the orbital plane. If the earth's rotational axis were straight up and down to a person standing upright on this orbital plane, then the earth's situation wouldn't change much as it followed its orbit and every day would be essentially the same.

But the earth's rotational axis is *not* straight up and down with respect to the orbital plane. The earth rotates about an axis that is tilted 23.45° away from straight up and down. And because the earth has so much angular momentum, the earth's rotational axis keeps pointing the same direction relative to the fixed stars, day after day, month after month, and year after year. It points almost directly toward a bright, solitary star—Polaris, the "north star."

For part of the year, the earth's northern hemisphere is tilted toward the sun (Fig. 4.3.4a) and for part of the year, the northern hemisphere is tilted away from the sun (Fig. 4.3.4c). When the northern hemisphere is tilted toward the sun, as it is around June 21 (Fig. 4.3.4a), the northern hemisphere experiences summer. The sun appears more nearly overhead at noon in the northern hemisphere and the day is much longer than the night. North of the Arctic Circle, a band located 23.45° south of the North Pole, the sun never sets at all during part of the summer. When the northern hemisphere is tilted away from the sun, as it is around December 22 (Fig. 4.3.4c), the northern hemisphere experiences winter. The sun never rises very high in the sky and the day is much shorter than the night. North of the Arctic Circle, the sun never rises at all for part of the winter.

Having the sun more directly overhead raises the local average temperature. To explain this effect, let's consider two possibilities. When the sun is low in the sky

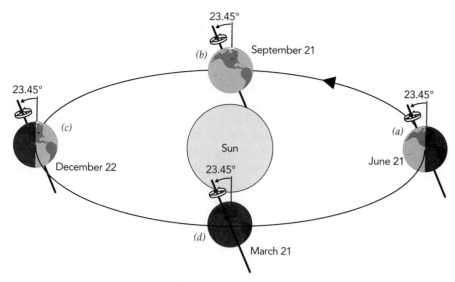

Fig. 4.3.4 The seasons arise because the earth's rotational axis is tilted with respect to its orbital plane. For half the year, the earth's northern hemisphere is tilted toward the sun and for half the year, it's tilted away from the sun. The more the northern hemisphere is tilted toward the sun, the longer are its days and the shorter its nights.

(Fig. 4.3.5*a*), solar radiation is spread out by its oblique incidence and its intensity on the ground is low. However, when the sun is directly overhead (Fig. 4.3.5*b*), the solar radiation incident on the ground is as concentrated as it can be and its intensity is high. The more intense the solar illumination on the ground, the warmer the local average temperature. The northern hemisphere is hottest during the summer months of June, July, and August because that's when the sun is highest in the sky and when the solar illumination of the ground is most intense.

In the southern hemisphere, the experiences are reversed. The southern hemisphere experiences the height of its summer around December 22 and the depths of its winter around June 21. South of the Antarctic circle, a band located 23.45° north of the South Pole, the sun never sets during part of its summer and never rises during part of its winter.

People living on a certain band around the earth, near the equator, find that the sun passes directly overhead in the middle of the day. The location of this band moves north as summer approaches and it move south as winter approaches. At the autumnal (September 21) and vernal (March 21) equinoxes, the earth's rotational axis is pointed neither toward nor away from the sun (Fig. 4.3.4*b,d*) and the sun appears directly overhead to people living on the earth's equator. On June 21, the earth's northern hemisphere is tilted as much as possible toward the sun, its summer solstice, and the sun appears overhead to people living on the Tropic of Cancer, a band located 23.45° north of the equator. On December 22, the northern hemisphere is tilted as much as possible away from the sun, its winter solstice, and the sun appears overhead to people living on the Tropic of Capricorn, located 23.45° south of the equator.

Actually, the observation that the earth's rotational axis always points in the same direction isn't exactly true. The sun and moon exert small torques on the earth that cause the earth's rotational axis to turn ever so slowly about an axis that would appear vertical to a person standing on the earth's orbital plane. This slow movement of the earth's rotational axis, called the precession of the equinoxes, completes one full cycle every 25,800 years. Because of it, the tropical or civil year (the time between successive vernal equinoxes) is a little shorter than the solar or sidereal year (the time in which the earth completes one full orbit of the sun relative to the fixed

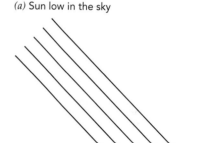

(a) Sun low in the sky

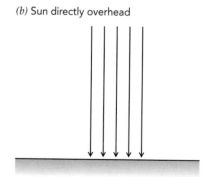

(b) Sun directly overhead

Fig. 4.3.5 *(a)* When the sun is low in the sky, sunlight strikes the ground at an oblique angle and the solar intensity is low. *(b)* When the sun is directly overhead, the sunlight reaching the ground is as intense and concentrated as possible.

stars). The stars overhead at midnight on any particular day of the year are slowly changing because of this precession, which is why the sections of the zodiac have drifted relative to the seasons since the days of Ptolomy.

There is one other interesting detail about the earth's orbit—the earth's orbit isn't a perfect circle, it's an ellipse. The earth is therefore not always the same distance from the sun. The earth makes its closest approach to the sun, its perihelion of 147,100,000 km (91,300,000 miles), on January 2 and reaches its farthest point from the sun, its aphelion of 152,100,000 km (94,400,000 miles), on July 4. So the northern hemisphere's summer has nothing to do with the closeness of the sun. The earth is actually farthest from the sun during the northern hemisphere's summer.

Other Objects in the Solar System

The solar system has many other objects that are too small to be called planets but that are visible from the earth. There are many more objects that can't be seen until their paths cross that of the earth and they appear in our atmosphere as meteors or "falling stars." All of these objects travel in elliptical orbits around the sun. The asteroids are a group of objects that orbit the sun between the orbit of Mars and the orbit of Jupiter. The asteroids have nearly circular orbits and they are sometimes called minor planets.

In contrast, comets are objects that orbit the sun in highly elongated elliptical orbits. They start very far from the sun and dive in close once per orbit. A typical comet is warmed by sunlight as it approaches the sun and becomes visible as gas evaporates from its surface and reflects the sun's light. A typical comet orbit appears in Fig. 4.3.6.

Let's follow the comet through one orbit to see how it behaves. The comet starts far from the sun, where it moves slowly and has relatively little kinetic energy. Since it's very small, it's essentially impossible to see even with a large telescope. However, its distance from the sun gives it a great deal of gravitational potential energy. The comet accelerates toward the sun but it doesn't have the sideways velocity needed to send it into a near-circular orbit, like the orbits of the planets. Instead, the comet dives almost directly toward the sun, picking up speed and kinetic energy at the expense of gravitational potential energy. As the comet approaches the sun, sunlight warms its surface and gas begins to evaporate from its nucleus. A steady spray of particles flowing away from the sun, the solar wind, blows this comet gas out and away from the sun like a gossamer streamer in the breeze from a fan. We see this gas because it's illuminated by sunlight.

The comet reaches its closest approach to the sun at a very high speed. It has converted most of its gravitational potential energy into kinetic energy and it sails quickly past the sun and begins to return to the distant reaches of the solar system. The streamer of gas recondenses on the cooling nucleus of the comet and the comet becomes harder and harder to see. It continues to accelerate toward the sun, but this time its acceleration slows its outward velocity. The comet is converting its kinetic energy back into potential energy. Eventually, it returns to its original position and the whole cycle begins again.

Fig. 4.3.6 A comet orbits the sun in a highly elongated or eccentric elliptical orbit. The comet moves slowly when it's far from the sun but speeds up as it approaches the sun. Its tail is actually gas that evaporates from its nucleus and that is driven back, away from the sun, by the solar wind.

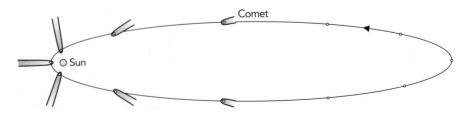

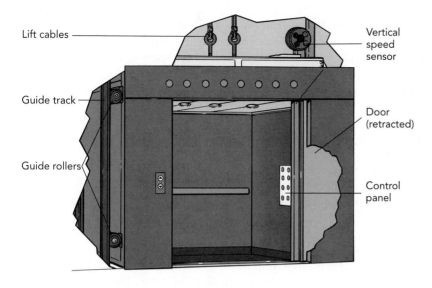

Lift cables

Vertical speed sensor

Guide track

Door (retracted)

Guide rollers

Control panel

Section 4.4 **Elevators**

While the invention of steel made skyscrapers possible, the invention of elevators made them practical. Imagine life in a big city without elevators. Business near the top of the Empire State Building would be limited to a few world-class athletes.

At the heart of an elevator is a very simple lifting machine. There are only a few different types of elevators and the techniques they use to raise or lower their cars have changed very little since Elisha Otis invented the safety elevator in 1853. What has changed is the source of power for operating the elevators and the sophistication of their control equipment. Electricity has long since replaced steam as the power source and elevator operators have been replaced by computers.

Jackscrew Elevators

One of the oldest and simplest lifting devices is the jackscrew—a screw used as a lifting mechanism. Jackscrews are used frequently in industry, construction, and maintenance to support or move heavy objects. Jackscrews are also used to level buildings and support sagging beams, and the repair jack that you have in your automobile is probably a jackscrew or a mechanism that incorporates a jackscrew. It's not surprising then that early elevators were based on the jackscrew. The elevator sat on top of a jackscrew and was raised or lowered by turning the screw into or out of a threaded hole. While jackscrews are no longer used in passenger elevators, they're still worth a few moments discussion.

A jackscrew elevator consisted of a lifting platform that was pushed upward from below by a jackscrew (Fig. 4.4.1). What made jackscrews so appealing for early elevators was that they were unlikely to fail catastrophically and that they exhibited mechanical advantage. The worry about catastrophic failure was very real before 1853—the cars in elevators built prior to that time were prone to dropping suddenly when the rope lifting them broke. Since the consequences of such a fall were awful, knowing that a thick metal jackscrew was pushing the car upward from beneath was very comforting to the passengers.

But what makes jackscrews so useful in lifting devices of all sorts is mechanical advantage. A modest torque exerted on the threaded cylinder in a jackscrew can lift a very heavy object. As I noted in Chapter 1, lifting a piano to the second floor requires a certain amount of work, regardless of how you get it there. In that chapter we used a ramp and here we use a jackscrew. Actually, the jackscrew is

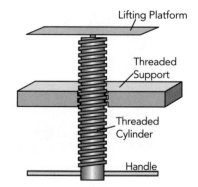

Fig. 4.4.1 A jackscrew uses the motion of a threaded cylinder—a screw—through a threaded hole to raise or lower a heavy object. A modest torque exerted on the screw's handles can keep the screw rotating and rising as it exerts a large upward force on the lifting platform above it.

just a rotating ramp, so the principle is exactly the same. The jackscrew allows the elevator operator to do the work required to lift the piano a little at a time. A modest torque exerted over many, many turns of the screw does the same amount of work as lifting the piano straight up to the second floor.

However, while a jackscrew provides a great deal of lifting force, it must be turned very rapidly in order to raise its platform at any reasonable rate. Unfortunately, a jackscrew encounters sliding friction and becomes extremely hot if it's turned too quickly. It may also wear out. Friction and wear severely limit the vertical speed of a jackscrew elevator and an elevator that takes minutes to move from floor to floor will lose most of its passengers to the staircase. Furthermore, the jackscrew itself must be as tall as the building it serves and buildings have become very tall. Jackscrew-based elevators long ago gave way to hydraulic and cable-lift elevators.

Pushing Up from Below: Hydraulic Elevators

A hydraulic elevator is lifted from below by a long metal shaft while a cable-lifted elevator is pulled up from above by a long metal cable. Let's begin by looking at hydraulic elevators.

The car of a hydraulic elevator is lifted from below by a hydraulic ram (Fig. 4.4.2). A hydraulic ram is a long piston that is driven into or out of a hollow cylinder by pressure in a hydraulic fluid. The hydraulic fluid, usually oil or water, exerts a force on any surface it touches, including the base of the piston. If the pressure in the hydraulic fluid is high enough, the force it exerts on the base of the piston will exceed the weight of the piston and elevator car and they will accelerate upward. (For a more complete discussion of fluids, pressure, and the basic concepts of hydraulics, flip ahead to Pascal's principle on p. 138.)

But as the piston rises, the hydraulic fluid has more volume to fill and its pressure drops. To keep the piston moving upward, something must continuously add high-pressure hydraulic fluid to the cylinder. That something is usually an electrically powered pump. This pump draws low-pressure hydraulic fluid from a reservoir and pumps it into the cylinder. The pump does work on the fluid and this work is what lifts the elevator car.

When the elevator car has reached the proper height, the pump stops and the piston rests on the high-pressure hydraulic fluid beneath it. As long as the amount of fluid in the cylinder doesn't change, the piston and car will stay where they are as the passengers get on and off.

To let the car descend, the elevator opens a valve and permits the high-pressure hydraulic fluid to return to the low-pressure reservoir. The fluid naturally accelerates toward the lower pressure and the cylinder begins to empty. The car descends. However, the descending car releases gravitational potential energy and that energy must go somewhere. It ends up in the swirling motion of hydraulic fluid returning to the reservoir and is quickly converted into thermal energy. When the swirling has stopped, the fluid in the reservoir is warmer than it was before the elevator made its trip up and down.

Because it lifts the car from below like a jackscrew, the hydraulic elevator is naturally very safe. Even if the cylinder springs a leak, the hydraulic fluid will probably not flow out of the cylinder fast enough for the car to descend at a dangerous speed. But unlike a jackscrew, a hydraulic ram encounters very little friction and wear, so its piston can move in or out of the cylinder rapidly. As a result, the car of a hydraulic elevator can be lifted as fast as the pump can deliver high-pressure hydraulic fluid. Of course, the pump has to do a great deal of work on that fluid in a short time, so it must be very powerful. Nonetheless, the speed of a hydraulic elevator is limited only by the power of the pump and the comfort of the passengers. Most passengers don't enjoy huge accelerations. While you could build a hydrau-

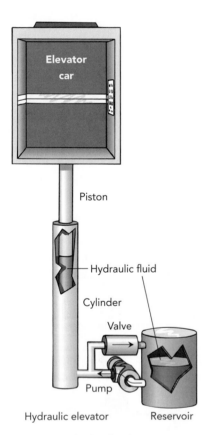

Fig. 4.4.2 A hydraulic elevator uses a hydraulic ram to support its car. The ram's piston rises as high-pressure hydraulic fluid is pushed into the hollow cylinder by a pump. The car is lowered by opening the valve and allowing the high-pressure hydraulic fluid to flow back into the storage reservoir.

lic elevator that would leap from one floor to another in the wink of an eye, it would require seat belts and airbags.

Although hydraulic elevators are wonderful in many situations, they do have at least two drawbacks. First, a hydraulic elevator is only as tall as its piston and cylinder. The piston has to reach all the way to the top floor and the equally tall cylinder must be hidden below the ground. Burying the cylinder is quite a procedure in a tall building. A deep hole must be drilled and the cylinder must be lowered into the hole with a crane. The difficulties involved in manufacturing the cylinder and piston and in assembling the completed hydraulic ram limit its height. It's extremely difficult to construct a hydraulic elevator over 30 stories tall. While some hydraulic elevators use telescoping pistons, pistons that fit inside one another like the segments of car radio antennas (Fig. 4.4.3), they still can't serve the tallest buildings.

The other deficiency of hydraulic elevators is that there is no mechanism for storing energy between trips. The energy expended in lifting people up 30 floors is not saved as those people descend. It becomes thermal energy in the hydraulic fluid as the hydraulic fluid returns to the reservoir. For a tall building with lots of up and down traffic, the elevator can turn a lot of electric energy into thermal energy in the hydraulic fluid.

Fig. 4.4.3 This elevator uses a telescoping hydraulic ram to lift its car to the top floor of a three-story building. There are actually two pistons in this telescoping system; the first piston rises out of a cylinder buried in the ground while the second, narrower piston rises out of the first piston. Because its pistons stack on top of one another when fully extended, the elevator car can rise almost twice as high as the height of the cylinder itself.

Pulling Up from Above: Cable-Lifted Elevators

To eliminate the need for long hydraulic rams, most elevators are lifted from above by cables. Introducing cable-lifted elevators was not easy because people were wary of any system that would drop disastrously if the rope broke. In 1853, the American inventor Elisha Graves Otis (1811–1861) demonstrated a "safety elevator" that would stop automatically if the rope broke. In a further improvement, the ropes used to lift early elevators were replaced with metal cables, which were less prone to wear and aging and made cable failure a rare event. With safety no longer an issue, cable-lifted elevators soon became the dominant form of elevator. But before we look at how a cable-lifted elevator actually works, we'll need to know how a rope lifts an object and how pulleys redirect forces exerted on a rope. Let's take a moment to look at ropes and pulleys.

Suppose that the elevator in your building is broken. You decide to lift the empty elevator car by hand with a light but sturdy rope (Fig. 4.4.4). The elevator is on the ground floor and you are pulling the rope up from the fifth floor, where your apartment is located. The empty elevator weighs 500 N (112 pounds), which is about all you can lift. If your arms were 5-stories long, you could pull the elevator up directly. The rope simply extends your reach so that you can exert an upward force on the elevator many meters below you.

Pulling on a rope produces tension throughout the rope. Tension means that every portion of the rope pulls inward on each of the two adjacent portions with a certain amount of force. To keep the empty elevator hanging motionless from the bottom of the rope, you must pull upward on the rope with 500 N of force. Each portion of rope then exerts 500 N of upward force on whatever is

Fig. 4.4.4 If you pull on a very light, stationary rope, you produce a uniform tension throughout the rope that is equal to the force you exert on it. Each portion of rope, for example the two end portions or the shaded portion near the middle, experiences an upward force from whatever is above it and a downward force from whatever is below it.

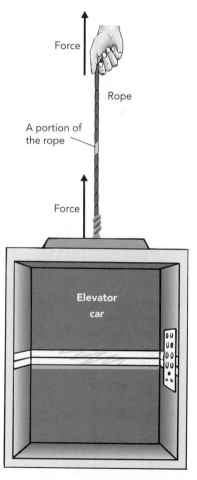

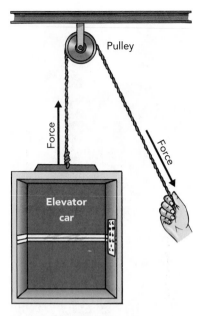

Fig. 4.4.5 With the rope drawn over a pulley, you can lift the elevator by pulling from almost anywhere. You exert the same force and the tension in the rope is the same, but the pulley redirects the force to make the job more convenient.

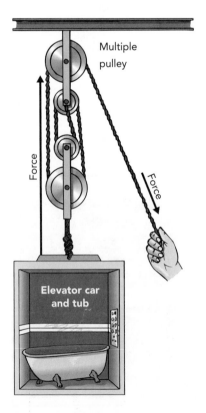

Fig. 4.4.6 A multiple-pulley is used to lift an elevator car containing a bathtub. Because there are four rope segments between the elevator and the support above it, the tension in the rope pulls upward on the elevator car four times and the upward force on the car is four times the tension in the rope.

below it and 500 N of downward force on whatever is above it. The bottom of the rope exerts 500 N of upward force on the elevator. Overall, your upward force of 500 N is conveyed meter by meter along the rope until it's exerted on the elevator far below. In effect, you are exerting an upward force of 500 N on the elevator and it's pulling back. As promised, the rope simply extends your reach.

Since the elevator weighs 500 N and you are exerting an upward force of 500 N on it, the net force on the elevator is zero and it doesn't accelerate. Because the elevator is initially stationary, it remains stationary. If you now exert a little more upward force on the rope, the elevator will experience a net upward force and will accelerate toward the fifth floor. Once the elevator has begun to move upward, you can reduce your force back to 500 N and the elevator will coast upward at constant velocity. You are now doing work on the elevator because you are pulling upward on it via the rope and it's moving upward.

Lifting the empty elevator to the fifth floor doesn't require an enormous amount of force, but that force must come from the middle of the elevator shaft. It would be nice to stand somewhere else as you pulled on the rope, so you suspend a pulley in the elevator shaft (Fig. 4.4.5). With the rope draped over the pulley, you can create tension in the rope from a different location. In fact, you can even pull downward on the rope. While each portion of rope continues to pull inward on its neighbors, the directions of these forces gradually change as the rope bends around the pulley. The pulley redirects the forces on the rope so that a downward force on one end of the rope can exert an upward force on the other end. This redirection makes it much easier to lift the elevator and even allows you to use the weight of some other object to help lift an elevator (see ❶).

Multiple Pulleys

The elevator, however, isn't always empty. Last week, the bathtub cracked and you and your friends pushed it off the fire escape. That was easy enough, although it ruined the flower garden next door. But the new bathtub weighs 1300 N (292 lbf), so how are you going to get it up to your apartment? It has to ride up in the elevator. You could rig up the same single pulley and get all of your friends to pull on the rope. But a better idea is to use a multiple-pulley, sometimes called a block-and-tackle. When you pull on a rope, you produce a tension all along that rope. If you could use that same tension several times, you could get mechanical advantage. That's what happens in a multiple-pulley.

In a multiple-pulley, the cord goes back and forth between a fixed set of pulleys and a moving set of pulleys (Fig. 4.4.5). The rope's far end is tied to one of the pulley sets. It's important that the cord pass easily over the pulleys. Now when you create tension in the cord, that same tension appears on every segment of cord between the two sets of pulleys. If you exert 500 N of force on the cord, each cord segment will have 500 N of tension. As a result, the two sets of pulleys will be pulled together with 500 N of force for each segment of cord connecting them. If there are 4 cord segments attached between the top of the elevator and the fifth floor, then the total lifting force on the elevator and bathtub will be 2000 N. Since the bathtub and elevator only weigh 1800 N, they will experience a net upward force and will accelerate upward.

While it takes less force on the cord to lift the bathtub and elevator with a multiple pulley than with a single pulley, you don't get something for nothing. To lift the elevator 1 m, you must shorten each segment of cord by 1 m. Since there are 4 segments, you will have to pull 4 m of cord through the system of pulleys. You are obtaining mechanical advantage, using a modest force exerted over a long distance to obtain a larger force exerted over a shorter distance. The amount of work required to lift the bathtub and elevator to your apartment is the same, whether you use a single or multiple pulley. The multiple pulley merely allows you to do this work more gradually, with a smaller force exerted over a longer distance.

Cable-Lifted Elevators and Counterweights

True cable-lifted elevators resemble the hand-powered one we have just discussed, except that machines pull the cables. In early cable-lifted elevators, the cables were pulled by steam-powered hydraulic rams. Steam was used to pump fluid into or out of the ram and the ram's movement was used to pull the cables. Usually, the ram was used to separate the two halves of a multiple-pulley. The cable coming out of this multiple pulley ran over a pulley at the top of the elevator shaft and down to the elevator car itself. As the two halves of the multiple pulleys were drawn apart, they drew in more cable and lifted the elevator car. As fluid was released from the hydraulic ram, the multiple pulley released cable and the elevator car descended.

The first improvement that appeared in cable-lifted elevators was the counterweight (Fig. 4.4.7). Lifting the elevator car by itself requires a considerable amount of work because the car's gravitational potential energy increases as it rises. It would be nice to get back this stored energy when the car descends. While it's hard to turn gravitational potential energy back into high-pressure steam, it's possible to use that energy to lift a counterweight.

The counterweight in an elevator descends when the car rises and rises when the car descends. Because the two objects have similar weights, the total amount of weight that's rising or falling as the elevator moves is almost zero. The overall gravitational potential energy of the elevator isn't changing much; it's simply moving around between the various parts of the machine. The counterweight balances the car so that it takes very little power to move the system. The elevator and counterweight resemble a balanced seesaw, which requires only a tiny push to make it move.

The counterweight on a typical elevator hangs from its own cable. That cable passes up and over a pulley at the top of the elevator shaft and then down to the top of the elevator car, where it's attached. The counterweight's weight is usually equal to the weight of the empty elevator car plus about 40% of the elevator's rated load. Thus, when the elevator is 40% filled, the counterweight will exactly balance the car and very little work will be done in raising or lowering the car.

Most modern elevators are driven by electric motors. The advantages of electric motors are their variable speeds of rotation, high torques, and reliability. While we won't discuss electric motors in detail until Section 11.3, what matters here is simply that they can provide mechanical power efficiently at many rotational speeds, torques, and overall power-levels. The output power of an electric motor is frequently rated in horsepower and the motors used in elevators may be as large as several hundred horsepower.

Because early electric motors couldn't provide much mechanical power, the first electric elevators used winches to lift their elevator cars. The cable from the elevator car was wound up on a drum at the top of the elevator shaft. The counterweight was attached to a second cable that was also wound on the drum. The two cables were arranged so that the counterweight cable unwound as the car cable wound up. An electric motor used gears to turn the drum.

This winch mechanism had a number of disadvantages. It raised or lowered the car relatively slowly because the gearing limited the drum's maximum angular speed.

❶ The development of safe elevators had an enormous effect on people's interest in tall buildings. Suddenly the upper floors became more desirable than the lower floors. Speed became very important. A "water balance" elevator was tried in the New York Western Union Building in 1873. The elevator car was drawn upward by the weight of an enormous bucket of water. To descend, the bucket of water was emptied. Controlled only by braking and without any automatic safety system, this elevator was too scary to be popular.

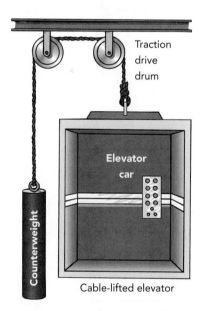

Cable-lifted elevator

Fig. 4.4.7 A cable-lifted elevator usually supports the car and a counterweight from opposite ends of its cable. A motor turns a traction drive that either raises or lowers the car. The counterweight moves in the other direction, assisting the motor in lifting the car or storing energy as the car descends.

Fig. 4.4.8 These three cable-lifted elevators are encased in glass so that all of their moving parts are visible. Each elevator is raised or lowered on four metal cables that rise into the top-floor penthouse, where most of the machinery is located.

The overall height of the elevator was limited because the drum had to be able to hold all of the cable when the elevator was at the top of its travel. The diameter of the drum was constrained by the need to keep torques low and only about 100 m of cable could be accommodated.

Instead of winding and unwinding cable from a drum, most modern elevators (Fig. 4.4.8) use traction to draw a cable over a drum. The cable rises from the elevator car, travels over the traction drum and then descends into the elevator shaft where it's attached to the counterweight. An electric motor turns the traction drum. When high speed isn't important, the drum can be turned by a small motor through the use of gears. However, in tall buildings, the drum is usually turned directly by a large motor. Elevators of this type can run at speeds as high as 10 m/s (22 mph) in buildings of any height.

The mechanical power required from the drive motor depends on how well balanced the car and counterweight are. If the elevator car is loaded to 40% of capacity so that the two weights are balanced, the motor will have little difficulty in moving the car up or down. If the car is particularly empty or particularly full, the motor will have to provide considerable mechanical power when lifting the heavy side of the system rapidly and various brakes will have to absorb power released by the elevator when the heavy side descends rapidly. The motor's maximum mechanical power, together with the strength of the cables, limits how much weight the elevator can lift and how quickly.

In many freight elevators, the car is lifted by a multiple pulley so that a single segment of cable doesn't have to support the entire load. Even when a single pulley is used, several separate cables support the car, both for safety and to reduce cable stretching. Cable stretching is a serious problem in tall elevators. Like most objects, a cable behaves as a spring when it's subject to tension: its length increases by an amount proportional to the tension it experiences. As people enter the elevator car and its total weight increases, the tension on its support cable increases and that cable stretches slightly. Modern elevators are equipped with automatic leveling systems that turn the traction drum to make up for the stretching of the cables. The passengers are unaware of this careful adjustment taking place as they step on or off the elevator. Nonetheless, you may be able to feel the cable stretch if you bounce up and down on a cable-lifted elevator.

Balance

Elevator cars must remain level no matter where the passengers choose to stand. To prevent the car from tilting, virtually all elevators employ tracks. While the piston of a hydraulic elevator could keep the car level on its own, the torques exerted on that piston by shifting passengers would probably damage it over time. And a cable lift elevator has no natural way of keeping its car level and has no choice but to use a track.

To see why the cable lift elevator can't do without a track, consider what happens as passengers move around in its car. The cable supports the top of the car and the car's center of gravity, the effective location of its overall weight, is located at a point near the car's bottom (Fig. 4.4.9a). Like anything else, the elevator car will accelerate in whatever direction reduces its total potential energy as quickly as possible. If the car can lower its center of gravity and thus its gravitational potential energy by tilting, it will accelerate toward that tilt.

If the passengers are carefully balanced around the middle of the car's floor, the car may settle into a level orientation. But if all the passengers stand on one side of the floor, that side of the floor will descend and the car will end up tilted (Fig. 4.4.9b). To make matters worse, the car will swing back and forth like a pendulum before settling down at this tilted equilibrium. The best way to prevent an elevator car

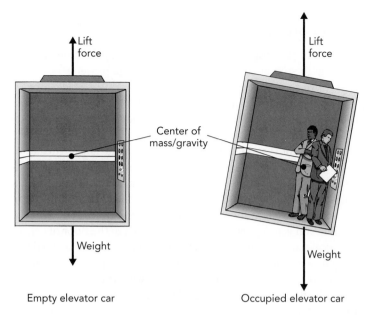

Center of mass/gravity

Empty elevator car

Occupied elevator car

Fig. 4.4.9 *(a)* An empty elevator car hangs level because its center of gravity is centered in the car and settles directly below the supporting cable. *(b)* An occupied car tilts because its center of gravity isn't centered in the car. The car again settles with its center of gravity directly below the support cable, but now the car is tilted. For the car to remain level as the passengers move around, the car must run in a track that can exert leveling torques on the car.

from tilting or swinging is to confine the car to a track. The rails of the track exert the torques needed to keep the car level.

Safety

All cable-lifted elevators have safety devices to keep them from falling if their cables break. Most modern elevators have more than one lifting cable, but they still require mechanisms to ensure that there are no accidents (see ❷).

The original safety device that Otis developed for his first elevators had jaws that would grab onto the rails of the elevator track if there were a loss of tension in the supporting cable. If the cable broke and its tension vanished, springs would force the jaws into the track.

Modern elevators use mechanisms that monitor the vertical speed of the elevator. If the speed exceeds a certain permissible value, brakes on the car grab the tracks. This speed control prevents a nearly empty elevator from moving upward too quickly just as it prevents a full elevator from falling.

One such speed-sensing device is the centrifugal governor, a mechanism that senses how quickly a shaft is turning (Fig. 4.4.10). When it's used with an elevator, the shaft is turned by a pulley on a special cable attached to the elevator car. The faster the elevator moves, the faster the shaft turns. The centrifugal governor swings several masses around in a circle. Since the masses travel in uniform circular motion, they need some centripetal force to accelerate them toward the center of the circle. In the centrifugal governor, this centripetal force is exerted by several rods that are held apart by a spring.

As long as the shaft is turning slowly, the spring can keep the rods from moving together. But when the shaft is turning quickly, the centripetal force becomes very large and the rods compress the spring. As the rods move, they push on a lever. In the case of the elevator, this lever activates brakes that slow the elevator down.

❷ The only time a safety elevator plummeted to the bottom of its shaft was in 1945, when a military airplane struck the Empire State Building. The plane lodged in the elevator shaft near the 79th floor, cutting all of the cables to the elevator car on the 38th floor. The car dropped to the basement, but its descent was cushioned by the increasing air pressure beneath it and by a mountain of severed cables and an emergency bumper at the bottom of the shaft. The only occupant of the car, a 20-year-old elevator operator, survived without serious injury.

(a)

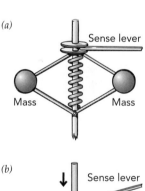

Sense lever

Mass   Mass

(b)

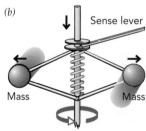

Sense lever

Mass   Mass

Fig. 4.4.10 A centrifugal governor uses the principle that a central force is required to accelerate masses around in a circle. As long as the shaft is stopped or spinning slowly *(a)*, the spring can keep the upper and lower rods apart. But once the shaft spins too quickly, the masses swing outward *(b)* and the sense lever is shifted.

Starting and Stopping

Simply moving the elevator car up or down isn't enough. To be useful, an elevator must be able to stop at the proper level, exchange passengers or freight, and then start to move to a new level. To be pleasant to ride, the elevator must start and stop slowly enough that it doesn't knock the passengers off their feet. To meet these added requirements on the motion of the elevator car, variable speed electric motors are used.

Whether the elevator is handled by an operator or is run automatically, the torque exerted on the traction drive drum is carefully controlled in order to avoid sudden accelerations. Whenever the elevator you are in accelerates upward, as it does when it starts moving upward or stops moving downward, you feel particularly heavy. Your apparent weight increases because of the upward acceleration. If the upward acceleration is too great, you may be thrown to the floor of car. Whenever the elevator accelerates downward, as its does when it starts moving downward or stops moving upward, you feel particularly light. Your apparent weight decreases because of the downward acceleration. If the downward acceleration is too great, you may leave the floor of the elevator and bump against its ceiling. Only after the elevator reaches constant velocity, either up or down, does your apparent weight return to your true weight.

A well-designed elevator accelerates and decelerates smoothly and gradually. This need for smooth deceleration means that the operator or the automatic mechanism must anticipate stops and begin to decelerate before reaching the stopping point. Operating an antique elevator, with no machinery to help anticipate the stop, required great skill. In manually operated elevators, the operator's ability to stop at the correct height limited the maximum vertical speed that could be used effectively. Modern elevators anticipate stops automatically and gradually reduce the speed of travel so as to come to a stop at exactly the right height. These elevators can move up or down extremely quickly and still stop properly.

CHAPTER 5
THINGS INVOLVING FLUIDS

So far all of the everyday objects we've examined have been solids. But since gases and liquids are also important parts of the world around us—as the air we breathe, the water we swim in, and even the blood our hearts pump through our veins—I'll now turn to objects that, unlike solids, don't have well-defined shapes. These objects are called fluids.

Fluids are generally given short shrift in the teaching of science. It's as though no one wants to take ownership or responsibility for them. After examining a few basic fluid issues, such as floating and pressure, most science instructors move on to other things. You can almost hear them saying: "pay no attention to that elephant in the corner."

This superficial treatment of fluids is unfortunate because it omits most of what makes fluids interesting in everyday life: pouring, swimming, sailing, flying, spraying, breathing, swirling, pumping, churning, and drinking. Life would be awfully dull if it weren't for fluids, even if you're not one who frequents coffee shops or bars. As you'll probably notice while reading this and the subsequent chapter, I find fluids amazing and never tire of learning new things about them. Just consider all the ways in which air and water influence your life and you'll begin to see why understanding how fluids behave might be valuable.

Despite being hidden from public view most of the time, the study of fluid behavior and motion is actually a broad field, extending across the sciences and engineering. Fluid dynamics, often called hydrodynamics, is as important to an oil-well engineer as to an animal physiologist or a stellar astrophysicist. The tools used to analyze fluids are somewhat more complicated than for solids because fluids themselves are more complicated: it's hard to exert a force directly on them, and, even if we could, they usually don't move as a single rigid object. In this chapter, we'll look at some of the concepts and tools needed to understand their complex behaviors.

Chapter Itinerary

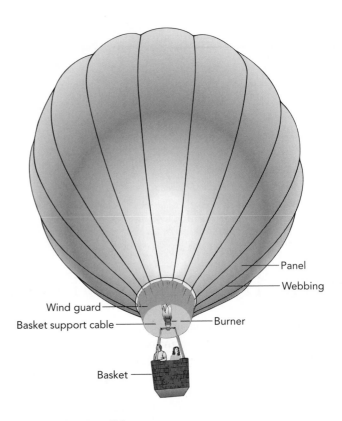

Panel

Webbing

Wind guard

Basket support cable

Burner

Basket

Section 5.1 **Balloons**

Because gravity gives every object near the earth's surface a weight proportional to its mass, objects fall when you drop them. Why then does a helium-filled balloon—which, after all, is just another object with a mass and a weight—sail upward into the sky when you let go of it? Does the balloon have a negative mass and a negative weight, or are we forgetting something?

We're forgetting air—specifically, the layer of air that sits atop the earth's surface and is held in place by gravity. Since this air is difficult to see and moves out of our way so easily, we often forget that it's there. But air sometimes makes itself noticeable. When you ride a bicycle, you feel its forces; when you blow up a beach ball, you see that it takes up space. And when you release a helium balloon, air lifts the balloon upward.

Air and Air Pressure

Hot-air and helium balloons are supported by the air around them. Although these balloons have positive masses and downward weights, the surrounding air pushes upward on them hard enough to balance their weights so that they float. To understand balloons, we must start by understanding air.

Like the objects we've already studied, air has mass and weight. Unlike those objects, however, it has no fixed shape or size. You can mold 1 kg of air into any form you like and it can occupy a wide range of volumes. Air is compressible, that is, you can squeeze a certain mass of it into almost any space. For example, 1 kg of air could fill a single scuba tank or a whole basketball arena.

This flexibility of size and shape originates in the microscopic nature of air. Air is a gas, a substance consisting of tiny, individual particles that travel around independently. These individual particles are atoms and molecules. An atom is the smallest portion of an element that retains all of the chemical characteristics of that element; a molecule, an assembly of two or more atoms, is the smallest portion of a chemical compound that retains all of the characteristics of that compound. A molecule's atoms are held together by chemical bonds, linkages formed by electromagnetic forces between the atoms.

Air particles are extremely small, less than a millionth of a millimeter in diameter. Most are nitrogen and oxygen molecules, but others include carbon dioxide, water, methane, and hydrogen molecules, and argon, neon, helium, krypton, and xenon atoms. Those atoms, which don't make strong chemical bonds and rarely form molecules, are called inert gases because of their chemical inactivity.

Like tiny marbles, these air particles have sizes, masses, and weights. But while marbles quickly settle to the ground when you spill them from a bag, air particles don't seem to fall at all. Why don't they pile up on the earth's surface?

The answer has to do with air's thermal energy, specifically the portion of that energy contained in the motions of the air particles. This internal kinetic energy keeps the tiny air particles moving, spinning, and away from the earth's surface. In contrast, real marbles are too massive and heavy to be moved noticeably by thermal energy. The air's internal kinetic energy per particle is measured as its temperature; the greater that energy per particle, the hotter the air. While air's thermal energy also includes a portion stored in the forces between particles, this internal potential energy is negligible because the average forces between air particles are so weak.

An expanded view of air would reveal countless individual particles in frenetic thermal motion (Fig. 5.1.1a). At room temperature, these particles travel at bulletlike speeds of roughly 500 m/s (1100 mph), but they collide so often that they make little progress in any particular direction. Between collisions, air particles travel in nearly straight-line paths because gravity has too little time to make them fall very far.

Let's ignore gravity for the moment and consider what happens in a box containing 1 kg of air. The air particles whiz around inside the box and each time a particle bounces off a wall of the box, it exerts a force on that wall. Although the individual forces are tiny, the number of particles is not, and together they produce a large average force. The size of this total force depends on the wall's surface area; the larger its surface area, the more average force it experiences. In order to characterize the air, however, we don't really need to know the wall's surface area; instead, we can refer to the average force the air exerts on each unit of surface area, a quantity called pressure.

Pressure is measured in units of force-per-area. Since the SI unit of surface area is the meter^2 (abbreviated m^2 and often referred to as the square meter), the SI unit of pressure is the newton-per-meter^2. This unit is also called the pascal (abbreviated Pa), after French mathematician and physicist Blaise Pascal. One pascal is a small pressure; the air around you has a pressure of about 100,000 Pa (2100 lbf/ft^2 or 15 lbf/inch^2), so that it exerts a force of about 100,000 N on a 1-m^2 surface. Since 100,000 N (22,500 lbf) is about the weight of a city bus, air pressure can exert enormous forces on large surfaces.

Besides pushing on the walls of our hypothetical box, air also pushes on any object immersed in it. Its particles bounce off the object's surfaces, pushing them inward. As long as the object can withstand these compressive forces, the air won't greatly affect it since the uniform air pressure ensures that the forces on all sides of the object cancel one another perfectly. A sheet of paper, for example, will experience zero net force because the forces exerted on its two sides will add to zero.

Air particles also bounce off one another, so that air pressure exerts forces on air, too. A cube of air inserted into the box experiences all of the inward forces that a cube of metal would experience. The air around the cube pushes inward on it, and the cube pushes outward on the air around it. Since the net force on the cube of air is zero, the cube doesn't accelerate.

(a)

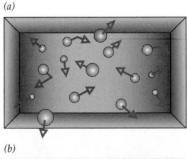

(b)

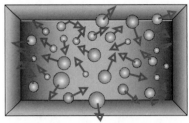

(c)

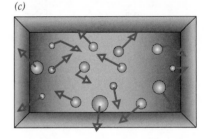

Fig. 5.1.1 (*a*) As air particles bounce off surfaces, they exert pressure on those surfaces; the amount of pressure depends on the air's temperature and on how densely its particles are packed. (*b*) Packing the air particles more densely increases the number of particles that hit the surfaces each second. (*c*) Increasing the temperature of the air increases the speed of the particles (shown by the arrows) so that they hit the surfaces harder and more frequently. Either change, in speed or collision frequency, increases the air's pressure.

Pressure, Density, and Temperature

Since air pressure is produced by bouncing air particles, it depends on how often, and how hard, those particles hit a particular region of surface. The more frequent or harder the impacts, the greater is the air pressure.

To increase the rate at which air particles hit a surface, we can pack them more tightly. If we add another 1 kg of air to our hypothetical box, we double the number of air particles in the same volume, which doubles the rate at which they hit each surface and therefore doubles the pressure (Fig. 5.1.1b). The air's pressure is thus proportional to its density, that is, its mass per unit of volume. Since the SI unit of volume is the meter^3 (abbreviated m^3 and often referred to as the cubic meter), the SI unit of density is the kilogram-per-meter^3 (abbreviated kg/m^3). The air around you has a density of about 1.25 kg/m^3 (0.078 lbm/ft^3). Water, in contrast, has a much greater density of about 1000 kg/m^3 (62.4 lbm/ft^3).

We can also increase the rate at which air particles hit a surface by speeding them up (Fig. 5.1.1c). If we double the internal kinetic energy of the air in our box, we double the average kinetic energy of each particle. Because a particle's kinetic energy depends on the square of its speed, doubling its kinetic energy increases its speed by a factor of $\sqrt{2}$. As a result, each particle hits the surface $\sqrt{2}$ times as often and exerts $\sqrt{2}$ times as much average force when it hits. With each particle exerting $\sqrt{2} \cdot \sqrt{2}$ or two times as much average force, the pressure doubles. Air's pressure is thus proportional to the average kinetic energy of its particles—to their average internal kinetic energies.

This average kinetic energy per particle is measured by the air's temperature; the hotter the air, the larger the average kinetic energy per particle and the higher the air's pressure. But the most convenient scale for relating the temperature of air to its pressure isn't the common Celsius (°C) or Fahrenheit (°F) scale; instead, it's a special absolute temperature scale. The SI scale of absolute temperature is the Kelvin scale (K). When the air's temperature is 0 K (−273.15 °C or −459.67 °F), it contains no internal kinetic energy at all and has no pressure; this temperature is called absolute zero. The Kelvin scale is identical to the Celsius scale, except that it's shifted so that 0 K is equal to −273.15 °C. In addition to associating the zero of temperature with the zero of internal kinetic energy, the Kelvin scale avoids the need for negative temperatures. Room temperature (20 °C or 68 °F) is about 293 K.

Since air pressure is proportional to both the air's density and its absolute temperature, we can express the relationship among these quantities as:

$$\text{pressure} \propto \text{density} \cdot \text{absolute temperature}. \tag{5.1.1}$$

The symbol "$\propto$" indicates proportionality rather than equality; although I'll refer to this expression as "Eq. 5.1.1," it's not actually an equation. This relationship is useful because it allows us to predict what will happen if we change the temperature or density of a specific gas, such as air. But it has its limitations; in particular, it doesn't work if we compare the pressures of two different gases, such as air and helium, which differ in their chemical compositions. To make such a comparison, we'll need to improve on Eq. 5.1.1. We'll do that later when we examine helium balloons.

Even in describing a specific gas, Eq. 5.1.1 has other shortcomings. The main problem is that real gas particles aren't completely independent of one another. If the temperature drops too low, the particles begin to stick together to form a liquid and Eq. 5.1.1 becomes invalid. But despite its limitations, this simple relationship between pressure, density, and temperature will prove useful in understanding how hot-air balloons float: it will help us understand the basic structure of the earth's atmosphere, the origins of the upward force that keeps a hot-air balloon aloft, and the reason why hot air rises.

Quantity	SI Unit	English Unit	SI → English	English → SI
Area	meter^2 (m^2)	foot^2 (ft^2)	1 m^2 = 10.764 ft^2	1 ft^2 = 0.092903 m^2
Volume	meter^3 (m^3)	foot^3 (ft^3)	1 m^3 = 35.315 ft^3	1 ft^3 = 0.028317 m^3
Pressure	pascal (Pa)	pound-force-per-foot^2 (lbf/ft^2)	1 Pa = 0.020885 lbf/ft^2	1 lbf/ft^2 = 47.880 Pa
Density	kilogram-per-meter^3 (kg/m^3)	pound-mass-per-foot^3 (lbm/ft^3)†	1 kg/m^3 = 0.062428 lbm/ft^3	1 lbm/ft^3 = 16.018 kg/m^3

The Earth's Atmosphere

Most of the mass of the earth's atmosphere is contained in a layer less than 6 km (4 miles) thick. Since the earth is 12,700 km (7,900 miles) in diameter, this layer is relatively thin—so thin that, if the earth were the size of a basketball, it would be no thicker than a sheet of paper.

The atmosphere stays on the earth's surface because of gravity. Every air particle, as we've seen, has a weight. Just as a marble thrown upward eventually falls back to the ground, so the particles of air keep returning toward the earth's surface. Although the particles are moving too fast for gravity to affect their motions significantly over the short term, gravity works slowly to keep them relatively near the earth's surface. An air particle, like a rapidly moving marble, may appear to travel in a straight line at first, but it will arc over and begin to fall downward eventually. Only the lightest and fastest moving particles in the atmosphere—hydrogen molecules and helium atoms—occasionally manage to escape from earth's gravity and drift off into interplanetary space.

While gravity pulls the atmosphere downward, air pressure pushes the atmosphere upward. As the air particles try to fall to the earth's surface, their density increases and so does their pressure. It's this air pressure that supports the atmosphere and prevents it from collapsing into a thin pile on the ground.

To understand how gravity and air pressure structure the atmosphere, picture a 1-m^2 column of the atmosphere as though it were a tall stack of 1-kg air blocks (Fig. 5.1.2). These blocks support one another with air pressure to form a stack of about 10,000 blocks. The bottom block must support the weight of all the blocks above it and is tightly compressed, with a height of about 0.8 m, a density of about 1.25 kg/m^3, and a pressure of about 100,000 Pa. A block farther up in the stack has less weight to support and is less tightly compressed. The higher in the stack you look, the lower the density of the air and the less the air pressure.

The atmosphere has essentially the same structure as this stack of blocks. The air near the ground supports the weight of several kilometers of air above it, giving it a density of about 1.25 kg/m^3 and a pressure of about 100,000 Pa; at higher altitudes, however, the air's density and pressure are reduced since there is less atmosphere overhead and the air doesn't have to support as much weight. High-altitude air is thus "thinner" than low-altitude air. Whatever the altitude, the pressure of the surrounding air is referred to as atmospheric pressure.

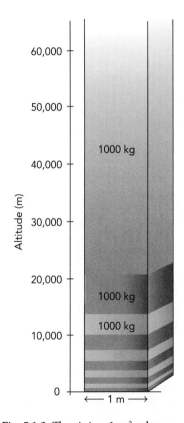

Fig. 5.1.2 The air in a 1-m^2 column of atmosphere has a mass of about 10,000 kg. The bottom 1000 kg is the most tightly compressed, because it supports the most weight above it. At higher altitudes, the air is less tightly compressed because it has less weight above it.

The Lifting Force on a Balloon: Buoyancy

So far we've examined air, air pressure, and the atmosphere. While it may seem that we've avoided dealing with balloons, these topics really are involved in keeping hot-air and helium balloons aloft. As we've seen, the air in the earth's atmosphere is a fluid, a shapeless substance with mass and weight. This air has a pressure and exerts forces on the surfaces it touches; that pressure is greatest near the ground and decreases with increasing altitude. Air pressure and its variation with altitude allow air to lift hot-air and helium balloons through an effect known as buoyancy.

Buoyancy was first described more than two thousand years ago by the Greek mathematician Archimedes (287–212 BC). Archimedes realized that an object partially or wholly immersed in a fluid is acted on by an upward buoyant force equal to the weight of the fluid it displaces. Archimedes' principle is actually very general and applies to objects floating or submerged in any fluid, including air, water, or oil. The buoyant force originates in the forces a fluid exerts on the surfaces of an object. We've seen that such forces can be quite large but tend to cancel one another. How then can pressure create a nonzero total force on an object, and why should that force be in the upward direction?

Archimedes' Principle
An object partially or wholly immersed in a fluid is acted on by an upward buoyant force equal to the weight of the fluid it displaces.

Without gravity the forces would cancel each other perfectly because the pressure of a stationary fluid would be uniform throughout. But gravity causes a stationary fluid's pressure to decrease with altitude. For example, when nothing is moving, the air pressure beneath an object is always greater than the air pressure above it. Thus air pushes upward on the object's bottom more strongly than it pushes downward on the object's top, and the object consequently experiences an upward overall force from the air—a buoyant force.

How large is the buoyant force on this object? It's equal in magnitude to the weight of the fluid that the object displaces. To understand this remarkable result, imagine replacing the object with a similarly shaped portion of the fluid itself (Fig. 5.1.3*a*). Since the buoyant force is exerted by the surrounding fluid, not the object, it doesn't depend on the object's composition. A balloon filled with helium will experience the same buoyant force as a similar balloon filled with water or lead or even air. So replacing the object with a similarly shaped portion of fluid will leave the buoyant force on it unchanged.

But a portion of fluid suspended in more of the same fluid doesn't accelerate anywhere; it just sits there, so the net force on it is clearly zero. It has a downward weight, but that weight must be canceled by some upward force that can only come from the surrounding fluid. This upward force is the buoyant force, and it's always equal in magnitude to the weight of the object-shaped portion of fluid displaced by the object.

This buoyant principle explains why some objects float while others sink. An object placed in a fluid experiences two forces: its downward weight and an upward buoyant force. If its weight is more than the buoyant force, it will accelerate

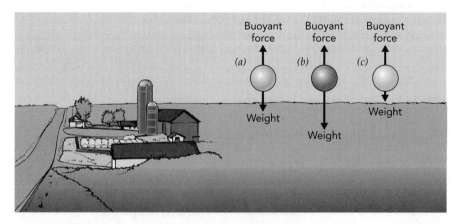

Fig. 5.1.3 (*a*) A portion of air immersed in that same air experiences an upward buoyant force equal to its weight and doesn't accelerate. (*b*) An object that is heavier than the air it displaces sinks, while (*c*) another object that is lighter than the air it displaces floats.

downward (Fig. 5.1.3b); if its weight is less than the buoyant force, it will accelerate upward (Fig. 5.1.3c). And if the two forces are equal, it won't accelerate at all and will maintain a constant velocity. If the balloon in this latter case starts motionless, it will remain motionless and will hover at a constant velocity of zero.

Whether or not an object will float in a fluid can also be viewed in terms of density. An object that has an average density greater than that of the surrounding fluid sinks, while one that has a lesser average density floats. A water-filled balloon, for example, will sink in air because water and rubber are more dense than air. If you double the volume of the balloon, you double both its weight and the buoyant force on it, so it still sinks. The total volume of an object is less important than how its density compares to that of the surrounding fluid.

Hot-Air Balloons

Since air is very light, with a density of only 1.25 kg/m³ (0.078 lbm/ft³), few objects float in it. One of these rare objects is a balloon with a vacuum inside it. Assuming that the balloon has a very thin outer shell or envelope, it will weigh almost nothing and have an average density near zero. Because its negligible weight is less than the upward buoyant force it experiences, the empty balloon will float upward nicely.

Unfortunately, this empty balloon won't last long. Because it's surrounded by atmospheric pressure air, each square meter of its envelope will experience an inward force of 100,000 N. With nothing inside the balloon to support its envelope against this crushing force, it will smash flat. A thick, rigid envelope might be able to withstand the pressure of the surrounding air, but then the balloon's average density would be large and it would sink. So an empty balloon won't work.

What will work is a balloon filled with something that exerts an outward pressure on the envelope equal to the inward pressure of the surrounding air. Then each portion of the envelope will experience zero net force and the balloon will not be crushed. We could fill the balloon with outside air, but that would make its average density too high. Instead, we need a gas that has the same pressure as the surrounding air but a lower density.

One gas that has a lower density at atmospheric pressure is hot air. Filling our balloon with hot air takes fewer particles than filling it with cold air, since each hot-air particle is moving faster and contributes more to the overall pressure than does a cold-air particle. A hot-air balloon contains fewer particles, has less mass, and weighs less than it would if it contained cold air. Now we have a practical balloon with an average density less than that of the surrounding air. The buoyant force it experiences is larger than its weight, and up it goes (Fig. 5.1.4).

Because the air pressure inside a hot-air balloon is the same as the air pressure outside the balloon, the air has no tendency to move in or out (an issue we will cover in the next section), and the balloon doesn't need to be sealed (Fig. 5.1.5). A large propane burner, located beneath the balloon's open end, heats the air that fills the envelope. The hotter the air in the envelope, the lower its density and the less the balloon weighs. The balloon's pilot controls the flame so that the balloon's weight is very nearly equal to the buoyant force on the balloon. If the pilot raises the air's temperature, particles leave the envelope, the balloon's weight decreases, and the balloon rises. If the pilot allows the air to cool, particles enter the envelope, the balloon's weight increases, and the balloon descends.

But even if the pilot heats the air to be very hot, the balloon won't rise upward forever. As the balloon ascends, the air becomes thinner and the pressure decreases both inside and outside the envelope. Although the balloon's weight decreases as the air thins out, the buoyant force on it decreases even more rapidly, and it becomes less effective at lifting its cargo. When the air becomes too thin to lift the balloon any higher, the balloon reaches a flight ceiling above which it can't rise, even

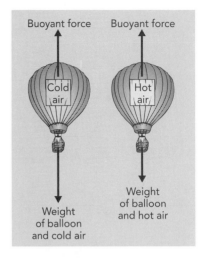

Fig. 5.1.4 Filling a balloon with hot air requires fewer air particles than filling it with cold air. That's because the average particle in hot air moves faster, collides more often, and effectively occupies more space than the average particle in cold air. A balloon filled with hot air therefore weighs less than a balloon filled with cold air. If the balloon's weight is small enough, the net force on the balloon will be in the upward direction and the balloon will accelerate upward.

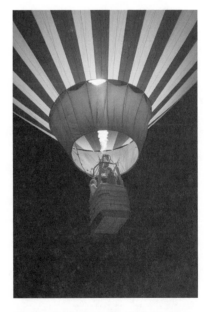

Fig. 5.1.5 The bottom of a hot-air balloon is open so that heated air can flow in and cold air can flow out. The heated air displaces more than its weight in cold air and makes the balloon lighter.

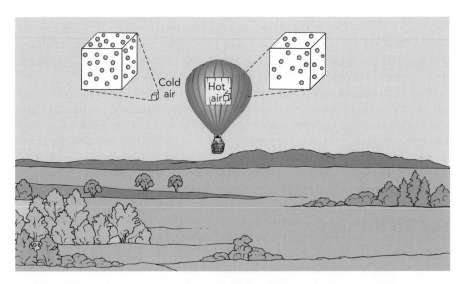

Fig. 5.1.6 A cube of hot air contains fewer air particles than a similar cube of cold air. Since it weighs less than the cold air it displaces, the hot air inside the balloon experiences an upward buoyant force that is greater than its weight.

if the pilot turns the flame on full blast. For each hot-air temperature, then, there is a cruising altitude at which the balloon will hover. When the balloon reaches that altitude, it's in a stable equilibrium. If the balloon shifts downward for some reason, the net force on it will be upward; if it shifts upward, the net force on it will be downward.

Helium Balloons

Although the particles in hot and cold air are similar, there are fewer of them in each cubic meter of hot air than in each cubic meter of cold air. We call the number of particles per unit of volume particle density, and hot air has a smaller particle density than cold air (Fig. 5.1.6). Because they contain similar particles, hot air also has a smaller density than cold air and is lifted upward by the buoyant force.

But there's another way to make one gas float in another: use a gas consisting of very light particles. Helium atoms, for example, are much lighter than air particles. When they have equal pressures and temperatures, helium gas and air also have equal particle densities. Since each helium atom weighs 14% as much as the average air particle, 1 m³ of helium weighs only 14% as much as 1 m³ of air. Thus a helium-filled balloon has only a fraction of the weight of the air it displaces, and the buoyant force carries it upward easily.

Why should air and helium have the same particle densities whenever their pressures and temperatures are equal? Because a gas particle's contribution to the pressure doesn't depend on its mass (or weight). At a particular temperature, each particle in a gas has the same average kinetic energy in its translational motion, regardless of its mass. Although a helium atom is much less massive than a typical air particle, the average helium atom moves much faster and bounces more often. As a result, lighter but faster-moving helium atoms are just as effective at creating pressure as heavier but slower-moving air particles.

Thus, if you allow the helium atoms inside a balloon to spread out until the pressures and temperatures inside and outside the balloon are equal, the particle densities inside and outside the balloon will also be equal (Fig. 5.1.7). Since the helium atoms inside the balloon are lighter than the air particles outside it, the balloon weighs less than the air it displaces, and it will be lifted upward by the buoyant force.

Fig. 5.1.7 A cube of helium gas contains the same number of particles as a similar cube of air, but each helium particle weighs less than the average air particle. Since it weighs less than the air it displaces, the helium inside the balloon experiences an upward buoyant force that is greater than its weight.

The pressure of a gas is proportional to the product of its particle density and its absolute temperature, as the following formula indicates:

$$\text{pressure} \propto \text{particle density} \cdot \text{absolute temperature.} \tag{5.1.2}$$

This proportionality holds regardless of the gas's chemical composition. Our previous proportionality, Eq. 5.1.1, worked only as long as the gas's composition didn't change, so that density and particle density remained proportional to one another. But now we have a relationship with a wider applicability.

Equation 5.1.2, with an associated constant of proportionality, is called the ideal gas law. This law relates pressure, particle density, and absolute temperature for a gas in which the particles are perfectly independent. It's also fairly accurate for real gases in which the particles do interact somewhat. The constant of proportionality is the Boltzmann constant, with a measured value of 1.381×10^{-23} Pa·m^3/(particle·K). Using the Boltzmann constant, the ideal gas law can be written:

$$\text{pressure} = \text{Boltzmann constant} \cdot \text{particle density} \cdot \text{absolute temperature.} \tag{5.1.3}$$

The Ideal Gas Law

The pressure of a gas is equal to the product of the Boltzmann constant times the particle density times the absolute temperature.

Helium isn't the only "lighter-than-air" gas. Hydrogen gas, which is half as dense as helium, is also used to make balloons float. But don't expect hydrogen to lift twice as much weight as helium. A balloon's lifting capacity is the difference between the upward buoyant force it experiences and its downward weight. Although the gas in a hydrogen balloon weighs half that in a similar helium balloon, the balloons experience the same buoyant force. Thus the hydrogen balloon's lifting capacity is only slightly more than that of the helium balloon. Hydrogen's main advantage is that it's cheap and plentiful, while helium is scarce (see ❶). But because hydrogen is also dangerously flammable (see ❷), it's avoided in situations where safety is important. However, even helium-filled airships can have problems (see ❸).

❶ Helium gas is obtained as a by-product of natural gas production from underground reservoirs in the United States, where it formed through the gradual radioactive decay of uranium and other unstable elements. While some of this gas is saved for industrial and commercial use, much of it is simply released into the atmosphere. The only other source of helium is the atmosphere, where helium is present at a level of 5 parts per million. Once the underground stores are consumed, helium will become a relatively rare and expensive gas.

❷ Rigid airships were appropriately named because they were truly ships that floated through the air. Unfortunately, the hydrogen gas that filled most airships, making them light enough that air could lift them, is very flammable. The *Hindenburg* burned on May 6, 1937, while trying to land at Lakehurst, New Jersey. Because hydrogen is so buoyant in air, most of the combustion occurred above the airship and many passengers survived.

❸ Even helium-filled airships were easily destroyed by bad weather. The *Shenandoah*, one of two U.S. airships based on the German designs, was destroyed by air turbulence on September 3, 1925, near Ava, Ohio. Crowds from a local fair immediately poured over the wreckage, collecting souvenirs.

Fig. 5.1.8 Helium balloons are sealed to keep helium from diffusing out and air from diffusing in.

Elastic Balloons

The air molecules in a hot air balloon with an open bottom are constantly being exchanged with air molecules from outside the balloon; some air molecules migrate out the opening and others migrate in. This drift of molecules is called diffusion, and it results from the thermal motions of the individual molecules. Diffusion is what carries odors around a room, even when the air is still. The odor molecules are bouncing about rapidly and, although they collide frequently with air molecules, they slowly drift throughout the room. Diffusion is not important to a hot air balloon, because any cold air molecules that migrate into the balloon are quickly heated by the gas flame.

But diffusion is a serious problem for a helium balloon. If a helium balloon is left open at the bottom, helium atoms will diffuse out and heavier air molecules will diffuse in; as a result, the balloon will slowly increase in weight until it can no longer float. To make matters worse, helium atoms move extremely rapidly, so that they diffuse very quickly. In fact, they move so quickly and are so tiny that they can actually diffuse right through certain solids—for example, the rubber envelope of a balloon.

To slow the outward diffusion of helium, helium balloons are usually sealed (Fig. 5.1.8). But a sealed helium balloon gradually deflates, since the helium atoms that diffuse out through the envelope's skin are rarely replaced by air molecules. Toy helium balloons made of rubber lose helium so quickly that they stop floating after a day or two. Those made of Mylar® plastic are far more resistant to diffusion and can retain their helium for a week or two (see ❹).

A rubber balloon, the type you normally blow up for parties, is an interesting elastic object. As it's inflated with helium, the rubber balloon is stretched away from its equilibrium shape, and it experiences restoring forces that try to return it to that equilibrium shape. As a result, each region of the balloon's surface experiences three forces: an inward force from the pressure of air outside, an outward force from the pressure of helium inside, and an inward force from tension in the elastic skin of the balloon itself. Since each region of surface is stationary, it must be experiencing zero net force; the outward force must balance the two inward forces. This balance requires that the pressure of the helium inside the balloon be somewhat greater than the pressure of the outside air.

Sure enough, the pressure of helium inside the balloon is higher than that of outside air. This difference of pressure explains why a balloon "pops" when you stick a pin in it. The rubber tears, abruptly releasing the pressurized helium gas. The noise you hear is mostly this gas being released (Fig. 5.1.9).

As a helium balloon rises upward in the atmosphere, the air pressure around it decreases. The pressure inside the balloon stretches the balloon's skin outward further until the skin exerts enough inward force to contain the helium. The balloon gets somewhat larger, so the pressure inside it diminishes. The higher the balloon travels, the less force the atmosphere exerts on the balloon's skin, and the more the skin itself must act to contain the helium. Eventually the skin can't withstand the tension and it tears, releasing the helium. Sadly, that toy helium balloon you let go of when you were a child probably suffered this fate.

Special helium balloons are sometimes used to lift objects almost to the top of the atmosphere (Fig. 5.1.10). These research balloons are specially built so that the helium inside them can expand to much greater volume as the balloon rises upward. The balloon's envelope is so light that even a small amount of helium can lift the envelope. As the balloon rises, the atmospheric pressure around it decreases, and the helium inside expands to become less dense and to match the atmospheric pressure. The envelope slowly grows in size until it is fully inflated. By that time, the balloon may be as much as 50 km (30 miles) above the ground.

Fig. 5.1.9 This balloon tears after being hit by a small pellet. The sudden release of high-pressure gas creates a loud pop.

⚏ Floating and Sinking in Water

Almost all water sports—swimming, surfing, scuba diving, and boating—ultimately depend on floating in water. If everything sank, there would be little to do at the seashore except lounge around in the sun or play in the sand. Floating in water depends on the buoyant force, the same force that supports a hot-air or helium balloon in the atmosphere. As always, the upward buoyant force on an object is equal in magnitude to the weight of the fluid the object displaces. An object that's less dense than water, such as a wooden ball, is lighter than the water it displaces and accelerates upward. An object that's denser than water, such as a metal coin, is heavier than the water it displaces and accelerates downward.

Just as in air, the buoyant force in water has its origin in pressure: the water pressure below an object is greater than that above the object, so the object experiences a larger upward force from the pressure below it than downward force from the pressure above it. However, the basis for pressure in water is somewhat different from that in air. While air pressure is caused by air molecules colliding with one another and with surfaces, water pressure is caused by the forces that water molecules exert on one another while they're touching. Although the water molecules are kept in contact by attractive forces that appear when they begin to separate, these molecules will also push one another away if they begin to approach too closely. This tendency to pull together when stretched and push apart when squeezed makes water almost incompressible—its density changes very little when you try to squeeze it or stretch it. Water's density is always about 1000 kg/m³.

In the absence of gravity, the pressure in a container of water would be uniform and would depend only on how hard the water was being squeezed by the walls of the container. However, gravity creates additional pressure near the bottom of the container because the water there must support the weight of the water above it. Just as the air pressure in the atmosphere increases as you descend toward the ground, so the water pressure of the ocean increases as you dive deeper below its surface. The pressure deep in the ocean is enormous because the water there must support the immense weight of the water above it.

Because water is nearly incompressible, its density is almost constant, regardless of depth. An object always displaces the same weight of water so it always experiences the same buoyant force. That's not the case for a balloon in air because the atmosphere becomes less dense as the altitude increases. A balloon in air experiences a flight ceiling above which the buoyant force is too small to accelerate the balloon upward. But there is no flight ceiling in water. If an object floats in water, it floats upward all the way to the surface of the water. If it sinks, it sinks all the way to the bottom.

One interesting aspect of swimming is that we can choose whether we sink or float. Our average densities are close to that of water. If we were made of stone we would always sink, and if we were made of wood we would always float. Our bodies, however, are mostly water. While the rest of our constituents vary—bones and muscle that are denser than water and fat that is less dense—our average densities are nearly that of water. Whether we sink or float depends in large part on how much air is in our lungs. A big breath may lower your average density enough for you to float and letting it out may raise your average density enough for you to sink.

Body type affects a person's ability to float. A person with extra fat will have a relatively low average density and should float easily while a person with less fat will tend to sink. A person with very little fat may be unable to float even after taking a huge breath of air. Since his downward weight exceeds the upward buoyant force, he can remain at the surface of the water only with the aid of an additional upward force. He may obtain this force by "treading water"—pushing downward on the water so that the water pushes upward on him.

Fig. 5.1.10 This scientific balloon is being inflated with helium and will soon begin its trip to the upper atmosphere. By the time it reaches its working altitude of almost 40 km (25 miles), it will occupy about 100 times its present volume.

❹ Mylar® is the same plastic (Polyethylene Terephthalate or PETE) used in most beverage bottles—those in recycling category 1 (⚠). Because PETE offers a fairly good barrier to diffusion, these bottles can keep their contents fresh about 1 year. Nonetheless, the carbon dioxide molecules that give soda its tangy, carbonated taste gradually escape through the plastic walls and the beverage goes flat. Moreover, oxygen molecules slowly diffuse into the beverage, where they may affect the taste. The shelf-life of soda in plastic bottles is therefore shorter than soda in glass or metal containers, which are both better diffusion barriers than PETE.

❺ A submarine is designed to have an average density near that of water. It can then choose to sink or float by making small adjustments to its average density. It makes those adjustments by varying the mixture of air and water in its ballast tanks. Located outside the ship's main pressure hull, these tanks can be filled with water to increase the submarine's average density so that it sinks or with air to decrease the submarine's average density so that it floats. If the mixture of air and water is just right, the submarine will be neutrally buoyant—experiencing zero net vertical force so that it can hover at a constant depth. Since the submarine can't rise to the surface unless it has enough air to blow the water out of its ballast tanks, it stores that air carefully in tanks inside its main hull.

❻ The water in a motionless bottle rests at the bottle's bottom while the air rests at its top. This situation is another case of floating; one involving a fluid (air) floating on a fluid (water). Situations in which fluids float on fluids are surprisingly common. For example, oil is less dense than vinegar so salad dressings tend to separate into an oil layer floating on a vinegar layer. Herbs and other ingredients that are less dense than vinegar but more dense than oil tend to float at the interface between the two fluids. In accordance with Archimedes' principle, those herbs displace just the right mixture of oil and vinegar to experience an upward buoyant force that exactly equals their weight.

Floating and sinking are easy compared to hovering at some particular depth. That's because hovering requires your average density to be exactly that of water. Such perfect equality is nearly impossible to achieve without continuous adjustment. Passive objects—ones that can't make adjustments in their average densities—either float or sink. People, fish, and submarines **❺** are able to hover between the surface and the bottom because they can carefully adjust their average densities to be almost exactly equal to that of water.

Since tissues and bones are denser than water and tend to sink, skin divers use the air in their lungs to lower their average densities to that of water. Because fish have no lungs, something else must keep them from sinking. In a remarkable evolutionary advance, bony fish developed an internal air-filled swim bladder that lowers their average densities to exactly that of water. More primitive cartilaginous fish such as sharks have no swim bladders and would sink if they didn't obtain additional upward forces. These dense fish "fly" through the water much as an airplane flies through air—the forces needed to keep them from sinking are obtained by pushing water downward.

This careful balancing act is more complicated for a scuba diver, since the scuba equipment and wet suit tend to decrease the scuba diver's average density. To increase the scuba diver's average density to that of water, the scuba diver must wear a heavy weight belt. In an emergency, the scuba diver can remove the weight belt and will then float upward to the water's surface.

An object that sinks eventually experiences enough upward support force from the sea bottom to bring it to rest. But why does a floating object remain at the surface of the water? It does this because, once the object floats upward far enough to project out of the water, it begins to displace air instead of water and the buoyant force on it decreases. As long as the object is denser than air and less dense than water, it will float partly in water and partly in air. It adopts a height at which the weight of the water and air it displaces is exactly equal to its weight. Archimedes' principle works perfectly, even in this mixed environment (see **❻**).

People, driftwood, rowboats, and battleships all remain at the surface of water for this same simple reason—they all weigh less than the water they would displace if they were fully submerged. Each one rises out of the water until it displaces a mixture of fluids, water and air, equal in weight to its own weight. While you might wonder how a metal battleship could possibly float, that ship contains vast amounts of air and has an average density much less than that of water. Its average density is so low that the battleship floats easily and towers above the water's surface. However, if a hole in the ship's bottom let water replace the air inside it, its average density would soon be greater than that of water and it would sink.

Before we leave the topic of floating, there are two more issues to discuss. First, there is a way to change the density of water: dissolve chemicals into it. Adding salt to water increases the water's density. Because ocean salt water is about 3% more dense than fresh water, it exerts a larger buoyant force on objects immersed in it. People find it easier to float in ocean water than in fresh water. Trapped inland seas can be even more salty than the ocean. The waters of the Dead Sea and the Great Salt Lake are so salty and dense that people float effortlessly in them.

Second, a floating or sinking object doesn't accelerate forever as it rises or falls through the water. That's because an object that's moving through water experiences a friction-like drag force. As I'll explain in Section 5.2, this drag force opposes the object's motion and converts its ordered energy into thermal energy. The faster the object moves through the water, the stronger the drag force becomes. A sinking object, for example, experiences an upward drag force that strengthens as its downward speed increases until finally the drag force stops it from accelerating. The object then descends at a constant velocity, its terminal velocity. Similarly, a floating object, such as an air bubble, eventually reaches an upward terminal velocity.

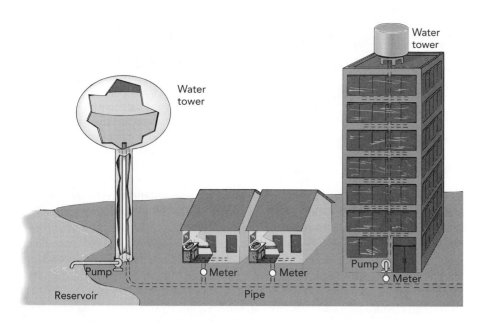

Section 5.2 **Water Distribution**

Now that we've explored the behavior of objects in fluids, let's turn to the behavior of fluids in objects. In this section, as we examine how plumbing distributes water, we'll see that pressure, density, and weight are just as important in plumbing as they are in ballooning. To keep things simple, we'll focus on the causes of water's motion through plumbing, leaving most of the complications associated with the motion itself for the next chapter. For example, we'll temporarily ignore drag and viscosity and the fascinating pressure changes that accompany fluid motion.

Water Pressure

Water distribution systems require two things: plumbing and water pressure. Plumbing is what delivers the water, and water pressure is what starts that water flowing. Water pressure is important because, like everything else, water has mass and accelerates only when pushed. If nothing pushed on the water when you opened a faucet, the water simply wouldn't budge. Since the pushes that send water through pipes come principally from differences in water pressure, we need to look carefully at how this pressure is created and controlled.

We'll begin our study of water distribution by ignoring gravity. As we've seen with the atmosphere, gravity creates pressure gradients in fluids—distributions of pressure that vary continuously with position. Pressure decreases with altitude and increases with depth, and these vertical pressure gradients complicate plumbing in hilly cities and skyscrapers. But if all of our plumbing is in a level region—for example, a single-story house in a very flat city—our job is much simpler. With no significant changes in height, we can safely ignore gravity, since no water is supporting the weight of water above it and gravity's effects are minimal.

In this simplified situation, water accelerates only in response to unbalanced pressures. Just as unbalanced forces make a solid object accelerate, so too do unbalanced pressures make a fluid accelerate. If the water pressure inside a pipe is uniform, then each portion of water feels no net force and doesn't accelerate; it either remains stationary or coasts at constant velocity (Fig. 5.2.1a). But if the pressure is out of balance, then each portion of water experiences a net force and accelerates toward the region of lowest pressure (Fig. 5.2.1b,c).

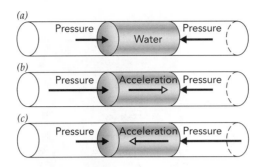

Fig. 5.2.1 (*a*) If the water in a horizontal pipe is exposed to a uniform pressure, then it will not accelerate. (*b, c*) If the pressure along the pipe is not uniform, however, the imbalance will produce a net force on each portion of the water and the water will accelerate toward the side with the lower pressure.

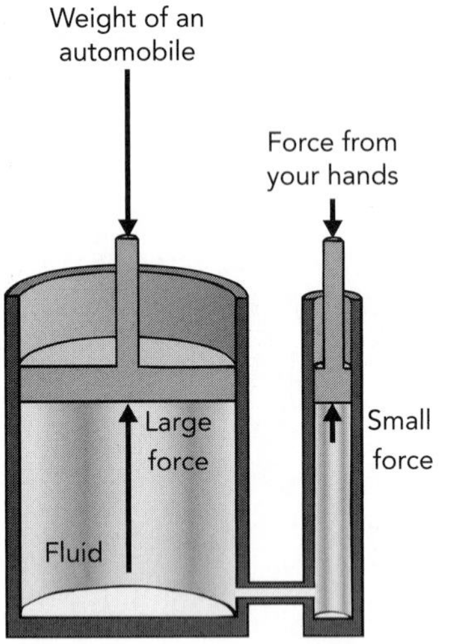

This acceleration doesn't mean that the water will instantly begin moving toward the lowest pressure. Because of its inertia, water changes velocity gradually: it speeds up, slows down, or turns to the side, depending on where the lowest pressure is located. A complicated arrangement of high and low pressures can steer water through an intricate maze of pipes, and that is exactly how water reaches your home from the city pumping station. Every change in its velocity during its trip through the plumbing is caused by a pressure imbalance.

You can create a pressure imbalance in water simply by squeezing parts of it. The pressure in a squeezed portion will rise and it will accelerate toward lower pressures elsewhere. Since this sort of pressure change isn't caused by the water's motion, it's a static variation in pressure. But the water's motion can also affect its pressure and such dynamic variations in pressure can be complicated and fascinating. As we'll see in the next chapter, they contribute to such diverse effects as the spray of a garden hose nozzle, the lift on an airplane's wing, and the curve of a curve ball.

Creating Water Pressure with Water Pumps

To start water flowing through the plumbing in a level house or city, you need a water pump—a device that uses mechanical work to deliver pressurized water through a pipe. At its most basic level, a water pump squeezes a portion of water to raise its local pressure so that it accelerates toward regions of lower pressure. The pump continues to squeeze that water as it flows out through the plumbing.

To understand how a pump works, picture a sealed plastic soda bottle full of water. When you don't squeeze the bottle, the pressure inside it is atmospheric and uniform (remember we're neglecting gravity). But when you squeeze the sides of the bottle and push inward on the water, the water responds by pushing outward on you—Newton's third law—and it does this by increasing its pressure. The harder you push, the harder the water pushes back and the greater its pressure becomes.

Because water, like all liquids, is incompressible and experiences almost no change in volume as its pressure increases, the bottle won't get smaller. But the rise in water pressure inside it can be substantial. It doesn't take much force, exerted with your thumb on a small area of the bottle, to increase the pressure in the bottle from atmospheric to twice that value or more.

The pressure increases uniformly throughout the water bottle, an observation known as Pascal's principle—a change in the pressure of an enclosed incompressible fluid is conveyed undiminished to every part of the fluid and to the surfaces of its container. This uniform pressure rise leads to a large upward force on the bottle's cap. If the cap were wider and had more surface area, the upward force on it might be large enough to blow it off the bottle. That effect is the basis for hydraulic systems and lifts, where pressure produced in an incompressible fluid by a small force exerted on a small area of the fluid's container results in a large force exerted on a large area of the fluid's container (Fig. 5.2.2). It also explains why plastic drinking bottles usually have small caps and why wide-mouth plastic containers are better suited for candies, cookies, and nuts.

Fig. 5.2.2 The force a pressurized fluid exerts on a piston is proportional to the surface area of that piston. This fact underlies hydraulic systems, in which a small force exerted on a small piston pressurizes a trapped liquid so that it exerts a large force on a large piston. In this figure, your gentle downward push on the small piston is balanced by a gentle upward force exerted on that small piston by the pressurized fluid. At the same time, the strong downward push of an automobile on the large piston is balanced by a strong upward force exerted on that large piston by the pressurized fluid. If the small piston descends quickly at constant velocity, the large piston will rise slowly at constant velocity and you'll be doing the work of lifting the heavy car a short distance by pushing the small piston down gently for a large distance. The hydraulic system is providing you with mechanical advantage and allowing you to raise something that's too heavy for you to lift unaided.

Pascal's Principle

A change in the pressure of an enclosed incompressible fluid is conveyed undiminished to every part of the fluid and to the surfaces of its container.

It's time to remove the cap from the water bottle. Now when you squeeze the bottle and increase the local water pressure, the water can move. As its pressure rises inside the bottle, the water begins to accelerate toward the lower pressure above the bottle's open top and the result is a fountain. You are pumping water!

You are also doing work: as the water flows out of the bottle, your hands move inward. Since you are pushing inward on the water and the water is moving inward, you are doing work on the water. Pumps do work when they deliver pressurized water, and pressurized water carries with it the energy associated with that work.

While a water bottle can act as a pump briefly, it soon runs out of water. A more practical pump appears in Fig. 5.2.3. In this pump, a piston slides back and forth in the open end of a hollow cylinder, making a watertight seal. Pushing inward on that piston squeezes any water in the cylinder and raises the local water pressure. Water begins to flow.

But in contrast to our simple bottle, this pump's cylinder is easy to refill. The cylinder actually has two openings, each with a valve that permits water to flow in only one direction. Water can leave the cylinder only through the top opening and can enter only through the bottom opening. As you push the pump's piston into the water-filled cylinder, the water pressure in the cylinder rises and water accelerates and flows out through the top valve. As you pull the pump's piston out of the water-filled cylinder, the water pressure inside the cylinder drops and water accelerates and flows in through the bottom valve. In fact, as you withdraw the piston, the pressure inside the cylinder drops below atmospheric pressure, so that even water in an open reservoir nearby will accelerate toward the partial vacuum in the cylinder and refill it.

Moving Water: Pressure and Energy

The pump of Fig. 5.2.3 can draw low-pressure water from a pond and fill a hose with high-pressure water. If the other end of the hose is open, the water will accelerate toward lower pressure at that end and will have considerable kinetic energy as it sprays out of the hose. From where does this kinetic energy come?

The energy comes from you and the pump. As you push inward on the piston, pressurizing the water and squeezing it out through the top valve, you're doing work on the water because you're exerting an inward force on the water's surface and the water is moving inward. The amount of work you do is equal to the product of the water pressure times the volume of water you pump. This simple relationship between work, pressure, and volume comes about because the inward *force* you exert on the water with the piston is equal to the water pressure times the surface area of the piston, and because the *distance* the water moves in the direction of that force is equal to the volume of water pumped divided by the surface area of the piston. *Force* times *distance* equals *work*.

As you pump the water, it sprays out of the open hose. The energy that makes the water accelerate out of the hose actually travels through the water directly from the pump to the end of the hose. Since water is incompressible, each time a liter of water leaves the pump, a liter of water also leaves the hose. While the water never really stores any energy, the pump gives each liter of water a certain amount of energy as it leaves the hose so we can imagine that this energy is associated with the water and not with the pump. We create a useful fiction: pressure potential energy. Water that's under pressure has a pressure potential energy equal to the product of the water's volume times its pressure.

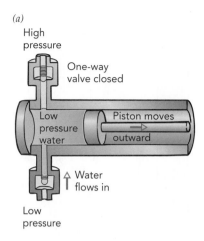

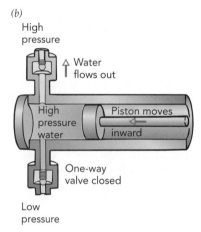

Fig. 5.2.3 Water is pumped from a region of low pressure to a region of high pressure by a reciprocating piston pump. (*a*) As the piston is drawn outward, water flows into the cylinder from the low-pressure region. (*b*) As the piston is pushed inward, the inlet one-way valve closes and water is driven out of the cylinder and into the high-pressure region.

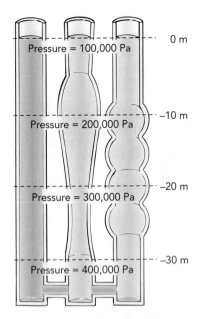

Fig. 5.2.4 The pressure of stationary water in pipes increases with depth by about 10,000 Pa per meter of depth. *The shape of the pipes doesn't matter.* For plumbing that's open on top and connected near the bottom, as shown here, water will tend to flow until its height is uniform throughout the plumbing.

❶ As a professor in Basel, Switzerland, Daniel Bernoulli (Swiss mathematician, 1700–1782) taught not only physics, but also botany, anatomy, and physiology. He correctly proposed that the pressure a gas exerts on the walls of its container results from the countless impacts of tiny particles that make up the gas. He also derived an important relationship between the pressure, motion, and height of a fluid—the relationship known as "Bernoulli's equation."

Because pressure potential energy actually comes from the pump, it vanishes as soon as you break the link between the water and the pump; you can't save a bottle of high-pressure water and expect it to retain this potential energy. The concept of pressure potential energy is only meaningful if the water is flowing freely so that water leaving the plumbing is immediately replaced; then, whatever energy leaves the plumbing as kinetic energy in the water is put back into the plumbing by the pump. Actually, the details of the pump aren't as important as the idea that any water moving through the plumbing is immediately replaced by more water with the same pressure. As long as the water is flowing steadily, you can safely use the concept of pressure potential energy, even if you don't know where the pump is or whether there actually is one.

Pressure potential energy is most meaningful in steady-state flow—a situation in which fluid flows continuously and steadily through a stationary environment, without starting or stopping or otherwise changing its characteristics anywhere. You can tell you're watching steady-state flow when you can't detect the passage of time in the fluid or its environment; if a video of the situation would look the same as a still photograph, then it's steady-state flow. With small allowances for the inevitable imperfections of real life, water spraying steadily out of a hose, wind blowing smoothly across your motionless face, and a gentle current flowing in a quiet river are all cases of steady-state flow in fluids.

Without gravity, the energy in a certain volume of water in steady-state flow is equal to the sum of its pressure potential energy and its kinetic energy. We've already seen that the pressure potential energy is the product of the water's volume times its pressure. The water's kinetic energy is given by Eq. 2.2.1 as one-half the product of its mass times the square of its speed. Since water's mass is its density times its volume, this sum is

$$\text{energy} = \text{pressure potential energy} + \text{kinetic energy}$$
$$= \text{pressure} \cdot \text{volume} + \tfrac{1}{2} \cdot \text{density} \cdot \text{volume} \cdot \text{speed}^2. \quad (5.2.1)$$

If we divide both sides of this expression by the volume involved, we can obtain another useful form of this relationship:

$$\frac{\text{energy}}{\text{volume}} = \frac{\text{pressure potential energy}}{\text{volume}} + \frac{\text{kinetic energy}}{\text{volume}}$$
$$= \text{pressure} + \tfrac{1}{2} \cdot \text{density} \cdot \text{speed}^2. \quad (5.2.2)$$

As each volume of water moves along with the flow, it is soon replaced by a new volume of water. Because the flow is steady-state, the energy in the new volume of water must be exactly the same as in the volume that preceded it; thus the energy in each volume of water that flows along a particular path must be identical. The particular path that a volume of water takes is called a streamline, and the energy-per-volume of fluid along a streamline is constant:

$$\frac{\text{energy}}{\text{volume}} = \frac{\text{pressure potential energy}}{\text{volume}} + \frac{\text{kinetic energy}}{\text{volume}}$$
$$= \text{pressure} + \tfrac{1}{2} \cdot \text{density} \cdot \text{speed}^2$$
$$= \text{a constant value } (\textit{along a streamline}). \quad (5.2.3)$$

Equation 5.2.3 is called Bernoulli's equation, after Swiss mathematician Daniel Bernoulli ❶ whose work led to its development, although Swiss mathematician Leonhard Euler (1707–1783) actually completed it.

Because energy is conserved, an incompressible fluid such as water that's in steady-state flow can exchange pressure for speed or speed for pressure as it flows along a streamline. As water accelerates out of a hose with a nozzle, for example, its pressure drops but its speed increases; it's converting pressure potential energy into kinetic energy. And as that moving water sprays against the car you're washing, it slows down but its pressure increases; it's converting kinetic energy back into pressure potential energy. In both cases, the water's total energy is constant.

Gravity and Water Pressure

Gravity creates a pressure gradient in water: the deeper the water, the more weight there is overhead and the greater the pressure. Since water is much denser than air, water pressure increases rapidly with depth. In a vertical pipe that's open on top, the water's surface is at atmospheric pressure (about 100,000 Pa), but only 10 m (33 feet) below the water's surface, the pressure has already doubled to 200,000 Pa. At that modest depth, the water overhead weighs as much as the air overhead, even though the atmosphere is several kilometers thick.

The shape of the pipe doesn't affect the relationship between pressure and depth. No matter how complicated the plumbing, the pressure of stationary water inside it increases with depth by 10,000 Pa per meter or 10,000 Pa/m (Fig 5.2.4). This uniform pressure gradient creates an upward buoyant force on anything immersed in the water. In fact, that buoyant force is what supports the water itself (Fig. 5.2.5).

The dependence of water pressure on depth has a number of important implications for water distribution. First, water pressure at the bottom of a tall pipe is substantially higher than at the top of that same pipe. Consequently, if only a single pipe supplies water to a skyscraper, then the water pressure on the ground floor will be dangerously high while the pressure in the penthouse will be barely enough for a decent shower. Tall buildings must therefore handle water pressure very carefully; they can't simply supply water to every floor directly from the same pipe.

Second, pressure in a city water main does more than just accelerate water out of a showerhead; it also supports water in the pipes of multistory buildings. Lifting water to the third floor against the downward force of gravity requires a large upward force, and that force is provided by water pressure. The higher you want to lift the water, the more water pressure you need at the bottom of the plumbing. Lifting the water also requires energy, which is often provided by a water pump.

Third, as water travels up and down the streets of a hilly city, its pressure varies with height. In the valleys the pressure can be very large, and at the tops of hills the pressure can be very small. Water mains in valleys must therefore be particularly strong to keep from bursting. The large pressure in a valley is quite useful because it helps push the water back uphill on the other side of the valley (Fig. 5.2.6). Nonetheless, a hilly city must have pumping stations and other water pressure control systems located throughout in order to provide reasonable water pressures to all the buildings, regardless of their altitudes.

Apart from those pressure-control systems, even a hilly city's plumbing often involves steady-state flow and can be explained using a version of Bernoulli's equation that includes gravity. But before we explore that topic, let's take a look at the non-steady-state flow that occurs when water in an isolated section of plumbing has free surfaces that can move up or down so that you see things change with time.

The simplest case is plumbing that's open on top, so that all the water's free surfaces are at atmospheric pressure. Like any object, water accelerates in the direction that lowers its total potential energy as quickly as possible. With neither the immovable plumbing beneath it nor the uniform air pressure above it contributing any potential energy, water's only potential energy is gravitational and it accelerates in the direction that lowers that gravitational potential as quickly as possible.

If the water's free surface is higher in one place than another, it can reduce its average height and therefore its gravitational potential energy by letting its highest

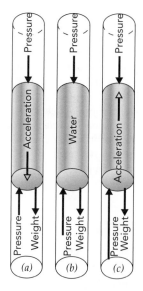

Fig. 5.2.5 When a pipe is oriented vertically, gravity affects the motion of water in the pipe. (*a*) If the water's pressure doesn't change with depth, the water will accelerate downward (i.e., fall) because of its weight. (*b*) If the water's pressure increases with depth by 10,000 Pa/m, the water won't accelerate. (*c*) If the water's pressure increases with depth by more than that amount, it will accelerate upward.

Fig. 5.2.6 Los Angeles receives much of its water from Owens Valley, 300 km north. The water negotiates the mountains and valleys in between, driven by gravity alone. Giant pipes allow pressure to build during downhill stretches in order to push the water back uphill later on. Parts of the 1913 aqueduct support so much pressure that the steel pipe used in them has to be more than an inch thick.

❷ The Romans used gravity to convey water to Rome from sources up to 90 km away. A very gradual slope in the aqueducts kept the water moving in spite of frictional effects that opposed the water's progress. Because lead was used extensively in these aqueducts, some historians have hypothesized that gradual poisoning by that lead contributed to the decline of the Roman Empire.

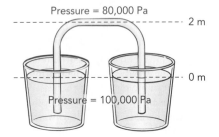

Fig. 5.2.7 The two open containers of water are connected by a siphon. This U-shaped tube permits water to flow until its level is equal in both containers. The sturdy tube permits the water pressure in the siphon to drop below atmospheric pressure.

water fill in its lowest valley. After some sloshing about, the water settles down in stable equilibrium with all of its free surfaces smooth and level at a single, uniform height. No matter how complicated the plumbing, water that is open to the air on top always "seeks its level." The natural flow associated with this leveling effect is often used in water delivery (see ❷).

But if you seal off part of the isolated plumbing and reduce the pressure above one of the water's free surfaces, that surface will rise higher than all of the others. It will rise until the added pressure produced by the taller column of water replaces the missing pressure above the water's free surface. The less pressure there is above that surface, the higher the water must rise to make up for the missing pressure. This effect lifts water in a drinking straw and allows it to travel between two open containers in order to "seek its level" through an elevated pipe known as a siphon (Fig. 5.2.7).

However, removing all of the air pressure above the water's free surface inside a long straw or siphon will raise its height only about 10 m (33 feet) above the level of the water elsewhere in an open container. Even with no pressure above it, this 10-m column of elevated water completely replaces the absent air pressure and thus prevents water in the rest of the plumbing from lifting it any higher. It's therefore impossible to draw water from a deep well simply by lowering a pipe into that well and reducing the air pressure in the pipe: the water will rise no more than 10 m upward. Instead, a pump must be attached to the bottom of the pipe to pressurize the water and push it all the way to the top of the pipe.

Moving Water Again: Gravity

As we've seen, it takes pressure and energy to lift water to the third floor of a building. We can now expand our statement of energy conservation in fluids undergoing steady-state flow to include gravity and gravitational potential energy.

Water's gravitational potential energy is equal to its weight times its height (the force required to lift it times the distance it has been lifted), and its gravitational potential energy-per-volume is its weight-per-volume times its height. Since its weight-per-volume is its density times the acceleration due to gravity, water's gravitational potential energy-per-volume is its density times the acceleration due to gravity times its height.

If we include gravitational potential energy in Eq. 5.2.2 and recognize that, for fluids in steady-state flow along a streamline, the energy-per-volume is constant, we obtain the relationship:

$$\frac{energy}{volume} = \frac{pressure\ potential\ energy}{volume} + \frac{kinetic\ energy}{volume} + \frac{gravitational\ potential\ energy}{volume}$$

$$= pressure + \tfrac{1}{2}\cdot density\cdot speed^2 + density\cdot acceleration\ due\ to\ gravity\cdot height$$

$$= a\ constant\ value\ (along\ a\ streamline). \tag{5.2.4}$$

This is a revised version of Bernoulli's equation, one that includes gravity. It correctly describes steady-state flow in streamlines that change height.

Bernoulli's Equation

For an incompressible fluid in steady-state flow, the sum of its pressure potential energy, its kinetic energy, and its gravitational potential energy is constant along a streamline. Equation 5.2.4 expresses this law as a formula.

Because energy is conserved, an incompressible fluid such as water that's in steady-state flow can exchange the energies associated with its speed, pressure, and height for one another. Thus as water flows downward, its speed or pressure or both increase; if it falls from an open faucet, its speed increases; and if it descends steadily inside a uniform pipe, its pressure increases. The reverse happens as water flows upward. Water rising from a fountain loses speed as it ascends while water rising steadily in a uniform pipe loses pressure on its way up.

This interchangeability of height, pressure, and speed makes it possible to pressurize plumbing by connecting a tall column of water to the pipes. That's why cities, communities, and even individual buildings have water towers. A water tower is built at a relatively high site within the region it serves. A pump fills the water tower with water, and then gravity maintains a constant high pressure throughout the plumbing that connects to it (Fig. 5.2.8). The water is at atmospheric pressure at the top of the water tower, but the pressure is much higher at the bottom; at the base of a 50-m-high water tower, for example, the pressure is about 600,000 Pa or six times atmospheric pressure.

In addition to providing a fairly steady pressure in the water mains, a water tower stores energy efficiently and can deliver that energy quickly (see ❸). When water is drawn out of the water tower, its gravitational potential energy at the top becomes pressure potential energy at the bottom. The water tower replaces a pump, supplying a steady flow of water at an almost constant high pressure. But unlike a pump, the water tower can supply this high-pressure water at an enormous rate. As long as the water level doesn't drop too far, high-pressure water keeps flowing.

Since the water leaving a water tower situated on a hill continues to flow downhill toward people's houses, the water pressure can become extremely high. It must be reduced by regulating devices near low-lying houses so as not to burst their pipes or hot water heaters. These regulating devices use friction-like effects to convert the water's pressure potential energy into thermal energy. We'll examine those effects in the next chapter.

Using a water tower to maintain pressure in the water mains eliminates the need to run a water pump continuously. Since a real water pump, because of frictional effects, turns ordered energy into thermal energy even if there is no water flowing, the pump is most energy efficient when it moves as much water as possible. A city saves ordered energy and money by allowing the pump to fill the water tower at night, when electricity is in low demand and can be purchased relatively inexpensively and when the pump can work at its full capacity and efficiency. The water level in the tower then drops slowly during the day as water is consumed by the houses. If the water level in the tower drops too far, reducing the pressure in the water mains, the pump turns on and refills the water tower.

If a pump maintains the water pressure, then the pump supplies the pressure potential energy. But the pump has a limited power capacity, typically rated in horsepower, and can transfer no more than a certain amount of energy to the water each second. If you draw water slowly from the pump, the pump will be able to deliver that water at high pressure. But if you open too many faucets at once, they will exceed the power capacity of the pump, and the pump will then deliver a large amount of water but at a lower pressure. Since people tend to use water in daily schedules, taking showers in the morning and washing dishes in the evening, a city served only by pumps can experience substantial drops in water pressure. Attaching a water tower to the water system nearly eliminates these pressure drops. (Compressed air can also maintain water pressure, see ❹.)

But skyscrapers present a serious problem for water delivery systems. It would take enormous water pressure to drive water all the way to the top of a skyscraper, and no city system can provide that much pressure. Instead, a skyscraper has pumps that lift the water up. It begins with water near the bottom of the building and pumps it to the top in several steps. The skyscraper actually has water towers inside it to maintain steady water pressures for its occupants. These internal tanks are also essential for fire safety in the event of electric power loss. A fire truck would

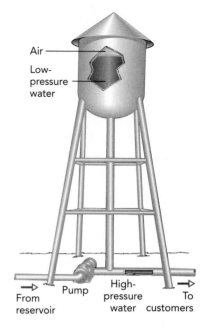

Fig. 5.2.8 A water tower uses the weight of water to create a large water pressure near the ground. The higher the tower, the greater the pressure near the ground. The water tower is able to maintain the water pressure passively and doesn't require constant pumping. Even during periods of peak water consumption, it maintains a fairly steady pressure. When the water level in the tower drops below a certain set point, a pump refills the tower.

❸ Energy stored in elevated water is the basis for hydroelectric power generation. Water descending from an elevated lake or river does work on a spinning turbine that does work on an electric generator. The water's gravitational potential energy is thus converted into electrical energy. The more water and the farther it descends, the more electrical energy the hydroelectric system can generate.

❹ The streams of water from some toy water guns are accelerated by high-pressure air. A small hand-pump compresses air inside the water container, a process requiring work. When the water is released, the pressure imbalance causes it to accelerate and leave the gun as a narrow spray. The stored energy becomes kinetic energy in the water. As the water leaves the container, the air gradually expands and its stored energy decreases.

have difficulty pumping water to the top of the skyscraper, but it can make good use of water already stored at the top.

Ⓜ Drinking Straws

Water in an open drinking straw inserted in a glass of water rises to the level of the water around it. But when you seal the straw's top with your mouth and remove air from within the straw, the water accelerates up the straw toward your mouth. By removing air from the straw, you reduce the pressure at the top of the water to less than atmospheric pressure; water then accelerates upward toward the lower pressure. Although it feels as though you are attracting the water, you are actually just permitting water pressure in the glass to push water toward your mouth.

Water at the surface of the glass of water is at atmospheric pressure; it can support the weight of either the air overhead or a 10 m column of water, but not both. By removing the air from above the water, you permit the water pressure to support a column of water instead of a column of air. But the tallest column of water that atmospheric pressure can support is 10 m. If you try to "suck" water up an 11 m straw, it won't reach your mouth. Even if you remove all the air from the straw, a feat that you can't actually do with your mouth, atmospheric pressure will only lift the water 10 m up the straw.

Similarly, a water pump at the surface can't "suck" water out of a well if the water level is more than 10 m below the pump. That is why most well pumps are actually inserted into the well hole itself. Gravity-produced water pressure pushes water into the submerged pump, and the pump then provides the enormous pressure needed to lift the water up and out of the well.

Ⓜ Scuba Diving

Let's suppose that you're on vacation in the Bahamas and have spent the morning floating leisurely on the ocean's surface. Bored with the sun and the gorgeous scenery, you now decide to explore the water below. You buy a very long soda straw, tie a rock to your waist, and descend 10 m to the sandy bottom. You try to breathe in through the straw and discover that you can't. Fortunately, you're good at untying knots. How do you breathe in and why did being underwater make it so difficult?

You breathe in by increasing the volume of your lungs with the muscles of your chest and diaphragm. As your lungs expand, the air inside them fills a larger volume and experiences a decrease in density and pressure. The air outside your mouth, which is at atmospheric pressure, then accelerates toward the lower pressure inside your lungs and your lungs fill with air. You breathe out by using your muscles to compress your lungs, packing the air inside them more tightly and increasing its density and pressure. Higher pressure air inside your lungs then accelerates toward the lower pressure outside your mouth and your lungs empty of air.

This scheme works as long as the pressure surrounding you is nearly uniform—when the pressures outside your mouth and chest are essentially the same. With minimal effort you can then vary the pressure in your lungs above and below the pressure outside your mouth so that air flows in and out and you breathe. But when you're trying to breathe atmospheric pressure air while your body is 10 m below the surface of the water, you've got pressure problems and can only breathe out.

Ten meters below the water's surface, your chest is experiencing twice the pressure it did on land. This pressure on your chest compresses the air in your lungs until its density and pressure are also twice those of the nearby atmosphere. You may not feel this compression happening in your lungs as you dive below the surface, but if you consider other body parts, namely your ears, you may well notice this compression happening.

Unfortunately, the air in your breathing straw is not supporting any water and is still at atmospheric pressure. With the air in your lungs at twice the pressure of the air in the straw, air will flow easily out of your lungs, but never into them. Since you're not strong enough to expand your lungs and reduce the pressure inside them below atmospheric pressure, you can't inhale. You can only exhale.

For you to be able to breathe air far below the surface of the sea, air entering your mouth must be compressed so that its pressure is about the same as the pressure your chest is experiencing. If you're 10 m below the surface, the air you breathe must be about twice atmospheric pressure or two "atmospheres." If you're 100 m below the surface, the air pressure must be about eleven atmospheres. This pressurized air can be sent to you through a pipe from the surface, like the old deep-sea divers, or you can carry it with you in scuba equipment.

The purpose of scuba equipment is to provide you with air at essentially the ambient pressure. A special pressure regulator ensures that the pressure of air delivered to your mouth is just equal to the water pressure surrounding you and your chest. When you expand your lungs with your chest muscles and diaphragm, this pressurized air gently accelerates inside. You exhale by contracting your lungs and pushing the air into the water as bubbles.

In scuba equipment, air is stored in steel tanks that can withstand enormous pressures. Air molecules are packed very tightly into the tanks so that each tank contains enough air for many minutes of breathing. Because the density of air in a scuba tank is enormous, its pressure is also enormous—typically 200 atmospheres (about 20,000,000 Pa). The scuba equipment reduces the air pressure with a pressure regulator. This regulator permits small amounts of the high-density air to expand into a separate volume and become low-density air. You breathe air directly from this second volume, which includes the mouthpiece.

The pressure regulator must determine just how many air molecules to allow into the second volume. If the pressure it provides is too low, you won't be able to breathe in, and if it's too high, you may blow up like a balloon. Adjusting the pressure is the delicate task of an ingenious regulating device.

The regulating device compares the pressure in the mouthpiece to the pressure in the water. The regulator uses a flexible membrane to control the flow of air from the high-pressure volume to the mouthpiece volume (Fig. 5.2.9). With water on one side of the membrane and mouthpiece air on the other, the membrane experiences zero net force only when the pressures on its two sides are exactly equal.

If the pressure in the mouthpiece is too low, the membrane experiences a net force and the regulator valve opens. More air from the tank flows to the mouthpiece. When the air pressure in the mouthpiece reaches the water pressure, the membrane no longer experiences a net force and the valve closes. As you breathe air from the mouthpiece, the regulator permits more air to flow in so that the air pressure in the mouthpiece is always very close to the water pressure.

Actually, the pressure regulator in scuba equipment has two stages of pressure regulation. The first stage reduces the enormous pressure in the storage tanks to about 10 atmospheres and the second stage reduces it to the ambient pressure. The first stage regulator is needed because the second stage regulator doesn't work well when the air it starts with has too high a pressure.

While the second stage regulator compares its outlet air pressure to the water pressure, the first stage regulator compares its outlet air pressure to a fixed reference—a spring. The air leaving the first stage regulator pushes on one side of a membrane while the spring pushes on the other side of that same membrane. If the air pressure on that membrane is too low, the spring dominates and the first stage regulator lets air flow through it. If the air pressure rises too high, the air dominates and the first stage regulator stops the airflow. Spring-operated pressure regulators are commonly used with propane tanks for home heating and cooking,

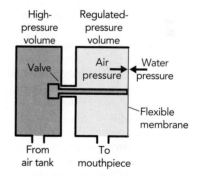

Fig. 5.2.9 The pressure regulator in scuba equipment uses a valve and a pressure-sensitive membrane to control the flow of high pressure air to the diver's mouthpiece. If the mouthpiece pressure is too low, the membrane opens the valve to allow more air molecules to flow into the mouthpiece region. The pressure inside the mouthpiece rises as a result. When the mouthpiece pressure is equal to the outside water pressure, the membrane closes the valve to prevent further flow.

and on carbon dioxide cylinders for soda machines. These regulators ensure that gas is delivered at a constant low pressure, substantially below the pressure that exists inside the storage tanks themselves.

The deeper a scuba diver goes, the more pressure the regulator must supply in order for him to be able to breathe in. This increased pressure has several consequences. First, since air's density increases with pressure, the regulator must put more air molecules into each breath. Since the air tanks contain a certain number of air molecules, the stored air is depleted faster at great depths than at shallow depths. The deeper the diver goes, the sooner his air supply will run out.

Second, increasing air's density affects its ability to flow. At its normal sea-level density, air flows easily and we rarely notice its movement into our lungs. At 10 times normal density, air is noticeably thicker and breathing becomes more difficult.

Third, pressure affects the solubility of air molecules in water, not to mention blood and tissues. A gas dissolves when its molecules become separated and dispersed among the molecules of water. The concentration of dissolved air molecules is related to the rates at which those molecules enter and leave the water. Anything that encourages entry tends to boost the concentration of dissolved molecules, while anything that encourages leaving tends to lessen that concentration. Raising the air's pressure, for example, increases its density and the rate at which its molecules collide with and enter the water; the concentration of dissolved air begins to rise.

That rise in dissolved air concentration doesn't continue indefinitely. As more air molecules enter the solution, the rate at which molecules encounter the water's surface and leave increases. The system tends to reach a balance where air molecules leave the water as often as they enter it and the concentration of dissolved air molecules then holds steady. When this balance is reached, the concentration of dissolved air molecules is roughly proportional to the air's density, although it also depends on temperature and on how strongly the specific gas molecules and water molecules cling to one another in the solution.

Gas molecules that bind strongly to water molecules (e.g., carbon dioxide) reach high concentrations before achieving the balance between entering and leaving (see ❺), but gas molecules that bind only weakly to water molecules (e.g., nitrogen) are much less soluble. Nonetheless, when the air density is high enough, even nitrogen dissolves moderately well in water and reaches a substantial concentration.

When a scuba diver is far below the water's surface and breathing dense, high-pressure air, his blood and tissues dissolve a considerable amount of nitrogen. As he returns to the water's surface, the reduction in pressure permits the nitrogen to come back out of solution. If he rises too quickly and experiences a sudden drop in pressure, bubbles of gaseous nitrogen will begin to form inside his body. It's just like opening a bottle of soda. While the nitrogen bubbles are quickly cleared from a diver's blood stream, the bubbles that form in his tissues are slow to dissipate and the diver experiences a painful case of "the bends" or decompression sickness.

To avoid trouble, a diver must return to the surface slowly so that the nitrogen will come out of solution gradually and be exhaled from his lungs. Nitrogen gas is slow to diffuse into and out of tissues, so that a diver who remains deep underwater for a long time must take a long time to return to the surface. With only one compressed air tank, a diver probably can't stay under water long enough to be at serious risk of the bends. However, two or three air tanks permit longer dives and increased concentrations of tissue-dissolved nitrogen. Only experienced divers, able to carry out gradual decompression, can safely use two or three air tanks.

At very great depths, so much nitrogen dissolves in the blood that it becomes toxic. At these depths, a diver must breathe air that contains little or no nitrogen. Since breathing pure oxygen is also not healthy, helium is used to dilute the oxygen. Helium, an inert gas, is so weakly attracted to other molecules that it's virtually insoluble in water, blood, or tissue, regardless of pressure.

❺ Carbonated beverages are made by exposing water to carbon dioxide gas under several atmospheres of pressure. Carbon dioxide is particularly soluble in water because carbon dioxide molecules feel a strong attraction to water molecules. The two merge to form carbonic acid, a chemical which gives soda much of its taste. But despite its wonderful solubility, carbon dioxide begins coming out of solution whenever a drop in pressure spoils the balance between entering and leaving. When you open a bottle of soda and release the pressure, the density of gaseous carbon dioxide plummets. Suddenly, far more carbon dioxide molecules are leaving the water than entering it and the soda begins releasing gas. Assisted by defects in the container or gas mixed into the soda by shaking it, carbon dioxide molecules deep in the soda join together to form bubbles of gas. The soda fizzes as these bubbles of carbon dioxide float to the surface of the beverage. The only way to curtail this loss of dissolved carbon dioxide is to repressurize the container with carbon dioxide gas. That procedure will boost the rate at which carbon dioxide molecules enter the water and thus restore the proper balance between entering and leaving. Pressurizing the container with ordinary air simply won't help, although there are plenty of gadgets on the market that claim otherwise.

CHAPTER 6

THINGS THAT MOVE WITH FLUIDS

Fluids are fascinating when they move. Stationary water and air may be essential to life, but they're also fairly simple; only their pressures vary from place to place and even these are determined primarily by gravity. Rushing rivers or gusts of wind, however, with their wonderful variety of simple and complicated behaviors are much more interesting. And the motions of fluids aren't just interesting, they're also important; our world is filled with objects and machines that work in whole or part because of the behaviors of moving fluids. In this chapter, we'll look at a variety of situations in which fluid motion contributes to the way things work.

Sometimes I have trouble knowing when to stop. My children tolerate my long answers to their short questions either because they share my belief that true understanding is more valuable than simplistic factoid collecting or because they're too good natured to tell me to put a sock in it. Given the nature of the writer/reader relationship, your informing me about what I can do with that sock is likely to come a bit too late. So I'm just going to have to curb my enthusiasm on my own. While I love observing how fluids behave in the world around us, I won't point out every fascinating feature of the river flowing behind your farm or the wind blowing past your umbrella. I'll save something for when I write *How Too Many Things Work*. To prove just how restrained I can be, I have left p. 196 completely blank. It looks so peaceful that way.

As you'll soon see, this chapter continues to develop the concept of energy conservation along a streamline that I introduced in Chapter 5. It also brings up several new types of forces that are present when fluids move past one another or past solid objects. These ideas are present in a vast array of commonplace activities, from washing windows with a hose to pumping water with a windmill.

Chapter Itinerary

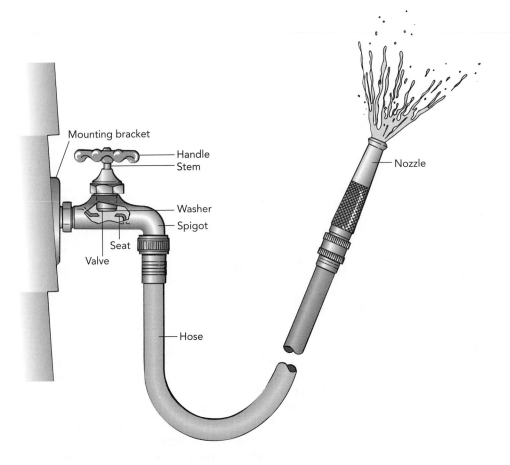

Mounting bracket

Handle
Stem

Washer
Spigot

Seat

Valve

Hose

Nozzle

Section 6.1 **Garden Watering**

Tending a flower garden often involves watering. While this once meant walking the garden's paths with a watering can, modern plumbing has made such effort unnecessary. With a hose and nozzle attached to a faucet, you can do your job without leaving your lawn chair. But while the tools involved—faucets, hoses, and nozzles—are simple and unsophisticated, the principles behind them are not. All three make elegant use of the laws of fluid flow, letting openings and channels control the delivery rate and speed of the water so that it arcs gracefully through the air to the farthest reaches of your garden.

Water's Viscosity

Having brought water to your home in the previous chapter, we're already well on our way to watering your garden. However, to reach the garden, water must first travel through a long stretch of hose that's lying straight on level ground. Does the length of this hose have any effect on the water delivery process?

The answer is yes, longer hoses generally deliver less water. But, according to what we learned in Chapter 5, water should coast through a straight, level hose at constant velocity and constant pressure, and the length of that hose shouldn't matter. We evidently overlooked something important: friction. Moving water doesn't slide freely through a stationary hose. In reality, it experiences frictional forces that oppose its motion relative to the hose.

But this friction is unusual because most of the water in the hose never actually touches the hose itself. If water deep inside the hose is going to experience any

❶ Your car's engine is protected by motor oil with a carefully chosen viscosity. If that oil were too thin, it would flow out from between surfaces and wouldn't keep them from rubbing against one another. If that oil were too thick, the engine would waste power moving its parts through the oil. Years ago, you had to change your motor oil for the season. Thick 40 weight motor oil was used in summer because hot weather made it thinner; thin 10 weight oil was used in winter because cold weather made it thicker. But a modern, multigrade oil maintains a nearly constant viscosity over a wide range of temperatures and need not be changed with the seasons. This oil contains tiny molecular chains that ball up when cold but straighten out when hot. These chains thicken hot oil so that 10W-40 oil resembles 10 weight oil in winter and 40 weight oil in summer.

forces due to relative motion, then those forces are going to have to occur *within* the water. Water must exert frictional forces on itself!

Sure enough, water does experience internal frictional forces. They're called viscous forces—forces that appear whenever one layer of a fluid tries to slide across another layer of that fluid. Viscous forces oppose relative motion and you can observe their effects easily when you pour honey out of a jar. The honey at the jar's surface is stuck there by chemical forces and remains stationary. But even honey that's far from the walls can't move easily; it experiences viscous forces as it tries to move relative to nearby honey. Since honey is a "thick" or viscous fluid, viscous forces act quite effectively to keep all the honey moving with nearly the same velocity. Since the honey at the walls can't move, viscous forces tend to prevent any of the honey from moving.

Water isn't as thick as honey (Table 6.1.1), so it's less resistant to relative motion. The measure of this resistance to relative motion within a fluid is called viscosity, and water's viscosity is less than that of honey. In fact, if you heat the water up it will become even less viscous and thus flow more easily. Typical of most liquids, this decrease in viscosity with increasing temperature reflects the molecular origins of viscous forces: the molecules in a liquid stick to one another, forming weak chemical bonds that require energy to break. In a hot liquid, the molecules have more thermal energy, so they break these bonds more easily in order to move past one another (see ❶).

❷ Water may flow easily and honey with difficulty, but these two liquids are hardly extremes of viscosity. The superfluid portion of ultracold liquid helium has exactly zero viscosity, while the viscosity of softened glass is extraordinarily high. Most other fluids, including air, water, honey, and shampoo, fall somewhere in between. However, not all liquids can be characterized by viscosity. Egg white is a strange liquid. You can pour it part way out of a cup and then, with a flick of your wrist, have it come back into the cup. Other exotic liquids include Silly Putty® and wet cornstarch. These unusual liquids are classified as "non-Newtonian liquids" because they seem to defy Newton's laws of motion as applied to fluids. They don't actually break any of Newton's laws, but they experience such complicated internal forces that it is often difficult to predict how they will behave in a given situation.

Table 6.1.1 Approximate Viscosities of a Variety of Fluids ❷

Fluid	Viscosity*
Helium (2 K)	0 Pa·s
Air (20 °C)	0.0000183 Pa·s
Water (20 °C)	0.00100 Pa·s
Olive oil (20 °C)	0.084 Pa·s
Shampoo (20 °C)	100 Pa·s
Honey (20 °C)	1000 Pa·s
Glass (540 °C)	10^{12} Pa·s

*The pascal-second (abbreviated Pa·s and synonymous with kg/m·s) is the SI unit of viscosity. Only the superfluid portion of ultracold liquid helium exhibits zero viscosity.

Flow in a Straight Hose: The Effect of Viscosity

Viscosity slows the flow of water through your hose. Chemical forces between the hose and the outmost layer of water hold that layer of water stationary, and this motionless layer exerts viscous forces on the layer of moving water inside it. As this second layer slows, it exerts viscous forces on yet another layer. Layer by layer, viscous forces hold back the moving water until even water at the center of the hose feels viscosity's slowing effects (Fig. 6.1.1). Although water at the center of the hose moves faster than water in any other layer, it's still affected by the stationary hose.

These viscous forces impede water delivery. Instead of coasting effortlessly through the straight, level hose, real water needs a pressure gradient to push it steadily forward. Like the file cabinet sliding on the sidewalk in Section 2.2, water must be pushed through the hose if it's to maintain a continuous flow. And like that file cabinet, the water becomes hotter as the work done pushing it forward is wasted and becomes thermal energy.

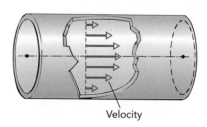

Velocity

Fig. 6.1.1 The speed of water flowing through a pipe is not constant across the pipe. The water near the walls is stationary, while the water at the center of the pipe moves the fastest. The differences in velocity are the results of viscous forces.

However, unlike the forces of ordinary sliding friction—which don't depend on relative velocities—viscous forces become larger as the relative velocities within a fluid increase. That's because as two layers of water slide past one another faster, their molecules collide harder and more frequently. Since it experiences stronger viscous forces, fast-moving water wastes more energy per meter and needs a larger pressure gradient to keep it moving steadily through a hose than does slow-moving water.

Because of viscous forces, the amount of water flowing steadily through a hose depends on four factors:

1. It's inversely proportional to the water's viscosity. The more viscous the water, the more difficulty it has flowing through the hose.
2. It's inversely proportional to the length of the hose. The longer the hose, the more opportunity viscous forces have to slow the water down.
3. It's proportional to the pressure difference between the hose's inlet and its outlet. This pressure difference determines the water's pressure gradient and thus how hard the water is pushed forward through the hose.
4. It's proportional to the fourth power of the diameter of the hose. Tripling the hose's diameter provides the water with nine times as much room and also allows water near the hose's center to move nine times faster.

We can turn all these proportional relationships into an equation by adding the correct numerical constant ($\pi/128$)—the odd-looking result of an arduous calculation. The final relationship is called Poiseuille's law and can be written:

$$\text{volume} = \frac{\pi \cdot \text{pressure difference} \cdot \text{pipe diameter}^4}{128 \cdot \text{pipe length} \cdot \text{fluid viscosity}}. \tag{6.1.1}$$

Poiseuille's Law

The volume of fluid flowing through a cylindrical pipe each second is equal to ($\pi/128$) times the pressure difference across that pipe times the pipe's diameter to the fourth power, divided by the pipe's length times the fluid's viscosity.

❸ To deliver large amounts of water at high pressure or velocity, fire hoses must have large diameters. When filled with high-pressure water, these wide hoses become stiff and heavy, making them difficult to handle. Chemicals are sometimes added to water to decrease its viscosity so that firefighters can use narrower, lighter, and more flexible hoses.

It's hardly surprising that the flow rate depends in this manner on the pressure difference, pipe length, and viscosity; we've all observed that low water pressure or a long hose lengthens the time needed to fill a bucket with water and that viscous syrup pours slowly from a bottle. But the dependence of the flow rate on the fourth power of pipe diameter may come as a surprise. Even a small change in the diameter of a hose significantly changes the amount of water that hose delivers each second. The two most common garden hoses in the United States have diameters of 5/8 inch and 3/4 inch, and while these hoses differ by a seemingly insignificant 20% or a factor of 1.2 in diameter, the 3/4 inch hose can carry about 1.2^4 or 2 times as much water as the 5/8 inch hose (see ❸❹).

We can also look at viscous forces in terms of total energy. By opposing the flow of water through a hose, viscous forces do negative work on it and reduce its total energy—the energy considered in Bernoulli's equation (Eq. 5.2.4), which doesn't include thermal energy. Just how much total energy the water retains depends on how fast it moves inside the hose. If you allow lots of water to leave the hose, water will move through it quickly and encounter large viscous forces. In the process, most of the water's total energy will be wasted as thermal energy and the water will pour gently out of the end of the hose.

❹ Very large diameter pipes are required to transport crude oil across the Alaskan wilderness. The distances are long and the fluid is viscous, although it is heated to reduce its viscosity and allow it to flow more easily.

But if you partially block the hose's opening with your thumb and reduce the flow, water will travel slowly through the hose and encounter smaller viscous forces. As a result, the water will retain most of its total energy and will still be at high pressure when it reaches your thumb. This high-pressure water will then accelerate to enormous speed as it passes through the narrow opening and sprays out into the air.

We can now explain why water delivery systems normally use the widest pipes that are practical and affordable. In contrast to a narrow hose, wide pipes can carry large amounts of water while letting that water travel slowly, experience weak viscous forces, and waste little of its total energy. In such energy efficient water delivery systems, friction is insignificant and Bernoulli's equation accurately predicts water's properties throughout its trip.

Flow in a Bent Hose: Dynamic Pressure Variations

On reaching your garden, let's suppose the hose bends toward the right and the flowing water bends with it. That water is accelerating as it turns and, as we observed in Chapter 5, water accelerates horizontally only in response to unbalanced pressures. Since the hose is motionless, the unbalanced pressures inside it must be caused by the water itself; the water is experiencing dynamic pressure variations.

To understand these dynamic pressure variations, let's follow the streamlines as water traverses this bend. Although we've just introduced viscous forces, we're going to ignore them here for clarity and simplicity. While viscous forces are certainly important in the long, narrow hose, the bend is so short that viscous forces have little effect on what happens to the water passing through it.

Neglecting viscous forces, the water's total energy is constant along each streamline and we can observe the interchanges of energy allowed by Bernoulli's equation. However, since the hose rests on level ground, water's gravitational potential energy can't vary and the only interchanges we'll see are between pressure potential energy and kinetic energy.

Figure 6.1.2 shows the water's steady-state flow pattern near the bend. We're looking down on the hose in this calculated drawing and, as indicated by the black streamlines, water that is initially flowing straight ahead arcs rightward at the bend and eventually continues directly toward the right.

Water approaches the bend through a straight section of hose in which it travels at constant velocity and has a uniform pressure. Its velocity is constant because the straight hose directs all the streamlines forward and because the water mov-

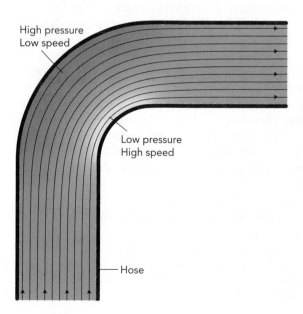

High pressure
Low speed

Low pressure
High speed

Hose

Fig. 6.1.2 Water in a bent hose experiences changes in speed and pressure. The black streamlines show the paths the water takes as it flows around the bend. The spacing between streamlines indicates flow speed (wider space is slower flow), and the background shading indicates pressure (dark is higher pressure; light is lower pressure).

ing along a given streamline can't change its speed; if it tried to speed up, it would leave an empty space behind it; if it tried to slow down, it would cause a "traffic jam." The water's pressure is uniform throughout this straight section because constant velocities mean no accelerations and thus no pressure differences.

The water's constant velocity and uniform pressure are represented visually in Fig. 6.1.2. You can see the local water velocity by looking at the direction and spacing of the streamlines. Streamlines always point in the direction of the local water velocity and their spacing varies inversely with the local water speed. Streamlines that become more widely spaced denote decreasing speed—water that slows down spreads out sideways and its streamlines separate from one another. Streamlines that become more narrowly spaced denote increasing speed—water that speeds up stretches out along its path and its streamlines draw toward one another. Since the streamlines leading up to the bend are straight and evenly spaced, we know that water moves along each streamline at a constant velocity.

You can see the local water pressure in Fig. 6.1.2 by looking at the shading. Dark shading denotes relatively high pressures while light shading denotes relatively low pressures. Since the straight section has uniform gray shading, the water there has a uniform, moderate pressure.

Once the water starts bending toward the right, its velocities and pressures begin to vary. Since the water is accelerating toward the inside of the bend, there must be a pressure imbalance pushing it in that direction. Sure enough, the turning stream of water develops higher local pressure (dark shading) near the outside of the bend and lower local pressure near the inside of the bend (light shading). A similar pressure imbalance accompanies any bend in a fluid's path: the pressure is always higher on the outside of the bend than it is on the inside of that bend. After all, that pressure imbalance is what causes the fluid to bend!

Bends and Pressure Imbalances
When the path of a fluid in steady-state flow bends, the pressure on the outside of the bend is always higher than the pressure on the inside of the bend.

To keep the total energy constant along a streamline, each decrease in the water's local pressure is accompanied by an increase in the water's local speed and vice versa. Water arcing around the outside of the bend slows down (the streamline spacing widens) as its pressure rises, while the water arcing around the inside of the bend speeds up (the streamline spacing narrows) as its pressure drops.

As the hose straightens out beyond the bend, water's pressures and speeds return to what they were before the bend. Water from the outside of the bend speeds up and its pressure drops, while water from the inside of the bend slows down and its pressure rises. In the straight section following the bend, water's velocity is once again constant along each streamline and its pressure is uniform.

Odd as these pressure and speed changes may seem, they are quite real and have real consequences. If your hose were clear and you could introduce thin threads of dye into the flowing water, you'd see these dyed streamlines arc around the bend just as they do in Fig. 6.1.2. And if the hose were weak and couldn't tolerate excessive pressure, it would be most likely to burst on the outside of the bend, where the local water pressure is highest.

You might wonder which causes which: does each pressure change cause a speed change or does each speed change cause a pressure change? The answer is that they occur together and are equally entitled to be called cause and effect. Once the steady-state flow pattern has established itself, water following a particular streamline experiences rises and falls in pressure at the same time that it experiences decreases and increases in speed. The two effects, pressure changes and speed changes, simply go hand in hand.

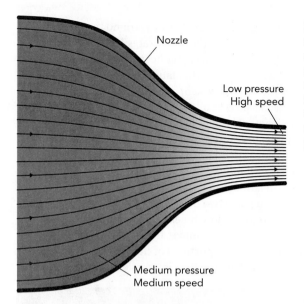

Fig. 6.1.3 Water flowing through a nozzle speeds up and its pressure drops. The narrowing spacing between streamlines indicates that the flow speed is increasing while the shading shift from dark toward light indicates that the pressure is dropping.

Flow through a Nozzle: From Pressure to Speed

When water finally flows through the nozzle at the end of your hose, it exchanges its remaining pressure potential energy for kinetic energy and sprays out into the garden. The nozzle's narrowing channel initiates this energy transformation so that low-speed, high-pressure water entering the nozzle becomes fast-moving, atmospheric-pressure water leaving the nozzle.

Figure 6.1.3 shows that as water passes through the nozzle, the narrowing channel herds all the streamlines together so that the water's local speed increases. The water following each streamline is speeding up in order to squirt through the bottleneck without causing a backup. And this increase in water's local speed is accompanied by a decrease in water's local pressure, as indicated by the shift toward lighter shading.

By the time the water leaves the hose nozzle, its pressure has dropped all the way to atmospheric pressure and it has turned all of its available pressure potential energy into kinetic energy. It emerges as a narrow stream of fast moving water and arcs gracefully through the air. No wonder you can reach the farthest parts of your garden with water when you use a nozzle.

Common Misconceptions: Speed and Pressure in Fluids

Misconception: A fast-moving fluid always has a low pressure.
Resolution: The pressure of a specific portion of fluid depends on its circumstances and can take any value, high or low. However, if a fluid speeds up without descending as it flows along a streamline in steady-state flow, its pressure will decrease. In that special context, the *faster* moving fluid has a *lower* pressure.

The Onset of Turbulence

As you direct the stream of water at the plants in your garden, you notice two interesting phenomena: first, the stream pushes on any surface that slows it down and, second, it tends to break up into fragments as it flows around obstacles. The pushing effect is another Bernoulli result: when the stream encounters a surface, it slows down and spreads out sideways. As the water slows close to the surface, its pressure there rises above atmospheric pressure and it is this elevated pressure that actually pushes the surface forward.

But the breakup effect is something new. In trying to go around the obstacle, the stream of water loses its orderly structure and disintegrates into a swirling, hissing froth. Actually, the hiss you hear is familiar; you heard it as you opened the faucet to start water flowing through the hose. That faucet uses a movable stopper to control the flow of water into the hose; as you opened the faucet, you gradually removed this stopper from the water pipe to allow water to flow more freely into the hose and the faucet hissed. Whether water encounters a plant or a faucet stopper, there is something about high speeds and obstacles that upsets the smoothly flowing fluid.

Up until now, we've discussed only laminar flow—smooth, silent flow that's characterized by simple streamlines. In laminar flow, adjacent regions of a fluid always remain adjacent. For example, if you place two drops of dye near one another in a smoothly flowing stream, they will remain close together indefinitely as they follow streamlines in the laminar flow (Fig. 6.1.4). Laminar flow is the orderly result of viscous forces, which tend to bring adjacent portions of fluid to the same velocity. When viscosity dominates a fluid's motion, the flow is usually laminar.

Fig. 6.1.4 Water flows slowly past rocks in the stream on the left, and its viscosity keeps it smooth and laminar. Water flows quickly past rocks in the stream on the right, and its inertia separates it into swirling, splashing pockets of turbulence.

But as the stream flows swiftly past rocks and obstacles, its streamlines break up into the eddies and churning "white water" that make rafting exciting. The dye is quickly dispersed in this frenzied turbulence. The stream is experiencing turbulent flow—roiling, noisy flow in which adjacent regions of fluid soon become separated from one another as they move independently in unpredictable directions. Turbulent flow is the disorderly consequence of inertia, which tends to propel each portion of fluid independently according to its own momentum. When inertia dominates a fluid's motion, the flow is usually turbulent.

The plants and faucet stopper are evidently initiating turbulence in what had been laminar flows; flows that were dominated by viscous forces are suddenly dominated by inertia instead. Whether a flow is laminar or turbulent depends on several characteristics of the fluid and its environment:

1. The fluid's viscosity. Viscous forces tend to keep nearby regions of fluid moving together, so high viscosity favors laminar flow (Fig. 6.1.5).
2. The fluid's speed past a stationary obstacle. The faster the fluid is moving, the more quickly two nearby regions of fluid can become separated and the harder it is for viscous forces to keep them together.
3. The size of the obstacle the fluid encounters. The larger the obstacle, the more likely that it will cause turbulence because viscous forces will be unable to keep the fluid ordered over such a long distance.
4. The fluid's density. The denser the fluid, the less it responds to viscous forces and the more likely it is to become turbulent.

Fig. 6.1.5 Honey's large viscosity keeps it flowing smoothly (laminar flow) when you pour it. Water's small viscosity allows it to splash about (turbulent flow) in a fountain.

Rather than keeping track of all four physical quantities independently, English mathematician and engineer Osborne Reynolds (1842–1912) found that they could be combined into a single number that permits a comparison of seemingly different flows. The Reynolds number is defined as

$$\text{Reynolds number} = \frac{\text{density} \cdot \text{obstacle length} \cdot \text{flow speed}}{\text{viscosity}}. \qquad (6.1.2)$$

The units on the right side of Eq. 6.1.2 cancel one another so that the Reynolds number is dimensionless; that is, it's just a simple number, such as 10 or 25,000. As the Reynolds number increases, the flow goes from viscous-dominated to inertia-dominated and therefore from laminar to turbulent. In his experiments, Reynolds found that turbulence usually appears when the Reynolds number exceeds roughly 2300. You can observe this transition by moving a 1-cm-thick (0.4-inch-thick) stick through still water. If you move the stick slowly, about 10 cm/s (4 inch/s), the Reynolds number will be about 1000 and the flow around the stick will be laminar. But if you speed the stick up to about 50 cm/s (20 inch/s), the Reynolds number will rise to about 5000 and the flow will become turbulent.

One of the most common features of turbulent flows is the vortex, a swirling region of fluid that moves in a circle around a central cavity. A vortex resembles a miniature tornado, with its cavity created by inertia as the fluid spins. Vortices are easily visible behind a canoe paddle or in a mixing bowl. Once an object moves fast enough through a fluid to create turbulence, these vortices begin to form. Each vortex builds up behind the object but is soon whisked away to form a wake of shed vortices—vortices that have broken free of the object that made them (Fig. 6.1.6).

While laminar flow is fully predictable, turbulent flow exhibits chaotic behavior or chaos: you can no longer predict exactly where any particular drop of water will go. The study of chaos is a relatively new field of science. Because a chaotic system—a system exhibiting chaos—is exquisitely sensitive to initial conditions, even the slightest change in those conditions may produce profound changes in its situation later on.

Even when you can't see turbulent water flow, you can usually hear it. The churning motion of turbulence converts some of the water's total energy into thermal energy and sound. The turbulence near the faucet slightly reduces the water's total energy as it enters the hose and therefore its speed as it emerges from the nozzle and sprays toward your garden.

A different sound occurs when you abruptly close the nozzle and stop the flow of water. Moving water has momentum, and stopping it suddenly requires an enormous backward force. Since the slowing flow is not steady state, Bernoulli's equa-

Fig. 6.1.6 When water flows rapidly around a cylinder, its flow becomes turbulent. A pattern of swirling vortices forms to the right of this cylinder.

tion doesn't apply and the pressure can surge to astronomical values near the front of the moving water. This pressure surge is what accelerates the water backward to slow it down and also what leads to the loud "thump" sound you hear as the water stops. Known as water hammer, the surging pressure in front of stopping water jerks the nozzle, swells the hose, and may even rattle the pipes in your home.

⤷ Faucets, Screws, and Seals

Watering your garden would be difficult if you couldn't turn the water on and off. You need a faucet. And even if you live in a studio apartment on the 30th floor, without so much as a window box, you probably still have faucets in your home.

Although there are a number of different faucet types, each using a somewhat different mechanism to control water flow, the type shown in Fig. 6.1.7 is probably the oldest and most common. This faucet uses a rubber washer to control the water flow. When the faucet is closed, it presses the washer against the water inlet pipe, preventing the water from flowing. When you open the faucet, the washer moves away from the inlet pipe's opening and water begins to flow. The rubber washer sits at the end of a shaft that's raised and lowered by a screw as you turn the faucet handle. But why does the faucet need this screw? Why not just push the washer down to block the flow of water and pull it up to unblock the flow of water?

That simple push-pull scheme has a pressure problem. When the washer is pressed against the inlet pipe so that no water flows, the pressure in the inlet pipe is city water pressure and the pressure in the outlet pipe is atmospheric pressure. There's a large pressure imbalance on the washer and thus a large upward force on it. As a result, you'd have to push down very hard on the shaft to keep the faucet closed and it would pop upward when you let go. To make things easier, therefore, most faucets include a screw.

As I noted in Section 4.4 on elevators, a screw is a rotating device that exhibits mechanical advantage. It's actually two ramps or inclined planes that are wrapped around two cylinders so that they slide across one another as the cylinders rotate relative to each other. When two ordinary ramps slide across one another, one goes up while the other goes down (Fig. 6.1.8a). However, the only parts of the ramps that are really important are their surfaces (Fig. 6.1.8b). If these surfaces are wrapped around cylinders, they become screws (Fig. 6.1.8c).

The cylindrical ramps themselves are called screw threads. In a screw, one of these screw threads attaches to an inner cylinder, while the other screw thread attaches to an outer cylinder. The threaded inner cylinder resembles a bolt; the threaded outer cylinder resembles a nut. When the inner cylinder of the screw turns rela-

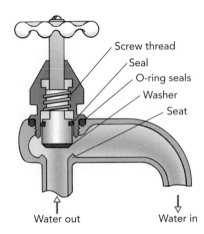

Fig. 6.1.7 A water faucet controls the flow of water by partially or completely blocking the inlet pipe with an elastic washer. A screw attached to the handle sets the position of the washer. Several other water seals prevent the faucet from leaking.

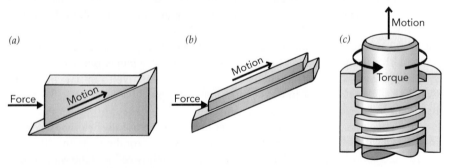

Fig. 6.1.8 (*a*) When you push a movable ramp across a second, fixed ramp, the movable ramp slides upward relative to the other. (*b*) The same holds for two inclined surfaces. (*c*) When the inclined surfaces are wrapped around cylinders to form cylindrical ramps, a screw is created. As the movable cylinder rotates, it moves upward or downward relative to the outer, fixed cylinder.

tive to the outer cylinder, the two screw threads slide across one another. The inner cylinder moves up or down in the outer cylinder, depending on the direction of their relative rotation.

When two ramps slide across one another, modest horizontal forces on the ramps can produce large vertical forces. The same holds true for a screw, where a modest torque on the screw can produce large vertical forces. Like the ramp, the screw exhibits mechanical advantage. In this case, the modest torque you exert as you rotate the handle several turns exerts a huge force as the shaft descends a small distance. While the work you do in closing the faucet is the same whether you push the shaft down directly or use a screw, the screw spreads out the work and lets you do it with smaller forces or torques exerted over longer distances or angles. Here, the screw easily exerts the large force needed to keep the washer pressed against the inlet pipe. Moreover, the screw's threads experience frictional forces that are strong enough to prevent the screw from turning on its own. With the help of this friction, the screw can keep the washer pressed against the inlet pipe even when you take your hand off the faucet handle.

A faucet isn't a solid object; it's built from several separate pieces with joints between them. Each joint presents a possible site for water leakage. The faucet doesn't leak, however, because seals block the water's path through each of these joints. The washer itself forms a seal when the faucet is closed, pressing down against the water inlet to stop the flow. But the faucet also contains at least two O-ring seals or their equivalent. An O-ring is a doughnut-shaped pieces of hard elastomer (rubberlike elastic material), that is confined in a groove and prevents liquid from flowing through joints between two pieces of metal, plastic, or ceramic (Fig. 6.1.9). Elastomers store energy when they're distorted and release that energy when they return to their equilibrium shapes. The faucet's O-rings keep water from flowing out around the handle. One of these O-rings is attached to the faucet's shaft and maintains its seal even as the shaft rotates and moves up or down.

An O-ring is particularly effective at blocking the flow of water because it's elastic. I'll discuss elastic materials, especially natural and synthetic rubbers, in Section 17.3 on plastics. What matters here is that an O-ring is springlike over a certain range of temperatures. The *Challenger* explosion in 1986 was caused by an O-ring that was too cold. It had lost its elastic character and therefore failed to seal well.

In a seal, where an O-ring is distorted away from its equilibrium shape and experiences a strong restoring force, it pushes against the surfaces that confine it, filling all gaps and blocking the passage of water. As long as it remains compressed in the seal, it will try to return to its equilibrium shape and keep the seal water-tight.

As the shaft O-ring turns and slides inside the faucet, it experiences sliding friction. Although the water lubricates its motion, the O-ring experiences some wear when you turn the faucet on or off, and with enough wear it begins to leak and must be replaced. In order to reduce this wear and prolong the O-ring's useful life, the faucet manufacturer polishes the metal surfaces against which the O-ring slides.

The washer that controls the flow of water in a faucet eventually wears out, too. Each time the washer presses against the water inlet, it distorts. Like any elastomer, it stores energy when it's distorted and releases that energy when it returns to its equilibrium shape. But sliding friction with the inlet pipe, long periods of distortion, and thermal damage in the hot water eventually change the washer's equilibrium shape forever. Since it no longer distorts from its new equilibrium shape when pressed against the inlet pipe, the washer fails to fill the gaps so the faucet leaks. To fix the faucet, you must unscrew the handle, shaft, and shaft holder from the faucet and replace the washer. During this repair, you can block water flow to the faucet using the cut-off valve—another faucet-like valve installed upstream of the faucet's water inlet and usually located underneath the kitchen sink, bathroom sink, or wherever water enters the home.

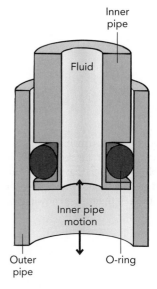

Fig. 6.1.9 An O-ring seals an inner pipe to an outer pipe. The inner pipe has a groove cut in its outer edge to hold the O-ring. Whenever the inner pipe moves through the outer pipe, that motion causes the O-ring to twist and thereby lubricate itself with the fluid contained within the pipes. The O-ring should continue to seal this joint between the two pipes for years, even if the two pipes move or rotate relative to one another occasionally. O-ring seals were invented during World War II and revolutionized the hydraulic plumbing fittings used in aircraft and other military equipment.

❺ A modest narrowing in a person's artery substantially reduces the flow of blood through that artery. That's why people should watch what they eat to keep their blood flowing freely.

Over time, the plumbing in a home narrows due to the buildup of minerals inside the pipes. As a result of this narrowing, filling a bathtub takes longer and longer. The pressure difference between where water enters the home and the bathtub spigot doesn't change, nor does the length of the pipes. But a small change in pipe diameter has an enormous effect on the rate of flow predicted by Poiseuille's law. Water struggling to get through the narrowed pipes loses total energy quickly and dribbles lethargically out of the spigot. Even a 15% narrowing in the pipes can cut the flow rate in half. (For the medical equivalent of this problem, see **❺**.)

A long wait for the tub to fill may irritate you, but things could be worse. You could be taking a shower when someone flushes a toilet nearby. If the circumstances are right (or maybe I should say wrong), you'll suddenly find yourself sprayed with scalding hot water. People in old buildings often learn to jump out of the spray the moment they hear a flushing toilet.

The sudden temperature surge is caused by an abrupt loss of pressure in the pipe delivering cold water to the showerhead. As long as your shower is the only consumer of cold water, the flow rate in this pipe is small and the water travels relatively slowly through the pipe. Since slow-moving water doesn't lose much of its total energy to thermal energy, the pressure is still high when the cold water arrives at the showerhead. Likewise, the hot water retains most of its total energy and enters the showerhead at high pressure. The two waters, cold and hot, mix easily and spray out at you at a comfortable temperature.

When someone flushes a nearby toilet, however, the toilet begins to consume large amounts of cold water. The flow rate in the cold water pipe increases dramatically and the now fast-moving cold water loses total energy quickly en route to the toilet and shower. The cold water now arrives at the showerhead at relatively low pressure and it no longer mixes well with the hot water. In fact, the shower may even allow the high-pressure hot water to flow backward into the low-pressure cold water pipe temporarily. But most importantly, the showerhead now sprays almost exclusively hot water. That's why the temperature of the spray rises and why you'd better leap to safety.

Another household nuisance is water hammer, a phenomenon I mentioned above briefly. The "clunk" of water hammer is the impact that occurs when moving water is suddenly brought to a stop. When a faucet is opened wide, the water runs freely through the pipe leading to that faucet. This water has momentum and the longer the pipe and the faster the water is moving, the greater that momentum. When you suddenly close the faucet, all of that moving water has to stop abruptly. The water, in effect, crashes into the closed faucet. The faucet must exert a large force on this long column of water and, since the only way to exert a force on a fluid is through pressure, the pressure in front of the moving water surges to an enormous value. The column of water then experiences a pressure imbalance that makes it accelerate backward and it quickly comes to a stop.

If the impact of water hammer is strong enough, you can hear the pipes move. The water suddenly transfers its momentum to the pipes and the faucet, just as a hammer does to the nail it hits. This effect is more than just a nuisance; it can actually break a pipe. Water hammer can be reduced, however, by putting an air shock absorber in the pipe (Fig. 6.1.10). The water gradually compresses the air in this device when it slams into the faucet. While the momentum transferred from the water to the pipe is still the same, that impulse occurs with a smaller force exerted over a longer period of time. A careful plumber may install sealed vertical tubes at water outlets for clothes washers and other machines that tend to stop the flow of water abruptly. These air pockets fill with air that naturally comes out of solution in the water and use that trapped air to cushion the water's impact.

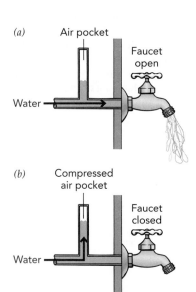

(a)

Air pocket

Faucet open

Water

(b)

Compressed air pocket

Faucet closed

Water

Fig. 6.1.10 Water flowing through a pipe has momentum. If the water is shut off suddenly, it exerts a large force on the pipe and faucet as it comes to a stop. To reduce the peak force, you can add an air shock absorber to the pipe. The water slows gradually as it compresses the air and therefore exerts a smaller force on the pipe and faucet.

Water hammer isn't limited to household plumbing; it can occur whenever an obstacle suddenly blocks the path of rapidly moving water. Both the pressure of the water and the force it exerts on the obstacle surge upward, particularly if the water can't find a path around the obstacle. A sudden heavy rain can send a flash flood coursing through a dry riverbed and a car caught by this flash flood can be spun around, knocked over, or washed away by the impact of water hammer. Waves crashing against cliffs also experience this effect. The pressure surge can squirt water high into the air through gaps or holes in the rocks.

◈ A Little More Chaos

A chaotic system is one that's exquisitely sensitive to initial conditions so that even the tiniest change in how the system starts off will radically alter its future. For example, your motion on a crowded highway is a chaotic system. If the movement of just one other car prevents you from switching lanes as you pass through a complicated interchange, you could soon find yourself in an entirely different city.

The classic example of a chaotic system is turbulent flow in a fluid. The slightest change in almost any aspect of turbulent water flow will dramatically alter its swirls and eddies. A grain of sand being carried along with the water will move unpredictably. Similarly, two grains of sand that enter the turbulent water side-by-side another will quickly become separated; the longer you watch them, the more independent their motions will become. This divergence of particles or paths that are initially near each other is one of the hallmarks of chaos.

The atmosphere is a chaotic system, so that slight changes in the air today will significantly affect the weather a day or a week from now. Surface winds are filled with vortices and eddies, common features in turbulent and chaotic flows. Tiny changes in the earth's surface can dramatically affect these turbulent flows and redirect the wind.

It's frequently hypothesized that the turbulence caused by a single butterfly flapping its wings can redirect the flow of the wind and will eventually affect the weather everywhere on earth. The butterfly changes the local winds, which in turn change other winds, and so on until the earth's entire atmosphere has been affected. Because of the atmosphere's chaotic nature, weather is difficult to predict. Weather prediction will never be completely reliable nor will it extend much further into the future than it does today. You simply can't record the movements of every butterfly on earth.

Spin
Lift
Turbulent
wake
Drag

Section 6.2 **Balls and Air**

Much of the subtlety and nuance in games such as baseball and golf come from the way balls interact with air. If baseball were played on the moon, which has no air or atmosphere, the sinking fastball would be the only interesting pitch. Moon golfers wouldn't have to worry about hooks or slices. In this section we will investigate how air affects the flight of balls and other related objects.

When a Ball Moves Slowly: Laminar Airflow

One of the first things you might notice if you joined a new baseball franchise on the moon is that pitched balls reach home plate faster than back at home. Since the moon has no atmosphere, there is no air resistance to slow a ball down. In the previous section, we saw how objects affect moving fluids. Now as we study aerodynamics, the science of air's dynamic interactions, we'll see how fluids affect moving objects. (For a discussion of the applicability of Bernoulli's equation to aerodynamic systems, see ❶.)

In air, a moving ball experiences aerodynamic forces, that is, forces exerted on it by the air because of their relative motion. These consist of drag forces that push the ball downwind and lift forces that push the ball to one side or the other (Fig. 6.2.1). We'll begin our study of ball aerodynamics with drag forces, commonly called "air resistance," and we'll start with a slow-moving ball. The reason for starting slow is that at low speeds, viscous forces are able to organize the air as it flows around the ball; viscosity dominates over inertia and the airflow around the slow-moving ball is laminar.

Figure 6.2.2 shows the pattern of laminar airflow around a slow-moving ball. Actually, the pattern is the same whether the ball moves slowly through the air or the air moves slowly past the ball. For simplicity, let's move along with the ball and study the airflow from the ball's inertial frame of reference. In that inertial frame, the ball appears stationary with the air flowing past it.

The slow-moving air separates neatly around the front of the ball and comes back together behind it. It produces a wake, an air trail behind the ball, that's smooth and free of turbulence. But the air's speed and pressure aren't uniform all the way

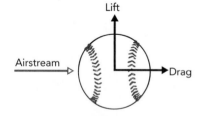

Lift

Airstream

Drag

Fig. 6.2.1 The two types of aerodynamic forces exerted on objects by air are drag and lift. Drag is exerted parallel to the onrushing airstream and slows the object's motion through the air. Lift is exerted perpendicular to that airstream and pushes the object to one side or the other. Lift is *not* necessarily in the upward direction.

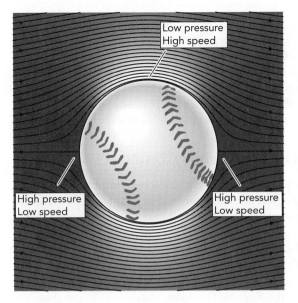

Low pressure High speed

High pressure Low speed

High pressure Low speed

Fig. 6.2.2 The airflow around a slowly moving ball is laminar. Air slows down in front of and behind the ball (widely spaced streamlines), and its pressure increases (shifts toward darker shading). Air speeds up at the sides of the ball (narrowly spaced streamlines), and its pressure decreases (shifts toward lighter shading). However, the pressure forces on the ball balance one another perfectly, and it experiences no pressure drag. Only viscous drag is present to affect the ball.

❶ Strictly speaking, we shouldn't be using Bernoulli's equation with air; it applies only to incompressible fluids in perfect steady-state flow, and air certainly isn't incompressible. But if certain conditions are met—if the air's velocity is less than about 300 km/h, and if there are no pressure differences of more than one tenth of an atmosphere—then we can *consider* air to be incompressible, since its density will remain fairly constant. That makes things much easier for us here as we discuss curve balls, airplanes, and vacuum cleaners. We can also ignore gravity. If the air were flowing up or down hundreds of meters, we would need to include gravity but here it's inconsequential because everything is happening at one altitude.

around the ball. The airflow bends several times as it travels around the ball and, as we saw in the previous section, such bends always involve pressure imbalances. Since the air pressure far from the ball is steadfastly atmospheric, those pressure imbalances are always caused by pressure variations near the ball's surface. Whenever air bends away from the ball, so that the ball is on the outside of a bend, the pressure near the ball must be higher than atmospheric. And whenever the air bends toward the ball, so that the ball is on the inside of a bend, the pressure near the ball must be lower than atmospheric.

With that introduction, let's examine the slow-moving airflow around the ball. Air heading toward the ball's front bends away from it, so the pressure near the front of the ball must be higher than atmospheric. This rise in air pressure is accompanied by a decrease in airspeed—the air's speed relative to the ball. Figure 6.2.2 indicates the pressure rise by a shift toward dark shading and the decrease in airspeed by the widening separation of the streamlines.

Air rounding the ball's sides bends toward it, so the pressure near those sides must be below atmospheric. This drop in air pressure is accompanied by an increase in airspeed. Figure 6.2.2 indicates the pressure drop by a shift toward light shading and the increase in airspeed by the narrowing separation of the streamlines.

The laminar airflow continues around to the back of the ball and then trails off behind it. Since the departing air again bends away from the ball, the pressure near the back of the ball must be higher than atmospheric. As before, the shift toward dark shading in Figure 6.2.2 indicates this pressure rise and the widening separation of the streamlines points out the accompanying decrease in airspeed.

It may seem strange that the air pressure can be different at different points on the ball, but that's what happens in a flowing stream of air (see ❷). It's particularly remarkable that low-pressure air at the sides of the ball is able to flow around to the back of the ball, where the pressure is higher. This air is experiencing a pressure imbalance that pushes it backward, opposite its direction of travel. But a pressure imbalance causes acceleration, not velocity, and the low-pressure air flowing past the sides of the ball has enough energy and forward momentum to carry it all the way to the back of the ball. Although this air slows as it flows into the rising pressure, it manages to complete its journey.

The airflow around the ball is symmetric, and the forces that air pressure exerts on the ball are also symmetric. These pressure forces cancel one another perfectly so that the ball experiences no overall force due to pressure. Most importantly, the high pressure in front of the ball is balanced by the high pressure behind it. As a

❷ Air pressure variations are common in the wind blowing around cars and buildings. Air pressure is highest on a car's front surfaces and lowest just outside its front side windows. Air passing the rear side windows has straightened out and returned to atmospheric pressure. When you open a front side window, air flows out of the car and the air pressure inside the car drops. If you then open a rear side window, air reenters the car through the rear side windows. This circulating pattern explains why objects thrown out the front side window frequently reenter the car through its rear side window. In a building, the air pressure is highest outside windward walls and can be quiet low outside convex side walls. Pressure changes on windy days can make doors hard to open or cause them to open spontaneously, depending on which way those doors swing. Moreover, peaked roofs experience low surface pressures on windy days and may be lifted off their buildings during windstorms.

❸ When the airflow around an object is laminar, the pressure forces on it cancel perfectly and it experiences no drag due to pressure imbalances—no *pressure* drag. The absence of pressure drag was a great puzzlement to early aerodynamicists, who knew that the airflow around dust is laminar and that it experiences a drag force. This mystery was named d'Alembert's paradox, after Jean Le Rond d'Alembert (1717–1783), the French mathematician who first recognized it. D'Alembert and his contemporaries didn't know about the viscous drag force, which is what really slows dust's motion through the air.

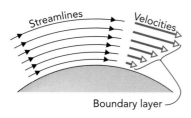

Fig. 6.2.3 As air flows past a surface, a thin layer of it is slowed by viscous drag forces. This boundary layer of air affected by the surface remains laminar at Reynolds numbers all the way up to about 100,000, above which it finally becomes turbulent. The freely flowing airstream, however, becomes turbulent at much lower Reynolds numbers, typically above about 2000.

❹ Among Ludwig Prandtl's (German engineer, 1875–1953) many pivotal contributions to aerodynamic theory is the concept of boundary layers in fluid motion. Prandtl was so engrossed in establishing Göttingen as the world's foremost aerodynamic research facility that he did not have time to court a wife. Deciding he should be married, Prandtl wrote his former advisor's wife, asking to marry one of her two daughters but not specifying which one. The family selected the eldest daughter, and the wedding took place.

result of this symmetric arrangement, the only aerodynamic force acting on the ball is viscous drag—the downstream frictional force caused by layers of viscous air sliding across the ball's surface (see **❸**).

Though we'll soon see that viscous drag is only a small fraction of the air resistance experienced by sports balls, it's the force that suspends dust in the air for hours and is an important issue for airplane wings. It's also a force that we've encountered before: viscous drag slowed the water in your garden hose in the previous section!

When a Ball Moves Fast: Turbulent Airflow

Balls don't always experience laminar airflow. Turbulence is common, particularly in sports, and brings with it a new type of drag force. When the air flowing around a ball is turbulent, the air pressure distribution is no longer symmetric and the ball experiences pressure drag—the downstream force exerted by unbalanced pressures in the moving air. These unbalanced pressures exert an overall force on the ball that slows its motion through the air.

A ball can experience turbulent airflow and pressure drag when the Reynolds number exceeds about 2000. The Reynolds number, introduced in the previous section, combines the ball's size and speed with the air's density and viscosity to give an indication of whether the airflow is dominated by viscosity or inertia. At low Reynolds numbers, the air's viscosity dominates over its inertia and the airflow is laminar. But at high Reynolds numbers, air's inertia dominates over its viscosity and the airflow tends to become turbulent. This turbulence, however, won't start until something triggers it, and viscosity provides that trigger.

To understand viscosity's role, we must look at the air near the ball's surface. Even in a strong wind, viscous forces slow down a thin boundary layer of air near the ball's surface (Fig. 6.2.3). Discovered by Ludwig Prandtl **❹** with help from Gustave Eiffel (Fig. 6.2.4), this boundary layer moves more slowly and has less total energy than the freely flowing air farther from the surface.

As air flows toward the back of the ball, it travels through an adverse pressure gradient—a region of rising pressure that pushes backward on the air and causes it to decelerate. While the freely flowing airstream outside the boundary layer has enough energy and forward momentum to continue onward and reach the back of the ball on its own, air in the boundary layer does not. It needs a forward push.

At low Reynolds numbers, the entire airstream helps to push that boundary layer all the way to the back of the ball and the airflow remains laminar. But at high Reynolds numbers, viscous forces between the freely flowing airstream and the boundary layer are too weak to keep the boundary layer moving forward into the rising pressure behind the ball.

Without adequate help, the boundary layer eventually stalls—it comes to a stop and thereby spoils steady-state flow. More horrible still, this stalled boundary layer air is pushed backward by the adverse pressure gradient and returns all the way to the ball's sides. As it does, it cuts like a wedge between the ball and the freely flowing airstream. The result is an aerodynamic catastrophe: the airstream separates from the ball, leaving a huge turbulent wake or air pocket behind the ball (Fig. 6.2.5).

Because of this turbulent wake, air no longer bends smoothly away from the back of the ball and there is no rise in pressure there. Instead, the pressure behind the ball is roughly atmospheric. The absence of a high-pressure region behind the ball spoils the symmetry of pressure forces on the ball and those forces no longer cancel. The ball experiences an overall pressure force downwind—the force of pressure drag. In effect, the ball is transferring forward momentum to the air in its turbulent wake and dragging that wake along with it.

Pressure drag slows the flight of almost any ball moving faster than a snail's pace. The pressure drag force is roughly proportional to the cross-sectional area of the

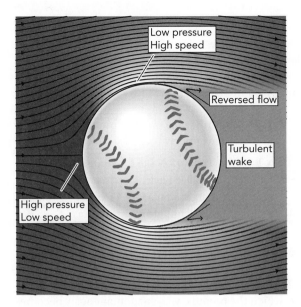

Fig. 6.2.5 When a ball's speed gives it a Reynolds number between about 2000 and 100,000, its laminar boundary layer stalls in the rising pressure behind the ball. The resulting reversed flow causes the main airflow to separate from the ball's surface, leaving a large, turbulent wake. The average pressure behind the ball remains low, and the ball experiences a large force of pressure drag.

Fig. 6.2.4 Early experiments in aerodynamics were performed by Gustave Eiffel (French engineer, 1832–1923), who designed the tower that bears his name. In the 1890s, Eiffel dropped objects of various sizes and shapes from his tower and measured the drag that they experienced. His work was used by Prandtl to explain the reduction in drag that accompanies the appearance of turbulent boundary layers.

turbulent air pocket and to the square of the ball's speed through the air. For a ball moving at a moderate speed, the air pocket is about as wide as the ball and the ball experiences a large pressure drag force.

The Dimples on a Golf Ball

If this were the whole story, you would never hit a home run at a baseball game or a 250-yard drive on the golf course. But inertia has yet another card to play.

At very high Reynolds numbers, the boundary layer itself becomes turbulent (Fig. 6.2.6). It loses its laminar streamlines and begins to mix rapidly within itself and with the freely flowing airstream nearby. This mixing brings additional energy and forward momentum into the boundary layer and makes it both harder to stop and more resistant to reversed flow. Although this turbulent boundary layer still stalls before reaching the back of the ball, the stalled air flows upstream only a short distance. While the freely flowing airstream still separates from the ball, that separation occurs far back on the ball and the resulting turbulent wake is relatively small (Fig. 6.2.7).

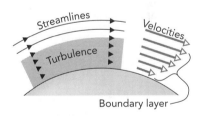

Fig. 6.2.6 When the Reynolds number exceeds about 100,000, the boundary layer of air flowing past a surface becomes turbulent. This whirling fluid brings in extra energy and momentum from the freely flowing airstream and can travel deep into a region of increasing pressure.

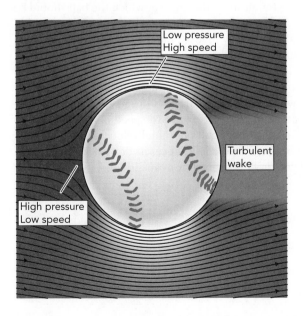

Fig. 6.2.7 When a ball travels fast enough that its Reynolds number exceeds 100,000, its boundary layer becomes turbulent. This turbulent layer travels much of the way around the back of the ball before it separates from the surface. The freely flowing air follows it, and the two leave a relatively small turbulent wake. The ball experiences only a modest force of pressure drag.

Fig. 6.2.8 Early golf balls (left) were handmade of leather and stuffed with feathers. Golf became popular when cheap balls made of a hard rubber called gutta-percha became available. But new, smooth "gutties" didn't travel very far; they flew better when they were nicked and worn. Manufacturers soon began to produce balls with various patterns of grooves on them (bottom and right), and those balls traveled dramatically farther than smooth ones. Modern golf balls (top) have dimples instead of grooves.

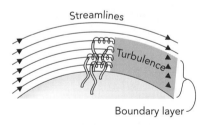

Fig. 6.2.9 The boundary layer can be made turbulent at Reynolds numbers below 100,000 by "tripping" it with obstacles such as fuzz or dimples.

As a result of this smaller air pocket, the pressure drag is reduced from what it would be without the turbulent boundary layer. The effect of replacing the laminar boundary layer with a turbulent one is enormous; it's the difference between a golf drive of 70 yards and one of 250 yards! The effects of the Reynolds number on the airflow around a ball are summarized in Table 6.2.1.

Table 6.2.1 Effects of Reynolds Number on the Airflow around a Ball or Other Object

Reynolds Number	Boundary Layer	Type of Wake	Main Drag Force
<2000	Laminar	Small laminar	Viscous
2000–100,000	Laminar	Large turbulent	Pressure
>100,000	Turbulent	Small turbulent	Pressure

Delaying the airflow separation behind the back of the ball is so important to distance and speed that the balls of many sports are designed to encourage a turbulent boundary layer (Fig. 6.2.8). Rather than waiting for the Reynolds number to exceed 100,000, the point near which the boundary layer spontaneously becomes turbulent, these balls "trip" the boundary layer deliberately (Fig. 6.2.9). They introduce some impediment to laminar flow, such as hair or surface irregularities, which causes the air near the ball's surface to tumble about and become turbulent. The drop in pressure drag more than makes up for the small increase in viscous drag. That's why a tennis ball has fuzz and a golf ball has dimples.

So how much does drag affect balls in various sports? For those that involve rapid movements through air or water, the answer is quite a bit. Drag forces increase dramatically with speed; as soon as a turbulent wake and pressure drag appear, the drag force increases as the square of a ball's speed. As a result, baseball pitches slow significantly during their flights to home plate, and the faster they're thrown, the more speed they lose. A 90-mph fastball loses about 8 mph en route, while a 70-mph curveball loses only about 6 mph.

A batted ball fares slightly better because it travels fast enough for the boundary layer around it to become turbulent, an effect that appears at around 160 km/h (100 mph). While the resulting reduction in drag explains why it's possible to hit a home run, the presence of air drag still shortens the distance the ball travels by as much as 50%. Without air drag, a routine fly ball would become an out-of-the-park home run. To compensate for air drag, the angle at which the ball should be hit for maximum distance isn't the theoretical 45° above horizontal discussed in Section 1.2. Because of the ball's tendency to lose downfield velocity, it should be hit at a little lower angle, about 35° above horizontal (Fig. 6.2.10).

Since the ball loses much of its horizontal component of velocity during its trip to the outfield, a long fly ball tends to drop almost straight down as you catch it. Gravity causes it to move downward, but drag almost stops its horizontal motion away from home plate. Drag also limits the downward speed of a falling ball to about 160 km/h (100 mph). That's the baseball's terminal velocity, the downward velocity at which the upward drag force exactly balances its downward weight and it stops accelerating. Even if you drop a baseball from an airplane, its velocity will not exceed this value ❺.

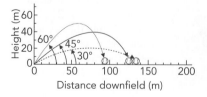

Fig. 6.2.10 Air drag slows the flight of a batted ball so that the ideal angle at which to hit it isn't the theoretical 45° of Fig. 1.2.7. An angle of roughly 35° above horizontal will achieve the maximum distance.

Badminton Birdies and Bullets

In addition to slowing balls down in flight, air can also exert torques on non-symmetric ones. Oblong balls such as American footballs and rugby balls frequently experience aerodynamic torques about their centers of mass. A quarterback throws the football with a spiral not only to minimize the drag it experiences, but also to avoid aerodynamic torques that would make the football tumble in flight and therefore hard to catch. But while aerodynamic torques are relatively small for balls, even oblong balls, they can be extremely large for the projectiles used in other sports.

The shuttlecock or "birdie" used in badminton has the peculiar property of always flying bumper first and feathers last (Fig. 6.2.11). This behavior is caused by the air, which exerts a large torque on the birdie about its center of mass whenever it's not flying bumper first. Air can exert that torque because the birdie's center of pressure—the point at which the overall pressure force on the birdie effectively acts—isn't located at the birdie's center of mass. Since an object's center of pressure tends to be located near any large surfaces that are pushed on by the passing air, the birdie's center of pressure is located in its broad, air-catching feathers. Since the birdie's center of mass is in its heavy bumper, air pressure can exert a torque on it about its center of mass and cause it to undergo angular acceleration. The birdie always rotates so that it flies bumper-first.

With its bumper flying ahead of its feathers, the shuttlecock has dynamic stability. If it turns clockwise, air pressure twists it counter-clockwise; if it turns counter-clockwise, air pressure twists it clockwise. This aerodynamic stabilizing effect is strong and reliable. It flips the shuttlecock around quickly after each hit and then keeps the shuttlecock flying bumper forward until the next hit. Other sports, including darts, archery, and javelin, use similar aerodynamic torques to orient and stabilize a moving object. The simple rockets I described in Section 4.2 use fins and sticks to achieve this same dynamic stability in air.

While aerodynamic torques stabilize those projectiles, they can cause trouble for others. A rifle bullet, for example, experiences tremendous pressure at its front as it rushes through the air. Its center of pressure is located in front of its center of mass, making it aerodynamically *unstable*. Without any compensating effect, the bullet would tumble wildly in flight and wouldn't be very accurate. To keep this from happening, a properly made rifle has rifling grooves that spin the bullet as it travels through the gun barrel. The bullet's large angular momentum keeps it from tumbling and its rapid rotation averages out the aerodynamic lift forces so that the bullet flies straight.

Curveballs and Knuckleballs

The drag forces on a ball push it downstream, parallel to the onrushing air. But in some cases, the ball may also experience lift forces—forces that are exerted perpendicular to the airflow (Fig. 6.2.1). To experience drag, the ball only has to slow the airflow down; to experience lift, the ball must deflect the airflow to one side or the other. Although its name implies an upward force, lift can also push the ball toward the side or even downward.

Curveballs and knuckleballs both use lift forces. In each of these famous baseball pitches, the ball deflects the airstream toward one side and the ball accelerates toward the other. Again we have action and reaction—the air and the ball push off one another. Getting the air to push the ball sideways is no small trick. Explaining it isn't easy either, but here we go.

A curveball is thrown by making the ball spin rapidly about an axis perpendicular to its direction of motion. The choice of this axis determines which way the ball curves. In Fig. 6.2.12, the ball is spinning clockwise, as viewed from above. With

❺ A skydiver who jumps out of a plane won't continue to accelerate downward indefinitely. After about 10 s, she'll have reached terminal velocity and will coast downward the rest of the way. Her terminal velocity is approximately 200 km/h (125 mph), depending on her body posture, and can rise to about 320 km/h (200 mph) if she makes herself as narrow and vertical as possible. Clearly, no one would want to hit the ground at any of these speeds. Once her parachute opens and begins to experience severe upward drag and/or lift forces, however, her terminal velocity drops to about 18 km/h (12 mph) and she makes a soft and graceful landing.

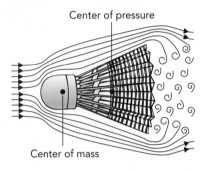

Fig. 6.2.11 A badminton shuttlecock or birdie flies into the wind bumper first because the overall force exerted by air pressure is located at its center of pressure, far from its center of mass. If the feathers try to overtake the bumper, air resistance exerts a torque about the shuttlecock's center of mass and returns the feathers to the rear.

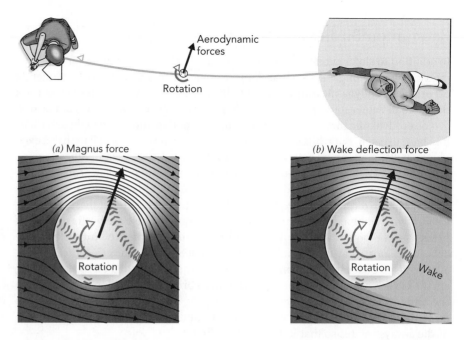

Fig. 6.2.12 A rapidly rotating baseball experiences two lift forces that cause it to curve in flight. (*a*) The Magnus force occurs because air flowing around the ball in the direction of its rotation bends mostly toward it, while air flowing opposite its rotation bends mostly away from it. (*b*) The wake deflection force occurs because air flowing around the ball in the direction of its rotation remains attached to the ball longer and the ball's wake is deflected.

this choice of rotation axis, the ball curves to the pitcher's right because the ball experiences two lift forces to the right. One is the Magnus force, named after the German physicist H. G. Magnus (1802–1870) who discovered it. The other is a force we will call the wake deflection force.

The Magnus force occurs because the spinning ball carries some of the viscous air around with it (Fig. 6.2.12*a*). The steady-state flow pattern that forms around this ball is asymmetric: the airstream that moves with the turning surface is much longer than the airstream that moves opposite that surface. Since the longer airstream bends mostly toward the baseball, the average pressure on that side of the ball must be below atmospheric. The shorter airstream bends mostly away from the ball, so the average pressure on that side of the ball must be above atmospheric. Because the pressure forces on the ball's sides don't balance one another, the ball experiences the Magnus force toward the low-pressure side—the side turning toward the pitcher—and deflects in that direction. The airflow deflects in the opposite direction.

In laminar flow, the Magnus force is the only lift force acting on a spinning object. But a pitched baseball has a turbulent wake behind it and is also acted on by the wake deflection force. This force appears when the ball's rapid rotation deforms the wide, symmetric wake (Fig. 6.2.5) that develops behind it at high Reynolds numbers. When the ball isn't spinning, the freely flowing airstream separates from the ball approximately at its side and that separation is symmetric all the way around the ball's middle. But when the ball is spinning (Fig. 6.2.12*b*), the moving surface pushes on the airstream with viscous forces. As a result, airstream separation is delayed on one side of the ball and hastened on the other. The overall wake of air behind the ball is thus deflected to one side and the ball experiences the wake deflection force toward the opposite side—the side turning toward the pitcher. The wake deflection force and the Magnus force both push the ball in the same direction.

Of these two forces, the wake deflection force is probably the more important for a curveball, although the Magnus force is usually given all the credit. A skillful pitcher can make a baseball curve about 0.3 m (12 inches) during its flight from the mound to home plate—the more spin, the more curve. The pitcher counts on this change in direction to confuse the batter. The pitcher can also choose the *direction* of the curve by selecting the axis of the ball's rotation. The ball will always curve toward the side of the ball that is turning toward the pitcher. When thrown by a right-handed pitcher, a proper curveball curves down and to the left, a slider curves horizontally to the left, and a screwball curves down and to the right.

When the pitcher throws a ball with backspin, so that the top of the ball turns toward the pitcher, the ball experiences an upward lift force. In baseball, this force isn't strong enough to overcome gravity, but it does make the pitch hang in the air unusually long and appear to "hop." Not surprisingly, a fastball thrown with strong backspin is called a hanging fastball. A fastball thrown with relatively little spin falls naturally and is called a sinking fastball.

In golf, where the club can give the ball tremendous backspin, the ball really does lift itself upward so that it flies down the fairway like a glider. Professional golfers use soft golf balls that grip the club heads strongly during impact and receive lots of backspin; the ball leaves the driver only about 10° above horizontal yet it remains aloft for a long time because of lift. For the less competent golfer, slices or hooks (horizontal curves toward or away from the golfer's dominant hand, respectively) are the results of various sidespins put on the ball by improper swings. A bad hook can curve the ball 15° in flight and send it into the woods. Novices, hoping to avoid these curves, use hard golf balls that don't grip the club heads well and don't receive much spin of any kind. This choice of ball sacrifices distance in exchange for better control over the direction of flight.

Other games in which spin and lift are crucial to the play include Ping-Pong, volleyball, and soccer. Sometimes hitting the ball straight and hard just isn't enough to get the ball past your opponent. It often helps to spin the ball so that it curves in flight and is harder to stop. The movie title *Bend it like Beckham* refers to the path of a spinning shot on goal in soccer, one that travels in an almost indefensible arc as the result of lift forces.

There are, however, some interesting cases when a ball's behavior stems from its *lack* of spin. In baseball, for example, a knuckleball is thrown by giving the ball almost no rotation. Its seams are then very important. As air passes over a seam, the flow is disturbed so that the ball experiences a sideways aerodynamic force: a lift force. The ball flutters about in a remarkably erratic manner. Releasing the ball without making it spin is difficult and requires great skill. Pitchers who are unable to throw a knuckleball legally sometimes resort to lubricating their fingers so that the ball slips out of their hands without spinning. Like its legal relative, this so-called spitball dithers about and is hard to hit. The same is true for a scuffed ball.

Other ball sports have their equivalents to the knuckleball. In volleyball, a spinless hit is called a floater and is particularly difficult to return. It flies in an unpredictable direction and, because of the volleyball's large surface to mass ratio, it zigzags more effectively than a baseball does.

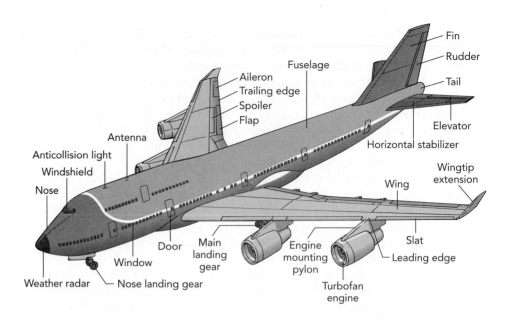

Fin
Rudder
Tail
Fuselage
Aileron
Elevator
Trailing edge
Spoiler
Horizontal stabilizer
Flap
Antenna
Wingtip
extension
Anticollision light
Wing
Windshield
Nose
Slat
Leading edge
Door
Main
landing
gear
Engine
mounting
pylon
Window
Turbofan
engine
Weather radar
Nose landing gear

Section 6.3 **Airplanes**

We have now set the stage for the ultimate aerodynamic machines: airplanes. Freed from contact with the ground, airplanes are affected only by aerodynamic forces and gravity, hopefully in that order. Despite their complex appearances, airplanes employ physical principles that we have already successfully examined. But while this section revisits many familiar concepts, it also explores new territory. For example, you may have already figured out what type of aerodynamic force holds an airplane up, but what type of aerodynamic force keeps it moving forward?

Airplane Wings: Streamlining

By now you've probably realized that an airplane is supported in flight by an upward lift force on its wings and that this lift force comes from deflecting the passing airflow downward. Each wing is an airfoil, an aerodynamically engineered surface that's designed to obtain particular lift and drag forces from the air flowing past it. More specifically, each wing is shaped and oriented so that, during flight, the airstream flowing over the wing bends downward toward its top surface while the airstream under the wing bends downward away from its bottom surface. These bends are associated with pressure changes near the wing itself and are responsible for the upward lift force that suspends the airplane in the sky.

However, to get a more complete understanding of how the wing develops this lift, let's go for a flight. Imagine yourself in an airplane that has just begun rolling down the runway. From your perspective, air is beginning to flow past each of the airplane's wings. When this moving air encounters the wing's leading edge, it separates into two airstreams: one traveling over the wing and the other under it (Fig. 6.3.1). These airstreams continue onward until they leave the wing's trailing edge. Since the airplane's nose is still on the ground, the wing is essentially horizontal and the airflow around it is simple and symmetric.

Since the wing isn't deflecting the airflow yet, it's experiencing no lift, only drag. But while this drag pushes the airplane downwind and thus opposite its forward motion along the runway, it's unusually weak. The wing produces almost no turbulent wake and thus experiences almost no pressure drag. What little drag it does experience is mostly viscous drag, essentially surface friction with the passing air.

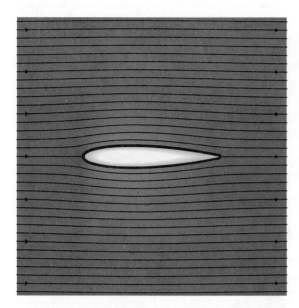

Fig. 6.3.1 An airplane wing is a streamlined airfoil and the airflow around it is laminar. This horizontal wing is symmetric, top and bottom, and the airflow splits evenly into airstreams above and below it. Since it doesn't deflect the airflow, it experiences no lift.

Although the wing's near lack of air resistance should surprise you, you probably take it for granted. That's because you've often observed that such "streamlined" objects cut through the air particularly well. Having a long, tapered tail allows the wing to avoid the flow separation and turbulent wake that occur behind an unstreamlined ball.

What makes the horizontal wing streamlined is the extremely gradual rise in air pressure after its widest point. While this gently rising pressure pushes the wing's boundary layer backward, opposite the direction of flow, the force it exerts is so weak that the layer doesn't stall. Driven onward by viscous forces from the freely flowing airstream, the wing's boundary layer manages to keep moving forward all the way to the wing's trailing edge and never triggers flow separation. The wing produces almost no turbulent wake and experiences almost no pressure drag.

Airplane Wings: Producing Lift

With so little air resistance, the airplane accelerates forward rapidly and soon reaches takeoff speed. The pilot then raises the airplane's nose so that its wings are no longer horizontal and they begin to experience upward lift forces. The airplane's total lift soon exceeds its weight and it begins to accelerate upward into the air. The airplane is flying!

But let's take a closer look at the moment of takeoff. If you could see the airflow and were paying close attention, you'd notice a remarkable sequence of events that begins when the wings tilt upward.

At first, the airflow around the tilted wings continues to travel horizontally on average, although it develops a peculiar shape (Fig. 6.3.2*a*). The two airstreams, one over the tilted wing and one under it, each bend twice—once up and once down. As we saw while studying balls, when an airstream bends toward the wing, the pressure near the wing is below atmospheric and when an airstream bends away from the wing, the pressure near the wing is above atmospheric. Since each airstream bends equally toward and away from the wing, it experiences no overall deflection or average pressure change and provides the wing with no overall lift.

But the lower airstream is making a sharp bend around the wing's trailing edge, essentially an upward kink. Air's inertia makes such a kink unstable and it soon blows away from the wing's trailing edge as a swirling horizontal vortex of air (Fig. 6.3.2*b*). After shedding that vortex, the wing establishes a new, stable flow pattern in which both airstreams pass smoothly away from the wing's trailing edge

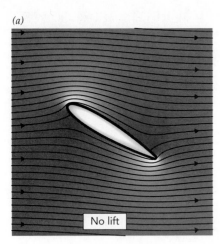

(a)

No lift

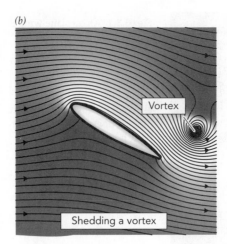

(b)

Vortex

Shedding a vortex

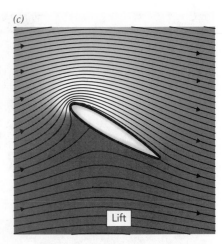

(c)

Lift

Fig. 6.3.2 (*a*) Although this wing's leading edge has been tipped upward, giving it a positive angle of attack, the airflow around it is relatively symmetric and produces no lift. (b) The kink at the trailing edge of the wing is unstable and is blown away or shed as a horizontal vortex. (c) The resulting airflow is deflected downward and the wing experiences an upward lift force.

(Fig. 6.3.2*c*), a situation named the Kutta condition after the German mathematician M. Wilhelm Kutta (1867–1944).

In this new pattern, the airstream flowing over the wing is longer than the airstream flowing under it and both bend downward (Fig. 6.3.3). The upper airstream bends primarily toward the wing, so the air's pressure just above the wing is below atmospheric (a shift toward light shading) and its speed is increased (narrowly spaced streamlines). In contrast, the lower airstream bends primarily away from the wing, so the air's pressure just below the wing is above atmospheric (a shift toward dark shading) and its speed is decreased (widely spaced streamlines). The air pressure is now higher under the wing than over it, so this new flow pattern produces upward lift. The air now supports your plane and up you go.

Another way to think about this lift is as a deflection of the airflow. Air approaches the wing horizontally but leaves it heading somewhat downward. To cause this deflection, the wing must push the airflow downward. In reaction, the airflow pushes the wing upward and produces lift. In other words, the wing transfers downward momentum to the air and is left with upward momentum as a result. These two explanations for lift—the Bernoullian view that lift is caused by a pressure difference above and below the wing and the Newtonian view that lift is caused by a transfer of momentum to the air—are perfectly equivalent and equally valid.

However, the overall aerodynamic force on the wing isn't quite perpendicular to the onrushing air; it tilts slightly downwind. The perpendicular component of this aerodynamic force is lift, but the downwind component is a new type of drag force—induced drag. Induced drag is a consequence of energy conservation: in addition to transferring momentum to the passing air, the wing also transfers some energy to it. The air extracts that energy from the wing by pushing the wing downwind with induced drag and thereby doing negative work on it. Since induced drag is undesirable, the airplane minimizes it by using as much air mass as possible to obtain its lift. A larger mass of air carries away the airplane's unwanted downward momentum while moving downward less quickly and with less kinetic energy. Since larger wings obtain their lift from larger air masses, they experience less induced drag.

Unfortunately, larger wings also have more surface area and experience more viscous drag, so bigger isn't always better. And because wing shape and airspeed affect aerodynamic forces, too, wings must be carefully matched to their airplanes. Small propeller airplanes that move slowly through the air need relatively large,

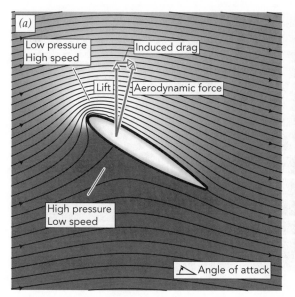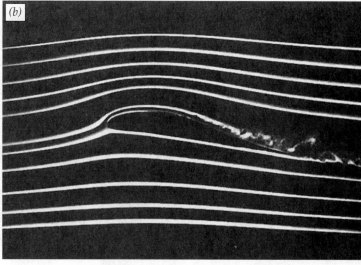

Fig. 6.3.3 (*a*) This airplane wing is shaped and oriented so that both airstreams, over it and under it, bend downward. The wing experiences a large aerodynamic force that points upward and slightly downstream. The upward component of this force is lift. The downstream component is induced drag. (*b*) Smoke trails in a wind tunnel show the airflow past a wing.

highly curved wings to support them. Those wings are often asymmetric—more curved on top than on bottom to make maximum use of the limited, low-speed air they encounter each second. Commercial and military jets fly faster and encounter far more high-speed air each second, so they can get by with relatively small, moderately curved wings. The faster a plane flies, the higher the maximum altitude at which it can support its weight with lift—its flight ceiling.

But even at constant airspeed, a wing's lift can be adjusted by varying its angle of attack—the angle at which it approaches the onrushing air. The larger the angle of attack, the more the two airstreams bend and the greater the wing's lift. Because the wings are rigidly attached to the plane, the pilot has no choice but to tip the entire plane to adjust its lift ❶. The pilot tips the nose of the plane upward to increase the lift and downward to reduce the lift. That's why raising the plane's nose during takeoff is what finally makes the plane leap up into the air.

Since lift depends so strongly on a wing's angle of attack, some planes can be flown upside down. As long as the inverted wing is tilted properly, it obtains upward lift and supports the plane. But this feat is easiest when a plane's wing has the same curvature, top and bottom. That's why stunt fliers who regularly fly upside down often use sport aircraft that have symmetric or nearly symmetric wings.

Lift Has Its Limits: Stalling a Wing

There's a limit to how much lift the pilot can obtain by increasing the wing's angle of attack because tilting the wing gradually transforms it from streamlined to blunt, that is, to having a rapid rise in air pressure after its widest point. As we saw for balls, blunt objects generally experience airflow separation and pressure drag. Indeed, beyond a certain angle of attack, the airstream over the top of a wing separates from its surface and the wing stalls. This separation starts when air in the upper boundary layer is brought to a standstill by the rapidly rising pressure beyond the wing's widest point. Once this boundary layer stalls, it shaves most of the airstream away from the wing's upper surface.

❶ While it may seem that the body of the plane must point roughly in the direction of acceleration, there's no reason why this must be so. If the wings could swivel without taking the body of the plane with them, the pilot would be able to maneuver the plane up and down while the passenger compartment remained perfectly level. Unfortunately, making sturdy wings that swivel without falling off is so hard that essentially all ordinary planes can only tilt their wings by tilting their bodies.

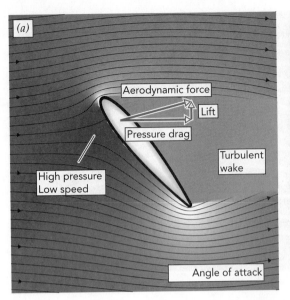

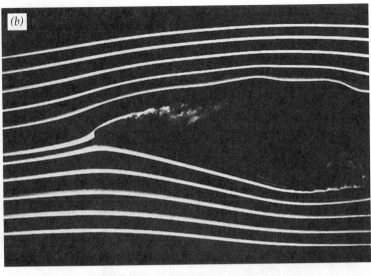

Fig. 6.3.4 (*a*) A wing stalls when the airstream over the top of the wing separates from its surface. A turbulent air pocket forms above the wing, making it much less efficient. The wing's lift decreases because the average pressure above the thickest part of the wing becomes higher, and the drag increases because the average pressure above the trailing edge becomes lower. (*b*) Smoke trails in a wind tunnel show that the air separates from the surface and becomes turbulent as it flows over a stalled wing.

The separated airstream over the top of the stalled wing leaves a billowing storm of turbulence beneath it (Fig. 6.3.4). This airstream separation is an aerodynamic catastrophe for the airplane. Because the average pressure above the wing increases, the wing loses much of its lift. And the appearance of a turbulent wake heralds the arrival of severe pressure drag. The plane slows dramatically and drops like a rock.

To avoid stalling, pilots keep the angle of attack within a safe range. But the possibility of stalling also limits the minimum speed at which the airplane will fly. As the airplane slows down, the pilot must increase its angle of attack to maintain adequate lift. Below a certain speed, the airplane can't obtain that lift without tilting its wings until they stall. It can no longer fly.

To avoid stalling, a plane must never fly slower than this minimum speed, particularly during landings and takeoffs. For a small, propeller-driven plane with highly curved wings, the minimum flight speed is so low that it's rarely an issue. For a commercial jet, however, the minimum airspeed is about 220 km/h (140 mph). Airplanes taking off or landing this fast would require very long runways on which to build up or get rid of speed. Instead, commercial jets have wings that can change shape during flight. Slats move forward and down from the leading edges of the wings, and flaps move back and down from the trailing edges (Fig. 6.3.5). With both slats and flaps extended, the wing becomes larger and more strongly curved, similar to the wings of a small propeller plane, and the minimum safe airspeed drops to a reasonable 150 km/h (95 mph). Vanes near the flaps also emerge during landings to direct high-energy air from beneath the wings onto the flaps. These jets of air keep the boundary layers moving downstream and help prevent stalling. (For another approach to stall prevention, see ❷).

Once a commercial jet lands, flat panels on the top surfaces of its wings are tilted upward and cause the airflow to separate from the tops of the wings. The resulting turbulence created by these spoilers reduces the lift of the wings and increases their drag, so that the plane doesn't accidentally start flying again. Even before landing,

❷ Airplane designers can reduce the dangers of stalling by adding special boundary layer control devices to their aircraft. Narrow metal strips called vortex generators, which stick up from the surfaces of the wings, introduce turbulence into the boundary layers over the wings. This turbulent flow allows higher energy air to mix with the boundary layers so that they can continue forward into rising pressure. This process helps keep the airstreams attached to the surfaces.

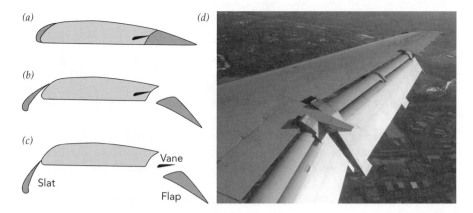

Fig. 6.3.5 At cruising speed, an airplane's wings are moderately curved airfoils (*a*). But during takeoffs (*b*) and landings (*c*), slats are extended from the leading edges and flaps from the trailing edges. The airfoils become much more highly curved, generating more lift at low speeds. During landing (*d*), a vane is also extended for boundary layer control to prevent stalling.

the spoilers are sometimes used to slow the plane and help it descend rapidly toward an airport (Fig. 6.3.6).

In flight, a wing does more than just push the passing air downward; it also twists the air near its tip. Since the air pressure below the wing is greater than the air pressure above it, air tends to flow around the wing's tip from bottom to top. The plane soon leaves this air behind, but not before the air has acquired lots of angular momentum and kinetic energy.

A swirling vortex thus emerges from each wingtip and trails behind the plane for several kilometers, like an invisible tornado. You can occasionally see them behind a plane that's landing or taking off in humid air. A wingtip vortex from a jumbo jet can flip over a small aircraft that flies through it or give passengers in a much larger plane an unexpected thrill. Entered from behind, one of these vortices feels like a horizontal blender; from the side, it feels like a speed hump that you might drive over in a car. I chuckle anxiously each time the plane I'm in encounters one.

For safety, air traffic controllers are careful to keep planes from flying through one another's wakes and schedule them at least 90 s apart on runways. Some modern airplanes have vertical wingtip extensions that reduce these vortices, both to save energy and to diminish the hazard (Fig. 6.3.7).

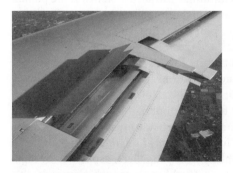

Fig. 6.3.6 This spoiler or air brake is producing a turbulent wake behind the wing. The resulting pressure drag extracts energy from the airplane to help it lose speed and altitude in preparation for landing.

Fig. 6.3.7 This vertical wingtip keeps air from flowing around the end of the wing, a motion that would otherwise leave a powerful vortex in the air trailing behind the plane. Such wingtip vortices waste energy and are hazardous for other aircraft.

Propellers

For a plane to obtain lift, it needs airspeed; air must flow across its wings. And since drag forces push it downwind, a plane in level flight can't maintain its airspeed unless something pushes it upwind. That's why a plane has propellers or jet engines: to push the air backward so that the air pushes the plane forward, action and reaction.

A propeller is an assembly of rotating wings. Extending from its central hub are two or more blades that together form a sophisticated fan (Fig. 6.3.8). These blades have airfoil cross sections and are designed to create forward lift forces when the propeller turns and the blades move through the air.

As a propeller blade slices through the air, the airstreams bending around that blade experience pressure variations (Fig. 6.3.9). The forward airstream bends toward the blade's front surface, so the pressure in front of the blade drops below atmospheric. And the rearward airstream bends away from the blade's rear surface, so the pressure behind the blade rises above atmospheric. The resulting pressure difference exerts a forward lift or thrust force on the propeller.

The propeller blades have all the features, good and bad, of airplane wings. Their thrust increases with size, front-surface curvature, pitch (i.e., angle of attack), and airspeed; in other words, the larger the propeller, the faster it turns, and the more its blades are angled into the wind, the more thrust it produces. The blades themselves have a twisted shape to accommodate the variations in airspeed along their lengths, from hub to tip.

And like a wing, a propeller stalls when the airflow separates from the front surfaces of its blades; it suddenly becomes more of an air-mixer than a propeller. This stalled-wing behavior was the standard operating condition for air and marine propellers before the work of Wilbur Wright in 1902 (see ❸). The Wrights were among the first people to study aerodynamics in a wind tunnel and their methodical and scientific approach to aeronautics allowed them to achieve the first powered flight. Since the Wrights' work, propellers have experienced almost no pressure drag.

A propeller does, however, experience induced drag. As the propeller's thrust pushes the plane through the air, induced drag extracts energy from the propeller. To keep the propeller turning steadily, an engine must do work on the propeller. Propellers are driven by high-performance reciprocating (piston-based) engines, like those found in automobiles, or the turbojet engines that we'll discuss later.

Propellers aren't perfect; they have three serious limitations. First, a propeller exerts a torque on the passing air, so that air exerts a torque on the propeller. This reaction torque can flip a small plane. To minimize torque problems, some planes use pairs of oppositely turning propellers and single-propeller planes usually locate their propellers in front, so that the spinning air can return angular momentum to them while passing over their wings.

A second problem with a propeller is that its thrust diminishes as the plane's forward speed increases. When the airplane is stationary, a propeller blade moves through motionless air (Fig. 6.3.10a). But when the airplane is traveling fast, the air approaches that same propeller blade from the front of the plane (Fig. 6.3.10b). To retain its thrust at higher airspeeds, the propeller blade must increase its pitch, that is, its angle of attack. It must swivel forward to meet the onrushing air.

The third and most discouraging problem with propellers, especially in high-speed aircraft, is drag. To keep up with the onrushing air at high airspeeds, the propellers must turn at phenomenal rates. The tips of the blades must travel so fast that they exceed the speed of sound—the fastest speed at which a fluid such as air can convey forces from one place to another. When the blade tip exceeds this speed, the air near the tip doesn't accelerate until the tip actually hits it. Instead of flowing smoothly around the tip, the air forms a shock wave—a narrow region of

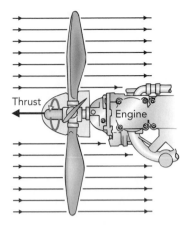

Fig. 6.3.8 A propeller behaves like a rotating wing. As the propeller turns, its blades create lift in the forward direction. This lift pushes the engine and the aircraft forward through the air, so that it's called thrust.

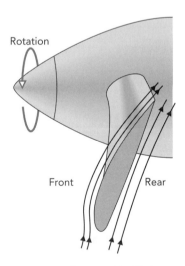

Fig. 6.3.9 As the propeller blade rotates, the flow of air around it creates a low pressure in front of it (left) and a high pressure behind it (right). The blade experiences a lift force that pushes the propeller and plane forward (toward the left). Induced drag extracts energy from the propeller and tends to slow its rotation.

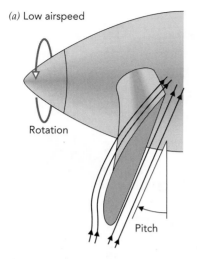

(a) Low airspeed

Rotation

Pitch

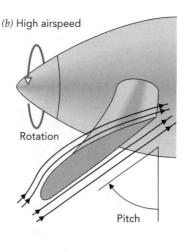

(b) High airspeed

Rotation

Pitch

❸ In addition to achieving the first self-propelled flight of an airplane in 1903, Orville (1871–1948) and Wilbur (1867–1912) Wright were exceptionally accomplished aerodynamicists. In 1902, Wilbur was the first person to recognize that a propeller is actually a rotating wing. Propellers up until his time were little more than rotating paddles, more effective at stirring the air than propelling the plane. Wilbur's aerodynamically redesigned propeller made flight possible and that propeller itself dominated aircraft design for a decade.

Fig. 6.3.10 At low airspeeds (*a*), the propeller blade approaches nearly stationary air as it rotates. At high airspeeds (*b*), air rushes past the propeller, so the blade must swivel forward to meet it. The blade's angle of attack is called its pitch.

high pressure and temperature caused by the supersonic impact—and the propeller stalls. That's why propellers aren't useful on high-speed aircraft.

Jet Engines

Unlike propellers, jet engines work well at high speeds. While a propeller tries to operate directly in the high-speed air approaching the plane, a jet engine first slows this air down to a manageable speed. To achieve this change in speeds, the jet engine makes wonderful use of Bernoulli's equation (p. 142).

Turbojet engines are depicted in Figs. 6.3.11 and 6.3.12. During flight, air rushes into the engine's inlet duct or diffuser at about 800 km/h (500 mph), the speed of the plane. Once inside that diffuser, the air slows down and its pressure increases, but its total energy is unchanged. The air then passes through a series of fanlike compressor blades that push it deeper into the engine, doing work on it and increasing both its pressure and its total energy. By the time the air arrives at the combustion chamber, its pressure is many times atmospheric.

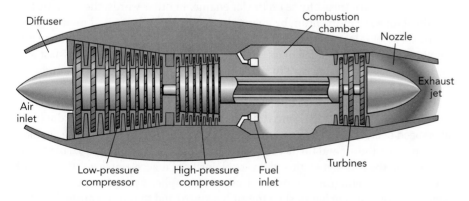

Diffuser

Combustion chamber

Nozzle

Exhaust jet

Air inlet

Low-pressure compressor

High-pressure compressor

Fuel inlet

Turbines

Fig. 6.3.12 This turbojet engine is mounted near the tail of a small commuter jet airplane.

Fig. 6.3.11 The turbojet engine operates by compressing incoming air with a series of fanlike blades. Fuel is mixed with the high-pressure air and the mixture is ignited. The high-energy, high-pressure air accelerates out of the rear of the jet, does work on the turbines, and leaves at a greater speed than when it arrived. The engine has accelerated the air backward and experiences a thrust forward.

Fig. 6.3.14 This turbofan engine hangs below the wing of a jumbo jet. The dark-colored fan duct ends well before the metallic-colored nozzle of the turbojet that powers the fan.

❹ Ramjets are jet engines that have no moving parts. Air that approaches the engine at supersonic speeds interacts with carefully tapered surfaces so that its own forward momentum compresses it to high density. The engine then adds fuel to this pressurized air, ignites the mixture, and allows the hot burned gas to expand out of a nozzle. The engine pushes this exhaust backward, and the exhaust propels the engine and airplane forward. Although the air enters the engine at supersonic speeds, it passes through the combustion chamber much more slowly. In a supersonic combustion ramjet or "scramjet," the fuel and air mixture flows through the combustion chamber at supersonic speeds. This motion makes it extremely difficult to keep the fuel burning because the flame tends to flow downstream and out of the engine. The flame can't advance through the mixture faster than the speed of sound, so it won't spread upstream fast enough to stay in the engine on its own.

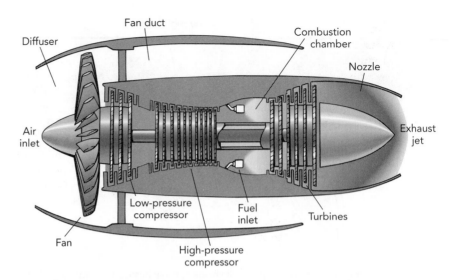

Fig. 6.3.13 The turbofan engine adds a giant fan to the shaft of a normal turbojet engine. Most of the air passing through the fan bypasses the turbojet and returns directly to the airstream around the engine. Because the fan does work on this air, it leaves the engine at a higher speed than it had when it arrived. The air has transferred forward momentum to the engine and the plane.

Now fuel is added to the air and the mixture is ignited. Since hot air is less dense than cold air, the hot exhaust gas takes up more space than it did before combustion. Furthermore, combustion subdivides the fuel molecules into smaller pieces that therefore take up still more volume. This hot exhaust gas pours out of the combustion chamber, traveling faster than when it entered.

The pressure of the exhaust gas is still very high as it streams through a windmill-like turbine. The air does work on that turbine and thereby spins the compressor for the incoming air. After the turbine, the high-pressure gas finally accelerates through the engine's outlet nozzle and emerges into the open sky at atmospheric pressure and extraordinarily high speed.

Overall, the engine slows the air down, adds energy to it, and then lets it accelerate back to high speed. Because the engine has added energy to the air, the air leaves the engine traveling faster than when it arrived. The air's increased backward velocity means that the jet engine has pushed it backward and the air has reacted by exerting a forward thrust force on the jet engine. In other words, the airplane has obtained forward momentum by giving the departing air backward momentum.

The turbojet is less energy efficient than it could be. Since it gives backward momentum to a relatively small mass of air, that air ends up traveling overly fast and with excessive kinetic energy. To make the engine more efficient, it should give backward momentum to a larger mass of air.

The turbofan engine solves this problem by using a turbojet engine to spin a huge fan (Figs. 6.3.13 and 6.3.14). Since this fan is located in the engine's inlet duct, the air's speed decreases and its pressure increases before it enters the fan. The fan then does work on the air and further increases its pressure. While about 5% of this air then enters the turbojet engine, the vast majority of it accelerates out the back of the fan duct and emerges into the open sky at atmospheric pressure and increased speed. Overall, the fan has pushed the air backward and the air has pushed the fan forward, producing forward thrust.

Like a turbojet, the turbofan slows air down, adds energy to it, and then lets it accelerate back to high speed. But because the turbofan engine moves more air than a turbojet engine, it gives that air less energy and uses less fuel. The huge fanlike engines on many jumbo jets are turbofans. (For another type of jet engine, see ❹.)

Stability and Steering

For an airplane to fly from one place to another, it must remain stable in flight and the pilot must be able to steer it. Stability means that the plane will continue to fly nose forward rather than spinning around wildly. As we saw with badminton birdies, aerodynamic stability depends on placing the airplane's center of aerodynamic pressure behind its center of mass. That way, whenever the airplane's orientation is disturbed so that its tail begins to overtake its nose, the air will exert a torque on the airplane to return its tail to the rear, where the tail belongs.

On most airplanes, the tail is a large, three-finned object that moves the center of pressure toward the rear of the airplane. Because the passengers can move about and the cargo can be rearranged, the airplane must be designed carefully so that it never becomes unstable. Seats are assigned with attention not only to weight and balance, but also to keeping the plane's center of mass forward of its center of pressure. Having passengers return to their seats and buckle themselves while the airplane is flying through turbulence also ensures that the airplane remains aerodynamically stable; that way, a severe bump can't toss everyone into the rear of the airplane.

On most airplanes, steering is performed by tilting one or more of five surfaces on the airplane's wings and tail. These movements change the airplane's orientation and the direction of the aerodynamic forces it experiences. The pilot controls the airplane's orientation by moving three types of control surfaces on the airplane: the two ailerons, the two elevators, and the rudder.

The ailerons are small horizontal panels located on the trailing edges of the wings, near their tips (Fig. 6.3.15). By turning the steering yoke as though it were the steering wheel of a car, the pilot tilts the two ailerons in opposite directions: the aileron on one wing tilts up so that it deflects the passing airstream upward while the aileron on the other wing tilts down so that it deflects the airstream downward. These airstreams push back on the wings and produce a torque on the airplane. The airplane undergoes angular acceleration and starts rotating so that one wing descends and the other rises. When the airplane has tilted far enough, the pilot reverses the ailerons and brings the rotation to a halt. To return the airplane to level, the pilot executes this procedure in the opposite order, first causing it to rotate the other way and then halting that rotation when the airplane becomes level.

Fig. 6.3.15 This wing has a wide aileron near the tip of its trailing edge. Although the aileron is presently level with the wing, it can be tilted up or down to exert aerodynamic torques on the plane. Small vortex generators project upward from the top of the wing just upwind of the aileron. By tripping the boundary layer to make it turbulent, these vortex generators ensure that the airstream stays attached to the aileron's upper surface. The small wires trailing off behind the wing are static dissipators that prevent the airplane from accumulating large amounts of electric charge.

❺ Planes that are very stable in flight are inherently hard to turn and maneuver. Their large tails also experience undesirable viscous drag forces. The Wright brothers' planes were designed to be unstable to increase their maneuverability. They tended to turn around or tumble in flight and required great skill to fly. Some modern fighter aircraft and recent commercial jets also obtain enhanced maneuverability and fuel efficiency by being aerodynamically unstable. These aircraft are flyable only with the aid of computers.

The elevators are small horizontal panels located on the trailing edges of the horizontal tail fins. By pushing or pulling on the steering yoke or stick, the pilot tilts the two elevators in the same direction. If the elevators tilt up, they deflect the airstream upward and the tail is pushed downward. The airplane rotates so that its nose becomes higher than its tail. The reverse happens when the elevators are tilted downward.

The rudder is a vertical panel on the trailing edge of the vertical tail fin and is controlled by foot pedals. As the pilot moves the pedals, the rudder swivels to the right or to the left and begins to deflect the airstream in that direction. The airstream pushes back, exerting a torque on the airplane so that it rotates about a vertical axis like an office chair.

An airplane isn't steered like an automobile or a boat. Simply swiveling the airplane's rudder will make its body rotate, but the airplane will end up coasting sideways through the air—not a safe situation. Instead, the pilot steers primarily by tilting one wing lower than the other. Dropping the airplane's inside wing—the wing nearest the center of the turn—and raising its outside wing tilts the airplane's overall lift force to give it a horizontal component. That horizontal lift component points toward the inside of the turn and causes the airplane to accelerate toward the inside of the turn and travel in an arc. The passengers are virtually unaware of this change in direction of flight.

But tilting the wings produces an imbalance in the induced drag forces that the wings experience; as the outside wing acts to obtain more lift in order to rise, it also experiences more induced drag. The rudder allows the pilot to keep the plane from twisting in response to this unbalanced drag forces on its wings. (For a discussion of steering in some modern aircraft, see **❺**.)

🔃 Cabin Pressurization

The earth's atmosphere is so thin at high altitudes that the passengers and crew of an unpressurized airplane flying 10 km above the ground would be sound asleep or worse. Because of this thin-air problem, all modern commercial aircraft are pressurized—air is pumped into the sealed cabin until the air density is about that found at 2500 m (8000 ft) above sea level. The sealed cabin also helps keep out noise, much of which originates in the turbulent airflows outside.

As the air's density increases, so does its pressure. With the pressure inside the cabin far greater than that outside, the cabin's walls experience large outward forces. While the pressurized air keeps everyone healthy, it requires the airplane to have a strong outer hull. Occasionally an aircraft loses pressurization, either because of a leak or because the pressurization system shuts down. When that happens, the passengers and crew find that breathing does them less and less good. Military and commercial aviators are trained to recognize the drunken stupor that sets in when an airplane depressurizes, to quickly don oxygen masks, and to descend to lower altitudes. Depressurization is generally a subtle process, far different from its Hollywood characterizations in movies such as *Goldfinger*.

Of course, the air used to pressurize the cabin has to come from somewhere. The only place in a jet aircraft where high-pressure air is normally found is in the compressors of its turbojet engines. A commercial jet extracts some of this dense, high-pressure air from the engines and lets it flow into the cabin. Unfortunately, the compression process warms the air so much that it must be refrigerated before it's sent into the cabin. That compression warming is an unavoidable result of thermodynamics, one that we'll examine in Chapter 8 when we look at why automobile engines sometimes experience knocking. While it may seem crazy to extract hot air from the engines and chill it to room temperature, there's actually no better alternative.

🡒 Navigation

People who only drive cars or bicycles are used to knowing exactly which way their vehicles are oriented and to following roads and reading signs in order to reach their destinations. But an airplane pilot knows a whole new world of uncertainty. When flying in the clouds, it's easy to lose track of which way you're heading, where you are relative to your destination, and even which way is up. Your sense of up and down doesn't work properly when you are accelerating and it's easy to make the plane accelerate by accident. Any pilot who has flown through a cloud knows that it's almost impossible to keep the plane flying straight and level without looking at the instruments. In the next few paragraphs, we'll examine five important instruments: the artificial horizon, the gyrocompass, the altimeter, the airspeed indicator, and the radio direction finder. There are many other interesting instruments, but I think these will be enough for us both.

For a pilot flying in the clouds or the dark, it's hard to tell when the plane is level or which way it's tilted. She uses an instrument called an artificial horizon to tell her how far she is from level flight. The artificial horizon displays an airplane symbol against a picture of the horizon in front of the plane. If one wing of the airplane is higher than the other, the artificial horizon shows the airplane symbol as tilted. If the airplane's nose is higher then its tail, the airplane symbol rises above the horizon.

Because the effects of gravity and acceleration are indistinguishable, the artificial horizon can't use a measurement of gravity to indicate which way is straight down. Instead, it maintains an accurate internal reference for true vertical, which it constantly compares with the airplane's present orientation. Its reference is a gyroscope—a disk that spins on a nearly frictionless axle with a huge amount of angular momentum. The gyroscope in an artificial horizon spins about a vertical axis. Since angular momentum is conserved, this axis will only change if something exerts a torque on the gyroscope. But the gyroscope is supported in gimbals, freely pivoting supports that isolate it from any outside torques. Once the gyroscope is spinning about a particular axis, it keeps spinning about that axis indefinitely. The gyroscope in an artificial horizon always spins about a vertical axis, even in the clouds. The artificial horizon compares the gyroscope's vertical axis with the airplane's orientation and moves the airplane symbol accordingly.

The pilot also has trouble knowing which direction the airplane is heading. Years ago, a pilot or a ship's captain might have used a magnetic compass to tell him which direction was north. But on modern airplanes, a gyrocompass is the preferred direction indicator. The gyrocompass is another gyroscope that spins about a north–south axis. It's supported in gimbals so that its axis of rotation never changes. A gyrocompass responds to the airplane's motion much more quickly than a magnetic compass and is also far more accurate.

Knowing how far above the ground you are is important when your airplane is traveling through the mountains, when you're trying not to hit another airplane, or when you're trying to land. To determine an airplane's height above sea level, the pilot uses an altimeter—a device that measures the air pressure outside the airplane. Since air pressure decreases with altitude, air pressure is a fairly accurate indication of altitude, although it must be calibrated according to the local weather. The altimeter compares the air pressure on one side of a metal diaphragm with a reference air pressure on the other side. The pressure imbalance exerts a force on the diaphragm. This force is measured, converted to an altitude, and displayed on the altimeter's dial.

The airplane's lift depends on how fast it's moving through the air, not on how fast it's moving relative to the ground. Knowing the airplane's airspeed is particularly important during landings or takeoffs. To determine airspeed, the pilot uses a pitot tube—a rod-shaped device that's mounted on the side of the plane and points forward, into the onrushing air. The pitot tube has two holes through which it can measure air pressure. One hole is located at the front of the pitot tube, where the onrushing air enters and stops. The air pressure in this hole is relatively high because the stopped air has converted all of its kinetic energy into pressure potential energy. The other hole is on the side of the pitot tube, where the onrushing air simply passes by at full speed. The air pressure in this hole is low because most of the air's total energy remains as kinetic energy. The difference in pressures measured in these two holes is directly related to the plane's airspeed. The two air pressures are compared with the help a diaphragm, much like that used in the altimeter. The difference in pressure is converted to airspeed and displayed on the dial of an airspeed indicator.

If a pilot can't see the ground or can't identify the ground he sees, he won't know where he is. To find out his location, he uses a radio direction finder—an instrument on the airplane that studies radio signals from special transmitters on the ground. From these signals, the radio direction finder determines the airplane's position relative to the transmitter.

One of these ground-based transmitters doesn't send its radio waves equally in all directions. Instead, it emits a group of rotating beacons, similar to the light beacon from a seaside lighthouse. Each radio beacon sweeps from north to east to south to west and back to north 15 times per second. The radio transmission also contains information to tell the airplane's radio direction finder when the beacons are aligned with north. By analyzing the pattern of beacons it observes, the radio direction finder can tell whether the airplane is to the north, east, south, or west of the ground-based transmitter. Actually, it can determine the direction from the transmitter to the airplane to about 1°.

By studying the airplane's direction from several ground-based transmitters, the pilot can determine where the airplane is on the map to within a kilometer or two. The special ground-based transmitters, called VOR and VORTAC transmitters, are situated near airports or along common aircraft flight paths and look like giant bowling pins on bowls. In many cases, an airplane flies from transmitter to transmitter as it goes from city to city. The airplane flies directly toward each transmitter, passes over it, and then flies directly away from it. Each radio direction finder tells the pilot that he is approaching the transmitter from the north or east or whatever angle is appropriate and the pilot simply maintains that same angle until the airplane arrives at the transmitter. The airplane leaves the transmitter from the opposite side. The pilot then tunes in a new transmitter and begins to head directly for it.

Alas, while the radio direction finder system has been the backbone of aircraft navigation for decades, it's being superseded by the satellite-based Global Positioning System (GPS). In GPS, a network of satellites in low orbit around the earth transmits precisely timed bursts of radio waves (or, more accurately, microwaves). By measuring the time it takes for those bursts to reach a GPS receiver, the receiver can determine how far it is from each of the satellites. The receiver can then use this information to calculate its position and altitude to within 20 m, anywhere on earth.

The accuracy of the public GPS system was initially limited to about 100 m by encryption instituted by the US Defense Department. This limitation was intended to prevent foreign countries from using GPS for military purposes. The encryption was removed in 2000 and the accuracy of public GPS increased to about 20 m. Its accuracy will continue to improve as more GPS features and radio frequencies are released to the public. Nonetheless, people have developed clever techniques that allow them to use the current public GPS system to measure a receiver's position to within a few millimeters.

⤴ Supersonic Flight

In air, sound is a density disturbance that travels at a speed of about 1193 km/h (331 m/s or 741 mph) in standard conditions (sea level and 0 °C). We'll look more carefully at sound in Section 9.2 on musical instruments, but for now you should know that it consists of density waves—patterns of compressions and rarefactions (i.e., de-compressions) that propagate outward from their source at the speed of sound. Sound is created when something upsets air's normally uniform density. Just as upsetting the uniform surface of a still lake sends water waves rippling off in all directions, so upsetting the uniform density of air sends sound waves rippling off in all directions. When these sound waves reach our ears, we hear them. And just as the water waves take some time to reach the shore, so sound waves take some time to travel from their source to our ears.

When the sound source is stationary, its sound spreads out evenly in all directions. But when the sound source is moving, the pattern of sound is much more complicated. To understand this pattern, we can look at how ripples travel on the surface of a lake. Imagine a stone thrown into the lake. Ripples on the surface head out as nice, concentric circles in all directions (Fig. 6.3.16a). But if the source of the ripples is moving, the ripples bunch up in one direction and spread out in the other (Fig. 6.3.16b). If the source of the ripples is moving very quickly, the ripples will bunch up further in one direction until finally the source exceeds the speed of the ripples. Now the source actually outruns the ripples it generated earlier (Fig. 6.3.16c,d). The pattern of ripples looks more like a triangle than a circle. The triangle's forward edge is a shock wave. Powerboats make shock waves like these whenever they move faster than about 15 km/h (9 mph) (Fig. 6.3.17).

The same kind of ripple behavior happens for sound waves in air. Because sound spreads in three dimensions, rather than the two dimensions available to surface waves on water, the ripples in air are spherical rather than circular. While an airplane is moving slower than the speed of sound, it emits simple sound waves in all directions. But once it travels faster than sound (see ❻), its sound expands in

❻ As they approached the speed of sound during dives, W.W.II aircraft encountered buffeting and stability problems that led people to believe that a plane could not fly faster than the speed of sound. This notion of a "sound barrier" was dispelled when, on October 14, 1947, Capt. Charles Yeager piloted his XS-1 rocket plane to 1.06 times the speed of sound (Mach 1.06). The flight was so uneventful that Yeager, who was flying with two broken ribs from a horse riding accident, could only tell that he had exceeded the speed of sound with the aid of instruments.

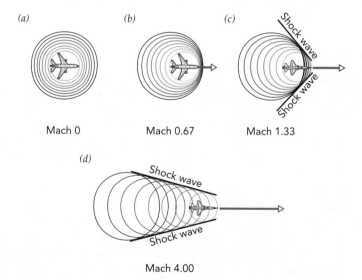

Fig. 6.3.16 (*a*) When a source of sound waves is not moving, the sound spreads out in a series of concentric spheres. (*b*) When the source begins to move, the spheres of sound bunch up in the direction of the source's motion. (*c,d*) Once the source travels faster than the speed of sound, the sphere pattern appears as a cone. The forward edge of the cone is a shock wave. All of the speeds are indicated in Mach (măk), where Mach 1 is the speed of sound. Thus, Mach 0.67 is 0.67 times the speed of sound or about 800 km/h (497 mph).

Fig. 6.3.17 This powerboat is generating a triangular shock wave on the water behind it. The shock wave forms because the boat is outrunning its own surface waves and they trail behind it in a triangular shape.

a cone. This cone extends outward behind the airplane and moves along with it. The outer surface of the cone is a shock wave. When this shock wave passes over observers on the ground, they hear a "sonic boom."

The edge of the cone, the shock wave, carries a great deal of energy. This energy is stored as a large difference in air pressure across the shock wave. The abrupt change in air pressure that occurs when the shock wave passes you can hurt your ears or break your windows. It can also heat objects it touches, particularly the surfaces of a supersonic airplane that are constantly being hit by this shock wave. Supersonic airplanes must be designed so that they don't fly through too many of their own shock waves and so that any surfaces that do fly through shock waves can withstand high temperatures. Since shock waves carry off energy from the airplane, it takes additional thrust to propel an airplane above the speed of sound. Supersonic aircraft are designed to minimize their energy losses due to the formation of shock waves.

⋔ Helicopters also Fly in Air

A helicopter is basically an aircraft body and an engine suspended from a spinning propeller. I previously described a propeller as an assembly of rotating wings, and a helicopter's rotor is exactly that. A basic helicopter uses an engine to exert a torque on the rotor and the rotor obtains lift as it spins through the air. Because the rotor is always moving, even when the helicopter itself is stationary, the rotor can always generate lift. That's why a helicopter can take off and land vertically, and why it can hover.

However, just as a single propeller exerts a torque on an airplane, so the rotor exerts a torque on the helicopter. If the helicopter didn't correct for this torque, its body would begin to rotate in the direction opposite the rotation of its rotor. That's why virtually all helicopters have two rotors. On simple helicopters, the body includes a tail boom with a small rotor attached to it. This second rotor spins about a horizontal axis and producing thrust and torque to counteract the torque from the rotor. Large helicopters support the body on two separate, counter-rotating rotors that cancel out their torque problems.

Supporting the helicopter in the air is pretty straightforward, but how does the helicopter steer or move about? A simple helicopter performs these tasks by changing the pitches of its rotor blades as its rotor turns. The lift experienced by each blade depends on its pitch as it travels through the air (Fig. 6.3.18). Using a

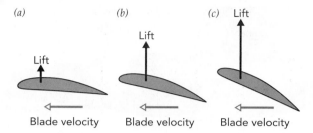

(a) (b) (c)

Lift

Lift

Lift

Blade velocity Blade velocity Blade velocity

Fig. 6.3.18 The blades of a helicopter rotor can be tilted to adjust their pitch. (*a*) When they are tilted to a low angle of attack, they produce a modest amount of lift. (*b,c*) As the angle of attack is increased, the pressure difference between the top and bottom of the blade increases and the blade produces more lift. By increasing the lift of each blade once per rotation as it passes over a certain point of the helicopter's body, the helicopter can be forced to tilt. The helicopter will then accelerate in the direction of the tilt.

mechanism called a swashplate, the helicopter varies the pitch of each blade once per rotation. If the helicopter increases the lift of each blade as it passes over the helicopter's tail, the helicopter's tail will accelerate upward and rise higher than its nose. With the helicopter tilted forward in this manner, part of the rotor's lift force will be directed in the forward direction so the helicopter will accelerate forward. Similarly, increasing the blades' lifts whenever they are in front or over one side will cause the helicopter to tilt and accelerate backward or sideways. And if the pilot wants the helicopter to pivot horizontally so that it faces a new direction, he uses its tail rotor to exert a torque on the body. To create this torque, the pilot changes the pitch of the blades on the tail rotor.

There is one final point worth noting about a helicopter's behavior in the air. When the helicopter is moving forward rapidly, the airspeeds of the various blades aren't always the same. A blade that is swinging forward, toward the front of the helicopter, moves faster through the air than one that is swinging backward, away from the front of the helicopter. Because of these differences in airspeeds, the helicopter tends to experience more lift on one side than it does on the other. This unbalanced lift exerts a torque on the helicopter and causes it to tilt.

The direction of this tilt is counterintuitive because it involves gyroscopic precession. The helicopter's blades together act as a giant gyroscope, with an angular momentum that points directly upward when they are horizontal and spinning counter-clockwise when viewed from above. The torque caused by the unbalanced lift forces points directly backward and causes the direction of the blades' angular momentum to tip toward the rear of the helicopter. This tipping of the angular momentum corresponds to a rise in the front of the helicopter. Fortunately, the pilot of the helicopter perceives this tilting effect and unconsciously corrects for it by adjusting the pitches of the helicopter blades.

⋒ Submarines Fly in Water

You might wonder what submarines are doing in a section on airplanes. I've put submarines here because they basically fly through water much the way airplanes fly through air. While it's true that submarines are supported primarily by buoyancy, they use lift forces to control height and orientation and they turn by tipping the way airplanes do. The last time I mentioned submarines (p. 136), it was to explain how they adjust the mixture of water and air in their ballast tanks in order to float, sink, or hover. This time, I'll explain how submarines orient themselves in the water and how they fly through it on their wings.

Orientation is just as important to a submarine as it is to an airplane; you don't want to be in a submarine that tips upside-down. Even if the submarine is in equi-

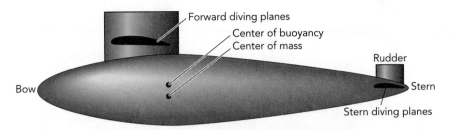

Fig. 6.3.19 This submarine is in stable rotational equilibrium when its center of mass is directly below its center of buoyancy. Its weight is then hanging directly below its support. As the submarine moves forward, its forward and stern diving planes and its rudder can deflect the passing water and obtain lift forces that guide the submarine's path.

❼ Because a ship floating on the water's surface displaces a mixture of air and water, its rotational stability is somewhat more complicated than that of a submarine. Like a submarine, a ship is in rotational equilibrium whenever its center of gravity (i.e., its weight) lies directly below its center of buoyancy (i.e., its support). The ship's center of buoyancy, however, can move as the ship tips and immerses different parts of its hull in the water. That movement of the center of buoyancy affects the ship's stability. A ship with a broad, flat hull has enhanced stability because its shifting center of buoyancy produces strong restoring torques whenever it begins to tip. That's why you can stand on basic surfboards and windsurfers without flipping them over. In contrast, a ship with a round hull has no shifting center of buoyancy and no enhanced stability. To be stable, this ship must have its center of gravity located well below its center of buoyancy; it will flip if it's top-heavy. Improperly loaded cargo ships occasionally do flip, particularly when their cargos aren't tied down well and shift in rough seas. The 19th century clipper ships used to bring back tea from China also brought back heavy clay pottery in the bottoms of their hulls to ensure that their centers of gravity were located well below their centers of buoyancy. And while round-bottomed canoes and kayaks are stable while you're seated in them, they become unstable when you stand up and raise their centers of gravity above their centers of buoyancy.

librium, it will still rotate about its center of mass if it's subjected to a torque. Since the submarine's weight acts at the center of mass (its center of gravity coincides with its center of mass), that weight has no lever arm with which to produce a torque on the submarine. But while gravity can't exert a torque on the submarine, the buoyant force can. The buoyant force can even flip the submarine upside-down.

Like other diffuse forces we have encountered (weight and pressure forces), the buoyant force can be thought of as acting at a single point in the submarine. In keeping with Archimedes' principle, this center of buoyancy is located where the center of mass would be if the submarine were replaced with a submarine-shaped portion of water. The water surrounding the submarine not only pushes up on the submarine with a force that would exactly support a submarine-shaped portion of water, it also acts on the submarine at a point that would exert exactly zero torque on that submarine-shaped portion of water: the portion's center of mass.

The submarine's actual center of mass, however, may be quite different from the center of mass of the water it displaces. The buoyant force can thus produce a torque on the submarine. That torque will cause the submarine to begin rotating and the submarine will eventually settle with its center of buoyancy directly above its center of mass (Fig. 6.3.19). Like a pendulum hanging with its support directly above its center of mass, the submarine will then be in a stable rotational equilibrium. If something tips it away from this rotational equilibrium, it will experience a restoring torque and accelerate back to the equilibrium.

Clearly, a submarine must be carefully balanced so that, when it's level, its center of buoyancy is above its center of mass ❼. But considering the mobility of its crew and the motions of its equipment, the submarine will never be perfectly balanced. With its engines off, a submarine will inevitably tip slowly one way or the other.

To remain level, the submarine needs to move. Once it's heading forward, the submarine behaves like an underwater airplane. It literally flies through the water on two pairs of small wings, assisted by a rudder. Its wings are called diving planes and are technically hydrofoils rather than airfoils. The two forward diving planes are located near the middle of the submarine and perform the same functions as an airplane's wings and ailerons. The two stern diving planes are located at the submarine's stern and perform the same functions as an airplane's horizontal tail fins and elevators. Finally, the rudder is a vertical hydrofoil located at the submarine's stern that performs the same function as an airplane's vertical tail fin and rudder.

Nearly everything I said about airplane stability also applies to submarines. To raise or lower the bow of the submarine, the captain tilts the stern diving planes so that they deflect the passing water up or down, respectively. To swivel the bow of the submarine toward the right or left, the captain rotates the rudder so that it deflects the passing water right or left, respectively. And to tip the submarine to its left side or right side, the captain rotates the forward diving planes in opposite directions so that the passing water twists the submarine in the desired direction.

The captain can make the submarine ascend or descend by changing the lift obtained by the forward diving planes. With both planes deflecting water downward, the submarine experiences a net force upward and it accelerates upward. Similarly, deflecting water upward causes the submarine to accelerate downward. To turn the submarine, the captain tips the submarine to lower the diving plane on the inside of the turn and raise the diving plane on the outside of the turn. The lift forces on these diving planes then have horizontal components that cause the submarine to accelerate in the direction of the turn. At the same time, the captain uses the rudder to keep the submarine pointing in the direction of the turn. (For other watercraft that obtain lift as they move through water, see ❽.)

♒ Sailboats Fly across Water

A sailboat propels itself across the water with the help of two wings, both of which are vertical and obtain horizontal lift forces. One wing is its sail which projects upward into the air and the other wing is its keel which projects downward into the water. While the keel is out of sight and easy to ignore, this underwater wing is surprisingly important to sailing. With only its sail, a sailboat would be able to maneuver only in a variety of downwind directions. With both its sail and its keel, however, the sailboat can actually travel upwind.

Because of its vertical orientation, the sail deflects the passing wind to one side and thereby obtains a horizontal lift force in the opposite direction. The keel or its equivalent is also oriented vertically and deflects the passing water to obtain another horizontal lift force. By carefully coordinating those two lift forces, the sailboat can propel itself in almost any direction.

Let's start with the sail. When wind encounters the sail, it separates into two airstreams in order to flow on both sides of the fabric (Fig. 6.3.20). Depending on the angle of the sail relative to the wind, these two airstreams bend toward or away from the sail and the two sides of the sail experience different air pressures. The sail bows outward toward the lower air pressure and takes on its familiar curved airfoil shape.

The airstream flowing across the sail's inside surface bends away from the sail and therefore develops high pressure (dark shading) and low speed (widely spaced streamlines) near the fabric. The airstream flowing across the sail's outside surface bends toward the sail and therefore develops low pressure (light shading) and high speed (tightly spaced streamlines) near the fabric.

The pressure difference across the fabric produces an aerodynamic force on the sail. That force includes a horizontal lift force pointing away from the convex outside of the sail and an induced drag force pointing directly downwind. Of course, the sail also experiences viscous drag due to surface friction between the wind and the fabric and it may experience pressure drag if it produces a turbulent wake.

A sail is most useful when it produces far more lift than drag. Getting blown directly downwind by drag forces is easy, but being propelled at a right angle to the wind by horizontal lift forces requires the sophistication of a good wing. A skilled sailor knows how to adjust the sail's orientation to maximize lift and minimize drag.

But the sail's lift and drag alone can't propel the sailboat upwind; drag is always downwind and lift is across the wind, so there is no upwind aerodynamic force available. To obtain a net force upwind, the sailboat must begin to deflect the water as well. When the boat is motionless in the water, there are no hydrodynamic forces on its keel. But once the boat begins to move, the keel can deflect water and obtain its own lift force. The direction of that hydrodynamic lift force depends on the keel's orientation and the direction in which the boat is moving. Amazingly, this lift force can point into the wind. By combining aerodynamic forces on its sail

❽ While stationary watercraft depend on buoyancy alone to keep them floating at the water's surface, moving watercraft often supplement buoyant forces with lift forces. Hydrofoil boats and planing-hull boats are designed to deflect the passing water downward to obtain upward lift. These boats rise high in the water when they're moving fast and thus experience less drag. Ultrafast racing boats barely skim the surface, supporting themselves almost entirely with lift forces. Water skiing and wakeboarding are also dependent on lift forces. The transition from floating chest high in the water to skimming the water's surface on water skis or a wakeboard is really the transition from buoyant support to lift support. Once you're moving fast on water skis or a wakeboard, buoyancy is no longer important. At sufficiently high speeds, people can even water ski barefoot; they obtain enough upward lift with their feet alone to support themselves. Lift forces are also important in many less extreme cases, including surfing, windsurfing, and jet skiing. Once these watercraft are moving fast enough, they obtain substantial upward lift forces and rise higher in the water because they no longer need much buoyant support.

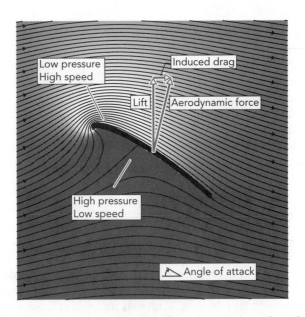

Fig. 6.3.20 Wind flows around a sail, as viewed from directly above the sail. The wind separates into two airstreams, one flowing against the sail's inside surface and the other arcing around the sail's outside surface. The inside airstream bends away from the sail, experiencing a rise in pressure (darker shading) and loss of speed (widening streamline spacing). The outside airstream bends toward the sail, experiencing a drop in pressure (lighter shading) and a rise in speed (narrowing streamline spacing). The pressure imbalance exerts an overall aerodynamic force on the sail, consisting of both a horizontal lift force and a downwind induced drag force.

and hydrodynamic forces on its keel, the sailboat can sail upwind!

If the sailboat's sail experienced only lift from the wind, that aerodynamic lift would push it exactly perpendicular to the wind and the boat would move through the water perpendicular to the wind. And if the boat's keel experienced only lift from the passing water, that hydrodynamic lift would push it exactly perpendicular to its motion through the water, which would be directly upwind. So an ideal sailboat could travel directly into the wind.

No real sailboat can travel directly into the wind because no sailboat can avoid drag forces. Both viscous drag and induced drag are unavoidable and pressure drag is present to some extent, too. The best sailboats can sail almost into the wind, but most have to settle for a few tens of degrees away from directly upwind.

♒ Frisbees®, Aerobies®, and Boomerangs

Even if there were no such thing as lift, baseball and golf would still look familiar. But sports involving flying disks wouldn't be the same at all. Frisbees® and Aerobies® are held aloft by aerodynamic lift and would fall like stones without it.

A Frisbee's lift arises from its shape and orientation as it flies through the air (Fig. 6.3.21). Like an airplane wing, a Frisbee is an airfoil, an aerodynamically engineered surface designed to obtain particular lift and drag forces from the air flowing around it. With its bowed top surface and its leading edge higher than its trailing edge, a thrown Frisbee develops an asymmetric pattern of airflow. Since the airstream passing over its top completes a long bend toward the top, the pressure just above the Frisbee must be below atmospheric pressure. In contrast, the airstream passing under the Frisbee completes a short bend away from its bottom surface so the pressure just below the Frisbee must be above atmospheric pressure.

Because of this difference in pressure on its two surfaces, the Frisbee experiences

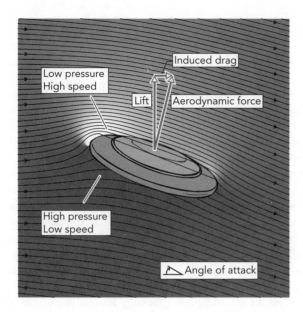

Fig. 6.3.21 Air separates into two airstreams as it flows around a Frisbee. The upper airstream bends toward the Frisbee so the pressure above the Frisbee is low. The lower airstream bends away from the Frisbee so the pressure below the Frisbee is high. The overall aerodynamic force on the Frisbee points up and slightly downwind: an upward lift force and a downwind induced drag force.

an overall aerodynamic force that pushes it upward and slightly toward the rear—a strong upward lift force and a weak rearward induced drag force. The lift force keeps the Frisbee aloft while the drag force gradually slows its forward motion.

That aerodynamic lift force keeps a Frisbee from falling, but what keeps it from tipping over or tumbling through the air? Its principal stabilizer is its rotation. A well-thrown Frisbee is spinning rapidly and has lots of angular momentum about its nearly vertical axis. As long as the air exerts no torque on the Frisbee, it will keep turning about that same axis and won't tip. In fact, the air exerts remarkably little torque on a genuine Frisbee. The Frisbee's center of pressure, the point at which all the aerodynamic forces act, is extremely close to its center of mass. With almost no lever arm between the center of pressure and the center of mass, air pressure can't exert much of a torque on the Frisbee.

When a Frisbee does experience an aerodynamic torque, it generally responds by precessing. As I have discussed before, precession is a phenomenon in which the direction of a spinning object's angular momentum gradually shifts toward the direction of the torque that's exerted on it.

Suppose that you throw a Frisbee so that its angular momentum initially points upward. You can obtain this direction of angular momentum by throwing the Frisbee backhand with your left hand or forehand with your right hand. Most people learn the backhand throw first, so that's probably the one you'd use. If the Frisbee now experiences a rightward torque, perhaps because the air pushes up a little too hard on its leading (or front) edge, its angular momentum will gradually tip from upward to rightward. From your vantage point, the Frisbee's right side will descend and it will veer off to your right.

If instead you throw the Frisbee so that its angular momentum points downward, by throwing it either backhand with your right hand or forehand with your left hand, it will precess the opposite way. Now a rightward aerodynamic torque will gradually tip the Frisbee's angular momentum from downward to rightward. You'll see the Frisbee's left side descend and it will veer off to your left.

Imitation Frisbees usually don't succeed in keeping their centers of pressure coincident with their centers of mass and they experience more serious aerodynamic torques. These ersatz Frisbees tip over quickly in flight as torques from the air rapidly change the directions of their angular momenta. You must spin them faster to keep them flying straight.

An exception to this rule is the Aerobie, a flat, ring-shaped flying disk that flies even better than a genuine Frisbee. Like the Frisbee, an Aerobie has its center of pressure aligned with its center of mass and it doesn't tip over. In a Frisbee, this near-perfect alignment is achieved with the help of turbulence in the bottom of the Frisbee. In an Aerobie, it's achieved with a tiny lip on the outside of the ring. The Frisbee, with its turbulence, experiences far more drag than the Aerobie and doesn't fly as far. If you play catch with an Aerobie, expect to do a lot of walking.

This brings us to boomerangs. Like Frisbees and Aerobies, boomerangs are supported by lift forces so that they don't fall. But while the flying discs try to avoid aerodynamic torques in order to fly straight, boomerangs deliberately obtain torques that cause them to precess. When a boomerang is properly made and thrown, it experiences an aerodynamic lift force that points up and to the side. It also experiences a small aerodynamic torque that causes the horizontal components of its angular momentum and its lift force to pivot gradually around in a circle. The horizontal component of the lift force then acts approximately as a centripetal force and the boomerang follows a circular path through the air. If you get things just right, you can play catch with yourself and not have to do any walking at all!

♒ Swimming: Propelling Yourself across the Water

On land, you propel yourself forward by pushing the ground backward. The ground responds by pushing you forward and it's the ground's force on you that causes you to accelerate forward. But suppose that you are floating motionless at the surface of a still lake. How can you propel yourself forward to reach the distant shore?

If you had a long pole, you could push backward on the lake bottom. The lake bottom would then push you forward and you would accelerate forward. But while this propulsion scheme is fine for some boats, notably trade and passenger boats on shallow rivers and canals (e.g., Venetian gondoliers), it's of no use to you now. With nothing around you but water, you're going to have to make the water itself push you forward. So you put your hand in the water and push it backward. Sure enough, the water pushes you forward and you accelerate forward toward the shore!

We just encountered this same action-reaction concept in the context of airplanes: an airplane propels itself forward by pushing the passing air backward. Although you don't have propellers or jet engines, you do have your hands and legs. As you push the water backward, you accelerate forward. Viewed in terms of momentum, you are propelling yourself forward by giving the water backward momentum. Since momentum is conserved, you are left with an equal amount of momentum in the forward direction and you are now moving across the water.

If the water exerted no further force on your body, you would coast at a steady pace across the lake to the shore. Unfortunately, once you begin to move through the water, you experience drag forces that act to slow you down. You'll have to continue pushing water backward the entire way or you'll come to a stop. To maintain a steady forward speed, the forward thrust force you obtain by pushing water backward with your hands and feet must exactly cancel the backward drag force you experience as your body moves forward through the water. Viewed in terms of momentum, drag keeps taking away your forward momentum so you have to keep replenishing it by pushing the water backwards.

Overall, we have good news and bad news. The good news is that you can propel yourself forward by pushing water backward. The bad news is that swimming requires a great deal of effort. One reason why swimming is so exhausting is that drag forces waste your energy quickly by turning it into thermal energy. To minimize this wasted energy, you need to reduce the drag forces you experience. To lessen pressure drag, you should adopt a streamlined shape and try not to produce much wake. To reduce viscous drag, wear a smooth bathing suit and cover your hair.

Sadly, drag isn't the only effect that saps your energy while swimming. The water that you push backward also carries away some of your energy. But while you must give that water backward momentum in order to obtain the forward momentum you need to keep moving forward, there is no value to giving it energy. Fortunately, you have some control over how much energy you give that water and you can try to give it as little energy as possible.

Drag slows you down by transferring a certain amount of backward momentum to you each second. You propel yourself forward by passing that backward momentum along to the water you push backward. To be an efficient swimmer, you should transfer that backward momentum to as much water as possible, so that the water ends up moving backward slowly and therefore with relatively little kinetic energy. A large mass of water moving backward slowly has less kinetic energy than a small mass of water moving backward rapidly, even though the two carry away the same backward momentum. That's because water's momentum is proportional to its speed while its kinetic energy is proportional to the square of its speed. For this reason, it's more energy efficient for you to push a large mass of water backward slowly than to push a small mass of water backward rapidly—you experience the same forward force but give the water less energy in the process.

Part of the art of swimming well is learning to push large masses of water directly backwards at modest speeds, rather than churning the water about at wild angles and high speeds. In this manner, you can give the water backward momentum without transferring much energy to it. And when you're snorkeling and want to propel yourself primarily with your feet, you do well to wear flippers. Flippers increase the surface areas of your feet so that you can move more water backward at lower speeds and save your energy.

I have always been astonished whenever a good swimmer cruises past me at twice my speed. To understand my amazement, you must realize several things about what this person is doing. First, the pressure drag force is roughly proportional to speed squared; if you travel twice as fast through water, you encounter twice as much water each second and you leave it moving twice as fast in your turbulent wake. Two times two is four, four times the momentum transferred to the water and four times the drag force. So a swimmer going twice as fast as me, all else being equal, is fighting four times the drag force. To make things even worse, that swimmer is traveling twice as far each second and therefore doing eight times as much work against the force of pressure drag. The swimmer is thus producing eight times as much swimming power as I am. I'm not a great athlete, but this exhibition of raw power still boggles my mind.

Actually, much of what that swimmer is doing is being efficient. While I flail around with imperfect strokes, the swimmer is pushing the water backward carefully and efficiently. The swimmer is also minimizing pressure drag by leaving a much smaller wake than I leave. One of these days I'll learn their tricks and become a better swimmer myself. In the meantime, I'll have to settle for writing about them.

Powerboats also propel themselves forward by pushing water backward. They, too, are most energy efficient when they move large masses of water directly backward. A typical powerboat pushes on the water with a propeller. As the propeller spins, the nearby water accelerates backward. The water in turn pushes the propeller forward and the boat accelerates forward. Modern boat propellers, like airplane propellers, are designed as rotating wings rather than paddles and are therefore efficient and effective ❾.

❾ Like powerboats, submarines use propellers to obtain forward thrust. Until recently, those propellers made so much noise that a submarine could be heard underwater from miles away. The problem was that conventional propellers produced both turbulence and cavitation—the formation of empty cavities in the water that subsequently collapse violently and noisily. To reduce this propeller noise, modern submarines have extremely sophisticated propellers that minimize turbulence and avoid cavitation. These propellers maintain smooth, laminar flow over their entire surfaces and exhibit no flow separations. Creating such a perfect propeller requires very complex surface structuring that can only be achieved using computer-controlled metal-shaping machines.

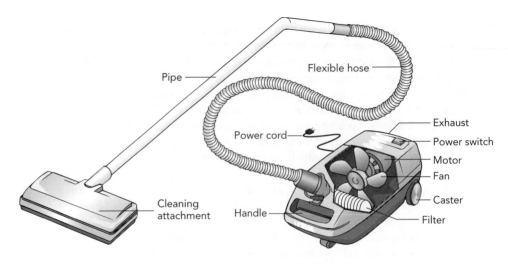

Pipe

Flexible hose

Power cord

Exhaust

Power switch

Motor

Fan

Caster

Filter

Cleaning
attachment

Handle

SECTION 6.4 **Vacuum Cleaners**

When we looked at garden watering in Section 6.1, we examined tools that permit a fluid to flow out of a hose. In this section, we'll look at the reverse, a device that draws a fluid into a hose. That device is a vacuum cleaner, and the fluid that it draws inward is air. This moving air gathers dust and debris as it rushes into the vacuum cleaner, which is why vacuum cleaners are useful.

Air Flowing into the Vacuum Cleaner

Vacuum cleaners use swiftly moving air to sweep up dust. To understand them, you'll need to know how they create that moving air and why dust is so easily carried along by it. But let's forget about the dust for now and look at how air itself flows into a vacuum cleaner. In particular, let's examine the airflow in a canister vacuum cleaner with a long hose.

For simplicity, I'm going to reduce the machine to its basics: a hose and a fan that draws air through that hose (Fig. 6.4.1). Outside the hose, the air is stationary at atmospheric pressure. When you turn the fan on, the pressure inside the hose drops and a partial vacuum is created (hence the name "vacuum cleaner"). Since air accelerates from higher pressure toward lower pressure, outside air accelerates toward the hose and rushes into its opening. After all, nature abhors a vacuum.

After a second or two, the air in the hose achieves steady-state flow and we can begin to use Bernoulli's equation to examine its pressures and speeds at various points along the streamlines. We can't, however, use Bernoulli's equation to compare the air before and after the fan because Bernoulli's equation requires that all obstacles be stationary and the fan is obviously moving. Most importantly, the fan does do work on the air and therefore increases the air's total energy. But upstream of the fan, Bernoulli's equation will give us some good insight into the airflow in the hose.

The vacuum cleaner operates at a constant altitude, so the air's gravitational potential energy isn't important and we should expect the sum of the air's pressure potential energy and kinetic energy to be constant along a streamline. As the air flowing along a streamline loses pressure, it gains speed; as it gains pressure, it loses speed. Since the pressure well inside the vacuum cleaner hose is low, air accelerates toward the hose's inlet and its speed increases as its pressure drops (Fig. 6.4.1). Even though it doesn't look like one, the inlet to the hose is behaving as a nozzle: the streamlines flowing into it are becoming more tightly spaced, so the air's speed is increasing. At the same time, its pressure is dropping.

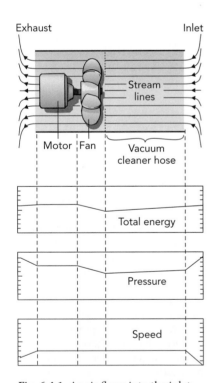

Exhaust

Inlet

Stream
lines

Motor

Fan

Vacuum
cleaner hose

Total energy

Pressure

Speed

Fig. 6.4.1 As air flows into the inlet of a vacuum cleaner hose, its pressure drops and its speed increases. The fan boosts the air's total energy, helping it to overcome viscous losses of total energy so that it can return to the outside air through the outlet.

Once inside the hose, the air continues at high speed and low pressure until it encounters the fan. The fan's job is to increase the air's total energy, not only so that it blows quickly out of the exhaust port but also to make up for energy lost to frictionlike effects in the hose. We'll look at those frictionlike effects and how the fan compensates for them later on.

Because the fan's inlet and outlet have the same diameter, air's speed can't change as it moves through the fan—the air would bunch up or tear apart if it slowed down or sped up, respectively. Instead, its pressure and pressure potential energy increase. Low-pressure air enters the fan through the hose and higher-pressure air leaves the fan to head out through the exhaust port. The fan thus maintains a pressure imbalance between the vacuum cleaner's hose and its exhaust port. The low pressure inside the hose is what got the air moving in the first place!

While the air leaving the fan may still be below atmospheric pressure, it has plenty of kinetic energy. As this air flows out of the exhaust port, its streamlines spread out (Fig. 6.4.1). The exhaust port is acting as a diffuser, allowing the streamlines to become more widely separated so that the air's speed drops and its pressure rises; the air is converting kinetic energy into pressure potential energy. Eventually, the air's pressure rises to atmospheric pressure and it reenters the room. The air has completed its trip through the vacuum cleaner.

But when you add a narrow cleaning attachment to the hose (Fig. 6.4.2), the airflow becomes more complicated. For the air to continue flowing through the fan at the same rate as before, it must rush rapidly through the narrow channel of the attachment. The streamlines bunch tightly together, indicating a dramatic rise in speed and a severe drop in pressure.

The dramatic increase in speed and drop in pressure that occur when a steady flow of fluid passes through a narrow channel is called the Venturi effect, after its discoverer, Italian physicist G. B. Venturi (1746–1822). The Venturi effect is a special case of the Bernoulli effect, which recognizes that any increase in a fluid's velocity along a streamline is accompanied by a drop in pressure.

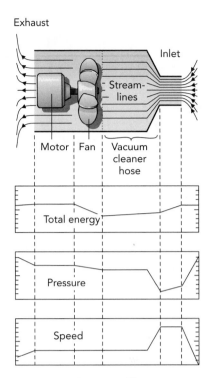

Fig. 6.4.2 When you add a narrow cleaning attachment to the end of the vacuum cleaner hose, the air in the inlet reaches very high speed and relatively low pressure. The attachment's narrow channel is kept relatively short to minimize the loss of total energy to viscous forces.

Dust and Drag Forces

Now that we understand speed and pressure in a vacuum cleaner, let's turn our attention to the dust. As air rushes toward the opening in the cleaning attachment, it carries dust with it. This phenomenon, in which a particle or portion of fluid is carried along in the flow of another fluid, is called entrainment. Dust entrainment is most effective in very high-speed air, which is why a narrow attachment cleans a carpet more thoroughly than a wide one.

Dust particles are entrained in air by viscous drag forces. Whether the dust moves through the air or the air moves past the dust, viscous drag acts to bring the dust to rest relative to the air. Pressure drag is absent because a dust particle is too small to produce a turbulent wake in the air; the Reynolds number for air flowing around a dust particle is simply too tiny and the flow is purely laminar.

The viscous drag force that the air exerts on a dust particle is proportional to the diameter of the particle and to the difference in velocities between the particle and the air. This relationship comes about because the amount of air the particle rubs against is proportional to its diameter (or girth) and to its velocity through the air. As always, the viscous drag force on the particle is directed so as to bring the particle to the same velocity as the air. Furthermore, the particle pushes back on the air with an equal but oppositely directed viscous drag force.

Because dust particles are often just tiny rocks, we might expect viscous drag to affect rocks and dust particles equally. But that isn't so. At issue is how large the drag force is in comparison to the object's weight and mass. Although a dust particle experiences a smaller viscous drag force than a rock does, it nonetheless responds more than the rock does. That's because the viscous drag force is proportional to diameter while mass and weight are proportional to diameter to the *third power*. Think of how many dust particles you can produce by grinding up even a small rock, and consider how light each of those dust particles must be. With so little weight to hold them in place yet so much surface area interacting with the air, it's not surprising that viscous drag forces easily blow dust particles about.

Dust particles are influenced so strongly by passing air that they are easily borne aloft by air currents. Dust swept into the air by winds, emitted by industrial smokestacks, or blown upward in volcanic eruptions can remain in the atmosphere for days, weeks, or even years. For example, volcanic ash from the 1980 explosion of Mount St. Helens in Washington State was carried through the Midwest and even to the eastern United States.

Although dust tries to fall, viscous drag keeps it from descending rapidly. Like any object falling through the atmosphere, dust has a terminal velocity—the velocity at which the upward drag force on it exactly cancels its downward weight. Because of its large surface to weight ratio, a dust particle's terminal velocity may be 1 mm per second or less. Any upward air current will therefore lift it back into the sky. In a calm, sunlit room, you can often see dust particles drifting about with the air currents, prevented from falling by the viscous drag force.

This same viscous drag force is what lets air carry dust particles into the vacuum cleaner. The force acts to reduce any difference in velocities between the air and the dust, so if the air rushes into the vacuum cleaner, the dust will, too. The faster the air moves, the larger the viscous drag force on a particle. This increase in force with air speed explains why a narrow attachment cleans better than the hose alone: the air speed is higher near the attachment and the viscous drag forces are larger.

Unfortunately, viscous drag also slows the air as it passes close to carpet fibers or the surface of the floor. It's hard to keep air moving quickly near surfaces because those surfaces develop boundary layers of slower-moving air. Removing really ground-in dirt from a carpet or floor requires a powerful fan and the high airspeed that comes from making air pass through a narrow opening. This need for high airspeed also explains why battery-powered or poor-quality vacuum cleaners don't clean well: their fans are too weak to move the air fast enough to remove the dirt effectively, particularly near surfaces. A beater brush, which uses inertia to jostle the dirt away from surfaces, helps to move the dirt out of the boundary layer and into the fast-moving air stream so that it can be carried into the vacuum cleaner.

All this explains how vacuum cleaners pick up tiny particles of dust, but what about larger objects? If you try to vacuum up a marble, you'll soon discover that the drag forces acting on it are too weak to pick it up. After sweeping over the marble a couple of times unsuccessfully, you'll probably pick up the marble by hand and drop it into the hose.

The marble simply weighs too much relative to its interactions with the air. Actually, the marble is large enough to experience pressure drag in the flowing airstream; the Reynolds number for a marble is much greater than for a dust particle and a turbulent wake forms in the air flowing behind the marble. But despite the presence of pressure drag, the airstream still can't carry the heavy marble into the vacuum cleaner. A vacuum cleaner is meant for collecting dust, not marbles.

The Fan

We've looked at how a vacuum cleaner picks up dust; now it's time to look at the fan. Without the fan, viscous drag forces would quickly stop air from flowing through the cleaning hose by converting its total energy into thermal energy. This effect is most severe when you use a narrow cleaning attachment because the high-speed air inside that attachment loses total energy particularly quickly. But even a long hose that's not wide enough will waste lots of energy; the excessively long hoses used by some truck-based cleaning services are not helping them clean effectively.

To keep air moving through the hose, the fan pumps the air from the low-pressure region in the cleaning hose to the high-pressure region at the exhaust port, against its natural direction of flow (Fig. 6.4.3). The fan does work on the air, replacing the energy that viscous drag has converted into thermal energy and adding some extra energy for increased speed at the exhaust port. Because the fan increases the air's total energy, the air passing through the fan is *not* in steady-state flow; the air's pressure increases as it flows through the fan without a corresponding decrease in speed.

In its basic form, the fan is just a rotating wing assembly, a propeller. Its moving wings do work on the air, using energy provided to an electric motor by the electric power company. What's interesting here is that work is being done by a rotational motion, not the translational motion that we normally associate with work. I've sidestepped this issue before a number of time, but now I'll take it on directly. We've seen that you do work when you exert a force on something and it moves a distance in the direction of that force. But you also do work when you exert a torque on something and it rotates through an angle in the direction of that torque.

The amount of work that you do in this manner is equal to the product of the torque you exert on the object times the angle through which it rotates in the direction of that torque, or:

$$\text{work} = \text{torque} \cdot \text{angle of rotation.} \qquad (6.4.1)$$

In this relationship, the angle of rotation is in the direction of the torque. If the object turns in the direction opposite your torque, it does work on you! To calculate the work you do on something when you twist it, you must measure the angle of rotation in the natural units of angle: radians. One radian is equal to $180/\pi$ degrees (about 57.3°). While degrees are probably more familiar to you than radians, using degrees in Eq. 6.4.1 will lead to incorrect results.

It's not hard to see why work is the product of a torque times angle (in radians). Think of pedaling a bicycle: the work you do on the pedal is the product of the force you exert on that pedal times the distance it travels in the direction of that force. Since the pedal follows a circular path, the distance it travels is the product of its lever arm times the angle through which it turns, measured in radians. But the force you exert on the pedal times the pedal's lever arm is actually the torque you exert on the pedal. So the work you do on the pedal can also be viewed as the product of the torque you exert on it times the angle through which it turns.

Eq. 6.4.1 is useful in devices such as bicycles and electric motors, where energy is transferred via rotational motion. At present, however, you need only recognize that the motor does work on the fan via a torque and that the fan uses this work to pump air from low pressure to high pressure.

We can now explain the high-pitched whine that most vacuum cleaners emit when you block their airflows. If air can't reach the fan's low-pressure side, the fan can't pump any air to its high-pressure side. Since the fan moves no air, it does no work and offers no resistance to the motor. The motor quickly exceeds its rated rotational speed and begins to whine. When you unblock the airflow, the fan and motor begin to do work on the air, the motor slows down, and the whine disappears.

Filtering the Dust

A vacuum cleaner's fan produces the moving stream of air that entrains the dust. But the vacuum cleaner eventually returns that air to the room. What prevents the dust from returning to the room along with the air?

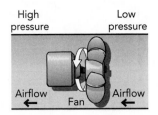

Fig. 6.4.3 As a vacuum cleaner's fan rotates, its blades transfer air from the low-pressure side to the high-pressure side, doing work on the air in the process and increasing its total energy.

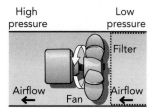

Fig. 6.4.4 In a canister vacuum cleaner, air flowing through the hose first passes through the filter and then through the fan. The fan pumps air from the low-pressure region inside the vacuum cleaner to the atmospheric pressure outside.

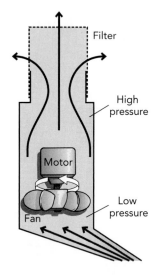

Fig. 6.4.5 In an upright vacuum cleaner, air flowing up from the carpet is pumped by the fan and flows out into the room through the filter.

The answer, in most cases, is a filter, a device that blocks the dust particles while permitting air molecules to pass. A typical filter is made of porous paper or cloth, with fibers that are loosely woven to create openings or pores large enough to pass air, but too small to pass dust.

This simple filtration is complicated by viscous drag. Air passing through the filter's pores experiences viscous drag and loses some of its total energy. The narrower and longer the pores, the more the air rubs against their surfaces and the worse the viscous energy loss. When the filter is new, the pores are wide, short, and abundant, so the air loses relatively little energy. The filter removes dust from the air without much effect on the air itself, so the vacuum cleaner works well. But as dust gradually accumulates on the filter, its pores become narrower, longer, and less abundant and the viscous energy loss rises. The air loses much of its energy as it struggles to pass through the clogged filter so the vacuum cleaner doesn't clean well.

To address this gradual "loss of suction" problem, the most recent form of vacuum cleaner abandons porous filters altogether and uses rapid acceleration to collect the dust. In a Dyson™ vacuum cleaner, dust carried into the unit is separated from air by whirling the mixture around in a tight spiral. Since such circular motion involves a centrally directed or centripetal acceleration, each substance needs a central force to bend its normally straight path into a spiral. The air negotiates this spiral easily. It spontaneously develops a pressure gradient—high pressure on the outside of the spiral and low pressure near the center of the spiral, like water following the bend in the hose shown in Fig 6.1.2. That pressure gradient pushes the air inward and steers the flow gracefully so that the air travels in the desired spiral motion.

The dust particles, however, are too dense to follow the spiraling air. Neither the force due to the air's pressure gradient nor the viscous drag force is strong enough to bend a dust particle's path into the required spiral. As the air and dust swirl around, the dust gradually drifts outward; its inertia wins out over the weak centripetal forces. The dust eventually leaves the spiraling airstream altogether and is collected for disposal. The air successfully completes its spiral trip and is then returned to the room.

Vacuum cleaners come in two main types: those that filter the air *before* the fan and those that filter the air *after* the fan. Conventional canister vacuum cleaners and the Dyson vacuum cleaners both filter the air before the fan (Fig. 6.4.4). In these devices, the fan's principal job is to increase the pressure of air leaving the filtering system so that it has enough energy to flow out into the room and the air leaves the fan near atmospheric pressure.

In contrast, upright vacuum cleaners place the filter after the fan. Air flowing up from the carpet passes through the fan, around the motor, and finally through the filter (Fig. 6.4.5). Since the fan in this device is responsible for blowing the air through the filtering system and then into the room, air leaves it well above atmospheric pressure.

Perhaps the biggest difference between filter-before-fan and fan-before-filter vacuum cleaners is in what happens when you vacuum up an object like a penny. In a filter-before-fan vacuum cleaner, the penny stops in the filter and never gets into the fan. In a fan-before-filter vacuum cleaner, the penny flies right through the fan on its way to the filter. Since a penny is fairly heavy and the vacuum cleaner has trouble moving it with drag forces, the penny tends to rattle around in the fan, making a terrific racket. If you use an upright vacuum cleaner, you probably recognize that sound all too well.

⚏ Hairdryers and Diffusers

A hairdryer resembles a short, stubby vacuum cleaner with a heating element just upstream from its outlet port. The hairdryer's fan creates a partial vacuum inside the unit, so that room air rushes into its inlet through a lint filter. This inflowing air exchanges pressure potential energy for kinetic energy and moves rapidly toward the fan. The fan does work on the air and increases its total energy. The air continues on through a heating element, which transfers thermal energy to it and increases its temperature. Finally, the heated air emerges from the outlet and sprays toward your hair.

As the air emerges from the hairdryer's outlet, it spreads out sideways so that its streamlines separate slightly. The air's speed therefore decreases and its pressure rises. Because of that final rise in pressure at the outlet port, the air can actually exit the hairdryer's fan at slightly less than atmospheric pressure and still manage to flow out into the room air. It sprays out warm and fast from the outlet port and blows your hair dry.

If you don't want such fast moving air, you can attach a diffuser to the normal outlet. The diffuser's reversed-nozzle shape allows the streamlines passing through it to separate widely so that the air flowing along each streamline experiences a dramatic loss of speed and increase in pressure. With a diffuser attached to the hairdryer, the air leaves the fan well below atmospheric pressure and its pressure rises to atmospheric pressure only near the end of the diffuser, when most of the air's kinetic energy has been transformed into pressure potential energy. Air emerges from the diffuser traveling relatively slowly so that the hairdryer doesn't blow your hair around as violently as it would without the diffuser.

⚑ Wind Power and Water Power

A vacuum cleaner's fan uses the rotational work provided by a motor to increase the total energy of the air passing through it. A windmill or water turbine decreases the total energy of the wind or water passing through it in order to provide rotational work to a generator. They're the reverse of one another!

It should come as no surprise then that fans, propellers, turbines, and windmills are all pretty much the same; they're all rotating wings. If you supply them with rotational work, they'll add energy to the fluid passing through them and increase that fluid's speed or pressure or both. And if you extract rotational work from them, they'll remove energy from the fluid passing through them and decrease that fluid's speed or pressure or both. At an abstract level then, I've said just about all there is to say about wind power and water power.

Practical windmills and water turbines, however, require careful engineering in order to be efficient. Since they use lift forces to extract energy from the passing fluid (they're the reverse of airplane propellers, p. 174, which use lift forces to add energy to the passing fluid), they do well to avoid drag forces whenever possible. If they experience strong viscous or pressure drag forces, they'll waste lots of energy heating the passing fluids.

They can limit viscous drag forces by reducing their surface areas and they can minimize pressure drag by maintaining laminar flow or at least minimal turbulence throughout their structures. They must also interact with and extract energy from as much fluid as possible, so they have to orient their wings carefully with respect to the onrushing wind or water. Much of the improvements in wind and water power systems have come from more efficient and effective wing assembles.

Chapter 7

Thermal Things

We can't see all of the motion that takes place around us. Some of it is hidden deep inside each object, where thermal energy keeps the individual atoms and molecules jiggling back and forth in an endless flurry of activity. We're usually aware of this thermal energy only because it determines an object's temperature; the more thermal energy an object contains, the higher its temperature and the hotter it feels.

However, thermal energy plays an important role in everyday life. In addition to moving from one place to another, thermal energy can transform a substance from a solid to a liquid to a gas. What you're feeling when you touch a hot object is actually its thermal energy flowing into your comparatively colder hand and raising the temperature of your skin. When thermal energy is flowing in this manner, from a hotter object to a colder one, we call it heat.

Heat and fire were great mysteries until the last few centuries. Though they're now well understood scientifically, they still retain much of their ancient mystique. We still gather around campfires and stare into them as though viewing some elemental part of nature. And it's hard not to feel a thrill while watching fireworks cast their hot sparks about the sky.

Like Prometheus, we want fire in our lives and can't always resist its appeal. Why else would we have the expression "playing with fire"? If you're like me, you've probably had a few close calls over the years; my parents were lucky they didn't know how often I burnt my fingers or singed my clothes with "scientific" experiments gone awry. In this chapter, we'll examine not only fire, but also temperature, heat, and the phases of matter in order to understand more about our hot and cold world.

Chapter Itinerary

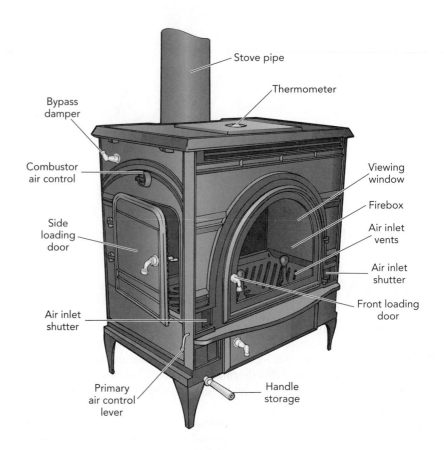

Stove pipe

Thermometer

Bypass damper

Combustor air control

Viewing window

Firebox

Air inlet vents

Side loading door

Air inlet shutter

Air inlet shutter

Front loading door

Primary air control lever

Handle storage

SECTION 7.1 **Woodstoves**

Winter would be pretty unpleasant for most of us were it not for heating. Heating keeps our rooms warm even when the weather outside is cold. One of the most fashionable types of heating is a woodstove, which burns logs in its firebox and sends thermal energy out into the room. In this section we'll look at how a woodstove produces thermal energy and how this thermal energy flows out of the stove to keep us warm.

A Burning Log: Thermal Energy

A woodstove produces thermal energy and distributes it as heat to the surrounding room (Fig. 7.1.1). We've encountered thermal energy before: in a file cabinet sliding along the sidewalk, in an old ball bouncing inefficiently off the floor, and in honey pouring slowly from a jar. In each case, ordered energy—energy that could easily be used to do work—became disordered thermal energy and the temperatures of the objects increased. But now that we're going to study devices that are intended to provide heat, let's reexamine thermal energy and temperature to see how thermal energy moves from one object to another.

When you burn a log in the fireplace or woodstove, you're turning the log's ordered chemical potential energy into thermal energy. Thermal energy is a disordered form of energy contained in the kinetic and potential energies of individual atoms and molecules. The presence of thermal energy in the log, the woodstove, or the room air is what gives it a temperature; the more thermal energy it has, the higher its temperature.

The nature of thermal energy depends somewhat on what it's in. In the hot, burning log, thermal energy is mostly in the wood's atoms and molecules, which

Fig. 7.1.1 This wood-stove transfers heat to the room by conduction through its metal walls, convection of air past its surfaces, and radiation from its black exterior.

jitter back and forth rapidly relative to one another. When each of these particles moves, it has kinetic energy. When it pushes or pulls on its neighbors, it has potential energy.

In the air near the burning log, thermal energy is again mostly in the atoms and molecules. But since those particles are essentially free and independent, most of this thermal energy is kinetic energy. The air particles store potential energy only during the brief moments when they collide with one another.

In the metal poker that you use to stir the fire, thermal energy is not only in the atoms and molecules, but also in the mobile electrons that move about the metal and allow it to conduct electricity.

While it's important to know what energy is thermal energy, it's also important to know what energy is not thermal energy. The log's thermal energy includes only its internal disordered energy and not the energy that's associated with the log as a whole. Moving the log with the poker to increase its kinetic energy, lifting it with tongs to increase its gravitational potential energy, and bending it with another log to increase its elastic potential energy all increase its energy as a whole and not its thermal energy.

Forces between Atoms: Chemical Bonds

To understand how a burning log produces thermal energy, let's take a look at bonds between atoms and the chemical potential energy that's stored in those bonds. Since both result from the forces between atoms, that's where we'll begin.

As you bring two atoms close together, they exert attractive forces on one another (Fig. 7.1.2a). These chemical forces are electromagnetic in origin and grow stronger as the atoms approach. But the attraction diminishes when the atoms start to touch and is eventually replaced by repulsion when the atoms are too close (Fig. 7.1.2b). The separation between atoms at which the attraction ends and the repulsion begins is their equilibrium separation, that is, the separation at which the atoms exert no forces on one another (Fig. 7.1.2c). Since atoms are tiny, this equilibrium separation is also tiny, typically only about a ten-billionth of a meter.

Imagine holding two atoms in tweezers and slowly bringing them together. They pull toward one another as they approach, doing work on you and increasing your energy. Since energy is conserved, their energy must be decreasing. They're giving up chemical potential energy—energy stored in the chemical forces between atoms.

Once the atoms reach their equilibrium separation, you can let go of them and they won't come apart. Like two balls attached by a spring, the atoms are in a stable equilibrium. Since they've given up some of their chemical potential energy, they can't separate unless that energy is returned to them. It takes work to pull them apart, so the atoms are held together by a chemical bond.

The bound atoms have become a molecule. The strength of their bond is equal to the amount of work the atoms did when they drew together or, equivalently, the work required to separate them. Bond strengths range from extremely strong in the case of two nitrogen atoms to extremely weak in the case of two neon atoms. I'll explain the different types of chemical bonds later in this book, particularly in Chapter 18 when we take a look at chemical physics.

If they have a little extra energy, bound atoms can vibrate back and forth about their equilibrium separation (Fig. 7.1.2d). Whenever the atoms are moving quickly toward or away from one another, most of their energy is kinetic. Whenever the atoms are traveling slowly while turning around, most of their energy is chemical potential. Overall, the molecule's total energy remains constant, and it vibrates back and forth until it transfers its extra energy elsewhere.

But many molecules have more than two atoms. In a large molecule, each pair of adjacent atoms has a chemical bond and an equilibrium separation. If you give

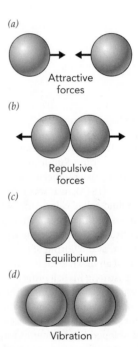

(a)

Attractive forces

(b)

Repulsive forces

(c)

Equilibrium

(d)

Vibration

Fig. 7.1.2 (a) Two atoms attract one another at moderate distances but (b) repel when they're too close. (c) In between is their equilibrium separation, at which they neither attract nor repel and are thus in equilibrium. (d) Pairs of atoms with excess energy tend to vibrate back and forth about their equilibrium separations.

this molecule excess energy, it will vibrate in a complicated manner as the energy moves among the various atoms and chemical bonds. The atoms in the molecule will continue to jiggle until something removes the excess energy from the molecule.

Like all liquids and solids, our burning log is just a huge assembly of atoms and molecules, held together by chemical bonds of various strengths. These particles push and pull on one another as they vibrate about their equilibrium separations. Their motion is thermal motion, and the energy involved in this disorderly jiggling is thermal energy. Because thermal energy is fragmented among the atoms and exchanged between them unpredictably, it can't be used directly to do work.

Heat and Temperature

Everything contains thermal energy, from the hot burning log to the cold metal poker you use to stir the fire. However, that doesn't mean that the thermal energy is equitably distributed. What happens to thermal energy when you push the log with the poker?

When they touch, the poker and the log begin to exchange thermal energy. In effect, the two objects become one larger object, and thermal energy that has been moving among the atoms in each individual object begins to flow across the junction between the two. Since each object starts with some amount of thermal energy, energy moves in both directions across this junction. Nonetheless, there may be some net flow from one object to the other. To allow us to predict the direction of this flow, we define a temperature for each of the objects.

Temperature is the quantity that indicates which way, if any, thermal energy will naturally flow between two objects. If no thermal energy flows when two objects touch, then those objects are in thermal equilibrium and their temperatures are equal. But if thermal energy flows from the first object to the second, then the first object is hotter than the second.

A temperature scale classifies objects according to which way thermal energy will flow between any pair. An object with a hotter temperature will always transfer thermal energy to an object with a colder temperature, and two objects with the same temperature will always be in thermal equilibrium. Thus the hot burning log will transfer thermal energy to the cold poker. We say that the burning log is hot because it tends to transfer thermal energy to most objects, while the poker is cold because most objects tend to transfer thermal energy to it.

Energy that flows from one object to another because of a difference in their temperatures is called heat. Heat is thermal energy on the move. Strictly speaking, the burning log doesn't *contain* heat; it contains thermal energy. However, when that log transfers energy to the cold poker because of their temperature difference, it's heat that *flows* from the log to the poker. (For a historical note about the understanding of heat, see ❶.)

Our present definition of temperature can order the objects around us from hottest to coldest, but it doesn't quantify temperature in any unique way. You could make your own temperature scale by comparing every pair of objects to see which way heat flows between them, but you probably wouldn't enjoy it. You do better to use a standard temperature scale such as Celsius, Fahrenheit, or Kelvin.

Standard temperature scales are based on an object's average thermal kinetic energy per atom. The more kinetic energy each atom has, on average, the more vigorous the object's thermal motion and the more thermal energy it transfers to the atoms in a second object by way of microscopic portions of work. Microscopic work is what actually passes heat between objects—a teeny shove here, a tiny yank there, all at the atomic scale. Since an object with more average thermal kinetic energy per atom will pass heat to an object with less, it makes sense to assign temperatures according to average thermal kinetic energies per atom.

❶ Before the time of Benjamin Thompson, Count Rumford (American-born British physicist and statesman, 1753–1814), heat was believed to be a fluid called caloric that was contained within objects. Thompson disproved the caloric theory by showing that the boring of cannons produced an inexhaustible supply of heat. Among his scientific and technological contributions, Thompson improved cooking and heating methods. He reshaped fireplaces and developed the damper as ways to reduce smoking and improve heat transfer to the room. Thompson also had a life of sensational escapades and great rises and falls in fortune. He fled New Hampshire in 1775 because he was a British loyalist, he fled London in 1782 under suspicion of being a French spy, and he was, at the time of his studies of heat, among the most powerful people in Bavaria.

The Celsius, Fahrenheit, and Kelvin scales all measure temperature in this manner. In each scale, a 1 degree or unit increase in temperature reflects a specific increase in average thermal kinetic energy per atom. The relationship between average thermal kinetic energy per atom and assigned temperature is based on several standard conditions: absolute zero, water's freezing temperature, and water's boiling temperature. (Recall from Section 5.1 that absolute zero is the temperature at which all thermal energy has been removed from an object.) Once specific temperatures have been assigned to two of these standard conditions, the whole temperature scale is fixed. For example, the Celsius scale is built around 0 °C being water's freezing temperature and 100 °C being water's boiling temperature. Temperatures for the three standard conditions appear in Table 7.1.1.

Table 7.1.1 Temperatures of Several Standard Conditions, as Measured in Three Temperature Scales: Celsius, Kelvin, and Fahrenheit

Standard Condition	Celsius (°C)	Kelvin (K)	Fahrenheit (°F)
Absolute zero	-273.15	0	-459.67
Freezing water	0	273.15	32
Boiling water	100	373.15	212

Open Fires and Woodstoves

Suppose you needed an easy way to heat your room. The oldest and simplest method would be to start a campfire in the middle of the floor (Fig. 7.1.3). The burning wood would produce thermal energy, which would flow as heat into the colder room. But how does burning wood produce thermal energy?

This thermal energy is released by a chemical reaction between molecules in the wood and oxygen in the air. Recall that atoms do work as they join together in a chemical bond and that the amount of work done depends on which atoms are being joined. For example, while carbon and hydrogen atoms can bond with one another to form hydrocarbon molecules, these atoms form much stronger bonds with oxygen atoms. Thus while it may take work to disassemble a hydrocarbon molecule, the work done by its hydrogen and carbon atoms as they bind to oxygen atoms more than makes up for that investment. As a hydrocarbon molecule burns in oxygen, new, more tightly bound molecules are formed and chemical potential energy is released as thermal energy. The reaction products produced by burning hydrocarbons in air are primarily water and carbon dioxide.

Wood is composed mostly of cellulose, a long carbohydrate molecule. Carbohydrates contain carbon, hydrogen, and oxygen atoms. Despite the presence of a few oxygen atoms, carbohydrates still burn nicely to form water and carbon dioxide. When you light the wood with a match, you're supplying the energy needed to break the old chemical bonds so that the new bonds can form. This starting energy is called activation energy—the energy needed to initiate the chemical reaction. Heat from the match flame gives the wood enough thermal energy to break chemical bonds between various atoms and start the reaction. (For a discussion of matches themselves, see ❷.)

Unfortunately, wood isn't pure cellulose. It also contains many complex resins that don't burn well and create smoke. If you plan to breathe the air in which you burn fuel, wood is an awful choice. You'd be better off with kerosene or natural gas, both of which are nearly pure hydrocarbons and burn cleanly. Actually, wood can be converted to a cleaner fuel by baking it in an airless oven to remove all of its volatile resins. This process converts the wood into charcoal, which burns to form nearly pure carbon dioxide, water vapor, and ash.

❷ The various chemicals in a new match head have excess chemical potential but can't release that energy without help. When you strike the match, you are using friction to supply the activation energy necessary to initiate its chemical reactions. Chemicals within the match and strike surface then begin to rearrange and release enough chemical potential energy to become extremely hot. The match head ignites first, then the match stick. The first chemical to burn in a match is phosphorus, which the frictional heating transforms from stable red phosphorus to its reactive white phosphorus. White phosphorus reacts easily with chemicals containing oxygen, including oxygen in the air, and the match bursts into flames. With safety matches, the red phosphorus is contained in the strike strip on the box or packet, so the match will only ignite when you strike it there. With strike-anywhere matches, the match tip itself contains the red phosphorus, so the match will ignite when you strike it on any rough surface.

Fig. 7.1.3 An open fire allows cellulose and other chemicals in wood to react with oxygen in the air and released their chemical potential energies as thermal energy. Although the fire must be lit with a match to provide the initial activation energy for the reactions, the fire is then self-sustaining.

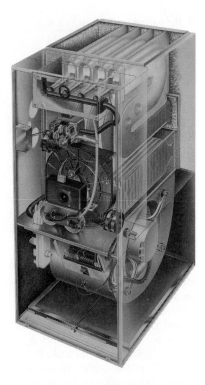

Fig. 7.1.4 This modern furnace burns natural gas in an S-shaped firebox. A fan at the bottom of the furnace blows fresh air past the hot outer surfaces of the firebox and then circulates the heated air amongst the rooms.

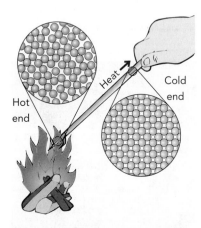

Fig. 7.1.5 When one end of a metal poker is hotter than the other, the atoms at the hot end vibrate more vigorously than those at the cold end. The poker then conducts heat from the hot end to the cold end. Some of this heat is conducted by interactions between adjacent atoms. In the metal poker, however, most of the heat is conducted by mobile electrons, which carry thermal energy long distances from one atom to another.

But even with clean burning fuels, the direct fire-in-the-room heating concept has its disadvantages: it consumes the room's oxygen and presents a safety hazard. Nonetheless, fires have heated dwellings for thousands of years. While fireplaces that burn wood or peat have chimneys that carry away their noxious fumes, the rising smoke takes with it much of the fire's thermal energy and some of the room's air. That's why a room that's heated by a fireplace often feels drafty away from the fireplace itself—cold outside air is seeping in through cracks to replace air drawn up the chimney. Even when clean burning fuels are used without a chimney, there are no simple solutions to the oxygen or safety problems.

Like a fireplace, a woodstove sends fumes from its burning wood up a chimney. But before its thermal energy can follow the fumes outside, a well-designed woodstove transfers most of that energy into the room. A woodstove is an example of a heat exchanger—a device that transfers heat without transferring the hot molecules themselves. Its smoke never enters the room but heat from that smoke does. The gas furnace in Fig. 7.1.4 also employs a heat exchanger.

The burning coals and hot gases inside the woodstove contain a great deal of thermal energy and are much hotter than the room air. Because of this temperature difference, heat tends to flow from the fire to the room. What is not so clear yet is how that heat is transferred.

There are three principal mechanisms by which heat moves from the fire to the room: conduction, convection, and radiation. The woodstove makes wonderful use of all three so that most of the thermal energy released by the burning wood is transferred to the room. Let's examine these three mechanisms of heat transport, beginning with conduction.

Heat Moving Through Metal: Conduction

Conduction occurs when heat flows through a stationary material. The heat moves from a hot region to a cold region but the atoms and molecules don't. For example, if you place the tip of a metal poker in the fire, the poker's handle will gradually become warm as the metal conducts heat.

Some of this heat is conducted by interactions between adjacent atoms. The vibrating atoms frequently push on one another, doing microscopic work in the process and exchanging miniscule amounts of thermal kinetic energy. In this fashion, thermal energy flows randomly from atom to neighboring atom.

But when the poker's tip is hotter than its handle, the flow is no longer completely random. The atoms at the hot tip have more thermal kinetic energy to exchange with their neighbors than atoms at the cold handle. The exchanges statistically favor the flow of thermal energy away from the hot tip and toward the cold handle. This flow of thermal energy from hot to cold through the poker is conduction (Fig. 7.1.5).

However, this atom-by-atom bucket brigade isn't the only way in which materials conduct heat. In a metal, the primary carriers of heat are actually mobile electrons—the tiny negatively charged particles that make up the outer portions of atoms. When atoms join together to form a metal, some of the electrons stop belonging to particular atoms and travel almost freely throughout the metal. These mobile electrons can carry electricity (as we'll discuss in Chapter 10) and are also good at transporting heat.

Mobile electrons participate in the bucket brigade of heat conduction because they, too, can push on vibrating atoms and exchange thermal kinetic energy with them. But while atoms can pass thermal energy only from one neighbor to the next, mobile electrons can travel great distances between exchange partners and can move thermal energy quickly from one place to another.

The ease with which electrons move heat about a metal explains why metals generally have higher thermal conductivities than nonmetals. Thermal conductivity is the measure of how rapidly heat flows through a material when it's exposed to a difference in temperatures. The best conductors of electricity—copper, silver, aluminum, and gold—are also the best conductors of heat. Poor conductors of electricity—stainless steel and insulators such as plastic, and glass—are also poor conductors of heat. There are a few exceptions to this rule. Diamonds, for example, are terrible conductors of electricity but wonderful conductors of heat. Of course, it would be silly to make a woodstove out of diamonds; after all, diamonds burn.

Conduction is what moves thermal energy from the woodstove's inside to its outside. No atoms move through the metal walls of the stove, just heat. So conduction serves as a filter, separating desirable thermal energy from the unwanted smoke and noxious gases that then go up the chimney.

Thus conduction makes the outside surface of the woodstove hot, allowing heat to flow from it to the colder room. But what carries heat into the room? If you touch the stove, conduction will immediately transfer a huge amount of heat to your skin and you'll be burned. But even without touching the stove, you're aware of its high temperature. It transfers heat into the room by convection and radiation.

Heat Moving with Air: Convection

Convection occurs when a moving fluid transports heat from a hotter object to a colder object. The heat moves as thermal energy in the fluid so that the two travel together. The fluid usually follows a circular path between the two objects, picking up heat from the hotter object, giving it to the colder object, and then returning to the hotter object to begin again.

This circulation often develops naturally. As the fluid warms near the hotter object, its density decreases and it floats upward, lifted by the buoyant force. When the fluid cools near the colder object, its density increases and it sinks downward.

Thus air heated by contact with the woodstove rises toward the ceiling and is replaced by colder air from the floor (Fig. 7.1.6). Eventually, this heated air cools and descends. Once it reaches the floor, it's drawn back toward the hot woodstove to start the cycle over. This moving air is a convection current, and the looping path that it follows is a convection cell. Within the room, convection currents carry heat up and out from the woodstove to the ceiling and walls. When you put your hand over the stove, you feel this convection current rising as it transfers heat to your hand.

Natural convection is good at heating the air above the woodstove, but most of that hot air ends up near the ceiling. While some of it will eventually drift downward to where you're standing, convection sometimes needs help. Adding a ceiling fan will help move the hot air around the room and make the woodstove more effective. This forced convection still transfers heat from the hot stove to the colder occupants of the room, but it doesn't rely on the buoyant force to keep the air circulating. The faster the air moves, the more heat it can transport from hot objects to cold objects.

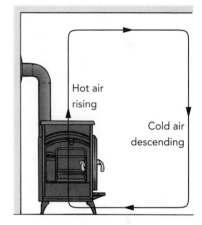

Hot air rising

Cold air descending

Fig. 7.1.6 When the woodstove is hot, convection carries heat from its surfaces to the ceiling and walls of the room. Warm air rises, supported by the buoyant force, and is replaced by cooler air from the floor. The warmed air eventually cools and descends. It then returns toward the stove to repeat the cycle.

Heat Moving as Light: Radiation

There is one more important mechanism of heat transfer: radiation. As the particles inside a material jitter about with thermal energy, they emit and absorb electromagnetic radiation. This radiation consists of electromagnetic waves, which include radio waves, microwaves, and infrared, visible, and ultraviolet light.

We'll study electromagnetic radiation in Chapters 13 and 14. For now, what's most important is that this radiation can carry thermal energy. When heat flows from a hot object to a cold object as electromagnetic radiation, we say that heat is being transferred by thermal radiation or simply radiation. Unlike conduction and convection, which depend on atoms, molecules, or electrons to carry the heat, radiation occurs directly through space. Radiative heat transfer happens even when two objects have nothing at all between them.

The types of electromagnetic waves in an object's thermal radiation depend on its temperature. While a colder object emits only radio waves, microwaves, and infrared light, a hotter object can also emit visible or even ultraviolet light. The red glow of a hot coal in the woodstove is that coal's thermal radiation.

Since our eyes are only sensitive to visible light, we can't see all of the thermal radiation emitted by an object, even when it's hot. But whether we see it or not, electromagnetic radiation contains energy and transfers heat to whatever absorbs it. While everything emits thermal radiation, the amount of that emission depends on temperature: the hotter an object, the more thermal radiation it emits. When two objects face one another, thermal radiation will travel in both directions between them. However, the hotter object will dominate this radiant exchange of thermal energy, resulting in a net transfer of thermal energy to the colder object. Exchanges of thermal energy via radiation always transfers heat from a hotter object to a colder one.

Radiation transfers a great deal of heat from the woodstove's surface to the surrounding objects. The stove bathes the room in infrared light, which warms everything it reaches. To encourage such radiative heat transfer, the woodstove and its chimney are often painted black. Black not only absorbs light well, but it's also particularly good at emitting thermal light (Fig. 7.1.7). If you heat a black poker red hot, it will glow much more brightly than one that's white, silvery, or transparent.

You can see that a white, silvery, or transparent surface is a poor absorber of light simply by shining a bright light on it; most of that light will bounce off it or pass through it. You can see that it is also a poor emitter of thermal light by looking at it when it's hot; it won't emit much light.

Common Misconceptions: Black Objects and Light
Misconception: that a black object never emits light.
Resolution: while a black object absorbs all light that strikes it, it still emits thermal radiation and can glow brightly if it's hot enough.

Even if air in the room is cold, you can usually feel the invisible infrared light from a woodstove on your face. When you block this light with your hands, your face suddenly feels colder because less heat is reaching your skin. This thermal radiation effect is even more pronounced with a fireplace or campfire, where thermal radiation from the hot coals and flames is the primary mechanism for heat transfer to the surroundings.

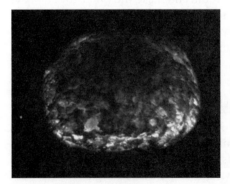

Fig. 7.1.7 A hot charcoal briquette glows brightly in the dark (*left*). However, a flash photograph reveals that its surface is actually gray and thus a partial absorber of light (*right*). If it weren't for the white ash, the briquette's black carbon would make it a nearly ideal emitter of blackbody radiation and absorber of light.

Overall, a modern woodstove is an excellent heat exchanger. As convection draws hot smoke up the long black chimney pipe, the smoke heats the stove and the pipe. These metal components conduct heat to their outer surfaces, which then distribute it around the room by convection and radiation. Although the stove consumes some room air, it controls the airflow with dampers so that it draws in only enough air to completely burn the wood. Overall, the stove extracts heat efficiently, cleanly, and safely from the burning wood.

Warming the Room

You light the woodstove and its heat begins flowing out into the cold room. Moving always from a hotter object to a colder object, heat enters each item in the room and gradually raises its temperature. For example, a once frigid brass bowl near the woodstove is soon pleasantly warm to the touch.

Let's take a look at the relationship between the heat added to that bowl and its temperature rise. Because the bowl's temperature increases steadily when heat is flowing into it steadily, its overall temperature rise must be proportional to the added heat. The constant of proportionality is called the bowl's heat capacity and is the amount of heat that must be added to the bowl to cause its temperature to rise by 1 unit. In effect, the bowl's heat capacity is the measure of its resistance to temperature changes.

But suppose that you have an assortment of bowls near the woodstove, each a different material. If you keep track of their temperatures, you'll find that some of them warm faster than others. Even when you take into account differences in their masses and in how much heat each one is receiving from the woodstove, you'll find that bowls made of different materials respond differently to added heat. Some materials are more thermally sluggish than others.

It makes sense to characterize each material by its heat capacity per unit mass, a quantity known as specific heat. The SI unit of specific heat is the joule-per-kilogram-kelvin (abbreviated J/kg·K). Each bowl's heat capacity is the product of its mass times the specific heat of the material from which it's made.

Table 7.1.2 Specific Heats Measured Near Room Temperature (293 K) and Atmospheric Pressure.

Material	Specific Heat
Lead	128 J/kg·K
Brass	380 J/kg·K
Copper	386 J/kg·K
Air (at constant volume)	715 J/kg·K
Glass	840 J/kg·K
Aluminum	900 J/kg·K
Air (at constant pressure)	1001 J/kg·K
Wood	~1100 J/kg·K
Plexiglas or Lucite	1349 J/kg·K
Steam (at constant pressure)	2027 J/kg·K
Ice	2220 J/kg·K
Water	4190 J/kg·K

Table 7.1.2 contains specific heats for a number of common materials. The wide range of values indicates that different materials respond quite differently to added heat. Each material's specific heat depends principally on the number of micro-

scopic ways it can store thermal energy per kilogram. Known as a degree of freedom, each independent way of handling thermal energy stores an average thermal energy equal to half the Boltzmann constant times the absolute temperature. The Boltzmann constant, which we first encountered in Section 5.1, has a measured value of 1.381×10^{-23} J/K and its units here are equivalent to those that appeared in Section 5.1.

Brass's relatively small specific heat explains why the brass bowl heats up so quickly when you place it directly on the woodstove; the bowl has relatively few degrees of freedom in which to store its thermal energy. But if you add even a modest amount of water to the bowl, the water's astonishingly large specific heat will dramatically slow the bowl's temperature rise. Water has a remarkable capacity for thermal energy.

Like brass or water, air also has a specific heat. However, air's specific heat depends on how you measure it. That's because gases tend to expand as they warm up. If you seal the air in a bottle, so that its volume doesn't change, it warms relatively easily; air's specific heat at constant volume is 715 J/kg·K. But if you allow the air to expand as its temperature rises so that its pressure doesn't change, it's harder to warm. That's because it needs extra energy to push the surrounding air out of its way as it expands; air's specific heat at constant pressure is 1001 J/kg·K.

⚒ Modern Combustion Furnaces

Unfortunately, woodstoves and their chimney pipes get extremely hot, so they can still burn you. True furnaces go one step farther, completely separating the firebox and chimney from the rooms they're heating. The furnace burns the fuel in the enclosed firebox and heat flows through a heat exchanger from the burned gas to air or water. This hot air or water then circulates throughout the building and heats the rooms. Since nothing in the rooms themselves gets extremely hot, there is less chance of being burned.

Most home furnaces now burn oil or gas, although coal was once popular. Regardless of the fuel, something must keep it flowing into the firebox at a controlled rate or the fire will go out. Shoveling coal into the furnace was hot, miserable work, so automated screw feeds were invented. Such feed systems are still used in sophisticated wood or pellet stoves.

Similarly, most modern oil or gas furnaces are automated. They use a pressure imbalance to push their fuels into the firebox. Propane gas is pressurized by evaporation in its storage tank while natural gas is pressurized by the gas company. These gases accelerate toward the lower pressure in the firebox and flow into it through many tiny holes in a burner assembly. Fuel oil is usually pressurized by a small electrically powered pump and sprays into the firebox through a nozzle, forming a fine mist of oil droplets in the air.

Since gas and oil furnaces turn on automatically whenever heat is needed, their flames must be reignited frequently. Something must supply the initial activation energy required to start the chemical reactions of burning. Many gas furnaces have a pilot light that burns all the time and ignites the main burner whenever fuel flows out of it. Other gas furnaces and nearly all oil furnaces use high voltage electricity to create a spark that lights the main burner. This ignition-on-demand is more complicated than a pilot light but wastes no gas or oil when heat isn't needed.

In all cases, a properly designed furnace has a temperature sensor that detects when the pilot light is out or when the main burner doesn't ignite. When one of these failures occurs, the sensor shuts off the fuel until the furnace is repaired or the pilot is relit. Unfortunately, some oil furnaces lack such a mechanism. These furnaces have been known to pump entire tanks of oil onto basement floors when they fail, creating hazardous waste spills.

The amount of air flowing through the firebox is controlled by a damper in the chimney. The hot, low-density burned gas is carried upward by buoyant forces and must flow past the damper. By closing the damper, you constrict the opening and impede the flow. Too much airflow lets excessive thermal energy flow out of the chimney while too little airflow leads to incomplete burning of the fuel. (For a discussion of the hazards of an overdamped furnace, see ❸.)

With the fuel ignited and the right amount of air flowing through the firebox, the furnace is ready to transfer heat to room air or to water. The hot burned gas passes through one side of a heat exchanger while room air or water passes through the other. The two sides are separated by a metal surface that conducts heat from the hot burned gas to the cooler air or water. The burned gas becomes colder and the air or water becomes hotter. By the time the burned gas reaches the chimney, it has cooled considerably and it carries relatively little thermal energy outside.

The best modern furnaces even let the moisture in the burned gases condense into liquid water before those gases leave through a chimney or exhaust pipe. As we'll see in the next section, condensing moisture releases its heat of evaporation—the chemical potential energy stored in separated water molecules. Water's heat of evaporation is enormous and by capturing this extra heat and transferring it to the room air, a condensing furnace greatly enhances its overall energy efficiency. A condensing furnace is the obvious choice if you're installing a new gas furnace and should be considered as well if you're installing a new oil furnace.

Heating Systems

There are several different styles of heating systems that may be connected to the furnace. In a house with warm-air heating, the heat exchanger passes the heat directly to indoor air and distributes that warm air to the rooms through a collection of air ducts. Warm-air heating systems differ, however, in how the warm air is propelled through the ducts.

In gravity warm-air heating, there is no active circulator for the warm air. Instead, the house makes use of natural convection to move the air around. The furnace is located in the basement so that buoyant forces lift the low-density warm air into the rooms. The hot air registers are located low in each room so that the warm air they deliver can float upward and raise the temperature of the whole room. High-density cold air accumulates near the floor, where it sinks into cold air registers at or near floor level and returns to the furnace to be heated again.

While gravity warm-air heating works adequately for some houses and was common long ago, it's not as flexible or effective as forced warm-air heating. Air's viscosity tends to slow its motion through ducts so that the natural circulation is weak and rooms far from the furnace receive little warmth.

In forced warm-air heating, room air is drawn into the furnace by a fan, passes through the heat exchanger, and blows out into the house through the ducts. The fan acts as a pump, increasing the air's total energy so that the air moves quickly and can easily overcome the viscous energy losses it experiences in the ductwork. Since forced warm-air heat doesn't depend on buoyant forces to circulate the air, the furnace can be located anywhere in the house.

Much like a house-wide hairdryer, the fan and furnace blow warm air easily to all of the rooms, wherever they're located. Moreover, the fan provides enough total energy to propel the air through a dust filter on the inlet side of the furnace. As with a vacuum cleaner, that filter extracts dust from the circulating air and cleans the house. It's important to clean the filter regularly because a clogged filter will extract too much total energy from the air and stop it from circulating properly.

Another common heating technique uses water-filled radiators. The furnace transfers thermal energy from hot burned gas to water and the hot water then flows

❸ When the flame in a combustion furnace receives too little air, the fuel won't burn completely to water and carbon dioxide. Instead, the furnace may produce carbon monoxide, a toxic and flammable gas. Carbon monoxide molecules bind tightly to hemoglobin, the oxygen-carrying component of blood, and can cause suffocation and death. Proper adjustment of the furnace's damper should prevent this problem, although a carbon monoxide detector is a sensible safety precaution in any home with a combustion furnace.

through pipes to the various rooms. In each room, the hot water passes through a radiator, where it transfers heat to the room via convection and radiation. Having given up most of its excess thermal energy, the lukewarm water returns through other pipes to the furnace and is reheated.

In gravity hot-water heating, the hot water is propelled by gravity alone. With the furnace in the basement, hot water naturally floats upward to the radiators and cold water sinks back to the furnace. But most hot-water heating systems actively circulate the hot water with a pump. Forced hot-water heating systems are more easily controlled than gravity driven ones and are often used to provide hot tap water as well. Because the radiator water itself isn't pure enough to drink, hot tap water is obtained by passing clean cold water through a heat exchanger with the hot circulating radiator water.

Sometimes the water-filled radiator panels are built directly into the walls and floor of a room. This heating concept is called radiant heat. Radiant heat is particularly useful in houses that have no basements and would otherwise have cold floors. Because there are no visible radiators, radiant heat doesn't waste room space. It's even under driveways to control ice and snow buildup. Mineral buildup or other damage to the radiator pipes, however, can impede the flow of hot water and prevent the radiant heating system from warming the rooms effectively. If those pipes are embedded in the concrete slab of the house, as is frequently the case, they won't be easy to clean or replace.

The other common heating system uses steam, the gaseous form of water. In steam heating, the basement furnace heats water in a boiler so that a reservoir fills with steam. Steam is much lighter than water, so it floats up through the house inside pipes and passes into radiators in every room. There the steam condenses back into water and transfers an enormous amount of heat to the radiators in the process. (I'll explain boiling and condensation in the next section.) The water sinks back to the furnace for reheating through the same pipes that delivered the steam.

Steam-heated radiators are much hotter (at 100 °C or 212 °F) than hot-water-heated radiators (less than 80 °C or 176 °F). Consequently smaller steam radiators are just as effective as much larger hot water ones. Steam is also less viscous than water so that it moves more rapidly through the pipes and heats the radiators faster.

But while steam heat works beautifully to heat rooms, it's far more dangerous than hot water heating. Steam radiators are hot enough to cause burns. I grew up with steam heat, both at home and at school, and I used to enjoy melting colored crayons on the steam radiators. Although I burned myself many times and occasionally got caught by my parents or teachers, it didn't seem to slow me down much. Steam is still widely used to supply heat to industrial and commercial buildings, but steam radiators themselves have almost entirely disappeared. Entire generations of melted wax artists have been lost as a result of this "progress."

None of these furnaces provides heating constantly. They are normally controlled by a thermostat, a device that turns the furnace or the circulator on only when the room temperature is below a desired value. I'll explain how thermostats work later in this chapter. While some homes use a single thermostat to control the entire heating system, others have separate thermostats for each room or region. When a separate thermostat is present in a room, it controls a valve, fan, or pump to turn on the flow of hot air, water, or steam whenever the room temperature is significantly below the desired value and to turn that flow off when the room reaches its goal.

Stoves and Cookware

Heating your home makes it comfortable on a cold day, but sometimes all you want to heat is your food. Since we've just covered most of the physics involved in stoves, ovens, and cookware, let's look at how they all work.

Although modern technology offers us a dozen or so variations on the classic stovetop burner, the physics of how most burners heat food hasn't changed all that much. When you activate a stovetop burner, its temperature rises and heat begins to flow from it to the colder objects around it. If there's a cold pot of food on the burner, heat flows from the burner to the pot and from the pot to the food. But while the pot may appear to be a trivial intermediary between the burner and the food, it actually has an important opportunity to make up for shortcomings in the burner. The best pots do exactly that.

How heat flows from the burner to the pot depends on the type of burner. A gas flame transfers heat to the pot's outer surface primarily via convection—heated gas produced by combustion rises up and around the pot convectively, conveying heat directly to the pot's outer surface. While a well-designed gas burner transfers heat relatively evenly to the pot's outer surface, the uniformity of that heating depends somewhat on the flame setting. A burner's lowest flame setting usually heats the center of the pot more than the pot's outer edges.

A conventional electric burner transfers heat to the pot via conduction, convection, and radiation. Although conduction would be the dominant heat transfer mechanism if the electric burner were in perfect contact with the pot, gaps inevitably appear between the two surfaces and heat must be conveyed across those gaps by convection and radiation. Since heat flow is strongest at the contact points between burner and pot, an electric burner heats the pot less uniformly than does a gas burner. The common spiral electric burner shown in Fig. 7.1.8 provides particularly non-uniform heating. A fancier electric burner may produce a hot surface rather than a hot spiral or it may effectively suspend the pot above the actual heating element so that convection and radiation transfer all the heat and can thereby smooth out the heating. Still, even the best electric burners don't provide perfectly uniform heating and gas burners aren't perfect, either.

If merely delivering heat to the pot were the whole story, all pots would cook equally well and you could choose pots exclusively according to look and feel. But cooking on the stovetop is complicated and different pots perform quite differently. The most important way in which pots differ is in their abilities to make up for imperfect burners. Unless you're simply heating water on the stovetop, you'll appreciate having a pot that heats food uniformly no matter what the burner does.

An ideal pot can receive heat anywhere on its outer surface and then distribute that heat quickly within itself so that its entire inner surface reaches a single, elevated temperature. Heat will then flow evenly from the pot to the food and the food will cook evenly. In short, a good pot should shield the food inside it from any unevenness in the heating it experiences from the burner. Even if only half the pot is on the burner, the food shouldn't be able to tell.

Pots range widely in how they distribute heat within themselves. Since heat flows naturally from hotter objects to colder objects, thermal conduction will always act to reduce temperature differences in the pot. Unfortunately, not all pots conduct heat well enough to achieve a uniform temperature and bad pots won't even come close to a single temperature. If you put a bad pot on a spiral electric burner, it will develop such severe hotspots that you'll probably find a spiral pattern of scorched food stuck to its bottom when you're done.

A good pot spreads heat quickly throughout its bottom surface and always approximates thermal equilibrium—it comes close to having a uniform temperature no matter how you heat it. The food inside the pot then has the best chance of cooking evenly and of not scorching on any hotspots. To facilitate heat flow, good pots use thick layers of copper or aluminum. Silver would be even better, but it's too expensive for common use. You'll find copper or aluminum at the heart of almost any good pot. One exception to that rule is cast iron; cast iron isn't as thermally conducting as aluminum, let alone copper, but if the cast iron is thick enough, it

Fig. 7.1.8 This electric burner produces extremely non-uniform heating and tends to produce hotspots in the cookware it heats.

can still provide reasonably uniform heating. That's one reason why heavy cast iron griddles and skillets remain popular.

In contrast, stainless steel is a terrible conductor of heat and a pot made exclusively of stainless steel can't make up for a burner's non-uniform heating. While a pure stainless steel pot may be serviceable in the uniform heat of a gas burner, it will develop awful hotspots when used with a typical electric burner.

Even if a pot uses aluminum or copper to spread its heat, those metals won't help much if they're too thin. A pot with a thin aluminum bottom or a stainless steel pot with a token layer of copper on its bottom surface won't provide the uniform heating of a truly great pot. And a glass pot is hopeless on the stovetop except for heating liquids or when used with a burner that provides extremely uniform heating.

Unfortunately, pots made only of pure copper or pure aluminum have their own problems. Like many elements, pure copper and pure aluminum are relatively soft and their surfaces damage easily. While they can be made sturdier by alloying them with other elements, that alloying also reduces their thermal conductivities. Moreover, copper forms toxic compounds in contact with certain foods so solid copper pots have limited uses. The best pots, therefore, are composite pots: pots that use pure copper or pure aluminum to distribute heat but encase those metals in materials that are tough, non-toxic, and easy to clean.

Anodizing aluminum is one possibility for protecting a pot that is otherwise pure aluminum. In anodization, an electrochemical process is used to grow a thick layer of aluminum oxide on the outside of pure aluminum. Aluminum oxide is an extremely hard mineral—the mineral on which both sapphire and ruby are based—and is almost impossible to scratch. Although pure aluminum spontaneously develops a thin layer of aluminum oxide when exposed to air, that layer isn't thick enough to provide much protection. Anodized aluminum, on the other hand, is extremely scratch resistant and thick, and so anodized aluminum pots are quite popular.

The other common type of composite pot is one that places a thick layer of pure copper or pure aluminum between two layers of protective stainless steel. The stainless steel then provides the scratch-resistant, easy-to-clean surface while the copper or aluminum layer ensures uniform heating of the food. Sometimes the inner stainless steel layer is replaced by a layer of non-stick plastic such as Silverstone®. Other times the outer stainless steel layer is omitted because keeping the outside of the pot looking nice doesn't matter. But a good pot is hard to make without a thick copper or aluminum layer.

Of course, you don't want the handle of the pot to be a good conductor of heat or you'll have to wear hot mitts whenever you cook. Tubular stainless steel or plastic handles are both terrible conductors of heat and therefore make wonderful pot handles. There is little excuse for putting an aluminum or copper handle on a pot, and the only "good" pots that have thermally conducting handles are cast iron skillets. People who favor cast iron skillets are usually more than willing to wear their cooking mitts and to put up with an occasional burn or two.

Conventional and Convection Ovens

When you put a pan of food in the oven, the cooking process isn't the same as it was on the stovetop. The food is still in contact with the pan, however, oven heat doesn't necessarily have to go through that pan to reach the food. Furthermore, conduction doesn't contribute significantly to the heat flowing from the burner to the food; all the cooking is done by convection and radiation. That change puts gas and electric ovens on a more even footing: the heating non-uniformities that plague some electric stovetops are largely absent in electric ovens.

But there are at least three different ways to cook food in an oven, not counting the microwave ovens that I'll discuss in Section 13.2. The three I have in mind are baking, broiling, and convection baking (or, more accurately, *forced* convection baking).

In baking, you place food in an oven with an active burner at its bottom. Heat flows from that burner to the food by natural convection and radiation. Convection is important because air heated by the burner floats upward to and around the food and transfers heat to it. Radiation is important because the hot burner and the heated walls of the oven bathe all surfaces of the food, particularly its bottom surface, with thermal radiation. Since the bottom of the food is exposed to the most direct convective heat and the brightest thermal radiation, the food tends to cook from the bottom up.

In broiling, you place food in an oven with an active burner at its top. Now heat flows from that burner to the food by radiation alone. Convection is absent because air heated by the burner floats to the top of the oven and stays there; it never encounters the food. The only way heat can flow from the burner to the food is as electromagnetic radiation and this radiation strikes primarily the top of the food. The food therefore tends to cook from the top down.

A brick pizza oven acts as a broiler, despite having wood burning on its bottom surface. Because the wood fire and pizza sit side-by-side on the same brick surface, conduction and convection are unable to carry much heat from the fire to the pizza. Bricks are terrible conductors of heat and hot air from the fire floats upward without ever touching the pizza. All that's left is radiative heat transfer. While some radiative heat travels directly from the fire to the pizza, most of the heat reaching the pizza comes from the bricks overhead. Heat flows from the fire to the bricks above it via both convection and radiation. The heated bricks, in turn, radiate heat at the pizza below them. So the pizza cooks from the top down, heated almost entirely by radiation. That's one reason brick oven pizza is often caramelized on top and rarely burns on the bottom.

Finally, there is convection baking. It should really be called *forced* convection baking because what distinguishes it from conventional baking is a fan. The hot air in a convection oven is circulated rapidly and deliberately by a fan. The food still sits in the middle the oven and it's still heated by convection and radiation. But because of the rapid airflow inside the oven, convection becomes the dominant heat transfer mechanism and it affects the top of the food just as much as it affects the bottom of the food. The food therefore cooks from the outside in. The food not only cooks faster, it also cooks more evenly.

Furthermore, you can bake more food in a convection oven than you can in a conventional oven. Since natural convection depends on weak buoyant forces to move the heated air, viscous or pressure drag forces can easily impede the airflow. An over-filled conventional baking oven won't cook uniformly. But in a forced convection oven, the flowing air has so much total energy that it can easily move through even relatively narrow openings, and all but the most tightly packed foods will cook uniformly.

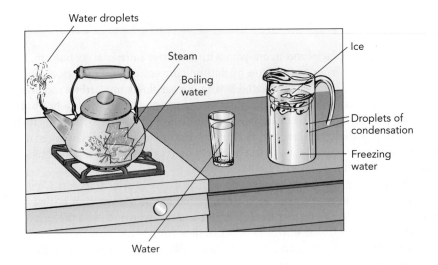

Water droplets

Steam

Boiling water

Ice

Droplets of condensation

Freezing water

Water

Section 7.2 **Water, Steam, and Ice**

Water is probably the most important chemical in our daily lives. It is so crucial to biology, climate, commerce, industry, and entertainment that it merits a whole section of its own. Moreover, it exhibits the three classic phases of matter, solid, liquid, and gas, and illustrates the role that heat plays in transforming one phase into another. While most of what we can learn from water is applicable to any material, there are a few aspects of water that are almost unique in nature. Water really is a remarkable substance.

Solid, Liquid, and Gas: the Phases of Matter

Like most substances, water exists in three distinct forms or phases: solid ice, liquid water, and gaseous steam (Fig. 7.2.1). These phases differ in how easily their shapes and volumes can change. Ice is solid—rigid and incompressible; you can't alter an ice cube's shape or its volume. Water is liquid—fluid but incompressible; you can reshape the water in a pitcher but you can't change its volume. Steam is gaseous—fluid and compressible; you can vary both the shape and the volume of the steam in a tea kettle.

These different characteristics reflect the different microscopic structures of steam, water, and ice. Steam or water vapor is a gas, a collection of independent molecules kept in motion by thermal energy. These water molecules bounce around their container, periodically colliding with one another or with the walls. The water molecules fill the container uniformly and can accommodate any changes in its shape or size. Enlarging the container simply decreases the steam's density and lowers its pressure.

When they are independent of one another as gaseous steam, water molecules have a substantial amount of chemical potential energy. They can release some of this energy by joining together to form ordinary water. Water is a liquid, a disorderly collection of molecules that cling to one another with chemical bonds. Because these bonds aren't very strong, the molecules in water can use thermal energy to break them temporarily and then change bonding partners. This rebonding process allows water to change shape so that it is a fluid. But despite their flexibility, these bonds manage to bunch the water molecules together so snuggly that even squeezing can't pack them much tighter. That's why water is incompressible.

The molecules in water can release still more chemical potential energy by linking together stiffly as ice. Ice is a solid, a rigid collection of chemically bound molecules.

Like most solids, ice is crystalline—its water molecules are arranged in an orderly latticework that extends over long distances and gives rise to the beautiful crystal facets seen on snowflakes and frost. Ice's crystalline structure is so constraining that its water molecules can't use thermal energy to change bonding partners and consequently ice can't change shape.

Just as an orderly stack of oranges at the grocery store takes up less volume than a disorderly heap, a crystalline solid almost always occupies less volume than its corresponding disorderly liquid. The solid phase of a typical substance is thus denser than the liquid phase of that same substance, so the solid phase sinks in the liquid phase.

There is only one common substance that violates that rule: water. Ice's crystalline structure is unusually open and its density is surprisingly low. Almost unique in nature, solid ice is slightly less dense than liquid water and ice therefore floats on water. That's why icebergs float on the open ocean and why ice cubes float in your drink. In fact, water reaches its greatest density at about 4 °C (39 °F).

Melting Ice and Freezing Water

Ice in a freezer is extremely cold, typically −18 °C (0 °F). When you place this ice on a warm countertop, heat flows into it and its temperature rises. The ice remains solid until its temperature reaches 0 °C (32 °F). At that point, the ice stops getting warmer and begins to melt. Melting is a phase transition, a transformation from the ordered solid phase to the disordered liquid phase. This transition occurs when heat breaks some of the chemical bonds between water molecules and permits the molecules to move past one another. The melting ice transforms into water, losing its rigid shape and crystalline structure.

Zero degrees Celsius is ice's melting temperature, the temperature at which heat added to ice goes into breaking its bonds and converting it into water, rather than into making it hotter. The ice-water mixture remains at 0 °C until all of the ice has melted. When only water remains, heat once again causes its temperature to rise.

The heat used to transform a certain mass of solid into liquid without changing its temperature is called the latent heat of melting or, more formally, latent heat of fusion. The bonds between the water molecules in ice are strong enough to give ice an enormous latent heat of melting: it takes about 333,000 J of heat to convert 1 kg of ice at 0 °C into 1 kg of water at 0 °C. Since water's specific heat is 4190 J/kg·K, that same amount of heat would raise the temperature of 1 kg of liquid water by about 80 °C. Thus it takes almost as much heat to melt an ice cube as it does to warm the resulting water all the way to boiling.

The latent heat of melting reappears when you cool the water back to its melting temperature and it starts to freeze. Freezing is another phase transition, a transformation from the disordered liquid phase to the ordered solid phase. As you remove heat from water at 0 °C, the water freezes into ice rather than becoming colder. Because the water molecules release energy as they bind together to form ice crystals, the water releases heat as it freezes. The heat released when transforming a certain mass of liquid into solid without changing its temperature is again the latent heat of melting. You must add a certain amount of heat to ice to melt it, and you must remove that same amount of heat from water to solidify it.

Phase Equilibrium: Leaving and Landing

We've seen that ice has a melting temperature, so now let's see *why* it has a melting temperature. To do that, we must look at the interface between solid ice and liquid water. Whenever both phases are in contact, they exchange water molecules across the interface between them. Water molecules regularly break free from the ice to enter the water, and they often drop out of the water to stick to the ice. In other words, water molecules are leaving the ice and landing on it all the time, like airplanes at a busy airport.

While you can't see the individual leavings and landings, you can observe their net effect. If leavings outpace landings, the ice will gradually transform into water. If the landings outpace leavings, the water will gradually transform into ice. And if the two processes balance one another, the ice and water will coexist indefinitely—a situation known as phase equilibrium.

Temperature plays a crucial role in this balance because it affects the rate at which water molecules leave the ice. The warmer the ice, the more often water molecules at its surface can gather enough thermal energy to break free and leave. Below ice's melting temperature, water molecules leave the ice too infrequently to balance the landing process and the water transforms completely into ice. Above ice's melting temperature, water molecules leave the ice so often that they outstrip the landing process and the ice transforms completely into water. Only at the melting temperature do the leaving and landing rates balance so that ice and water can coexist in phase equilibrium.

Ice's huge latent heat of melting has a stabilizing effect on the phase equilibrium between ice and water. Whenever ice and water are mixed together, the mixture's temperature will shift rapidly toward 0 °C. That's because if the mixture's temperature is above 0 °C, the ice will melt—a phase transformation which absorbs the heat of melting and thereby lowers the mixture's temperature toward 0 °C. And if the mixture's temperature is below 0 °C, the water will freeze—a phase transformation which releases the heat of melting and thereby raises the mixture's temperature toward 0 °C.

As long as the mixture doesn't run out of ice or water, its temperature will soon reach 0 °C and remain there even as you add or remove heat from it. Any heat you add to the mixture goes into melting more ice, not into raising its temperature. Any heat you remove from the mixture comes from freezing more water, not from lowering its temperature. That's why your glass of ice water remains at 0 °C, even in the hottest or coldest weather (Fig. 7.2.2). (For more about melting ice and why ice is slippery, see ❶.)

Evaporating Water and Condensing Steam

Water's open surface is another active interface between phases, but this time the liquid water is exchanging molecules with gaseous steam. Water molecules are feverishly leaving the water for the steam and landing on the water from the steam, once again like planes at a busy airport.

While this frantic exchange of molecules is interesting, what matters most is its net effect. If more molecules leave the water than return to it, then water gradually evaporates into steam. Evaporation is the phase transition from liquid to gas.

Fig. 7.2.2 Ice and water can coexist only at 0 °C, ice's melting temperature.

On the other hand, if more molecules are landing on the water than leaving it, the steam gradually condenses into water. Condensation is the phase transition from gas to liquid. And if landing and leaving are balanced, water and steam coexist in phase equilibrium.

These two phase transitions have enormous thermal consequences. Since the molecules in water cling to one another with chemical bonds, it takes energy to separate them. Although the bonds *between* water molecules are weaker than the bonds *within* water molecules, it still takes a great deal of energy to transform water into steam.

The heat needed to transform a certain mass of liquid into gas, without changing its temperature, is called the latent heat of evaporation or, more formally, latent heat of vaporization. Water's latent heat of evaporation is truly enormous because water molecules are surprisingly hard to separate. About 2,300,000 J of heat is needed to convert 1 kg of water at 100 °C into 1 kg of steam at 100 °C. That same amount of heat would raise the temperature of 1 kg of water by more than 500 °C!

You are most aware of this latent heat of evaporation on hot summer days, when perspiration that evaporates from your skin draws heat from you and lowers your temperature. As each water molecule leaves your skin to become steam, it gathers up more than its fair share of thermal energy from its environment and carries it away as chemical potential energy. The departing water molecules thus leave you bereft of thermal energy and you grow colder.

The latent heat of evaporation reappears when steam condenses and the gathering water molecules release their chemical potential energy as heat. The heat released when transforming a certain mass of gas into liquid without changing its temperature is again the latent heat of evaporation. You must add a certain amount of heat to water to evaporate it and you must remove that same amount of heat from steam to condense it.

The huge amount of heat released by condensing steam is often used to cook food or warm radiators in older buildings. When you steam vegetables, you are allowing steam to condense on the vegetables and transfer heat to them. A double-boiler uses condensing steam to transfer heat from a burner to a cooking container in a controlled manner.

Relative Humidity

Although we've examined the consequences of evaporating and condensing, we haven't yet seen what conditions determine when they'll occur. It all comes down to water molecules leaving and landing, so let's take a look at why one process wins out over the other.

The basic indicator of whether water will evaporate or steam will condense is relative humidity. Relative humidity measures the water molecule landing rate as a percentage of the leaving rate. When the relative humidity is 100%, the two rates are equal and water and steam are in phase equilibrium. If the relative humidity is less than 100%, the landing rate is less than the leaving rate and the water evaporates. Finally, if the relative humidity is more than 100%, the landing rate is more than the leaving rate and the steam condenses.

Relative humidity depends on the temperature and on the density of the steam. Temperature affects the leaving rate. The warmer the water, the more thermal energy it contains and the more frequently water molecules leave its surface to become gas. By itself, an increase in temperature will boost the leaving rate and thereby decrease the relative humidity. Rising temperatures thus favor evaporation.

The density of water molecules in the steam affects the landing rate. The denser the steam, the more often water molecules land on the water to become liquid. By itself, an increase in steam density will boost the landing rate and thereby increase

❶ Solid ice is remarkably slippery because it easily develops a thin layer of liquid water that lubricates motion and nearly eliminates traction. The origins of this water layer have been a matter of debate for more than a century and it is only recently that scientists have established a complete explanation for ice's slipperiness. There are three factors at work. First, ice is almost unique in nature in that increased pressure destabilizes solid ice relative to liquid water; squeezing ice at 0 °C causes it to melt. Ice's decreasing melting temperature with increasing pressure stems from the fact that solid ice is less dense than liquid water. It takes great pressure to depress ice's melting temperature to −1 °C, however, so this factor is only important in special circumstances. When the narrow blade of an ice skate presses sharply on ice near 0 °C, this pressure melting effect contributes to the ice's slipperiness. Second, sliding friction can easily heat a thin surface layer of ice hot enough to melt it, even when the ice's overall temperature is as low as −35 °C. This frictional melting effect explains the sudden and dramatic loss of traction that occurs just after your feet begin to slide on ice. Skis and sleds depend on the loss of traction that occurs once sliding begins, while automobile tires try to avoid it. Third, the molecular surface of crystalline ice is surprisingly noncrystalline and begins to have liquid-like properties even at −33 °C. The liquidlike layer thickens as the ice warms toward 0 °C and contributes to ice's increasing slipperiness as it approaches its melting temperature.

Fig. 7.2.3 You can determine the air's relative humidity using two thermometers—one of which has a wet cloth wrapped around its bulb (right). Evaporation cools the wet bulb thermometer by an amount related to the air's relative humidity. The dryer the air, the colder the wet bulb thermometer becomes.

Fig. 7.2.4 This frost formed when moisture in the air deposited directly as ice crystals on the cold surface of a car window.

the relative humidity. Rising steam densities thus favor condensation. Even when that steam is mixed with air, as it often is, the air molecules act as passive bystanders. The density of water molecules alone determines the air's relative humidity.

Relative humidity plays an important role in countless experiences of everyday life. When the relative humidity is low, water evaporates quickly and the air feels dry. Perspiration cools you effectively. When the relative humidity is high (near 100%), water barely evaporates at all and the air feels damp. Perspiration clings to your skin and doesn't cool you much.

And when the relative humidity exceeds 100%, perhaps because of a sudden temperature drop, steam begins to condense everywhere. Water droplets grow on surfaces as dew or form directly in the air as fog, mist, or clouds. If the humidity remains high, these droplets grow larger and eventually fall as rain.

You can measure relative humidity by observing the cooling that accompanies evaporation. The most common scheme involves two thermometers (Fig. 7.2.3), one of which has a wet cloth wrapped around its bulb (right). Water evaporating from the wet cloth cools the thermometer until that water reaches phase equilibrium with steam in the air and evaporation ceases. The dryer the air, the colder the thermometer gets. The temperatures of the two thermometers can then be used to determine the relative humidity, usually with the help of tabulated values.

Subliming Ice and Depositing Steam

We've examined phase transitions between ice and water and between water and steam. That brings us to the phase transitions between ice and steam. Oddly enough, water molecules can leave ice to become steam and can land from steam to become ice. In fact, ice and steam regularly exchange water molecules even when there is no liquid water present at all.

As usual, this exchange of water molecules occurs at the surface of the ice, the interface between ice and steam. Since we're most interested in the net movement of molecules, it comes down to leaving and landing rates on the ice. If molecules leave the ice more often than they land, the ice sublimes. Sublimation is the phase transition from solid to gas. And if molecules land on the ice more often than they leave, the steam deposits. Deposition is the phase transition from gas to solid. Once again, relative humidity measures the landing rate as a percentage of the leaving rate. At 100% relative humidity, ice and steam are in phase equilibrium.

When the relative humidity is below 100%, ice sublimes. This effect gives rise to a number of familiar phenomena. When the weather is cold and dry, snow gradually disappears from the ground without ever melting. In the low relative humidity of a frostless freezer, the ice cubes slowly shrink to miniature size. And when you leave food unprotected in that same frostless freezer, it eventually dries out. While this "freezer burn" is a nuisance at home, sublimation from frozen food is used commercially to prepare freeze-dried foods.

And when the relative humidity exceeds 100%, steam deposits. This process yields several more familiar effects. Frost forms on cold windows and lawns that are exposed to humid air (Fig. 7.2.4). In the high relative humidity of a non-frostless freezer, frost and ice accumulate on the walls and require periodic defrosting. And snowflakes grow in clouds and then descend gracefully to the ground.

Boiling Water

With three phases and six phase transitions, we seem to be out of possibilities. So where does boiling fit into this picture? Boiling is simply an accelerated form of evaporation in which bubbles of pure steam grow by evaporation inside the water itself. To understand boiling, let's look at the interplay between water and steam.

Suppose we seal some water inside an airless container and keep it at a constant temperature. The water will evaporate as steam until the relative humidity inside the container reaches 100%. At that point, the water and steam will have reached phase equilibrium; the steam will have just the right density so that its water molecules will land on the water as often as they leave. Steam at its equilibrium density is said to be saturated.

Of course, the density of that saturated steam depends on the temperature of the container. If you warm up the container, water molecules will leave the water more frequently and the density of the steam will have to increase in order to match the landing rate to the leaving rate. The density of saturated steam is thus an increasing function of the temperature.

The saturated steam's density, together with its temperature, determine its pressure—the pressure inside our container. If we warm the container, the density of the saturated steam will increase and so will its pressure. Near room temperature, the pressure of saturated steam is a few percent of atmospheric pressure. As the temperature approaches 100 °C (212 °F), however, the pressure of saturated steam approaches atmospheric pressure.

With that background, suppose that we place a bubble of pure saturated steam in room temperature water. Because the pressure inside that bubble is much lower than atmospheric pressure, the surrounding water will rush inward and compress it. As the steam bubble's volume shrinks, the steam will exceed its saturated density and begin to condense. In almost no time, the bubble will be smashed out of existence.

Now suppose we begin warming the water on the stove. As the water temperature increases, steam's saturated density and pressure both increase. At first, saturated steam bubbles remain unstable; they're crushed quickly by atmospheric pressure. But when the water temperature is near 100 °C, something remarkable happens: saturated steam bubbles suddenly become stable and the water can begin to boil (Fig. 7.2.5). At that temperature, water's boiling temperature, the pressure of saturated steam reaches atmospheric pressure and bubbles of saturated steam can survive indefinitely within the water. Even more remarkably, these bubbles can grow by evaporation; each bubble's surface is an interface between water and steam, so when heat is added to the water, water can transform into steam and enlarge the bubble. Although the bubbles quickly float to the water's surface and pop, new bubbles can promptly take their place.

Boiling converts water to steam so rapidly that it consumes almost any amount of heat you add to the water. That's why it's so difficult to warm water above its boiling temperature. An open pot of water on the stove warms to water's boiling temperature and then remains at that temperature until all of the water has transformed into steam. Only then can the pot's temperature again begin to increase.

The constant, well-defined temperature of boiling water allows you to cook vegetables or an egg at a particular rate. When you place an egg in boiling water, it cooks in 3 minutes because it's in contact with water at its boiling temperature.

Fig. 7.2.5 Water boils at 100 °C, when atmospheric pressure can no longer crush (or implode) the bubbles of water vapor.

Changing Water's Boiling Temperature

Water's boiling temperature depends on the ambient pressure. For an open pot or pan, that pressure is atmospheric pressure. However, atmospheric pressure decreases with altitude and depends somewhat on the weather. Water boils at 100 °C (212 °F) near sea level but at only 90 °C (194 °F) at an altitude of 3000 m (9800 ft). This reduction in water's boiling temperature with altitude explains why many recipes must be adapted for use at higher elevations. At 3000 m, an egg cooks slowly in boiling water because it's surrounded by 90 °C water, not 100 °C water. The same problem slows the cooking of rice, beans, and many other foods at high altitudes.

Fig. 7.2.6 The water in this glass measuring cup was superheated in a microwave oven and then disturbed with a fork. Explosive boiling blew all of the water out of the cup in a fraction of second.

Fig. 7.2.7 The beverage in this glass is supersaturated with carbon dioxide (CO_2) and is gradually losing CO_2 molecules to the room air through its top surface. Bubbles of CO_2 are also streaming upward from the bottom of the glass, where they start as seed bubbles formed at nucleation sites. Although those seed bubbles are microscopic, they grow quickly as more CO_2 molecules enter them from the carbonated beverage. Each bubble stream originates from its own nucleation site. Those nucleation sites are rare, which is why this glass has only a handful of bubble streams in it.

At sufficiently low pressures, water boils even at room temperature or below. At the other extreme, high pressures can prevent water from boiling until it's extremely hot. The boilers in steam engines and power plants often operate at such high pressures that water's boiling temperature inside them may exceed 300 °C (572 °F).

One way to decrease cooking times is to use a pressure cooker, a pot that seals in steam so that the pressure inside it can exceed atmospheric pressure. This increased pressure prevents boiling until the water temperature is well above 100 °C. If you subject water to twice sea-level atmospheric pressure, it won't boil until 121 °C (250 °F). An egg cooks very quickly at that temperature, as do vegetables and other foods.

However, just because water *can* boil, doesn't mean that it *will* boil. Boiling depends on tiny seed bubbles that subsequently grow by evaporation. Without seed bubbles, the water won't boil. Because nucleation or seed-bubble formation almost never occurs spontaneously in water below 300 °C, something else must create those seeds. Most nucleation occurs at defects or hotspots on the container or at contaminants in the liquid. Those nucleation sites usually trap air or other permanent gases and then serve as nurseries for seed bubbles of steam. Reliance on these specific nucleation sites explains why the bubbles of boiling water, like those in soda or champagne, often stream upward from specific spots on their containers.

When you heat water uniformly in a clean, glass container, it may not boil properly at its boiling temperature. Glass has a liquid-like smoothness even at the atomic scale and rarely aids seed bubble formation. Without any long-lived nucleation sites, the water may stop forming seed bubbles and cease boiling. Once boiling stops, the water's temperature can rise above the boiling temperature so that it becomes superheated.

Superheated water, which forms easily and often in a microwave oven, can be extremely dangerous. Touching it with a fork, adding sugar or salt, or even just tapping its container can initiate violent or even explosive boiling (Fig. 7.2.6). The more the water's temperature exceeds its boiling temperature, the more energy it can release suddenly if it abruptly boils. Be careful when you heat water in a microwave oven, particularly in a glass or glazed container. If it doesn't appear to be boiling properly despite being very hot, recognize that it may be superheated. Your safest bet is to stay away from it until it has cooled down.

There is one other interesting way to change water's boiling temperature: dissolve chemicals in it. A dissolved chemical keeps the water molecules busy so that they are less likely to leave the water to become steam, or ice for that matter. Since dissolved chemicals discourage water molecules from leaving water's liquid phase, they suppress any phase transitions that reduce the amount of liquid phase water. That's why dissolving sugar or salt in water slows its evaporation and raises its boiling temperature. It's also why saltwater freezes at a lower temperature than freshwater and why salt tends to melt ice. In contrast, sand doesn't melt ice because it doesn't dissolve in water.

When water has too much of a substance dissolved in it to be in phase equilibrium, it is supersaturated and should undergo phase separation—some of the substance should go elsewhere to leave the water saturated but not supersaturated. Initiating that phase separation can be difficult, however, and often requires nucleation similar to that involved in boiling.

For example, a carbonated beverage that you've just poured into a glass is supersaturated with carbon dioxide (CO_2) molecules. In the bottle, the beverage was in phase equilibrium with high-density CO_2 gas and its equilibrium concentration of dissolved CO_2 molecules was enormous. When the high-density CO_2 was released, that phase equilibrium was spoiled, and the beverage became supersaturated. In your glass, it loses CO_2 molecules from its open surface and gradually goes "flat." If there are nucleation sites on the glass or in the beverage to provide seed bubbles, it can also lose CO_2 into those bubbles as they float upward (Fig. 7.2.7). Shaking the bottle before opening it fills the beverage with such seed bubbles, and these bubbles grow explosively when you open the bottle.

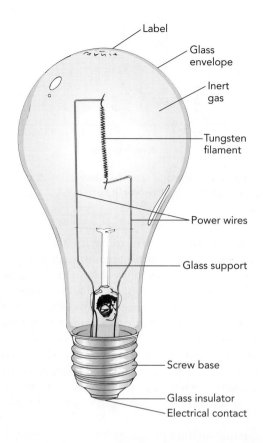

Label
Glass envelope
Inert gas
Tungsten filament
Power wires
Glass support
Screw base
Glass insulator
Electrical contact

Section 7.3 Incandescent Lightbulbs

For more than a century, incandescent lightbulbs have provided light at the flip of a switch. Their invention brought to a close the era of candles and gaslights and spurred the development of electric power. While the variety of incandescent bulbs has grown over the years to include everything from heat lamps to halogen headlights, all incandescent bulbs have at their hearts one simple object: an extremely hot filament.

Light, Temperature, and Color

Light from an incandescent lightbulb is part of the thermal radiation emitted by its hot filament. While most types of electromagnetic waves are invisible, our eyes are sensitive to a narrow range of waves that we call visible light. Any object that's hotter than about 400 °C (750 °F) emits enough visible light for us to see it in a dark room. At higher temperatures, that visible light brightens and shifts in color from red to orange to yellow to white. At 500 °C (930 °F), an object glows a dull red. At 1700 °C (3100 °F), it emits the orange light of a candle. And at 5800 °C (10,500 °F), the temperature of the sun's surface, it gives off the brilliant white light of the sun.

To reproduce pure white sunlight, the bulb's filament should be heated to 5800 °C. Unfortunately, nothing is solid at that high temperature. Even tungsten metal, the best filament material known, readily sublimes at temperatures above 2500 °C (4500 °F). Since incandescent lightbulbs must operate at lower temperatures, they can't really reproduce sunlight. Most give off the warm, yellow-white light that's characteristic of tungsten metal at about 2500 °C.

The filament's brightness and color both depend on its temperature (Fig. 7.3.1). Since light carries energy, we can measure its brightness as the number of watts

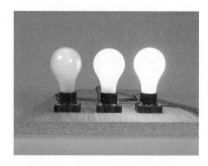

Fig. 7.3.1 As you increase the power to an incandescent lightbulb, its filament becomes hotter and emits a brighter and whiter light. The cooler filament on the left is dim and red while the hotter one on the right is bright and yellow-white. Because these bulbs are frosted, you can't see their filaments directly.

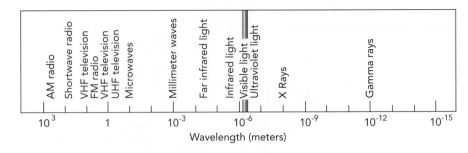

Fig. 7.3.2 The spectrum of electromagnetic radiation, arranged by wavelength. The scale here is logarithmic, meaning that the wavelength decreases by a factor of 10 with each tick mark to the right.

of visible light it emits. But how do we characterize its color? Moreover, what distinguishes visible light from the invisible types of electromagnetic radiation? Although the full answers to those questions will have to wait until Chapters 13 and 14, we can make a few essential observations about them now.

Visible light is part of a continuous spectrum of electromagnetic radiation that extends from radio waves at one extreme to gamma rays at the other (Fig. 7.3.2). Different types of electromagnetic radiation are distinguished by their wavelengths, that is, the distance between their wave crests. Wavelength is easy to see in the waves on a lake or sea, where the crests are visible and you can directly measure the distance from one to the next. But while the wave crests of electromagnetic waves aren't so easy to observe, they exist and it's possible to measure the distance between them.

The electromagnetic radiation produced by a hot filament is mostly infrared, visible, and ultraviolet light. Although this light is just a tiny portion of the overall electromagnetic spectrum, it's particularly important to our everyday world. Figure 7.3.3 gives an expanded view of the visible portion of the electromagnetic spectrum. Various colors that we see correspond to specific wavelength ranges. For example, light with a wavelength of 530 nanometers (billionths of a meter, abbreviated nm) appears green to our eyes.

But the thermal radiation emitted by a filament isn't a single electromagnetic wave with one specific wavelength. Instead, it's many individual waves that cover a broad range of wavelengths. Some of these waves are red light, some green, some blue, and some are invisible.

The distribution of wavelengths emitted by the filament depends on its temperature and surface properties, particularly its emissivity—the efficiency with which it emits and absorbs light. Emissivity is measured on a scale from 0 to 1, with 1 being ideal efficiency. A perfectly black object has an emissivity of 1; it absorbs all light that strikes it and emits thermal light as efficiently as possible. Although tungsten's emissivity is only 0.43, the filament wire is wound in such a way that it has many dark nooks and crannies. The filament is thus so nearly black that its emissivity is essentially 1.

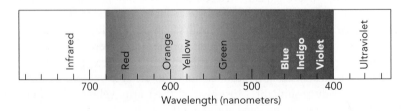

Fig. 7.3.3 A portion of the electromagnetic radiation spectrum around visible light. Wavelengths are measured in nanometers (nm or billionths of a meter).

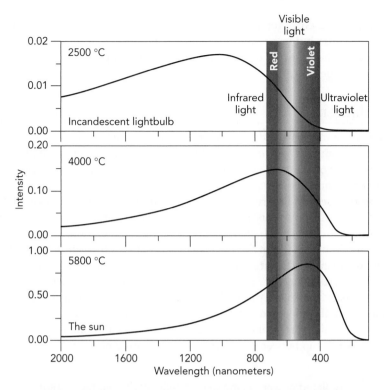

Fig. 7.3.4 The distributions of light emitted by black objects at 2500 °C (*top*), 4000 °C (*middle*), and 5800 °C (*bottom*). In addition to containing a larger fraction of visible light, the 5800 °C object is much brighter than the 2500 °C object (note the different intensity scales on the left).

The distribution of wavelengths emitted by a black object is determined by its temperature alone and is called a blackbody spectrum. As you can see from the examples in Fig. 7.3.4, the spectrum of a blackbody brightens and shifts toward shorter wavelengths as its temperature increases. An object that isn't black emits somewhat less thermal radiation, but that radiation still brightens and shifts toward shorter wavelengths as the object becomes hotter.

Our eyes make an average assessment of the distribution of wavelengths emitted by a black object, and we observe reddish, orangish, yellowish, whitish, or bluish light, depending on the object's temperature (Table 7.3.1). The temperature associated with a particular distribution of wavelengths is the color temperature of that light.

Table 7.3.1 Temperatures and Colors of Light Emitted by Hot Objects

Object	Temperature	Color
Heat lamp	500 °C (930 °F)	Dull red
Candle flame	1700 °C (3100 °F)	Dim orange
Bulb filament	2500 °C (4500 °F)	Bright yellow-white
Sun's surface	5800 °C (10500 °F)	Brilliant white
Blue star	6000 °C (10800 °F)	Dazzling blue-white

We can already see two of the principal shortcomings of an incandescent lightbulb: its poor efficiency at converting electric energy into visible light and its low color temperature. At 2500 °C, only about 12% of its thermal radiation is visible light; the rest is invisible infrared light. Its filament would have to reach 5000 °C before the infrared fraction of its thermal radiation would drop below 50%. Moreover, its 2500 °C color temperature makes it look yellowish when compared to sunlight because it doesn't emit enough blue light. Most of the developments in lighting over the past half century have focused on improving energy efficiency and color temperature.

The Filament

The bulb's filament is heated by an electric current, which provides it with thermal power in the amount specified on the bulb. For example, the filament of a 60-watt bulb produces 60 watts of thermal power from the 60 watts of electric power it consumes. Although the filament accumulates thermal energy at first, its temperature quickly rises until thermal energy flows out of it as heat as quickly as it's produced from electricity. Thus a 60-watt bulb sends 60 watts of thermal power into its environment as heat. Much of that heat is thermal radiation.

The temperature at which this heat balance occurs is surprisingly specific. That's because the filament's thermal radiation increases so rapidly with temperature that even a small excursion above its normal operating temperature will cause the filament to radiate away more thermal energy than it produces from electricity, and it will quickly cool back down to normal. Like any hot object, the power radiated by the filament is proportional to the fourth power of its absolute temperature. The precise relationship between temperature and emitted power is:

$$\text{radiated power} = \text{emissivity} \cdot \text{Stefan-Boltzmann constant}$$
$$\cdot \text{temperature}^4 \cdot \text{surface area}. \tag{7.3.1}$$

This relationship is called the Stefan–Boltzmann law, and the Stefan–Boltzmann constant that appears in it has a measured value of 5.67×10^{-8} J/(s·m²·K⁴). Remember that the temperature must be measured in kelvin.

Increasing the filament's temperature therefore increases both its color temperature and its brightness. Each of these characteristics is important to you, so it's fortunate that you can adjust them somewhat independently. Unless you're enjoying a romantic dinner and want the reddish glow of candlelight, you usually prefer higher color temperatures. For an incandescent bulb, that normally means a hotter filament. But once you've chosen the filament temperature, you can still make the filament brighter by increasing its surface area.

While the concept of an incandescent bulb is simple, finding a material that can tolerate extremely high temperatures is not. Early filaments were made of carbon and platinum. Of these materials, carbon showed the most promise. In 1879, Thomas Edison developed an incandescent lamp with a carbon filament that operated for several hundred hours. His wasn't the first incandescent bulb ever made but rather the first practical one. (For more about the development of carbon filaments, see ❶.)

However, while carbon has the highest melting temperature of any element (3550 °C or 6422 °F), it sublimes relatively quickly even at much lower temperatures. Thus a carbon filament that is heated close to its melting temperature quickly disappears as a gas. When a gap appears in the filament, it stops carrying electricity and "burns out." Carbon is also flammable, so it must be enclosed in an airtight glass bulb that contains either inert gases or a vacuum.

A better choice for filaments, now used in virtually all incandescent bulbs, is tungsten metal. Tungsten melts at 3410 °C (6170 °F) and sublimes extremely slowly below that temperature. Tungsten filaments can thus operate at higher temperatures than carbon filaments, producing richer, whiter light. Like carbon, however, hot tungsten burns in air and must be protected in a glass bulb.

To ensure that most of the electric power passing through the filament is converted into thermal power, the filament must be long and thin. A typical 60-W lightbulb has about 0.5 m (20 inches) of 25-micron (0.001-inch) tungsten wire, coiled up into a filament only about 2 cm (0.8 inches) long. To minimize the filament's length, it's wound into a double spiral. First it's wound into a thin springlike coil about 0.25 mm wide. Then this coil is wound into another coil to form the actual filament (Fig. 7.3.5). Fabricating such a complicated tungsten filament is so difficult that it wasn't accomplished until 1937.

The Glass Bulb

To keep the white-hot filament from burning, it's surrounded by a glass bulb that usually contains oxygen-free inert gas. This gas, typically nitrogen and argon, slows sublimation by bouncing some of the escaping tungsten atoms back onto the filament. Although the gas extends the filament's life, it has at least two drawbacks. First, it allows conduction and convection to carry some heat away from the filament (see ❷). Second, tiny tungsten particles that form in this gas rise with convection currents to produce a dark spot on the top of the bulb.

The glass bulb presents another challenge: operating a filament inside it involves passing wires right through the glass. This step isn't so easy because the glass and metal must seal to one another perfectly. Complicating this sealing requirement is the fact that materials expand as their temperatures increase. If the glass and metal don't expand equally as the bulb warms up, the bulb may leak or even break.

A material's thermal expansion is caused by atomic vibrations. Because of thermal energy, adjacent atoms vibrate back and forth about their equilibrium separations (Fig. 7.3.6). This vibrational motion isn't symmetric; the repulsive force the atoms experience when they're too close together is stiffer than the attractive force they experience when they're too far apart. As a result of this asymmetry, they push apart more quickly than they draw together and thus spend most of their time at more than their equilibrium separation. On average, their actual separation is larger than their equilibrium separation, and the material containing them is bigger than it would be without thermal energy.

As an object's temperature rises, its atoms move farther apart on average and the object grows larger in all directions. The extent to which an object expands with increasing temperature is normally described by its coefficient of volume expansion: the fractional change in the object's volume per unit of temperature increase. Fractional change in volume is the net change in volume divided by the total volume. Since most materials expand only a small amount when becoming 1 °C (or 1 K) hotter, coefficients of volume expansion are small, typically about 10^{-5} K^{-1} for metals, about 10^{-6} K^{-1} for special low-expansion glasses, and about 10^{-4} K^{-1} for liquids. In a lightbulb, the metal wires and the glass are carefully selected to have similar coefficients of volume expansion. As the bulb warms up, the wires and glass expand together and the seals remain intact.

Extended Life, Halogen, and Three-Way Bulbs

One way to prolong the life of the filament is to increase its surface area while providing it with the same electric power. With more surface area to radiate away thermal power, the filament doesn't get quite as hot and doesn't sublime as quickly. The result is an extended life bulb. Unfortunately, extended life bulbs are redder than conventional bulbs and also less energy efficient. Because an extended life bulb emits a smaller fraction of its input power as visible light, it must have a higher wattage to give equivalent lighting. As a result, extended life bulbs aren't always a bargain; the money you save on replacement bulbs may well be spent on

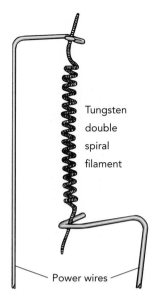

Fig. 7.3.5 The tungsten filament of a modern incandescent lightbulb is a double spiral—a coil wound from a smaller coil of extremely fine tungsten wire. The double spiral allows the manufacturers to put a long length of wire in a small space.

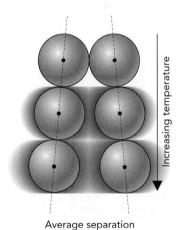

Fig. 7.3.6 The thermal kinetic energy of a solid increases with temperature, causing its atoms to bounce against one another more vigorously. As they vibrate, the atoms repel more strongly than they attract so their average separation increases slightly.

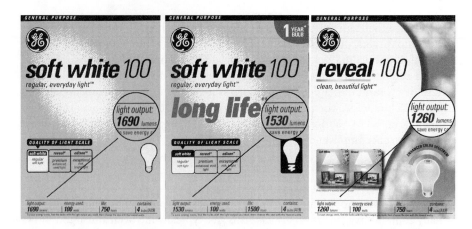

Fig. 7.3.7 Different 100-W lightbulbs produce different amounts of useful light, as revealed by the lumens listed on their packaging.

increased energy costs. When you're choosing bulbs, it's worth looking at how many lumens—the measure of useful illumination—they produce per watt of electric power consumed (Fig. 7.3.7).

You're much better off buying a halogen bulb, which is both longer-lived and more energy efficient than a conventional bulb. A halogen bulb uses a chemical trick to rebuild its filament continuously during operation. This filament is enclosed in a small tube of quartz or aluminosilicate glass, which can tolerate high temperatures and reactive chemicals (Fig. 7.3.8). The tube contains molecules of the halogen elements bromine and/or iodine. During operation, the tube becomes extremely hot and the halogen reacts with any tungsten atoms on its inside surface. They form a gas of tungsten–halogen molecules that drift about the tube until they encounter the white-hot filament. The molecules then break apart and the tungsten atoms stick to the filament.

Fig. 7.3.8 These halogen bulbs operate at higher temperatures than normal incandescent bulbs and produce whiter light. The large glass envelope of the upper bulb protects a smaller lamp inside.

The halogens act as recycling agents, seeking out tungsten atoms that have sublimed from the filament and returning them to it. Unfortunately, this recycling process slowly changes the structure of the filament. The returning tungsten atoms deposit unevenly, so that the filament gradually develops thin spots and eventually burns out. Nonetheless, the filament lives more than 2000 hours, even when it runs several hundred degrees hotter than in a conventional bulb. Its higher filament temperature allows a halogen bulb to produce whiter light than a conventional bulb and increases its energy efficiency.

Halogen bulbs do have some drawbacks. The quartz glass tube is small, extremely hot, and filled with toxic halogen gas. It's a fire and safety hazard, which is why it's often protected by an outer glass tube. The quartz tube is sensitive to fingerprints, which discolor and damage it when it becomes hot. Finally, the tungsten recycling system only functions at high temperatures. When a halogen lamp operates on a dimmer at less than full power, tungsten can accumulate on the quartz tube, darkening the bulb and reducing its life. The bulb needs to operate at full power periodically to clean off the quartz tube and return the tungsten to the filament.

Bulbs with different power ratings have filaments with different amounts of surface area. The filament in a 100-W bulb has four times the surface area of the filament in a 25-W bulb and thus emits four times as much light. To support this fourfold increase in thermal power, the 100-W filament also consumes four times as much electric power. Both filaments operate at the same temperature and emit light with the same color temperature.

One way to make an incandescent bulb with a variable light output is to put several independent filaments in it. A "three-way bulb" has two filaments (Figs. 7.3.9 and 7.3.10) that can be turned on and off separately. In a 50-100-150 W bulb, one filament uses 50 W of electric power and the other uses 100 W. If only the low-power filament is on, the bulb appears to be a 50-W bulb. If only the high-power filament is on, it appears to be a 100-W bulb. But when both filaments are on, the bulb appears to be a 150-W bulb. Since one of the filaments invariably burns out before the other, the bulb fails by going from having three light levels to only one.

〰 Toasters and Space Heaters

If the filament of a lightbulb never operated above red-hot, it wouldn't need to be made of expensive tungsten. It could be made instead of an inexpensive nickel-chromium alloy called nichrome. Although nichrome will melt if heated beyond red-hot, it has an advantage over tungsten at lower temperatures: it doesn't burn in air. So while nichrome filaments aren't suitable for lightbulbs that are meant to emit lots of visible light, they're great for "lightbulbs" that emit red and infrared light. In fact, nichrome is wonderful for all sorts of electrical heating elements!

Nichrome-based filaments and heating elements are common in homes, where they appear in toasters, space heaters, hot-water heaters, clothes irons, and many other appliances. They're also standard in electric stovetops and electric ovens. You can often see these nichrome filaments directly in devices that can protect them from contact. And in other devices, the nichrome filaments are typically encased in electrically insulating tubes to form sturdy, reliable electrical heating elements.

A toaster offers a good starting point because you can see its red-hot nichrome filament at work. That filament is relatively large because it operates cooler and radiates far less thermal power per unit of surface area than an ordinary lightbulb filament. As long as bread never touches the toaster's filament directly, that filament transfers heat to the bread only by convection and radiation. In most cases, the bread is placed beside or below the filament, so natural convection isn't effective at transferring heat to the bread. Instead, the toaster heats the bread primarily by radiation and the bread experiences the high surface temperatures that are necessary for caramelizing and browning. In a pinch, you can make toast in a broiler because it also cooks with radiation. Baked toast usually doesn't brown as well and doesn't taste the same.

If you turn the toaster inside-out, you'll have a space heater. While a real space heater may have a fan to circulate heated air or a fluid-filled radiator to diffuse the heat, it's still a first cousin to the toaster. A space heater uses a nichrome filament to convert electric power into thermal power and that thermal power flows out into the room. With space heaters that allow you to see their hot filaments directly, you can feel their radiant heat bathing your skin. Even in cold room air, this radiative heat transfer can warm your skin significantly and make you feel comfortable.

Other space heaters hide their hot filaments to reduce the radiant heat transfer and depend instead on forced convection to carry heat from their filaments to the room air. These space heaters use fans to blow warm air into the room. Hairdryers work exactly the same way, except that they're specialized to blowing warm air at

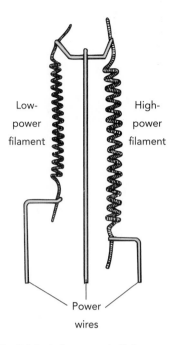

Low-power filament   High-power filament

Power wires

Fig. 7.3.9 A three-way bulb has two independent filaments. The filament on the left is shorter and thinner than the one on the right and emits about half as much light. The three different light levels correspond to having the left filament on, the right filament on, and both filaments on.

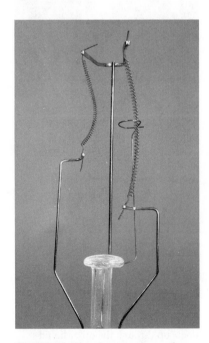

Fig. 7.3.10 The glass envelope of this three-way lightbulb has been removed to expose its two filaments. The shorter, thinner filament (*left*) uses 50 W, while the longer, thicker one (*right*) uses 100 W.

your hair. You can usually see the nichrome heating element inside a hairdryer, although it never gets hot enough to emit visible light. The filaments in electric heat guns, however, do glow red-hot and should obviously never be used with hair!

Finally, there are space heaters in which heat from a nichrome heating element flows into a fluid that's sealed inside a radiator. This fluid undergoes natural convection inside the radiator and the entire radiator warms up simultaneously. The large heated surface of that radiator then transfers heat to the room by convection and radiation. The surface temperature of this radiator is warm but not overly hot, which makes it safe for use around children. Touching it won't burn them and there's no chance that they'll start a fire by poking things into it.

Hot Water Heaters and Lava Lamps

An electric hot water heater works like one of the radiator space heaters. The water in a large tank is heated by an electric element that's immersed in the water. Since the heated water rises convectively, all of the water near or above the heating element warms up simultaneously to the desired temperature. When that temperature is reached, a thermostat turns off the heating element. Hot water is extracted through a pipe at the top of the tank and cold water is introduced into the bottom of the tank to replace it. Since hot water floats on cold water, the entering cold water sits at the bottom of the tank and you can continue to extract hot water from the top of the tank until there is no hot water left.

As soon as the cold water level rises above the heating element, the thermostat senses the temperature drop and turns the heating element back on. As long as you don't remove too much hot water too quickly, the hot water heater will successfully replace it and you'll never run out. But if you do run out of hot water, the system will take some time to recover. That's because the heating element heats all of the cold water above it at the same time—the convection cell that forms inside the hot water heater then reaches all the way from the heating element up to the top of the tank.

To speed the recovery process, many electric hot water heaters have two heating elements: one located near the bottom of the tank and one located near the middle of the tank. Activating both heating elements simultaneously would overload the hot water heater's electrical supply, so it can only turn on one heating element at a time. When you've depleted all the hot water, it turns on the middle heating element and heats only the water from the middle of the tank on up. With only half as much water to heat, the hot water heater is able to start delivering hot water in half the time.

Once the top half of the tank is hot, the middle heating element turns off and the bottom heating element turns on. A new convection cell forms that reaches from that heating element up to the bottom of the existing hot water. Water in the lower half of the tank now heats up. If you only use a little hot water at a time, it is always the bottom heating element that warms up the incoming cool water.

Since the bottom heating element does most of the work in a typical hot water heater, it's usually the first to fail. When it does, you must turn off and drain the hot water heater, swap in a replacement heating element, refill the hot water heater with water, and then turn it back on. If you accidentally turn the hot water heater on before filling it with water, the heating element will have no water around it to carry away its thermal energy and that element will overheat and melt in a matter of seconds.

A Lava® Lamp is charming toy in which heat from an incandescent lightbulb causes blobs of softened wax to float up through another liquid in a glass bottle and then cool and descend to repeat the trip again (Fig. 7.3.11). The convection cell that carries the wax up and down in the bottle would be easy to understand

Fig. 7.3.11 The wax in this Lava Lamp melts when heated from below by an incandescent lightbulb. The molten wax has a low density, so it floats up through the other liquid until it reaches the top of the bottle. There the wax cools and begins to solidify, experiencing a rise in density so that it sinks back to the bottom of the bottle. The wax repeats this journey over and over again.

except that there are two liquids in the bottle and they seem to circulate in opposite directions: the heated wax rises through the other liquid and the cooled wax descends through it.

The phenomenon that makes this system work is that the wax is close to its melting temperature and it becomes more liquid-like when it's heated by the incandescent lightbulb at the bottom of the bottle and more solid-like when it cools at the top of the bottle. Like most materials, the wax is significantly denser as a solid than it is as a liquid. When the lightbulb heats the wax, the wax becomes more fully liquid. The liquid or melting wax is buoyant in the other liquid and floats toward the top of the bottle. But once the wax reaches the top of the bottle, it cools and begins to transform into its solid phase. Solid or solidifying wax is relatively dense and sinks through the other liquid toward the bottom of the bottle. This cycle then repeats over and over again.

Knit acrylic fibers

Woven nylon shell

Fiber insulation and trapped air

Closed-pore foam insulation

SECTION 7.4 **Clothing and Insulation**

When you sit in front of a fire on a cold winter day, your skin is warmed by heat from the hot embers. But when you walk through the snow on your way to the store, the last thing you want is heat transfer. As the hottest object around, you will become colder, not warmer. Instead, you do your best to avoid heat transfer. So you bundle up tight in your new down coat. Its thermal insulation keeps you warm in your frigid environment. In this section, we'll examine thermal insulation and see how it keeps heat from moving between objects.

The Importance of Body Temperature

Thermal insulation slows the heat transfer between objects and keeps your home warm, your refrigerator cold, and your fingers comfortable when you pick up a cup of hot coffee. One of the most important examples of thermal insulation is your clothing. The principal non-aesthetic purpose of clothing is to control the rate at which heat flows into or out of your body. Clothing helps you maintain your proper body temperature.

The goal of keeping body temperature in homeostasis is unique to mammals and birds. Cold-blooded animals such as reptiles, amphibians, and fish make no attempts to control their body temperatures. Instead, they exchange heat freely with their surroundings and are generally in thermal equilibrium with their environments.

Unfortunately, the chemical processes that are responsible for life are very sensitive to temperature. That sensitivity is due in part to thermal energy's role in initiating chemical reactions; it provides the activation energies that many chemical reactions need in order to proceed. As a cold-blooded animal's temperature decreases, there is less thermal energy per molecule and these chemical reactions occur less frequently. The animal's whole metabolism slows down and it becomes sluggish, dimwitted, and vulnerable to predators.

In contrast, warm-blooded animals have temperature regulation systems that allow them to maintain constant, optimal body temperatures. Regardless of its environment, a mammal or bird keeps the core of its body at a specific temperature so that it functions the same way in winter as in summer. The advantages of uniform temperature are enormous. On a cold day, a warm-blooded predator can easily catch and devour its slower-moving cold-blooded prey.

But there is a cost to being warm-blooded. The thermal energy associated with an animal's temperature must come from somewhere and the animal must struggle against its environment to maintain its body temperature. Without realizing it, many of our behaviors are governed by our need to maintain body temperature. Our bodies are careful about how much thermal energy they produce and we work hard to control the rate at which we exchange heat with our surroundings.

A resting person converts chemical potential energy into thermal energy at the rate of about 80 Calories-per-hour. Our bodies use that much ordered energy even when we are doing no work on the outside world. Our hearts keep pumping, we keep synthesizing useful chemicals and building cells, and we keep thinking. Since the chemical energy isn't doing outside work or creating much potential energy anywhere, most of it ends up as thermal energy.

Eighty Calories-per-hour is a measure of power, equal to about 100 W. A resting person is using about as much power as a 100 W light bulb and, as with the light bulb, most of that power ends up as thermal energy. If a person is more active, he or she will produce more thermal energy. This steady production of thermal energy is why a room filled with people can get pretty warm. 100 W may not seem like very much power, but when a hundred people are packed into a tight space, they act like a 10,000 W space heater and the whole room becomes unpleasantly hot.

If you had no way to get rid of this thermal energy of metabolism, you would become hotter and hotter. To maintain a constant temperature, you must transfer heat to your surroundings. Since heat flows naturally from a hotter object to a colder object, your body temperature must generally be hotter than your surroundings. This requirement is one reason why human body temperature is approximately 37 °C (98.6 °F). This temperature is higher than all but the hottest locations on earth so that heat flows naturally from your body to your surroundings. You produce thermal energy as a by-product of your activities and transfer this thermal energy as heat to your colder surroundings.

Since the rate at which your resting body generates thermal energy is fairly constant, the principal way in which you maintain your temperature is by controlling heat loss. You and other warm-blooded animals have developed a number of physiological and behavioral techniques for controlling heat loss. Let's examine those techniques in terms of the three mechanisms of heat transfer: conduction, convection, and radiation.

Retaining Body Heat: Thermal Conductivity

Overall, you must lose thermal energy at the same rate as you produce it; about 100 joules each second. This modest rate is relatively easy to achieve. Except on hot days or when you are exercising hard, your body must struggle to avoid losing heat too quickly. Since all three heat-transfer mechanisms are involved in this heat loss, you must control them all in order to keep warm.

One way in which your body retains heat is by impeding conductive heat loss. Some materials are better conductors of heat than others; they have different thermal conductivities. Thermal conductivity measures how rapidly heat flows through a material that is exposed to a difference in temperatures. Skin has a particularly low thermal conductivity, meaning that it conducts relatively little heat compared to materials such as glass or copper.

Thermal conductivity is a characteristic of a material itself, so it's defined for a small cube of that material with a temperature difference of one degree across it. Your skin has a specific thermal conductivity. But to determine how much heat will flow through your skin you must consider not only its thermal conductivity, but also its size and shape and the temperature difference across it. The more skin surface you have and the greater the temperature difference across it, the more heat your skin will conduct. However, thickening your skin reduces the temperature difference across each cube of it and lessens the heat conduction through it.

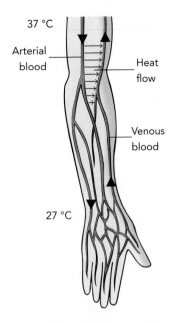

37 °C

Arterial
blood

Heat
flow

Venous
blood

27 °C

Fig. 7.4.1 Blood flowing toward your hand through arteries exchanges heat with blood returning to your heart through veins. In this fashion, your blood is able to carry oxygen and food to your fingers without warming them all the way up to core body temperature. This adaptation reduces the rate at which you lose heat in cold weather.

Thus the amount of heat flowing through your skin depends on its thermal conductivity, its surface area, the temperature difference across it, and its thickness. Your body controls all of these factors in trying to minimize heat loss:

1. It uses materials with very low thermal conductivities in your skin.
2. It makes your skin as thick as is practical.
3. It minimizes the surface area of your skin.
4. It minimizes the temperature difference across your skin.

Your skin and the layers immediately beneath it contain fats and other thermal insulators. Fat's thermal conductivity is about 20% that of water and only about 0.03% that of copper metal. Your body uses fat for energy storage anyway, but by locating the fat in and beneath your skin, your body improves its heat retention. Furthermore, the presence of a fatty layer beneath your skin effectively thickens your skin and reduces the temperature difference across each unit of thickness. "Thick-skinned" people retain body heat better than those who are "thin-skinned."

Minimizing surface area means that your body is relatively compact, shaped more like a ball than a sheet of paper. Many other adaptive influences have led to the evolution of arms, legs, and fingers that increase your total surface area. However, you have little superfluous surface through which to lose heat.

Finally, your body tries to lessen conductive heat loss by reducing the temperature difference between your skin and the surrounding air. It does this by letting your skin temperature drop well below your core body temperature. On a cold day, your hands and feet feel cold because they are cold. The colder they get, the less heat they lose to the cold air they touch.

Allowing your hands to become cold would be simple were it not for your circulating blood. Your blood must cool down from core body temperature as it approaches your cold fingers and must warm back up to core body temperature as it returns to your heart. This change in blood temperature occurs via a mechanism called countercurrent exchange. As the warm blood flows through arteries toward your cold fingers, it transfers heat to the blood returning to your heart through nearby veins (Fig. 7.4.1). The blood heading toward your fingers becomes colder while the blood returning to your heart becomes warmer.

Retaining Body Heat: Convection

On a cold day, heat leaving your skin warms the nearby air. The amount of heat required to warm this air depends on how much air is involved, on that air's specific heat (see p. 205), and on how much the air's temperature must rise to reach body temperature. Since air is a very poor conductor of heat, however, your skin warms only a thin layer of it. If this layer of warmed air never moved, you would only have to heat it once and that would be that. Protected by this warmed air, your skin would no longer have to transport any heat and the temperature difference across your skin would shrink to zero. You would feel comfortably warm.

But air moves all too easily. In reality, each time your skin manages to warm the nearby air, convection gently removes that warmed air and replaces it with fresh cold air. The temperature difference across your skin thus remains large and heat flows quickly out through your skin. You feel cold. Wind worsens this heat loss because it blows away warmed air near your skin even faster than does natural convection. Just as a forced convection oven cooks food faster, so a forced convection freezer (i.e., a cold, windy day) chills people faster. The enhanced heat loss caused by moving air is called wind chill—you feel even colder on a windy day.

To combat convective heat loss and wind chill, warm-blooded animals are covered with hair or feathers. Hair is itself a poor conductor of heat but its main purpose is to block airflow. Air passing through hair experiences large drag forces that slow its motion. In the dense tangle of a sheep's wool, air is trapped and can barely move at all. Since convection requires airflow, the sheep can only lose heat via conduction through the hair and the air. Since both are terrible conductors of heat, the sheep stays warm.

We humans have relatively little hair and are thus poorly adapted to living in cold, windy climates. Our lack of natural insulation is one of the reasons we wear clothing. Like hair and feathers, our clothing traps the air and reduces convection. Finely divided strands or filaments are particularly effective at stopping the flow of air. Not surprisingly, the best insulating clothing is made of hair (natural or synthetic) and feathers (also natural or synthetic). Since motionless air has a lower thermal conductivity than the hair or feathers that trap it, the ideal coat uses only enough material to keep a thick layer of air from moving.

This discussion also applies to water and swimming. If the water around you didn't move, you would soon be nice and warm. That's why some swimmers wear wet suits. The spongy material in a wet suit keeps the layer of water near the swimmer's skin from moving. As long it remains motionless, water is a respectable thermal insulator. This is evident in Fig. 7.4.2, where heating the top of a tube of water inhibits convection.

Retaining Body Heat: Radiation

You can also lose body heat via radiation. Your skin emits thermal radiation toward your surroundings and they emit thermal radiation toward you. The amount of heat transferred by thermal radiation depends on the temperature of each surface and on their emissivities, that is, on how well they absorb and emit light. As I explained while introducing the Stefan–Boltzmann law on p. 222, the amount of heat radiated by a surface depends roughly on the fourth power of its temperature, measured in an absolute temperature scale, so that hotter objects radiate far more heat than colder objects.

As always, heat flows from the hotter object to the colder object. However, while conduction and convection transfer heat in proportion to the temperature difference between objects, radiation transfers heat in proportion to the difference between the *fourth powers* of their temperatures. That is why radiative heat transfer to or from your skin is most noticeable when you're exposed to an unusually hot or cold object.

The sun warms your skin quickly because it radiates more heat at you than the rest of your surroundings combined. Measured on an absolute temperature scale, the sun's surface temperature (6000 K) is about 20 times that of your skin (310 K). Though it's very distant and appears small to your eye, the sun radiates about 20^4 or 160,000 times as much heat toward you as you radiate toward it.

In contrast, the dark night sky cools you quickly because it has an extremely low temperature. The nearly empty space beyond the earth's atmosphere is only a few degrees above absolute zero. When you stand in an open field on a dark, clear night, you radiate about a hundred watts of thermal power toward space but it radiates very little back toward you. Since you lose heat quickly, you feel cold.

You can improve your situation by standing under a leafy tree. Even in cold weather, the tree is much hotter than space and emits far more thermal radiation toward you. While the tree can't replace a crackling campfire, it will still help to keep you warm.

Since the air overhead has the same temperature as the tree, you might wonder why the air itself doesn't keep you warm. The answer is that air is reasonably transparent to infrared light, absorbing and emitting just a small fraction of it. Only water vapor, carbon dioxide, and a few other gases in the air interact with infrared light,

Fig. 7.4.2 Convection only occurs if the hotter object is below or next to the colder object so that the heated fluid is able to rise. If you heat a tube of water near its top, the hot water stays near the top and the cold water remains at the bottom. Because water itself is a poor conductor of heat, the water at the top of the tube can boil while the bottom of the tube is cool enough to hold in your hand.

so air's emissivity is small. When there's nothing above you but dark, clear sky, you don't exchange much thermal radiation with the atmosphere; you exchange thermal radiation mostly with the empty space beyond it. Unlike the atmosphere, space is a true blackbody with an emissivity of 1 and a temperature of about 2.725 K. Its dim thermal glow is a cooled relic of the universe's hot early years, dating from about 400,000 years after the Big Bang.

Emissivities are clearly important when you're feeling cold or feeling hot. If you're feeling cold, you want to exchange thermal radiation with warmer objects but not with colder objects. If you're feeling hot, you want to do the reverse. We've seen that black surfaces are ideal whenever you want to exchange thermal radiation with another object, but what about when you don't want to exchange thermal energy? That's where shiny, white, and transparent surfaces come into play.

A perfectly shiny or perfectly white surface has an emissivity of zero. The two are closely related to one another: a shiny surface reflects thermal radiation back in a mirrorlike fashion while a white surface scatters thermal radiation back in all direction. Though they have different appearances, both surfaces inhibit radiative heat transfer. When you face a perfectly shiny or perfectly white surface, it sends your thermal energy back at you and prevents you from exchanging thermal energy with whatever lies beyond it.

Shiny and white surfaces can be used as thermal insulators. If you put a shiny or white surface between yourself and a cold manikin, the two of you won't be able to exchange thermal radiation. Instead, each of you will see your own thermal radiation coming back at you. This insulating effect works even if one of you wears the shiny or white surface as clothing: whoever is wearing the shiny or white clothing won't emit thermal radiation at all and the other will see his/her own thermal radiation sent back by that clothing.

When it comes to using shiny or white surfaces as insulation, however, I must warn you about an important complication. A surface's emissivity varies with temperature, so that it may have one emissivity at 1000 °C, but a rather different emissivity at room temperature. Because room-temperature thermal radiation is infrared light, which we can't see, it's hard to guess an object's room-temperature emissivity just by looking at it. An object that appears white or shiny to visible light may look quite different to the infrared light of room temperature thermal radiation. In fact, it may well absorb infrared light and therefore be black to room-temperature thermal radiation!

Clothing is a case in point. Nearly all clothing materials, regardless of what color they appear to your eye, are black in the infrared and have emissivities near 1 for room-temperature thermal radiation. That means that your clothing is probably emitting thermal radiation just about as brightly as it can, given its temperature. The main exceptions to that rule are metallic fabrics and lamé fabrics—fabrics with metal threads woven into them. These special fabrics have the reduced room-temperature emissivities associated with their shiny metal surfaces. Even then, the choice of metal affects their room-temperature emissivities in ways that may not be visible to your eye.

Metals often look different in the infrared than they do in visible light. For example, the lunar module shown in Fig. 7.4.3 was insulated with gold foil rather than aluminum foil because gold foil is shinier in the infrared than is aluminum foil. Despite its yellow hue, gold's room-temperature emissivity is about 0.02 while that of aluminum is about 0.05. The gold foil was therefore particularly effective at keeping the shadowed side of a lunar module from exchanging thermal radiation with empty space.

Transparent surfaces also have emissivities near zero, but they avoid exchanging thermal radiation by letting it pass through them. When you put a transparent surface between two objects, one hot and the other cold, the surface doesn't block

Fig. 7.4.3 This Lunar Lander is wrapped in reflective foil to reduce its emissivity. As a result, it emits and absorbs relatively little thermal radiation.

any light and the two objects exchange thermal radiation as though the surface weren't there.

Truly transparent clothing would provide no radiative insulation, but since few clothing materials are transparent to room-temperature thermal radiation, I won't dwell on this topic. Nonetheless, materials that are transparent to high-temperature thermal radiation (i.e., visible light) are common. An artist who works with molten glass will know from painful experience that this transparent material provides almost no visible warning that it's hot. Its high-temperature emissivity is so small that you can barely see it glow.

These issues of radiative heat transfer explain why we wear certain colors and why we are careful about exposing ourselves to the sun. On hot, sunny days, it makes sense to wear light colors and to sit in the shade. Both actions reduce the amount of heat transferred to you by the sun. Since the sun's surface temperature is about 6000 K, a garment's ability to absorb sunlight depends on its high-temperature emissivity. Light colored clothes have small high-temperature emissivities, so they don't absorb much sunlight. Sitting in the shade prevents the sun from exchanging heat with you directly. If you are feeling cold, however, you do well to switch to dark clothes and to sit in the direct sun.

While low-emissivity clothing would help you retain body heat in a cold, sunless environment, most clothes radiate body heat all too efficiently. With room-temperature emissivities near 1, they are effectively black and glow brightly with thermal radiation. Only fabrics that incorporate metals, such as the metallic and lamé fabrics I mentioned above, can use low emissivity to slow the loss of body heat. That's why metal-coated plastic blankets are included in many emergency rescue kits; wrapping yourself in one of these blankets shiny-side out keeps you from exchanging thermal radiation with your surroundings (see ❶). And lamé fabrics may well help to keep their typically underdressed wearers warm.

Keeping Cool When It's Hot Outside

Slowing heat loss isn't always a good idea. If you retain heat too well, you'll overheat. When exercising or on a very hot day, it may be necessary to encourage heat transfer to your surroundings by enhancing conduction, convection, or radiation.

You can increase conductive heat loss by moving into cold air or, even better, cold water. With a larger temperature difference across your skin, the rate of heat conduction through it will increase. You can increase convective heat loss by actively circulating the air or water with a fan or pump. The more cold air or water that directly touches your skin, the more heat you'll lose. And you can increases radiative heat loss by standing in front of an open freezer. Regardless of what clothes you're wearing, they're probably black to body-temperature thermal radiation and you'll radiate far more heat toward the freezer than it radiates toward you.

But what happens when you are stuck in an environment that's hotter than body temperature? If you are the coldest object around, you are going to get hotter and hotter. For a minute or two, insulating clothing can slow the rate at which your temperature rises so that you can pull a casserole from a hot oven or rescue a person from a fire. But even when you are perfectly insulated from your surroundings, your metabolism will cause your body temperature to rise. What does your body do to keep from overheating?

It sweats. By covering your skin with water, your body is using a new trick to eliminate heat. For water to evaporate, changing from a liquid to a gas, it needs energy. It needs its latent heat of evaporation. As I discussed in Section 7.2, the molecules in liquid water are held together by chemical bonds that must be broken in order for it to transform into a gas. The energy to break those bonds and evaporate your sweat is drawn from your body as heat. The faster your sweat evaporates,

❶ Smoke jumpers who battle forest fires occasionally get trapped by the fires they're fighting and must try to survive as the fires burn over them. Their chances improve significantly if they use small personal survival shelters that have shiny metallic surfaces. Huddled against the cool ground underneath one of these tent-like shelters, a firefighter is relatively insulated from the fire overhead. Assuming the firefighter is in a low spot with nothing to burn nearby and makes no contact with the shelter itself, conduction and convection can't convey much heat to the firefighter. And with the shelter's shiny, low-emissivity surface reflecting most of the fire's thermal radiation, radiative heat transfer is also greatly diminished. On August 29, 1985, 73 firefighters were trapped for several hours by a fire in the Salmon National Forrest near Salmon, Idaho and survived only with the aid of their personal fire shelters.

the more heat it draws out of your skin. As long as you can keep producing sweat and the air is dry enough to allow it all to evaporate, you can maintain your usual body temperature in surprisingly hot air.

Animals with hair can't sweat directly because they don't have enough air circulation near their skins to carry away the water vapor. Instead, these animals pant. Evaporation from their mouths and lungs draws heat from their bodies. On a hot day or when a dog has been exercising hard, its tongue will dangle out of its mouth to evaporate saliva and cool the dog.

Insulating Houses

The same techniques that keep people and animals warm are used to control heat flow in houses and household objects. However, because houses and their contents don't move much, they can make use of insulating methods that are heavy, bulky, rigid, or fragile. Let's take a look at some of these insulating schemes.

The goal of housing insulation is to render a house's internal temperature effectively independent of the outside temperature. When it's cold outside, you want as little heat as possible to flow out of your warm house. When it's hot outside, you want as little heat as possible to flow into your cool house. So you fill its walls with insulating materials.

While there are many solid materials that are poor conductors of heat, including glass, plastic, hair, sand, and clay, the best insulator used in normal construction is air. Most modern buildings use air insulation. Unfortunately, air tends to undergo convection so that it can't be used by itself. To prevent convection, air is trapped in porous or fibrous materials such as glass wool, sawdust, plastic foam, or narrow channels.

Glass wool or fiberglass is made by spinning glass into very long, thin fibers that are then matted together like cotton candy. Solid glass is already a poor conductor of heat but reducing it to fibers makes it even more insulating. The path that heat must take as it's conducted through the tangled fibers is very long and circuitous and very little heat gets through. Most of the volume in glass wool is taken up by trapped air. The glass fibers keep the air from undergoing convection so the air must carry heat by conduction.

Overall, glass wool and the air trapped in it are excellent insulators. They also have the advantage of being nonflammable. In addition to its use in buildings, glass wool serves as insulation in ovens, hot water heaters, and many other machines that require nonflammable insulation. Most modern houses have about 10–20 cm (4–8 in) of glass wool insulation built into their outside walls, along with a vapor barrier to keep the wind from blowing air directly through the insulation. (For a discussion of older insulating techniques, see Fig. 7.4.4.)

Because hot air rises and cold air sinks, the temperature difference between the hot air below the ceiling of the top floor and the cold air above that ceiling can become quite large. That ceiling is thus a very important site of unwanted heat transfer and requires heavy insulation. Glass wool inserted above the ceiling of the top floor in a modern house may be more than 30 cm (12 in) thick (Fig. 7.4.5).

While glass wool is an excellent insulator, other materials are used in certain situations. Urethane and polystyrene foam sheets are both waterproof and better insulators than glass wool. Unfortunately, they are also flammable and relatively difficult to work with. Nonetheless, they are used in construction and are particularly well suited for refrigerators and coffee cups, where rigidity and flammability aren't problems.

In older houses that weren't insulated properly during construction, insulation can be blown into the walls or ceilings through holes drilled into the surfaces. As always, these insulators are porous or fibrous materials so that the main insulator

Fig. 7.4.4 While stone is not a good conductor of heat, it's not nearly as good an insulator as air trapped in a fibrous mat. Medieval stone castles were notoriously cold in winter because heat flowed too easily out of them through their stone walls. This tapestry slows the flow of heat to the outside air and helps to keep the room warm.

Fig. 7.4.5 (*left*) The space between the sloped ceiling of this building and its actual roof is insulated with a thick mat of fiberglass insulation. (*right*) The importance of that insulation is clearly visible on a snowy day, when the snow melts first where the fiberglass insulation is compressed by the beams and is therefore thinnest.

is trapped air. Urea-formaldehyde foams are convenient for filling walls and ceilings because they can be pumped into cavities before they harden. However, concerns that they release toxic chemicals have reduced their appeal. Vermiculite and fire-proofed cellulose chips are among the most common loose fill insulations.

Bringing Fresh Air into a House

While it might seem ideal to block all movement of air into or out of a house so as to prevent all heat transfer, a truly sealed house isn't very pleasant or healthy. Every smell will linger for weeks because it can't get out of the house. Older houses are drafty enough that the air inside is exchanged with outside air many times a day. But modern, energy efficient houses are nearly sealed and exchange air with the outside only a few times a day.

One way to deliberately exchange air but not heat with the outside is to use countercurrent exchange. Special ventilators are available in which entering air and leaving air pass near one another on opposite sides of thin metal ducts. In such a ventilator, air entering a house is allowed to exchange heat with air leaving the house. By the time it reaches the inside, the entering air is almost at room temperature. In principle, very little heat should be exchanged with the outside by a countercurrent exchange ventilator. Unfortunately, this concept is hard to implement effectively and operating these ventilators often requires more energy than they save.

Windows

While most household insulation revolves around air trapped in pores or around fibers, there are special circumstances in which finely divided materials just won't do. Windows have a special requirement that they must be transparent. They can't be filled with foam or fiberglass, and solid glass is just not a good enough insulator.

The most common way to insulate windows is to use two or more panes of glass separated by narrow gaps. These vertical gaps are filled with gases such as nitrogen or argon and prevent the easy conduction of heat from one side of the window to the other. While convection can occur inside the gaps, the convection cells that form are tall and thin and therefore relatively ineffective at carrying heat from one side of the window to the other.

If they're not mounted vertically, however, these multiple-pane windows can form convection cells that carry heat fairly efficiently between the panes so they don't insulate as well. And even when they're mounted vertically, multiple-pane

❷ Most metals have plasma frequencies so high that they can be expected to reflect the entire visible spectrum and even part of the ultraviolet. Aluminum, for example, reflects all the colors of the rainbow nearly perfectly. But the quantum physics of metals that we'll encounter on p. 405 introduces a complication that affects primarily the coinage metals. Gold, silver, and copper each has only a narrow range of empty conduction levels available and thus has difficulty interacting with and reflecting photons at the violet end of the visible spectrum. Silver provides a warm reflection because it doesn't reflect violet well. Gold appears yellowish because it reflects neither blue nor violet well. And copper is orange because it has trouble reflecting even green photons.

Fig. 7.4.6 This silver coffee urn may not seem to be insulated, but it is. Its polished silver surface is so shiny that it is an almost perfect mirror for infrared and visible light. As a result, it neither emits nor absorbs room-temperature thermal radiation well. Moreover, the air near the urn's vertical sides forms tall convection cells that aren't very efficient at carrying away heat from the urn. Overall, the urn loses heat much more slowly than it would if it were black and had fewer vertical surfaces.

windows provide far less insulation than an actual wall. That's one reason to use shades and curtains: those extra surfaces not only block the view, they also reduce heat transfer through the window.

Perhaps the biggest failing of conventional multiple-pane windows is that they can't control radiative heat transfer. Although ordinary glass is clear to visible light, it's nearly jet black to room-temperature thermal radiation—it has a room-temperature emissivity of about 0.92. That means that the panes of a conventional multiple-pane window are exchanging thermal radiation with nearly maximum efficiency all the time. When there's a big temperature difference between the inside and outside panes, heat will flow via thermal radiation from the hotter pane to the cooler pane and spoil the window's insulating behavior.

Sophisticated multiple-pane windows reduce radiative heat transfer by using low-emissivity glass. A low-emissivity (Low-E) window has a thin coating on the inside surface of one of its glass panes. This coating makes the glass surface behave like a shiny mirror for infrared light: the coated surface doesn't emit any room-temperature thermal radiation and it reflects back room-temperature thermal radiation emitted by the glass surface opposite it. But because the coating reflects only infrared light, the window still looks transparent to your eye. The coating is a "heat mirror"—it reflects infrared thermal radiation while transmitting visible light. Although you can still see through the window, its hidden internal heat mirror dramatically slows the passage of heat through the window.

A typical low-emissivity window coating has a room-temperature emissivity of about 0.10, so it's almost as good as a real metal mirror. Made of metals and/or metal-oxides, these coatings use various physical phenomena to reflect infrared light while transmitting visible light. New concepts come on the market every year or two.

One of the most commonly used coatings is indium-tin-oxide, a transparent electrical conductor that's also widely used in electronic displays. Indium-tin-oxide controls the digits of most digital watches, so you're probably looking through it whenever you read a watch! Indium-tin-oxide is transparent in the visible because it has a low plasma frequency—the natural resonant frequency of the mobile charges in an electrical conductor. Electrical conductors don't conduct electricity well at frequencies above their plasma frequencies. As we'll see later in this book, visible light has frequencies that are higher than those of infrared light. While indium-tin-oxide can respond to and reflect low-frequency infrared light, its plasma frequency is too low for it to respond to and reflect high-frequency visible light (For more about plasma frequencies, see **❷**.)

Unfortunately, multiple-pane windows eventually leak and allow their moisture-free low-thermal-conductivity gases to be replaced by ordinary air. When a conventional multiple-pane window leaks, it usually fogs up. When a Low-E window leaks, the moisture-laden air often attacks the coating chemically and the coating changes colors. In most cases, your only option is to replace the entire window assembly: all the glass panes and their housing.

Food Insulation

Food storage also depends on thermal insulations such as plastic foam and fiber mats. For example, Styrofoam® cups block heat flow so well that they can keep coffee warm for an entire commute, especially when you prevent the coffee from losing heat via evaporation by putting a lid on your cup. But sometime food insulation takes unusual forms: the polished silver coffee urn in Fig. 7.4.6 is insulated by silver's high reflectivity. Polished silver is such a poor emitter of thermal radiation that the urn loses heat only via convection. And since the urn's surfaces are primarily vertical, even convection isn't all that effective at drawing heat out of the urn.

But if you try to keep food hot or cold for a very long time, you'll find that even a fairly thick blanket of foam or fiber insulation won't be a sufficient barrier against heat transport. You do better with a glass or metal Thermos® bottle, which makes use of a completely different technique of insulation: a vacuum.

A Thermos bottle is a consumer version of a Dewar flask, named after Sir James Dewar who invented it in the late 1800's. Instead of using air trapped in fibers or foam as insulation, a Thermos surrounds the food with a region that contains nothing at all, not even air (Fig. 7.4.7). In order to withstand the crushing effects of atmospheric pressure, the Thermos has two strong walls. One wall surrounds the food and the other surrounds the first wall at a small distance. Since there is nothing between the two walls, there is no conduction and no convection. The two walls have mirror finishes so that they reflect thermal radiation and have very low emissivities. This mirroring dramatically reduces the radiative heat exchange between the walls. Since the only way heat can flow to or from the food is through its narrow mouth, a properly made Thermos bottle can keep food hot or cold for a remarkably long time.

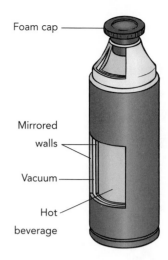

Fig. 7.4.7 A Dewar flask or Thermos bottle uses a vacuum to insulate its inner volume. The vacuum can't conduct heat or undergo convection and mirrored walls reduce the role of radiative heat transport as well. The only significant heat transfer occurs through the narrow mouth of the vessel, where the two walls meet.

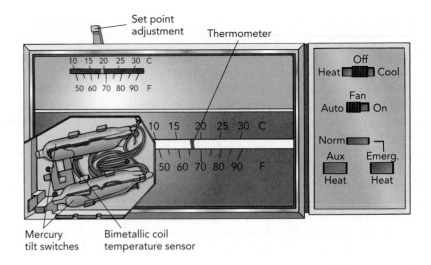

Set point adjustment

Thermometer

10 15 20 25 30 C

50 60 70 80 90 F

10 15 20 25 30 C

50 60 70 80 90 F

Off
Heat ▮▮▮ Cool

Fan
Auto ▮▮▮ On

Norm ▮▮▯
Aux   Emerg.
Heat   Heat

Mercury
tilt switches

Bimetallic coil
temperature sensor

SECTION 7.5 Thermometers and Thermostats

Knowing the temperature is important when you are going on a picnic, baking bread, or lying in bed with the flu. You use a thermometer to measure how hot it is outside, in the oven, or on your forehead. A thermometer is able to measure temperature because the characteristics of its components change with temperature. In this section, we will examine some of those changes to see how thermometers work.

There are also times when you want to control temperature. You don't just want to know how hot your house is, you want to maintain it at a certain level. In such cases, you need a thermostat, a thermometer that uses its temperature measurements to control other equipment.

Glass Thermometers and Liquid Thermostats

Temperature plays such an important role in everyday life that we frequently measure it and often control it. Whenever you're choosing what to wear, baking a cake, or trying to keep the milk from spoiling, you're probably paying some attention to temperature. Temperature, fortunately, is relatively easy to measure. Materials change with temperature in a great many ways and thermometers and thermostats are based on those changes.

The classic glass thermometer measures temperature using thermal expansion, a concept I introduced on p. 223 while discussing lightbulbs. In a lightbulb, the goal is to avoid differences in thermal expansion so that the lightbulb won't break as it warms or cools. In a glass thermometer, the goal is to use differences in thermal expansion to measure temperature.

Let me remind you that the extent to which a material expands with increasing temperature can be described by its coefficient of volume expansion, the fractional change in the material's volume caused by a temperature increase of 1 °C. Coefficients of volume expansion for a number of common materials are listed in Table 7.5.1.

It's not hard to understand how a glass thermometer works. The liquid inside the thermometer expands much faster with temperature than the glass tube around it, so the liquid flows up the tube in order to find room for itself and indicates the temperature increase.

Table 7.5.1 Coefficients of Volume Expansion (CVE) for Common Materials near 20 °C

Material	CVE (× 10⁻⁶)
Alcohol (ethanol)	1400
Alcohol (methanol)	1490
Aluminum	69
Brass	57
Cobalt	41
Copper	50
Glass (common)	26
Glass (Pyrex® and Kimax®)	9.6
Glass (quartz)	1.6
Glass-ceramics	<1.0
Iron	35
Lead	86
Mercury	181
Nickel	38
Silver	57
Stainless Steel	48
Tin	66
Titanium	26
Zinc	90

❶ A Galileo thermometer is a set of weighted glass spheres that float or sink in a liquid-filled vessel. The liquid expands more rapidly with temperature than the glass and its decreasing density reduces the buoyancy of the spheres. The higher the temperature, the harder it is for the spheres to float in the expanding liquid and the more of them sink to the bottom of the vessel. Each sphere is labelled to indicate the temperature above which it will sink. You can determine the current temperature by reading the label of the lowest-temperature sphere that is still floating.

More specifically, the thermometer's hollow body contains just enough colored alcohol or mercury to completely fill the cavity in its bulb, plus a little extra (Fig. 7.5.1). That extra liquid projects part way up a fine, hollow capillary connected to the bulb. As the thermometer's temperature increases, both the liquid and the glass enclosure expand but the liquid expands faster with temperature than the glass does. Although the cavity in the bulb becomes slightly larger and can therefore accommodate slightly more liquid, the liquid's greater expansion squeezes some of it out of the bulb and into the capillary. The column of liquid in the capillary becomes longer and you see a taller red or silver bar. The top of the capillary could be left open in principle, though its generally sealed to protect the liquid. (For an interesting glass-in-liquid thermometer, see ❶.)

Although mercury was once widely used in thermometers, it has been replaced by alcohol in most common thermometers. While mercury has a wider range of operating temperatures than the alcohols, mercury is an insidious and cumulative poison, and broken mercury thermometers present a lingering environmental hazard. Alcohols do have one important advantage over mercury: they expand so rapidly with temperature that an alcohol thermometer can have a wide capillary that's relatively easy to see.

Alas, even mercury fever thermometers have vanished from the market. Each of these elegant thermometers had a constriction in the base of its capillary so that the mercury column could expand up the capillary as the thermometer warmed in a patient's mouth, but wouldn't return down the capillary when the thermometer was allowed to cool. The weak cohesive forces within the narrow thread of mercury weren't strong enough to pull the mercury back through the constriction. Instead, an empty gap would appear between the constriction and the mercury-filled bulb. Since the column of mercury remained in place for a long time, you could read the thermometer at your leisure. To send the mercury back down the capillary and fill in the gap, you had to shake the thermometer vigorously. Inertial effects then flung the mercury through the constriction so that the thermometer was ready for reuse.

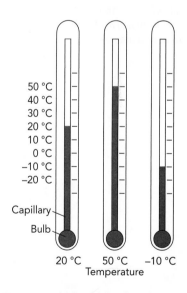

Fig. 7.5.1 The liquid inside a glass thermometer has a larger coefficient of volume expansion than the glass. As the temperature increases, excess liquid is driven out of the bulb at the base of the thermometer, and the thin column of liquid in the capillary rises. As the temperature decreases, liquid returns to the bulb and the column in the capillary descends.

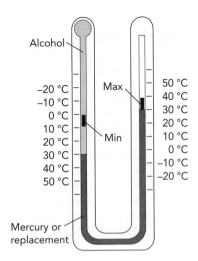

Alcohol

Max

Min

−20 °C
−10 °C
0 °C
10 °C
20 °C
30 °C
40 °C
50 °C

50 °C
40 °C
30 °C
20 °C
10 °C
0 °C
−10 °C
−20 °C

Mercury or
replacement

Fig. 7.5.2 This glass thermometer records the maximum and minimum temperatures. The expanding liquid's volume changes with temperature and it pushes the indicating liquid back and forth around the bottom of the U-shaped capillary. The indicating objects can rise when the indicating liquid pushes them up, but friction keeps them from descending when the indicating liquid moves back down. The bottoms of the indicating objects report minimum temperature on the bulb arm and maximum temperature on the open arm. The maximum and minimum are reset magnetically.

Another elegant type of glass thermometer—still in common use—is one that records both maximum and minimum temperatures. As shown in Fig. 7.5.2, this thermometer has a U-shaped capillary and its liquid-filled bulb is located atop one arm of the U. It contains two different liquids, one that exhibits strong thermal expansion and one that serves merely as an indicator of the present temperature. The expanding liquid is usually alcohol and it fills the bulb and part of the bulb side of the capillary. The indicating liquid is mercury or a mercury-replacement and it occupies the lower portion of the U. Since there is no gap between the two liquids, they meet somewhere on the bulb side of the capillary. Although the indicating liquid also expands somewhat with temperature, for simplicity I'll neglect that small expansion. The liquids are immiscible: like oil and water, they do not mix.

When the temperature rises, the expanding liquid flows out of the bulb and down the bulb side of the capillary. It pushes the indicating liquid ahead of it, down the bulb side of the U and up the open side of the U. You can read the present temperature by examining the height of the indicating liquid on either side of the U. The higher the indicating liquid rises in the open side of the capillary, the higher the temperature. That's why there is an ordinary temperature scale written on the open side of the thermometer. The higher the indicator fluid rises in the bulb side of the capillary, however, the colder the temperature. That's why there is an inverted temperature scale on the bulb side of the thermometer.

To this odd, double-reading thermometer there have been added two tiny indicating objects, one inside each arm of the capillary. These indicating objects have difficulty sliding through the capillaries, typically because they're magnetic and are held against the capillary walls by their attractions to a magnetic surface located behind the capillary. But while these objects normally remain in place when not touching the indicating liquid, the indicating liquid can push them up their respective sides of the capillary. The objects either float on the indicating liquid (for example, iron will float on mercury) or they won't wet in the indicating liquid (for example, wax won't wet in water). Regardless of why the indicating objects won't enter the indicating liquid, an indicating object will rise as the surface of the indicating liquid rises beneath it but it won't descend as that surface descends. The indicating object gets left behind.

As the temperature fluctuates up and down, the indicating liquid flows back and forth around the bottom of the U-shaped capillary and its upper surfaces occasionally push against and lift the tiny indicating objects. Since those objects can rise but not descend, the one on the bulb side of the capillary remains at the height to which it was pushed when the temperature was at its lowest, and the one on the open side of the capillary remains at the height to which it was pushed when the temperature was highest. Those minimum and maximum temperatures are indicated by the bottoms of the two indicating objects.

To reset the maximum and minimum readings, you must return the indicating objects to the top of the indicating liquid. In some thermometers, you drag the indicating objects down with separate magnets. In others, pushing a button moves a magnetic strip away from the capillaries so that the indicating objects are free to descend and they are then pulled downward by their own weights.

Differences in thermal expansion are also used in thermostats—devices that control equipment based on temperature measurements. Your home contains numerous thermostats, many of them hidden from view inside thermal appliances such as toasters, clothes irons, clothes dryers, and coffee pots and others built into gadgets such as computers and power adapters that don't seem to have much to do with temperature. Your body has its own collection of thermostats, which is how you maintain a nearly constant body temperature.

One of the simplest thermostats works much like a glass thermometer. In this scheme, a liquid expands out of a container as the temperature increases and operates a switch. This type of thermostat is used in many ovens, where it controls the flow of electricity or gas to the oven's burner. If the oven temperature rises above the desired value, the thermostat's switch turns the burner off. When the oven temperature falls back below the desired value, the switch turns the burner back on. In this manner, the thermostat keeps the oven temperature within a narrow range.

An oven that uses this type of thermostat will have a liquid-filled metal bulb inside it, usually located just below the top of the oven. A thin tube connects that bulb to the oven's control unit, where the tube connects to a small cylinder with a movable knob at its end. As the oven temperature rises, the expanding liquid flows from the bulb through the tube to the cylinder and pushes the knob outward. The hotter the oven, the farther the knob extends outward. Setting the desired oven temperature is a matter of choosing how far the knob must move to reach the switch and turn off the burner. The higher the temperature setting, the farther the knob is from the switch and the hotter the oven will become by the time the knob finally shuts off the burner. One nice feature of this simple thermostat scheme is that it can control a gas oven without any need for electricity.

Metal Thermometers: Bimetallic Strips

Not all thermometers are based on expanding liquids. There are many other thermometers that use metal strips to measure temperature. However, because solids have much smaller coefficients of volume expansion than liquids, it's difficult to make sensitive metal thermometers. This distinction between liquids and solids is caused by fundamental differences in their microscopic structures. I'm going to take a moment to explain those structures.

The atoms and molecules of most solids are held together rigidly and form orderly structures called crystals (Fig. 7.5.3). Crystals are familiar to most of us as the beautiful, faceted minerals in geology exhibits and museum gift shops. The natural faceting that appears on these crystals reflects the extraordinary order present in crystalline solids at the atomic and molecular scale. The particles in a crystal—its atoms and molecules—arrange themselves in nearly perfect lattices, repetitive and uniform arrangements that resemble stacks of oranges or food containers at the grocery store. Crystalline order isn't limited to faceted minerals. Most solids, including metals, are actually crystalline and their particles are arranged in regular lattices.

In contrast to solids, normal liquids aren't crystalline at all (Fig. 7.5.4). The particles in a normal liquid don't form any orderly lattice and don't even maintain their relative positions for long. Although they are weakly bound to their neighbors most of the time, they let go and change neighbors frequently. A liquid has much more microscopic activity than a solid and it is this added level of flexibility and action that gives liquids both their large specific heats and their large coefficients of volume expansion.

When you add heat to a crystalline solid, most of that heat goes into making the solid's particles vibrate more vigorously and thus into increasing the solid's temperature. That's why solids have relatively small specific heats: it doesn't take much thermal energy to raise a solid's temperatures. The increased vigor of its vibrations also cause the solid's volume to increase, but only a little. That's why solids have relatively small coefficients of volume expansion; they don't undergo any dramatic structural changes as they get hotter—they just vibrate harder.

When you add heat to a liquid, however, not all of that heat goes into making the particles vibrate more vigorously. A large fraction of it goes instead into breaking chemical bonds and decoupling the particles from one another. This alternative use of thermal energy increases the amount of heat you must add to the liquid to raise its temperature, which is why liquids have relatively large specific heats.

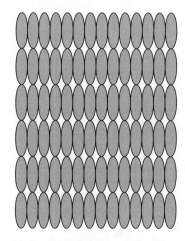

Fig. 7.5.3 In a crystalline solid, the atoms or molecules are arranged in a highly ordered lattice.

Fig. 7.5.4 In a normal liquid, the atoms or molecules are disordered. They touch one another but don't form an ordered lattice.

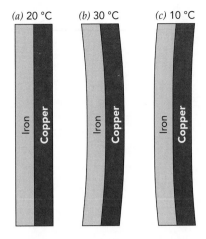

(a) 20 °C   *(b)* 30 °C   *(c)* 10 °C

Fig. 7.5.5 A copper and an iron strip are bonded together to form a bimetallic strip. Since copper has a larger coefficient of volume expansion than iron, the strip curls in response to changes in temperature. (*a*) This strip is straight at 20 °C, however, (*b*) the expanded copper curls it toward the left at 30 °C and (*c*) the shrunken copper curls it toward the right at 10 °C.

❷ Light blinkers used in automobiles and holiday lights contain a bimetallic strip thermostat. Electricity flowing through the blinker heats its filament. When the temperature of the filament becomes high enough, the thermostat stops the current flow and turns off the filament. When it cools down, the thermostat again permits electricity to flow and the filament heats up. This process repeats over and over and causes any lights attached to the blinker to wink on and off endlessly.

Fig. 7.5.6 This refrigerator thermometer uses a coiled bimetallic strip to measure temperature. The coil unwinds as it cools.

There are two important consequences to the breaking of bonds in a liquid when you heat it. First, the liquid's viscosity decreases and it flows more easily. That's why pancake syrup is easier to pour if you heat it first. Second, the liquid's volume increases substantially because its particles become even more loosely packed as they break free of one another and they take up additional space. This unpacking effect is absent in solids, where the particles remain in their lattices during heating. That's why liquids generally have larger coefficients of volume expansion than solids.

Despite their small coefficients of volume expansion, metals are often used in thermometers. The most common type of metal thermometer is based on a bimetallic strip. In this design, narrow sheets of two different metals, such as copper and iron, are permanently bonded together to make a thin metal sandwich (Fig. 7.5.5). Since these two metals have different coefficients of volume expansion, the bimetallic strip deforms as its temperature changes. Copper has the larger coefficient of volume expansion, so the strip's copper layer expands more when the strip is heated and shrinks more when the strip is cooled.

As the two layers of the bimetallic strip expand or shrink, the strip curls to one side or the other. There is only one temperature at which the strip is straight (Fig. 7.5.5*a*). Above that temperature, the strip curls so that the longer copper layer is outside the iron (Fig. 7.5.5*b*) and below that temperature, the strip curls so that the shorter copper layer is inside the iron (Fig. 7.5.5*c*). Since the strip's shape depends on its current temperature, it makes a good thermometer.

Most dial thermometers, including meat and candy thermometers, are based on bimetallic strips. To increase the sensitivities of these thermometers, their bimetallic strips are wound into small coils (Fig. 7.5.6) or spirals (Fig. 7.5.7) that curl or uncurl with temperature. One end of the coil is fixed to the thermometer's frame while the other end is attached to the pointer. As the thermometer's temperature changes, the curling bimetallic coil moves the pointer to indicate the temperature.

Nearly all home-heating thermostats from the pre-electronic era use bimetallic strips. Switches attached to the bimetallic coil in such a thermostat control the furnace. When the temperature becomes too high, the coil and switch turn the furnace off. When the temperature becomes too low, the coil and switch turn the furnace back on.

The furnace switch is usually a glass tube that's partially filled with liquid mercury metal (Fig. 7.5.8). This tube is attached to the movable end of the bimetallic coil, the end that normally turns the pointer of a dial thermometer. As the coil winds or unwinds, it tips the mercury from one end of the glass tube to the other. Two electric contacts are embedded in the wall of the tube. When the mercury is at one side of the tube, it connects these two contacts so that electricity can flow from one to the other. When the mercury tips to the other side of the tube, the contacts are not connected and no electricity flows. This mercury tilt switch allows the shape of the bimetallic coil to control the furnace.

When you set the temperature of the thermostat, you are actually changing the orientation of the bimetallic coil. This coil is attached to the temperature control knob so that turning that knob tilts both the coil and the mercury switch. When you turn up the thermostat, you are tilting the coil so that it won't operate the mercury switch until the room becomes hotter. When you turn down the thermostat, you are tilting the coil the other way so that it operates the switch at a relatively low temperature. Once the room reaches the desired temperature, the thermostat turns the furnace on and off as needed to maintain a nearly constant temperature.

Bimetallic strip thermostats are also used in clothes irons, toasters, coffee pots, and portable space heaters, where they directly control the flow of electricity through heating elements. As its temperature drops, the bimetallic strip in one of these simple thermostats bends until it makes electrical contact with a second piece of metal. Once that contact is made, electricity flows from the strip to its

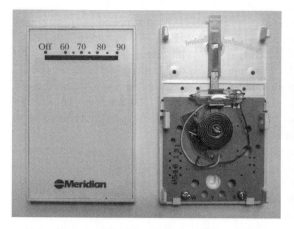

Fig. 7.5.8 The coiled bimetallic strip in this home thermostat (shown disassembled) controls the orientation of a mercury tilt switch. As the ball of liquid mercury tips from one end of the glass switch to the other, it turns the furnace on or off.

contact and then through the heating element. That way, whenever the thermostat becomes too cold, it turns on the heating element. (For an interesting application of bimetallic strips, see ❷.)

Direct contact thermostats aren't as sensitive, precise, or durable as those that use mercury tilt switches, but they operate well in any orientation. Tilt switch thermostats are sensitive to orientation because they use gravity to move the mercury around. To keep them operating at the right temperatures, tilt switch thermostats must be permanently mounted so that they always remain level.

Plastic Strip Thermometers: Liquid Crystals

The plastic strip thermometer is a thin flexible strip that displays the current temperature as a brightly colored number (Fig. 7.5.9). The strip has a whole range of temperatures printed on it, but only one number is easily visible at any given temperature. As the temperature changes, that number vanishes into the background and another number becomes visible. A plastic strip thermometer is particularly useful for measuring the temperature of a surface, such as the glass wall of a tropical fish aquarium or a child's fevered brow.

The strip isn't really a solid piece of plastic. It's a multi-layer sandwich containing a remarkable material phase called a liquid crystal. A liquid crystal has properties intermediate between a solid and a liquid. Behind each number on the strip is its own carefully formulated drop of liquid crystal that reflects colored light only at the temperature associated with that number.

To understand liquid crystals, we must return again to the microscopic structures of solids and liquids. Crystalline solids are highly ordered materials. The spacings between the particles in a crystal—its atoms and molecules—are so regular that, once you know exactly where a few of those particles are located, you can predict exactly where millions of other nearby particles are located. This regularity is called positional order and the crystal shown in Fig. 7.5.3 exhibits positional order. The particles in a crystal are also highly oriented, so that if you know the orientations of a few particles, you can also predict exactly how millions of other nearby particles are oriented. This second regularity is called orientational order and the crystal in Fig. 7.5.3 also exhibits orientational order. In contrast, normal liquids have neither positional nor orientational order (see Fig. 7.5.4). Knowing the positions and orientations of a few particles in a liquid tells you next to nothing about the positions and orientations of nearby particles.

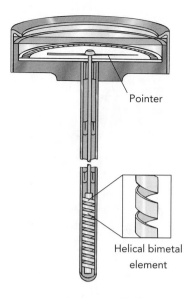

Fig. 7.5.7 A kitchen thermometer is based on a bimetallic strip that has been wound into a helix. As the temperature changes, the helical element coils or uncoils and turns the pointer.

Fig. 7.5.9 This plastic strip thermometer contains 12 different liquid crystals, each reflecting light only over a narrow range of temperatures. At present, the liquid crystal behind the 74 °F (23 °C) mark is reflecting light strongly, indicating that the room's temperature is near that value.

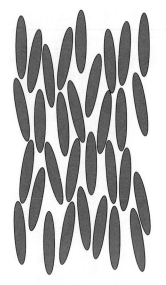

Fig. 7.5.10 In a liquid crystal, the rod-like or disk-like molecules don't have positional order but they do have orientational order. Here all the rod-like molecules point in more or less the same direction, a behavior characteristic of a nematic liquid crystal.

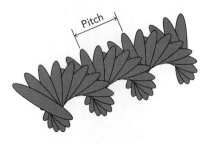

Fig. 7.5.11 The molecules in a chiral nematic liquid crystal have an orientation that rotates in a smooth spiral along one direction through the liquid. The spacing between adjacent regions of upward pointing molecules is called the pitch.

❸ Some insects obtain their striking colorations from chiral nematic liquid crystals. These liquid crystals contain oriented molecules that selectively reflect light of certain colors. In insects, these liquid crystal secretions harden to form solids that retain both the special spiral molecular order and the unusual optical effects.

Liquid crystals lie in between solids and liquids. Like normal liquids, liquid crystals have little positional order. Knowing where some of the particles in the liquid crystal are won't help you predict the locations of other nearby particles. But liquid crystals do have substantial orientational order. Liquid crystals are composed of rod-like or disk-like molecules that align themselves with one another, even though their positions are free to change (Fig. 7.5.10). Because these molecules move about like those in a normal liquid yet remain highly oriented like those in a crystalline solid, they are called liquid crystals.

Liquid crystals are actually quite common in biological systems. For example, cell membranes in animals are liquid crystals. Among the most familiar liquid crystals are the pearly, iridescent liquid hand soaps and shampoos that have become popular in recent years. Their striking optical properties come about because of their remarkable orientational order. Liquid crystals interact strangely with light, a feature that makes them useful in electronic watch displays and computer screens. It's a strange interaction between a liquid crystal and light that makes a plastic film thermometer work.

The liquid crystal used in a thermometer isn't quite as simple as that depicted in Fig. 7.5.10. That figure shows a nematic liquid crystal, which has molecules that can be located anywhere, but that all point in roughly the same direction throughout the material. The liquid crystal used in a thermometer is a chiral nematic liquid crystal, which has a natural twist to it so that the preferred molecular orientation spirals around like a corkscrew as you look across the liquid in one particular direction (Fig. 7.5.11). A chiral nematic liquid crystal still has only orientation order, but that orientational order is a complicated spiral one.

The spiral orientation of its molecules gives the chiral nematic liquid crystal a twisting, wave-like appearance. It even has "crests," where the molecules all tend to point up and down rather than to the side. The spacing between adjacent crests is called the pitch, the same word used to describe the distance between adjacent threads on a screw. The pitch of a chiral nematic liquid crystal is responsible for its remarkable optical properties.

This pitch may be only a few tens of nanometers or many microns, depending on the liquid crystal's chemical composition and *on its temperature*. Increasing the temperature shortens the pitch, a behavior that makes liquid crystal thermometers possible. In a particular chiral nematic liquid crystal, there may be a narrow range of temperatures over which the pitch of the spiral is equal to the wavelengths of visible light in that liquid. When the liquid crystal's temperature is in this range, it suddenly begins to reflect colored light!

For example, if at 28 °C the pitch of a particular liquid crystal is equal to the wavelength of blue light, then that liquid will appear brilliantly blue when illuminated by white light because it will reflect blue light back toward your eyes. If at 26 °C, the pitch is longer and is equal to the wavelength of red light, then the liquid will appear red. If at 24 °C, the pitch is longer than the wavelength of any visible light, then the liquid will reflect only infrared light and will appear transparent. This phenomenon is called selective reflection and is caused by constructive interference, a wave behavior that we'll examine in Chapter 14. (This effect also appears in nature—see ❸.)

A plastic strip thermometer contains a series of different chiral nematic liquid crystals that are viewed through clear number-shaped openings in the otherwise opaque strip. Behind each temperature number is a liquid crystal that reflects visible light only at the temperature represented by that number. Because the liquid crystals have black plastic behind them, they appear black unless they are selectively reflecting light. For any given temperature, only one patch of liquid crystal is selectively reflecting light and it illuminates the number corresponding to the strip's temperature.

Electronic Thermometers and Thermostats

As part of the explosion in modern consumer electronics, digital electronic thermometers and thermostats are gradually replacing traditional thermometers and thermostats. We no longer have the patience to read a dial or the height of a column of liquid—we expect a digital display. Reporting the temperature numerically is something a computer can do easily and controlling a furnace or oven based on that numerical value is no problem either. However, the computer still needs a sensor that measures the temperature in electronic form before it can display that temperature or control a furnace. The most common electronic temperature sensors are thermocouples and thermistors. These two devices operate on different principles but both are based on electronic properties that change with temperature.

Thermocouples are based on the Seebeck effect. As I noted in Section 7.1, the principal carriers of heat in a metal are its mobile electrons—the electrons associated with the metal's electrical conductivity. The mobile electrons at the hot end of a metal rod have extra kinetic energy and are faster-moving than those at the rod's cold end. On average, mobile electrons tend to carry heat from the rod's hot end toward its cold end. The added vigor with which mobile electrons leave the hot end of the rod for the cold end also creates a small imbalance of electric charge. The cold end of the rod ends up with a few too many electrons that are now missing from the hot end. This redistribution of electric charges is the Seebeck effect.

While charge imbalances due to the Seebeck effect are extremely small, they can be measured by delicate electronic equipment. However, it's tricky to measure a charge imbalance between a hot metal and a cold metal. It's much easier to make such a measurement between two metals that are at the same temperature. For this reason, a thermocouple is constructed by joining two wires made of different metals together at one point. When the junction between the wires is heated, each wire experiences the Seebeck effect but the amount of charge imbalance that occurs in each wire depends on the type of metal.

Because electrons can flow freely across the heated junction, the different charge imbalances in the two metal wires create an overall charge imbalance between their cold ends. For example, one type of standard thermocouple is made by joining a platinum wire with a platinum-rhodium alloy wire. When the junction between the wires is heated, electrons flow away from that junction through both wires, but more electrons flow up the platinum wire than up the platinum-rhodium alloy wire. As a result, there are more electrons on the room temperature end of the platinum wire than on the room temperature end of the platinum-rhodium wire. Measuring charge imbalances between two room-temperature wires is relatively easy.

Thermocouples appear frequently in temperature control units for furnaces and manufacturing equipment. They can measure very high temperatures with ease, even in a blast furnace. Thermocouples can also measure very low temperatures because the Seebeck effect works in reverse if you make the junction colder than the free ends of the wires. Just which two metals are used in a thermocouple depends on the desired temperature range and on the chemical environment that these metals must endure. Platinum and platinum-rhodium are wonderfully inert metals so that they tolerate almost any environment up to 1769 °C—the temperature at which platinum melts. Unfortunately, both metals are extremely expensive. Less costly thermocouple metals include iron, copper, and a variety of copper, nickel, chromium, and aluminum alloys.

A thermistor is quite different from a thermocouple. It indicates its temperature by changing its electrical conductivity. Thermistors are made out of semiconductors, materials that are neither good electrical conductors like metals nor good electrical insulators like glasses or plastics. I'll examine semiconductors in detail in Chapter 12, but what matters here is easy enough to explain.

A true semiconductor has no mobile electrons at all at very low temperature, so that it behaves like an insulator. Without any mobile electrons, a piece of semiconductor can't carry an electric current from one side to the other. But a semiconductor stops being a good insulator as it warms up. Thermal energy affects not only the semiconductor's atoms and molecules, it also affects the semiconductor's electrons. As the temperature rises, electrons begin to break free of the forces that hold them in place and they begin to move about the material. The semiconductor begins to conduct electricity. At low temperatures, semiconductors are still very poor conductors of electricity but at high temperatures, semiconductors conduct electricity pretty well.

With the help of electronics that can measure electrical conductivity, a semiconductor does an excellent job of measuring temperature. Commercial thermistors, such as those found in electronic fever thermometers and other household electronic thermometers, are built out of specialized semiconductors. These semiconductors are designed so that their resistances change dramatically over the ranges of temperatures they are supposed to measure. Properly made thermistors can be extremely accurate over a considerable range of temperatures. Some thermistors can even measure temperatures close to absolute zero. However, thermistors can't be used to measure very high temperatures without suffering permanent damage because semiconductor crystals are just not as robust as the metals that are used to make thermocouples.

Optical Thermometers

I've written so frequently about thermal radiation in this chapter that you probably knew it was only a matter of time before I brought it up with thermometers. That time has come. An optical thermometer observes the thermal radiation emitted by an object and uses it to determine that object's temperature. Optical thermometers are common in doctor's offices, where a two-second peek inside your ear is all they need to measure your body temperature. They're also widely used in industrial, commercial, and even household applications, where they can measure the temperatures of engines, machinery, and just about anything else from a distance and without any direct contact.

An optical thermometer (Fig. 7.5.12) has a lens system that collects the thermal radiation coming toward it from a narrow range of angles. You point it at the object you're studying and it collects that object's thermal glow. Making a detailed study of the object's entire thermal spectrum would be too complicated for an inexpensive device, so it normally makes a simple intensity measurement in a selected range of the infrared spectrum. That intensity measurement is enough to determine the object's temperature with fairly good accuracy.

But while being able to measure temperature almost instantly and from a distance is wonderful, optical thermometry has its complications. The thermometer can't distinguish the object's own thermal radiation from any thermal radiation that the object is reflecting or transmitting. The thermometer therefore works best if the object is black in the observed portion of the thermal spectrum. If the object is shiny or transparent in that portion of the spectrum, the thermometer may not read correctly.

Optical fever thermometers are highly accurate because the ear canal is almost perfectly black. Light entering your ear canal bounces repeatedly on its way in and has many opportunities to be absorbed. Almost none of the light is reflected back out of your ear. Since your ear canal is almost perfect at absorbing light that enters it, it is also almost perfect at emitting thermal radiation characteristic of its temperature. By looking into your ear canal, an optical thermometer is viewing thermal radiation that almost perfectly characterizes your body temperature.

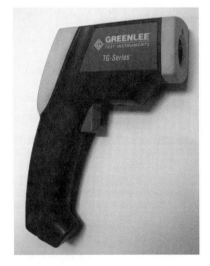

Fig. 7.5.12 This optical thermometer observes the thermal radiation coming from a surface and uses the intensity of that radiation to determine the surface's temperature. The device assumes that the surface has a high emissivity, but can be adjusted to work with other emissivities. It projects a ring of laser spots to identify the surface it's measuring.

When you use an optical thermometer to measure the temperature of a surface, however, you have to be careful about the surface's emissivity. If the surface isn't black, you might end up measuring the temperature of reflected or transmitted thermal radiation. Skin has a body-temperature emissivity of about 0.98, so measuring its temperature optically is almost foolproof. Thermal imaging cameras, which photograph infrared thermal radiation rather than visible light, can map out skin temperature with great precision because skin is a nearly perfect blackbody. Thermal imaging is used frequently as a medical diagnostic tool because a hot or cold region on a person's skin can indicate underlying health problems.

Most organic materials, fabrics, paints, glasses, and ceramics are also almost black to room temperature thermal radiation, with emissivities exceeding 0.90. You can measure their temperatures optically without worrying too much about making a mistake. But when you measure the temperatures of metals, you have to be extremely careful. Unless those metals are heavily oxidized so that they have large room-temperature emissivities, they won't emit thermal radiation efficiently and you may accidentally measure the temperature of reflected thermal radiation instead. You do best to measure the temperature in a cavity or crevice of the metal, where the multiple reflection effects increase the effective emissivity toward 1. And never try to measure the temperature of a cool object that's bathed in the thermal glow of much hotter objects around it; even a weak reflection of that intense thermal glow will fool the optical thermometer so that it reports too high a temperature.

♒ The Problem with Thermal Expansion

Thermal expansion in liquids rarely causes trouble because liquids don't break (but see ❹). The same can't be said for solids. Differences in coefficients of volume expansion or in the rates at which different parts of the same object are heated or cooled can cause damage when temperatures change. Concrete pavement, bridges, and railroad tracks all expand differently than the ground on which they rest. Without careful design, concrete pavement will crack or buckle, a bridge will tear itself away from the roads at either end, and railroad tracks will bend into so much steel spaghetti.

To avoid these potential disasters, special joints are introduced in each case. Concrete pavement is poured as individual slabs that are joined by soft materials so that the slabs can expand or contract without buckling or breaking. A bridge is separated from the roadways at either end by special expansion joints so that thermal changes in the bridge's length don't cause damage to its surface. And train rails are interrupted periodically by expansion joints, which permit the rails to expand or contract safely.

Even objects that are made of a single material may be damaged by non-uniform temperatures. A metal pan may bend and contort when you put it on the stove or in the oven because non-uniform heating causes different parts of the pan to expand by different amounts. Only metal's flexibility keeps it from breaking.

Glass isn't flexible so it's particularly susceptible to thermal damage. As shown in Table 7.5.1, common glass has a large coefficient of volume expansion and will shatter if you heat it non-uniformly—the rigid, brittle glass literally tears itself apart. One of the most important developments in glass fabrication over the past century was the formulation of heat-resistant glasses—glasses that expand very little when heated. Pyrex® and Kimax® glasses have relatively small coefficients of thermal expansion and can tolerate fairly non-uniform temperatures without shattering. That's why most glassware used in cooking is made from these glasses.

Even cookware made of Pyrex or Kimax will crumble, however, if you move it straight from the oven to cold water or put it directly on the red-hot spiral burner of an electric stove. To survive such sudden thermal shocks or severely non-uni-

❹ Thermal expansion in liquids presents a problem for sealed food containers. If beverage bottles, jars, or cans are filled to the top, they'll burst or leak when the temperature rises. That's because the foods' liquids expand more rapidly with temperature than do the solid containers. Most commercial food containers therefore include small air or vacuum spaces. Since home canned foods are normally sealed at high temperatures, they naturally develop empty spaces between their lids and their food as they cool.

form heating, the cookware must have a remarkably low coefficient of volume expansion; it must be made of a quartz glass or a glass-ceramic. Though more expensive than Pyrex or Kimax, quartz glasses and glass-ceramics are so strong and so resistant to thermal fracture that they're frequently used as the cooking surfaces on modern electric stoves. They're even used in stovetop cookware. Although pots made of quartz glass or glass-ceramic don't conduct heat well and therefore won't necessarily cook evenly, they'll take amazing thermal and mechanical abuse without breaking.

CHAPTER 8

THINGS THAT WORK WITH HEAT

Heat normally flows from a hotter object to a colder object, which is why the hot sun warms your skin as you sit on a beach and why a cold winter breeze cools it as you sled down a mountain. But not everything in nature permits heat to flow passively. Our technological world includes many devices that actively transfer heat from colder objects to hotter objects or that use the flow of heat to do useful work. In this chapter we will examine the rules governing the movement of heat, a field of physics known as thermodynamics.

More broadly, however, the laws of thermodynamics deal with the relationships between order and disorder. Yes, physics has its own version of Law and Order. In this case the plot focuses on the relentless and unavoidable loss of order that dominates our physical universe and many aspects of our lives. As I'll describe in many different ways in this chapter, turning order into disorder is far easier than doing the reverse.

The natural loss of order in most systems allows you to look at a collection of photographs and put them in time sequence. The more disorder you see in a particular photograph, the later it occurred in the sequence. That's certainly true of most party photographs, at least until someone cleans up and literally throws out the disorder along with the trash. You can export disorder but you can't make it go away completely.

There is no shortage of energy in our world; thermal energy is everywhere you look. What is in short supply is order. Fortunately, our sun will provide us with a fresh, ongoing supply of order for as long as we continue to exist on this planet. What we must learn to do before long is to use the order we get from the sun efficiently and to become far more careful and conservative in how we consume the precious stores of order already on earth—the mineral wealth beneath the ground and the plant and animal wealth on the land and in the water. That existing order will have to last us an awfully long time and, once it's gone, we'll have trouble undoing the disorder.

Chapter Itinerary

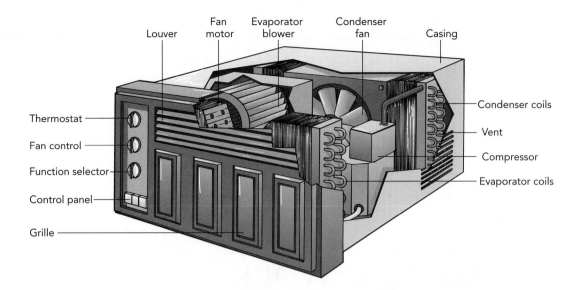

Louver | Fan motor | Evaporator blower | Condenser fan | Casing

Thermostat
Fan control
Function selector
Control panel
Grille

Condenser coils
Vent
Compressor
Evaporator coils

SECTION 8.1 **Air Conditioners**

On a summer day your problem isn't staying warm; it's keeping cool. Instead of looking for something to burn in your woodstove you turn on your air conditioner. An air conditioner is a device that cools room air by removing some of its thermal energy. But the air conditioner can't make thermal energy disappear. Instead, it transfers thermal energy from the cooler room air to the warmer air outside. Since the air conditioner transfers heat against its natural direction of flow, the air conditioner is a "heat pump." It's also a classic illustration of the laws of thermodynamics in action.

Moving Heat Around: Thermodynamics

On a sweltering summer day, the air in your home becomes unpleasantly hot. Heat enters your home from outdoors and doesn't stop flowing until it's as hot inside as it is outside. You can make your home more comfortable by getting rid of some of its thermal energy. But while we've already looked at ways to *add* thermal energy to room air, we haven't yet learned how to *remove* it. At present, the only cooling method we've discussed is contact with a colder object. Unless you have an icehouse nearby, you need another scheme for eliminating thermal energy. You need an air conditioner.

An air conditioner transfers heat against its natural direction of flow. Heat moves from the colder air in your home to the hotter air outside, so that your home gets colder while the outdoor air gets hotter. There's a cost to transferring heat in this manner. The air conditioner requires ordered energy to operate and typically consumes large amounts of electric energy. It's a type of heat pump—a device that uses ordered energy to transfer heat from a colder object to a hotter object and against its natural direction of flow.

Before learning how an air conditioner pumps heat, we should first show that pumping is necessary. There are a number of seemingly reasonable cooling alternatives that we should consider before turning to air-conditioning. Three such alternatives are

1. Letting heat flow from your home to your neighbor's home.
2. Destroying some of your home's thermal energy.
3. Converting some of your home's thermal energy into electric energy.

Unfortunately, these three alternatives can't be done. Still, it will be useful for us to examine them more closely because in doing so, we'll learn about the laws governing the movement of thermal energy, the laws of thermodynamics.

The first alternative raises an interesting issue. Your home is in thermal equilibrium with the outdoor air, meaning that no heat flows from one to the other and they're at the same temperature. Your neighbor's home is also in thermal equilibrium with the outdoor air. What will happen if you permit heat to flow between your home and your neighbor's home? Nothing. Since both homes are simultaneously in thermal equilibrium with the outdoor air, they're also in thermal equilibrium with one another. All three are at the same temperature.

This observation is an example of the zeroth law of thermodynamics, which says that two objects that are each in thermal equilibrium with a third object are also in thermal equilibrium with one another. This seemingly obvious law is the basis for a meaningful system of temperatures. If you had a roomful of objects at 35 °C (95 °F) and some were in thermal equilibrium with one another while others were not, then "being at 35 °C" wouldn't mean much. However, every object that has a temperature of 35 °C is in thermal equilibrium with every other object at 35 °C. The zeroth law is observed to be true in nature so that temperature does have meaning. And since your neighbor's home is just as hot as yours, they can relax because you're not going to be sending them any extra heat.

The Zeroth Law of Thermodynamics
Two objects that are each in thermal equilibrium with a third object are also in thermal equilibrium with one another.

The second alternative sounds unlikely from the outset. We've known since the first chapter that energy is special, that it's a conserved quantity. You can't cool your home by destroying thermal energy because energy can't be destroyed. To eliminate thermal energy, you must convert it to another form or transfer it elsewhere.

This concept of energy conservation is the basis for the first law of thermodynamics, which states that the change in a stationary object's internal energy is equal to the heat transferred into that object minus the work that object does on its surroundings. An object's internal energy is the sum of its thermal energy and any additional potential energy stored entirely within the object. This law says that heat added to the object increases its internal energy while work done by the object decreases its internal energy. In other words, since energy is conserved, the only way the object's internal energy can change is by transferring energy as heat or work. The first law of thermodynamics can be written as:

$$\text{change in object's internal energy} =$$
$$\text{heat added to object} - \text{work done by object.} \quad (8.1.1)$$

The First Law of Thermodynamics
The change in a stationary object's internal energy is equal to the heat transferred into that object minus the work that object does on its surroundings.

Disorder, Entropy, and the Second Law

The third alternative looks much more promising than the first two. It seems as though you should be able to convert thermal energy into electricity (or some other ordered form of energy). You could then sell it back to the electric company and get credit on your bill. Wouldn't that be great?

But there's a problem. Ordered energy and thermal energy aren't equivalent. You can easily convert ordered energy into thermal energy but the reverse is much harder. For example, you can burn a log to convert its chemical potential energy into thermal energy, but you'll have trouble converting that thermal energy back into chemical potential energy to recreate the log.

The basic laws of motion are silent on this issue. It isn't that the smoke doesn't have the energy to recreate the log. It's that the individual smoke particles must pool their thermal energies together to carry out the reassembly, a remarkably unlikely event. The particles would all have to move in just the right ways to turn the burned gases back into wood and oxygen, an incredible coincidence that simply never happens. Similarly, all of the air particles in your home would have to act together to convert their thermal energy into electricity. Since that coordinated behavior is ridiculously improbable, you're not going to be selling thermal-energy power to the electric company any time soon.

Once energy has been scattered randomly among the individual air particles, you can't collect that energy back together again. Creating disorder out of order is easy, but recovering order from disorder is nearly impossible. As a result, systems that begin with some amount of order gradually become more and more disordered, never the other way around. The best they can do is to stay the same for a while so that their disorder doesn't change. From these observations, we can state that the disorder of an isolated system *never decreases*.

This notion of never-decreasing disorder is one of the central concepts of thermal physics. There is even a formal measure of the total disorder in an object: entropy. All disorder contributes to an object's entropy, including its thermal energy and its structural defects. Breaking a window or heating it both increase its entropy. Although its name sounds similar to energy, don't confuse energy and entropy. Energy is a conserved quantity, while entropy is a quantity that can and generally does increase. It's easy to make more entropy.

Because disorder never decreases, the third cooling alternative is impossible. Turning your home's thermal energy into electric energy would reduce its disorder and decrease its entropy. But our observations about entropy aren't yet complete. There's one way to decrease your home's entropy: you can export that entropy somewhere else. In fact, you export entropy every time you take out the garbage, though that action also changes the contents of your home. You can also export entropy without modifying your home's contents by transferring heat somewhere else. Heat carries disorder and entropy with it, so getting rid of heat also gets rid of entropy.

Our rule about entropy never decreasing is weakened by the possibility of exchanging heat and entropy between objects. Before asserting that an object or system of objects can't decrease its entropy, we must ensure that it's thermally isolated from its surroundings so that it can't export its entropy. With that in mind, the strongest statement that we can make concerning entropy is that the entropy of a thermally isolated system of objects never decreases. This observation is the second law of thermodynamics.

The Second Law of Thermodynamics
The entropy of a thermally isolated system of objects never decreases.

Because of the second law, the only way to cool your home is to export its thermal energy and entropy elsewhere. Such a transfer would be easy if you had a cold object nearby to receive the heat. But lacking a cold object, you must use an air conditioner. Like all heat pumps, an air conditioner transfers heat and entropy in such a way that the second law of thermodynamics is never violated and the entropy of each thermally isolated system of objects never decreases. As we'll see, the air conditioner lowers the entropy of your home but raises the entropy of the outdoor air even more so that, overall, the entropy of the world actually increases.

There's a limit to how much entropy an air conditioner can remove from your home. As it exports thermal energy and entropy, the air conditioner lowers your home's temperature. In principle, your home will eventually approach absolute zero, where all thermal motion ceases. As it does, its entropy will approach zero. This relationship between the zero of temperature and the zero of entropy is the third law of thermodynamics, which states that as an object's temperature approaches absolute zero, its entropy approaches zero. The third law establishes absolute zero as a destination with no disorder left, but the second law ultimately makes it impossible to extract all the disorder from an object. Because absolute zero is unattainable, the third law refers to *approaching* it rather than to *arriving* at it.

The Third Law of Thermodynamics
As an object's temperature approaches absolute zero, its entropy approaches zero.

Pumping Heat Against Its Natural Flow

While the second law of thermodynamics doesn't allow the entropy of a thermally isolated system to decrease, it does permit the objects in that system to redistribute their individual entropies. One object's entropy can decrease as long as the entropy of the rest of the system increases by at least as much. Such entropy redistribution allows part of the system to become colder if the rest of the system becomes hotter.

For example, suppose that there's a pond of cold water behind your home. You pump that water through your bathtub and let it draw heat out of the room air. Your home becomes colder while the pond becomes warmer. This transfer of heat from the hot air in your home to the cold water in the pond satisfies the second law of thermodynamics. The entropy of the combined system—your home and the pool of water—doesn't decrease. In fact, it actually increases!

This entropy increase occurs because heat is more disordering to cold objects than it is to hot objects. Each joule of heat that flows from your home to the pool creates more disorder in the pool than it creates order in your home.

A useful analog for this effect involves two parties taking place simultaneously: the garden society's annual tea party and a 4-year old's birthday party. The orderly tea party represents the cold pool while the disorderly birthday party represents your hot home. The analogy to letting heat flow from your hot home to the cold pool is to trade one lively 4-year old from the disorderly birthday party for one quiet octogenarian from the orderly tea party. This exchange will reduce the birthday party's disorder only slightly, but it will dramatically increase the disorder of the tea party. The attendance at each party will be unchanged but their total disorder will increase.

When heat flows from your home to the pool, the overall entropy increases and the second law is more than satisfied. A similar increase in entropy occurs whenever heat flows from a hot object to a cold object, which is why heat normally flows in that direction.

But an air conditioner does the seemingly impossible: it transfers heat from a cold object, your home, to a hot object, the outdoor air. This heat flows in the wrong direction and the disorder it creates by entering the hot outdoor air is less than the disorder it removes by leaving the cold indoor air. It's like returning the tea party's lone 4 year old to the birthday party in exchange for the elderly garden fancier—the birthday party becomes only a tiny bit more disorderly while the garden party becomes much more orderly, so the net disorder of the two gather-

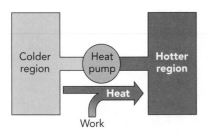

Fig. 8.1.1 A heat pump transfers heat from a colder region to a hotter region. In doing so, it converts some work (ordered energy) into heat (thermal energy in the hotter region). The larger the temperature difference between the two regions, the more work is required to transfer each joule of heat.

ings decreases substantially. Similarly, if nothing else happened when the air conditioner moved heat from the cold indoor air to the hot outdoor air, the entropy of the combined system would decrease and the second law of thermodynamics would be violated!

However, we've omitted an important feature of the air conditioner's operation: the electric energy it consumes. The air conditioner converts this ordered energy into thermal energy and delivers it as additional heat to the outdoor air (Fig. 8.1.1). In doing so, the air conditioner creates enough extra entropy to ensure that the overall entropy of the combined system increases. The second law is satisfied after all.

The amount of ordered energy the air conditioner consumes depends on the indoor and outdoor temperatures. If the two are close in temperature, the transfer of heat reduces the entropy only slightly so the air conditioner doesn't need to convert much ordered energy into thermal energy. But if they are far apart in temperature, the air conditioner must create lots of extra entropy to make up for the entropy lost in the transfer.

This requirement that entropy not decrease explains why an air conditioner works best when it's cooling your home the least. The greater the temperature difference between the indoor air and the outdoor air, the more electric energy or other form of work the air conditioner must consume to transfer each joule of heat. For an ideally efficient air conditioner or other heat pump, the relationships between the work consumed, the heat removed from the cold object, and the heat added to the hot object are:

$$\text{heat removed from cold object} =$$
$$\text{work consumed} \cdot \frac{\text{temperature}_{cold}}{\text{temperature}_{hot} - \text{temperature}_{cold}}$$
$$\text{heat added to hot object} =$$
$$\text{heat removed from cold object} + \text{work consumed}. \quad (8.1.2)$$

The hot object receives not only the heat removed from the cold object, but also an amount of heat equal to the work consumed by the transfer. Note also that the work needed to remove heat from your home approaches infinity as its temperature approaches absolute zero; that's why absolute zero is unattainable.

Sadly, a practical air conditioner never reaches ideal efficiency, so it moves less heat than promised by Eqs. 8.1.1. Moreover, heat leaks back into your home through its walls at a rate that's roughly proportional to the temperature difference. No wonder your electric bill soars when you turn down the thermostat too far!

How an Air Conditioner Cools the Indoor Air

Having determined the air conditioner's goals, it's time for us to look at how a real air conditioner meets them. In most cases, the air conditioner uses a fluid to transfer heat from the colder indoor air to the hotter outdoor air. Known as the working fluid, this substance absorbs heat from the indoor air and releases that heat to the outdoor air.

The working fluid flows in a looping path through the air conditioner's three main components: an evaporator, a condenser, and a compressor (Fig. 8.1.2). The evaporator is located indoors, where it transfers heat from the indoor air to the working fluid (Fig. 8.1.3). The condenser is located outdoors, where it transfers heat from the working fluid to the outdoor air. And the compressor is also located outdoors, where it squeezes the working fluid and does the work needed to move heat against its natural flow. To see how these three components pump heat out of your home, let's look at them individually.

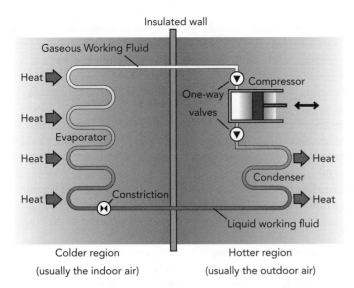

Insulated wall

Gaseous Working Fluid

Heat

Heat

Evaporator

Heat

Heat

Constriction

One-way valves

Compressor

Heat

Condenser

Heat

Liquid working fluid

Colder region
(usually the indoor air)

Hotter region
(usually the outdoor air)

Fig. 8.1.2 A typical air conditioner transfers heat from colder indoor air to hotter outdoor air by evaporating a liquid to a gas in the indoor air and condensing that gas to a liquid in the outdoor air. The air conditioner's working fluid travels endlessly around a loop of sealed plumbing, carrying heat with it as it moves. As the liquid working fluid enters the indoor evaporator through a constriction, its pressure drops and it begins to evaporate into a gas by absorbing heat from the indoor air. The low-pressure gaseous working fluid then flows to the outdoor compressor, which does work on the gas and increases its pressure. The high-pressure gaseous working fluid enters the outdoor condenser and it begins to condense into a liquid and release heat to the outdoor air. The high-pressure liquid working fluid than flows to the indoor constriction and evaporator to begin its journey all over again. The compressor provides the input of ordered energy that is necessary to move heat from a colder region to a hotter region. The compressor's work is transformed into thermal energy and released as heat into the outdoor air.

We'll begin with the evaporator, a long metal pipe that's decorated with thin metal fins. The evaporator is a heat exchanger that allows heat to flow from the warm air around it to the cool working fluid inside it. Its fins provide additional surface area to facilitate that heat flow and a fan blows the indoor air rapidly past the fins so that heat moves quickly into the working fluid.

True to its name, the evaporator allows the working fluid inside it to evaporate. That's why the working fluid becomes so cool and absorbs so much heat. Like any evaporating substance, the liquid working fluid needs its latent heat of evapora-

Fig. 8.1.3 The evaporator of this central air-conditioning unit extracts heat from the indoor air. That heat is transferred outside, where the compressor and condenser release it into the outdoor air.

tion to separate its molecules from one another and become a gas. The working fluid obtains that latent heat of evaporation from its own thermal energy, so its temperature drops and heat then flows into it through the walls of the evaporator. By the time the gaseous working fluid leaves the evaporator, it has absorbed a great deal of the indoor air's thermal energy and carries that energy with it as chemical potential energy.

To make the liquid working fluid vaporize in the evaporator, the air conditioner abruptly reduces the fluid's pressure. Recall from Chapter 7 that phase transitions such as evaporation depend on molecular landing and leaving rates. When both the liquid and the gaseous phases are present, molecules are continually leaving the liquid for the gas and landing on the liquid from the gas. If leaving dominates, the liquid evaporates and the gas is stable. If landing dominates, gas condenses and the liquid is stable. And if the two balance one another, the liquid and gas can coexist in phase equilibrium.

As it flows through a pipe toward the evaporator, the working fluid's pressure is high and it's stable as a liquid. At this high pressure, any gaseous working fluid would be so dense that landing would dominate leaving and the gas would condense. Thus the pipe carries only high-pressure liquid working fluid to the evaporator.

But just before reaching the evaporator, the liquid working fluid passes through a narrow constriction in the pipe and its pressure drops dramatically. The low-pressure working fluid that results isn't stable as a liquid. At low pressure, any gaseous working fluid is so dilute that leaving dominates landing and the liquid evaporates. Thus as the low-pressure liquid working fluid emerges from the constriction and pours into the evaporator, it's evaporating rapidly. It continues to evaporate even though its temperature decreases as it absorbs its latent heat of evaporation.

By the time the working fluid emerges from the evaporator, it has evaporated completely and has absorbed considerable thermal energy from the indoor air. It leaves the evaporator as a cool low-pressure gas and travels through a pipe toward the compressor.

Half the air conditioner's job is done: it has removed heat from the indoor air. But the remaining half of its job is more complicated: it must add heat to the outdoor air while ensuring that the total entropy of the combined system doesn't decrease. After all, there's no getting around the second law of thermodynamics.

How an Air Conditioner Warms the Outdoor Air

Satisfying the second law is the task of the compressor. The compressor receives low-pressure gaseous working fluid from the evaporator, compresses it to much higher density, and delivers it as a high-pressure gas to the condenser. The compressor may use a piston and one-way valves, like the water pump in Fig. 5.2.3, or it may use a rotary pumping mechanism. But regardless of how it functions, the result is the same: the gaseous working fluid undergoes a dramatic increase in density and pressure as it passes through the compressor.

Compressing a gas requires work because the compressor must squeeze the gas inward as it moves inward—force times distance. Since work transfers energy, the compressor increases the energy of the gas. The air conditioner usually obtains this energy from the electric company and converts it into mechanical work with an electric motor.

In accordance with the first law of thermodynamics, this work increases the internal energy of the working fluid. The only way that the gaseous fluid can store this additional energy is as thermal energy in its individual particles. These particles begin to move about more rapidly so that the gaseous working fluid leaves the compressor much hotter than when it arrived. There is no getting around that temperature rise; compressing the working fluid unavoidably raises its temperature.

This hot, high-pressure working fluid then flows into the condenser. Like the evaporator, the condenser is a long metal pipe with fins. It acts as a heat exchanger and its metal fins provide extra surface area to speed the flow of heat from the hotter working fluid inside it to the less hot outdoor air. There may also be a fan to move outdoor air quickly past the condenser and speed up the heat transfer.

As its name suggests, the condenser allows the gaseous working fluid inside it to condense into a liquid. Compression prompts that condensation. While it flows through a pipe toward the compressor, the low-pressure working fluid is stable as a gas. But in the high-pressure, high-density working fluid that emerges from the compressor, the landing rate far outpaces the leaving rate and the gas condenses.

Like any condensing substance, the gaseous working fluid releases its latent heat of evaporation as its molecules bind together and it becomes a liquid. This latent heat of evaporation becomes thermal energy in the working fluid, so its temperature rises still further and heat flows out of it through the walls of the condenser.

By the time the liquid working fluid leaves the condenser, it has transformed a great deal of chemical potential energy into thermal energy and released that energy into the outdoor air. The outdoor air receives as heat not only the thermal energy extracted from the indoor air, but also the electric energy consumed by the compressor. The working fluid leaves the condenser as a warm, high-pressure liquid and travels through a pipe toward the evaporator.

The second half of the air conditioner's job is now complete: it has released heat into the outdoor air and, in the process, converted ordered energy into thermal energy. From here, the working fluid returns to the evaporator to begin the cycle all over again. The working fluid passes endlessly around the loop, extracting heat from the indoor air in the evaporator and releasing it to the outdoor air in the condenser. The compressor drives the whole process and thereby satisfies the second law of thermodynamics. The same technique is used to extract heat from the air inside a refrigerator and to release that heat to the room air (Fig. 8.1.4).

In fact, heat pumps of exactly this type appear in many common appliances, including drinking fountains (Fig. 8.1.5) and home heating units. Many homes in moderate climates are heated by effectively running their air conditioners backward. Called heat pumps rather than air conditioners, these systems are capable of pumping heat against its natural direction of flow in winter as well is in summer.

In summer, a heat pump moves heat from indoor air to outdoor air to cool the home. But in winter, it moves heat from outdoor air to indoor air to heat the home. Rather than turning electricity directly into thermal energy to heat the home, it leverages that electrical energy by using it to gather the abundant thermal energy from outdoors and to carry that indoors.

Before leaving air conditioners, we should take a moment to look at the working fluid itself. This fluid must become a gas at low pressure and a liquid at high pressure, over most of the temperature range encountered by the air conditioner. For decades, the standard working fluids were chlorofluorocarbons such as the various Freons. These compounds replaced ammonia, a toxic and corrosive gas used in early refrigeration.

Chlorofluorocarbons are ideally suited to air conditioners because they easily transform from gas to liquid and back again over a broad range of temperatures. They're also chemically inert and inexpensive. Unfortunately, chlorofluorocarbon molecules contain chlorine atoms and, when released into the air, can carry those chlorine atoms to the upper atmosphere. There they promote the destruction of ozone molecules, essential atmospheric constituents that absorb portions of the sun's ultraviolet radiation. Recently, chlorine-free hydrofluorocarbons have replaced chlorofluorocarbons as the working fluids in most air conditioners. Though not as energy efficient and chemically inert as the materials they replace, hydrofluorocarbons appear to be safe for the environment.

Fig. 8.1.4 The compressor (bottom) and condenser coils (top) are visible on the back of this refrigerator. The compressor squeezes the working fluid into a hot, dense gas and delivers it to the condenser. There it gives up heat to the room air and condenses into a liquid. The working fluid evaporates inside the refrigerator, extracting heat from the food.

Fig. 8.1.5 These drinking fountains have heat pumps built into them to cool the water. Heat removed from the water, along with heat produced by the heat-pumping process itself, is released into the room air through the louvers visible on the sides of the drinking fountains.

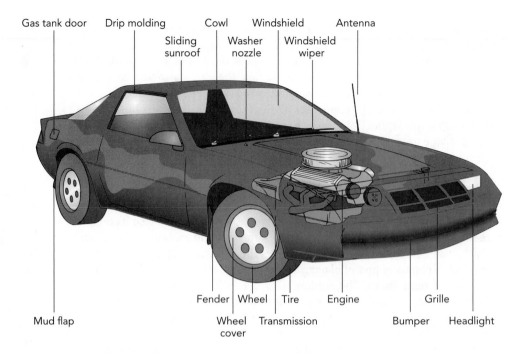

Gas tank door Drip molding Cowl Windshield Antenna
Sliding sunroof Washer nozzle Windshield wiper

Fender Wheel Tire Engine Grille
Mud flap Wheel cover Transmission Bumper Headlight

Section 8.2 **Automobiles**

Nothing is more symbolic of freedom and personal independence than an automobile. With its keys in your hand you can go almost anywhere at a moment's notice. The mechanism that makes this instant transportation possible is the internal combustion engine. Though it has been refined over the years, this engine's basic design has changed little since it was invented more than a century ago. It uses thermal energy released by burning fuel to do the work needed to propel the car forward. That thermal energy can do work at all is one of the marvels of thermal physics and the primary focus of this section.

Using Thermal Energy: Heat Engines

The light turns green and you step on the accelerator pedal. The engine of your car roars into action and, in a moment, you're cruising down the road a mile a minute. The engine noise gradually diminishes to a soft purr and vanishes beneath the sound of the radio and the passing wind.

The engine is the heart of the automobile, pushing the car forward at the light and keeping it moving against the forces of gravity, friction, and air resistance. It's not simply a miracle of engineering. It's also a wonder of thermal physics because it performs the seemingly impossible task of converting thermal energy into ordered energy. But the second law of thermodynamics forbids the direct conversion of thermal energy into ordered energy, so how can a car engine use burning fuel to propel the car forward?

The car engine avoids conflict with the second law by being a heat engine—a device that converts thermal energy into ordered energy *as heat flows from a hot object to a cold object* (Fig. 8.2.1). While thermal energy in a single object can't be converted into work, that restriction doesn't apply to a system of two objects *at different temperatures*. Because heat flowing from the hot object to the cold object increases the overall entropy of the system, a small amount of thermal energy can be converted into work without decreasing the system's overall entropy and without violating the second law of thermodynamics.

Another way to look at a heat engine is through the contributions of the two objects. The hot object provides the thermal energy that's converted into work. The cold object provides the order needed to carry out that conversion. As the heat engine operates the hot object loses some of its thermal energy and the cold object loses some of its order. The heat engine has used them to produce ordered energy. Since the heat engine needs both thermal energy and order, it can't operate if either the hot or the cold object is missing.

In a car engine, the hot object is burning fuel and the cold object is outdoor air. Some of the heat passing from the burning fuel to the outdoor air is diverted and becomes the ordered energy that propels the car. But what limits the amount of thermal energy the engine can convert into ordered energy?

To answer that question, let's examine a simplified car engine. We'll treat the burning fuel and outdoor air as a single, thermally isolated system and look at what happens to their total entropy as the engine operates. In accordance with the second law of thermodynamics, this total entropy can't decrease while the engine is transforming some thermal energy into ordered energy.

When the car is idling at a stoplight its engine is doing no work, and heat is simply flowing from the hot burning fuel to the cold outdoor air. The system's total entropy increases because this heat is more disordering to the cold air it enters than to the hot burning fuel it leaves. In fact, the system's entropy increases dramatically because the burning fuel is extremely hot compared to the cold outdoor air.

This increase in the system's entropy is unnecessary and wasteful. The second law of thermodynamics only requires that the engine add as much entropy to the cold outdoor air as it removes from the hot burning fuel. Since a little heat is quite disordering to cold air, the car engine can deliver much less heat to the outdoor air than it removes from the burning fuel and still not cause the system's total entropy to decrease. As long as the engine delivers enough heat to the outdoor air to keep the total entropy from decreasing, there's nothing to prevent it from converting the remaining heat into ordered energy!

This conversion starts as soon as you remove your foot from the brake and begin to accelerate forward. Instead of transferring all of the thermal energy in the burning fuel to the outdoor air, your car then extracts some of it as ordered energy and uses it to power the wheels. The car engine can convert thermal energy into ordered energy, as long as it passes along enough heat from the hot object to the cold object to satisfy the second law of thermodynamics.

Obeying the second law becomes easier as the temperature difference between the two objects increases. When the temperature difference is huge, as it is in an automobile engine, a large fraction of the thermal energy leaving the hot object can be converted into ordered energy—at least in theory. For an ideally efficient automobile engine or other heat engine, the relationships between the heat removed from the hot object, the heat added to the cold object, and the work provided are:

work provided =

$$\text{heat removed from hot object} \cdot \frac{\text{temperature}_{\text{hot}} - \text{temperature}_{\text{cold}}}{\text{temperature}_{\text{hot}}}$$

heat added to cold object =

$$\text{heat removed from hot object} - \text{work provided.} \qquad (8.2.1)$$

Unfortunately, theoretical limits are often hard to realize in actual machines, and the best automobile engines extract only about half the ordered energy specified by Eq. 8.2.1. Still, obtaining even that amount is a remarkable feat and a tribute to scientists and engineers who, in recent years, have labored to make automobile engines as energy efficient as possible.

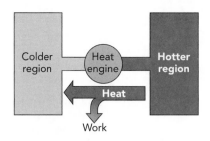

Fig. 8.2.1 A heat engine converts heat (thermal energy from the hotter region) into work (ordered energy) as heat flows from a hotter region to a colder region. The larger the temperature difference between the two regions, the larger the fraction of heat that can be converted into work.

The Internal Combustion Engine

Invented by the German engineer Nikolaus August Otto in 1867, an internal combustion engine burns fuel directly in the engine itself. Gasoline and air are mixed and ignited in an enclosed chamber. The resulting temperature rise increases the pressure of the gas and allows it to perform work on a movable surface.

To extract work from the fuel, the internal combustion engine must perform four tasks in sequence:

1. It must introduce a fuel–air mixture into an enclosed volume.
2. It must ignite that mixture.
3. It must allow the hot burned gas to do work on the car.
4. It must get rid of the exhaust gas.

In the standard four-stroke fuel-injected engine found in modern gasoline automobiles, this sequence of events takes place inside a hollow cylinder (Fig. 8.2.2). It's called a "four-stroke" engine because it operates in four distinct steps or strokes: induction, compression, power, and exhaust. "Fuel-injected" refers to the technique used to mix the fuel and air as they're introduced into the cylinder.

Automobile engines usually have four or more of these cylinders. Each cylinder is a separate energy source, closed at one end and equipped with a movable piston, several valves, a fuel injector, and a spark plug. The piston slides up and down in the cylinder, shrinking or enlarging the cavity inside. The valves, located at the closed end of the cylinder, open to introduce fuel and air into the cavity or to permit burned exhaust gas to escape from the cavity. The fuel injector adds fuel to the air as it enters the cylinder. And the spark plug, also located at the closed end of the cylinder, ignites the fuel–air mixture to release its chemical potential energy as thermal energy.

The fuel–air mixture is introduced into each cylinder during its induction stroke. In this stroke, the engine pulls the piston away from the cylinder's closed end so that its cavity expands to create a partial vacuum. At the same time, the cylinder's inlet valves open so that atmospheric pressure can push fresh air into the cylinder. The cylinder's fuel injector adds a mist of fuel droplets to this air so that the cylinder fills with a flammable fuel–air mixture. Because it takes work to move air out of the way and create a partial vacuum, the engine does work on the cylinder during the induction stroke.

At the end of the induction stroke, the inlet valves close to prevent the fuel–air mixture from flowing back out of the cylinder. Now the compression stroke begins. The engine pushes the piston toward the cylinder's closed end so that its cavity shrinks and the fuel–air mixture becomes denser. Because it takes work to compress a gas, the engine does work on the mixture during the compression stroke. In accordance with the first law of thermodynamics, this work increases the internal energy of the fuel-air mixture. That gaseous mixture can't store the added energy in potential form, so its thermal energy rises and it becomes hotter. Since increases in a gas's density and temperature both increase its pressure, the pressure in the cylinder rises rapidly as the piston approaches the spark plug.

At the end of the compression stroke, the engine applies a high-voltage pulse to the spark plug and ignites the fuel–air mixture. The mixture burns quickly to produce hot, high-pressure burned gas, which then does work on the car during the cylinder's power stroke. In that stroke, the gas pushes the piston away from the cylinder's closed end so that its cavity expands and the burned gas becomes less dense. Since the hot gas exerts a huge pressure force on the piston as it moves outward, it does work on the piston and ultimately propels the car. As it does work, the burned gas gives up thermal energy and cools in accordance with the

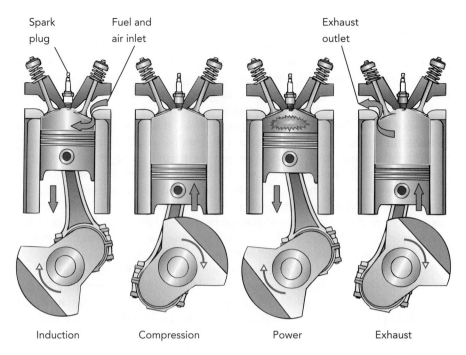

Spark plug   Fuel and air inlet   Exhaust outlet

Induction   Compression   Power   Exhaust

Fig. 8.2.2 A four-stroke engine cylinder. During the induction stroke, fuel and air enter the cylinder. The compression stroke squeezes that mixture into a small volume. The spark plug ignites the mixture and the power stroke allows the hot gas to do work on the automobile. Finally, the exhaust stroke ejects the exhaust gas from the cylinder.

first law of thermodynamics. Its density and pressure also decrease. At the end of the power stroke, the exhaust gas has cooled significantly and its pressure is only a few times atmospheric pressure. The cylinder has extracted much of the fuel's chemical energy as work.

The cylinder gets rid of the exhaust gas during its exhaust stroke. In this stroke, the engine pushes the piston toward the closed end of the cylinder while the cylinder's outlet valves are open. Because the burned gas trapped inside the cylinder at the end of the power stroke is well above atmospheric pressure, it accelerates out of the cylinder the moment the outlet valves open. These sudden bursts of gas leaving the cylinders create the "poof-poof-poof" sound of a running engine. Without a muffler on its exhaust pipes, the engine would be loud and unpleasant.

Just opening the outlet valves releases most of the exhaust gas, but the rest is squeezed out as the piston moves toward the cylinder's closed end. The engine again does work on the cylinder as it squeezes out the exhaust gas. At the end of the exhaust stroke, the cylinder is empty and the outlet valves close. The cylinder is ready to begin a new induction stroke.

Engine Efficiency

The goal of an internal combustion engine is to extract as much work as possible from a given amount of fuel. In principle, all of the fuel's chemical potential energy can be converted into work because both are ordered energies. But it's difficult to convert chemical potential energy directly into work, so the engine burns the fuel instead. This step is unfortunate, for in burning the fuel, the engine converts the fuel's chemical potential energy directly into thermal energy and produces lots of unnecessary entropy.

But all is not lost. Since the burned fuel is extremely hot, a good fraction of its thermal energy can be converted into ordered energy by diverting some of the heat

❶ A gun uses hot, high-pressure gas to propel its bullet through its barrel. That hot, high-density gas forms abruptly when energetic powder inside the bullet cartridge undergoes a sudden chemical reaction. The pressure of this gas is so great that the brass cartridge stretches outward and its thin, elastic wall presses tightly against the inside of the steel barrel, forming a leak-tight seal. With this trapped gas behind it and atmospheric pressure ahead of it, the bullet then experiences an enormous pressure imbalance and it accelerates rapidly toward the open end of the barrel. As the bullet moves through the barrel, the trapped gas does work on the bullet, so the gas's temperature decreases along with its density and pressure. By the time the bullet leaves the barrel, much of the gas's thermal energy has been converted into kinetic energy in the bullet, although this conversion is more complete for a long-barrelled rifle than for a short-barrelled handgun. As the bullet emerges from the barrel, the sudden release of pressurized gas into the air creates a shock wave that's responsible for most of the gun's noise. If the bullet is supersonic, however, it will produce a sonic boom as its own shock wave passes your ears. A silenced gun uses baffles to prolong the release of gas from the barrel and thereby suppress the barrel shock wave. It also uses subsonic bullets to avoid sonic booms.

❷ Older cars use a device called a carburetor to mix fuel and air before they enter the cylinders. The carburetor sits on top of the engine and adjusts the balance between the fuel and air, normally mixing about 1 unit of gasoline with 15 units of air—as measured by mass. It performs this mixing using the Venturi effect: as air flows through a narrow constriction in the carburetor, and on its way to the cylinders, the air's speed increases and its pressure drops. Gasoline accelerates through a jet nozzle toward this low-pressure region and is atomized as it sprays out into the rapidly moving airstream. This fuel and air mixture then flows out of the carburetor and into the cylinders, where it's burned.

that flows from the burned fuel to the outdoor air. As we noted earlier, the hotter the burned fuel and the colder the outdoor air, the more ordered energy the engine can extract. To maximize its fuel efficiency, an internal combustion engine obtains the hottest possible burned gas, lets that gas do as much work as it can on the piston, and releases the gas at the coldest possible temperature. (For another system that extracts work from an expanding and cooling gas, see **❶**.)

It would be wonderful if, during the power stroke, the burned gas expanded and cooled until it reached the temperature of the outdoor air. The exhaust gas would then leave the engine with the same amount of thermal energy it had when it arrived, and the engine would have extracted all of the fuel's chemical potential energy as work. Unfortunately, that would violate the second law of thermodynamics by converting thermal energy completely into ordered energy. As Eqs. 8.2.1 indicate, an operating heat engine always adds some heat to its cold object. In this case, the engine releases the burned gas before it cools to the temperature of the outdoor air. It has no choice; the engine's exhaust must be hot!

But a real internal combustion engine wastes energy and extracts less work than the second law allows. For example, some heat leaks from the burned gas to the cylinder walls and is removed by the car's cooling system. This wasted heat isn't available to produce work. Similarly, sliding friction in the engine wastes mechanical energy and necessitates an oil-filled lubricating system. Overall, a real internal combustion engine converts only about 20% to 30% of the fuel's chemical potential energy into work.

Improving Engine Efficiency

To obtain the hottest possible burned gas, the compression stroke should squeeze the fuel-air mixture into as small a volume as possible. The more tightly the piston compresses the mixture, the higher its density, pressure, and temperature will be before ignition and the hotter the burned gases will be after ignition. Since the efficiency of any heat engine increases as the temperature of its hot object increases and since the hot burned gas is the automobile engine's "hot object," its high temperature after ignition is good for efficiency.

The extent to which the cylinder's volume decreases during the compression stroke is measured by its compression ratio—its volume at the start of the compression stroke divided by its volume at the end of the compression stroke. The larger this compression ratio, the hotter the burned gas and the more energy efficient the engine. While normal compression ratios are between 8 : 1 and 12 : 1, those in high-compression engines may be as much as 15 : 1.

Unfortunately, the compression ratio can't be made arbitrarily large. If the engine compresses the fuel–air mixture too much, the flammable mixture will become so hot that it will ignite all by itself. This spontaneous ignition due to overcompression is called preignition or knocking. When an automobile knocks, the gasoline burns before the engine is ready to extract work from it and much of the energy is wasted.

There are two ways to reduce knocking. First, you can mix the fuel and air more uniformly. In a non-uniform mixture, there may be small regions of gas that get hotter or are more susceptible to ignition than others. The fuel-injection technique used in all modern cars provides excellent mixing and also allows a car's computer to adjust the fuel–air mixture for complete combustion and minimal pollution. So unless a car is seriously out of tune, there isn't much room for improvement as far as mixture uniformity is concerned. (For an earlier mixing system, see **❷**.)

Second, you can use the most appropriate fuel. Not all fuels ignite at the same temperature, so you should select a fuel that is able to tolerate your car's compression process without igniting spontaneously. That's exactly what you do when you purchase the proper grade of gasoline. Regular gasoline ignites at a relatively low temperature and is most susceptible to knocking. Premium gasoline ignites at a relatively high temperature and is most resistant to knocking.

Fuels that are more difficult to ignite and more resistant to knocking are assigned higher "octane numbers." Regular gasoline has an octane number of about 87 while premium has an octane number of about 93 (Table 8.2.1). Choosing the proper fuel is simply a matter of finding the lowest octane gasoline that your car can use without excessive knocking. A little knocking in the most demanding circumstances is quite acceptable. Most modern well-tuned automobiles work beautifully on regular gasoline. Since only high-performance cars with high-compression engines need premium gasoline, putting anything other than regular gasoline in a normal car is usually a waste of money.

Table 8.2.1 Approximate Ignition Temperatures for the Three Standard Grades of Gasoline during Compression.

Octane Number	Approximate Ignition Temperature
87 (regular)	750 °C (1382 °F)
90 (plus)	800 °C (1472 °F)
93 (premium)	850 °C (1562 °F)

Diesel Engines and Turbochargers

Since knocking sets the limit for compression ratio, it also sets the limit for efficiency in a gasoline engine. However, diesel engines avoid the knocking problem by separating the fuel and air during the compression stroke (Fig. 8.2.3). Invented by German engineer Rudolph Christian Karl Diesel (1858–1913) in 1896, the diesel engine has no spark plug to ignite the fuel. Instead, it compresses pure air with

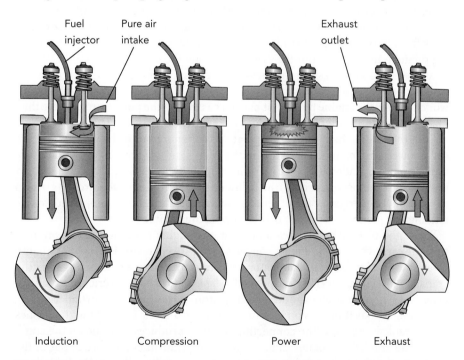

Fuel injector   Pure air intake   Exhaust outlet

Induction   Compression   Power   Exhaust

Fig. 8.2.3 A diesel engine cylinder contains pure air during the compression stroke. As the piston does work on it, this air becomes extremely hot. At the start of the power stroke, diesel fuel is injected into the cylinder. The fuel ignites spontaneously, and the hot burned gas does work on the piston and engine during the power stroke.

an extremely high compression ratio of perhaps 20 : 1 and then injects diesel fuel directly into the cylinder just as the power stroke begins. The fuel ignites spontaneously as it enters the hot, compressed air.

Because of its higher compression ratio, a diesel engine burns its fuel at a higher temperature than a standard gasoline engine and can therefore be more energy efficient. It effectively has a hotter "hot object" and can convert a larger fraction of heat into work. Unfortunately, a diesel engine is also harder to start than a gasoline engine and requires carefully timed fuel injection.

Some gasoline or diesel engines combine fuel injection with a turbocharger. A turbocharger is essentially a fan that pumps outdoor air into the cylinder during the induction stroke. By squeezing more fuel–air mixture into the cylinder, a turbocharger increases the engine's power output. The engine burns more fuel each power stroke and behaves like a larger engine. The fan of a turbocharger is powered by pressure in the engine's exhaust system. A nearly identical device called a supercharger is driven directly by the engine's output power.

The downside of a turbocharger, other than being expensive and wearing out rather quickly, is that it encourages knocking. As it squeezes air into the cylinder, it does work on that air and the air becomes hot. Since the fuel–air mixture enters the engine hot, it may ignite spontaneously during the compression stroke. To avoid knocking in a car equipped with a turbocharger, you may need to use premium gasoline. Some turbocharged cars are equipped with an intercooler, a device that removes heat from the air passing through the turbocharger. By providing cool, high-density air to the cylinders, the intercooler reduces the peak temperature of the compression stroke and avoids knocking.

Multicylinder Engines

Since the purpose of the engine is to extract work from the fuel–air mixture, it's important that each cylinder do more work than it consumes. Three of the strokes require the engine to do work on various gases, and only one of the strokes extracts work from the burned gas. During the induction stroke, the engine does work drawing the fuel-air mixture into the cylinder. During the compression stroke, the engine does work compressing the fuel–air mixture. During the exhaust stroke, the engine does work squeezing the exhaust gas out of the cylinder. Fortunately, the work done on the engine by the hot burned gas during the power stroke more than makes up for the work the engine does during the other three strokes.

Still, the engine has to invest a great deal of energy into the cylinder before each power stroke. To provide this initial energy, most four-stroke engines have four or more cylinders (Fig. 8.2.4), timed so that there is always one cylinder going through the power stroke. The cylinder that is in the power stroke provides the work needed to carry the other cylinders through the three nonpower strokes, and there is plenty of work left over to propel the car itself.

While the pistons move back and forth, the engine needs a rotary motion to turn the car's wheels. The engine converts each piston's reciprocating motion into rotary motion by coupling that piston to a crankshaft with a connecting rod. The crankshaft is a thick steel bar, suspended in bearings, that has a series of pedal-like extensions, one for each cylinder. As the piston moves out of the cylinder during the power stroke, it pushes on the connecting rod and the connecting rod pushes on its crankshaft pedal. The connecting rod thus produces a torque on the crankshaft. The crankshaft rotates in its bearings and transmits this torque out of the engine so that it can be used to propel the car. So, while each cylinder initially exerts a force, the crankshaft uses that force to produce a torque.

The spinning crankshaft conveys its rotary power to the car's transmission, and from there the power moves on to the wheels. Overall, a significant portion of the

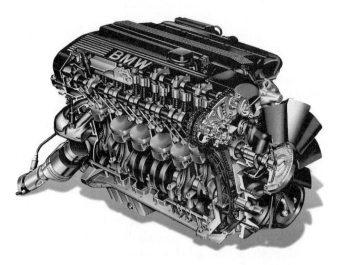

Fig. 8.2.4 This cutaway view of a BMW automobile engine shows the six pistons arranged in a single row. The cylinders have been omitted from the drawing.

heat flowing out of the burning fuel-air mixture is being converted into work and used to spin the car's wheels. Assisted by friction with the pavement, the wheels push the car forward, and you cruise down the highway toward your destination.

The Ignition System

In each cylinder of a gasoline engine, a spark plug is responsible for igniting the fuel and air mixture exactly at the beginning of the power stroke. Ignition occurs when an electric current suddenly leaps across a narrow gap at the end of the spark plug. But where does this electric current come from?

In older automobiles, the current is created by the coil. The coil stores energy in the form of magnetic potential energy—something we'll examine in Chapter 11. This potential energy is initially transferred to the coil by the car's electric system. The electric system, which includes the car's battery, passes an electric current through a switch called the points and through one section of the coil. This current magnetizes the coil and gives it its magnetic potential energy. At the start of the power stroke, a mechanical device opens the points and interrupts the current flowing through first part of the coil. The coil abruptly loses its magnetism and its stored energy is released as a current that suddenly appears in the second part of the coil. This current through the second part of the coil passes through the spark plug, creating the spark. Actually, the engine has many spark plugs (one for each cylinder), so the current from the coil is directed to the proper spark plug by a rotating mechanical switch called the distributor.

However, the fuel and air mixture doesn't burn instantly. The flame propagates through the mixture at less than the speed of sound, taking roughly a thousandth of a second to move from the spark plug to the piston's surface. When the engine is turning slowly, this burning time isn't very important. But at high engine speeds, the mixture must be ignited a little early to obtain the maximum work from the burned gas. The timing system advances the opening time of the points when the engine is turning quickly.

With the advent of the electronic ignition, techniques for producing a well-timed spark at the cylinder have become more sophisticated. Electronic ignition systems have replaced the points and coil with more complicated electronic devices. Nonetheless, electronic ignition systems still look to the engine for timing information, they still use a mechanical distributor, and they still produce the high-energy electric pulse with a magnetic device resembling a coil. But now a computer is used to adjust the spark timing subtly so that the engine works extremely effectively at all speeds and in any environment.

Pollution Control

One way in which an internal combustion engine can fail to achieve peak energy efficiency is by not burning all of the fuel it consumes. It's difficult to mix the fuel and air perfectly, combining just the right number of fuel and oxygen molecules to allow complete combustion. If there are too many fuel molecules, some will be unable to burn completely. If there are too many air molecules, the whole mixture will have trouble burning at all. But even if the mixture is perfectly balanced, not every fuel molecule will react completely with oxygen molecules to form water and carbon dioxide molecules. Instead, some fuel molecules will pass through the cylinder more or less intact. These imperfectly burned molecules don't just waste energy—they also pollute the air.

Modern engines scavenge the unburned molecules in at least two ways. First, an engine tries to reburn fuel that has escaped from the cylinders. During the induction stroke, the pressure inside a cylinder drops below atmospheric pressure. This low pressure is used to suck fuel vapors out of the engine and into the cylinders so that it can be burned. The positive crankcase ventilation system (PCV) collects fuel vapor that has slipped past the pistons and into the crankcase surrounding the crankshaft. This vapor flows from the higher pressure in the crankcase toward the lower pressure inside a cylinder, passing through the PCV valve on its way.

The other way in which an engine eliminates unburned fuel molecules in the exhaust gas is in a catalytic converter. This converter uses a catalyst to facilitate the reaction of fuel molecules with oxygen. A catalyst is a chemical surface that helps a reaction to proceed by reducing the activation energy needed to carry out that reaction. At relatively low temperatures, a fuel molecule and an oxygen molecule won't react because they lack the initial energy needed to break several old chemical bonds so that new, stronger bonds can begin to form. The catalyst helps to break those old bonds so that the new bonds form easily.

In an automobile, the catalyst on which the residual fuel molecules oxidize is usually platinum. Microscopic particles of that metal are suspended on a porous ceramic honeycomb through which the exhaust gases must flow to reach the muffler and tailpipe. The platinum particles are so small that most of their atoms are on their surfaces and together they expose a phenomenal amount of platinum surface to the exhaust gases. The imperfectly burned fuel molecules cling briefly to the catalyst particles, react with oxygen, and leave the catalytic converter as water and carbon dioxide molecules.

Joining the platinum particles in the catalytic converter are particles of rhodium. Rhodium acts as a catalyst to "unburn" nitrogen oxide molecules that formed accidentally during combustion in the cylinders. Rhodium converts those nitrogen oxide molecules into ordinary nitrogen and oxygen molecules, which are no longer pollutants. The car's computer uses data from oxygen sensors in the exhaust system to adjust the fuel and air mixture to minimize the emissions of both unburned fuel and nitrogen oxides in the car's exhaust.

The catalytic converter must be heated by the engine exhaust before it can begin to function, so it doesn't work for the first minute or two after you start the car. Once it's hot enough, however, the reactions proceed and they heat the converter still further. The hotter they get, the better the catalysts work. However, the converter assembly itself can be damaged by overheating if it receives too much unburned fuel in too short a time (see also ❸). The catalytic converter can also be "poisoned" by sending it materials, such as lead, that coat the noble metal particles and stop them from assisting reactions. One tank of leaded gasoline is enough to ruin a car's catalytic converter. During the era when both leaded and unleaded fuels were available simultaneously, their delivery spouts were different sizes to prevent people from putting leaded fuels in cars that required unleaded fuels.

❸ A catalytic converter is usually located below the engine as part of the exhaust system. Because it becomes so hot while operating, it is protected by a heat shield. Nonetheless, it is extremely dangerous to park a car over flammable materials after the catalytic converter has become hot. Every autumn, a few cars go up in flames after their drivers unwittingly park them on piles of fallen leaves.

Manual Transmission: Friction and Gears

Just as a bicycle has gears and a freewheel between its pedals and rear wheel, an automobile has a transmission between its engine and wheels. This transmission receives torque from the engine and delivers it to the wheels, though it generally provides mechanical advantage along the way and can also decouple the engine from the wheels when necessary. Mechanical advantage allows the car to climb a hill and decoupling permits it to stop at a traffic light.

There are two types of transmissions: manual and automatic. In a car with manual transmission, the driver is responsible for decoupling the engine from the wheels and for adjusting the mechanical advantage. An automatic transmission accomplishes both tasks on its own.

In a manual transmission, the decoupling is performed by the clutch. The clutch consists of two disks, one of which is turned by the engine and one of which ultimately turns the wheels. Normally, a spring presses the two disks together hard, so that friction between them forces them to turn together. However, when you depress the clutch pedal inside the car, a mechanism separates the two disks so that they no longer rub together and the engine can turn independently of the wheels. As you release the clutch pedal, the two plates come back together and sliding friction begins to equalize their angular velocities. Once the plates have reached the same angular velocity, static friction keeps them turning together. Among other things, the clutch permits you to start a car moving gradually without having to slow the engine's rotation so much that it stops turning.

Because the two clutch plates experience sliding friction as you step on or off the clutch pedal, they convert work into thermal energy during those moments. Normally, you spend only a moment turning the clutch on or off so that these periods of sliding friction are brief and infrequent. However, if you "ride the clutch" and allow the plates to slip against one another for a long time, sliding friction will overheat the clutch and it will burn out prematurely.

The clutch transmits torque from the engine to the gearbox. In the gearbox, incoming torque turns a set of gears. Each turning gear will transmit torque to any other gear that engages with it. The wheels are connected to another set of gears that are normally not engaged with the turning gears. However, a lever by the driver's seat allows him or her to engage a particular pair of gears so that the engine begins to turn the wheels.

In a typical car with five forward gears and one reverse gear, the driver can bring together five different combinations of gears for forward motion. The lowest gear is one in which the engine turns much more rapidly than the wheels. Although the wheels turn slowly, they receive a large torque from the gearbox. You can use first gear to start a car moving or to climb a very steep hill. In the highest gear, the wheels turn as fast or even faster than the engine. While the wheels turn rapidly, they receive only a small torque from the gearbox. You can use fifth gear to drive at high speed on a fairly level road or down a hill.

A mechanical gearbox also has a reverse gear. In reverse, the wheels turn the opposite direction from the forward gears. The gearbox makes this reversal simply by adding one extra gear between the engine and the wheels. Each time one gear turns a second gear, the direction of rotation reverses. So the reverse gear just involves one extra reversal.

A slight variation on the gearbox idea is the synchromesh gearbox. In such a gearbox, the driver doesn't choose which gears are meshing and which are not. Instead, the gears are always meshing. However, the gears aren't directly coupled to the various drive shafts in the gearbox. To connect a gear to a drive shaft, the gearshift lever moves a toothed sleeve so that it couples a toothed cylinder on the gear to a toothed cylinder on the driveshaft. As the sleeve moves, a friction cone is used to bring the shaft and the gear to about the same angular velocity. This is the origin of the word "synchromesh." Because the gears themselves are always engaged, they can have curved teeth that make them quieter than straight-toothed gears. Teeth in straight-toothed gears click every time they meet while those in curve-toothed gears meet gradually as the gears turn and don't click.

Automatic Transmission: Fluids and Gears

In a car with an automatic transmission, one or more fluid couplings replace the clutch, and the gearbox is designed so that the car can operate it automatically. In a fluid coupling, torque is transmitted from the engine to the wheels by a fluid that circulates between two fan-like impellers. The engine turns the first impeller, which pushes transmission fluid toward the second impeller. This moving fluid turns the second impeller, which ultimately turns the car's wheels. The fluid transfers power from the engine to the wheels because the first impeller does work on the fluid and the fluid does work on the second impeller.

What makes this peculiar, indirect coupling so useful is that it doesn't waste power, even if the two impellers are turning at very different speeds. When the second impeller is stopped while your car is at a red light, the transmission fluid can't do any work on it. The torque on the impeller may be large, but it doesn't turn (see p. 193 for a reminder that work can be done by a torque exerted over an angle). The transmission fluid stops circulating through the transmission and the first impeller spins easily so that it extracts very little power from the engine.

When you permit the car to move, the second impeller begins turning and gradually picks up speed until it's turning at roughly the same angular velocity as the first impeller. Throughout this process, power flows efficiently from the first impeller to the second impeller and the car accelerates forward smoothly. An automatic transmission produces fewer jerks and bumps than a manual transmission.

The gearbox in an automatic transmission operates hydraulically. Hydraulic fluid pushes friction plates together so that they stop various parts of the transmission from turning. Complicated collections of gears are used so that stopping one gear from turning transfers torque to other gears. Mechanisms in the car sense its speed and choose which friction plates should be opened and which should be closed. These plates determine the relative rate of rotation between the shaft that carries torque into the automatic transmission and the shaft that carries torque out of it.

Some automatic transmissions combine the functions of the fluid coupling and part of the gearbox into one hydraulic mechanism called a torque converter. A torque converter is a fluid coupling that contains two differently shaped impellers. The first impeller turns rapidly, at the rotational speed of the engine. The second impeller turns more slowly but delivers more torque to the wheels.

Starting the Engine

Up until now, we've considered only an engine that's already running. But how does that engine begin running? The car starts the engine using an electric motor called the starter. The starter motor is bolted to the engine next to a toothed disk called the flywheel. The flywheel turns when the engine turns and visa versa. The starter motor's only purpose is in starting the engine; once the engine is running, the starter motor just sits there idly.

When you start the engine, an electromagnet called the solenoid moves a toothed gear on the starter motor so that it engages with the flywheel. The motor and the flywheel must then turn together. Electric power from the car's battery is directed to the starter motor and it begins to turn (I'll explain electric motors in Section 11.3). The starter motor exerts a torque on the flywheel and it begins to turn, too.

The gear on the starter motor is small, while the flywheel disk is large. As a result, the starter motor has a great deal of mechanical advantage. The starter motor turns rapidly and exerts a modest torque on the flywheel, while the flywheel turns slowly but exerts an enormous torque on the engine. Although the starter motor does all the work needed to turn the engine the gears allow it to do that work with a modest torque exerted over a large angle of rotation.

As the cylinders pass through the power strokes they begin to function normally and pretty soon the whole engine is operating on its own. You can then turn off the starter. As you do, the solenoid moves the starting motor's gear away from the flywheel and the starter returns to its normal idle state.

The Differential

So far we've followed the torque from the engine through the transmission to the drive shaft, and we're almost at the wheels. While it might seem that all the automobile has to do is couple that drive shaft to a single axle attached to both powered wheels, there is a serious problem with attaching both wheels rigidly to the same axle. When you turn a corner, the inside wheel travels a shorter distance than the outside wheel and the inner wheel must turn more slowly than the outside wheel. If both wheels share a common axle, they will have to turn together and will skid on the pavement, squandering traction, tread, and energy.

To solve this problem, automobiles use a complicated gear assembly called a differential. The wheels are attached to two separate halves of the axle and these two half-axles and the drive shaft from the transmission meet in the differential. Torque from the drive shaft turns the four gears of the differential, and two of these gears then drive the half-axles. As the drive shaft rotates, the differential causes one or both of the two half-axles to turn. However, the differential allows one half-axle to turn faster than the other—it only requires that the *sum* of the two angular velocities be proportional to the drive shaft's angular velocity. Overall, the differential exerts equal torques on each wheel rather than making both wheels turn at the same angular velocity.

During a turn, the differential permits one of the wheels to turn more slowly than the other. It exerts the same torque on each wheel and doesn't waste any power. Without the differential, a car would be hard to handle during a turn. Unfortunately, there is also a dark side to the differential. If one of your car's wheels is resting on ice and has no traction, that wheel will do all of the turning while the other wheel remains stationary. The tractionless wheel will require almost no torque to rotate so the differential will give it all of the rotation. On some vehicles, such as tractors, the driver can lock the differential or limit its operation so that both wheels are forced to rotate. These mechanisms make it easier to drive the vehicle out of snow or mud but should be turned off on clear pavement.

Wheels, Suspension, and Steering

The story doesn't end when torque from the engine reaches the wheels. The driver still needs to be able to steer the automobile and doesn't want to sit directly on the wheels, feeling every bump in the road. So the wheels are connected to a complicated support structure. This support structure suspends the body of the car off the wheel axles and permits at least two of the wheels to swivel for steering.

In all cars, the goal of the suspension is the same: to support the body on elastic devices so that passengers don't feel the road. The suspension is also responsible for damping out and thereby eliminating oscillations—a car that bounces up and down for a minute or two after hitting a pothole is uncomfortable to ride in and unsafe to drive.

Elastic devices used in car suspensions include coil springs, leaf springs, air springs, and torsion bars. The idea is to have the body of the car rest on springy devices rather than directly on the wheels. The wheels support the springs and the springs support the car body. During assembly, the car body's weight compresses the springs somewhat. After a few moments, the system reaches equilibrium, with the springs exerting just enough upward force on the car body to exactly balance the body's weight. The car is then in a stable equilibrium.

Whenever the wheel of a moving car hits a rock, the spring attached to that wheel compresses. The spring then begins to exert an upward force on the car body and the body begins to accelerate upward. However, the spring transfers vertical momentum to the car slowly, with modest forces, so that the passengers don't accelerate upward rapidly. The wheel is much less massive than the car body and accelerates much more rapidly in response to forces from the spring. Often, the wheel returns to its original position before the body of the car can respond significantly. In such a situation, the passengers barely notice the rock.

If the bump is broader than a rock, such as a rise in the height of the roadway, then the car's body must eventually move up, too. The car body continues to accelerate upward until it has returned to the correct height above the wheels. But by that time, the car body has a substantial upward velocity and overshoots its proper height. As a result, the car body oscillates up and down, just as a spring scale in a grocery store does when you suddenly throw a heavy bag of vegetables into it.

If the only component of the suspension were the springs, your car really would bounce down the road. However, your car's suspension also contains shock absorbers. These mechanical devices use sliding friction between surfaces or viscous effects in a fluid to damp out the oscillations of the car body. Each time the car body tries to move relative to the wheels, the shock absorber resists the motion. It wastes the car's excess energy—the energy in the bounce—and quickly brings the car back to stable equilibrium. When your car's shock absorbers are functioning properly, you will not be able to make the car bounce. When they have failed, your car will bounce repeatedly each time you hit a broad bump or dip in the road. The bouncing can reduce the wheels' traction and your ability to control the direction your car is heading.

You control the direction your car is heading by turning the steering wheel. This action shifts the axis about which the front wheels rotate. The car will always try to follow a path that lets the wheels turn without skidding. As you steer the car to the left, a gear mechanism turns the front wheels so that the front end of each wheel is to the left of the rear end of that wheel. As the wheel spins, it first makes contact with the road ahead and to the left of the middle of the wheel. This contact point then shifts backward and toward the right as the wheel continues to spin. In order that the wheel not slip from the surface at that contact point, the whole wheel and the car as well must accelerate and move toward the left. Static friction at that contact point provides the force that causes the car to accelerate the car to the left and thereby travel around the turn.

Of course, if you turn too quickly, the contact point will begin to slip. Static friction can provide only so much force before slipping occurs. The maximum amount of static friction between the tires and the road determines just how quickly you can turn the car. On an icy road, that may not be very quickly at all. Once the wheels begin to slip, you will have only sliding friction between the wheels and the road. Because sliding friction doesn't depend on which way the wheels are pointing, you will have little control over the direction the car is heading. The car will tend to travel in a straight line while friction tries to slow its forward motion. You should steer toward the direction in which you are skidding so that the wheels begin to turn and sliding friction is replaced by static friction. Once static friction returns, you will again be able to steer the automobile.

The Brakes

After all the discussions about getting the car moving, we should know how to stop it. The last major component of an automobile is its brakes. Just as in a bicycle, a car uses sliding friction to stop. When the driver steps on the brake pedal of the car, she pushes a piston into the master cylinder of the brake system. Brake fluid is squeezed out of that cylinder and passes through four separate tubes to the four wheels. At each wheel, the pressurized brake fluid pushes a piston out of a slave cylinder. The piston's motion brings a brake pad into contact with a metal surface that is turning with the wheel.

The harder the driver pushes on the brake pedal, the higher the pressure in the brake fluid and the more force the slave cylinder exerts on the brake pad. Since the forces of sliding friction are roughly proportional to the support forces between the pad and metal surface, the brake's stopping force is proportional to the force the driver exerts on the brake pedal. This mechanism, where fluid squeezed out of one cylinder is used to push the piston out of a second cylinder, is typical of hydraulic controls. Airplanes, construction equipment, and industrial machines contain many such mechanisms. For safety's sake, most master cylinders actually contain two separate cylinders—one that controls the brakes for the front wheels and one that controls the brakes for the rear wheels. With this scheme, even if one of the cylinders fails, half of the brakes will continue to function.

The actual shape of a brake varies. In some cases, the brake pad is pushed against the inside of a drum that's connected to the wheel. In other cases, the brake pad is pushed against a disk that's connected to the wheel. Both brakes turn the kinetic energy of the car into thermal energy in the metal and pad.

Getting rid of that thermal energy is important to the continued function of the brakes. If the brakes overheat, the brake fluid can boil and the brakes will stop functioning. Disk brakes have better exposure to the air than drum brakes and do a better job of transferring heat to the air. As a result, disk brakes are less likely to stop working due to overheating than drum brakes. Nonetheless, on long downhill stretches, it's best to dissipate the car's excess energy—the gravitational potential energy it releases as it descends—through the engine rather than through the brakes. By switching to a low gear and letting the spinning engine turn the extra energy into heat in the radiator and exhaust systems, you save wear on the brakes and avoid the possibility of overheating them to the point of failure.

Many recent cars come equipped with anti-lock brakes. These brakes are designed to avoid skidding by ensuring that the wheels continue to turn during braking. With traditional brakes, the frictional forces between the pads and the metal surfaces can become so high that the wheels stop turning and begin to skid along the road. The wheels suddenly experience sliding friction with the road, leading to a loss of steering, damage to the tire surfaces, and reduction in stopping force.

As long as the wheels continue to turn, static friction supplies the stopping force, and static friction can exert a larger stopping force than sliding friction can. This change in stopping force reflects the fact that static friction will exert far more force to prevent a wheel from skidding than sliding friction will exert to slow down a wheel that *is* skidding.

Anti-lock brakes analyze wheel rotation and release the brakes whenever they sense that a wheel isn't turning properly. The wheels continue to turn throughout a fast stop so that only static friction occurs between the tires and the pavement. All of the car's kinetic energy is turned into thermal energy by sliding friction in the brakes, not sliding friction between the roadway and the tires. When the anti-lock brakes activate, you can feel them adjusting the slowing forces in the brakes so that the wheels continue turning. It's worth experimenting with anti-lock brakes occasionally, so that you'll know what to expect when you use them in earnest.

Automobile Aerodynamics

Even on a level road, a car can't coast for long. Its engine must continue to propel it forward against the slowing effects of friction and air resistance. The principal form of air resistance that a car experiences is pressure drag. Like the ball moving through the air in Fig. 6.2.5, a forward-moving automobile leaves a turbulent wake in the air behind it and it gives that wake air a considerable amount of its forward momentum.

Minimizing the size of a car's turbulent wake is an important part of automobile design. While sleek, aerodynamically streamline shapes are more critical for cars that travel fast on highways or race courses (Fig. 8.2.5) than for those that travel primarily in stop-and-go urban traffic (Fig. 8.2.6), virtually all modern cars are streamlined wherever possible. You can usually look up a car's drag coefficient: the ratio of its wake cross section to its physical cross section. For example, a car with a drag coefficient of 0.3 produces a wake that's only 30% as large in cross section as the car itself.

Fig. 8.2.5 This automobile has been meticulously streamlined to minimize its drag coefficient so that it experiences relatively little pressure drag even at high speeds.

Fig. 8.2.6 This automobile is designed for stop-and-go traffic and to be able to park in tiny spaces. It has a relatively large drag coefficient.

Minimizing drag isn't the only goal of aerodynamic car design. Because the air flowing around a car tends to arc over the top of the car in a long, downward curve, the car typically experiences a relatively low air pressure on its top surface. With nearly atmospheric pressure below the car and less than atmospheric pressure above it, the car experiences an upward lift force. Although that force isn't enough to pick the car up off the ground, it supports part of the car's weight and therefore reduces the support forces that the pavement exerts on the car's tires. The reduced forces between the car's tires and the pavement reduce its traction. At high speeds, a car may lose enough traction to have trouble with skidding during turns or other rapid accelerations.

To counter this lift force and improve their traction, some cars have spoilers (Fig. 8.2.7). An automobile spoiler is a wing that's designed to obtain a downward lift force from the passing air. By deflecting air upward, the spoiler obtains a downward force that pushes the automobile harder against the ground and improves its traction.

In fact, many race cars have overall aerodynamic designs that produce tremendous downward lift forces. They are often pushed against the ground with downward lift forces that exceed their weights. These racing cars grip the pavement so tightly with the help of aerodynamic lift forces that they could actually drive on the ceiling if they could figure out some way to start and stop!

Fig. 8.2.7 The spoiler on this racing car is a carefully designed airfoil that deflects the passing airstream upward while experiencing no unnecessary drag forces. The entire vehicle is designed with aerodynamics in mind. It produces only a tiny turbulent wake and develops large downward lift forces to improve its traction.

In contrast, many ordinary cars have relatively ineffectual "spoilers" mounted above their trunks. These spoilers are often so carelessly designed or oriented that they provide little downward lift and considerable backward drag. Instead of helping the car stay on the road, some of the worst models act as spoilers for the fuel efficiency and the view out of the rear window.

Steam Engines

The internal combustion engines used in modern automobiles developed out of earlier external combustion engines during the latter half of the 19th century. An external combustion engine is a heat engine in which fuel is burned outside the engine itself. The classic external combustion engine is the steam engine, first made practical by Scottish inventor James Watt (1736–1819) in 1763.

In a steam engine, a fire boils water to produce high-pressure steam. This hot steam carries heat toward the cold outside air and does work on a moving surface along the way. By the time the steam is released from the engine, some of its thermal energy has been converted into ordered energy. In addition to powering early automobiles, steam engines were used to propel boats and locomotives and are still commonly used in electric power plants. Of course, modern cars have long since abandoned steam engines in favor of internal combustion engines.

Perpetual Motion Machines

A perpetual motion machine is one that continues in steady motion without slowing down and usually while providing some output of ordered energy. It does this without consuming any external source of order and without being allowed to transfer heat from a hotter region to a colder region. Those last two restrictions are necessary because it's easy to make a perpetually moving system if you're allowed to supply it with electric power or if you can run a steam engine based on heat transfer to power it.

While people keep trying to invent perpetual motion machines, particularly those that would offer an inexhaustible supply of free ordered energy, they'll never succeed. These devices would have to produce work either out of nothing at all or out of thermal energy. Creating work out of nothing is really impossible; energy is a conserved quantity and cannot be created or destroyed. For a machine to do work, it has to obtain energy from somewhere or something else. That's the first law of thermodynamics. If anyone tries to sell you an engine that never needs fuel—that creates work out of nothing—don't buy it! It's a fraud.

As for machines that convert thermal energy completely into work, they're also impossible, but for a different reason. While they don't violate energy conservation, they do violate the laws of thermodynamics. Thermal energy is disordered energy—it is energy that has been distributed randomly among the individual atoms and molecules in an object so that it cannot be easily reassembled to do useful work. When you burn a candle, all of the energy the candle once had is still in the room, but it's much harder to use. Just as a coffee cup is much more useful before you drop it than after you drop it, so energy is much more useful before you disorder it than after you disorder it.

The difficulty with reassembling thermal energy to do useful work is a statistical one: it's unlikely that this energy will spontaneously reassemble itself in a useful manner, just as it's unlikely that a dropped coffee cup will spontaneously reassemble itself in a useful manner. The laws of mechanics don't prevent either of those reassemblies from occurring, but both reassemblies are extraordinarily unlikely to occur. That's the second law of thermodynamics. How often have you dropped a broken cup and had it fall together rather than apart? So if someone tries to sell you an engine that uses the thermal energy in the surrounding air as "fuel"—that turns thermal energy completely into work—don't buy it! It's also a fraud.

That said, you should understand that the world has caches of order scattered everywhere and an engine that consumes that order is permitted under the laws of thermodynamics—it is *not* a perpetual motion machine. A tank of gasoline, a new battery, and a freshly baked jelly donut are obvious examples of order—they can easily be used to run engines (where I'm using the term "engine" broadly to include anything that can consume order and use it to perform work). But even a full glass of water on a dry day, a warm rock on a cold day, and a pile of leaves on a raked yard have order to them and can, in principle, be used to operate engines and perform work. Living things, particularly bacteria, are fantastically adept at consuming whatever small tidbits of order they can find. If there's order to be had, you can pretty much count on some creature having figured out how to live on that order. We humans are terrible, wanton consumers of order, devouring precious resources and leaving heaps of disordered, high-entropy refuse in our wakes. But many other creatures carefully husband the order they find around them and consume it sparingly.

SECTION 8.3 The Atmosphere

Our atmosphere does more than just provide the oxygen we breathe. This layer of gas helps to maintain the earth's surface temperature and shields us from both interplanetary debris and some of the sun's ultraviolet light. The atmosphere also contributes to the dynamic character of the earth's surface—it forms the weather, moves water about the globe, and creates the winds. In this section, we'll look at the origins and characteristics of these atmospheric phenomena.

Earth's Temperature and the Greenhouse Effect

Despite variations with time, place, and season, the earth's surface temperature maintains a fairly constant average value of about 15 °C (59 °F). When America is experiencing night, Asia is experiencing day and vice versa. When the northern hemisphere is experiencing winter, the southern hemisphere is experiencing summer and vice versa. It all averages out.

To maintain this constant average temperature, the earth's heat flow must be balanced. If instead there were a net flow of heat to the earth, its average temperature would increase. And if there were a net flow of heat away from the earth, its average temperature would decrease. The earth maintains its constant average temperature of 15 °C by getting rid of heat just as quickly as that heat arrives. That way the earth's store of thermal energy never changes.

The earth's main source of heat is the sun, and this heat reaches the earth mostly as electromagnetic radiation (Fig. 8.3.1). Because the sun's surface temperature is about 5500 °C, solar radiation is primarily visible light, although it also includes a substantial amount of infrared and ultraviolet light. The total solar power reaching the earth is about 1.73×10^{17} W, or 1.73×10^{17} J of heat each second. For comparison, the world's total electric generating capacity is roughly 3×10^{12} W.

While about 34% of the sunlight is simply reflected or scattered back from the earth's surface and atmosphere, the rest is absorbed by the earth's surface and atmosphere and must be eliminated by radiating it into the dark, empty space surrounding the earth. Because earth's surface and atmosphere are much colder than the sun, their thermal radiation is mostly infrared light (Fig. 8.3.1). For the earth to radiate 1.73×10^{17} W of thermal power, the surface that emits this infrared radiation must have a temperature of about −18 °C (0 °F).

If there were no atmosphere, the source of earth's thermal radiation would be the ground and the earth's average surface temperature would therefore be about −18 °C. But the earth does have an atmosphere and while that atmosphere is nearly transparent to visible light, it absorbs and emits infrared light fairly well. As a result, the earth's principal source of the thermal radiation isn't its surface but its atmosphere! In fact, the effective source of the earth's thermal radiation is located about 5 km (3 miles) above the earth's surface, and it is the atmosphere at that altitude that has the −18 °C average temperature.

This displacement of earth's effective radiating surface from ground level to an elevated layer in the atmosphere gives rise to the greenhouse effect. Since the earth's surface isn't responsible for radiating away the earth's heat, its temperature doesn't have to be −18 °C. And because the atmosphere naturally develops a temperature gradient of about −6.6 °C per kilometer of altitude, the ground 5 km below the radiating surface is about 33 °C hotter, or roughly 15 °C.

The atmosphere's decreasing temperature with altitude, which we'll examine more carefully later in this section, reflects the decrease in temperature that occurs when air is allowed to expand into a region of lower pressure. Rising air encounters decreasing air pressure and expands, so its temperature drops. The presence

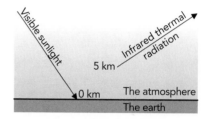

Fig. 8.3.1 Most thermal radiation from the 5500 °C sun is visible light and passes easily through the atmosphere to the ground. The earth's thermal radiation, however, is infrared light and is emitted primarily by the atmosphere. The earth's overall thermal radiation is effectively emitted from an altitude of about 5 km.

of moisture in the air weakens this effect, but the temperature of the earth's atmosphere still decreases by about 6.6 °C per kilometer.

Since life would be difficult at −18 °C, we are fortunate to have the greenhouse effect. However, too large a greenhouse effect could be a problem. If the atmosphere were to become even more effective at absorbing and emitting infrared light, the average altitude from which the earth's thermal radiation is emitted would move upward and the temperature at the earth's surface would increase. We want that altitude to stay about 5 km and not to rise to 6 km or more.

Just how effective the earth's atmosphere is at absorbing and emitting infrared light depends on its chemical makeup. Nitrogen and oxygen molecules, though extremely common in the atmosphere, are remarkably transparent to both infrared and visible lights. It's the less common gas molecules that allow air to absorb and emit infrared light and are thus greenhouse gases. While the principal greenhouse gas is water vapor, other gases such as carbon dioxide and methane (natural gas) are also important. The more of these gases there are in the atmosphere, the higher will be the average altitude from which the earth's thermal radiation is emitted and the warmer the earth's surface will become.

It is now fairly certain that the overproduction of greenhouse gases, notably carbon dioxide, is causing a rapid warming of the earth's surface and a change in its climate. But the greenhouse effect is only one of many influences on climate and, in the absence of overwhelming scientific evidence either way, politics and self-interest continue to dominate the national and global discussions. Nonetheless, many scientists are alarmed by the existing evidence for global warming and believe that ongoing human production of greenhouse gases is a major threat to our future wellbeing. (To see how changes in the greenhouse gases are studied, see ❶.)

Warming the Air and Creating Wind

The air's tendency to absorb infrared radiation does more than simply raise the earth's average surface temperature. It also allows the earth's surface to transfer heat to the atmosphere. Because air is a poor conductor of heat, little heat flows from the ground to the air by contact between the two. Instead, the main mechanism for warming the air is radiative heating from the earth's surface—infrared radiation from the ground is absorbed by greenhouse gases in the air and the air becomes hotter. Since water vapor is the main greenhouse gas and it is concentrated near the ground, the lower atmosphere is heated more effectively than the upper atmosphere.

Because the air is heated from below, the earth's surface temperature has a profound impact on the air above it. Moist air above a hot region of land or water absorbs a great deal of heat while dry air above a cold region absorbs relatively little heat. These variations in heating rates are what cause winds.

To see how wind works, consider the air above a level region of the earth's surface. For starters, let's imagine that the surface temperature is uniform everywhere (Fig. 8.3.2a). In this case, the air on the east side of the region is identical to the air on the west side. In each location, the air is hottest near the ground and its temperature decreases slowly with increasing altitude. The air pressure also begins at a certain value near the ground and decreases slowly with increasing altitude.

So far, there is no wind because there are no horizontal imbalances in pressure. There is a vertical pressure imbalance but it serves only to support the air's weight—without a vertical pressure imbalance, the atmosphere would fall. But a horizontal pressure imbalance would cause the air to accelerate horizontally toward the lower pressure. It would create wind.

Now, let's imagine that the rising sun begins to warm the ground only on the east side of the region. This hotter ground quickly warms the air above it. The hot-

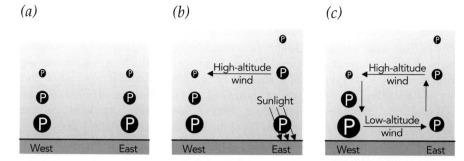

Fig. 8.3.2 (*a*) When surface temperatures are equal, there are no winds. (*b*) But when the eastern ground becomes hot, the air above it expands upward. A high-altitude wind begins to blow toward the colder western side. (*c*) The additional weight of air on the western side increases the pressure near the ground and a low-altitude wind begins to blow toward the hotter eastern side. A convection cell is established.

ter air expands, just as it does in a hot-air balloon, and the column of air over the east side becomes taller. Because the weight of the air column doesn't change, the air pressure at the ground remains the same. But the upward expansion of the air column slows the rate at which the air pressure decreases as the altitude increases. Since the air and its weight move upward, the pressure gradient associated with increasing altitude also spreads upward.

This upward movement of air creates a horizontal pressure imbalance (Fig. 8.3.2*b*). Air far above the hotter eastern side has a higher pressure than corresponding air above the colder western side. High altitude air begins to accelerate toward the colder western side. In effect, the taller air column over the hotter ground acts to reduce its height by flowing toward the shorter air column over the colder ground. A high altitude wind begins to blow from the hotter eastern side to the colder western side.

But this high altitude wind reduces the weight of air above the hotter eastern side and increases the weight of air above the colder western side. As a result, air pressures near the ground begin to change. As the weight of air over the colder western side increases, so does its ground-level air pressure. A higher air pressure is needed to support the increased weight of the air overhead. Similarly, the air pressure at ground level in the hotter eastern side decreases because there is less weight to support.

Now a second wind begins to flow near the earth's surface. This surface wind flows from the colder western side to the hotter eastern side, accelerated in that direction by the ground-level pressure imbalance. Overall, there are two winds: one at high altitude that blows from the hotter side to the colder side and one at low altitude that blows from the colder side to the hotter side. At some intermediate altitude, there is no wind at all.

These two winds create an upward flow of air above the hotter eastern side and a downward flow of air above the colder western side, so the air circulates continuously. An air molecule starting near the ground on the eastern side travels upward, westward, downward, and eastward in an endless cycle. The winds are just the horizontal motions of a huge convection cell.

We have rediscovered convection, as discussed in Section 7.1, but on an atmospheric scale. Air rising above hot ground flows upward and then outward, carrying heat away from the hot ground and delivering it to the colder ground somewhere else. The air that rises upward above the hot ground is replaced by a steady inward flow of colder air from surrounding areas. As usual, heat flows from the hotter region of earth's surface to the colder region and the mechanism of heat transfer is convection.

The winds themselves often contain large amounts of kinetic energy and this energy can be used to do useful tasks, including grinding grain and generating electricity. This wind energy originates as heat in the hotter region and is converted to work by the convection process. The atmosphere acts as a heat engine, moving heat from the hotter region of ground to the colder region of ground while converting a small portion of that heat into work. As usual, the second law of thermodynamics limits the amount of heat that can be converted into work. At best, the earth's winds convert only about 3% of the heat they carry into ordered energy. Nonetheless, there is a huge amount of energy in wind and it has only begun to be exploited for useful purposes.

The pressure differences that cause surface winds can be measured with barometers. By determining the air pressure near the ground, a barometer helps predict which way the wind will blow and even what weather it will bring. As we've seen, the expanded column of air above a hot portion of the earth's surface generally has a lower-than-normal air pressure at ground level and is traditionally called a "low." In contrast, the compressed air above a cold portion of the earth's surface usually has a higher-than-normal pressure at ground level and is called a "high." Knowing where highs and lows are located helps to predict wind direction because surface winds accelerate away from highs and toward lows.

As I noted while discussing the greenhouse effect, air's temperature changes as it rises or falls through the atmosphere. A portion of air descending to the ground is compressed by the increasing weight of the air above it. Just as in the automobile engine (Section 8.2), it takes work to compress air and the result is hotter air. As work is done on the descending portion of air, its temperature increases. In our atmosphere, the temperature of dry air increases by about 10 °C for every kilometer of its descent. Similarly, a portion of dry air that rises away from the ground expands and does work on the air above it. Its temperature decreases by about 10 °C for every kilometer of its ascent. Moist air experiences smaller temperature changes of about 6 or 7 °C per kilometer. These changes in temperature are particularly important for winds that blow down out of mountains and thus experience rises in temperature because of their descent.

Daily or diurnal temperature changes give rise to repetitive winds. For example, daily heating and cooling cycles cause the land and sea breezes that occur near the seashore. Land heats up more rapidly than water so that in the morning, a surface sea breeze blows from the cooler water toward the warmer land. At night, the land cools off first and a surface land breeze blows from the cooler land toward the warmer water. Valley and mountain breezes are caused by a similar effect. As the sun rises, the mountainside warms first and a valley wind blows up the side of the mountain. At sunset, the mountainside cools first and a mountain breezes blows down the side of the mountain toward the warmer valley below. A similar seasonal effect produces monsoons (see ❷).

Relative Humidity and Weather

Relative humidity is a crucial factor in the weather. Rain occurs when warm, moist air cools to the point that its relative humidity exceeds 100%. Air can cool by contact with colder air or by moving upward and expanding. Thus it often rains when hot and cold air masses collide or when warm, moist air blows upward into the mountains and cools. Rain is often found at the edges of hot and cold weather fronts and on the windward slopes of mountains.

In contrast, air that blows downward out of the mountains becomes warmer and its relative humidity drops. Downslope winds, called katabatic winds, are very dry and warm. Examples of these katabatic winds include the chinook winds of the eastern Rocky Mountains and the Santa Ana winds of southern California.

❷ Monsoons are caused by a seasonal variation of the sea breeze. Summer heating of the land in Eastern Asia gives rise to giant sea breezes that bring water-laden air far into land from the surrounding oceans. The resulting rains occupy much of the summer months.

The chinook winds are famous for melting and evaporating snow at a furious pace, while the Santa Ana winds have fanned many disastrous fires in the chaparral between Los Angeles and San Diego.

A final way in which water vapor and humidity influence the weather is by storing energy. Water vapor contains an enormous amount of chemical potential energy, energy that is released when the gaseous water molecules bind together to form liquid water. This latent heat of vaporization is an essential source of energy that powers thunderstorms and hurricanes. When water vapor condenses inside one of these storms, it releases its latent heat of vaporization and warms the surrounding air. The air expands and floats upward. This heating due to water's condensation produces the ferocious updrafts that are present inside the storm.

Global Wind Patterns and the Coriolis Effect

The global wind patterns would be simple if the earth weren't rotating. Because the sun warms land and water near the earth's equators more than that near the earth's poles, there would be steady thermal winds between the equator and the two poles (Fig. 8.3.3). High altitude winds would blow from the hotter equator toward the colder poles while low altitude winds would return from the colder poles toward the hotter equator. There would always be a low over the equator and highs over both poles.

But the earth does rotate and this rotation complicates the global wind patterns. The problem is caused by the Coriolis effect—because the earth rotates, an object moving freely across its surface appears to curve. Nothing is actually exerting a horizontal force on the object and it's actually traveling in a straight line. However, the turning earth is an accelerating frame of reference and straight paths may not appear straight from an accelerating reference frame.

To illustrate the Coriolis effect, imagine playing catch with your friend on a carousel (Fig. 8.3.4). Your friend is closer than you to the center of the carousel. If the carousel is stopped, you can easily toss the ball back and forth (Fig. 8.3.4a). You simply aim at your friend and throw. However, if the carousel is turning counterclockwise, you will have a much harder time throwing the ball so your friend can catch it. If you aim directly at your friend, the ball will miss (Fig. 8.3.4b). Because you are moving sideways at the moment you let go of the ball, you inadvertently give the ball a sideways component of velocity. Your friend is also moving sideways, but more slowly than you are because you are farther from the center of the carousel. As a result, the ball heads to the right of your friend. From your perspective on the turning carousel, you see the ball curve toward the right (Fig. 8.3.4c).

The ball's curved path is the result of your own accelerating frame of reference. The ball really travels in a straight line and you are the one who is actually curving. Still, it really looks like the ball is curving. You are observing the Coriolis effect.

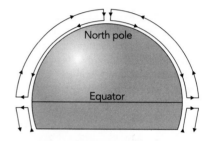

Fig. 8.3.3 If the earth weren't rotating, thermal winds would extend from the hot equator to each of the colder poles. High altitude winds would blow toward the poles while surface winds would blow toward the equator.

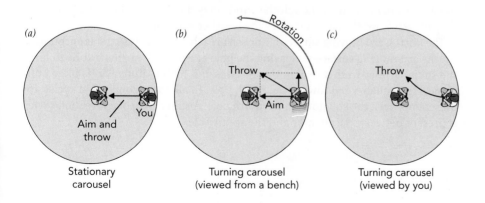

Fig. 8.3.4 (*a*) If you throw a ball to a friend near the center of a stopped carousel, the ball travels in a straight line and your friend can catch it. (*b*) But if the carousel is turning counterclockwise, the ball leaves your hand with a sideways component to its velocity and travels to the right of your friend. (*c*) Viewed from your accelerating reference frame on the carousel, the ball appears to curve toward your right and miss your friend.

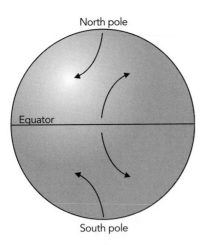

Fig. 8.3.5 The earth rotates toward the east. As a result, objects in the northern hemisphere appear to curve toward the right and objects in the southern hemisphere appear to curve toward the left.

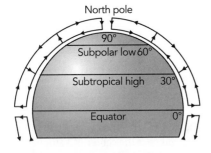

Fig. 8.3.6 Because of the earth's rotation, the winds in the northern (and southern) hemisphere break up into three sections. Convection currents occur between the equator and the subtropical high at about 30° north latitude, between the subtropical high and the subpolar low at about 60° north latitude, and between the subpolar low and the north pole.

The Coriolis effect is also present for objects moving about the surface of the earth. Because the earth is rotating toward the east at about 15°/hour, the ground near the equator has an enormous eastward velocity of about 1670 km/hour. As you head north or south from the equator, you move closer to the earth's rotational axis and this eastward velocity decreases. Near the poles, very close to the earth's rotational axis, this eastward velocity is almost zero.

Just as in the carousel example, an object thrown from the equator directly toward the north pole will miss it. The object starts off with a 1670 km/hour eastward component to its velocity and this eastward motion causes it to drift eastward relative to the more slowly moving ground as it heads north toward the pole. According to someone standing on the ground, the object curves toward the right (Fig. 8.3.5).

Similarly, an object thrown from the north pole toward a point on the equator will also miss it. The object starts off without any sideways velocity, so the moving earth rotates out from under it. The object drifts westward relative to the more rapidly moving ground as it heads south toward the equator. According to someone standing on the ground, the object curves toward the right.

In fact, north of the equator, freely moving objects traveling horizontally always appear to curve to the right. South of the equator, they always appear to curve to the left. Exactly on the equator, there is no Coriolis effect.

We can now look at how the Coriolis effect contributes to the global wind patterns. Over the equator, the air warms and rises upward, causing high-altitude winds to begin blowing toward the poles (Fig. 8.3.6). As the high altitude wind flows north from the equator, it curves toward the right. Instead of traveling directly toward the north pole, this air current curves toward the east. By the time it reaches about 30° north latitude (a third of the way to the north pole), the air is traveling almost due east. This eastward high-altitude flow is called the northern subtropical jet stream.

Unable to make further progress toward the pole, the air piles up over this region and creates the northern subtropical high. The air in this high-pressure region cools by radiating its heat into space and descends toward the ground. From there, it spreads south toward the equator and north toward the pole. The air returning toward the equator again curves toward the right and begins to head westward. These westward surface winds between 30° north latitude and the equator are called the trade winds (Fig. 8.3.7). These winds were important during the era of sailing ships because they helped to carry ships from Europe to the Americas. At the equator itself, the air is rising and there is little surface wind. This region is called the doldrums.

There is also little surface wind at the subtropical high, a calm region around 30° north latitude that is sometimes called the horse latitudes. But the surface winds again pick up as low-altitude air flows northward toward the pole. It, too, curves to the right and creates west-to-east surface winds called the westerlies. These winds, which in North America extend from Texas up into Canada, cause an overall eastward movement of air. Although the winds may fluctuate locally, on the average they blow toward the east and carry the weather with them.

The northward moving surface air doesn't reach the pole. Instead, it encounters southward moving surface air leaving the pole at about 60° north latitude to form the subpolar low. In this region, the air rises and high altitude winds blow both toward the equator and toward the north pole. In all, three convection cycles are present between the equator and the north pole. A similar trio of convection cycles is present between the equator and the south pole.

Hurricanes and Cyclones

On the surface of our rotating earth, the Coriolis effect prevents a surface wind from flowing directly into a region of low pressure. The wind curves relative to the ground and misses the low. In the northern hemisphere, the wind curves toward the right and ends up circling around the low in a counter-clockwise direction, as viewed from above (Fig. 8.3.8). The pressure imbalance causes the wind to accelerate leftward, toward the low, but the Coriolis effect acts to make the wind curve rightward, away from the low. These two competing effects balance one another and the wind circles around and around the low. This circulation of air around a region of low surface pressure is called a cyclone.

When cold air passes over a warm ocean, the air is heated and expands. High altitude winds begin to blow outward and a region of low surface pressure forms. As more cold air rushes in across the water's surface to replace the rising air, the Coriolis effect deflects this inbound air and causes it to circulate around the low pressure region as a cyclone.

In certain special cases the circulating winds can reach very high speeds, creating a powerful storm we call a hurricane. Hurricanes only form when the conditions are exactly right, requiring several unusual circumstances to occur simultaneously. First, a hurricane will only form above a very warm ocean, one that is warmer than about 26 or 27 °C (80 °F) to a depth of at least 50 m. The hurricane is a giant heat engine and its source of heat is the hot water. For the storm to have a great deal of energy, the water must be prepared to deliver a very large amount of heat. That's why hurricanes only occur in late summer.

Second, the air above the warm ocean must be relatively cool so air that is heated by the water's surface can rise upward to very great heights. Air that is warm enough to rise upward through colder air above it is called unstable air. As it rises, the air expands and cools, so it must be quite warm in order to continue its ascent.

Third, the air rising over the warm ocean must carry a great deal of moisture with it. As is discussed in Section 7.2, moist air releases heat when its moisture condenses to form raindrops, and this extra heat helps to propel the air still further upward. Much of the ferocity of hurricanes and thunderstorms comes from energy released by water's changes in phase, from gaseous water vapor to liquid water or solid ice.

Finally, the low and high altitude winds must be favorable or the hurricane will rip itself apart before it starts. At low altitude, the cool winds must converge on the warm ocean water. At high altitude, the winds must help to carry away the air that rises through the center of the storm.

When all of these requirements are met, a region of unusually low surface pressure forms, producing a hurricane. Air accelerates to enormous velocities as it approaches the low, but the Coriolis effect causes it to circulate endlessly around the region of lowest pressure. A calm "eye" forms at the center of the hurricane, protected from the circulating winds by the Coriolis effect. The eye is typically about 60 km in diameter. Just outside the eye, the most violent winds in the storm form a narrow ring only a few kilometers thick. The eye itself has little wind, very low air pressure, and may even be dry and only lightly overcast.

Near the center of the hurricane but outside the eye, air rises upward and carries heat away from the water. This rising air eventually travels outward as high altitude winds. These outward moving high altitude winds also experience the Coriolis effect and they curve weakly in the opposite direction from the surface winds.

In the northern hemisphere, surface winds always spiral counter-clockwise in a hurricane. In the southern hemisphere, the spiral is clockwise because the Coriolis effect is reversed. Because of these opposite spirals, a hurricane spawned in one hemisphere can never cross the equator into the other hemisphere.

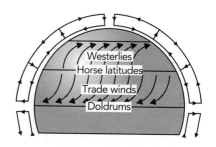

Fig. 8.3.7 The surface winds over the earth generally flow toward the west between the subtropical high and the equator (the trade winds) and toward the east between the subtropical high and the subpolar low (the westerlies). Regions of relative calm are found at the equator and the subtropical high (the doldrums and horse latitudes).

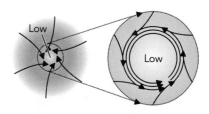

Fig. 8.3.8 Wind accelerated toward a region of low pressure in the northern hemisphere is deflected toward the right by the Coriolis effect. The wind ends up circling around the low.

The Atmosphere's Oxygen Content

The earth's oxygen content is maintained by plants. Plants use a process called photosynthesis to convert carbon dioxide and water molecules into carbohydrate and oxygen molecules. Carbohydrates are a broad class of molecules ranging from the cellulose that gives the plant its rigidity to the sugars that provide energy for the plant and for the animals that eat it.

Since the carbohydrate and oxygen molecules contain more chemical potential energy than the carbon dioxide and water molecules from which they're produced, the plant needs an input of energy to carry out the conversion. In the process of photosynthesis, this energy is provided by sunlight.

Plants absorb light in brightly colored photosynthetic pigments, such as the green chemical chlorophyll, and these chemicals use the light's energy to perform chemical reactions. In effect, photosynthesis is the opposite of combustion. While it's easy to burn wood in oxygen to form water and carbon dioxide, it's much harder to use light energy to turn carbon dioxide and water into wood and oxygen. Nonetheless, plants have developed very capable mechanisms for performing this reverse-combustion process. Without plants, the earth's atmosphere would quickly become depleted of oxygen. Much of the effort to reduce deforestation throughout the world is motivated by a desire to keep the atmosphere's oxygen level high and its carbon dioxide level low.

Plants are able to use light energy to induce chemical reactions in part because light is emitted and absorbed in discrete packets of energy. Each time it absorbs a packet of light energy, a photon, the photosynthetic chemical takes a step toward converting carbon dioxide and water into carbohydrate and oxygen. The energy in a photon is related to the wavelength of the light—long wavelength infrared light is absorbed or emitted as small packets of energy while short wavelength ultraviolet light is absorbed or emitted as large packets of energy. Visible light falls somewhere in between and has the right range of photon energies to support photosynthesis.

Atmospheric oxygen goes on to support life and many other important processes in our world. But atmospheric oxygen also has an important light-filtering role. Some ultraviolet photons in sunlight contain so much energy that they can cause chemical damage to plants and animals. These photons cause permanent photochemical changes to the molecules that absorb them, denaturing proteins—making them non-functional—and causing such biological injuries as sunburns and cataracts.

Ultraviolet light can even damage DNA and RNA, the molecules of genetic information that cells use to synthesize proteins and to reproduce. When ultraviolet light causes sufficient damage to the genetic information in a cell, it may kill that cell. It can also cause changes to a cell that don't kill the cell, but instead render it defective. One possible defect is cancer.

Tanning is your skin's response to this photochemical threat and it provides a modest amount of protection. Nonetheless, ultraviolet light continues to damage the skin molecules that absorb it. To truly protect your skin from ultraviolet light, you must wear sunscreen. Sunscreen molecules actually absorb ultraviolet light and convert its energy into thermal energy. Very little of the ultraviolet light penetrates the chemical barrier and reaches your skin.

Fortunately, the atmosphere also acts as a sunscreen. Atmospheric oxygen molecules absorb most of the sun's highest-energy ultraviolet photons so that very little of this extreme ultraviolet light reaches the earth's surface. Once an oxygen molecule absorbs an extreme ultraviolet photon, it can fall apart into two oxygen atoms. One of these oxygen atoms can then combine with another oxygen molecule to form an ozone molecule. Ozone is a moderately toxic gas that's highly reactive chemically, smelling and behaving much like chlorine. Near the earth's surface, ozone is a pollutant and part of photochemical smog. Fortunately, it breaks down quickly into normal oxygen and causes no long-term contamination. Because of its similarity to chlorine and its environmental advantages, ozone is often used instead of chlorine as a disinfectant for swimming pools and a bleaching agent for paper mills.

But in the upper atmosphere, ozone is very important. Oxygen molecules only absorb extreme ultraviolet photons, leaving the earth unprotected from less extreme ultraviolet light. Ozone molecules, however, absorb this less energetic ultraviolet light. Although the ultraviolet photon may break up the ozone molecule that absorbs it, at least that photon doesn't reach the earth's surface. Without ozone in the upper atmosphere, we would be exposed to far more ultraviolet light and would suffer far more photochemical damage from sunlight.

Ozone molecules are continuously being created and destroyed by ultraviolet light. There are just about the right number of them in the upper atmosphere at any one time to ensure that we don't get too much ultraviolet exposure. However, chlorine-containing molecules released into the atmosphere since the industrial revolution have upset that delicate balance. Chlorine atoms act as catalysts, facilitating the conversion of ozone back into normal oxygen. Ozone molecules have extra chemical potential energy, and two ozone molecules tend to become three normal oxygen molecules. The only thing preventing this conversion is the activation energy needed to break apart the two ozone molecules so that they can begin to form the three normal oxygen molecules. By reducing the activation energy required, chlorine atoms ease the conversion. The more chlorine atoms we put into the upper atmosphere, the faster ozone reverts to normal oxygen and the less ozone the atmosphere contains.

Recognizing the threat to life posed by a decrease in the atmospheric ozone content, countries have curtailed or banned the production of chlorine-containing gases such as chlorofluorocarbons. Still, chlorine atoms have a long lifetime in the upper atmosphere and a single chlorine atom can aid in the destruction of countless ozone molecules. It will take a long time for the ozone balance to be fully reestablished.

SECTION 8.4 **Water Purification**

Fresh water is an essential ingredient of modern life. Though it's often available as the result of natural processes, there are times when it must be extracted from impure water, typically salt water. In some countries where rain water is scarce, desalinated sea water is the main source of drinking water. Any extraction process that purifies water must separate water molecules from contaminating liquids, solids, or gases. This section examines some of the techniques that make this molecular separation possible.

Distilling Water from Salt Water

One way to purify water is by distillation. Distillation is a general technique for separating various chemicals from one another. The chemicals are heated to form a vapor, and that vapor is condensed to form a new mixture of chemicals. Because the various chemicals have different tendencies to form vapors at a particular temperature, the newly formed mixture has a different balance of the chemicals from the original mixture. In some cases, the condensed liquid contains primarily a single chemical—all of the other chemicals are left behind in the original liquid.

To understand how distillation can purify water, let's look at the phases of water. At any temperature above absolute zero, there's a possibility of finding gaseous water molecules above the surface of ice or water. These water molecules have acquired enough thermal energy to break free of the solid or liquid and become a gas.

If you place some water in an enclosed container, water molecules will evaporate until there are enough of them in the gas phase that they return to the liquid's surface as often as they leave it. At that point the two phases, liquid and gas, are in phase equilibrium. Although molecules constantly shift back and forth between the two phases, neither phase grows at the expense of the other. Overall, there's no net movement of molecules from one phase to the other.

At this phase equilibrium, the relative humidity is 100%—the water vapor has reached its saturated density. Associated with that density is a pressure known as the water's saturated vapor pressure. Saturated or not, water's vapor pressure is the effective pressure produced by the water molecules alone. For example, if air at atmospheric pressure (101,300 Pa) contains 1% water molecules, then water's vapor pressure in that air is 1% of the total pressure or 1013 Pa. And if the relative humidity is 100%, then the water's saturated vapor pressure is 1013 Pa.

But we have forgotten to pay attention to temperature. Since water's saturated density and saturated vapor pressure depend on temperature, the present balance of gas and liquid in the container is ideal only at its current temperature. If you warm up the container, more water molecules will enter the gas phase, and the amount of liquid water will decrease. If you cool down the container, more water molecules will enter the liquid phase, and the amount of gaseous water will decrease. This connection between temperature, saturated density, and saturated vapor pressure is the central principle behind distillation.

What happens if there is air inside the container, along with the water and water vapor? Surprisingly, the air doesn't matter. The density of water molecules in the gas phase and the water's vapor pressure are the same, whether the air is there or not. This interesting observation makes it possible to perform distillation with or without air around, although air's presence affects the total pressure on the water and thus its boiling temperature. The desalination schemes that we'll examine in the next few pages operate most effectively at less than atmospheric pressure, so many distillation plants remove air from their equipment in order to reduce the gas pressure.

How much water vapor is there in the container? Water's saturated vapor pressure at 20 °C, typical room temperature, is about 2,300 Pa or about 2% of atmospheric pressure. That means that when the humidity is 100% at room temperature, about 2% of the molecules in the air are water molecules. At 0 °C, the melting temperature of ice, water's saturated vapor pressure is only about 600 Pa or about 0.6% of atmospheric pressure. That means that there is still quite a bit of moisture in the air even at freezing. And at 100 °C, water's vapor pressure is about 101,300 Pa or 100% of atmospheric pressure at sea level. That's why water boils at 100 °C at sea level.

The simultaneous presence of both water and water vapor in the container means that there is a phase separation. The two phases, liquid and gas, appear in separate regions of the container. The denser water phase sinks to the bottom of the container while the less dense water vapor phase floats to the top. These two phases can exist in phase equilibrium over a wide range of temperatures. There is actually one special temperature and pressure, water's triple point, at which water, water vapor, and ice can all exist together in a single container and in phase equilibrium with one another. This triple point occurs at about 0.01 °C.

Before going further into water distillation, let's consider the behaviors of a few other chemicals to see how they differ from water. After all, if they behaved exactly as water does, distillation wouldn't separate them from water.

Table salt, sodium chloride, is a solid at room temperature. It doesn't even melt until 801 °C and its boiling temperature is about 1450 °C at atmospheric pressure. Salt's saturated vapor pressure is almost negligible at any temperature below about 500 °C. So if you put a block of salt in a container at less than 500 °C, it will reach a phase equilibrium with almost all the salt molecules in the solid phase and only a tiny number in the gas phase. (For some interesting cases of materials in different phases, see ❶❷.)

The situation is quite different for ethyl alcohol (grain alcohol). Ethyl alcohol melts at −112 °C and boils at only 78 °C near sea level. Although ethyl alcohol molecules are larger than water molecules, they don't form many hydrogen bonds and are relatively easy to separate from one another. As a result, ethyl alcohol is more likely than water to form a gas.

Distillation uses these differences in vapor pressures to separate chemicals. When you heat a mixture of chemicals to a particular temperature, the whole mixture tries to establish phase equilibrium. The chemicals that tend to be gaseous at that temperature accumulate in the system's gas phase. The chemicals that tend to be a liquid at that temperature accumulate in its liquid phase. When the phase equilibrium is finally reached after a minute or two, the balance of chemicals in the gas phase may be very different from that in the liquid phase.

Distillation starts when you transport the vapor from this hot region to a second region with a colder temperature. There are now two different regions sharing the same vapor and each of these regions tries to maintain its own phase equilibrium. Some vapor molecules condense to form a liquid in the cold region and are replaced by evaporation from the liquid in the hot region.

The condensed liquid contains primarily those molecules that evaporated readily in the hot region and condensed readily in the cold region. Those molecules are the ones that experienced the largest changes in saturated vapor pressures between the two temperatures. They tended to become gaseous at the temperature of the hot region and liquid at the temperature of the cold region.

We can now look at how to desalinate salt water by distillation. A simple distillation system appears in Fig. 8.4.1. Two separate liquid containers, an evaporator containing the original salt water and a condenser containing fresh water, share the same gas, and that gas moves freely between them. Since salt molecules rarely enter the gas phase near room temperature, the gas is virtually pure steam (air is typically removed at the start). The salt water evaporator is kept hot so that its

❶ Although gaseous salt molecules are rarely found in air near room temperature, breaking waves fill the air with tiny salt water droplets. These droplets evaporate to leave minute salt grains, which are carried aloft by drag forces and account for the salty air near the ocean.

❷ Nitrogen, oxygen, and argon are all gases at room temperature, regardless of pressure. Compressing them simply pushes their molecules closer together, without creating a separate liquid phase. They must be cooled to very low temperatures before they will liquefy.

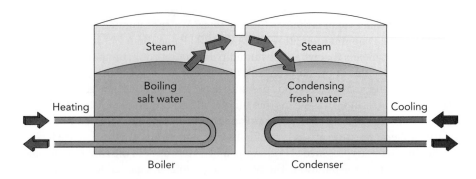

Fig. 8.4.1 Fresh water can be obtained by distilling salt water. In this process, salt water in boiled in one region to form a vapor that is mostly pure steam. This steam is then condensed in a second, cooler region and becomes fresh water.

water molecules tend to evaporate into steam. The fresh water condenser is kept cold so that its water molecules tend to condense into liquid water. Overall, water molecules tend to move from the salt water to the fresh water.

With enough patience, only a small temperature difference is needed to separate fresh water from salt water. This temperature difference just has to be large enough to make sure that, on the average, water molecules leave the salt water as gas and arrive at the fresh water as liquid. Nature is very patient and it uses small temperature differences to produce fresh water. Rain, dew, and frost are all created by natural distillation. Salt water evaporates in warmer weather and the resulting water vapor condenses in colder weather.

It might seem as if even the tiniest temperature difference can perform distillation, but that's not true. When you mix salt with water, you create disorder and entropy. Separating salt from water decreases entropy and the second law of thermodynamics comes into play. If you're going to separate the two chemicals and reduce their entropy, you must create extra entropy elsewhere. In distillation, that extra entropy comes from letting heat flow from the hotter region to the colder region. The temperature difference between the two regions must be large enough so that the total entropy doesn't decrease.

In real water distillation systems, the temperature difference is usually quite large. A large temperature difference doesn't make purer water but it does speed up the process. Most systems heat the salt water to its boiling temperature, a step which dramatically increases the evaporation rate. They also condense the water vapor rapidly by bringing it into contact with a very cold surface. The water molecules move swiftly from the salt water side to the fresh water side so the facility produces fresh water much more rapidly than nature itself. In exchange, the facility consumes far more ordered energy per liter of fresh water than nature does.

To avoid wasting too much energy, large distillation-based desalination plants are rather sophisticated. For the distillation to proceed quickly, they do have to boil the water, but they don't boil it at atmospheric pressure. Instead, they pump the air out of the distilling chambers so that water boils at a much lower temperature. Everything proceeds as above, except they don't have to heat the salt water as much. While the facility can't produce quite as much fresh water per hour because the density of water molecules in the vapor is lower, a little patience saves a lot of energy.

However, distillation still uses more energy than anyone would like. The problem lies in water's huge latent heat of vaporization. Water vapor carries away lots of heat from the salt water as it evaporates, and it gives that heat to the fresh water when it condenses. The salt water gets colder, and the fresh water gets hotter, reducing the temperature difference and slowing the distillation process. To keep everything working quickly, the distillation plant must continuously add heat to the salt water and remove it from the fresh water. It's this heating and cooling that makes distillation so expensive.

The transfer of heat from the boiler to the condenser, due only to water's latent heat of vaporization, is unfortunate. The second law of thermodynamics doesn't require such a transfer in order to separate water from salt; it's just a side-effect of using the liquid/gas phase transition to separate two chemicals. Fortunately there are tricks that a desalination plant can use to reduce this heat transfer.

The best trick is to reuse the heat. Most distillation plants use the same heat over and over again, operating several separate distillation systems with it. Because distillation units that operate at different pressures also operate at different temperatures, the waste heat leaving the condenser of a higher pressure distillation unit can be used to heat the boiler of a lower pressure distillation unit. Since water's boiling temperature depends on pressure, the two distillation units function properly at very different temperatures.

An example of such heat reuse appears in Fig. 8.4.2. The heat first distills water in a high-pressure distillation unit and then in a low-pressure distillation unit. In the high-pressure unit, the heat travels from the boiler to the condenser in the steam and then leaves the condenser in its cooling water. This cooling water leaving the high-pressure distillation unit is actually hot enough to be the heating water for the low-pressure distillation unit. There, the heat travels from the boiler to the condenser and finally leaves the plant in a second cooling water system. This waste heat is dispersed into the great outdoors.

A desalination plant may reuse the same heat five or more times before sending it out into the ocean or the atmosphere. The heat may originate as solar energy or it may come from burning fuel or from a nuclear reactor. In some cases, it comes from waste heat released by an electric power plant.

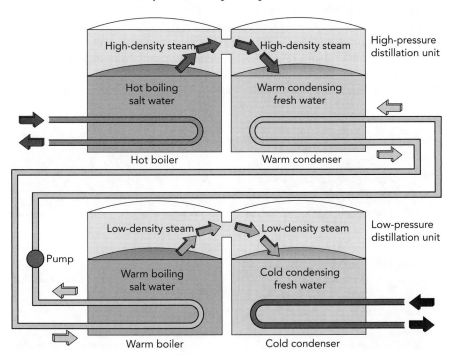

Fig. 8.4.2 Distilling water at atmospheric pressure requires a large amount of heat. This heat is used to raise the temperature of the water to 100° C and then to separate the molecules into a vapor. When the steam condenses in the condenser, this heat is released and becomes waste heat in the cooling water. However, a more sophisticated distillation plant lowers the pressure in a second distillation unit and reuses heat from the first unit to operate the second unit.

This same process of distillation is used to create liquor. Natural fermentation can't produce liquids that are more than about 20% ethyl or grain alcohol because too much alcohol kills the yeast that causes the fermentation. However, the alcohol and water mixture can be distilled to create much more concentrated alcohol and water mixtures. Near room temperature, alcohol has a much higher vapor pressure than water, and it boils at a lower temperature. When alcohol and water are heated together at atmospheric pressure until the mixture boils, the vapor above the mixture will be mostly alcohol. If this vapor is condensed, the new liquid is as much as 90% alcohol.

Freezing Salt Water to Produce Fresh Water

Boiling isn't the only change of phase that's used to purify water. The ice that forms when sea water freezes is essentially pure fresh water, a phenomenon that Eskimos have used to obtain potable water for thousands of years. This purification effect results from a balance between energy and disorder.

Physical systems tend to minimize their potential energies. For example, a ball tends to roll down a hill to minimize its gravitational potential energy. Similarly, a mixture of ice and water tends to minimize its potential energy by excluding contaminants from the ice and leaving them in the water.

Water is disordered already and having solids, liquids, or gases dissolved in it isn't a problem. Ice, on the other hand, is a highly ordered crystal that is seriously disrupted by the presence of dissolved chemicals. The water molecules in an ice crystal pack more neatly and obtain a lower overall potential energy when the ice contains no contaminants. So a contaminated mixture of water and ice reaches its lowest overall potential energy when all of the contaminants stay in the water and the ice contains only water molecules.

As long as the water freezes slowly, the ice crystal it forms will have very little contamination in it, and the unfrozen water will end up with a relatively high concentration of contaminants (Fig. 8.4.3). By separating pure from impure water, this process reduces the water's disorder so that, to avoid violating the second law of thermodynamics, additional disorder must be created somewhere else. This addi-

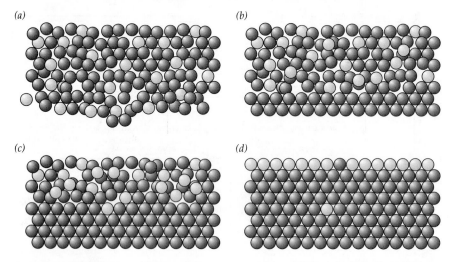

Fig. 8.4.3 While water can incorporate many dissolved ions and molecules into its disordered structure (*a*), ice is too ordered to include such contaminants. As water freezes from the bottom up (*b–d*), the interface between solid ice and liquid water moves slowly upward through the container. The ice incorporates more and more of the water molecules (dark) and rejects almost all of the contaminant molecules (light). Eventually the contaminant molecules form solid, liquid, or gaseous regions of their own (*d*).

tional disorder appears in the cold region that freezes the salt water. The salt water doesn't freeze until the temperature in that cold region drops well below water's ordinary freezing temperature. As it freezes, the salt water releases heat into this very cold region and thereby introduces considerable disorder. Adding heat to a something that's very cold creates more disorder than adding the same heat to something less cold. The increased entropy in the cold region more than makes up for any decrease in entropy in the water as it freezes.

However, disorder appears whenever it can, and so the ice crystals that form always contain imperfections. Even if there are no impurities around, the crystals will probably have minor defects. These defects include flaws in the stacking of molecules or empty spots in the otherwise orderly arrays of molecules. Truly perfect crystals are extremely hard or even impossible to grow.

When you freeze a bucket of salt water, the ice that forms first contains very little salt. The salt is in the remaining salt water, which becomes more and more concentrated as additional water molecules are bound up in the ice. By the time there is only a small amount of water remaining, that salt water has a very high concentration of salt and salt crystals begin to form. These salt crystals can easily become trapped in the ice, so care must be taken to remove the ice from the concentrated salt solution before salt crystals begin to form. In sea ice, the salt is carried away by the sea water, so only pure ice is formed.

When you freeze ice cubes, the outer surface freezes first and the impurities become concentrated near the middle of the cube. One of the main impurities is dissolved air, which eventually comes out of solution and forms tiny air bubbles in the ice. These air bubbles appear as a white cloudy region inside the ice cube. One way to reduce this clouding is to boil the water before you freeze it. Boiling the water drives most of the air out of solution so that few air bubbles form in the resulting ice.

Freezing salt water to form pure ice works best in cold climates where low temperatures are available directly. Active refrigeration can also freeze salt water to obtain fresh water, but it's expensive. Because of water's latent heat of melting, you must remove a large amount of heat from salt water to freeze it. Although refrigerated water desalination plants have been built, they have proven to be less economical than distillation plants. (For another case of purification through freezing and melting, see ❸.)

❸ Some of the purest materials on earth are made by zone refining. In this technique, a rod of semi-pure material passes slowly through a heater that creates a moving zone of molten material. As each portion of the rod melts and then solidifies, its impurities are carried away in the molten zone. The final rod is remarkably pure, with as few as one atom in a billion being an impurity.

Osmosis

Another way to desalinate water is by reverse osmosis, a process resembling filtration except that it takes place at the molecular scale. In effect, salt water is converted to fresh water by filtering the impurities out of it with an incredibly fine filter. Because it operates at such a tiny size scale, however, reverse osmosis encounters some peculiar pressure effects that you don't see with larger filters. In order to describe these effects in reverse osmosis, let's first examine osmosis itself.

Osmosis occurs whenever two different fluids are placed on opposite sides of a semipermeable membrane. A semipermeable membrane is a surface that only allows certain molecules to pass through it. The molecules in the fluid jiggle about rapidly because they have thermal energy. They bounce and push off one another and off the membrane. Those molecules that can pass through the membrane often do, flowing back and forth until the two fluids reach phase equilibrium. There is then no net flow of the mobile molecules through the membrane and the two fluids stop changing with time.

If the two fluids have the same pressure, phase equilibrium occurs when they have equal concentrations of the immobile molecules—equal numbers of those molecules per volume. At that point, it's just as likely for a mobile molecule to cross

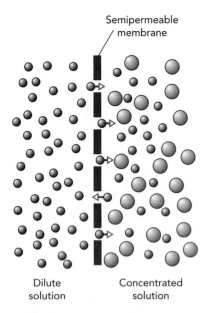

Semipermeable
membrane

Dilute
solution

Concentrated
solution

Fig. 8.4.4 Osmosis occurs when two different fluids sit on opposite sides of a semipermeable membrane. Only the smaller, mobile molecules can flow through the pores in the membrane. On the average, these mobile molecules flow toward the side that has the largest concentration of immobile molecules. Since the most concentrated solution is on the right side here, there is a net flow of mobile molecules toward the right.

the membrane from one side as from the other. If you put a relatively concentrated solution of the immobile molecules on one side of the semipermeable membrane, it will have relatively few mobile molecules in it, so mobile molecules will tend to flow through the membrane and into that concentrated solution (Fig. 8.4.4). This flow will dilute the concentrated solution and make it less concentrated. The fluid that loses its mobile molecules becomes more concentrated.

Some semipermeable membranes allow water to pass through them but don't pass salt. In solution, the salt exists as separated ions, wrapped up in water solvation shells. Although the ions themselves are very small, the solvation shells are huge and can't pass through the membranes. Since the salt ions are immobile, osmosis occurs between salt water and fresh water. If you put salt and fresh water on opposite sides of a semipermeable membrane, water molecules from the fresh water flow through the membrane to dilute the salt water. The total mass of fluid on the fresh water side decreases as water molecules flow over to the salt water side. Water molecules will continue to flow until the concentrations of immobile molecules on each side of the membrane are equal.

The only way to stop osmosis from diluting concentrated salt water is to increase the pressure of the salt water. Increasing that pressure tends to drive water molecules back out of the salt water and through the semipermeable membrane. The two fluids will still reach phase equilibrium, but the salt water will retain a higher concentration of immobile molecules than the fresh water. Maintaining a large difference in concentrations between the two fluids requires a huge pressure difference of many tens or hundreds of atmospheres.

Osmosis shows up in a great many biological systems because most biological membranes are semipermeable. The mobile molecules, usually water, flow through these membranes and try to equalize the concentrations of immobile molecules in the two fluids. For example, your skin tends to lose moisture when it comes into contact with a concentrated solution of salt or other chemicals. Water molecules flow out of your skin as they act to dilute the concentrated solution. At the other extreme, your skin tends to absorb moisture from fresh water. Your skin cells contain immobile molecules and fresh water enters them to dilute their contents.

Water flow of this sort, into or out of cells, is a very important issue for fresh and salt water animals, particularly microorganisms. These microorganisms must control osmosis or else suffer severe consequences. One way in which they handle osmosis is to maintain concentrations of immobile molecules inside their cells that are similar to those in the water around them. To do this, they must adapt specifically for salt or fresh water. Salt water can extract the moisture out of a fresh water microorganism while fresh water can swell and explode a salt water microorganism.

Plants use osmosis to attract water into their root hairs. The root hairs contain a relatively concentrated solution of immobile molecules and ground water flows into the root hairs to dilute the concentrated solution. In fact, water flows into the root hairs so aggressively that it raises the pressure inside the root hairs to many times atmospheric pressure. This high osmotic pressure is partly responsible for pushing water up toward the tops of trees.

Reverse Osmosis

We noted that you can use pressure to stop the flow of water molecules into a concentrated salt solution. Osmosis will only equalize the concentrations of immobile molecules when the pressures are equal on both sides of the membrane. If you exert extra pressure on the salt-water side, it will remain more concentrated than the fresh-water side. If the pressure on the salt-water side becomes high enough, it can actually make osmosis run backwards. Water molecules will flow from the salt-water side to the fresh-water side!

This reverse osmosis can be used to extract fresh water from salt water but it requires a lot of pressure. A typical reverse osmosis cell operates at about 1000 times atmospheric pressure. The semipermeable membrane must be able to withstand this imbalance of pressure between the low-pressure fresh-water side and the high-pressure salt-water side. Osmosis is also very slow, so that a desalination plant must use membranes that have very large surface areas.

To obtain such large surface areas, desalination is done by making a stack of several different materials and then rolling the stack into a long cylinder. The stack consists of a layer that carries in the salt water, a semipermeable membrane, and a layer that carries out the fresh water. The roll is arranged so that the membrane is well supported and doesn't explode when exposed to the high-pressure salt water.

Salt water flows into this rolled sandwich and fresh water flows out. Actually, the salt water also flows out of the sandwich through a separate tube. If the desalination plant tried to squeeze too much fresh water out of the same salt water, that salt water would become very concentrated. This increase in concentration would make further reverse osmosis even more difficult. To ease the extraction, only a small fraction of fresh water is extracted from any volume of salt water. The slightly concentrated salt water is then returned to the ocean and is replaced by new salt water.

Reverse osmosis is fairly energy efficient because it doesn't involve any change of material phase so there is no latent heat to provide or absorb. The only energy needed is that required to pump the salt water through the system at very high pressure. This energy, however, is substantial and it comes in the form of mechanical work. Since heat is much easier to obtain than work, multi-stage distillation plants are still very competitive with reverse osmosis plants when producing large quantities of fresh water from salt water.

Drinking Water Filters

Despite what they're called, most filters for drinking water don't produce chemically pure water. That's actually a good thing because chemically pure water doesn't taste all that pleasant. Instead, the activated carbon and ion exchange resins in these filters selectively remove unwanted ions and molecules from the water passing through them while leaving various other constituents that contribute to the water's refreshing taste.

Being selective in this manner also prolongs the water filter's life. Its capacity for absorbing chemicals from the water is limited by the second law of thermodynamics: as the water becomes more pure and orderly, the filter must become more impure and disorderly. Overall, entropy cannot decrease and will actually increase in this case. The disorder that accumulates in the filter gradually lessens its effectiveness so that it must be replaced eventually. Leaving the innocuous and even desirable chemicals in the water permits the filter to accumulate disorder more slowly than it would otherwise and handle more water before replacement.

One of the filter's key components is activated carbon: a highly porous form of carbon that acts as a sponge for certain types of unwanted molecules. These contaminant molecules become permanently trapped within the carbon's extensive network of microscopic pores. A single gram of activated carbon can have 1,000 m^2 or more of surface area inside it—nearly as much surface area as a football or soccer field has. Because of its broad cavity structure, the filter's activated carbon can bind a phenomenal number of molecules in its pores before running out of space.

Activated carbon is particularly efficient at ridding water of hydrophobic organic molecules—oil-like molecules that bind poorly to water. These hydrophobic molecules cling tightly to the activated carbon whenever they encounter it. The carbon also clears the water of many unpleasant odor and taste molecules.

Perhaps most importantly, activated carbon reacts with the aggressive "free chlorines"—chlorine molecules (Cl_2) and hypochlorous acid (HOCl)—that are put in municipal water to keep it sanitary. It converts those molecules into non-toxic chloride (Cl^-) and hydrogen (H^-) ions.

The ion exchange resins used in drinking water filters are high-tech plastics that replace less-soluble, toxic metal ions such as lead (Pb^{++}), copper (Cu^{++}), mercury (Hg^{++}), and cadmium (Cd^{++}) in the water with more-soluble hydrogen (H^+) ions. These resins also remove enough calcium (Ca^{++}) and magnesium (Mg^{++}) ions from hard water to stop it from forming mineral deposits in kettles and tea cups. Calcium and magnesium ions, however, are important to the tastes of many waters so the best filters don't remove too many of them. They are examples of the desirable constituents that the filters do well to leave in the water.

CHAPTER 9

THINGS WITH RESONANCES AND MECHANICAL WAVES

Many fascinating motions in the world around us are repetitive ones. Our lives are filled with cycles, from the sun's daily passage overhead to a pond's undulating ripples on a rainy day. These cyclic motions are governed by the physical laws and steadily mark our journey through time and space. Some of these cycles structure our lives out of necessity or tradition, while others are simply there to be observed. Still other cycles have become part of our everyday world because they're useful or enjoyable. In this chapter, I'll cover those three possibilities by examining cyclic motions in three contexts: in clocks, in musical instruments, and at the seashore.

Part of developing a good understanding of science is learning to generalize. Just as nature doesn't keep reinventing physical mechanisms for each new situation, a person who is willing to think carefully doesn't have to rediscover everything when they encounter yet another example of a familiar physics concept. The proverbial expression, "if you've seen one, you've seen them all" applies to many scientific issues.

Despite their disparate appearances, the three things I'll discuss in this chapter share many of the same basic physics principles. Those conceptual similarities are so exciting to most physicists that we're tempted to begin leaping back and forth between specific examples in a manner that lay listeners find disorienting and confusing. On the other hand, someone who steadfastly avoids making any connections based on shared physics concepts will never begin to see the forest for the trees.

In writing this chapter on one of the most important of all physics concepts, the harmonic oscillator, I have tried to balance specificity with generality. I hope that I've taken a sufficiently methodical passage through this fabulous forest of resonances and mechanical waves to give you a good sense for how each tree works, but also an appreciation for the forest as a whole.

Chapter Itinerary

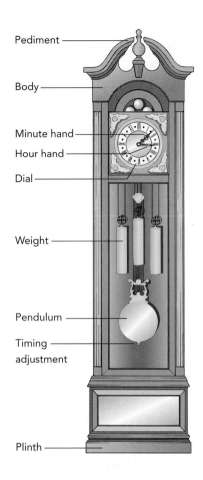

Pediment

Body

Minute hand

Hour hand

Dial

Weight

Pendulum

Timing
adjustment

Plinth

Section 9.1 **Clocks**

People measure their lives according to the sky, dividing existence into days, months, and years according to the celestial motions of the sun, moon, and stars. But on the less romantic scale of daily life, the sky offers little help. Since it provides no easy way to measure short periods of time, people invented clocks.

Early clocks were based on the time it took to complete simple processes—the flow of sand or water, or the burning of candles. However, these clocks had limited accuracy and required constant attention. Better clocks measure time with repetitive motions such as swinging or rocking. In this section, we'll examine the workings of modern clocks based on repetitive motions. As we do, we'll see that repetitive motions are interesting in their own right and appear throughout nature in countless objects besides clocks.

Time

Before examining clocks we should take a brief look at time itself. Scientists treat time as a dimension, similar but not identical to the three spatial dimensions that we perceive in the world around us. In total, our universe has four dimensions: three spatial dimensions and one temporal dimension. Thus it takes four numbers to completely specify when and where an event occurs; three numbers identify the event's location and one identifies its moment in time.

An obvious difference between space and time is that, while we can see space stretched out around us, we can only observe the *passage* of time. Though we occupy only one location in space at a given moment, we are somehow more aware of the expanse of space around us. It's much harder to sense the whole framework of time stretching off into the past and future; you must use your imagination.

Our perception of space is ultimately based on the need for forces, accelerations, and velocities to travel from one place to another. A city seems far away because we know that traveling there with reasonable forces, accelerations, and velocities would take a long time. Our perception of time is based on the same mechanical principles. If two moments are separated by a long time, then reasonable forces, accelerations, and velocities will permit us to travel large distances between the two moments. In short, our perceptions of space and time are interrelated, and measurements of time and space are connected as well.

We measure space with rulers and time with clocks. But how would you make a ruler? You could construct a rather large ruler by driving a car at constant velocity and marking the pavement with paint once each second. Your ruler wouldn't be very practical, but it would fit the definition of a ruler as having spatial markings at uniform distances. You would be using your movement through time to measure space.

How would you make a clock? You could make a rather strange clock by driving a car at constant velocity down your giant ruler and counting each time you saw one of your marks go by. You would then be using your movement through space to measure time. Most clocks really do use motion to measure time. As we are about to see, however, they use motions that are a bit more compact than a car ride.

Natural Resonances

An ideal timekeeping motion should offer both accuracy and convenience. That rules out some of the obvious choices. The sun, moon, and stars keep excellent time but fail the convenience test. Sure, conservation of energy, momentum, and angular momentum so dominate their motions that these celestial bodies move steadily and predictably through the heavens, century after century, but what do you do on a cloudy day? And while simple interval timers like sandglasses and burning candles are easy to make and use, they're not very accurate. Besides, who's going to stay up all night lighting fresh candles just to keep the "clock" running? (For an interesting astronomical clock, see ❶.)

Instead, practical clocks are based on a particular type of repetitive motion called a natural resonance. In a natural resonance, the energy in an isolated object or system of objects causes it to perform a certain motion over and over again. Many objects in our world exhibit natural resonances, from tipping rocking chairs, to sloshing basins of water, to waving flagpoles, and those natural resonances usually involve motion about a stable equilibrium. Like the bouncing spring scale in Section 3.1, an object that has been displaced from its stable equilibrium accelerates toward that equilibrium but then overshoots; it coasts right through equilibrium and must turn around to try again. As long as it has excess energy, this object continues to glide back and forth through its equilibrium and thus exhibits a natural resonance.

Some resonances, such as those of bouncing balls and teetering bottles, don't maintain a steady beat and aren't suitable for clocks. But we are about to encounter a group of resonances that are extremely regular and can be used to measure the passage of time with remarkable accuracy. Those resonances belong to an important class of mechanical systems known as harmonic oscillators.

Pendulums and Harmonic Oscillators

One of the first natural resonances to find its way into clocks is the swing of a pendulum—a weight hanging from a pivot (Fig. 9.1.1). When the pendulum's center of gravity is directly below its pivot, it's in a stable equilibrium. Its center of gravity is then as low as possible, so displacing it raises its gravitational potential energy and a restoring force begins pushing it back toward that equilibrium position (Fig. 9.1.2). For geometrical reasons, this restoring force is almost exactly proportional to how far the pendulum is from equilibrium. As you displace the pendulum steadily from equilibrium, the restoring force on it also increases steadily.

When you release the displaced pendulum, its restoring force accelerates it back toward equilibrium. But instead of stopping, the pendulum swings back and forth about its equilibrium position in a repetitive motion called an oscillation. As it swings, its energy alternates between potential and kinetic forms. When it swings rapidly through its equilibrium position in the middle of a swing, its energy is all kinetic. When it stops momentarily at the end of a swing, its energy is all gravita-

❶ The daughter of a planetarium architect, Jocelyn Bell Burnell (British astronomer, 1943–) acquired an early interest in radio astronomy. Advised to study physics first, she became the only woman in a class of 50 at Glasgow University. While working on her Ph.D. at Cambridge, Bell discovered an extraterrestrial source of radio bursts, occurring precisely 1.33730113 seconds apart. She had discovered the first pulsar, a collapsed star whose angular momentum keeps it turning at an extraordinarily uniform rate. Each burst coincided with one rotation of the star remnant.

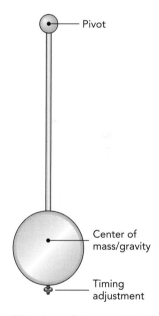

Fig. 9.1.2 If you tilt a pendulum's center of gravity away from its equilibrium position, it experiences a restoring force proportional to its distance from that equilibrium position.

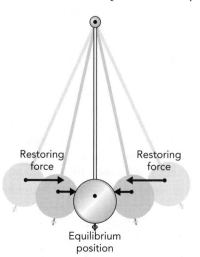

Fig. 9.1.1 A pendulum consists of a weight hanging from a pivot. The pendulum is in a stable equilibrium when its center of gravity is directly below the pivot.

tional potential. This repetitive transformation of excess energy from one form to another is part of any oscillation and keeps the oscillator—the system experiencing the oscillation—moving back and forth until that excess energy is either converted into thermal energy or transferred elsewhere.

But the pendulum isn't just any oscillator. Because its restoring force is proportional to its displacement from equilibrium, the pendulum is a harmonic oscillator—the simplest and best understood mechanical system in nature. As a harmonic oscillator, the pendulum undergoes simple harmonic motion, a regular and predictable oscillation that makes it a superb timekeeper.

The period of any harmonic oscillator—the time it takes to complete one full cycle of its motion—depends only on how stiffly its restoring force pushes it back and forth and on how stubbornly its mass resists that back-and-forth motion. Stiffness is the measure of how sharply the restoring force strengthens as the oscillator is displaced from equilibrium; stiff restoring forces are associated with firm or hard objects, while less stiff restoring forces are associated with soft objects. The stiffer the restoring force, the more forcefully it pushes the oscillator back and forth and the shorter the oscillator's period. On the other hand, the larger the oscillator's mass, the less it accelerates and the longer its period.

However, the most remarkable and important characteristic of a harmonic oscillator is not that its period depends on stiffness and mass, but that its period *doesn't* depend on amplitude—its furthest displacement from equilibrium. Whether that amplitude is large or small, the harmonic oscillator's period remains exactly the same. This insensitivity to amplitude is a consequence of its special restoring force, a restoring that is proportional to displacement from equilibrium. At larger amplitudes, the oscillator travels farther each cycle, but the forces accelerating it through that cycle are stronger as well. Overall, the harmonic oscillator completes a large cycle of motion just as quickly as it completes a small cycle of motion.

Any harmonic oscillator can be thought of as having a restoring force component that drives the motion and an inertial component that resists the motion. Their competition determines the oscillator's period. Harmonic oscillators with stiff restoring forces and little inertia have short periods, while those with soft restoring forces and great inertia have long periods. Their amplitudes of oscillation simply don't affect their periods, which is why harmonic oscillators are so ideal for timekeeping. Because practical clocks can't control the amplitudes of their timekeeping oscillators perfectly, virtually all of them are based on harmonic oscillators.

Harmonic Oscillators

A harmonic oscillator is an oscillator with a restoring force proportional to its displacement from equilibrium. Its period of oscillation depends only on the stiffness of that restoring force and on its mass, not on its amplitude of oscillation.

Actually, a pendulum is an unusual harmonic oscillator because its period is independent of its mass. That's because increasing the pendulum's mass also increases its weight and therefore stiffens its restoring force. These two changes compensate for one another perfectly so that the pendulum's period is unchanged.

A pendulum's period does, however, depend on its length and on gravity. When you reduce the pendulum's length—the distance from its pivot to its center of mass—you stiffen its restoring force and shorten its period. Similarly, when you strengthen gravity (perhaps by traveling to Jupiter), you increase the pendulum's weight, stiffen its restoring force, and reduce its period. Though we won't try to prove it, the period of a pendulum is

$$\text{period of pendulum} = 2\pi \sqrt{\frac{\text{length of pendulum}}{\text{acceleration due to gravity}}}.$$

Thus a short pendulum swings more often than a long one, and any pendulum swings more often on the earth than it would on the moon.

On the earth's surface, a 0.248-m (10-inch) pendulum has a period of 1 s, making it suitable for a wall clock that advances its second hand by 1 second each time the pendulum completes a cycle. Since a pendulum's period increases as the square root of its length, a 0.992-m (40-inch) pendulum (four times as tall as a 0.248-m pendulum) takes 2 s to complete its cycle and is appropriate for a floor clock that advances its second hand by 2 seconds per cycle (Fig. 9.1.3).

Because a pendulum's period depends on its length and gravity, a change in either one causes trouble. As we learned in Chapter 7, materials expand with increasing temperature, so a simple pendulum slows down as it heats up. A more accurate pendulum is thermally compensated by using several different materials with different coefficients of volume expansion to ensure that its center of mass remains at a fixed distance from its pivot.

While gravity doesn't change with time, it does vary slightly from place to place. To correct for differences in gravity between the factory and a clock's final destination, its pendulum has a threaded adjustment knob. This knob allows you to change the pendulum's length to fine-tune its period.

Pendulum Clocks

While a pendulum maintains a steady beat, it's not a complete clock. Something must keep the pendulum swinging and use that swing to determine the time. A pendulum clock does both. It sustains the pendulum's motion with gentle pushes, and it uses that motion to advance its hands at a steady rate.

The top of the pendulum has a two-pointed anchor that controls the rotation of a toothed wheel (Fig. 9.1.4). This mechanism is called an escapement. A weighted cord wrapped around the toothed wheel's shaft exerts a torque on that wheel, so that the wheel would spin if the anchor weren't holding it in place. Each time the pendulum reaches the end of a swing, one point of the anchor releases the toothed wheel while the other point catches it. The wheel turns slowly as the pendulum rocks back and forth, advancing by one tooth for each full cycle of the pendulum. This wheel turns a series of gears, which slowly advance the clock's hands. Although these hands are actually counting the number of pendulum swings since midnight, their movement is calibrated so that their positions indicate the current time.

Fig. 9.1.3 The swinging pendulum controls the movement of this grandfather clock's hands. The pendulum is 0.992 m long, from pivot to center of mass/gravity, so each cycle takes 2 s to complete and advances the hands by 2 seconds.

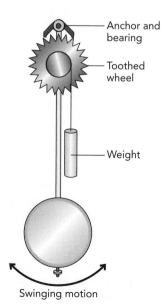

Anchor and bearing

Toothed wheel

Weight

Swinging motion

Fig. 9.1.4 A pendulum clock uses a swinging pendulum to determine how quickly a toothed wheel turns and advances a series of gears that control the hands of the clock. The anchor permits the toothed wheel to advance by one tooth each time the pendulum completes a full cycle. A weight exerts the torque that causes the toothed wheel to advance when the anchor releases it and provides the energy that keeps the pendulum swinging despite the slowing effects of friction and air resistance.

The toothed wheel also keeps the pendulum moving by giving the anchor a tiny forward push each time the pendulum completes a swing. Since the anchor moves in the direction of the push, the wheel does work on the anchor and pendulum, and replaces energy lost to friction and air resistance. This energy comes from the weighted cord, which releases gravitational potential energy as its weight descends. When you wind the clock, you rewind this cord around the shaft, lifting the weight and replenishing its potential energy.

While these pushes from the toothed wheel can keep even the clumsiest pendulum swinging, a clock works best when its pendulum swings with almost perfect freedom. That's because any outside force—even the push from the toothed wheel—will influence the pendulum's period. The most accurate timekeepers are those that can oscillate without any assistance or energy replacement for thousands or millions of cycles. These precision timekeepers need only the slightest pushes to keep them moving and thus have extremely precise periods. That's why a good pendulum clock uses an aerodynamic pendulum and low-friction bearings.

Finally, the clock must keep the oscillation amplitude of its pendulum relatively constant. From a practical perspective, drastic changes in that amplitude will make the toothed wheel turn erratically. But there is a more fundamental reason to keep the pendulum's amplitude steady: it's not really a perfect harmonic oscillator. If you displace the pendulum too far, it becomes an anharmonic oscillator—its restoring force ceases to be proportional to its displacement from equilibrium, and its period begins to depend on its amplitude. Since a change in period would spoil the clock's accuracy, the pendulum's amplitude must be kept small and steady. That way, the amplitude has almost no effect on the pendulum's period.

Balance Clocks

Because it relies on gravity for its restoring force, a swinging pendulum mustn't be tilted or moved. That's why there are so few pendulum-based wristwatches. To make use of the excellent timekeeping characteristics of a harmonic oscillator, a portable clock needs some other restoring force that's proportional to displacement but independent of gravity. It needs a spring!

As we saw in Section 3.1, the force a spring exerts is proportional to its distortion. The more you stretch, compress, or bend a spring, the harder it pushes back toward its equilibrium shape. Attach a block of wood to the free end of a spring, stretch it gently, and let go, and you'll find you have a harmonic oscillator with a period determined only by the stiffness of the spring and the block's mass (Fig. 9.1.5). Since the period of a harmonic oscillator doesn't depend on the amplitude of its motion, the block oscillates steadily about its equilibrium position and makes an excellent timekeeper.

Unfortunately, gravity complicates this simple system. Although gravity doesn't alter the block's period, it does shift the block's equilibrium position downward. That shift is a problem for a clock that might be tilted sometimes. However, there's another spring-based timekeeper that marks time accurately in any orientation or location. This ingenious device, used in most mechanical clocks and watches, is called a balance ring or simply a balance.

A balance ring resembles a tiny metal bicycle wheel, supported at its center of mass/gravity by an axle and a pair of bearings (Fig. 9.1.6). Any friction in the bearings is exerted so close to the ring's axis of rotation that it produces little torque and the ring turns extremely easily. Moreover, the ring pivots about its own center of gravity so that its weight produces no torque on it.

The only thing exerting a torque on the balance ring is a tiny coil spring. One end of this spring is attached to the ring while the other is fixed to the body of the clock. When the spring is undistorted, it exerts no torque on the ring, and the ring

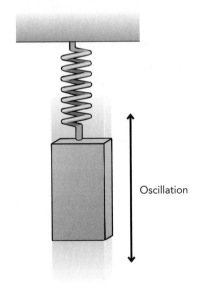

Fig. 9.1.5 A block attached to a spring is a harmonic oscillator. The oscillator's period is determined only by the stiffness of the spring and the mass of the block.

Oscillation

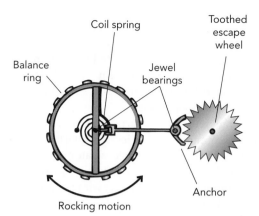

Fig. 9.1.6 A balance clock uses a rocking balance ring to control the rotation of a toothed wheel and the gears that advance the clock's hands. The anchor permits the toothed wheel to advance by one tooth each time the ring completes a full cycle. The clock's energy comes from a main spring (not shown) that exerts a steady torque on the toothed wheel.

is in equilibrium. But if you rotate the ring either way, torque from the distorted spring will act to restore it to its equilibrium orientation. Since this restoring torque is proportional to the ring's rotation away from a stable equilibrium, the balance ring and coil spring form a harmonic oscillator!

Because of the rotational character of this harmonic oscillator, its period depends on the *torsional* stiffness of the coil spring—how rapidly the spring's torque increases as you twist it—and on the balance ring's *rotational* mass. Since the balance ring's period doesn't depend on the amplitude of its motion, it keeps excellent time. And because gravity exerts no torque on the balance ring, this timekeeper works anywhere and in any orientation.

The rest of a balance clock is similar to a pendulum clock (Fig. 9.1.7). As the balance ring rocks back and forth, it tips a lever that controls the rotation of a toothed wheel. An anchor attached to the lever allows the toothed wheel to advance one tooth for each complete cycle of the balance ring's motion. Gears connect the toothed wheel to the clock's hands, which slowly advance as the wheel turns.

Because the balance clock is portable, it can't draw energy from a weighted cord. Instead, it has a main spring that exerts a torque on the toothed wheel. This main spring is a coil of elastic metal that stores energy when you wind the clock. Its energy keeps the balance ring rocking steadily back and forth and also turns the clock's hands. Since the main spring unwinds as the toothed wheel turns, the clock occasionally needs winding. (For an interesting example of a balance clock, see ❷.)

❷ The son and grandson of freed slaves, Benjamin Banneker (African-American mathematician, astronomer, and writer, 1731–1806) grew up on a Maryland tobacco farm. He was fascinated by mathematics and science, and supplemented his limited schooling with borrowed books. Though he is best remembered for his work in astronomy and for compiling six almanacs, he also produced one of the first clocks made entirely in America. With only a borrowed pocket watch as a guide, Banneker built his wooden balance clock by hand, using a knife to shape the parts. The clock kept accurate time for half a century and even struck the hours.

Fig. 9.1.7 (*a*) The balance ring in this modern transparent watch twists back and forth rhythmically under the influence of the spiral spring near its center. (*b*) The watch even has tiny jewels on its anchor to reduce friction and wear as the anchor allows the toothed escapement wheel in the lower right to advance tick by tick. (*c*) The balance ring in this antique French carriage clock is supported by tiny ruby bearings, visible at the center of the image, to minimize friction and permit this clock to keep very accurate time.

Electronic Clocks

The potential accuracy of pendulum and balance clocks is limited by friction, air resistance, and thermal expansion to about ten seconds per year. To do better, a clock's timekeeper must avoid these mechanical shortcomings. That's why so many modern clocks use quartz oscillators as their timekeepers.

A quartz oscillator is made from a single crystal of quartz, the same mineral found in most white sand. Like many hard and brittle objects, a quartz crystal oscillates strongly after being struck. In fact, it's a harmonic oscillator because it acts like a spring with a block at each end (Fig. 9.1.8a,b). The two blocks oscillate in and out symmetrically about their combined center of mass, with a period determined only by the blocks' masses and the spring's stiffness. In a quartz crystal, the spring is the crystal itself and the blocks are its two halves (Fig. 9.1.8c,d). Since the forces on the blocks are proportional to their displacements from equilibrium, they're harmonic oscillators.

Because of its exceptional hardness, a quartz crystal's restoring force is extremely stiff. Even a tiny distortion leads to a huge restoring force. Since the period of a harmonic oscillator decreases as its spring becomes stiffer, a typical quartz oscillator has an extremely short period. Its motion is usually called a vibration rather than an oscillation because vibration implies a fast oscillation in a mechanical system. Oscillation itself is a more general term for any repetitive process and can even apply to such nonmechanical processes as electric or thermal oscillations.

Because of its rapid vibration, a quartz oscillator's period is a small fraction of a second. We normally characterize such a fast oscillator by its frequency—the number of cycles it completes in a certain amount of time. The SI unit of frequency is the cycle-per-second, also called the hertz (abbreviated Hz) after German physicist Heinrich Rudolph Hertz. Period and frequency are reciprocals of one another (period equals 1/frequency and vice versa) so an oscillator with a period of 0.001 s has a frequency of 1000 Hz.

Because the vibrating crystal isn't sliding across anything or moving quickly through the air, it loses energy slowly and vibrates for a long, long time. And because quartz's coefficient of thermal expansion is extremely small, the crystal's period is nearly independent of its temperature. With its exceptionally steady period, a quartz oscillator can serve as the timekeeper for a highly accurate clock, one that loses or gains less than a tenth of a second per year.

Of course, a quartz crystal isn't a complete clock. Like the pendulum and balance, it needs something to keep it vibrating and to use that vibration to determine the time. While these tasks could conceivably be done mechanically, quartz clocks are normally electronic. There are two reasons for this choice. First, the crystal's vibrations are too fast and too small for most mechanical devices to follow. Second, a quartz crystal is intrinsically electronic itself; it responds mechanically to electrical stress and electrically to mechanical stress. Because of this coupling between its mechanical and electrical behaviors, crystalline quartz is known as a piezoelectric material and is ideal for electronic clocks.

Fig. 9.1.8 A quartz crystal acts like a spring with masses at each end. Just as the two masses alternately accelerate (a) toward and (b) away from one another, so the two halves of the vibrating crystal alternately accelerate (c) together and (d) apart.

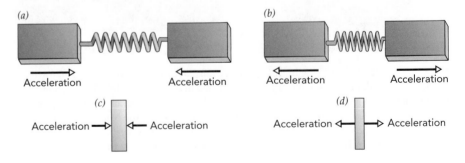

Fig. 9.1.9 The quartz crystal in this wristwatch is located inside the silver cylinder at the bottom right, behind the second hand. Carefully polished to vibrate at a precise frequency, the crystal keeps the watch accurate to about one second per month.

The clock's circuitry uses electrical stresses to keep the quartz crystal vibrating (Fig. 9.1.9). Just as carefully timed pushes keep a child swinging endlessly on a playground swing, carefully timed electrical stresses keep the quartz crystal vibrating endlessly in its holder. Because the crystal loses so little energy with each vibration, only a tiny amount of work is required each cycle to maintain its vibration.

The clock also detects the crystal's vibrations electrically. Each time its halves move in or out, the crystal experiences mechanical stress and emits a pulse of electricity. These pulses may control an electric motor that advances clock hands or may serve as input to an electronic chip that measures time by counting the pulses.

The quartz crystals used in clocks and watches are carefully cut and polished to vibrate at specific frequencies. The thinner the crystal, the faster it vibrates—less mass and a stiffer restoring force. In effect, these crystals are tuned like musical instruments to match the requirements of their clocks.

While most tiny quartz crystals vibrate millions of times each second, common watch crystals vibrate at 2^{15} Hz or 32,768 Hz. This low frequency prolongs a watch's battery life because counting each pulse consumes some of the battery's energy. To make a small crystal vibrate this slowly, the manufacturer cuts away most of the center of the crystal to weaken its restoring force and slow its oscillations (Fig 9.1.10). The resulting quartz "tuning fork" oscillator is carefully metal-

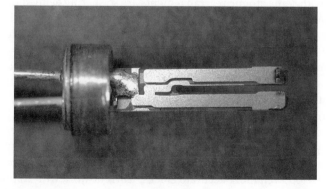

Fig. 9.1.10 Shaped like a tiny tuning fork, this watch crystal vibrates almost exactly 32,768 times per second. Burn marks at its tips were created when its vibrational frequency was tuned by a laser beam.

ized to permit the watch to interact with it electrically, and it's then tuned to exactly 32,768 Hz by burning away some of the metal mass with a laser beam. While crystal tuning forks that oscillated half as often (16,384 Hz) would make watches more energy efficient, they would be larger and would vibrate at a frequency that some people, particularly children, would be able to hear.

Atomic Clocks

Quartz clocks are accurate enough for all but the most demanding measurements of time. Still, there are occasions when an even greater accuracy is needed from a clock. The most accurate clocks in existence make use of atoms and atomic structure. Just as astronomical clocks are based on the motions of celestial objects orbiting one another in the heavens, atomic clocks are based on the motions of electrons and nuclei orbiting one another in atoms.

Because an atom is so small, the motions and behaviors of the particles inside it are dominated by the strange world of quantum physics. While the planets in the solar system can travel in any orbit about the sun, the electrons in an atom can travel only in a few specific orbits about the atom's positively charged center, its nucleus. These allowed orbits depend only on the numbers of electrons and other particles in the atom so that two atoms with identical constituents also have identical allowed orbits. Thus one atom of cesium has the same allowed orbits as every other atom of cesium.

The quantum structure of an atom governs the spectrum of light wavelengths that it can absorb and emit. As we'll see in Chapter 12, an atom can only absorb or emit light when that light has the right wavelength to shift the atom from one arrangement of allowed orbits to another such arrangement. As a result, atoms absorb and emit only specific wavelengths of light.

Light is effectively an electromagnetic oscillation so that, in addition to its wavelength, it also has a frequency. This frequency is inversely related to its wavelength. Thus atoms are selective not only about which wavelengths they absorb and emit, but also about which frequencies they absorb and emit.

This selective absorption and emission of light makes atoms useful as timekeepers. In suitable conditions, atoms will emit one of their permitted frequencies of light and this emission can be used as the timekeeper for an electronic clock. Alternatively, atoms can be exposed to light from another source that is then carefully adjusted in frequency until the atoms begin to absorb the light. Both of these schemes are used in atomic clocks.

Because atoms are completely free of friction and most of the problems affecting other timekeepers, they produce exquisitely accurate clocks. The frequencies associated with some atomic rearrangements are so sharply defined that the atom can distinguish between electromagnetic waves differing in frequency by only 1 Hz per 10^{15} Hz or even less. This precise behavior has made it possible to define the second as 9,192,631,770 periods of a certain electromagnetic wave absorbed by cesium atoms and the meter as the distance traveled by light during 1,650,763.73 periods of a certain light wave emitted by krypton atoms. In effect, the meter (a unit of length) is now officially defined in terms of time and the speed of light. Time and space really are closely related.

Modern atomic clocks use cesium, rubidium, or hydrogen as their timekeeper atoms. They are so accurate that they gain or lose as little as 1 s every 10 million years. But even this accuracy is inadequate for some purposes, so scientists are trying to develop even more accurate clocks. Current work with trapped and isolated atoms is intended to reach timing errors of perhaps 1 Hz per 10^{20} Hz. A clock with this accuracy would have gained or lost no more than a few thousandths of a second since the creation of the universe.

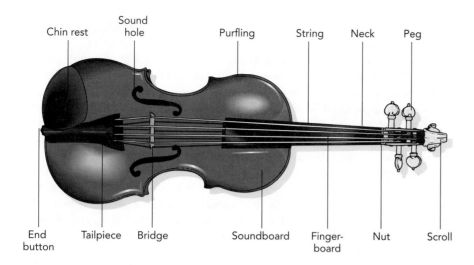

Chin rest · Sound hole · Purfling · String · Neck · Peg

End button · Tailpiece · Bridge · Soundboard · Finger-board · Nut · Scroll

Section 9.2 Musical Instruments

Music is an important part of human expression. While what qualifies as music is a matter of taste, it always involves sound and often involves instruments. In this section we'll examine sound, music, and several instruments: violins, pipe organs, and drums. As examples of the three most common types of instruments, strings, winds, and surfaces, this trio will help us to understand many other instruments as well.

Sound and Music

To understand how instruments work we'll need to know a bit more about sound and music. In air, sound consists of density waves—patterns of compressions and rarefactions that travel outward rapidly from their source. When a sound passes by, the air pressure in your ear fluctuates up and down about normal atmospheric pressure. Even when these fluctuations have amplitudes less than a millionth of atmospheric pressure, you hear them as sound.

When the fluctuations are repetitive you hear a tone with a pitch equal to the fluctuation's frequency. Pitch is the frequency of a sound. A bass singer's pitch range extends from 80 Hz to 300 Hz, while that of a soprano singer extends from 300 Hz to 1100 Hz. Musical instruments can produce tones over a much wider range of pitches, but we can only hear those between about 30 Hz and 20,000 Hz, and that range narrows as we get older.

Most music is constructed around intervals—the frequency ratio between two different tones. This ratio is found by dividing one tone's frequency by that of the other. Our hearing is particularly sensitive to intervals, with pairs of tones at equal intervals sounding quite similar to one another. For example, a pair of tones at 440 Hz and 660 Hz sounds similar to a pair at 330 Hz and 495 Hz because they both have the interval 3/2.

The interval 3/2 is pleasing to most ears and is common in Western music, where it's called a fifth. A fifth is the interval between the two "twinkles" at the beginning of "Twinkle, Twinkle, Little Star." If your ear is good, you can start with any tone for the first twinkle and will easily find the second tone, located at 3/2 the frequency of the first. Your ear hears that factor of 3/2 between the two frequencies.

The most important interval in virtually all music is 2/1 or an octave. Tones that differ by a factor of 2 in frequency sound so similar to our ears that we often think of them as being the same. When men and women sing together "in unison," they

❶ In addition to his contributions to mathematics, geometry, and astronomy, the Greek mathematician Pythagoras (ca 580–500 B.C.) was perhaps the first person to use mathematics to relate intervals, pitches, and the lengths of vibrating strings. He and his followers laid the groundwork for the scale used in most Western music.

often sing an octave or two apart and the differences in the tones, always factors of 2 or 4 in frequency, are only barely noticeable.

The octave is so important that it structures the entire range of audible pitches. Most of the subtle interplay of tones in music occurs in intervals of less than an octave, less than a factor of 2 in frequency. Thus most traditions build their music around the intervals that lie within a single octave, such as 5/4 and 3/2. They pick a particular standard pitch and then assign notes at specific intervals from this standard pitch. This arrangement repeats at octaves above and below the standard pitch to create a complete scale of notes. (For a history of scales, see ❶.)

The scale used in Western music is constructed around a note called A_4, which has a standard pitch of 440 Hz. At intervals of 9/8, 5/4, 4/3, 3/2, 5/3, and 15/8 above A_4 lie the six notes B_4, $C_5^\#$, D_5, E_5, $F_5^\#$, and $G_5^\#$. Similar collections of six notes are built above A_5 (880 Hz), which has a frequency twice that of A_4, and above A_3 (220 Hz), which has a frequency half that of A_4. In fact, this pattern repeats above A_1 (55 Hz) through A_8 (7040 Hz).

Actually, Western music is built around 12 notes and 11 intervals that lie within a single octave. Five more intervals account for five additional notes, B_4^b, C_5, $D_5^\#$, F_5, and G_5. It's also not quite true that every note is based exclusively on its interval from A_4. While A_4 remains at 440 Hz, the pitches of the other 11 notes have been modified slightly so that they're at interesting and pleasing intervals from one another as well as from A_4. This adjustment of the pitches led to the well-tempered scale that has been the basis for Western music for the last several centuries.

A Violin's Vibrating String

The tones produced by a violin begin as vibrations in its strings. But these strings are limp and shapeless on their own and rely on the violin's rigid body and neck for structure. The violin subjects its strings to tension—outward forces that act to stretch it—and this tension gives each string an equilibrium shape: a straight line.

To see that a straight violin string is in equilibrium, think of it as composed of many individual pieces that are connected together in a chain (Fig. 9.2.1). Tension exerts a pair of outward forces on each piece of the string; its neighboring pieces are pulling that piece toward them. Since the string's tension is uniform, these two outward forces sum to zero; they have equal magnitudes but point in opposite directions. With zero net force on each of its pieces, the straight string is in equilibrium.

When the string is curved, however, the pairs of outward forces no longer sum to zero (Fig. 9.2.2). Although those outward forces still have equal magnitudes, they now point in slightly different directions. As a result, each piece experiences a small net force.

The net forces on its pieces are restoring forces because they act to straighten the string. If you distort the string and release it, these restoring forces will cause the string to vibrate about its straight equilibrium shape in a natural resonance. But the string's restoring forces are special: the more you curve the string, the stronger the restoring forces on its pieces become. In fact, the restoring forces are spring-like forces—they increase in proportion to the string's distortion—so the string is a form of harmonic oscillator!

Actually, the string is much more complicated than a pendulum or a balance ring. It can bend and vibrate in many distinct modes or basic patterns of distortion, each with its own period of vibration. Nonetheless, the string retains the most important feature of a harmonic oscillator: the period of each vibrational mode is independent of its amplitude. Thus a violin string's pitch doesn't depend on how hard it's vibrating. Think how tricky it would be to play a violin if its pitch depended on its volume!

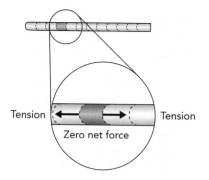

Fig. 9.2.1 A taut violin string can be viewed as composed of many individual pieces. When the string is straight, the two forces exerted on a given piece by its neighbors cancel perfectly, and it experiences zero net force.

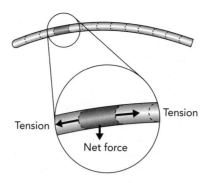

Fig. 9.2.2 When a violin string is curved, the two forces exerted on a given piece by its neighbors don't point in exactly opposite directions and don't balance one another. The piece experiences a net force.

A violin string has one simplest vibration: its fundamental vibrational mode. In this mode, the entire string arcs alternately one way then the other (Fig. 9.2.3). Its kinetic energy peaks as it rushes through its straight equilibrium shape, and its potential energy (elastic potential energy in the string) peaks as it stops to turn around. The string's midpoint travels the farthest (the vibrational antinode) while its ends remain fixed (the vibrational nodes). At each moment its shape is the gradual curve of the trigonometric sine function.

In this fundamental mode, the violin string behaves as a single harmonic oscillator. As with any harmonic oscillator, its vibrational period depends only on the stiffness of its restoring forces and on its inertia. Stiffening the violin string or reducing its mass both quicken its fundamental vibration and increase its fundamental pitch.

A violin has four strings, each with its own stiffness and mass and therefore its own fundamental pitch. In a tuned violin, the notes produced by these strings are G_3 (196 Hz), D_4 (294 Hz), A_4 (440 Hz), and E_5 (660 Hz). The G_3 string, which vibrates rather slowly, is the most massive. It's usually made of gut wrapped in a coil of heavy metal wire. The E_5 string, on the other hand, must vibrate quite rapidly and needs to have a low mass. It's usually a thin steel wire.

You tune a violin by adjusting the tension in its strings, using pegs in its neck and tension adjusters on the tailpiece. Tightening the string stiffens it by increasing both the outward forces on its pieces and the net forces they experience during a distortion. Since temperature and time can alter a string's tension, you should always tune your violin just before a concert.

A string's fundamental pitch also depends on its length. Shortening the string both stiffens it and reduces its mass, so its pitch increases. That stiffening occurs because a shorter string curves more sharply when it's displaced from equilibrium and therefore subjects its pieces to larger net forces. This dependence on length allows you to raise a string's pitch by pressing it against the fingerboard in the violin's neck and effectively shortening it. Part of a violinist's skill involves knowing exactly where on the string to press it against the fingerboard in order to produce a particular note.

If the arc of a string vibrating in its fundamental mode reminds you of a wave, that's because it is one. It's a mechanical wave—the natural motions of an extended object about its stable equilibrium shape or situation. An extended object is one like a string, stick, or lake surface that has many parts that move with limited independence. Since its parts influence one another, an extended object with a stable equilibrium exhibits fascinating natural motions that involve many parts moving at once; it exhibits mechanical waves.

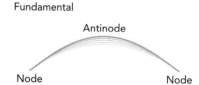

Fig. 9.2.3 This string is vibrating between two fixed points in its fundamental vibrational mode. The whole string moves together, traveling up and down as a single harmonic oscillator. It is exhibiting a standing wave.

With its innumerable linked pieces and its stable equilibrium shape, the violin string exhibits such waves. And the string's fundamental mode is a particularly simple wave, a standing wave—a wave in which all the nodes and antinodes remain in place. A standing wave's basic shape doesn't change with time, it merely scales up and down rhythmically at a particular frequency and amplitude—its peak extent of motion. Most importantly, the standing wave doesn't travel along the string.

Although this wave extends along the string, its associated oscillation is *perpendicular* to the string and therefore *perpendicular* to the wave itself. A wave in which the underlying oscillation is perpendicular to the wave itself is called a transverse wave. Waves on strings, drums, and the surface of water are all transverse waves.

The Violin String's Harmonics

The fundamental vibrational mode isn't the only way in which a violin string can vibrate. The string also has higher-order vibrational modes in which the string vibrates as a chain of shorter strings arcing in alternate directions (Fig. 9.2.4). Each of these higher-order vibrational modes is another standing wave, with a fixed shape that scales up and down rhythmically at its own frequency and amplitude.

For example, the string can vibrate as two half-strings arcing in opposite directions and separated by a motionless vibrational node. In this mode the violin string not only vibrates as half-strings; it has the pitch of half-strings as well. Remarkably, that half-string pitch is exactly twice the whole-string (i.e., fundamental) pitch! In general, a string's vibrational frequency is inversely proportional to its length, so halving its length doubles its frequency. Frequencies that are integer multiples of the fundamental pitch are called harmonics, so this half-string vibration occurs at the second harmonic pitch and is called the second harmonic mode.

A violin string can also vibrate as three third-strings, with a frequency that's three times the fundamental. The interval between this third harmonic pitch and the fundamental pitch is an octave and a fifth (2/1 times 3/2). Overall, the fundamental and its second and third harmonics sound very pleasant together.

While the violin string can vibrate in even higher harmonics, what's more important is that the string often vibrates in more than one mode at the same time. For example, a violin string vibrating in its fundamental mode can also vibrate in its second harmonic and emit two tones at once.

Harmonics are important because bowing a violin excites many of its vibrational modes. The violin's sound is thus a rich mixture of the fundamental tone and the harmonics. Known as timbre, this mixture of tones is characteristic of a violin, which is why an instrument producing a different mixture doesn't sound like a violin.

When a violin string is vibrating in several modes at once, its shape and motion are complicated. The individual standing waves add on top of one another, a process known as superposition. Each vibrational mode has its own amplitude and therefore its own volume contribution to the string's timbre.

While these individual waves coexist beautifully on the string, with virtually no effect on one another, the string's overall distorted shape is now the superposition of the individual waveshapes. Not only is that overall shape quite complicated, it actually changes substantially with time. That's because the different harmonic waves vibrate at different frequencies and their superposition changes as they change. The string's overall wave is not a standing wave and its features can even move along the string!

Bowing and Plucking the Violin String

You play a violin by drawing a bow across its strings. The bow consists of horsehair, pulled taut by a wooden stick. Horsehair is rough and exerts frictional forces on

Second harmonic

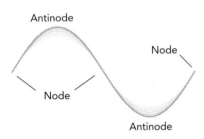

Third harmonic

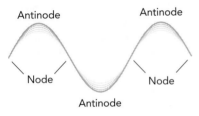

Fig. 9.2.4 These strings are vibrating between two fixed points in their second and third harmonic modes. The strings vibrate as two or three segments, completing cycles at two or three times the fundamental frequency, respectively. Both strings are exhibiting standing waves.

the strings as it moves across them. But most importantly, horsehair exerts much larger static frictional forces than sliding ones.

As the bow hairs rub across a string, they grab the string and push it forward with static friction. Eventually the string's restoring force overpowers static friction, and the string suddenly starts sliding backward across the hairs. Because the hairs exert little sliding friction, the string completes half a vibrational cycle with ease. But as it stops to reverse direction, the hairs grab the string again and begin pushing it forward. This process repeats over and over.

Each time the bow pushes the string forward, it does work on the string and adds energy to the string's vibrational modes. This process is an example of resonant energy transfer, in which a modest force doing work in synchrony with a natural resonance can transfer a large amount of energy to that resonance. Just as gentle, carefully timed pushes can get a child swinging high on a playground swing, so to can gentle, carefully timed pushes from a bow get a string vibrating vigorously on a violin. Similar rhythmic pushes can cause other objects to vibrate strongly, notably a crystal wineglass and the Tacoma Narrows Bridge near Seattle, Washington ❷, but including many others ❸❹. The wineglass's response to a certain tone is also an example of sympathetic vibration—the transfer of vibrational energy between two systems that share a common vibrational frequency.

The amount of energy the bow adds to each vibrational mode depends on where it crosses the violin string. When you bow the string at the usual position, you produce a strong fundamental vibration and a moderate amount of each harmonic. Bowing the string nearer its middle reduces the string's curvature, weakening its harmonic vibrations and giving it a more mellow sound. Bowing the string nearer its end increases the string's curvature, strengthening its harmonic vibrations and giving it a brighter sound.

The sound of a plucked violin string also depends on harmonic content and thus on where that string is plucked. But this sound is quite different from that of a bowed string. The difference lies in the sound's envelope—the way the sound evolves with time. This envelope can be viewed as having three time periods: an initial attack, an intermediate sustain, and a final decay. The envelope of a plucked string is an abrupt attack followed immediately by a gradual decay. In contrast, the envelope of a bowed string is a gradual attack, a steady sustain, and then a gradual decay. We learn to recognize individual instruments not only by their harmonic content but also by their sound envelopes.

An Organ Pipe's Vibrating Air

Like a violin, a pipe organ uses vibrations to create sound. However, its vibrations take place in the air itself. An organ pipe is essentially a hollow cylinder, open at each end and filled with air. Because that air is protected by the rigid walls of the pipe, its pressure can fluctuate up and down relative to atmospheric pressure and it can exhibit natural resonances.

In its fundamental vibrational mode, air moves alternately toward and away from the pipe's center (Fig. 9.2.5), like two blocks on a spring. As air moves toward the pipe's center, the density there rises and a pressure imbalance develops. Since the pressure at the pipe's center is higher than at its ends, air accelerates *away* from the center. The air eventually stops moving inward and begins to move outward. As air moves away from the pipe's center, the density there drops and a reversed pressure imbalance occurs. Since the pressure at the pipe's center is lower than at its ends, air now accelerates *toward* the center. It eventually stops moving outward and begins to move inward, and the cycle repeats. The air's kinetic energy peaks each time it rushes through that equilibrium and its potential energy (pressure potential energy in the air) peaks each time it stops to turn around.

❷ The Tacoma Narrows Bridge collapsed in November 1940, as the result of resonant energy transfer between the wind and the bridge surface. Shortly after construction, the automobile bridge began to exhibit an unusual natural resonance in which its surface twisted slowly back and forth so that one lane rose as the other fell. During a storm, the wind slowly added energy to this resonance until the bridge ripped itself apart.

❸ The coffee in a typical cup sloshes back and forth resonantly several times per second. Unfortunately, it's easy to shake the cup at this resonant frequency accidentally as you walk. When you do, resonant energy transfer causes the coffee to slosh so vigorously that it leaves the cup. When you're trying not to slosh the coffee out of the cup, you're actually trying to avoid shaking it at its resonant frequency or even to damp out whatever sloshing motion you observe.

❹ Many parts of the human anatomy can oscillate about stable equilibriums and exhibit natural resonances. Walking or running can cause these body parts to begin oscillating because of resonant energy transfer—if the body part's natural resonant frequency matches the frequency of a stride, energy will shift from the striding motion to the body part's oscillation. Woman often find running uncomfortable because their breasts are driven into vigorous oscillation by the unfortunate coincidence between the frequency of their strides and the frequency of their breasts' fundamental vibrational modes. The advent of sports bras made running more practical by stiffening the restoring forces experienced by each breast and thereby increasing its resonant frequency greatly. The resonant frequencies of bra-supported breasts are so much higher than the frequencies involved in running that there is little resonant energy transfer and thus far less discomfort.

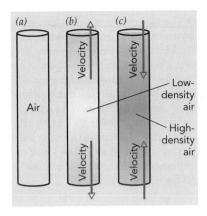

Fig. 9.2.5 In a pipe that's open at both ends (*a*), the air vibrates in and out about the middle of the pipe. (*b*) For half a cycle, the air moves outward and creates a low-pressure region in the middle, and (*c*) for half a cycle the air moves inward and creates a high-pressure region there.

This air is vibrating about a stable equilibrium of uniform atmospheric density and pressure, and is clearly experiencing restoring forces. It should come as no surprise that those restoring forces are springlike and that the air column is yet another harmonic oscillator. As such, its vibrational frequency depends only on the stiffness of its restoring forces and on its inertia. Stiffening the air column or reducing its mass both quicken its vibration and increase its pitch.

These characteristics depend on the length of the organ pipe. A shorter pipe not only holds less air mass than a longer pipe; it also offers stiffer opposition to any movements of air in and out of that pipe. With less room in the shorter pipe, the pressure inside it rises and falls more abruptly, leading to stiffer restoring forces on the moving air. Together, these effects make the air in a shorter pipe vibrate faster than that in a longer pipe. In general, an organ pipe's vibrational frequency is inversely proportional to its length.

Unfortunately, the mass of vibrating air in a pipe also increases with the air's average density, so that even a modest change in temperature or weather will alter the pipe's pitch. Fortunately, all of the pipes shift together so that an organ continues to sound in tune. Nonetheless, this shift may be noticeable when the organ is part of an orchestra.

As you may suspect, the fundamental vibrational mode of air in the organ pipe is another standing wave. Air in the pipe is an extended object with a stable equilibrium and the disturbance associated with its fundamental vibrational mode has a basic shape that doesn't change with time; it merely scales up and down rhythmically.

However, the shape of the wave in the pipe's air now has to do with back-and-forth compressions and rarefaction, not with side-to-side displacements as it did in the violin string. In fact, all of the wave's associated oscillation is *along* the pipe and therefore *along* the wave itself. A wave in which the underlying oscillation is parallel to the wave itself is called a longitudinal wave. Waves in the air, including those inside organ pipes and other wind instruments and sound waves in the open air, are all longitudinal waves.

Playing an Organ Pipe

The organ uses resonant energy transfer to make the air in a pipe vibrate. It starts this transfer by blowing air across the pipe's lower opening (Fig. 9.2.6*a*), although for practical reasons that lower opening is usually found on the pipe's side (Fig. 9.2.6*b*). As the air flows across the opening, it's easily deflected to one side or the other and tends to follow any air that's already moving into or out of the pipe. If the air inside the pipe is vibrating, the new air will follow it in perfect synchrony and strengthen the vibration.

This following process is so effective at enhancing vibrations that it can even initiate a vibration from the random noise that's always present in a pipe. That's how the sound starts when the organ's pump first blows air across the pipe. Once the vibration has started, it grows quickly in amplitude. That amplitude increases until energy leaves the pipe as sound and heat just as quickly as it arrives via compressed air. The more air the organ blows across the pipe each second, the more power it delivers to the pipe and the louder the vibration.

Like a violin string, an organ pipe can support more than one mode of vibration. In its fundamental vibrational mode, the pipe's entire column of air vibrates together. In the higher-order vibrational modes, this air column vibrates as a chain of shorter air columns moving in alternate directions. If the pipe has a constant width, these vibrations occur at harmonics of the fundamental. When the air column vibrates as two half-columns, its pitch is exactly twice that of the fundamental mode. When it vibrates as three third-columns, its pitch is exactly three times that of the fundamental. And so on.

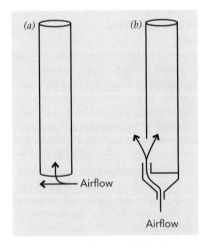

Fig. 9.2.6 (*a*) Air blown across the bottom of an open pipe will follow any other air that's moving into the pipe. If the air in the pipe is vibrating, this effect will add energy to that vibration. (*b*) The lower opening in an organ pipe is cut in its side for practical reasons.

But the air column inside a pipe can vibrate in more than one mode simultaneously. As with a violin string, the standing waves superpose and the fundamental and harmonic tones are produced together. The shape of the organ pipe and the place where air is blown across it determine the pipe's harmonic content and thus its timbre. Different pipes can imitate different instruments. To sound like a flute, the pipe should emit mostly the fundamental tone and keep the harmonics fairly quiet. To sound like a clarinet, its harmonics should be much louder. An organ pipe's volume always builds slowly during the attack, so it can't pretend to be a plucked string. However, a clever designer can make the organ imitate a surprising range of instruments.

A Drum's Vibrating Surface

After examining violin strings and organ pipes, it might seem that drums offer little new. But while a drumhead is yet another extended object with a stable equilibrium and springlike restoring forces, its overtone vibrations have an important difference: they *aren't* harmonics.

Violin strings and organ pipes are effectively one-dimensional or linelike objects, dividing easily into half-objects or third-objects that then vibrate at second or third harmonic pitches. Together with the many other one-dimensional instruments in an orchestra or band, they blend seamlessly when they're playing the same fundamental pitch because they share the same harmonics.

But because a drumhead is effectively two-dimensional or surfacelike, it doesn't divide easily into pieces that resemble the entire drumhead. As a result, the pitches of its overtone vibrations have no simple relationship to its fundamental pitch. A timpani stands out relative to other instruments in part because of the unique overtone pitches.

Figure 9.2.7 illustrates the fundamental (*a*) and five lowest-pitched overtone vibrational modes (*b–f*) for a drumhead. Each vibrational mode is a standing wave, but with vibrational nodes that are curves or lines rather than points. The fundamental mode (*a*) has only one node on its outer edge, while the overtone modes have additional nodes within the surface. In each vibrational mode, these nodes remain motionless as the rest of the surface vibrates up and down, its peaks and valleys interchanging alternately. The pitches of the overtone vibrations are indicated relative to that of the fundamental vibration. (For a historical note on the understanding of surface modes, see ❺.)

❺ In 1809, the French Academy of Sciences announced a competition to explain the intricate patterns observed on vibrating surface plates. The only respondent was French mathematician Sophie Germain [1776–1831]. As a woman, Germain was barred from formal education in mathematics and struggled to learn the subject from books and via correspondences with leading mathematicians, which she conducted under the pseudonym Antoine-August Le Blanc. It took her three tries, but in 1816 she was awarded the prize. Because she was a woman, however, she didn't attend the ceremony. Her analysis of surface vibrations, though imperfect, was a visionary effort, made all the more extraordinary by her circumstances. Although her mentor, Carl Fredrick Gauss, managed to convince the University of Göttingen to award her an honorary degree, she died of breast cancer before she received it.

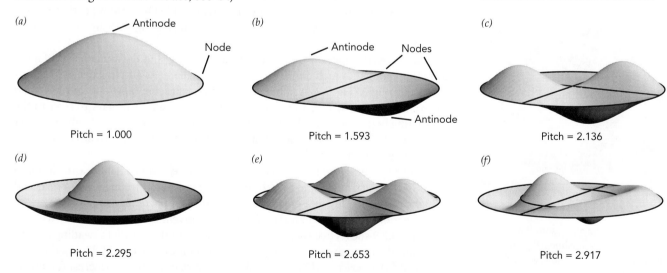

(a) Antinode Node Pitch = 1.000

(b) Antinode Nodes Antinode Pitch = 1.593

(c) Pitch = 2.136

(d) Pitch = 2.295

(e) Pitch = 2.653

(f) Pitch = 2.917

Fig. 9.2.7 The six lowest-pitch vibrational modes of a drumhead, including (*a*) the fundamental vibrational mode and (*b–f*) overtone modes. Pitches are shown relative to the fundamental pitch.

Because striking a drumhead causes it to vibrate in several modes at once, the drum emits several pitches simultaneously. The amplitude of each mode, and consequently its volume, depends not only on *how hard* you hit the drumhead, but also on *where* you hit it. If you hit it at its center, it vibrates primarily in circular modes like (*a*) and (*d*). If you hit it nearer its edge, it also vibrates in noncircular modes like (*b*), (*c*), (*e*), and (*f*).

A timpani sounds most musical when it's struck off-center in such a way that the amplitude of its fundamental vibrational mode is nearly zero and its overtones, particularly (*b*), dominate its sound. That's because the fundamental vibrational mode emits sound so well that its vibrational energy dissipates before it can produce a discernible tone. Unless all you want is a loud "thump," you must hit the timpani off-center so that its long-lived overtone vibrations receive most of the energy and emit most of the sound. The dominant pitch of a properly played timpani is that of its first overtone vibration and it is tuned with that pitch in mind.

In truth, the pitches shown in Fig. 9.2.7 neglect the effects of air's inertia on the drumhead's vibrations. Since air adds inertia to the drumhead, it lowers the pitches of all the vibrational modes, some more than others. Because of air's influence on pitch, a drum must be tuned to accommodate changes in temperature and weather.

Sound in Air

All these vibrations would serve little purpose if we couldn't hear them, so it's time to look at how instruments produce sound. We'll start by looking at sound itself.

We noted at the beginning of this section that sound in air consists of density waves—patterns of compressions and rarefactions that travel outward rapidly from their source. While that observation was mysterious at the time, we can now understand those waves as vibrations in an extended object with a stable equilibrium. That extended object is air.

Neglecting gravity, air is in a stable equilibrium when its density is uniform. If we disturb it from equilibrium, the resulting pressure imbalances will provide spring-like restoring forces. These forces, together with air's inertia, lead to rhythmic vibrations—the vibrations of harmonic oscillators. In open air, the most basic vibrations are waves that move steadily in a particular direction and are therefore called traveling waves. Like the standing waves inside an organ pipe, these traveling waves in open air are longitudinal—air vibrates along the same direction as the sound wave travels.

As it moves through the open air, a basic traveling sound wave consists of an alternating pattern of high density regions we'll call crests and low density regions we'll call troughs (Fig. 9.2.8). While those names will seem more appropriate when we examine water surface waves in the next section, it's customary to refer to the alternating highs and lows of any wave as crests and troughs, respectively. Whether a wave is standing or traveling, the shortest distance between two adjacent crests is known as the wavelength.

While a standing wave's crests and troughs merely flip back and forth in place, crests becoming troughs and troughs becoming crests, a traveling wave's crests and troughs move steadily in a particular direction at a particular speed. That speed and direction of travel together constitute the traveling wave's wave velocity.

Figure 9.2.9 shows five snapshots of a simple sound wave that's heading toward the right. If we watch air's density at the same point in space (solid line), it begins as a crest (*a*), decreases (*b*) to a trough (*c*), then rises (*d*) back to a crest (*e*) during one complete vibration cycle. However, if we follow the same crest (dashed line) over time, it travels one wavelength to the right during one complete vibration cycle (*a–e*). Since a crest moves one wavelength per vibration cycle, and frequency

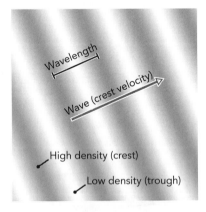

Fig. 9.2.8 A traveling sound wave in air consists of a washboard pattern of high density (dark) and low density (light) regions. The distance separating adjacent crests is the wavelength, and the speed and direction of crest motion are the wave velocity.

is the number of vibration cycles per second, the speed at which the crest moves is the product of the wavelength times the frequency:

$$\text{wave speed} = \text{wavelength} \cdot \text{frequency}. \qquad (9.2.1)$$

Remarkably enough, all sound waves travel at the same speed through air, regardless of wavelength or frequency. That's because a sound's wavelength is always inversely proportional to its frequency. We saw this same inverse relationship for the standing waves in organ pipes: if you double the length of a pipe, and therefore double the wavelength of its fundamental vibrational mode, you halve the frequency of that vibration. Even when it's not confined by a pipe, vibrating air has a wavelength that's inversely proportional to the frequency of that vibration.

Because of this extraordinary inverse relationship between a traveling sound wave's wavelength and its frequency, Eq. 9.2.1 yields the same wave speed for any sound wave. Known as the speed of sound in air, it's about 331 m/s (1086 ft/s) in standard conditions at sea level (0 °C, 101,325 Pa pressure). While that's fast, there is still a noticeable delay between when a percussionist strikes the cymbals and when you hear them from across the concert hall. Fortunately, because the speed of sound doesn't depend on frequency, when the entire orchestra plays in unison you hear all its different pitches simultaneously.

This discussion of sound assumes that the instruments and listener maintain a constant separation, as they usually do at an orchestra concert. But when a marching band dashes toward or away from the listener, something odd happens: the listener hears its music shifted up or down in pitch. Known as the Doppler effect, this frequency shift occurs because the listener encounters sound wave crests at a rate that's different from that at which those crests were created. If an instrument and the listener are approaching one another, the listener encounters the crests at an increased rate and the pitch increases. If the two are separating from one another, the listener encounters the crests at a decreased rate and the pitch decreases. Fortunately, the Doppler effect is subtle at speeds that are small compared to the speed of sound, so you can listen to parades without them sounding flat or sharp.

Turning Vibrations into Sound

Anything that disturbs air's otherwise uniform density can produce traveling sound waves. Instruments emit sound by compressing and rarefying the nearby air in synch with their own vibrations. How they accomplish this task differs from instrument to instrument, so we'll have to look at them individually. As we'll see, some instruments find it easier to produce sound than others.

A drum produces sound when its vibrating drumhead alternately compresses and rarefies the nearby air. As portions of that drumhead rise and fall, they upset the air's uniform density and thereby produce sound waves. But whenever it can, air simply flows silently out of the drumhead's way, leading to smaller density fluctuations and less intense sound. For example, when the drumhead is experiencing one of the five overtone vibrations shown in Fig. 9.2.7, air flows away from each rising peak in the undulating surface and toward each falling valley. The overtone vibrations still manage to produce sound, but it's less intense and the vibrational energy in the drumhead transforms relatively slowly into sound energy.

Air's partial success in dodging the drumhead's overtone vibrations allows those overtones to complete many vibrational cycles before running out of vibrational energy. Their vibrations are therefore long-lived and have distinct pitches. In contrast, air has difficulty dodging the drumhead's fundamental vibrational mode, which alternately compresses and rarefies the air so effectively that it transfers all of its vibrational energy to the air in just a few cycles. That's why the fundamental

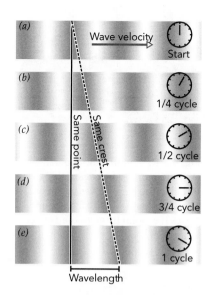

Fig. 9.2.9 A sound wave at five evenly spaced times (*a–e*) showing one complete cycle of oscillation. During that cycle, the pressure at a specific point in space goes from high to low to high (the solid line) and a specific crest moves one wavelength to the right (the dashed line).

Fig. 9.2.10 A violin's bridge transfers energy from its vibrating strings to its belly. The belly moves in and out, emitting sound. Some of this sound leaves the violin through the *f*-holes in its body.

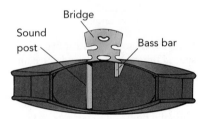

Fig. 9.2.11 The bridge is supported by the bass bar on one side and the sound post on the other. As the strings vibrate back and forth, the bridge experiences a torque that causes the belly of the violin to move in and out and emit sound.

vibrational mode of a drumhead produces the intense and nearly pitchless "thump" sound that we associate with a drum.

If air can dodge a vibrating surface, it can certainly dodge a vibrating string. Little of a violin's sound comes directly from its vibrating strings; the air simply skirts around them. Instead, the violin creates sound with its top plate or belly (Fig. 9.2.10). The strings transfer their vibrational motions to the belly, and the belly pushes on the air to create sound.

Most of this vibrational energy flows into the belly through the violin's bridge, which holds the strings away from the violin's body (Fig. 9.2.11). Beneath the G_3 string side of the bridge is the bass bar, a long wooden strip that stiffens the belly. Beneath the E_5 side of the bridge is the sound post, a shaft that extends from the violin's belly to its back.

As a bowed string vibrates across the violin's belly, it exerts a torque on the bridge about the sound post. The bridge rotates back and forth, causing the bass bar and belly to move in and out. The belly's motion produces most of the violin's sound. Some of this sound comes directly from the belly's outer surface, and the rest comes from its inner surface and must emerge through its *f*-shaped holes.

While all violin bodies project sound, some exhibit resonances of their own that complicate a violinist's job. If a violin body doesn't respond equally to different pitches, the violinist must compensate by adjusting the volume for every pitch. It takes a great violinist to make a simple violin sound good. A great violin, such as the fabled instruments built near the beginning of the 18th century by Antonio Stradivarius (1644–1737) and his colleagues in Cremona, Italy (Fig. 9.2.12), simplifies the violinist's task by responding with equal brilliance over the instrument's entire range.

Fig. 9.2.12 Although this violin was built 300 years ago by Antonio Stradivarius for the Royal Court of Spain, it looks much like a modern instrument. That's because Stradivarius and his colleagues in Cremona, Italy so nearly perfected the instrument that few significant improvements have been developed since their time.

An organ pipe doesn't have to produce sound because that sound already exists. In effect, the pipe's vibrating column of air is a standing sound wave that gradually leaks out of the pipe as a traveling one. Trapped sound is escaping from its container.

This conversion of a standing wave into a traveling wave isn't so remarkable because the two types of waves are closely related. The pipe's standing wave can be thought of as a reflected traveling wave, a traveling wave that's bouncing back and forth between the two ends of the pipe. Because of the reflections, the traveling wave is superposed with itself heading in the opposite direction and the sum of two equal but oppositely directed traveling waves *is* a standing wave!

The fact that sound reflects from the open end of an organ pipe is rather surprising. If that end were closed, you'd probably expect a reflection. After all, sound echos from cliffs and other rigid surfaces. But sound partially reflects from a surprising range of other transitions, including the transition from inside a pipe to outside it. If you don't believe that, clap your hands inside a long pipe and listen for the decaying echos.

The reflections at the organ pipe's open ends aren't perfect, so the trapped sound wave gradually leaks out and becomes the sound you hear. This process of letting a standing sound wave emerge slowly as a traveling wave is typical of woodwind and brass instruments. The reflection at an open pipe end depends on the shape of that end. Flaring it into the horn shape common in brass instruments reduces the reflection and eases the transition from standing wave to traveling wave. That's why horns project sound so well.

Music Boxes

The heart of a music box (Fig. 9.2.13) is its comb of metal teeth. Each tooth vibrates at a particular pitch, which it maintains even as its volume decays away. The tooth is able to hold that constant pitch because, like most musical instruments, it's a harmonic oscillator. Equivalent to an object bouncing on a leaf spring, the tooth bends back and forth rhythmically with a pitch determined only by its stiffness and its inertia, and not by the amplitude of its motion. The tooth's stiffness acts to straighten it, while its inertia at first delays that straightening and then causes it to overshoot.

Since shorter teeth are stiffer and have less inertia, they also have higher pitches. To avoid having large variations in the lengths of their teeth, most modern music boxes thicken the ends of the low-pitched teeth to increase their masses without increasing their stiffnesses. The back of the comb is hard to view without disassembly, but the back side clearly shows this thickening (Fig. 9.2.14).

Fig. 9.2.13 This music box has a comb of 18 musical teeth that vibrate at different pitches when they're plucked by pins on the rotating metal drum. A mainspring in the upper left stores the energy that powers this device and the fan-fly in the lower left controls the tempo with which the music box plays its tune. Because its box has been removed, this music box projects relatively little sound.

Fig. 9.2.14 The back side of the metal comb reveals that the low-pitched teeth are thicker at their ends than the high-pitched teeth. That added mass is primarily responsible for the lower pitches, since increasing the inertial aspect of a harmonic oscillator without altering the stiffness of its restoring forces reduces the oscillator's vibrational frequency.

Fig. 9.2.15 The music box's spinning paddle or "fan-fly" uses air drag to set the tempo of the tune. Since the force of pressure drag increases rapidly with speed, this fan-fly turns only fast enough to balance the forces exerted on it by the mainspring.

Pins on the turning music cylinder pluck the teeth to make them vibrate. But the teeth alone don't emit much sound. The teeth are so thin that the air simply flows around them as they vibrate and they have little effect on its density. To project sound, a complete music box couples its vibrating teeth to a surface. Since air can't easily flow around a surface, a surface is much more effective at producing sound.

The energy needed to generate the sound is stored in a large coil spring. You wind this spring with a key and it gradually unwinds as the music plays. To wind the spring, you push each key handle in the direction of its motion and thereby do mechanical work on it—you convey energy to the key and ultimately to the spring. As it unwinds, the spring does work on the gears, and the gears then do work on the music cylinder and the teeth. Energy from your body is thereby becoming musical sound energy.

The spring alone, however, can't control the tempo of the music. Without additional help, the spring would unwind so quickly that the music box would play its entire piece in a fraction of a second. To prolong the piece and slow the release of energy from the spring, the music box spins a small paddle called the fan-fly in the open air (Fig. 9.2.15). This fan-fly experiences air resistance that opposes its rotation. What makes air resistance so useful here is that it is speed dependent—the faster the fan-fly spins, the stronger the air resistance it experiences. The fan-fly's speed increases only until the forward push from the spring is balanced by the backward push of air resistance. From then on, the fan-fly and music cylinder turn steadily until the spring has nearly run out of stored energy.

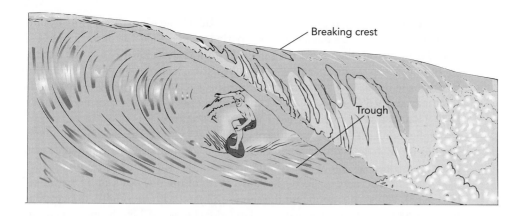

Breaking crest

Trough

SECTION 9.3 **The Sea**

The sea is never still. If you've visited the seashore, you've probably noticed two of the sea's most important motions: tides and surface waves. In this section, we'll examine the cycle of tides and look at how surface waves travel across water. These water waves can help us understand other wave phenomena, including the electromagnetic waves that are responsible for light and the density waves that are the basis for sound.

The Tides

If you watch the sea for a few days, you'll notice the tides. In a cycle as old as the oceans themselves, the water level rises for about 6¼ hours to reach high tide, drops for about 6¼ hours to arrive at low tide, and then begins rising again. Once a wonderful mystery, we now know that the tides are caused by the earth's rotation, the moon's gravity and, to a lesser extent, the gravity of the sun.

On earth, the moon's gravity is so weak that we normally don't notice it. The moon is far away and, as we learned in Section 4.2, gravity diminishes with distance. But this dependence on distance also means that the moon's gravity is stronger on one side of the earth than the other; you experience a stronger pull when you're on the side of the earth nearest the moon than you do when you're on the side opposite it (Fig. 9.3.1). While you can't feel these variations in moon gravity yourself, the earth's oceans respond to them. The oceans are deformed by the moon's gravity (Fig. 9.3.2), and this deformation produces the tides.

The differences between the moon's gravity at particular locations on earth and its average strength for the entire earth give rise to tidal forces—residual gravitational forces that act to displace those locations relative to the earth as a whole. The near side of the earth is pulled toward the moon more strongly than average, so it experiences a tidal force toward the moon. The far side of the earth is pulled less strongly than average, so it experiences a tidal force away from the moon.

If the earth were less rigid, these tidal forces would stretch it into an egg shape. The near side of the earth would bulge outward toward the moon, while the far side of the earth would bulge outward away from the moon. But while the earth itself is too stiff to deform much, the oceans are not, and they bulge outward in response to the tidal forces. Two tidal bulges appear: one closest to the moon and one farthest from the moon (Fig. 9.3.2). The near-side bulge develops because the water there tries to fall toward the moon faster than the earth as a whole. The far side bulge forms because water there tries to fall toward the moon slower than the earth as a whole. A beach located in one of these tidal bulges experiences high tide, while one in the ring of ocean between the bulges experiences low tide.

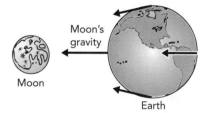

Fig. 9.3.1 The moon's gravity varies over the surface of the earth. The closer a point is to the moon, the stronger the gravity it experiences. This variation in the moon's gravity produces the tides.

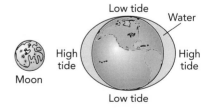

Fig. 9.3.2 Variations in the moon's gravity produce tidal forces on the earth's surface and cause the earth's oceans to bulge outward in two places. These bulges, which are located nearest and farthest from the moon, move over the earth's surface as it rotates.

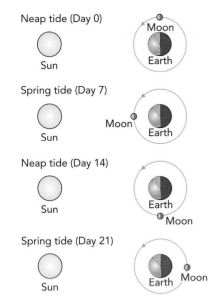

Fig. 9.3.3 The tides vary over a lunar month. They're strongest when the sun and the moon are aligned (spring tides) and weakest when they are at 90° from one another (neap tides).

As the earth rotates, the locations of the two tidal bulges move westward around the equator. Since a particular beach experiences high tide whenever it's closest to or farthest from the moon, the full cycle from high to low to high tide occurs about once every 12 hours and 24.4 minutes. The extra 24.4 minutes reflects the fact that the moon isn't stationary; it orbits the earth and completes a lunar solar month every 29.53 days (see p. 114). The moon thus passes overhead once every 24 hours and 48.8 minutes, rather than every 24 hours.

But the moon isn't the only source of tidal forces on the earth's oceans. Although the sun is much farther away than the moon, it's so massive that the tidal forces it exerts are almost half as large as those exerted by the moon. The sun's principal effect is to increase or decrease the strength of the tides caused by the moon (Fig. 9.3.3). When the moon and the sun are aligned with one another, their tidal forces add together and produce extra large tidal bulges. When the moon and the sun are at right angles to one another, their tidal forces partially cancel and produce tidal bulges that are unusually small.

Twice each lunar month, the time it takes for the moon to orbit the earth, the tides are particularly strong. These spring tides occur whenever the moon and sun are aligned with one another (full moon and new moon). Twice each lunar month the tides are particularly weak. These neap tides occur whenever the moon and sun are at right angles to one another (half moon).

Because of this interplay between lunar and solar tidal effects, the cycle of tides varies slightly from day to day. While the average cycle is 12 hours and 24.4 minutes, it fluctuates over a lunar month. Moreover, the exact moment of high or low tide at a particular location is influenced by water's inertia, the earth's rotation, and the environment through which water must flow to form the tidal bulge. That's why shore areas often publish tables of the local tides.

Tidal Resonances

The sizes of the tides depend on where you are, but high tide is typically a meter or two above low tide. Because the tidal bulges are located near the equator, tides far to the north or south are smaller than that; tides in isolated lakes or seas are smaller still because water can't flow in to create the bulges. However, there are a few special places that have enormous tides. For example, tides in the Bay of Fundy, an estuary between New Brunswick and Nova Scotia, can change the water level by as much as 15 m. How can tides ever get this large?

Giant tides result from natural resonances in channels and estuaries. Just as air in an organ pipe can be made to vibrate strongly by a series of carefully timed pushes from a pump, water in a channel or estuary can be made to oscillate strongly by a series of carefully timed pushes from the tides. The water in that channel is another extended object with a stable equilibrium, so it will oscillate about its equilibrium after being disturbed. Giant tides occur through resonant energy transfer when the cycle of the tides gradually builds up the amplitude of a suitable standing wave in a channel.

However, while standing waves in an organ pipe involve the column of air as a whole, standing waves in a channel involve primarily the water's open surface. That water is at equilibrium when its surface is smooth and horizontal, and it experiences springlike restoring forces whenever its surface is disturbed. For the broad waves we're considering in this section, those restoring forces are due to gravity, and they are known as gravity waves. For the miniature waves in a drinking glass, however, water's springy, elastic surface contributes significantly to the restoring forces. Waves that involve this surface tension are known as capillary waves.

You can observe standing gravity waves on the surface of water in a large basin. If you begin pushing the water back and forth rhythmically with your hand, you'll produce waves on its surface. And if you time your pushes carefully so that they're synchronized with the natural rhythm of a particular standing wave, you'll build up the amplitude of that wave by resonant energy transfer. Countless children have discovered this phenomenon as a way to amuse themselves at bath time—to the dismay of their parents and the delight of people who repair water-damaged floors and ceilings.

Like standing waves on violin strings and drumheads, standing surface waves on water are transverse—the water's vertical surface vibrations are perpendicular to the horizontal waves themselves. In a basin the fundamental vibrational mode (Fig. 9.3.4) has a node along the middle of the basin and antinodes at either end. At one antinode water arcs up to a crest—a maximum upward displacement from equilibrium. At the other antinodes water arcs down to a trough—a maximum downward displacement from equilibrium. Over time the crest drops to become a trough while the trough rises to become a crest. That process reverses and then repeats, over and over again until the sloshing water turns all of its vibrational energy into thermal energy or transfers it elsewhere.

The giant tides at the end of an estuary are simply the antinode of a standing wave fluctuating up and down between a crest and a trough. While the standing waves in an ordinary washbasin have periods measured in seconds or less, large bodies of water can sustain standing waves known as seiches that have periods of minutes or even hours.

Water in the Bay of Fundy has a fundamental seiche mode with a period of roughly 13.3 hours. Since this period nearly matches the 12.5 hour cycle of the tides, there's a resonant transfer of energy from the moon to the water oscillating in the estuary. The tides drive water back and forth in this estuary until, after many cycles, that water is moving so strongly that its height varies dramatically with time.

Traveling Waves on the Surface of Water

As you sit at the seashore on a cloudless day, enjoying a warm, steady breeze, you can't help but notice that the sea in front of you is covered with ridges. These ridges move steadily toward land and finally crash on the beach. Though it's customary to think of each breaking swell as a separate wave, we'll find it useful to view the entire moving pattern of evenly spaced ridges as a single wave—a traveling surface wave on water.

Traveling surface waves are the basic modes of oscillation on the open ocean, the simplest waves on that effectively *limitless* surface. Despite their steady progress across the water, these traveling waves actually involve oscillation. You can see this oscillation by watching a fixed point on the water's surface. That point fluctuates up and down as the crests and troughs of a traveling wave pass through it. The period of this oscillation is the time required for one full cycle of rise and fall, and the frequency is the number of crests passing through that fixed point each second.

The ocean's surface can host an incredible variety of traveling waves, each moving in its own direction with its own period and frequency. Moreover, these basic waves can coexist, adding together on the ocean's surface to create ever more complicated patterns. Like primary colors, which when blended in proper proportions can produce any possible color, traveling waves can be superposed in proper proportions to produce any possible surface pattern or wave. When the ocean is rough and its surface features ripples layered upon swells layered upon broad undulations, you're seeing this superposition of "primary" traveling waves in all its glory.

In contrast, the basic modes of oscillation on a channel or lake are standing surface waves—the simplest waves on that *limited* surface. In a standing wave the water's surface oscillates up and down vertically, with its crests and troughs interchanging periodically: crests become troughs and troughs become crests. The standing

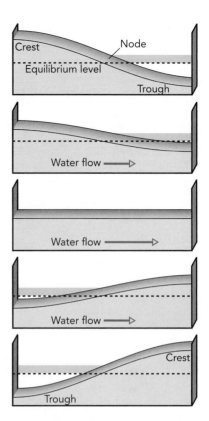

Fig. 9.3.4 Water sloshing in a basin in its fundamental mode. Its surface oscillates up and down, its crest becoming a trough and vice versa, over and over again.

wave's pattern of crests and troughs doesn't move anywhere; it simply flips up and down in place at a certain frequency.

On their limited surface, these standing waves can be superposed to produce any possible surface pattern or wave, so they, too, are like primary colors. Overall, traveling waves constitute the primary palette of waves for a limitless ocean, and standing waves constitute the primary palette of waves for a limited channel or lake.

Standing and Traveling Waves

The most basic waves on an extended object of limited dimensions are standing waves. With their different periods and/or patterns, these standing waves can be superposed to form any possible wave on that limited object.

The most basic waves on an extended object of limitless dimensions are traveling waves. With their different periods and/or directions of travel, these traveling waves can be superposed to form any possible wave on that limitless object.

Actually, we've seen these ideas before in the context of musical instruments and sound. Since instruments are limited objects, their basic vibrations are standing waves—the fundamental and overtone vibrations. And because air is effectively limitless, its basic vibrations are traveling waves—the sound waves it carries. The timbre of an instrument reveals the superposition of its many standing waves, while the full sound of an orchestra or band displays the superposition of its many traveling waves.

Both standing and traveling water surface waves carry energy, energy which they typically obtain from the wind, the tide, or occasionally seismic activity. Each wave's energy consists of kinetic energy in moving water and gravitational potential energy in water that has been displaced from level.

In a standing wave the energy of the entire wave fluctuates back and forth between kinetic and gravitational potential; the wave's kinetic energy peaks as the surface rushes through its level equilibrium, and its potential energy peaks as the surface stops to turn around at its maximum displacement from equilibrium.

A traveling wave's crests and troughs move steadily forward, so the water is never level or motionless. The energy in a traveling wave is therefore always an even mixture of kinetic and potential energies. And because of its directed motion, a traveling wave carries momentum pointing in the same direction as the wave velocity.

The Structure of a Water Wave

We've seen that a traveling surface wave moves across the open ocean with a certain velocity, wavelength, and frequency. But what is the water itself doing as the wave passes?

You can begin to answer that question by watching a bottle floating on the water's surface (Fig. 9.3.5). As a wave passes it, that bottle rises and falls with the crests and troughs, but it makes no overall progress in any direction. Instead, the bottle travels in a circle. Like the bottle, the water itself doesn't actually move along with the passing wave. Although this water bunches up to create each crest and spreads out to create each trough, it returns to its starting point once the wave has left.

Like the bottle in Fig. 9.3.5, a patch of water on the ocean's surface moves in a circular pattern (Fig. 9.3.6) as a traveling wave passes. Water that starts out on top of a crest moves down and forward as the crest departs. It moves down and backward as the trough arrives, then up and backward as the trough departs, and finally up and forward as the next crest arrives. When it reaches the top of the arriving crest, this water is back where it started on the ocean's surface. Which way the water circles depends on the wave velocity's direction—the wave's direction

Wave (crest) velocity ⟶▷

Fig. 9.3.5 You can see that water doesn't move with a wave by watching a bottle floating on the water. The bottle moves in a circle as each crest passes by. Starting at (*a*) a crest, the bottle moves (*b*) down and right, (*c*) down and left, (*d*) up and left, and then (*e*) up and right. It returns to its original position just as the next crest arrives.

of travel. The water at the top of a crest always moves in the same direction as the wave itself.

It isn't only the surface water that moves; water beneath the surface also circles. However, the circles diminish in radius gradually with depth and become negligible at a depth roughly equal to half the wave's wavelength. Thus while it's called a *surface* wave, it has a *depth* to it and a sensitivity to shallow water, as we'll soon see.

These surface waves have another interesting characteristic: their wave speed increases with wavelength. As you may have noticed, long-wavelength swells travel faster than short-wavelength ripples. That's quite different from sound waves, which all have the same wave speed regardless of wavelength.

Such dependence of wave speed on wavelength is known as dispersion. It occurs in this case because water's surface is surprisingly stiff when carrying long-wavelength waves. Unlike a tense string, which opposes short-wavelength distortions much more stiffly than long-wavelength ones, water's surface uses its weight to oppose long-wavelength disturbances almost as vigorously as it opposes short-wavelength ones. That heightened stiffness for long-wavelength waves boosts their frequencies and therefore increases their wave speeds (see Eq. 9.2.1).

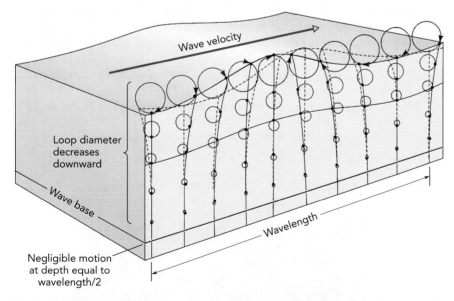

Fig. 9.3.6 Surface water moves in a circular motion as a wave passes. The water currently located at each dark dot will follow the circular path outlined around it as time passes. The circles are largest at the surface and become relatively insignificant once you look more than half a wavelength below the surface. The sense of the circular motion (clockwise or counterclockwise) determines the direction in which the wave travels. This wave travels toward the right.

❶ The Indian Ocean tsunami of December 26, 2004 was initiated by a sudden rise in the ocean floor off the northern coast of Sumatra. This long-wavelength traveling wave moved so quickly through the surrounding Indian Ocean that most shore inhabitants received no warning and were caught unprepared. Moreover, the wave's phenomenal troughs exposed vast stretches of seabed, attracting inquisitive people offshore where they were then unprotected from the destructive crests that followed. Roughly a quarter-million people perished.

Little ripples have short wavelengths and travel slowly, while large ocean swells have long wavelengths and travel much more rapidly. Giant waves produced by earthquakes and volcanic eruptions, known as tsunamis, have extremely long wavelengths and can travel at hundreds of kilometers-per-hour. Because these giant waves travel so fast and move water so deep in the ocean, they carry enormous amounts of energy and momentum and are potentially disastrous to shore areas ❶.

Waves at the Shore

As a wave approaches the shore, it begins to travel through shallow water. Since the water's circular motion extends below the surface, there comes a point at which the wave begins to encounter the seabed. Once the water is shallower than half the wavelength of the wave, the seabed distorts the water's circular motion so that it becomes elliptical.

That change has a number of interesting effects on the wave. First, its wave speed gradually decreases so that the crests begin to bunch together. Second, its amplitude—the height of its crests and depth of its troughs—increases as the wave acts to keep its overall forward momentum constant despite its decrease in speed. These two effects explain why waves that look broad and gradual on the open ocean look quite steep and dangerous nearer the beach. Their crests have bunched together and grown taller, so the slopes between crests and troughs really have become steeper.

A third effect of the shallowing water is a gradual change in the wave's direction of travel. Known as refraction, this bending occurs whenever a wave's speed changes as it passes from one environment to another. Since a water surface wave slows as it approaches the shore, that wave refracts—it bends—so as to head more directly toward the shore (Fig. 9.3.7). Because of refraction, waves approach the beach almost perpendicular to it, even if they were traveling at relatively oblique angles far from shore.

A fourth and final effect is the destruction of the wave—it eventually runs out of water and crashes onto the beach. The wave builds each of its crests using local water. When the crest enters very shallow water, there isn't enough water in front of it to construct its forward side. The crest becomes incomplete and begins to "break."

The form of its demise depends on the slope of the seabed. If that slope is gradual, the wave breaks slowly to form a smooth, rolling, "boiling" surf (Fig. 9.3.8). But if the slope of the seabed is steep, the wave breaks quickly by having the top of its

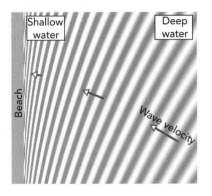

Fig. 9.3.7 When a traveling water surface wave encounters shallow water, it slows down, and its direction of travel changes. This refraction process bends the wave velocity so that it points more directly toward the beach.

Fig. 9.3.8 When the slope of the sea bottom is very gradual, the wave crests crumble gently into rolling surf.

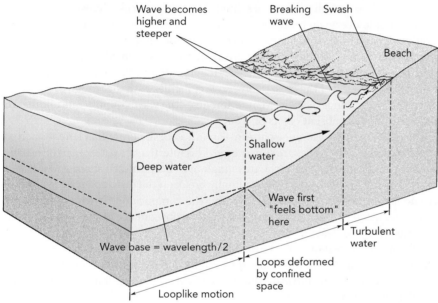

Fig. 9.3.9 When the water becomes too shallow to form a complete crest for the wave, the wave "breaks." If the slope of the shore is steep enough, the crest will be quite incomplete on its shore side and will plunge forward over the trough in front of it.

crest plunge forward over the trough in front of it (Fig. 9.3.9). The steep slope essentially prevents the forward half of the crest from forming. The rearward half continues through its normal circular motion and dives over the missing half-crest in front of it.

However, the wave can avoid this violent end by colliding with a seawall or cliff instead of a beach. Rather than breaking, the wave will then bounce off the wall and continue on in a new direction. Known as reflection, this bouncing effect occurs whenever certain dynamical properties of a wave, particularly its speed, change abruptly as it passes from one environment to another. Reflection is a surprisingly general phenomenon, affecting light waves just as much as water waves. In this case, the water surface wave would have to change so radically to enter the seawall or cliff that it instead reflects almost perfectly. But even less severe changes in environment can make the wave reflect, if only partially. Thus when a water surface wave passes over a sandbar or coral reef and its wave speed changes, it may experience both reflection and refraction. These effects contribute to the complicated dynamics of waves near the shore.

The Rhythm of the Surf: Wave Interference

If the ocean were carrying only one pure traveling wave toward shore, every breaking swell would look and sound the same. However, there is often a complicated rhythm to the crashing surf; its volume fluctuates with an overall pattern known as surf beat. Surf beat is a sign that the ocean's surface is a busy place: it's actually carrying more than one traveling wave at a time and these various waves all contribute to the surf.

To understand how multiple traveling waves produce surf beat, let's consider a simple case. Suppose that two traveling waves are heading toward shore and that they have equal amplitudes but different wavelengths (Fig. 9.3.10). Such a situation can easily arise when winds over two portions of the ocean produce two different traveling waves that later overlap. Since they are sharing the ocean's surface, they are superposed on one another.

Because these traveling waves are different, their patterns of crests and troughs can't coincide everywhere. Instead, these waves experience interference—their overlying patterns enhance one another at some locations and cancel one another at other locations. Wherever their crests or troughs coincide, they experience constructive interference—the waves act together to produce enormous crests or troughs. But wherever the crest of one wave coincides with the trough of the other wave, they experience destructive interference—the waves oppose one another to produce muted or absent crests or troughs.

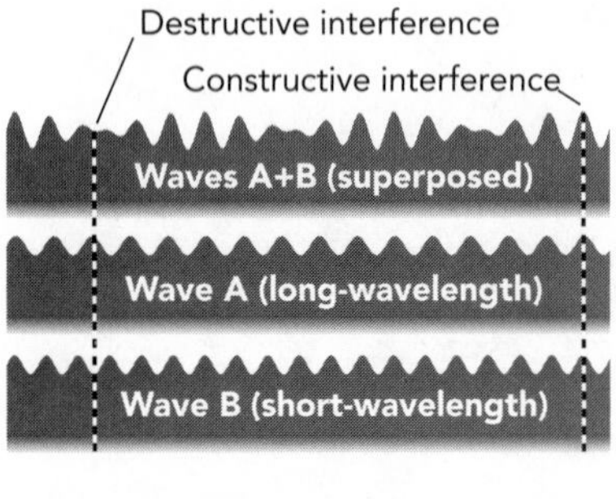

The result is an interference pattern—an intricate structure that spreads across space and time when waves are superposed. This interference pattern on the ocean's surface moves and evolves as the traveling waves head toward shore and it leaves its impression on the crests that eventually break on the beach. Since these crests are no longer equal in height, they exhibit the complicated rhythm of surf beat. When you listen to that beat, you're hearing the consequence of superposition and the interference of waves.

Of course, the real ocean carries many traveling waves, each with its own amplitude, wavelength, and direction of travel. But no matter how complicated the ocean's surface or how intricate the surface beat, you are still just observing the interference of waves.

Fig. 9.3.10 When two traveling waves A and B are superposed on the surface of the ocean, they produce interference pattern A+B. As this moving pattern encounters the shore, its varying crest heights give rise to surf beat.

Summary of Important Wave Phenomena

Reflection: the complete or partial mirror redirection of a wave that occurs when certain dynamic properties of that wave, particularly its speed, change abruptly as it passes from one environment to another.

Refraction: the bending of a wave that occurs when that wave's speed changes as it passes from one environment to another.

Dispersion: the dependence of a wave's speed on its wavelength.

Interference: when two or more waves are superposed, their crests and troughs can enhance one another (constructive interference) or cancel one another (destructive interference) and produce an interference pattern.

CHAPTER 10

ELECTRIC THINGS

While it's hard to see the electric charges that are responsible for electricity, it's easy to see their effects. They're all around us, in the sparks and shocks of a cold winter day, the imaging process of a xerographic copier, and the illumination of a flashlight when you turn on its switch. Although we often take electricity for granted, it clearly underlies many aspects of our modern world.

Just imagine what life would be like if there were no electric charges and no electricity. For starters, we'd probably be sitting around campfires at night, trying to think of things to do without television, cell phones, or computer games. But before you remark on just how peaceful such a pre-electronic-age existence would be, let me add one more sobering thought: we wouldn't exist either. Whether it's motionless as static charge or moving as electric current, electricity really does make the world go 'round.

The abstract nature of electric charge and our inability to sense it directly complicate my task in this chapter. I hope that you'll recognize many of the situations I describe and find them helpful in understanding electricity. I also hope that you'll try to observe still other effects of electricity as I point them out.

In fact, I've always loved electricity but I can't recommend that you follow the intensely hands-on approach I took while learning about it. Though I had a number of great successes among my childhood electrical experiments, most notably a 200,000 volt Tesla coil I built with some help from my father when I was 12, I had many terrifying failures as well. In addition to burning my fingers and hair almost daily, I endured a lifetime's worth of nasty shocks before finishing high school. The folly of youth and all that.

Although this chapter concentrates on electricity and its charges, we'll see in Chapter 11 that magnetism and its poles are closely related. While I'll leave the relationships between electricity and magnetism for that chapter, you may already begin seeing similarities between those two seemingly separate phenomena as you read Chapter 10.

Chapter Itinerary

Section 10.1 **Static Electricity**

Electricity may be difficult to see, but you can easily observe its effects. How often have you found socks clinging to a shirt as you remove them from a hot dryer or struggled to throw away a piece of plastic packaging that just won't leave your hand or stay in the trash can? The forces behind these familiar effects are electric in nature and stem from what we commonly call "static electricity." But static electricity does more than just push things around, as you've probably noticed while reaching for a doorknob or a friend's hand on a cold, dry day. In this section, we'll examine static electricity and the physics behind its intriguing forces and often painful shocks.

Electric Charge and Freshly Laundered Clothes

Unless you have always lived in a damp climate and avoided synthetic materials, you have experienced the effects of static electricity. Seemingly ordinary objects have pushed or pulled on one another mysteriously and you've received shocks while reaching for light switches, car doors, or friends' hands. But static electricity is more than an interesting nuisance; it's a simple window into the inner workings of our universe and worthy of a serious look. It will take some time to lay the groundwork, but soon you'll be able to explain most of the effects of static electricity and even to control it to some extent.

The existence of static electricity has been known for several thousand years. About 600 B.C., the Greek philosopher Thales of Miletus (ca 624–546 B.C.) observed that when amber is rubbed vigorously with fur, it attracts light objects such as straw and feathers. Known in Greek as *elektron*, amber is a fossil tree resin with properties similar to those of modern plastics. The term "static electricity," like many others in this chapter, derives from that Greek root.

Static electricity begins with electric charge, an intrinsic property of matter. Electric charge is present in many of the subatomic particles from which matter is constructed, and these particles incorporate their charges into nearly everything. No one knows why charge exists; it's simply one of the basic features of our universe and something that people discovered through observation and experiment. Because electric charge has so much influence on objects that contain it, we'll sometimes refer to those objects as electric charges or simply as charges.

Charges exert forces on one another and it is these forces that you observe with static electricity. Next time you're doing laundry, experiment with your clothes as they come out of the dryer. You'll find that some electrically charged garments attract one another while others repel. Evidently, there are two different types of charge. But while this dichotomy has been known since 1733, when it was discovered by French chemist Charles-François de Cisternay du Fay (1698–1739), it was Benjamin Franklin ❶ who finally gave the two charges their present names. Franklin called what appears on glass when it's rubbed with silk "positive charge" and what appears on hard rubber when it's rubbed with animal fur "negative charge."

Two like charges (both positive or both negative) push apart, each experiencing a repulsive force that pushes it directly away from the other (Fig. 10.1.1a,b). Two opposite charges (one positive and one negative) pull together, each experiencing an attractive force that pulls it directly toward the other (Fig. 10.1.1c). These forces between stationary electric charges are called electrostatic forces.

When you find that two freshly laundered socks push apart, it's because they both have the same type of charge. Whether that charge is positive or negative depends on the fabrics involved (more on that later), so let's just suppose that the dryer has given each sock a negative charge. Since like charges repel, the socks push apart. But what does it mean for the dryer to give each sock a negative charge?

The answer to that question has several parts. First, the dryer didn't create the negative charge that it gave to a sock. Like momentum, angular momentum, and energy, electric charge is a conserved physical quantity—it cannot be created or destroyed, only transferred. The negative charge that the dryer gave to the sock must have come from something else, perhaps a shirt.

Second, positive charge and negative charge aren't actually separate entities—they're just positive and negative amounts of the same physical quantity: electric charge. Positive charges have positive amounts of electric charge, while negative charges have negative amounts. Like most physical quantities, we measure charge in standard units. The SI unit of electric charge is the coulomb (abbreviated C). Small objects rarely have a whole coulomb of charge and your sock's charge is only about −0.0000001 C.

Third, the sock's negative charge refers to the sock as a whole, not to its internal pieces. As with all ordinary matter, the sock contains an enormous number of positively and negatively charged particles. Each of the sock's atoms consists of a dense central core or nucleus, containing positively charged protons and uncharged neutrons, surrounded by a diffuse cloud of negatively charged electrons. The electrostatic forces between those tiny charged particles hold together not only the atoms, but also the entire sock. However, in giving the sock a negative charge, the dryer saw to it that the sock's net electric charge—the sum of all its positive and negative amounts of charge—is negative. With its negative net charge, the sock behaves much like a simple, negatively charged object.

Finally, the sock became negatively charged when it contained more electrons than protons. Underlying that seemingly simple statement is a great deal of painstaking scientific study. To begin with, experiments have shown that electric charge is quantized: charge always appears in integer multiples of the elementary unit of electric charge. This elementary unit of charge is extremely small, only about 1.6×10^{-19} C, and is the magnitude of the charge found on most subatomic parti-

❶ Although best remembered for his political activities, American statesman and philosopher Benjamin Franklin (1706–1790) was also the preeminent scientist in the American colonies during the mid-1700s. His experiments, both at home and in Europe, contributed significantly to the understanding of electricity and electric charge. In addition to demonstrating that lightning is a form of electric discharge, Franklin invented a number of useful devices, including the Franklin stove, lightning rods, and bifocal eyeglasses.

(a)

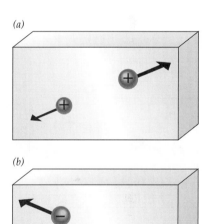

(b)

(c)

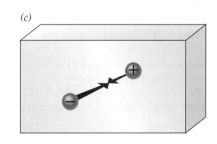

Fig. 10.1.1 (a) Two positive charges experience equal but oppositely directed forces exactly away from one another. (b) The same effect occurs for two negative charges. (c) Two opposite charges experience equal but oppositely directed forces exactly toward one another.

❷ In 1781, after a career as a military engineer in the West Indies, French physicist Charles-Augustin de Coulomb (1736–1806) returned to his native Paris in poor health. There he conducted scientific investigations into the nature of forces between electric charges and published a series of memoirs on the subject between 1785 and 1789. His research came to a close in 1789 when he left Paris because of the French Revolution.

cles. An electron has −1 elementary unit of charge, while a proton has +1 elementary unit of charge. Since the only charged subatomic particles in normal matter are electrons and protons, the sock becomes negatively charged simply by having more electrons than protons.

Returning to the original question, we now know what the dryer did in order to give a sock a negative charge. Assuming the sock started electrically neutral—it had zero net charge—the dryer must have added electrons to the sock or removed protons from the sock or both. These transfers of charge upset the sock's charge balance and gave it a negative net charge.

In keeping with our convention regarding conserved quantities, all unsigned references to charge in this book imply a positive amount. For example, if the dryer gave charge to a jacket, we mean it gave a positive amount of charge to that jacket. We follow this same convention with money: when you say that you gave money to a charity, we assume that you gave a positive amount.

Finally, Franklin's charge-naming scheme was brilliant in concept but unlucky in execution. While it reduced the calculation of net charge to a simple addition problem, it required Franklin to choose which type of charge to call "positive" and which to call "negative." Unfortunately, his seemingly arbitrary choice made electrons, the primary constituents of electric current in wires, negatively charged. By the time physicists had recognized the mistake, it was too late to fix. Scientists and engineers have had to deal with negative amounts of charge flowing through wires ever since. Imagine the awkwardness of having to carry out business using currency printed only in negative denominations!

Coulomb's Law and Static Cling

Although your sock and shirt pull together strongly when they're only inches apart, you can put on your shirt and go to the movies without fear of being attacked by your sock from the other side of town. Evidently, the forces between charges weaken with distance.

Over two centuries ago, French physicist Charles-Augustin de Coulomb ❷ studied electrostatic forces experimentally and determined that the forces between two electric charges are inversely proportional to the square of their separation (Fig. 10.1.2). For example, doubling the separation between your shirt and sock reduces their attraction by a factor of four, which explains your uneventful night out on the town.

Coulomb's experiments also showed that the forces between electric charges are proportional to the amount of each charge. That means that doubling the charge on either your shirt or your sock doubles the force each garment exerts on the other. Finally, changing the sign of either charge turns attractive forces into repulsive ones or vice versa. If both garments were either positively charged or negatively charged, they'd repel instead of attracting.

These ideas can be combined to describe the forces acting on two charges and written as:

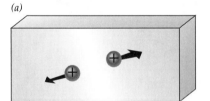

(a)

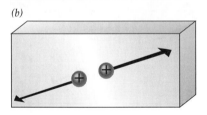

(b)

Fig. 10.1.2 The electrostatic forces between two charges increase dramatically as they become closer. As the distance separating two positive charges decreases by a factor of 2 between (a) and (b), the forces those two charges experience increase by a factor of 4.

$$\text{force} = \frac{\text{Coulomb constant} \cdot \text{charge}_1 \cdot \text{charge}_2}{(\text{distance between charges})^2}. \qquad (10.1.1)$$

The force on charge_1 is directed toward or away from charge_2, and the force on charge_2 is directed toward or away from charge_1.

This relationship is called Coulomb's law, after its discoverer. The Coulomb constant is about 8.988×10^9 N·m²/C² and is one of the physical constants found in nature. Consistent with Newton's third law, the force that charge_1 exerts on charge_2 is equal in amount but oppositely directed to the force that charge_2 exerts on charge_1.

Coulomb's Law

The magnitudes of the electrostatic forces between two objects are equal to the Coulomb constant times the product of their two electric charges divided by the square of the distance separating them. If the charges are like, then the forces are repulsive. If the charges are opposite, then the forces are attractive.

In addition to protecting you from distant charged socks, this relationship between electrostatic forces and distance gives rise to another intriguing feature of laundry static: charged clothes can cling to objects that are electrically neutral! For example, a negatively charged sock can stick to a neutral wall.

The origin of this attraction is a subtle rearrangement of charges within the wall. Even though the wall has zero net charge, it still contains both positively and negatively charged particles. When the negatively charged sock is near the wall, it pulls the wall's positive charges a little closer and pushes the wall's negative charges a little farther away (Fig. 10.1.3). Although each individual charge shifts just a tiny distance, the wall contains so many charges that together they produce a dramatic result. The wall develops an electric polarization—it remains neutral overall, but has a positively charged region nearest the sock and a negatively charged one farthest from the sock.

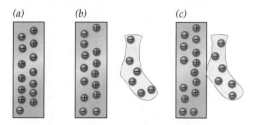

Fig. 10.1.3 (*a*) A neutral wall contains countless positive and negative charges. (*b*) As a negatively charged sock approaches the wall, the positive charges move toward it and the negative charges move away. (*c*) The polarized wall continues to attract the sock and holds it in place.

The wall's positive region attracts the sock while its negative region repels the sock. Though you might expect those two opposing forces to balance, Coulomb's law says otherwise. Since electrostatic forces grow weaker with distance, the sock is attracted more strongly to the nearer positive region than it is repelled by the more distant negative region. Overall, there is a net electrostatic attraction between the charged sock and the polarized wall, so the sock clings to the wall!

Transferring Charge: Sliding Friction or Contact?

While it's clear that the dryer transfers charge between the clothes, why does that charge move and what determines which garments gain charge and which lose it?

You might suppose that sliding friction is responsible for the transfer—that the dryer rubs the clothes together and somehow wipes charge from one garment to the other. After all, friction seems to help you charge a balloon as you rub it through your hair or against a wool sweater. But be careful—there are other cases of charge transfer that don't involve rubbing at all. For example, the plastic wrap you remove from a store package can acquire a charge no matter how careful you are not to rub it against its contents. And an antique car can build up enough charge to give you a nasty shock even when its white rubber tires never skid across the pavement.

Charge transfer is less the result of rubbing than it is of contact between dissimilar surfaces. When two different materials touch one another, a few electrons normally shift from one surface to the other. That transfer results from the chemical differences between the two touching surfaces and the associated change in an electron's potential energy when it shifts. In effect, some surfaces are "hungrier" for electrons than others and whenever two dissimilar surfaces touch, the hungrier surface steals a few electrons from its less hungry partner.

The physics behind this theft has to do with chemical potential energy—energy stored in the chemical forces that bind together a material's constituent atoms and electrons. To hold onto its electrons, a surface reduces their chemical energies to less than zero, meaning that it would take additional energy to free those electrons from the surface. However, some surfaces reduce the electron chemical potential energies more than others and thus bind their electrons more tightly. If an electron on one surface can reduce its chemical potential energy by shifting to the other surface, it will accelerate toward that "hungrier" surface and eventually stick there. You can picture the electron as "rolling downhill" from a chemical "valley" on one surface to an even deeper valley on the other surface.

This transfer of electrons is self-limiting. As electrons accumulate on the lower energy surface, they begin to repel any electrons that try to follow and the transfer process soon grinds to a halt. It stops altogether when the electrons reach equilibrium—when the forward chemical force an electron experiences is exactly balanced by the backward electrostatic force. The transfer won't resume until you bring fresh, uncharged surface regions into contact.

That's where rubbing enters the picture. Rubbing involves lots of surface contact and almost endless opportunities for charge transfer between those surfaces. As clothes tumble about in the dryer, touching one another and often rubbing, some fabrics steal electrons and become negatively charged while other fabrics lose electrons and become positively charged.

That said, you should be aware that the details of contact charging are messy. For starters, the surfaces that actually touch one another are neither chemically pure nor free of microscopic defects. While it's generally true that whichever fabric binds electrons most tightly is the one most likely to develop a negative net charge, surface contamination and defects can change the outcome radically. Even your choice of laundry detergent may affect the fabric's surface chemistry and thus how it charges. Furthermore, water molecules cling to most surfaces and influence the contact charging process. Finally, while we've concentrated on the exchange of electrons, it's also possible for certain surfaces to exchange ions—that is, electrically charged atoms, molecules, or small particles—along with electrons and acquire net charges as a result.

Separating Your Clothes: Producing High Voltages

The dryer stops and you take out your favorite shirt. It has several socks clinging to it, so you begin to remove them. As you separate the garments, they crackle and spark. Their attraction is obviously due to opposite charges, but why does separating them make them spark?

To answer that question, let's think about energy as you pull the negatively charged sock steadily away from the positively charged shirt. Since the sock would accelerate toward the shirt if you let go, you are clearly exerting a force on the sock. And because that force and the sock's movement are in the same direction, you are also doing work on the sock. You are transferring energy to it.

That energy is stored in the electrostatic forces—the shirt and sock accumulate electrostatic potential energy. Electrostatic potential energy is present whenever opposite charges have been pulled apart or like charges have been pushed together. With the negatively charged sock now far from the positively charged shirt, both attraction and repulsion contribute to the electrostatic potential energy: opposite charges are separated on the two garments and like charges are assembled together on each garment.

The total electrostatic potential energy in the shirt and sock is the work you did to separate them. But that potential energy isn't divided equally among the individual charges on these garments. Depending on their locations, some charges have more electrostatic potential energy than others and are therefore more important when it comes to sparks. In recognition of those differences, we need a proper way to characterize the electrostatic potential energy available to a charge at a particular location. The measure we're seeking is voltage—the electrostatic potential energy available per unit of electric charge at a given location.

Voltage is a difficult quantity to conceptualize because you can't see charge or sense its stored energy. To help you understand voltage, we'll use a simple analogy. In this analogy, the role of charge will be played by water and the role of voltage will be played by altitude. Where voltage is high, visualize water far above you. Where voltage is low, picture water at a lesser height. And just as water tends to flow from higher altitude to lower altitude, so charge tends to flow from higher voltage to lower voltage.

This analogy works well because both voltage and altitude measure the energy in a unit of something. Voltage is the electrostatic potential energy per unit of charge and altitude can be construed as the gravitational potential energy per unit of weight. Though thinking of altitude this way may seem strange, both water at high altitude and charge at high voltage are loaded with energy per unit and likely to cause trouble!

Since the SI unit of energy is the joule and the SI unit of electric charge is the coulomb, the SI unit of voltage is the joule-per-coulomb, more commonly called the volt (abbreviated V). Where the voltage is positive, (positive) charge can release electrostatic potential energy by escaping to a distant neutral place. Charge at positive voltage is analogous to water on a hill, which can release gravitational potential energy by flowing down to a distant level place. Where the voltage is negative, charge needs energy to escape to a distant neutral place. Charge at negative voltage is analogous to water in a valley, which needs energy in order to flow up to a distant level place.

As you can see, this voltage–altitude analogy is quite a boon. But while you should find it helpful now and throughout this book, please remember that the ups and down in altitude that you're using to visualize voltage occur only in your mind's eye and not in the real world. Your clothes don't necessary move up or down as their voltages change!

Returning to those clothes, you'll find that each point on the shirt or sock has its own voltage. You can determine that voltage by taking a tiny amount of (positive) charge at that point—a "test charge"—and moving it to a distant neutral place. The point's voltage is simply the electrostatic potential energy the test charge releases during that trip divided by the amount of its charge. If the point you examine is on the positively charged shirt, you'll obtain a large positive voltage—probably several thousand volts. If it's on the negatively charged sock, you'll obtain a negative voltage of similar magnitude. Whether positive or negative, these high voltages tend to cause sparks.

We'll look at the physics of sparks and discharges soon, but you can already see why oppositely charged clothes spark as you separate them: that's when the high voltages develop. As long as your sock is clinging tightly to your shirt, there isn't much electrostatic potential energy available. But as soon as you begin to separate them, watch out!

Accumulating Huge Static Charges

We've seen that touching two different materials together causes a small transfer of charge from one surface to the other and that separating those oppositely charged surfaces produces elevated voltages and perhaps sparks. However, the quiet crackling and snapping of items in your laundry basket is nothing compared to the miniature lightning bolts you can unleash after walking across a carpet on a dry winter day, stepping out of an antique car, or playing with a static generator. To get a really big spark, you need to separate lots of charge and that usually requires repeated effort.

Walking across a carpet is just such a repetitive process. Each time your rubber-soled shoe lands on an acrylic carpet, some (positive) charge shifts from the carpet to your shoe. Although the transfer is brief and self-limiting, you now have a little extra charge on your shoe. When you lift that shoe off the carpet, you do work on its newfound charge and your shoe's voltage surges to a high positive value. High voltage charge tends to leak from one place to another and the shoe's charge quickly spreads to the rest of your body. By the time your foot lands again on a fresh patch of carpet, the shoe has given away most of its charge and is ready to begin the process all over again.

Each time your foot lands on the carpet, it picks up some charge. And each time it lifts off the carpet, that charge spreads out on your body. By the time you finally reach for the doorknob, you are covered with charge and have an enormous positive voltage. As your hand draws close to the doorknob, it begins to influence the doorknob's charges—pulling the doorknob's negative charges closer and pushing its positive charges away. You are polarizing the doorknob.

As we saw while separating your freshly laundered sock from your shirt, oppositely charged objects that are close but not touching can have both large electrostatic potential energies and strong electrostatic forces. That's the situation here. The closer your hand gets to the doorknob, the stronger the electrostatic forces become until finally the air itself cannot tolerate the forces and a spark forms. In an instant, most of your accumulated electrostatic potential energy is released as light, heat, and sound. And that doesn't include any screams.

But as good as walking is at building up charge, an antique car is even better. Its pale rubber tires gather negative charge when they touch the pavement and develop large negative voltages as they roll away from it. This charge migrates onto the car body so that after a few seconds of driving, the car accumulates enough charge to give anyone who touches it a painful shock. Collecting tolls used to be hazardous work! Fortunately, modern tires are formulated to allow this negative charge to return safely to the pavement, so that cars rarely accumulate much charge. Instead, most shocks associated with cars now come from sliding across the seat as you step in or out.

While cars try to avoid static charging, there are machines that deliberately accumulate separated charge to produce extraordinarily high voltages. The most famous of these static machines is the Van de Graaff generator (Fig. 10.1.4). It uses a rubber belt to lift positive or negative charges onto a metal sphere until the magnitude of that sphere's voltage reaches hundreds of thousands or even millions of volts.

A typical classroom Van de Graaff uses a motor-driven rubber belt to carry negative charges from its base to its spherical metal top. Once inside the sphere, the belt's negative charges flow outward onto the sphere's surface, where they can be as far apart as possible. There they remain until something releases them.

Suspended at the top of a tall, insulating column, the Van de Graaff generator's sphere can accumulate an enormous negative charge. You may hear the motor struggling as it pushes the belt's negative charges up to the sphere, a reflection of how much negative voltage the sphere is developing. Eventually it releases its negative charge via an immense spark.

But even without sparks, the Van de Graaff is an interesting novelty. If you isolate yourself from the ground and touch the metal sphere while it's accumulating negative charges, some of those negative charges will spread onto you as well. If your hair is long and flexible, and permits the negative charges to distribute themselves along its length, it may stand up, lifted by the fierce repulsions between those like charges.

Fig. 10.1.4 Static electricity can be produced by mechanical processes. In this Van de Graaff generator, a moving rubber belt transfers negative charges from the base to the shiny metal sphere. This negative charge creates dramatic sparks as it returns through the air toward the positive charge it left behind.

Controlling Static Electricity: Fabric Softeners and Conditioners

Now that we've seen what static electricity is and how to produce it, we're ready to see how to tame it. Static cling, flyaway hair, and electrifying handshakes aren't everyone's cup of tea. The basic solution to static charge is mobility: if charges can move freely, they'll eliminate static electricity all by themselves. Opposite charges attract, so any separated positive and negative charges will join up as soon as they're allowed to move.

Materials such as metals that permit free charge movement are called electrical conductors. Those such as plastic, hair, and rubber that prevent free charge movement are called electrical insulators. Since charge movement eliminates static electricity, our troubles with static electricity stem mostly from insulators. If you wore metal clothing, you wouldn't have static problems with your laundry.

The simplest way to reduce static electricity is to turn the insulators into conductors. Even slight conductors, ones that just barely let charges move, will gradually get rid of any accumulations of separated charge. That's one of the main goals of fabric softeners, dryer sheets, and hair conditioners. They all turn insulating materials—fabrics and hair—into slight electrical conductors. The result is the near disappearance of static electricity and all its fashion inconveniences.

How these three items work is an interesting tale. They all employ roughly the same chemical: a positively charged detergent molecule. A detergent molecule is a long molecule that is electrically charged at one end and electrically neutral at the other end. Its charged end clings electrostatically to opposite charges and is chemically "at home" in water. Its neutral end is oil-like, slippery, and "at home" in oils and greases. This dual citizenship is what makes detergents so good for cleaning.

But while it might seem that positively and negatively charged detergent molecules would clean equally well, that's not the case. Since cleaning agents shouldn't cling to the materials they're cleaning, it's important that the two not have opposite charges. Fabrics and hair generally become negatively charged when wet—another example of a charge shift when two different materials touch—so negatively charged detergent molecules clean much better than positively charged ones.

Positively charged detergents are still useful, however, although you mustn't apply them until after you've cleaned your clothes or hair. Because they cling so well to wet fibers, these slippery detergent molecules will remain in place long after washing and give fabrics and hair a soft, silky feel. And they'll allow those materials to conduct electricity, albeit poorly, so as to virtually eliminate static electricity!

This conductivity is due principally to their tendency to attract moisture. Water is a slight electrical conductor and damp surfaces allow charges to move around. That's why moist air decreases static electricity. By making fabrics and hair almost imperceptibly damp, the positively charged detergents allow separated charges to get back together and do away with static hair problems and laundry cling. That's why they're the main ingredients in fabric softeners, dryer sheets, hair conditioners, and even many antistatic sprays.

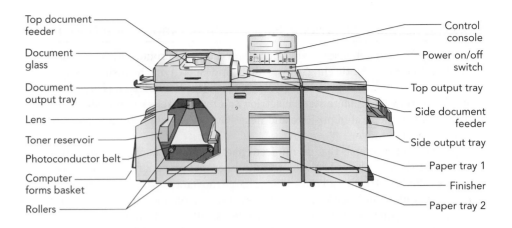

Top document feeder

Document glass

Document output tray

Lens

Toner reservoir

Photoconductor belt

Computer forms basket

Rollers

Control console

Power on/off switch

Top output tray

Side document feeder

Side output tray

Paper tray 1

Finisher

Paper tray 2

SECTION 10.2 **Xerographic Copiers**

The days of carbon paper and mimeograph machines are long gone. What modern office could operate without a xerographic copier? Advertisements for copiers are everywhere, and while each manufacturer claims to make the best copiers, that's mostly just salesmanship. In reality, all xerographic copiers are based on the same principles, discovered in 1938 by Chester Carlson. In this section, we'll examine xerographic copiers and the ideas that make them possible.

Xerography: Using Light to Print Copies

The image that a xerographic copier prints on a sheet of paper begins as a pattern of tiny black particles or toner on a smooth, light-sensitive surface. The copier uses static electricity and light reflected from the original document to arrange this toner on the surface and then carefully transfers the toner to the paper (Fig. 10.2.1). Invented in 1938 by Chester Carlson ❶, this process is basically our old friend static electricity doing something useful.

At the heart of the xerographic copier is a thin, light-sensitive surface made from a photoconductor—a normally insulating material that becomes a conductor while exposed to light. Although the darkened photoconductor can keep positive and negative charges apart, these charges quickly draw together when light hits the photoconductor (Fig. 10.2.2). That flexibility allows light from the original document to determine the pattern of static electricity on the photoconducting surface and consequently the placement of toner on the piece of paper.

Each copying cycle begins in the dark with the copier spraying negative charges onto its photoconductor. On the other side of the photoconductor is a grounded metal surface—grounded in the sense that it's electrically connected to the earth so that charges are free to flow between the two. As negative charges land on the open surface of the photoconductor, they attract positive charges onto the metal surface beneath it. When the charge-spraying process is complete, the open surface of the photoconductor is uniformly coated with negative charges while the underlying metal surface is uniformly coated with positive charges (Fig. 10.2.3a).

After this precharging, the copier uses a lens to cast a sharp image of the original document onto the photoconducting surface. We'll examine lenses and the formation of images when we study cameras in Chapter 15. For now, what matters is that light hits the photoconductor only in certain places, corresponding to white parts of the original document.

There are two standard techniques for exposing the photoconductor to light. Some copiers illuminate the whole original document with the brilliant light of a

❶ Impoverished as a youth, American inventor Chester F. Carlson (1906–1968) supported his family by washing windows and cleaning offices after school. His work in a print shop as a teenager started him thinking about copying and he began to experiment with electrophotography. After attending Caltech, he worked for Bell Laboratories but was laid off in the Depression. While attending law school, he continued his experiments and invented the xerographic copying process in 1937–1938. Development of commercial copiers was slow and it wasn't until 1960 that Haloid Xerox Corporation produced its first successful copier, Model 914. Carlson became extremely wealthy but gave most of his money away anonymously.

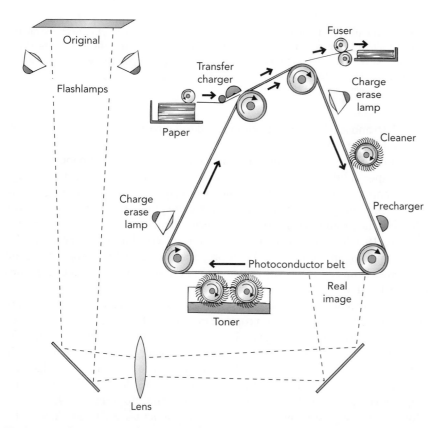

Fig. 10.2.1 This xerographic copying machine uses a photoconductor belt to form black-and-white images of an original document. The copying process begins with the pre-charger, which coats the photoconductor with charge. The optical system then forms a real image on a flat region of the photoconductor belt, producing a charge image. After the charge image picks up toner particles, the first charge erase lamp eliminates the charge image and weakens the toner's attachment to the belt. The toner is then transferred and fused to the paper.

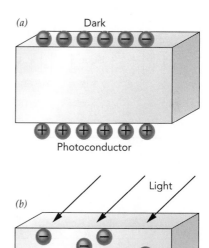

Fig. 10.2.2 (*a*) In the dark, a photoconductor is an electrical insulator so that separated electric charges on its surfaces remain there indefinitely. (*b*) When the photoconductor is exposed to light, it becomes an electrical conductor and the opposite electric charges soon join one another.

flash lamp and cast a complete image onto a flattened portion of a photoconductor belt. In other copiers, a moving lamp or mirror illuminates the original a little at a time and the image is cast as a moving stripe on a rotating photoconductor drum. Either way, charges move through any regions of the photoconductor that are exposed to light, leaving these regions electrically neutral (Fig. 10.2.3*b*). The result is a charge image—a pattern of electric charge on the photoconductor's surface that exactly matches the pattern of ink on the original document (Fig. 10.2.3*c*).

To develop this charge image into a visible one, the xerographic copier exposes the photoconductor to positively charged toner particles (Fig. 10.2.3*d*). This toner is a fine, insulating plastic powder containing a colored pigment, usually black. Applying toner to the photoconductor must be done gently and it's often accomplished with the help of Teflon-coated iron balls. These tiny balls are held together in long filaments by a rotating magnetic shaft, so that the shaft resembles a spinning brush with extraordinarily soft bristles. These bristles wipe toner particles out of their storage tray and onto the photoconductor. Contact with the Teflon leaves the toner particles positively charged, so they stick to the negatively charged portions of the photoconductor (Fig. 10.2.3*e*).

The photoconductor now carries a black image of the original document, an image that the copier must transfer to the paper. Before attempting that transfer, the copier first weakens the photoconductor's grip on the toner by exposing it to light from a charge erase lamp. This light eliminates the photoconductor's charge

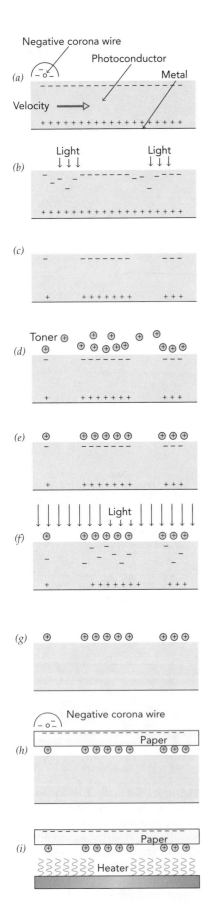

Fig. 10.2.3 The photoconductor is first coated (*a*) with a uniform layer of negative charge. Exposure to light (*b*) erases some charge to form a charge image (*c*). The charge image attracts (*d*) positively charged toner particles (*e*). The charge image is erased (*f*) to release the toner particles (*g*). The toner is transferred to the negatively charged paper (*h*) and fused to the paper with heat (*i*).

(Fig. 10.2.3*f*) and leaves the positively charged toner particles clinging only loosely to its surface (Fig. 10.2.3*g*).

The copier then transfers the toner image to a blank sheet of paper by pressing that paper lightly against the photoconductor while spraying negative charge onto the paper's back (Fig. 10.2.3*h*). The positively charged toner is attracted to the negatively charged paper and the two leave the photoconductor together. The copier then heats and presses the copy, permanently fusing the toner onto the paper (Fig. 10.2.3*i*). Once the image has been transferred to the paper, the copier cleans its photoconducting surface in preparation for the next copy: a second charge erase lamp eliminates any remaining charge and a brush or squeegee mops up any residual toner.

With that introduction to xerography, you can already explain many things about copiers. For example, while fixing a copier jam, you may find that you have removed unfinished copies—ones bearing toner images that haven't yet been fused onto the paper. The toner of an unfused copy comes off on your hand because it's held in place only by electrostatic forces. And when you replace the "toner cartridge" in a personal copier, in addition to adding new toner, you're also installing a new precharge system, photoconductor drum, and toner applicator (Fig. 10.2.4).

However, we've glossed over three important physics issues. Two we'll leave for later chapters: why a photoconductor becomes conducting when exposed to light (Chapter 12 on Electronics) and how a lens projects an image of the document onto the photoconductor (Chapter 15 on Optics). But the third issue is relevant now and so we'll examine it carefully: how the copier sprays charges onto surfaces.

Fig. 10.2.4 This xerographic copier places the photoconductor drum, toner supply, and a corona wire inside a disposable cartridge. After the paper passes through the cartridge, toner is fused onto its surface and it leaves the copier.

Discharges and Electric Fields

At the start of the copy cycle, the xerographic copier coats its photoconducting surface uniformly with electric charges. Because this precharging process is done in the dark, while the surface is an electrical insulator, the charges must be sprayed onto it like paint. The copier's charge sprayer is a corona discharge—a gentle, sustained spark that forms in the air near a needle or fine wire that's maintained at high voltage.

It's a type of discharge—a flow of electric charge through a gas. Air is normally an insulator because its atoms and molecules are neutral and can't convey charge from one place to another. However, by seeding air liberally with individual charged particles, the copier manages to turn that air into a conductor and then to produce a discharge in it. But how does the copier seed the air with charges and produce its discharge? And how does it use that discharge to coat its photoconducting surface? To answer those questions, we'll need to know more about electrostatic forces and voltages, and about a related concept: electric fields.

Because free charges are hard to come by in the air, the copier begins with just a few charged particles and uses them to generate more. The idea is simple: the copier uses electrostatic forces to accelerate those initial charges to enormous speeds and lets them smash into air's neutral particles. When hit hard enough, a neutral air particle breaks into oppositely charged fragments and thus adds two more free charges to the air. These new charges join the mix, accelerating, colliding, and breaking up still more air particles. A cascade of collisions ensues and the air "breaks down"—transforming from an insulator to a conductor. The copier then uses this conducting air to spray the photoconductor with charges.

So where do those initial charges come from? Surprisingly, they're already there, the products of cosmic rays and natural radioactivity! Every cubic centimeter of ordinary air contains almost 2000 charged particles, roughly half positive and half negative. Considering that this same volume of air contains almost 3×10^{19} neutral particles, that's not many charges. But it's enough to get the discharge started.

To parlay those initial charges into the vast numbers it needs, the copier must accelerate them aggressively. The neutral air particles are so densely packed that it's difficult for the charged ones to pick up much speed before they hit something and slow down. To give each initial charge a good shot at breaking the first neutral particle it hits, the copier must accelerate that charge very quickly.

The copier accelerates its free charges using electrostatic forces. Up until now, we've associated electrostatic forces with pairs of charges, each charge pushing or pulling on the other. Since the individual forces on an object sum to give the net force on that object, it's easy to figure out how three charges affect one another, or four, or five.... But in the copier's wires and discharge, there are so many individual charges that adding up their forces is virtually impossible. We need some other way to characterize the overall electrostatic force on a particular charge.

Instead of thinking about the many interactions between that charge and all the other charges around it (Fig. 10.2.5a), we can view the electrostatic force on our charge as the result of its one interaction with something local: an electric field—an attribute of space that exerts an electrostatic force on a charge (Fig. 10.2.5b). From this new perspective, our charge accelerates forward because it interacts with the local electric field, a field that's created by all of the surrounding electric charges.

This electric field appears to be nothing more than an intermediary: the surrounding charges produce the electric field and this electric field pushes on our charge. But in later sections, we'll see that an electric field is more than that, more than a seemingly unnecessary fiction. That's because an electric field truly exists in space, independent of the charges that produce it. In fact, electric fields are often created by things other than charges and can influence things other than charges as well.

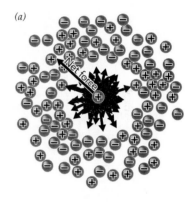

(a)

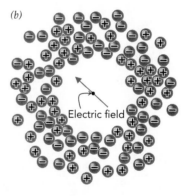

(b)

Fig. 10.2.5 (*a*) When a charge interacts with many other charges, adding up the individual electrostatic forces on that charge to obtain the net force on it becomes a daunting task. (*b*) It is often simpler to introduce an electric field as an intermediary. The other charges produce this electric field, which in turn exerts an electrostatic force on the charge (omitted for clarity). The electric field arrow passing through the center dot indicates the magnitude and direction of the force a positive test charge would experience at the dot's location.

The copier's electric field varies with location, meaning that the electrostatic force on our charge depends on where it is. This force is equal to the product of the charge times the electric field and points in the direction of that electric field. We can write this relationship as:

$$\text{force} = \text{charge} \cdot \text{electric field}, \tag{10.2.1}$$

where the force is in the direction of the electric field. Note, however, that a particle carrying a negative amount of charge (e.g., an electron) experiences a force opposite the electric field. The SI unit of electric field is the newton-per-coulomb (abbreviated N/C).

The copier employs a very strong electric field to "break down" the air so that it can operate its discharge. That field accelerates charges so rapidly that collision cascades occur and fill the air with free charges. Unfortunately, you can't sense electric fields directly, so it's hard to visualize a strong one. We'll work on that problem, but for now just remember that strong electric fields can initiate discharges in air. That's how thunderstorms produce lightning!

Conductors and Voltage Gradients

A copier's precharging system uses the gentle corona discharge that develops in the strong electric field just outside a fine, high-voltage wire. This discharge ferries charges to the photoconductor surface and coats it uniformly. But to understand why a strong electric field exists outside a fine, high-voltage wire and why the discharge it produces is "gentle," we need some background. Let's start by looking at electric fields inside and outside electrical conductors.

Consider the simplest conducting object: a solid metal ball. If you release some positive charges inside that ball (Fig. 10.2.6a), what happens to them? Because they repel one another, those charges accelerate outward and move apart. In fact, they'd leave the ball altogether if they weren't chemically bound to its metal. After

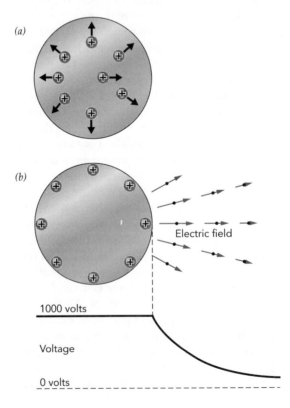

Fig. 10.2.6 (a) When like charges are placed inside a conducting sphere, they repel one another and accelerate toward the sphere's surface. (b) When those charges have reached equilibrium on the sphere's surface, the sphere has a single, uniform voltage and zero electric field inside it. Outside the sphere, the voltage decreases toward zero and there is an electric field.

spending a moment or two ridding themselves of extra electrostatic potential energy, principally as heat, the charges settle down in stable equilibria on the ball's surface (Fig. 10.2.6b). At its equilibrium point, the outward electrostatic force each charge experiences from its fellow charges is perfectly balanced by the inward chemical force it experiences from the metal. The net force on it is zero.

At equilibrium, each charge has also minimized its total potential energy. After all, it can't stop accelerating until there is no direction in which it can move to lower its potential energy further. But what's amazing about how the charges arrange themselves on the ball's surface is that each one ends up with the *same* total potential energy. That's because if one of them had less total potential energy than the rest, the other charges would accelerate toward it to lower their total potential energies as well!

Since the only potential energy that significantly affects charges in our small, homogeneous ball is electrostatic potential energy, every charge in our ball has essentially the same electrostatic potential energy. Because voltage is the electrostatic potential energy per unit of charge, equal potential energies on equal charges means equal voltages—the entire ball has a single, uniform voltage! In our voltage–altitude analogy, this observation is analogous to the fact that, at equilibrium, the water level in a swimming pool has a single, uniform altitude.

Because of the ball's perfect symmetry, charges in equilibrium are spread evenly on its surface. Had we chosen a less symmetric conducting object, such as the copier's fine metal wire, charges in equilibrium would not be spread so evenly. Nonetheless, those charges would still be on the outside of the object and it would still have a single, uniform voltage.

Voltage and Charge on a Conducting Object

With its charges at equilibrium, a homogeneous conducting object has a single, uniform voltage and the net charge anywhere in its interior is zero.

But while the voltage is uniform *in* and *on* the copier's fine conducting wire, it varies rapidly with position *outside* that wire (Fig. 10.2.6b). Accompanying this large spatial variation in voltage is a strong electric field. Officially called a voltage gradient, you can think of a spatial variation in voltage as a "slope" in the voltage. In our voltage–altitude analogy, a voltage gradient is analogous to an "altitude gradient"—the slope of an ordinary hill. And just as water accelerates swiftly down a steep slope toward lower altitude, so charge accelerates swiftly down a large voltage gradient toward lower voltage.

Since both electric fields and voltage gradients cause charges to accelerate, it shouldn't surprise you to learn that a voltage gradient *is* an electric field. Although we'll uncover a second source of electric fields in the next chapter, we'll treat a voltage gradient and an electric field as equivalent for now. Their relationship can be written as:

$$\text{electric field} = \text{voltage gradient} = \frac{\text{voltage drop}}{\text{distance}}, \qquad (10.2.2)$$

where the electric field points in the direction of most rapid voltage decrease.

This relationship gives us a second way to look at an electric field. In addition to being the electrostatic force exerted per unit of charge, electric field is also the voltage drop per unit of distance. The SI unit of electric field therefore has a second form: the volt-per-meter (abbreviated V/m). The V/m is exactly the same unit as the N/C. As an example of an electric field produced by a voltage drop, consider the top of a normal 9 V battery. With its two terminals separated by just 0.005 m and a voltage drop of 9 V between them, the space between those terminals contains an electric field of about 1800 V/m, pointing toward the negative terminal.

Fine Wires and High Voltages: Corona Discharges

Ordinary air breaks down in an electric field of about 3×10^6 volts-per-meter or, in customary units, about 30,000 volts-per-centimeter. At that field, free charges accelerate so rapidly that a cascade of charge-freeing collisions suddenly transforms air from a nearly perfect insulator into a reasonably good conductor.

You can produce such a strong field all by yourself. On a dry winter day, you can coat yourself with positive charges and raise your voltage to about 30,000 volts simply by scuffing your rubber-soled shoes across an acrylic carpet. As you then approach a grounded doorknob at 0 volts, the voltage difference between the doorknob and your hand will be 30,000 volts. When your hand is about 1 cm from the doorknob, the electric field will reach 30,000 volts-per-centimeter and the air will break down with a brilliant spark (Fig. 10.2.7).

Because your hand and the doorknob are similar in size and shape, the voltage changes smoothly between them (Fig. 10.2.8a). It varies steadily from 0 volts on the doorknob to 30,000 volts on your hand, so the voltage gradient or electric field is nearly uniform. But when two objects differ significantly in size, the larger object dominates voltages in the space between them. For example, if you hold a long pin in your hand as you approach the doorknob, the doorknob will control the voltage most of the way to the pin and nearly all the increase in voltage will occur just outside the pin's point (Fig. 10.2.8b). Rather than being uniform, the voltage gradient or electric field will be strongest near that point.

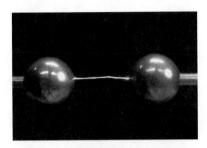

Fig. 10.2.7 These two metal spheres are 1 cm apart. When their difference in voltage is about 30,000 V, the air between them breaks down and forms a spark.

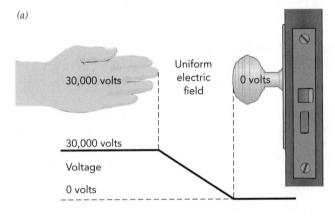

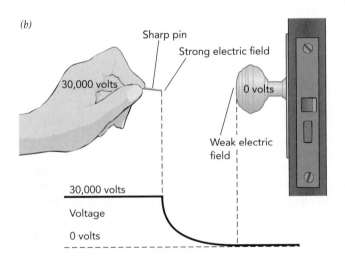

Fig. 10.2.8 Your voltage is 30,000 V when you reach for the 0 V doorknob. (*a*) Since your hand and the doorknob are similar in size, the voltage decreases steadily between them and the electric field is uniform. (*b*) But when you hold out a pin, the voltage plummets near its sharp point and the electric field there is extremely strong.

The copier makes good use of this nonuniform field. Its fine, high-voltage wire is nearly surrounded by a much larger metal shroud. The wire is so thin that its influence fades just a hair's breadth from its surface and the grounded shroud dominates voltage almost all the way to the wire. Although the wire's negative voltage is only −3000 volts and it's about 1 cm from the shroud, the voltage changes so rapidly in the air just outside this wire that the electric field there easily exceeds 30,000 volts-per-centimeter and breaks down the air.

The discharge that forms near the fine wire is a special, self-regulating one—a corona discharge (Fig. 10.2.9). While most discharges can't control how many free charges they produce, a corona discharge automatically maintains a steady production. Because free charges form only in the strong electric field near its thin conductor, their production rate is very sensitive to changes in that conductor's effective thickness. If there are too many free charges in the air near the conductor, their ability to conduct electricity effectively thickens the conductor, weakens the electric field, and slows the production of free charges. The discharge is correcting its own mistake.

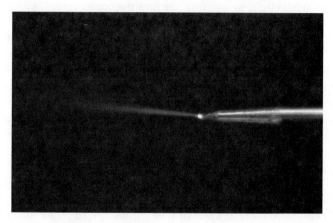

Fig. 10.2.9 The electric field near this sharp, high-voltage pin is so strong that it breaks down the air and forms a corona discharge. The resulting glow is produced by air particles that receive energy from the discharge.

Because of this stabilizing effect, the air in a corona discharge maintains a steady electrical conductivity that's ideal for charging a copier's photoconductor. However, corona discharges were common long before copiers. They often occur spontaneously near sharp points or fine wires at high voltages, leading to charge leakage from power transmission lines and occasionally producing a glow called St. Elmo's fire on the masts and rigging of sailing ships (see ❷).

Getting Ready to Copy: Charging by Induction

A corona discharge does more than just turn air into a conductor; it also produces an outward spray of electric charges. Those charges are pushed outward by the electric field surrounding the corona wire. Since the copier's corona wire has a negative voltage, the surrounding electric field points toward that wire. And because negative charges accelerate opposite an electric field, the copier's corona produces a shower of outgoing negative charges. They spray onto the photoconducting surface as it moves steadily past the corona and the photoconductor thus acquires a uniform coating of negative charges.

As each negative charge lands, it draws a positive charge onto the grounded metal surface beneath the photoconductor and the attraction between those two opposite charges holds them firmly in place. While the photoconductor's open surface

❷ Contrary to popular belief, lightning rods don't simply attract lightning strikes so as to protect the surrounding roof. Instead, they produce corona discharges that diminish any local buildups of electric charge. By neutralizing the local electric charge, the lightning rod reduces the chances that lightning will strike the house. Similar devices, called static dissipaters, are found near the tips of airplane wings and protect planes from lightning strikes.

is acquiring its uniform negative charge, the metal layer underneath is acquiring an equivalent positive charge (Fig. 10.2.3a). This process, whereby a grounded conductor acquires a charge through the attraction of nearby opposite charge, is called "charging by induction."

The induced positive charge on the metal side of the photoconductor is important to the xerographic process for several reasons. First, it lowers the electrostatic potential energy of the negative charge so that the surface's negative voltage isn't so enormous. Second, without that positive layer nearby, repulsion between like charges would tend to push negative charges on the open surface toward the edges of the photoconductor and distort the resulting images.

But most significantly, the positive charge layer gives the negative charge layer somewhere to go when the photoconductor is exposed to light! Wherever light from the original document turns a patch of photoconductor into a conductor, the negative and positive charge layers rush together and cancel. The resulting uncharged portion of photoconductor subsequently attracts no toner and produces a white patch on the finished copy.

Having come full circle, we can now see how the copier achieves its goal. It uses a corona discharge to coat a photoconducting surface with negative charge and then selectively erases portions of that charge layer with light from the original document. The remaining charged patches on the photoconductor attract positively charged black toner, which is then transferred permanently to the paper.

The only additional detail worth mentioning is that, for technical reasons, some copiers precoat their photoconductors with positive rather than negative charges and then use that charge to attract negatively charged toner. These copiers put high positive voltages on their fine wires so that their coronas spray positive charges.

⤷ Color Copiers and Laser Printers

Xerographic copying isn't limited to black and white. The toner itself can be any color, so it's easy to make red copies or blue copies. But to create full color copies, the copier needs to work with 3 or 4 different toners. Three of these toners are the primary colors of pigment: yellow, cyan, and magenta, while the fourth is black. As I'll explain in Section 14.2 on discharge lamps, yellow, magenta, and cyan pigments can be combined to make our eyes perceive any color imaginable. The black toner is optional but it improves the image contrast.

To make a full color copy, the color copier makes 3 or 4 different charge images and uses each of these images to transfer a particular toner to the paper. It forms these charge images through colored filters, so that the charge image that controls the yellow toner is formed by blue light (yellow toner absorbs blue light, just as black toner absorbs white light). The yellow, cyan, and magenta toner images are carefully assembled and transferred to the paper. Finally, a charge image is made with white light and used to control the placement of black toner. It, too, is transferred to the paper and the assembled image is fused in place. The result is a full color copy that looks just like the original document.

A laser printer is also a xerographic device, but it uses a laser beam to write a charge image directly onto its photoconductor drum. This rotating drum is precharged by a corotron and then a spinning mirror scans laser light rapidly across its surface. Wherever laser light hits the drum, charge flows through the photoconductor. A computer in the printer turns the laser on and off as it systematically constructs the charge image, one dot at a time.

The printer then sticks toner to the charge image and transfers it to the paper. However, most laser printers use electrostatic tricks to reverse the charge image, so that toner sticks only to portions of the drum exposed to laser light. This reversal makes it easier for the printer to produce pure white backgrounds for its prints.

▥ Electronic Air Cleaners

Dust doesn't float in air. Like the rocks, dirt, soot, ash, and pollen from which it's made, dust is denser than air and belongs on the ground. But air resistance makes it difficult for dust to fall. With so much surface relative to its tiny volume, a dust grain experiences severe viscous drag forces as it moves through air. Even at millimeter-per-second speeds, the drag force on a dust grain may exceed its weight. Thus dust grains have extremely slow terminal velocities and take minutes or hours to drop through a few meters of still air to the ground. Moreover, moving air carries dust along with it. Since drag forces oppose relative motion between dust and air, the slightest upward breeze can keep dust aloft indefinitely. Gravity is clearly too weak a force to pluck dust from the air quickly.

But suppose that we put negative charge on the dust grains in the air and then let that air flow past positively charged collecting plates. Near a plate's positive charges, a negatively charged dust grain experiences strong electrostatic forces. Although gravity is too weak to clear the air stream, electrostatic forces can easily overwhelm viscous drag and pull the dust grains speedily from the air (Fig. 10.2.10). In a properly designed electronic air cleaner, those grains collect on the charged plate and the air continues on without them.

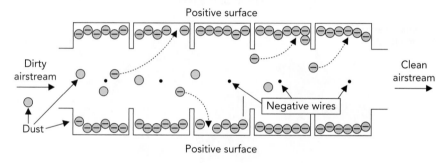

Fig. 10.2.10 In an electrostatic precipitator, dust grains in dirty air are given negative charges so that they'll be attracted to positively charged surfaces.

Because it uses corona discharges to charge the passing dust and then uses electrostatic forces to precipitate that dust in clumps on its collecting plates, this type of electronic air cleaner is called an electrostatic precipitator. Its chief advantage over a porous paper or fiber air filter is that huge amounts of dust can accumulate on its collecting plates without blocking the airflow. It's also easy to clean. Just tap or rinse.

But an electrostatic precipitator can't be a passive device. If it were, the negatively charged dust grains arriving on its positively charged collecting plates would quickly neutralize those plates and it would stop cleaning the air. To continue working, the precipitator must steadily transfer electric charge from the incoming dust to the collecting plates. That's why an electronic air cleaner has a high-voltage power supply that delivers negative charge to its corona discharges and positive charge to its collecting plates.

Actually, this cleaning scheme only requires that the dust and the collecting plates have opposite charges, so not all electrostatic precipitators put negative charge on the dust. Some put positive charge on the dust and attract it to negatively charged surfaces. While negative corona discharges are less susceptible to accidental sparking and are easier to maintain, they also produce ozone, an irritating and reactive form of oxygen that is considered to be a pollutant at ground level. Electronic air cleaners that need to minimize ozone production may opt for positive corona discharges and therefore positively charged dust.

❸ In 1906, while an instructor at the University of California, Berkeley, Frederick Gardner Cottrell (American scientist and inventor, 1877–1948) became involved with local industries that were attempting to combat air pollution in the San Francisco Bay area. Within the year he had invented the electrostatic precipitator. Cottrell donated the patents for the precipitator to a nonprofit corporation he founded. For almost a century, Research Corporation has supported scientific work in the United States. Cottrell gave up a personal fortune to help build the careers of a great many young scientists.

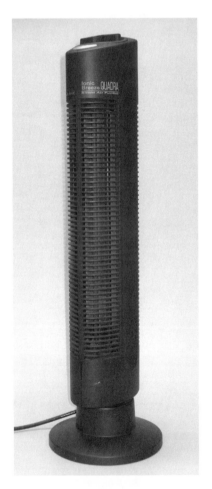

Fig. 10.2.12 This household electronic air cleaner has two corona wires and three collecting plates inside its plastic housing. The flow of charges from the corona wires to the plates also causes air to flow through the cleaner, so it requires no separate fan.

Industrial electrostatic precipitators (Fig. 10.2.11) are essential to pollution control in nearly all smokestack industries and have been in widespread use since they were invented in the early 20th century ❸. Household electronic air cleaners have also become relatively popular in recent years. Some of the most sophisticated consumer models don't even need fans; they use the attraction between charged particles in the air and their collecting plates to circulate air through themselves (Fig. 10.2.12).

Household ion generators also remove dust and smoke from room air. These devices resemble electrostatic precipitators but have no internal collecting plates. While they still use corona discharges to charge passing dust grains, they don't attempt to remove the dust from the air. Instead, they let those electrically charged particles, or ions, drift around the room on their own.

Those ions don't drift for long because they're attracted to neutral surfaces. A charged dust grain electrically polarizes any surface it approaches and the dust grain and that polarized surface then attract one another and stick. You can see this effect by charging a toy balloon in your hair and touching it to a wall. The balloon's charge will polarize the wall and the two will cling to one another. Thus an ion generator removes particles from the air by sticking them to surfaces in the room. This method is cheap and effective, but the dust ends up decorating the walls and furniture, a problem unless you're fond of the color gray.

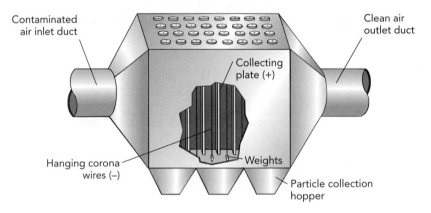

Fig. 10.2.11 In an industrial electrostatic precipitator, hanging corona wires transfer negative charge to particles in the air and those particles are then collected on positively charged surfaces. Rapping the surfaces vigorously dislodges the accumulated particles, which then fall in the collection hopper for disposal.

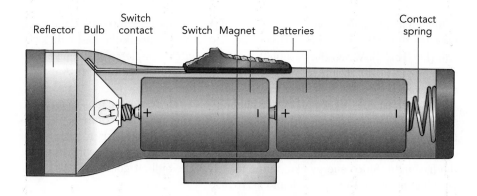

Reflector Bulb Switch contact Switch Magnet Batteries Contact spring

Section 10.3 **Flashlights**

There isn't much to a typical flashlight; you can see what few parts it has when you open it to replace its batteries. But a flashlight isn't a mechanical device, it's an electrical one: it contains an electric circuit and most of its components are involved in the flow of electricity. Understanding how a flashlight works means understanding how an electric circuit works and how electricity carries power from batteries to a bulb. As we'll see, flashlights aren't as simple as they appear.

Electricity and the Flashlight's Electric Circuit

A basic flashlight has just three components—a battery, a bulb, and a switch—connected together by metal strips. When the switch is on, the strips transfer energy from the batteries to the bulb. But how does energy move through the strips and why does the switch start or stop that energy transfer? To answer these questions, we must first understand electricity and electric circuits, so that's where we'll begin.

When you turn on a flashlight, electricity conveys energy from the batteries to the bulb. An electric current—a current of electric charges—flows through these components, carrying the energy with it. Though we'll examine the exact nature of this current soon, you can picture it as a steady stream of tiny positive charges following a circular route that takes them through the batteries, through the bulb, then back to the batteries for another trip (Fig. 10.3.1). As long as the flashlight is on, charges flow around this loop, receiving energy from the batteries and delivering it to the bulb, over and over again. En route, the charges carry this energy mostly as electrostatic potential energy.

The looping path taken by charges in a flashlight is called an electric circuit. Because a circuit has no beginning or end, charges can't accumulate in one place, where their mutual repulsion would eventually stop them from flowing. Circuits are present in virtually all electric devices and explain the need for at least two wires in the power cord of any home appliance: one wire carries charges to the appliance to deliver energy and the other wire carries those charges back to the power company to receive some more.

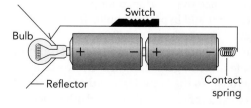

Bulb
Switch
Reflector
Contact spring

Fig. 10.3.1 A flashlight contains one or more batteries, a lightbulb, a switch, and several metal strips to connect them all together. When the switch is on (as shown), the components in the flashlight form a continuous loop of conducting materials. Electrons flow around this loop counterclockwise.

But what role does the switch play in all of this? As part of one conducting path between the batteries and bulb, the switch can make or break the flashlight's circuit (Fig. 10.3.2). When the flashlight is on, the switch completes the loop so that charges can flow continuously around the closed circuit (Fig. 10.3.2a). A closed circuit appears on the left side of Fig. 10.3.3.

However, when you turn off the flashlight, the switch breaks the loop to form an open circuit (Fig. 10.3.2b). Although one conducting path still connects the batteries and bulb, the loop now has a gap in it and can no longer carry a continuous current. Instead, charges accumulate at the gap and current stops flowing through the flashlight. Since energy can no longer reach the bulb, it goes dark. An open circuit appears on the right side of Fig. 10.3.3.

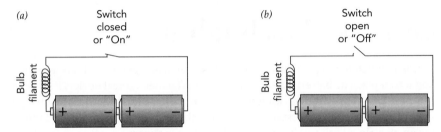

Fig. 10.3.2 (*a*) When the flashlight's switch is on, it closes the circuit so that current can flow continuously from the batteries, through the bulb's filament, and back through the batteries. It follows this circuit over and over again. (*b*) When the flashlight's switch is off, it opens the circuit so that current stops flowing.

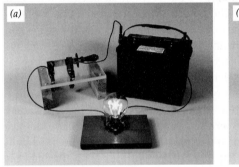

Fig. 10.3.3 (*a*) When the large knife switch at the left is on, current flows around the closed circuit and carries power from the battery to the light bulb. (*b*) But when the knife switch is off, no current flows through the open circuit.

There's one other type of circuit worth mentioning. A short circuit forms when the two separate paths connecting the batteries to the bulb accidentally touch one another (Fig. 10.3.4). This unintended contact creates a new, shorter loop around which the charges can flow. Because the bulb is supposed to extract energy from the charges, it's designed to impede their flow and to convert their electrostatic potential energy into thermal energy and light. This opposition to the flow of electricity is called electrical resistance. Since the shortened loop offers little resistance, most of the charges flow through it and bypass the bulb. The bulb dims or goes out altogether.

Since the bulb is the only part of the flashlight that's designed to get hot, a short circuit leaves the charges without a safe place to get rid of their electrostatic potential energies. They deposit it dangerously in the batteries and the metal paths, making them hot. Since short circuits can start fires, flashlights and other electric equipment try to avoid them.

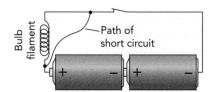

Fig. 10.3.4 When an unwanted conducting path allows current to bypass the flashlight's filament, it forms a short circuit. Because it has no proper place for electrons to deposit their energy, the short circuit becomes hot.

The Electric Current in the Flashlight

Each of the tiny charged particles flowing through the flashlight's circuit carries with it just a single elementary unit of electric charge and a miniscule amount of electrostatic potential energy. However, because those charges flow in astonishing numbers, they convey a considerable amount of energy per second—the quantity we know as power (see Section 2.2) and measure in watts (abbreviated W). The bulb needs a certain amount of power to keep its filament glowing brightly and you can determine how much power is reaching the bulb by multiplying the number of elementary charges passing through the bulb each second by the amount of energy each one delivers.

But there are too many elementary charges to count. You'll do much better to measure the circuit's current, that is, the amount of charge passing a particular point in the circuit per unit of time. The SI unit of current is the ampere (abbreviated A) and corresponds to 1 C (1 coulomb) of charge passing by the designated point each second. One coulomb is roughly 6.25×10^{18} or 6,250,000,000,000,000,000 elementary charges, so even a 1-A current involves a tremendous flow of elementary charges.

Using electric current instead of counting charges, you can determine how much power is reaching the bulb by multiplying that current by its electrostatic energy per coulomb—the quantity we already know as voltage. For example, a current of 2 amperes (2 coulombs-per-second) at a voltage of 3 volts (3 joules-per-coulomb) would bring 6 watts of power (6 joules-per-second) to the bulb. Brighter flashlights involve larger currents, greater voltages, or both.

Current has a direction, pointing along the route of positive charge flow. When you turn on the flashlight in Fig. 10.3.1, charge flows around the circuit clockwise—from the battery chain's positive terminal, through the bulb's filament, through the switch, and into the battery chain's negative terminal. However, it's time to address an awkward issue: the positive charges that flow clockwise around this circuit are fictitious. In reality, the electric current is actually carried by negatively charged electrons heading in the opposite direction!

This issue dates back to Franklin's unfortunate choice of which charge to call positive and which to call negative. By the time scientists discovered the electron and realized that these negatively charged particles carry currents in wires, current had already been defined as pointing in the direction of positive charge flow. Since it was far too late to make current and electron-flow point in the same direction, scientists and engineers simply pretend that current is carried by fictitious positive charges heading in the current's direction.

This fiction works extremely well, as illustrated by a simple example. When negatively charged electrons flow to the right through a neutral piece of wire, the wire's right end becomes negatively charged and its left end becomes positively charged (Fig. 10.3.5a). But exactly the same thing would happen if a current of fictitious positively charged particles were to flow to the left through that same piece of wire (Fig. 10.3.5b). Without sophisticated equipment, you can't tell whether negative charges are flowing to the right or positive charges are flowing to the left because the end results are essentially indistinguishable.

We, too, adopt this fiction and pretend that current is the flow of positively charged particles. In this and subsequent chapters, we'll stop thinking about electrons and

(a)

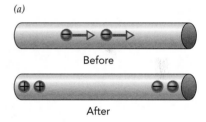

(b)

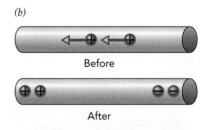

Fig. 10.3.5 A current of negatively charged particles flowing to the right through a piece of wire (*a*) can't easily be distinguished from a current of positively charged particles flowing to the left (*b*). The end result of both processes is an accumulation of positive charge on the left end of the wire and negative charge on its right end.

imagine that electricity is carried by positive charges moving in the direction of the current. There are only a few special cases in which the electrons themselves are important, and we'll consider those situations separately when they arise.

Batteries, ...

While a battery is basically a portable source of electric power, here are two other interesting ways to think of it. The first is rather abstract: a battery is a type of pump. It "pumps" charge from low voltage to high voltage, much as a water pump pumps water from low altitude to high altitude. Once again, our voltage–altitude analogy is helpful. Each of these pumps moves something against its natural direction of flow, pushing it forward and doing work on it in the process. The battery increases a charge's electrostatic potential energy by pushing it up a voltage gradient, while the water pump increases water's gravitational potential energy by pushing it up an altitude gradient.

The second perspective on batteries is more mechanical: a battery is a chemically powered machine. It uses chemical forces to transfer charges from its negative terminal to its positive terminal. As positive charges accumulate on the battery's positive terminal, the voltage there rises and as negative charges accumulate on the battery's negative terminal, the voltage there drops. Since the battery does work transferring charges from low voltage to high voltage, it is converting its chemical potential energy into electrostatic potential energy in these separated charges.

A battery's rated voltage reflects its chemistry, specifically the amount of chemical potential energy it has available for each charge transfer. As the voltage difference between its terminals increases, so does the energy required for each charge transfer. Eventually, the chemicals can't do enough work on a charge to pull it away from the negative terminal and push it onto the positive terminal, so the transfers stop. The battery is then in equilibrium—the electrostatic forces opposing the next charge transfer exactly balance the chemical forces promoting it. A typical alkaline battery reaches this equilibrium when the voltage of its positive terminal is 1.5 V above the voltage of its negative terminal. Lithium batteries, with their more energetic chemistries, can achieve voltage differences of 3 V or more.

When you turn on the flashlight, you upset its equilibrium by allowing charges to leave the battery's positive terminal for its negative terminal. With fewer separated charges now on its terminals, the battery's voltage difference decreases slightly and it begins pumping charges again. That renewed charge transport replenishes the terminals' separated charges and opposes any further decrease in the battery's voltage. In this manner, a 1.5-V alkaline battery maintains a nearly steady voltage difference of 1.5 V between its terminals, whether its flashlight is on or off.

That alkaline battery is powered by an electrochemical reaction in which powdered zinc at its negative terminal reacts with manganese dioxide paste at its positive terminal. This reaction resembles controlled combustion. In effect, the battery "burns" zinc to obtain the energy it needs to pump charges from its negative terminal to its positive terminal. However, as the battery consumes its chemical potential energy, its ability to pump charges diminishes. When its chemicals are nearly exhausted, the battery's increasing disorder reduces its voltage. An aging battery can pump less current than a fresh one and provides that current with less voltage. Ultimately, less power reaches the flashlight's bulb and it goes dim.

Most flashlights use more than one battery. When two alkaline batteries are connected together in a chain, so that the positive terminal of one battery touches the negative terminal of the other, the two batteries work together to pump charges from the chain's negative terminal to its positive terminal (Fig. 10.3.6). Each battery pumps charges until its positive terminal is 1.5 V above its negative terminal, so the chain's positive terminal is 3.0 V above its negative terminal. Because charges never leave the flashlight's circuit, only relative voltages matter in that circuit. We'll find

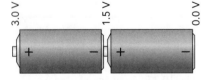

Fig. 10.3.6 When two 1.5-V batteries are connected in a chain, their voltages add so that the chain's positive terminal has a voltage that is 3.0 V higher than the chain's negative terminal. If the chain's negative terminal is chosen to be at 0 V, then the chain's positive terminal is at 3.0 V.

it convenient to ignore the flashlight's absolute voltages and define the voltage of the battery chain's negative terminal to be 0 V (Fig. 10.3.6). With that choice, the voltage of its positive terminal becomes 3.0 V.

The more batteries in the flashlight's chain, the more energy a charge receives overall and the more the voltage will increase from the chain's negative terminal to its positive terminal. A flashlight that uses six alkaline batteries in its chain has a positive terminal that is 9 V above its negative terminal. A typical 9-V battery actually contains a chain of six miniature 1.5-V batteries, arranged so that their voltages add up to 9 V (Fig. 10.3.7).

If you reverse one of the batteries in a chain, the reversed battery will extract energy from any charge passing through it (Fig. 10.3.8). While the chain may still pump charge from its negative terminal to its positive terminal, its overall voltage will be reduced because instead of adding 1.5 V to the chain's overall voltage, the reversed battery will subtract that amount. If the chain has three batteries, two will add energy to the charge while one will subtract it and the chain's overall voltage will be only 1.5 V.

As the reversed battery extracts energy from the charges passing through it, at least some of that extracted energy is converted into chemical potential energy. The reversed battery is recharging! Battery chargers follow that concept, pushing current backward through a battery—from its positive terminal to its negative terminal—to restore the chemical potential energy in a rechargeable battery. However, normal alkaline batteries are "nonrechargeable," meaning that they turn most of the recharging current's energy into thermal energy instead of chemical potential energy. Nonrechargeable batteries may overheat and explode during recharging.

... Bulbs, and Metal Strips

While a battery gives charges electrostatic potential energy by pushing them *up* a voltage gradient, a bulb releases that electrostatic potential energy by letting charges slide *down* another voltage gradient. Those two devices make a perfect pair: the battery provides electric power and the bulb consumes it. We saw back in Chapter 7 that the bulb uses this power to heat its tungsten filament to incandescence, thereby producing light. Now it's time to look at how electricity heats the filament.

We'll consider a flashlight with two alkaline batteries (Fig. 10.3.9*a*). The bulb's filament is a fine wire and its two ends are electrically connected to battery chain's terminals. With one end at 3.0 V and the other at 0.0 V, the filament has a voltage gradient across it and therefore an electric field. But how is that possible? While discussing copiers, we observed that a conductor has a uniform voltage throughout. Isn't the filament violating that rule?

No. While a conductor has a uniform voltage *when its charges are in equilibrium*, the charges in the bulb are only in equilibrium when the flashlight is off. When you switch the flashlight on, you impose a 3.0-V difference between the two ends of the filament and the filament's charges immediately began accelerating down the voltage gradient toward the 0.0-V end.

In our voltage–altitude analogy, it's as though you suddenly tilted a level field to create a hill and water that was lying motionless on that field now accelerates downhill. However, a better analog to the individual charges that we're considering at the moment would be bicyclists: picture hundreds of bicyclists on a level field that suddenly tilts to form a hill. All the bicyclists that were at equilibrium on the level field now accelerate downhill.

If the filament were a perfect conductor of electricity, each charge would accelerate steadily down the voltage gradient and convert its electrostatic potential energy into kinetic energy. But the filament has a large electrical resistance and significantly impedes the flow of electric current. Each charge bounces its way down the voltage gradient, colliding frequently with the filament's tungsten atoms and

Fig. 10.3.7 A 9-V battery actually contains six small 1.5-V cells, connected in a chain. Positive charges that enter the chain at the battery's negative terminal pass through all six cells before arriving at the battery's positive terminal.

Fig. 10.3.8 When one battery in a chain of three is reversed, the reversed battery's voltage is subtracted from the sum of the others. The chain's positive terminal has a voltage only 1.5 V higher than its negative terminal. The reversed battery recharges.

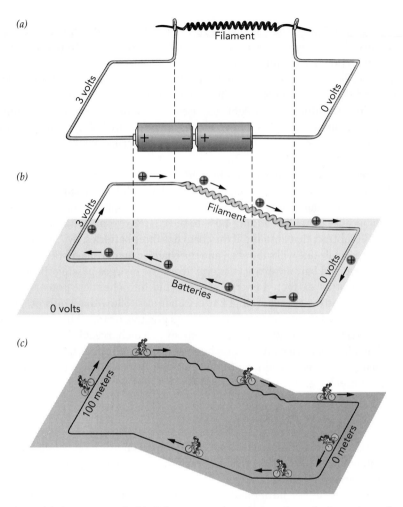

Fig. 10.3.9 (*a*) Current in a flashlight's circuit conveys power from the batteries and to the filament. (*b*) Its voltage rises in the batteries and decreases in the filament. Despite an electric field inside the filament, pushing the charges forward, they travel at constant velocity because of collisions. (*c*) This behavior is analogous to bicyclists pedaling up a smooth hill and then rolling down a rough one at constant velocity.

giving up kinetic energy with each collision (Fig. 10.3.9*b*). What began as ordered electrostatic potential energy in the charges becomes thermal energy in the tungsten atoms, and the filament glows brightly. Referring again to our voltage–altitude analogy, picture the bicyclists riding down a rough hill strewn with rocks and trees (Fig. 10.3.9*c*). They pick up bruises instead of speed.

What about the metal strips connecting the flashlight's batteries and bulb? These thick conductors have small electrical resistances and carry current easily. Charges emerge from a strip with nearly as much electrostatic potential energy as they had when they entered it, so the voltages at strip's ends are almost equal. In general, the less electrical resistance in the wires carrying current to and from the bulb, the less power is wasted en route and the more power reaches the bulb. That's why it's so important to use thick metal strips or even the flashlight's metal case in the connections.

But a poor connection anywhere in the circuit can spoil this efficient transfer of power. If there is dirt or grease on a battery terminal or worn materials in the switch, the current will have to pass through a large electrical resistance and waste power. Improving that connection, either by shaking the flashlight or by cleaning the metal surfaces, will increase the current flow through the circuit, reduce the wasted power, and brighten the flashlight.

Voltage, Current, and Power in Flashlights

When you turn on the flashlight, an electric current carries power from its two alkaline batteries to its bulb. Let's suppose that a current of 1 A is flowing through the flashlight's circuit and take a look at how much power is being transferred.

A bulb consumes electric power because the current passing through it slides down the voltage gradient and experiences an overall drop in voltage. This voltage drop measures the electrostatic potential energy each unit of charge loses while struggling through the filament. Multiplying the voltage drop by the current passing through the bulb gives you the power consumed by the bulb:

$$\text{power consumed} = \text{voltage drop} \cdot \text{current}. \qquad (10.3.1)$$

Since the voltage drop across the bulb is 3.0 V and the current passing through it is 1.0 A, it's consuming 3.0 W of power.

A battery chain produces electric power because the current passing through it is pushed up a voltage gradient and experiences an overall rise in voltage. This voltage rise measures the electrostatic potential energy each unit of charge gains while being pumped through the batteries. Multiplying the voltage gain by the current passing through the batteries gives you the power provided by the batteries:

$$\text{power provided} = \text{voltage rise} \cdot \text{current}. \qquad (10.3.2)$$

Since the voltage rise across the chain is 3.0 V and the current passing through it is 1.0 A, it's providing 3.0 W of power.

Choosing the Bulb: Ohm's Law

Our flashlight's bulb is designed to operate properly with a voltage drop of 3.0 V. Subjected to that voltage drop, it will carry a current of 1 A and thus consume 3 W of electric power: just enough to make it glow properly. If you were to use the wrong bulb in this flashlight, one designed for a different voltage drop, its filament would carry the wrong amount of current and receive the wrong amount of power. Too much power would quickly burn out its filament while too little power would make the filament glow dimly.

The bulb's filament must clearly match the flashlight, particularly the voltage of flashlight's battery chain. For example, flashlights that use many batteries require bulb filaments that are designed to operate with large voltage drops. But why is the current carried by a particular bulb filament related to the voltage drop across it and why do different bulbs respond differently to a particular voltage drop?

The relationship between current and voltage drop is the result of collisions. Charges effectively stop each time they crash into tungsten atoms, so they need the push of an electric field to keep them moving forward (Fig. 10.3.10). Doubling that electric field doubles each charge's average speed and, because the number of mobile charges in the filament is fixed, also doubles the overall current flowing through the filament. Since the electric field that propels this current is the filament's voltage gradient, doubling the voltage drop through the filament doubles the current as well.

In our voltage–altitude analogy, picture bicyclists riding on extremely rocky terrain without pedaling. These lazy bicyclists effectively stop each time they crash into rocks, so they need the push of a slope to keep them moving forward. Doubling the slope's altitude gradient—the altitude drop per meter of downhill travel—doubles each bicyclist's average speed and, because the number of bicyclists who can fit on the hill at once is fixed, also doubles the overall current of bicyclists rolling down

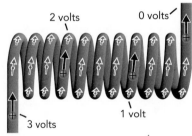

Fig. 10.3.10 Charges moving through this filament experience a voltage drop of 3 V and are pushed forward by the resulting electric field. They maintain a constant speed despite frequent collisions with tungsten atoms.

the hill. Since the slope that propels this bicyclist-current is the hill's altitude gradient, doubling the hill height doubles the current of bicyclists as well.

The influence of filament choice on current flow reflects the different electrical resistances of those filaments. Anything that increases the number of mobile electric charges across the filament's width or allows those charges to maintain a higher average speed for a given voltage drop will decrease the filament's electrical resistance and increase the current flowing through it. In fact, electrical resistance is defined as the voltage drop through the filament divided by the current that arises as a result. Making the filament thicker or shorter will lower its resistance, as will changing its composition to make collisions less frequent.

Our voltage–altitude analogy with bicyclists on a hill is again helpful. Anything that increases the number of bicyclists across the hill's width or allows those bicyclists to maintain a higher average speed for a given hill height will decrease the hill's "bicycle resistance" and increase the current of bicyclists rolling down it. In fact, "bicycle resistance" is defined as the hill height divided by the current of bicyclists it produces. Making the hill wider or shorter will lower its bicycle resistance, as will changing its rockiness to make collisions less frequent.

Combining these observations, we see that the current flowing through the filament is proportional to the voltage drop through it and inversely proportional to the filament's electrical resistance, which can be written:

$$\text{current} = \frac{\text{voltage drop}}{\text{electrical resistance}}. \tag{10.3.3}$$

This relationship is called Ohm's law, after its discoverer Georg Simon Ohm ❶. Structuring it this way separates the causes (voltage drop and electrical resistance) from their effect (current flow). However, this equation is often rearranged to eliminate the division. The relationship then takes its customary form:

$$\text{voltage drop} = \text{current} \cdot \text{electrical resistance}. \tag{10.3.4}$$

The SI unit of electrical resistance, the volt-per-ampere, is called the ohm (abbreviated Ω, the Greek letter omega). Despite its simplicity, Ohm's law is extremely useful in physics and electrical engineering. It applies to so many systems that nearly everything can be characterized by an electrical resistance. Once an object's electrical resistance is known, the current flowing through it can be calculated from its voltage drop or its voltage drop can be calculated from the current flowing through it. An object that obeys Ohm's law is often described as ohmic.

Ohm's Law
The voltage drop through a wire is equal to the current flowing through that wire times the wire's electrical resistance.

Finally, an object's electrical resistance is typically temperature-dependent. Rising temperature increases the number of mobile charges in an object but also makes them collide more frequently with the jiggling atoms. If the increasing collision frequency dominates, as it does in metals, an object's resistance increases with temperature. For example, a filament carries less and less current as it approaches operating temperature, a behavior that helps it avoid overheating. However, if the increase in mobile charges dominates, as it does in semiconductors, an object's resistance decreases with temperature. This explains why semiconductor-based computer chips carry more and more current as they get hotter and will self-destruct at excessive temperatures.

❶ German physicist Georg Simon Ohm (1787–1854) served as a professor of mathematics, first at the Jesuits' college of Cologne and then at the polytechnic school of Nuremberg. His numerous publications were undistinguished, with the exception of one pamphlet on the relationship between current and voltage. This extraordinary document, written in 1827, was initially dismissed by other physicists, even though it was based on good experimental evidence and explained many previous observations by others. In despair, Ohm resigned his position at Cologne, and it was not until the 1840s that his work was accepted. He was finally appointed professor of physics at Munich only two years before his death.

CHAPTER 11

MAGNETIC AND ELECTROMAGNETIC THINGS

Like electricity, magnetism is an important part of daily life. We use it to post notes on the refrigerator, to figure out which way is north, and to store information on credit cards and computer hard drives. The attraction between opposite magnetic poles is so familiar that it, probably more than the attraction between opposite electric charges, is responsible for the familiar phrase "opposites attract."

Thankfully, magnets are common so tinkering with them is relatively safe and easy. If you haven't done so lately, it's worth taking a minute or two to feel the forces between magnets and to observe how those forces change as you move or rotate the magnets. Their ability to push on one another even while on opposite sides of your hand is always wonderfully mysterious, although you should be careful with some of the latest and strongest magnets; they can really pinch hard if you catch your skin between them.

But the story of magnetism wouldn't be complete without including electricity. As we'll see, these two topics are related to one another through change and motion. For example, moving electric charges give rise to magnetism, and changing magnetism gives rise to electricity. I couldn't limit this chapter to magnetism alone because some of the most interesting magnets in our world involve electricity.

In this chapter, we'll examine magnetism itself, as well as several objects that use the relationships between electricity and magnetism to perform useful tasks. Since the word "dynamics" covers change and motion, these relationships are part of a field of physics known as "electrodynamics." Brevity isn't the only reason for omitting reference to magnetism in the title of that field; the other reason is that most magnetism is actually produced by electricity. In other words, most magnetism is actually electromagnetism.

Chapter Itinerary

Section 11.1 Household Magnets

How would a family stay organized without refrigerator magnets? How would the doorbell ring if it couldn't use a magnet to strike its bells? How would a scout navigate in the woods without a compass? And how would you get cash or charge purchases without magnetic strips on plastic cards?

We're so used to having magnets around that we take them for granted. But along with being useful, household magnets let us experiment with another of the basic forces in nature. Though we'll see that magnetism is so intimately related to electricity that the two are ultimately a single, unified whole, we'll find it helpful to begin our study of magnetism as a separate phenomenon and bring electricity into the picture gradually.

Button-Shaped Refrigerator Magnets

Refrigerator magnets come in all shapes and sizes, and some are more magnetically complicated than others. It's always best to start simple, so we'll begin with button-shaped magnets.

As you bring two button magnets together, they'll begin exerting forces on one another. You'll find that those forces can be either attractive or repulsive, depending on how the magnets are oriented, but always grow weaker with distance. Such magnetic forces resemble the electric ones you encounter while removing clothes from a hot dryer, but there are at least two important differences. First, reorienting two electrically charged garments won't turn their attraction into repulsion or vice versa. Second, no matter how you arrange two button magnets, you can't get a magnetic spark to jump from one to the other. Electricity and magnetism are evidently similar yet different. What's the story?

Magnetism is a phenomenon that closely resembles electricity. Just as there are two types of electric charges which exert electrostatic forces on one another, so there are two types of magnetic poles which exert magnetostatic forces on one another. The word "pole" serves to distinguish magnetism from electricity; poles are magnetic while charges are electric.

The two types of poles are called north and south, respectively, and in keeping with this geographical naming, they're exact opposites of one another. Both types of poles carry just one physical quantity: magnetic pole. North poles carry positive amounts of magnetic pole while south poles carry negative amounts. It should come as no surprise that like poles repel while opposite poles attract. Furthermore, the magnetostatic forces between two poles grow weaker as they move apart and are inversely proportional to the square of the distance between them. So far, the similarities between electricity and magnetism are striking.

However, we now come to a crucial difference between electricity and magnetism: while subatomic particles that carry pure positive or negative electric charges are common, particles that carry pure north or south magnetic poles have never been found. Called magnetic monopoles, such pure magnetic particles may not even exist in our universe. That cosmic omission explains why there are no magnetic sparks: without monopoles, there is no magnetic equivalent of an electric charge that can leap from one place to another as a magnetic current, let alone a magnetic spark.

But while isolated magnetic poles aren't available in nature, *pairs* of magnetic poles are. These pairs consist of equal north and south poles, spatially separated from one another in an arrangement called a magnetic dipole. Since the two opposite poles have equal magnitudes, they sum to zero and the magnetic dipole has zero net magnetic pole.

A simple button magnet has both a north pole *and* a south pole, usually on opposite faces of the button (Fig. 11.1.1a). There are no purely north buttons or purely south buttons. Amazingly enough, even slicing a button magnet in half won't yield separated north and south poles (Fig. 11.1.1b). Instead, new poles will appear at the cut edges and each piece of the original magnet will end up with zero net pole! Breaking the button magnet in half (Fig. 11.1.1c) will also produce pieces with zero net pole.

We can now explain why two of these magnets sometimes attract and sometimes repel. With two poles on each magnet, we have to consider four interactions: two repulsive interactions between like poles (north-north and south-south) and two attractive interactions between opposite poles (north-south and south-north). While it might seem that all these forces should cancel, the distances separating the various poles and therefore the forces between them depend on the magnets' orientations. Since the closest poles experience the strongest forces, they dominate. If you turn two like poles toward one another, the two magnets will push apart (Fig. 11.1.2a). If you turn their opposite poles toward one another, they'll pull together (Fig. 11.1.2b). And if you tip them at an angle, they'll experience torques which tend to twist opposite poles together and like poles apart.

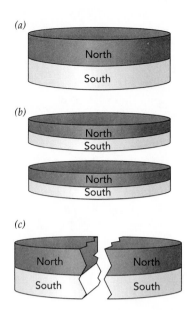

Fig. 11.1.1 (*a*) A typical button magnet has a north pole on one face and a south pole on the other. Its net pole is zero. (*b*) Slicing it between its poles or (*c*) breaking it through its poles always yields a pair of magnets, each with zero net pole.

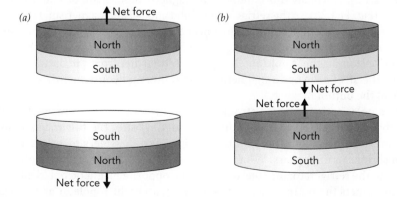

Fig. 11.1.2 (*a*) When like poles of two button magnets are turned toward one another, the magnets repel. (*b*) When opposite poles are turned toward one another, they attract.

Without monopoles, we're going to need some imagination to understand magnetism well. Let's start with units. Even though we can't collect a unit of pure north pole, we can still define such a unit and understand its behavior. The SI unit of magnetic pole is the ampere-meter (abbreviated A·m). That astonishing choice, an *electric* unit appearing in a *magnetic* unit, foreshadows the profound connections between electricity and magnetism that we'll soon encounter.

Just as there is a Coulomb's law for electric charges, there is a Coulomb's law for magnetic poles. Coulomb's magnetic experiments, which were complicated by the fact that he had to work with magnetic dipoles rather than individual poles, showed that the forces between magnetic poles are proportional to the amount of each pole and inversely proportional to the square of their separation. The exact relationship can be written:

$$\text{force} = \frac{\text{permeability of free space} \cdot \text{pole}_1 \cdot \text{pole}_2}{4\pi \cdot (\text{distance between poles})^2}. \tag{11.1.1}$$

The force on pole_1 is directed toward or away from pole_2, and the force on pole_2 is directed toward or away from pole_1. The permeability of free space is $4\pi \times 10^{-7}$ N/A². Consistent with Newton's third law, the force exerted on pole_1 by pole_2 is equal in amount but oppositely directed from the force exerted on pole_2 by pole_1.

Coulomb's Law for Magnetism

The magnitudes of the magnetostatic forces between two magnetic poles are equal to the permeability of free space times the product of the two poles divided by 4π times the square of the distance separating them. If the poles are like, then the forces they experience are repulsive. If the poles are opposite, then the forces are attractive.

The Refrigerator: Iron and Steel

While two button magnets can push or pull on one another, what if you have only one? The easiest way to observe magnetic forces is then to hold that single magnet near your refrigerator or another piece of iron or steel. The magnet is attracted to the refrigerator. But if you flip the button magnet over, thinking that it will now be repelled by the refrigerator, you'll be disappointed. Although the refrigerator is clearly magnetic, its magnetism somehow responds to the button magnet so that the two always attract.

Actually, the refrigerator's behavior isn't all that mysterious. Its steel is composed of countless microscopic magnets, each with a matched north pole and south pole (Fig. 11.1.3). Normally those individual magnetic dipoles are oriented semi-randomly (Fig. 11.1.3*a*), so the refrigerator exhibits no overall magnetism. However, as you bring one pole of a button magnet near the refrigerator, its tiny magnets evolve in size, shape, and orientation (Fig. 11.1.3*b*). Overall, opposite poles shift closer to the button magnet's pole and like poles shift farther from the button magnet's pole. The steel develops a magnetic polarization and consequently attracts the pole of the button magnet.

This polarization remains strong only as long as the button magnet's pole is nearby. When you remove the button magnet, most of the tiny magnets in the steel resume their semi-random orientations and the steel's magnetic polarization shrinks or disappears. When you then bring the button magnet's other pole close to the refrigerator, its steel develops the opposite magnetic polarization and again attracts the button magnet's pole. No matter which pole or assortment of poles you bring near the refrigerator, its steel will polarize in just the right way to attract those poles.

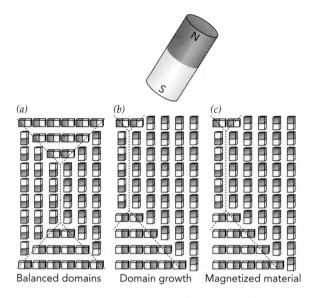

Fig. 11.1.3 (*a*) The countless microscopic magnets in iron or steel are normally oriented somewhat randomly. (*b*) But when a strong magnetic pole is nearby, those tiny magnets reorient to attract it. In soft magnetic materials, this reorientation is only temporary. (*c*) Hard magnetic materials, however, remain magnetized long after the external pole has departed.

(*a*) Balanced domains   (*b*) Domain growth   (*c*) Magnetized material

If you try this trick with a plastic or aluminum surface, the button magnet won't stick. What's special about steel that allows it to develop such a strong magnetic polarization? The answer is that ordinary steel, like its constituent iron, is a ferromagnetic material—it is actively and unavoidably magnetic on the size scale of atoms.

To understand ferromagnetism, we must start by looking at atoms and the subatomic particles from which they're constructed: electrons, protons, and neutrons. For complicated reasons, all of those subatomic particles have magnetic dipoles, particularly the electrons, and the atoms they form often display this magnetism. Despite a tendency for the subatomic particles to pair up with opposite orientations so that their magnetic dipoles cancel one another, most isolated atoms have significant magnetic dipoles.

But while most atoms are intrinsically magnetic, most materials are not. That's because another round of pairing and canceling occurs when atoms assemble into materials. This second round of cancellation is usually so effective that it completely eliminates magnetism at the atomic scale. Materials such as glass, plastic, skin, copper, or aluminum retain no atomic-scale magnetism at all and your button magnet won't stick to them. Even most stainless steels are nonmagnetic.

However, there are a few materials that avoid this total cancellation and thus manage to remain magnetic at the atomic scale. The most important of these are the ferromagnets, a class of magnetic materials that includes ordinary steel and iron. If you examine a small region of ferromagnetic steel, you'll find that it is composed of many microscopic regions or magnetic domains that are naturally magnetic and cannot be demagnetized (Fig. 11.1.3*a*). Within a single domain, all of the atomic scale magnetic dipoles are aligned and together they give the overall domain a substantial net magnetic dipole.

While common steel always has these magnetic domains, magnetic interactions orient nearby domains so that their magnetic dipoles oppose one another and cancel. The microscopic magnets balance one another so well that the steel appears non-magnetic. That's too bad; the appliance showroom would be a much more exciting place to visit if the cancellation weren't so good.

But when you bring a strong magnetic pole near steel (Fig. 11.1.3*b*), the individual domains grow or shrink, depending on which way they're oriented magnetically, and the steel becomes magnetized (Fig. 11.1.3*c*). The atoms themselves don't move during this process; the change is purely a reorientation of the atomic-scale magnetic dipoles. Domains that attract your button magnet's pole grow while those that repel it shrink and the button magnet sticks to the refrigerator.

Plastic Sheet Magnets and Credit Cards

When you remove your button magnet from the refrigerator, the steel returns to its original nonmagnetic state—it becomes demagnetized. Well, *almost* demagnetized. The demagnetization process isn't perfect because some of the domains get stuck. While magnetic forces within the steel favor complete return to apparent nonmagnetism, chemical forces can make it hard for the domains to grow or shrink. Adjacent domains are separated by domain walls—boundary surfaces between one direction of magnetic orientation and another. These domain walls must move if the domains are to change size. However, flaws and impurities in the steel can interact with a domain wall and keep it from moving. When that happens, the steel fails to demagnetize itself completely (Fig. 11.1.4). To remove the last bit of residual magnetism from steel, you must help the domain walls move, typically with heat or mechanical shock.

A soft magnetic material is one that demagnetizes itself easily when all nearby poles are removed. Chemically pure iron, which has few flaws or impurities, is a soft magnetic material—easy to magnetize and easy to demagnetize. A hard magnetic material is one that does not demagnetize itself easily and that tends to retain whatever domain structure is imposed on it by its most recent exposure to strong nearby poles (Fig. 11.1.3c). Your button magnet is made from a hard magnetic material!

Like steel, the material in your button magnet is ferromagnetic (or closely related to ferromagnetic). But unlike steel, your button magnet's domains do not shrink or grow easily. During its manufacture, the button magnet was magnetized by exposing it to such strong magnetic influences that its domains rearranged to give it permanent magnetic poles. It now has a north pole on one face and a south pole on

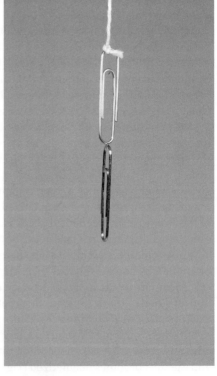

Fig. 11.1.4 (*a*) Although these paper clips were initially unmagnetized, the pole of a strong permanent magnet magnetizes them as a chain. (*b*) After the magnetizing magnet is removed, the clips retain some of their magnetization.

the other. Unless you expose the button to extremely strong magnetic influences, heat it, or pound it, it will retain its present magnetization almost indefinitely. In that respect, the button is a permanent magnet.

Not all permanent magnets are as simple as button magnets. Depending on how they were magnetized, they can have their north and south poles located in unexpected places and or even have more than one pair of poles. Plastic sheet magnets are a good example of multiple-pole magnets: each has a repeating pattern of alternate poles along its length. The exact patterns vary, but most have poles that form alternating parallel stripes. You can find these stripes by letting them polarize and attract iron powder (Fig. 11.1.5), or by sliding two identical sheet magnets across one another. The sheets will attract and bind together most strongly when opposite poles are aligned across from each other. And they'll repel when you shift one of the magnets so that like poles are aligned.

A hard magnetic material's ability to "remember" its magnetization can be useful for saving information. Once magnetized in a particular manner so as to represent a piece of information, the material will retain its magnetization and the associated information until something magnetizes it differently. Information retention in hard magnetic materials forms the basis for most magnetic recording and storage, including the magnetic stripes on credit cards, magnetic tapes, computer disks, and magnetic random access memory (MRAM) (Fig. 11.1.6).

Compasses

If you've spent time hiking, you may well own a magnetic compass. Like a button magnet, the needle of that compass is a simple permanent magnet with one north magnetic pole and one south magnetic pole. This needle aids navigation because the earth itself has a magnetic dipole and that dipole affects the orientation of the needle: the needle's north magnetic pole tends to point northward.

Already, we can see what must be located near the earth's *north* geographic pole: a *south* magnetic pole. Attraction from that south magnetic pole is what draws the compass's north magnetic pole toward the north. However, the full story is more complicated. To begin with, the earth's magnetic poles are actually located far beneath its surface and aren't perfectly aligned with the geographic poles. To make matters worse, magnetically active materials in everything from distant mountains to nearby buildings assert their own magnetic influences on the compass needle. Overall, the compass needle is responding to the influences of countless magnetic poles, both near and far. Given how difficult it would be to sum up all those separate influences, we do better to view the compass needle as interacting with something local: a magnetic field—an attribute of space that exerts a magnetostatic force on a pole. According to this new perspective, the compass needle responds to the local magnetic field, a field that's created by all of the surrounding magnetic poles.

As with an electric field, the magnetic field here appears to be acting merely as an intermediary: various poles produce the magnetic field and this magnetic field affects our compass needle. But as we'll see, a magnetic field is more than just an intermediary or fiction. It is quite real and can exist in space, independent of the poles that produce it. And just as electric fields can be created by things other than charge, so magnetic fields can be created by things other than pole.

The magnetic field at a given location measures the magnetostatic force that a unit of pure north pole would experience if it were placed at that point. More specifically, the magnetostatic force is equal to the product of the pole times the magnetic field and points in the direction of that magnetic field. We can write this relationship:

$$\text{force} = \text{pole} \cdot \text{magnetic field}, \tag{11.1.2}$$

Fig. 11.1.5 Iron powder forms bridges between the magnetic poles of this plastic sheet magnet.

Fig. 11.1.6 Iron powder discloses the locations of magnetic poles on this credit card's magnetic strip. The powder sticks between each pair of opposite poles. Information is stored as the locations of those poles.

Earth's magnetic field

Fig. 11.1.7 A compass needle aligns with the local magnetic field. Its north pole experiences a magnetostatic force in the direction of the field and its south pole experiences a force opposite the field.

where the force is in the direction of the magnetic field. Note, however, that a negative amount of pole (a south pole) experiences a force opposite the magnetic field. The SI unit of magnetic field is the newton-per-ampere-meter, also called the tesla (abbreviated T).

The earth's magnetic field is relatively weak, about 0.00005 T in a roughly northward direction. (For comparison, the field near your button magnet may be 0.1 T or more.) The earth's field pushes the compass needle's north pole northward and south pole southward (Fig. 11.1.7). Unless the compass needle is perfectly aligned with that field, it experiences a torque and undergoes angular acceleration. Since its mount only allows the needle to rotate horizontally and it experiences mild friction as it does, the needle soon settles down with its north pole pointing roughly northward. If its mount allowed it to rotate vertically as well as horizontally, the needle's north pole would dip downward in the northern hemisphere and upward in the southern hemisphere. In general, the needle minimizes its magnetostatic potential energy by pointing along the direction of the local magnetic field and is thus in a stable equilibrium when orientated that way. After a few swings back and forth, your compass needle points along the local magnetic field, which hopefully points northward.

Because the earth's magnetic field is so uniform in the vicinity of your compass, its northward push on the needle's north pole exactly balances its southward push on the needle's south pole and the needle experiences zero net force. But if you bring your compass near a button magnet, the local magnetic field will not be uniform and the needle may experience a net force. The magnetic field gets stronger near one of the button's poles and the compass needle will experience a net force toward or away from that pole, depending on which way it's orientated.

When the needle is aligned with a nonuniform field—its north pole pointing in the same direction as the local field—the forces on its two opposite poles won't balance and it will experience a net force in the direction of increasing field. If it is aligned against the field, it will experience a net force in the direction of decreasing field. In practice, as you bring the compass near your button magnet, its needle will first pivot into alignment with the local field and then find itself pulled toward increasing field, toward the nearest pole of the button magnet. The same thing happens when you bring two button magnets together: each pivots into alignment with the other's magnetic field and the two then leap at each other. Watch out for your fingers!

A piece of steel exhibits similar behavior when you hold it near a button magnet: it becomes magnetized along the direction of the local magnetic field and then finds itself pulled toward increasing field, toward the button magnet's nearest pole. That's how the button magnet holds your notes to the refrigerator!

Iron Filings and Magnetic Flux Lines

Magnetic fields seem so abstract; it would be helpful if you could see them. Remarkably enough, you can: just sprinkle iron filings into the field! Though you'll need to support their weight with paper or a liquid, an interesting pattern will form. Like tiny compass needles, iron particles magnetize along the local magnetic field and then stick together, north pole to south pole, in long strands that delineate the magnetic field (Fig. 11.1.8)!

These strands map the magnetic field in an interesting way. First, at each point on a strand, the strand points along the local magnetic field. Second, the strands are most tightly packed where the local magnetic field is strongest. In other words, the strands follow along the local magnetic field direction and have a density proportional to that local field. The lines highlighted by these strands are so useful that they have their own names: magnetic flux lines.

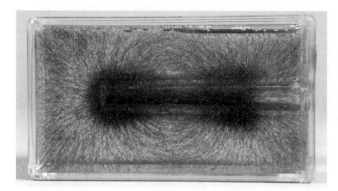

Fig. 11.1.8 Supported by a liquid, this iron powder shows the magnetic flux lines around a small bar magnet.

Flux lines are often helpful when exploring a magnetic field. If you're studying the magnetic field in a large area, you probably don't want to use iron filings. Instead, you can hold a compass in your hand and walk in the direction its needle is pointing—the direction of the magnetic field. The path you'll follow in this compass-guided walk is a magnetic flux line. If you repeat this trip from many different starting points, you'll explore the whole magnetic field, flux line by flux line. Since a magnetic field tends to point away from north poles and toward south poles, these tours will typically take you from north poles to south poles. In fact, for our permanent magnets, every magnetic flux line begins at a north pole and ends at a south pole.

That last observation about flux lines is quite general: they never start at or end on anything other than a magnetic pole. While flux lines emerge in all directions from a north pole and converge from all directions on a south pole, that's it; flux lines never begin or end in empty space. If you're following a magnetic flux line with your compass, you will either reach a south pole or walk forever!

The possibility of that endless walk is somewhat disconcerting; if the flux line you're following doesn't end at a pole, what created its magnetic field? The answer reveals a deep connection between magnetism and electricity: some magnetic fields aren't produced by magnetic poles at all, they're produced by electricity! To see how that's possible, let's take a look at another common household magnet: the electromagnet in an ordinary doorbell.

Electric Doorbells and Electromagnets

A classic electric doorbell uses a magnet and a spring to drive a piece of iron into two chimes, "ding-dong." When you press the doorbell button, you close an electric circuit and the resulting electric current pushes the iron *magnetically* into the first chime, "ding." When you release the button, you open the circuit, stopping the current and its magnetism so that the spring can push the iron back into the second chime, "dong."

The big news here is that electric currents can produce magnetic forces. In fact, there is nothing optional about this connection: electric currents *are* magnetic. More specifically, moving electric charge produces a magnetic field.

First Connection between Electricity and Magnetism
Moving electric charge produces a magnetic field.

Imagine the surprise of Danish physicist Hans Christian Oersted (1777–1851) when he observed in 1820 that current in a wire caused a nearby compass needle to rotate. Until that moment, electricity and magnetism had appeared as independent phenomena. Inspired by Oersted's experiment, French physicist André-Marie Ampère ❶ undertook a 7-year study of the relationships between electricity and

❶ Self-educated before the French Revolution, during which his father was executed, French physicist André-Marie Ampère (1775–1836) became a science teacher in 1796. He served as a professor of physics or mathematics in several cities before settling at the University of Paris system in 1804. In 1820, only a week after learning of Oersted's experiment showing that an electric current causes a compass needle to deflect, Ampère published an extensive treatment of the subject. Evidently, he had been thinking about these ideas for a long time.

Fig. 11.1.9 Iron powder shows that flux lines around a current-carrying wire form concentric rings around that wire.

magnetism, and started the revolution that eventually unified them within a single overarching conceptual framework.

When we use iron powder to disclose the magnetic flux lines surrounding a long, straight current-carrying wire, we too are in for a surprise (Fig. 11.1.9). Those flux lines circle the wire like concentric rings, growing more widely separated as the distance from the wire increases. The wire is an electromagnet—a device that becomes magnetic when it carries an electric current. But because an electromagnet has no true magnetic poles, the magnetic flux lines can't stretch from north pole to south pole. Instead, each flux line of an electromagnet is a closed loop. If you took a compass-guided walk along one of these flux lines, you'd retrace your steps over and over.

Since the flux lines are packed tightest near the surface of the current-carrying wire, that's where the magnetic field is strongest. Recalling that a piece of iron is pulled toward increasing magnetic field, we see that the wire will attract iron to it whenever it's carrying a current.

The magnetic field around a current-carrying wire is fairly weak, however, and a practical doorbell winds that wire into a coil to concentrate and strengthen its field. While the magnetic field around a current-carrying coil is complicated, we can use iron powder to make it visible (Fig. 11.1.10). Remarkably enough, the flux lines outside the coil resemble those outside a button magnet of similar dimensions (Fig. 11.1.11). It's as though the coil has a north pole at one end and a south pole at the other! But because there are no true poles present, the flux lines don't end anywhere. Instead, they continue right through the middle of the coil and form complete loops.

When current flows through the coil, nearby iron finds itself magnetized along the local magnetic field and then pulled toward increasing field—toward the tightly packed flux lines at the coil's end. But why stop there? Since the flux lines continue right into the coil and grow even more tightly packed inside, the iron will be pulled inward toward the very center of the coil!

That's how the doorbell works. When you press the doorbell button, current flows through a coil of wire and the resulting magnetic field yanks an iron rod into the center of that coil. About the time the rod reaches the center, part of it hits the first chime. When you then open the switch, stopping the current and its magnetism, a spring pushes the iron rod back out of the coil and it hits the second chime. These two chimes make the familiar ding-dong!

While current is flowing through the coil and the iron rod is inside it, the two objects act as a single, powerful electromagnet. The magnetic field surrounding the pair is the sum of the coil's modest magnetic field and the magnetized iron's much stronger field. In effect, the current in the coil magnetizes the iron and the iron then creates most of the surrounding magnetic field. Practical electromagnets, which

Fig. 11.1.10 Iron powder shows that flux lines pass straight through a current-carrying coil and return outside it, much like the flux lines around a similarly shaped bar magnet. In case you're wondering, you're looking at an open spiral-bound notebook with current passing through its spiral coil. The steel wire is almost red hot and has begun to burn the paper!

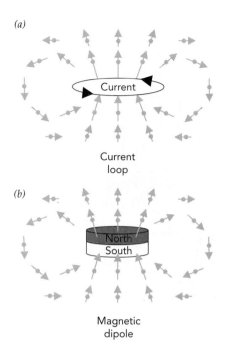

(a)

Current loop

(b)

North
South

Magnetic dipole

Fig. 11.1.11 (a) The magnetic field around a loop of current-carrying wire points up through the loop and down around the outside of the loop. The magnetic field arrow passing through each green dot indicates the magnitude and direction of the force a north test pole would experience at the dot's location. (b) The field produced by a two-pole button magnet is almost identical to that of the loop.

control switches and valves in your furnace or air conditioner and can lift cars at the scrap yard, generally use iron or related materials to dramatically enhance the magnetic field produced by a current in a coil of wire (Fig. 11.1.12).

Fig. 11.1.12 This electric switch is controlled by an electromagnet and is called a relay.

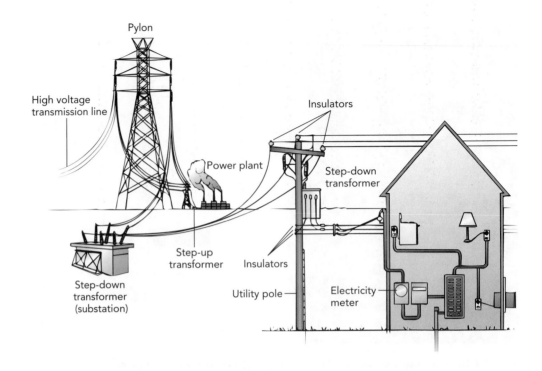

Pylon

High voltage
transmission line

Insulators

Power plant

Step-down
transformer

Step-up
transformer

Insulators

Step-down
transformer
(substation)

Utility pole

Electricity
meter

SECTION 11.2 Electric Power Distribution

Electricity is a particularly useful and convenient form of ordered energy. Because it's delivered to our homes and offices as a utility, we barely think about it except to pay the bills. The wires that bring it to us never plug up or need cleaning and work continuously except when there's a power failure, blown fuse, or tripped circuit breaker.

Just how does electricity get to our homes? In this section, we'll look at the problems associated with distributing electricity far from the power plant at which it's generated. To understand these problems, we'll examine the ways in which wires affect electricity and see how electric power is transferred and rearranged by devices called transformers.

WARNING

__Electricity is dangerous, particularly when it involves high voltages__. The principal hazard is that an electric current will pass through your body in the vicinity of your heart and disrupt its normal rhythm. While very little current is needed to cause trouble, your skin is such a poor conductor that it normally keeps harmful currents from passing through you. However, large voltages can propel dangerous currents through your skin and put you at risk. While your body usually has to be part of a closed circuit for you to receive a shock, don't count on the absence of an identifiable circuit to protect you from injury—circuits have a tendency to form in surprising ways whenever you touch an electric wire. Be especially careful whenever you're near voltages of more than 50 V, even in batteries, or when you're near any voltages if you're wet or your skin is broken.

Direct Current Power Distribution

Batteries may be fine for powering flashlights, but they're not very practical for lighting homes. Early experiments that placed batteries in basements were disappointing because the batteries ran out of energy quickly and needed service and fresh chemicals all too frequently.

A more cost-effective source for electricity was coal- or oil-powered electric generators. Like batteries, generators do work on the electric currents flowing through them and can provide the electric power needed to illuminate homes. But while generators produce electricity more cheaply than batteries, early ones were large machines that required fresh air and attention. These generators had to be built centrally, with people to tend them and chimneys to get rid of the smoke.

This was the approach taken by the American inventor Thomas Alva Edison (1847–1931) in 1882 when he began to electrify New York City. Each of the Edison Electric Light Company's generators acted like a mechanical battery, producing direct current that always left the generator through one wire and returned to it through another. Edison placed his generators in central locations and conducted the current to and from the homes he served through copper wires. However, the farther a building was from the generator, the thicker the copper wires had to be. That's because wires impede the flow of current and making them thicker allows them to carry current more easily.

Wire thickness is important because, like the filament of the flashlight bulb we studied in the previous chapter, wires have electrical resistance. In accordance with Ohm's law (Eq. 10.3.4), the voltage drop through a wire is equal to its electrical resistance times the current passing through it. In the case of a wire conducting current from a generating plant to a home, our primary concern is how much power the wire wastes as thermal power. We can determine this wasted power by combining Ohm's law with the equation for power consumed by a device (Eq. 10.3.1):

$$\text{power consumed} = \text{voltage drop} \cdot \text{current}$$
$$= (\text{current} \cdot \text{electrical resistance}) \cdot \text{current}$$
$$= \text{current}^2 \cdot \text{electrical resistance.} \qquad (11.2.1)$$

The wire's wasted power is proportional to the square of the current passing through it! This relationship became all too clear to Edison when he tried to expand his power distribution systems. The more current he tried to deliver over a particular wire, the more power it lost as heat. Doubling the current in the wire quadrupled the power it wasted.

Edison tried to combat this loss by lowering the electrical resistances of the wires. He used copper because only silver is a better conductor of current. He used thick wires to increase the number of moving charges. And he kept the wires short so that they didn't have much chance to waste power. This length requirement forced Edison to build his generating plants within the cities he served. Even New York City contained many local power plants. (For an interesting tale about the early days of electric power, see ❶.)

Edison also tried to avoid waste by delivering smaller currents at higher voltages. Since the delivered power is equal to the voltage drop times the current (see Eq. 10.3.1), Edison could reduce the current flowing through the wires by raising the voltage. Although less current flowed through each home, it underwent a larger voltage drop so the power delivered was unchanged. However, high voltages are dangerous because they tend to create sparks as current jumps through the air. They also produce nasty shocks when current flows through your body. While high voltages could be handled safely outside a home, they could not be brought inside. Edison used the highest voltages that safety would allow.

❶ Love Canal is the United States' most famous toxic waste dump. The dump was created in the 1920s at an abandoned section of canal constructed in 1892 by William T. Love. Love intended his canal to connect the upper and lower Niagara Rivers, so that the descending water could be used to generate DC electric power for the citizens of Niagara Falls, New York. The advent of AC power transmission systems in 1896 made the canal less useful, and it was never finished.

Although scientists have discovered a number of materials that lose their electrical resistance at extremely low temperatures and become perfect electrical conductors or superconductors (Fig. 11.2.1), these superconductors are still too impractical for power distribution systems. Their use is limited to local applications such as large electromagnets and specialized electronic devices.

Fig. 11.2.1 A magnetic cylinder floats above the surface of a superconductor at 78 K. Currents flowing freely in the superconductor make it magnetic and cause it to repel the magnetic cylinder. Its expulsion of magnetic flux lines is known as the Meisner effect.

Introducing Alternating Current

The real problem with distributing electric power via direct current (DC) is that there's no easy way to transfer power from one DC circuit to another. Because the generator and the lightbulbs must be part of the same circuit, safety requires that that entire circuit use low voltages and large currents. DC power distribution therefore wastes much of its power in the wires connecting everything together.

However, as we'll soon see, alternating current (AC) makes it easy to transfer power from one AC circuit to another, so that different parts of the alternating current power distribution system can operate at different voltages with different currents. Most significantly, the wires that carry the power long distances are part of a high-voltage, low-current circuit and therefore waste little power.

An alternating current is one that periodically reverses direction—it alternates. For example, when you plug your desk lamp into an AC electrical outlet and switch it on, the current that flows through the lamp's filament reverses its direction of travel many times each second.

The power company propels this alternating current through the lamp's filament by subjecting it to an alternating voltage drop—a voltage drop that periodically reverses direction. As you may recall from Section 10.3 on flashlights, current in an ohmic filament flows down a voltage gradient from positive to negative, much as bicyclists roll down an altitude gradient from high to low. While a flashlight's battery subjects the flashlight's filament to a steady voltage drop and obtains a direct current, the power company subjects your lamp's filament to an alternating voltage drop and obtains an alternating current.

Alternating voltages are present at any AC electrical outlet. In the United States, an ordinary AC outlet offers three connections: *hot*, *neutral*, and *ground* (Fig. 11.2.2). In a properly installed outlet, the absolute voltage of *neutral* remains near 0 V (0 volts) while the absolute voltage of *hot* alternates above and below 0 V. *Ground*, which is an optional safety connection that we'll discuss later, also remains near 0 V absolute.

One end of your lamp's filament is connected to *hot* and the other to *neutral*. Since current always flows through the filament from higher voltage to lower volt-

Fig. 11.2.2 This electric outlet follows the U.S. standard for 120-V AC, 15-A service. The wide slots (left) are neutral, the narrow slots (right) are hot, and the curved holes (centered) are ground. This outlet provides ground-fault circuit interruption (GFCI) protection: if any current leaving hot fails to return to neutral, or vice versa, the outlet shuts off instantly until it is reset. The test button simulates a current leak and will shut off the outlet if its protection is functioning properly.

age, it flows from *hot* to *neutral* when *hot* has a positive voltage and from *neutral* to *hot* when *hot* has a negative voltage.

In normal AC electric power, the *hot* voltage varies sinusoidally—it's proportional to the trigonometric sine function with respect to time (Fig. 11.2.3). This smoothly alternating voltage propels a smoothly alternating current through the lamp. During each reversal, the current in the filament gradually slows to a stop before gathering strength in the opposite direction. In the United States, AC voltages reverse every 120th of a second, yielding 60 full cycles of reversal (back and forth) each second (60 Hz). In Europe, the reversals occur every 100th of a second, so AC voltages complete 50 full cycles of reversal each second (50 Hz).

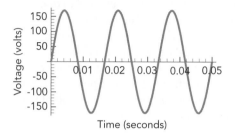

Fig. 11.2.3 The voltage of the hot wire of a U.S. 120-V AC outlet varies sinusoidally in time and completes 60 full cycles per second. While it peaks at ±170 V, its effective time average or RMS voltage is 120 V. The voltage of the neutral wire is always 0 V.

Fortunately, these reversals have little effect on many household devices. Lamps and toasters (Fig. 11.2.4) consume power because of their electrical resistances and don't care which way current passes through them. In fact, power consumption in such simple ohmic devices is used to define an effective voltage for AC electric power. An outlet's nominal AC voltage—technically, its root mean square or RMS voltage—is defined to be equal to the DC voltage that would cause the same average power consumption in an ohmic device. Thus 120-V AC power delivers the same average power to a toaster as 120-V DC power.

However, the reversals of AC power aren't without consequence. First, some electric and most electronic devices are sensitive to the direction of current flow and must handle the reversals carefully. Second, the power available from an ordinary AC outlet rises and falls with each voltage reversal and is momentarily zero at the reversal itself. Your lamp actually flickers slightly because of these power fluctuations and devices that can't tolerate even an instant without power must store energy to avoid shutting down during the reversals.

Finally, AC power's sinusoidally varying voltages peak well above their nominal values, exceeding those values by a factor of the square root of 2 (about 1.414). For example, the voltage of the *hot* connection in an ordinary 120-V AC power outlet actually swings between +170 V and -170 V. Those higher peak voltages are important for insulation and electrical safety.

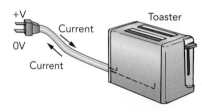

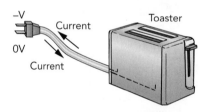

Fig. 11.2.4 The current passing through this toaster reverses directions periodically as the voltage of the toaster's hot terminal (top) reverses. The toaster's neutral terminal (bottom) remains at 0 V.

Magnetic Induction

Edison was adamantly opposed to alternating current, which he viewed as dangerous and exotic. Indeed, its fluctuating voltages and moments without power don't make alternating current look attractive at all.

The champion of alternating current was Nikola Tesla (1856–1943), a Serbian-American inventor, who was backed financially by the American inventor and industrialist George Westinghouse (1846–1914). The advantage that Tesla and Westinghouse saw in alternating current was that its power could be transformed—it could be passed via electromagnetic action from one circuit to another by a device called a transformer.

A transformer uses two important connections between electricity and magnetism to convey power from one AC circuit to another. The first is familiar: moving

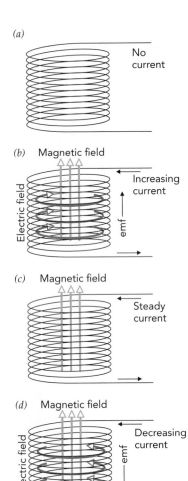

(a) No current

(b) Magnetic field — Increasing current — Electric field — emf

(c) Magnetic field — Steady current

(d) Magnetic field — Decreasing current — Electric field — emf

Fig. 11.2.5 (*a*) Without current, this inductor has neither electric nor magnetic fields. (*b*) An increasing current produces an increasing magnetic field in the inductor, which in turn produces an electric field. The emf resulting from that electric field opposes the current increase. (*c*) A steady current produces only a steady magnetic field. (*d*) A decreasing current produces a decreasing magnetic field in the inductor, which in turn produces another electric field. However, the resulting emf now opposes the current decrease.

electric charge creates magnetic fields. This connection allows electricity to produce magnetism. The second connection, however, is something new: magnetic fields that change with time create electric fields. Discovered in 1831 by Michael Faraday ❷, this relationship allows magnetism to produce electricity!

Second Connection between Electricity and Magnetism
Magnetic fields that change with time produce electric fields.

Whether you wave a permanent magnet back and forth, or switch an electromagnet on and off, you are changing a magnetic field with time and thereby producing an electric field. If there are mobile electric charges around to respond to that electric field, they'll accelerate and you'll have created or altered an electric current and possibly done work on it as well. This process, whereby a time-changing magnetic field initiates or influences an electric current is called magnetic induction.

A transformer combines these two connections in sequence—electricity produces magnetism produces electricity. But rather than returning electric power to where it started, the transformer moves that power from the current in one coil of wire through a magnetic field to the current in a second coil of wire.

Alternating Current and a Coil of Wire

To help us understand the power transfer that takes place in a transformer, let's start with a simpler case. What happens when you send an alternating current (AC) through a single coil of wire?

Because currents are magnetic, the coil becomes an electromagnet. However, since the current passing through it reverses periodically, so does its magnetic field. And because a magnetic field that changes with time produces an electric field, the coil's alternating magnetic field produces an alternating electric field.

This electric field has a remarkable effect: it pushes on the very alternating current that produces it! While it's not obvious how this electric field should affect that current, the result turns out to be simple (Fig. 11.2.5). As the coil's current increases, the induced electric field pushes that current backward and thereby opposes its increase (Fig 11.2.5*b*). And as the coil's current decreases, the induced electric field pushes that current forward and thereby opposes its decrease (Fig 11.2.5*d*). Amazingly, however the coil's current changes, the induced electric field always opposes that change!

This opposition to change is universal in magnetic induction, where it's known as Lenz's law: the effects of magnetic induction oppose the changes that produce them. In the present case, self-directed magnetic induction or "self-induction" leads our coil to oppose its own changes in current. A wire coil's natural opposition to current change makes it quite useful in electrical equipment and electronics, where it's called an inductor.

Lenz's Law
The effects of magnetic induction oppose the changes that produce them.

However, magnetic induction does more than just push currents around; it can also transfer energy. Its induced electric field does work on any charge that moves with its push and negative work on any charge that moves opposite its push.

When induction does work on a charge that goes through our coil, that charge experiences a rise in voltage. The overall voltage rise, from the coil's start to its finish, is called the induced emf (short for electromotive force). But because the coil's current and induced electric field alternate, so does the induced emf—it swings between positive and negative voltages. And for related reasons, energy alternately leaves the current and returns. But where does that energy reside when it's not in the current?

The missing energy is in the coil's magnetic field! Magnetic fields contain energy. The amount of energy in a uniform magnetic field is half the square of the field strength times the volume of the field divided by the permeability of free space. We can write this relationship as:

$$\text{energy} = \frac{\text{magnetic field}^2 \cdot \text{volume}}{2 \cdot \text{permeability of free space}}. \tag{11.2.2}$$

Common Misconceptions: Magnets as Limitless Sources of Energy

Misconception: Magnets are infinite sources of energy that could provide electric or mechanical power forever!

Resolution: While a magnet's field does contain energy, that energy is limited and was invested in it during its magnetization. To extract that energy, you'd have to demagnetize and thus destroy the magnet.

In effect, our coil is playing with the alternating current's energy, storing it briefly in the magnetic field and then returning it to the current. The coil stores energy while the magnitude of the current increases—the field strengthens and the current loses voltage. The coil returns energy while the magnitude of the current decreases—the field weakens and the current gains voltage. Because the coil's self-induced emf is responsible for bouncing this energy back to the current, it's frequently called a back emf.

The coil's self-induction and back emf allow it to handle alternating currents and alternating voltages with astonishing grace. You can actually plug the two ends of a properly designed coil into an AC electrical outlet without any trouble at all; the coil will rhythmically store energy and return it. That's not a stunt you'd want to try with an ordinary wire!

Unlike an ordinary wire, which can't safely receive current from the outlet at one voltage and return it to the outlet at a different voltage, the coil can use its back emf to "ride" the outlet's alternating voltage like a bottle riding ocean waves. Pushed forward or backward by the induced emf, current can enter this coil at one voltage and leave it at a completely different voltage. In fact, the coil's back emf always has just the right voltage so that current passing through the coil's *hot* end at the *hot* voltage passes through the coil's *neutral* end at the neutral voltage (0 V). For example, when *hot* is +170 V, the back emf is −170 V; when *hot* is −50 V, the back emf is +50 V; and when *hot* is 0 V, the back emf is 0 V.

Two Coils Together: A Transformer

In a single coil, energy that's transferred from the current to the magnetic field must eventually return to the current. It has nowhere else to go. But what if there are two coils and two currents? In that case, energy transferred from one current to the magnetic field can move on to the second current!

That possibility is the basis for a transformer. In its simplest form, a transformer consists of two separate coils that share the same electromagnetic environment. Some or all of the energy invested in the magnetic field by current in the first coil can be withdrawn from the magnetic field by current in the second coil. Although the two currents never touch and don't exchange a single charge, power can move from one current to the other with ease.

We'll illustrate this energy transfer by examining an ordinary halogen desk lamp (Fig. 11.2.6). This device consists of a two-coil transformer and a halogen lightbulb. One coil of the transformer, its "primary" coil, is plugged directly into an electrical

❷ With only a primary education, English chemist and physicist Michael Faraday (1791–1867) apprenticed with a bookbinder at 14. At 21, he became a laboratory assistant to Humphry Davy, a renowned chemist. Faraday's experiments with electrochemistry and his knowledge of work by Oersted and Ampère led him to think that, if electricity can cause magnetism, then magnetism should be able to cause electricity. Through careful experimentation, he found just such an effect. Toward the end of his career, Faraday became a popular lecturer on science and made a particular effort to reach children.

Fig. 11.2.6 The bulb in this halogen desk lamp operates from low voltages provided with the help of a transformer. The lamp's two support rods also carry 12-V AC current to and from the bulb.

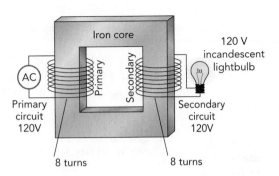

Fig. 11.2.7 This transformer conveys power from current in its primary circuit to current in its secondary circuit. The iron core guides magnetic flux lines so that the two coils share the same electromagnetic environment. With equal turns in its primary and secondary coils, this transformer supplies the same AC voltage from its secondary coil as it receives at its primary coil.

outlet and completes a circuit with the power company (Fig. 11.2.7). The power company pushes an alternating current through this primary circuit. The other coil of the transformer, its "secondary" coil, is connected to the lightbulb and completes another circuit—the secondary circuit. To ensure that the two coils share the same electromagnetic environment, both are wound around a ring-shaped magnetizable core. We'll discuss how that core works later in this section.

By itself, the primary coil acts as an inductor, alternately storing energy in its magnetic field and then returning that energy to the primary current by way of its back emf. Since this back emf mirrors the supply voltage, which we'll suppose is 120 volts AC, the back emf is also 120 volts AC.

However, because the secondary coil shares the primary coil's electromagnetic environment, the secondary coil also experiences an induced emf and a voltage difference appears between its two ends. Since the secondary coil forms a circuit with the lamp's filament, this voltage difference imposes a voltage drop on the filament and propels a current through it. That current alternates because the emf alternates. In short, power is moving via electromagnetic action from an alternating current in the primary circuit to an alternating current in the secondary circuit and lighting up the bulb.

If the transformer is providing power to current in its secondary circuit, it must be removing the same amount of power from current in its primary circuit. Sure enough, it's using magnetic induction to do just that. However, this time the induction is reversed: the current in the secondary circuit is inducing an emf in the primary coil and that emf is removing power from the primary current!

This removal happens in an interesting way. The emf induced in the primary coil increases the primary current whenever it's investing energy in the transformer's magnetic field and decreases that current whenever it's withdrawing energy from the field. With more investment than withdrawal, the primary current is leaving energy behind in the magnetic field and the secondary current is carrying that energy away!

Remarkably, the power transfer process responds automatically to any changes in the secondary circuit's power consumption. For example, if you replace the desk lamp's bulb with one that consumes more power, more current will flow through the secondary coil so induction will transfer more power from the primary current to the secondary current. And if you remove the bulb completely, the secondary current will vanish and the primary current's energy investment and withdrawal will balance perfectly.

Although that last observation implies that a transformer with nothing attached to its secondary coil consumes no power, real transformers aren't quite perfect. You'll save energy by unplugging transformers that aren't providing any power.

Changing Voltages

Transformers may be interesting, but why does the desk lamp need one? Why not just connect the bulb directly to the power outlet to form a single circuit with the power company?

In the desk lamp, the transformer's job is to provide the bulb with low-voltage AC power. Like the bulb in a flashlight, the desk lamp's bulb is designed to operate on small voltages. This low-voltage bulb derives its heating power from a large current experiencing a small voltage drop, so its filament has a small electrical resistance and is thick, short, and sturdy. In contrast, a high-voltage bulb derives its heating power from a small current experiencing a large voltage drop, so its filament needs a large electrical resistance and must be thin, long, and fragile. The shorter low-voltage filament is also a more concentrated light source, making it ideal for a desk lamp. To provide this low-voltage AC power, the transformer's secondary coil is wound differently from its primary coil.

In any transformer, the secondary coil experiences an induced emf that depends on its number of turns—the number of times its wire encircles the core. The more loops the secondary current makes around the core, the more work the transformer's electric field does on that current and the larger the induced emf. Since the amount of work done is proportional to the number of turns, so is the secondary coil's induced emf.

But what is the actual induced emf in a specific transformer? Suppose that we have a simple transformer in which the two coils, primary and secondary, are identical—they have equal numbers of turns—and that the primary coil of that transformer is plugged into a 120-V AC outlet.

Since the transformer's primary coil acts as an inductor, its back emf mirrors the AC voltage applied to it and is therefore 120 V AC. But because the two coils are identical and share the same electromagnetic environment, that same induced emf appears in the secondary coil: 120 V AC. If we connect the secondary coil to an appropriate bulb to form a complete circuit, the secondary coil will act as a source of 120-V AC electric power and light up the bulb.

This simple device is known as an isolation transformer. When you plug its primary coil into an AC outlet, its secondary coil acts as a source of AC power at the outlet's voltage. Though its secondary coil merely mimics the power outlet, an isolation transformer provides an important measure of electrical safety. Since its primary and secondary circuits are electrically isolated, charge can't move between those circuits and cause trouble. For example, when lightning strikes the power company's wires, the resulting burst of charge on the primary circuit can't pass to any appliances that are part of the secondary circuit. Not surprisingly, hospitals often employ isolation transformers to protect patients from shocks.

But most transformers have unequal coils and therefore different emfs in their coils. Since the secondary coil's induced emf is proportional to its number of turns, it acts as a source of AC power with a voltage equal to the voltage applied to its primary coil times the ratio of secondary turns to primary turns or

$$\text{secondary voltage} = \text{primary voltage} \cdot \frac{\text{secondary turns}}{\text{primary turns}}. \quad (11.2.3)$$

An isolation transformer is simply the special case where the turn numbers are equal and their ratio is 1.

The transformer in our desk lamp is called a step-down transformer because it has fewer secondary turns than primary turns and provides a secondary voltage that is less than the primary voltage (Fig. 11.2.8). If we suppose that the ratio of secondary turns to primary turns is only 0.1, the secondary coil will act as a source of 12-V AC power.

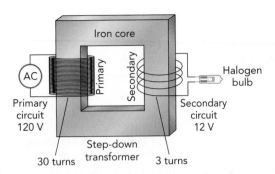

Fig. 11.2.8 With 10 times as many turns in its primary coil as in its secondary coil, this step-down transformer transforms 120-V AC power to 12-V AC power. To conserve energy, the current in the secondary circuit is 10 times that in the primary circuit. The total power being delivered to the transformer by the primary circuit is therefore the same as the total power being delivered to the bulb by the secondary circuit.

Not surprisingly, there are also step-up transformers that have more secondary turns than primary turns and that provide secondary voltages that are greater than their primary voltages (Fig. 11.2.9). The transformer that powers a neon sign typically has 100 times as many turns in its secondary coil as in its primary coil. When its primary coil is supplied by 120-V AC power, its secondary coil provides the 12,000-V AC power needed to illuminate the neon tube.

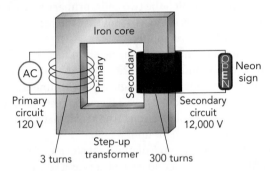

Fig. 11.2.9 With 100 times as many turns in its secondary coil as in its primary coil, this step-up transformer transforms 120-V AC power to 12,000-V AC power. To conserve energy, the current in the secondary circuit is 0.01 times that in the primary circuit. The total power being delivered to the transformer by the primary circuit is therefore the same as the total power being delivered to the neon sign by the secondary circuit.

Even when its primary and secondary voltages are different, a transformer still manages to conserve energy. While each additional secondary coil turn increases the secondary coil's voltage, it also increases the rate at which the secondary current withdraws energy from the transformer's magnetic field. If you scale up the number of turns in the secondary coil to increase its voltage, you must scale down the current flowing through that coil in order to leave the amount of energy it withdraws from the magnetic field unchanged. As a result, the secondary current is equal to the primary current times the ratio of primary turns to secondary turns or

$$\text{secondary current} = \text{primary current} \cdot \frac{\text{primary turns}}{\text{secondary turns}}. \quad (11.2.4)$$

In our desk lamp's step-down transformer, the primary coil has 10 times as many turns as the secondary coil. According to Eq. 11.2.3 the secondary voltage is 0.1 times the primary voltage, and according to Eq. 11.2.4 the secondary current is 10 times the primary current. If the 12-V bulb that you install in the desk lamp has been designed to consume 24 W of power, a current of 2 A will flow through the secondary circuit. To provide this power, the transformer's primary coil will carry 0.2 amperes of current supplied at 120 V AC. In all, 24 watts of power are flowing from the transformer's primary circuit to its secondary circuit.

Real Transformers: Not Quite Perfect

Although we've been pretending that inductors and transformers are flawless and that their wires conduct electricity perfectly, that's not quite true. In reality, the wires used in those devices have electrical resistances and waste power in proportion to the squares of the currents they carry. To minimize this wasted power, real inductors and transformers are designed so as to minimize their resistances. To the extent practical, they employ thick wires made of highly conducting metals and those wires are kept as short as possible.

Unfortunately, inductors and transformers built only from wires can't develop the strong magnetic fields they need to store large amounts of energy unless they use long, many-turn coils. To avoid long coils, many inductors and virtually all transformers wrap their coils around magnetizable cores. Such cores respond magnetically to the alternating currents around them, boosting the resulting magnetic fields and making it easier to store large amounts of energy. Aided by those magnetizable materials—typically iron or iron alloys—cored inductors and transformers work well even with short, few-turn coils.

A core provides another crucial benefit to a transformer: it guides the transformer's magnetic flux lines so that nearly all of them pass through both coils, even when those coils are somewhat separated in space. Sharing their flux lines in that manner gives the coils a common electromagnetic environment and permits them to exchange electric power easily.

Making two separate coils share their flux lines isn't easy. Since a coil has no net magnetic pole, each flux line that emerges from it must ultimately return to it. But without a core, most flux lines leaving a coil return to it almost directly and remain nearby throughout their trip. Those unadventurous flux lines are unlikely to pass through a second, separate coil. Not surprisingly, a coreless transformer only works well when its two coils are wound so closely together that they can't help but share the same flux lines.

Winding both coils around a ring-shaped magnetic core makes it easy for the flux lines to pass through both coils because those flux lines are drawn into the core's soft magnetic material and follow it as if in a pipe. Although the flux lines leaving a coil must still return to it eventually, most of them complete that trip by way of the core—a journey that then takes them through the other coil. With nearly all the flux lines channeled by the core through both coils, power can flow easily from one coil to the other.

A core thus provides a transformer with great flexibility: its coils can be practically anywhere as long they encircle that core. However, cores aren't quite perfect pipes for flux; they leak slightly. Therefore, the most efficient transformers have coils that are wound nearby or on top of one another.

But while magnetizable cores make small, efficient transformers practical, they also introduce a few problems. First, the cores must magnetize and demagnetize easily in order to keep up with the energy investment and withdrawal processes. If they lag behind, they'll waste power as thermal power. Sadly, perfect magnetic softness is unobtainable and all cores waste at least a little power through delays in their magnetizations.

Second, because these cores are subject to the same electric fields that push currents around in the coils, they shouldn't conduct electricity. If they do, they'll develop useless internal currents known as eddy currents and thereby waste power

❸ Most metal detectors are essentially complicated transformers with missing coils and cores. The metal that is being detected effectively provides those missing parts. A simple metal detector, such as that commonly used in beachcombing, supplies its detecting coil with a high-frequency alternating current. This device detects the presence of nearby metal when it senses the change in the coil's self-inductance caused by its electromagnetic interactions with the metal. More sophisticated metal detectors, such as those used in airport security checkpoints, often use two separate coils to detect nearby metal—one to induce currents and magnetizations in the metal and the second to measure emfs produced by the changing magnetic fields around that metal.

heating themselves up. Since most soft magnetic materials are electrical conductors, transformer cores are frequently divided up into insulated particles or sheets so that little or no current can flow through them. But despite best efforts at minimizing resistive heating in their coils, and magnetization and eddy current losses in their cores, all transformers still waste some power. Even the best transformers are only about 99% energy efficient. (For an interesting device that resembles a transformer, see ❸.)

Alternating Current Power Distribution

We're finally prepared to deal with the basic conflicts of power transmission. To minimize resistive heating in the power lines connecting a power plant with a distant city, electric power should travel through those lines as small currents at very high voltages. But to be practical, as well as to avoid shock and fire hazard, electric power should be delivered to homes as large currents at modest voltages.

While there is no simple way to meet both requirements simultaneously with direct current, transformers make it easy to satisfy them both with alternating current. We can use a step-up transformer to produce the very-high-voltage current suitable for cross-country transmission and a step-down transformer to produce the low-voltage current that's appropriate for delivery to communities (Fig. 11.2.10).

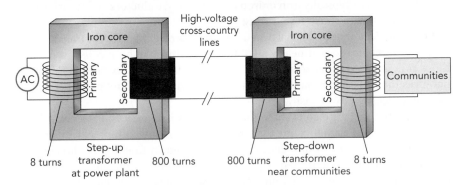

Fig. 11.2.10 Power is transmitted cross-country by stepping it up to very high voltage near the power plant, transmitting it as a small current at very high voltage, and stepping it back down to low voltage near the communities that are to be served. The secondary circuit for the step-up transformer is the primary circuit for the step-down transformer.

At the power plant, the generator pushes a huge alternating current through the primary circuit of a step-up transformer at a supply voltage of about 5000 V. The current flowing through the secondary circuit is only about 1/100th that in the primary circuit, but the voltage supplied by the secondary coil is much higher, typically about 500,000 V.

This transformer's secondary circuit is extremely long, extending all the way to the city where the power is to be used. Since the current in this circuit is modest, the power wasted in heating the wires is within tolerable limits.

Once it arrives in the city, this very-high-voltage current passes through the primary coil of a step-down transformer (Fig. 11.2.11). The voltage provided by the secondary coil of this transformer is only about 1/100th that supplied to its primary coil, but the current flowing through the secondary circuit is about 100 times that in its primary circuit.

Now the voltage is reasonable for use in a city. Before entering homes, this voltage is reduced still further by other transformers. The final step-down transformers can frequently be seen as oil-drum sized metal cans hanging from utility poles (Fig. 11.2.12) or as green metal boxes on the ground (Fig. 11.2.13). Current enters

Fig. 11.2.11 This giant transformer transfers millions of watts of power from the very-high-voltage cross-country circuits above it to the medium-high-voltage neighborhood circuits to its left. Fans keep the transformer from overheating.

the buildings at between 110 V and 240 V, depending on the local standards. While 240-V electricity wastes less power in the home wiring, it's more dangerous than 110-V power. The United States has adopted a 120-V standard while Europe has a 230-V standard.

Fig. 11.2.12 The three metal cans on this utility pole are transformers. They transfer power from the medium-high-voltage neighborhood circuits that run above them to the low-voltage household circuits at the lower right.

Fig. 11.2.13 This transformer transfers power from a medium-voltage underground circuit to a low-voltage underground circuit used by nearby homes. It handles 50 kV·A or 50,000 W of power.

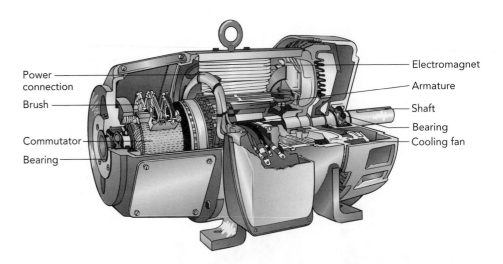

Power connection

Brush

Commutator

Bearing

Electromagnet

Armature

Shaft

Bearing

Cooling fan

SECTION 11.3 Electric Generators and Motors

We've seen what electricity is and how it's distributed. Now we'll look at how it's generated—how mechanical motion can be used to produce electric power. We'll also examine how electric power can be used to produce mechanical motion by studying electric motors. In fact, the symmetry of those two remarks foreshadows an amazing result: generators and motors are often the same devices!

Although generators are essential to electric power plants, they aren't common in your home. You encounter them mostly in cars and in emergency power equipment. However, electric motors are everywhere, spinning the parts of countless household machines. Sometimes this rotary motion is obvious, as in a fan or a mixer, but often it's disguised, as in the agitation of a washing machine or the vibration of a cell phone. Motors come in many shapes and sizes, each appropriate to its task. No matter how much torque or power a motor must provide, you can probably find one that's suitable. Some motors operate from direct current and can be used with batteries, while others require alternating current. There are even motors that work on either type of current.

AC Electric Generators: Mechanical Power Becomes Electric

A generator converts mechanical power into electrical power. Its spinning magnet produces an alternating magnetic field within a coil of wire and thereby induces an alternating emf in it. If you connect a lamp to the generator's coil and complete a circuit, that coil will act as a source of AC electric power and the lamp will light up.

If this arrangement sounds familiar, that's because it's almost the same as in a transformer. In fact, the only important difference between a transformer and a generator is in what produces the alternating magnetic field!

Figure 11.3.1 shows a simple generator, one that looks strikingly like the transformer in Fig. 11.2.7. Both devices have a (secondary) coil wrapped around a magnetizable core. But in place of the transformer's primary coil, the generator has a spinning magnet or rotor. As that magnetic rotor spins, it produces a sinusoidally alternating magnetic field in the coil. This alternating magnetic field, in turn, produces an alternating electric field and induces an alternating emf in the coil. That emf lights the lamp.

This brief outline accurately explains how a generator provides AC electric power. But how is energy conserved in the process and what determines the frequency

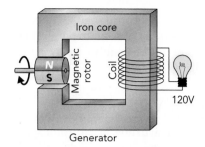

Fig. 11.3.1 This generator resembles the transformer in Fig. 11.2.7, except that power reaches it through the motion of its spinning magnetic rotor, rather than through an electric current in a primary coil.

and voltage of the generator's alternating current? Since the generator can't create energy, whatever electric power it delivers to the current in its coil must originate as mechanical power in its rotor. Let's begin by looking at how the act of generating electric power extracts mechanical power from the rotor.

For clarity and simplicity, we're mostly going to ignore forces due to the magnetizable core itself. The one aspect of the core we won't overlook is its guidance of magnetic flux lines. The core forms a magnetic bridge between rotor and coil, conveying the rotor's magnetic field into the coil even when they're separated in space. As the rotor spins, its magnetic field sweeps through the coil as though the two objects were almost touching one another. Thanks to the core's help, the rotor and coil share the same electromagnetic environment.

With the lamp unplugged, there is no circuit and no current flows through the generator's coil. Since the generator isn't producing electric power, it shouldn't extract mechanical power from the rotor. Sure enough, the rotor turns freely; the current-free coil is nonmagnetic and the rotor's magnetic field sweeps through it effortlessly!

But with the lamp plugged into the generator, a circuit forms and current can flow. Now as the rotor's magnetic field sweeps through the coil, it induces an alternating current in the coil and lights the lamp. This electric power generation has serious consequences for the rotor; because the current-carrying coil is now magnetic, the coil interacts with the rotor and extracts mechanical power from it!

That interaction between the rotor and the coil starts the moment the rotor's magnetic field begins to sweep into the coil. The arriving magnetic field induces a current in the coil, rendering it magnetic with a field oriented *opposite* the rotor's field. Consistent with Lenz's law, the coil's induced magnetism is opposing the rotor's incoming magnetic field—the change that produced it. In effect, the coil becomes an electromagnet with its like poles turned to repel the approaching poles of the rotor. The rotor must therefore do mechanical work in order to align its poles with the coil.

The rotor keeps turning and soon begins sweeping its magnetic field back out of the coil. This withdrawal again induces current in the coil, but this time in the opposite direction. The coil becomes magnetic with a field oriented along the rotor's field. The coil's induced magnetism is again opposing the change that produced it—the rotor's *departing* magnetic field. In effect, the coil becomes an electromagnet with its opposite poles turned to attract the departing poles of the rotor. The rotor must now do mechanical work to turn itself out of alignment with the coil.

The rotor can't get a break; it must do work to align its poles with the coil and more work to undo that alignment. All of the electric power consumed by the lamp is being extracted from the generator's rotor as mechanical power.

Nothing makes this power transfer more evident than turning the rotor of a small generator by hand; you can feel the lamp draw power from you as you spin the rotor. And the more power the lamp consumes, the stronger the coil's induced current and magnetic field become and the more mechanical power you must provide to keep the rotor turning.

Of course, most generators are driven by things other than people. Industrial generators typically obtain their mechanical work from steam turbines, using steam produced by fossil fuels such as coal, oil, and natural gas or uranium. Other industrial generators use renewable resources such as water or wind to power their turbines. And smaller commercial or home generators frequently employ gas or diesel engines to keep their rotors turning.

A Generator's Frequency and Output Voltage

The frequency of a generator's AC power is proportional to how fast its rotor spins. That's because the generator's alternating output voltage—the emf induced in its coil—is the result of the rhythmic sweep of its rotor's magnetic poles past its coil. This emf reverses every time a pair of poles aligns with the coil, so it takes two alignments to produce one full cycle of alternation in the generator's output voltage. For a rotor with a single pair of poles (e.g., Fig 11.3.1), the emf alternates once per rotation of the rotor.

The generator in Fig. 11.3.1 thus provides 60-Hz AC power when its rotor spins 60 times per second. If its rotor were a more complicated magnet with two pairs of poles rather than one, it would only have to spin 30 times per second to produce 60-Hz AC power. This relationship between rotation rate and frequency explains why every generator of the 60-Hz U.S. power grid turns at either 60 rotations per second or some integer fraction of that rate. In the 50-Hz European power grid, the basic rate is 50 rotations per second. And because their output voltages must alternate together as one, all the generators of a power grid spin their rotors in perfect synchrony. Each power grid resembles a well-choreographed ballet in which every dancer is forever in step with the rest of the company.

In addition to maintaining the correct frequency, a generator must produce the right output voltage; its rotor must induce the proper emf in its coil. That induced emf depends on three different factors: the number of turns in the coil, the magnetic field strength, and the frequency at which that magnetic field alternates. Increasing any of those factors boosts the coil's emf and therefore the generator's output voltage.

Increasing the number of turns in a generator's coil is like increasing the number of turns in a transformer's secondary coil: the more times the wire coil encircles the generator's alternating magnetic field, the more work the resulting electric field can do on charges in that wire and the larger the coil's induced emf. As with a transformer, the generator's output voltage is proportional to the number of turns in its coil.

The magnetic field's strength and alternation frequency affect the generator's output voltage because both factors influence the electric field in the coil. Stronger or faster-changing magnetic fields produce stronger electric fields and thus larger emfs in the coil.

A good AC generator always provides power at its specified voltage and frequency. Since the rotation rate of its rotor affects both those characteristics, the generator has a control system to keep its rotor spinning steadily no matter how much electric power is being consumed. When you connect more lamps to the generator, so that the current extracts more mechanical work from its rotor, this control system boosts the flow of steam or fuel so that the rotor maintains its rotation rate. You can hear this automatic response in a gasoline- or diesel-powered generator; when you plug in more equipment, the control system powers up the engine.

With its rotor spinning steadily, a generator's output voltage will be relatively constant. However, that voltage can be further regulated by adjusting either the number of turns in the generator's coil or the strength of the magnetic field. A generator that automatically regulates its output voltage typically uses an electromagnet in its rotor, making that electromagnet stronger or weaker to raise or lower the generator's output voltage, respectively.

It's also common for a generator to allow you to use just part of its coil to obtain a smaller emf and therefore a smaller output voltage. For example, a typical household emergency generator provides 240 V AC from its full coil and 120 V AC from either half of its coil. The 120-volt outlets on this generator connect to half of the coil while the 240-volt outlet connects to the entire coil. You can even use the two voltages from the same generator at the same time!

AC Electric Motors: Electric Power Becomes Mechanical

A motor converts electrical power into mechanical power. If you're wondering what motors are doing in the same section as generators, recall that a generator converts mechanical power into electrical power. As you can see, these two devices are the reverse of one another.

You might therefore expect motors and generators to be reverse in structure. Not quite. Motors and generators are actually the same in structure; what's reversed about them is their direction of current flow. Although they're usually specialized to one purpose or the other, you can often make a generator act as a motor or vice versa. Just reverse the current!

The effect of reversing the current in a generator shouldn't come as a total surprise. We have already seen that the mechanical work a generator extracts from its rotor is zero when the current is zero and rises in proportion to the current the generator propels through a lamp. What we hadn't considered was what happens if that current drops below zero—if it travels opposite its normal direction of flow through a lamp. In that case, the mechanical work extracted from the rotor should also drop below zero: the generator should *provide* mechanical work to its rotor and thereby become a motor!

However, the current that we're reversing is an *alternating* current, so how do we "reverse" it? Since a generator normally pushes current out of whichever end of its coil is momentarily at higher voltage, reversal means pushing current *into* that end of the coil. The effect of this reversal is to make the "generator" consume electric power rather than produce it. It acts as a motor.

This transition from generator to motor can also be explained in terms of a reversal of magnetic forces. Reversing the alternating current in the coil interchanges its magnetic poles so that it pushes on the rotor when it used to pull and pulls when it used to push. Now as the rotor sweeps into alignment with the coil, the coil *attracts* the approaching pole and does work on it. And as the rotor continues on, sweeping back out of alignment, the coil *repels* the departing pole and again does work on it. Overall, electric power in the coil is becoming mechanical power in the rotor—the hallmark of a motor.

It's an AC synchronous motor—a type of motor whose rotor turns in perfect synchrony with current from an AC electric power source (Fig. 11.3.2). When this type of motor is plugged into a 60-Hz electrical outlet, its rotor spins exactly 60 times per second, or at an integer fraction of that rate if the rotor has multiple pairs of poles.

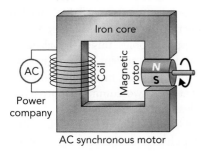

Fig. 11.3.2 This AC synchronous motor resembles the transformer in Fig. 11.2.7, except that power leaves it through the motion of its spinning magnetic rotor, rather than through an electric current in a secondary coil.

Its rigid adherence to the AC power frequency makes an AC synchronous motor steady and precise, but also hard to start. If its rotor isn't turning or is turning at the wrong speed, the coil's alternating magnetic poles will push or pull on the rotor erratically and the rotor may never spin properly. That's why most practical AC synchronous motors have extra components and often multiple coils to help their rotors start turning in the desired direction and reach the proper rotation rates.

The similarity between Figs. 11.3.1 and 11.3.2 raises an important question: if a generator and a motor are the same device and you plug one into the AC power grid, what determines whether it acts as a generator or as a motor? The basic answer to that question is surprisingly simple: it acts as a generator when you do work on its rotor and as a motor when you extract work from its rotor!

The device responds to you this way in order to keep its rotor turning in synch with the AC power grid. If you attempt to speed the rotor up by twisting it forward and doing work on it, the device acts as a generator: it extracts mechanical power from you and provides electric power to the grid. Your forward twist will have advanced the rotor's orientation to the point where its poles will be repelled as they sweep into alignment with the coil and attracted as they continue on out of alignment—the forces in a generator.

On the other hand, if you attempt to slow the rotor down by twisting it backward and having it do work on you, it acts as a motor: it extracts electric power from the grid and provides you with mechanical power. Your backward twist will have delayed the rotor's orientation to the point where its poles will be attracted as they sweep into alignment with the coil and repelled as they continue on out of alignment—the forces in a motor.

And if you simply leave the spinning rotor alone and let it coast freely, it will orient itself halfway between these two angular extremes so that the electromagnetic coil does zero average work on it.

DC Electric Motors

Since most portable devices are powered by batteries and direct current, they can't use AC synchronous motors to turn their components. If you send direct current through the coil of an AC synchronous motor, that coil will act like a permanent magnet and the rotor won't spin. Instead, attraction between opposite poles will snap the rotor into alignment with the coil and it will never move again. With its poles as close as possible to the coil's opposite poles, the rotor will be in a stable equilibrium.

To keep the rotor spinning, something must reverse the coil's magnetism each time the rotor reaches this stable equilibrium. That reversal will turn attraction into repulsion and the rotor will suddenly find itself in an *unstable* equilibrium. Instead of stopping, the rotor will spin onward in search of a new stable equilibrium. But when it gets there, the coil's magnetism will reverse again. The rotor thus spins forever, seeking to align its poles with the coil's opposite poles but never quite succeeding. It's the myth of Sisyphus—rolling a huge stone up a hill, only to have it roll down again as it nears the top—realized in an electromechanical device.

Flipping the coil's magnetism is as simple as reversing its current. All that's needed is a switch that interchanges the two wires connecting the coil to the battery. Whenever the rotor reaches alignment with the coil's opposite poles, this switch abruptly swaps the connections and thereby reverses the coil's current and magnetism.

Such a switch can be found in virtually every DC electric motor, but different DC motors implement it differently. There are at least two common approaches, used in at least two types of DC motors: brushless and brushed.

A brushless DC motor is just a synchronous AC motor plus a high-tech switch that reverses the current whenever the rotor's poles have aligned with opposite

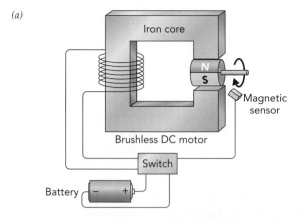

(a)

Iron core

N
S

Magnetic sensor

Brushless DC motor

Switch

Battery – +

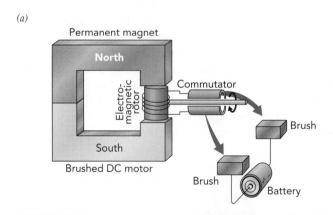

(b)

Fig. 11.3.3 (*a*) A brushless DC motor uses an electromagnet to spin its magnetic rotor. A sensor monitors the orientation of the rotor and reverses the current in the electromagnet each time the rotor aligns with the magnetic field. (*b*) This computer fan uses a brushless DC motor. Its stationary four-pole electromagnet fits inside the rotor's ring-shaped permanent magnet. The black magnetic sensor is below the electromagnets.

poles on the stationary coil (Fig. 11.3.3*a*). Although their high-tech switches make them relatively expensive, these brushless motors spin silently and safely for years without service. The fans that cool your computer are spun by brushless motors (Fig. 11.3.3*b*).

A brushed DC motor puts its electromagnetic coil on the rotor and pushes on it with a stationary permanent magnet (Fig. 11.3.4*a*). That change is inconsequential except that now it's the rotor that interchanges its poles each time they align with

(a)

Permanent magnet

North

Electro-magnetic rotor

Commutator

Brush

South

Brushed DC motor

Brush

Battery

(b)

Fig. 11.3.4 (*a*) A brushed DC motor spins its electromagnet rotor in the field of a permanent magnet. Each time the rotor aligns with the magnetic field, its commutator reverses the current in the electromagnet. (*b*) This toy train uses a brushed DC motor. Its electromagnetic rotor spins about a vertical axis above its commutator and brushes.

opposite poles of the permanent magnet. The value of this swap is that it allows the motor to use a simple and inexpensive switch called a commutator to control the direction of current flow in its coil.

In its simplest form, a commutator consists of two curved plates that are fixed to the rotor and connected to opposite ends of the rotor's coil. Electric current flows into the rotor through a conducting brush that touches one of these plates and leaves the rotor through a second brush that touches the other plate. As the rotor turns, each brush makes contact first with one plate and then with the other. Every time the rotor completes half a turn, these brushes trade plates and the rotor's magnetic poles flip. The plates are arranged so that this reversal occurs just when opposite poles have aligned, so the rotor spins forever.

While brushed DC motors are cheaper than brushless ones, friction in their commutators gradually wears them out (Fig. 11.3.4b). Moreover, imperfect electrical contact between brushes and pads leads to sparking, so that brushed motors are generally unsuitable near flammable gases or liquids. Brushless DC motors are better for tasks requiring long and continuous service or where sparking is hazardous.

The single-coil DC motors that we've just discussed often have trouble starting and may spin in either direction when they do. To start reliably and to spin in predictable directions, more practical DC motors have more than one coil and correspondingly more complicated switches. However, this improvement makes these motors sensitive to the overall direction of current flow. When you reverse the batteries powering a good DC motor, the current everywhere in that motor reverses and so do all the poles of its electromagnets. Since the motor's permanent magnets are unchanged, the forces affecting the rotor reverse—attraction becomes repulsion and vice versa. As a result, the motor spins backward.

Rotation Speed and DC Electric Generators

We've seen what makes the rotor of a DC motor spin, but not what determines how fast it spins. Since the motor effectively makes its own alternating current, won't its rotor spin faster and faster forever? The answer is no. Its rotor spins at a specific rate that's proportional to the voltage supplied to the coil. The origin of this natural rate is induction: as the rotor spins, the coil experiences an induced emf and that induced emf limits the rotor's rotation rate.

When you first connect the motor to a battery and its rotor is stationary, the battery-imposed voltage drop pushes current through its coil from higher voltage to lower voltage. But once the rotor starts spinning, an induced emf appears in the coil and opposes that current flow. That emf grows larger as the rotor spins faster and eventually becomes large enough to stop the current flow altogether. The induced emf is then equal to the battery-imposed voltage drop and the rotor is spinning at its natural rate.

The harder it is for the motor's emf to equal the battery voltage, the faster the rotor must spin to reach its natural rate. Therefore, increasing the battery voltage, weakening the motor's permanent magnet, or reducing the number of turns in its coil(s) will all increase the motor's natural rotation rate.

Once its rotor is spinning at that natural rate, the motor opposes any further change in rotation rate. If you try to slow the rotor down by twisting it backward, the motor will begin drawing current from the battery again. It will turn DC electric power into mechanical power to prevent the rotor from slowing down.

If you try to speed the rotor up by twisting it forward, however, the motor's emf will begin pushing current in the opposite direction. It will turn mechanical power into DC electric power to prevent the rotor from speeding up. By doing work on the rotor in this manner, you will cause current to flow backward through the battery and recharge it. The DC motor will be acting as a DC generator!

After seeing that AC motors and AC generators are the same, it's not so surprising that DC motors and DC generators are often the same as well. If you do work on the rotor of a brushed DC motor, it will act as a DC generator. Edison used such DC generators in his New York City power plants. Brushless DC motors are more finicky because of their high-tech switches, but some can act as DC generators as well.

Universal Motors

If you plug a brushed DC motor into an AC electrical outlet, its rotor will hum rather than spin. That's because the rotor tries to reverse each time the AC outlet voltage reverses and it just can't make any progress either way.

But if you replace the stationary magnet in a brushed DC motor with an electromagnet and connect that electromagnet to the same power source as the rest of the motor, you'll have made a universal motor (Fig. 11.3.5a). As indicated by its name, this motor will spin properly when powered by either direct or alternating current.

To understand its flexibility, let's consider what happens when you supply a universal motor with DC power. The stationary electromagnet will magnetize in one direction like a permanent magnet and the rotor will spin as if it were in a DC motor. However, there's an important difference: the rotor of a universal motor won't spin backward when you reverse the supply voltage. It will continue turning in the same direction because the current reversal reverses every pole in the entire motor and therefore leaves the motor's magnetic forces unchanged. If you want to make a universal motor turn backward, you must interchange the wires of its stationary electromagnet.

Since its rotor turns in a fixed direction, regardless of which way current flows, a universal motor works fine on AC power. And like the DC motors on which it's

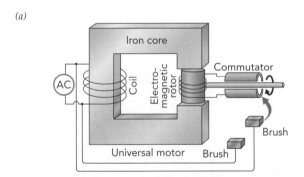

(a)

(b)

Fig. 11.3.5 (a) A universal motor resembles a DC motor, except that it uses only electromagnets. Since it is unaffected by the direction of current flow in its wires, it can operate on either DC or AC power. (b) This mixer uses a universal motor. Its electromagnetic rotor (center) spins inside a stationary electromagnet, while its commutator and brushes (right) control the direction of current flow through the rotor.

modeled, a universal motor's speed is determined by voltages and currents rather than by the AC power frequency. Flexible and reliable, universal motors are standard in such household appliance as kitchen mixers, blenders, and vacuum cleaners (Fig. 11.3.5b).

Induction Motors

Our final type of motor is the most conceptually sophisticated: the induction motor. Its rotor is neither a permanent magnet nor an ordinary electromagnet; it's a collection of conducting loops that are magnetized by induction alone. When the motor exposes this rotor to a rotating magnetic field, it experiences electromagnetic forces that make it spin along with that field.

To understand the forces acting on this rotor, let's start by looking at a loop of wire in a stationary magnetic field (Fig. 11.3.6). As long as that loop is motionless, it carries no current and experiences no forces. But when the loop begins to rotate through the field (Fig. 11.3.6a), a new force acts on its mobile electric charges: the Lorentz force.

Named after its discoverer, Dutch physicist Hendrik Antoon Lorentz [1853–1928], the Lorentz force affects a charge that is moving through a magnetic field. This force pushes the charge at right angles to both its velocity and the magnetic field (Fig. 11.3.7). The strength of the Lorentz force is proportional to the charge, to the velocity, to the magnetic field, and to the sine of the angle between the velocity and the magnetic field. Finally, the direction of the Lorentz force on a positive charge follows a right-hand rule: when the extended index finger of your right hand points along the charge's velocity and your bent middle finger points along the magnetic field, the force on the charge points along your extended thumb. A negative charge experiences a force in the opposite direction. This relationship can be written:

$$\text{force} = \text{charge} \cdot \text{velocity} \cdot \text{magnetic field} \cdot \text{sine of angle}, \qquad (11.3.1)$$

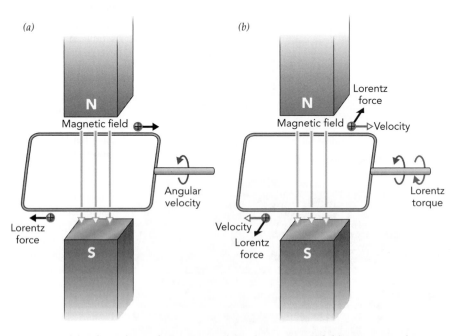

Fig. 11.3.6 (a) When a loop of wire rotates through a magnetic field, its moving charges experience Lorentz forces along the wire and begin to move as a current around the loop. (b) The circulating charges experience additional Lorentz forces perpendicular to the wire itself and produce a torque on the loop opposing its rotation.

where the angle involved is between the velocity and the magnetic field, and the direction of the Lorentz force follows the right-hand rule.

When our loop of wire rotates, it carries its mobile charges through the stationary magnetic field and they experience Lorentz forces. Since the two sides of the loop move opposite one another, the charges in those sides are pushed in opposite directions along the wire and begin to circulate around the loop as a current—an induced current!

With charges now moving around the wire loop, the Lorentz force acts yet again (Fig. 11.3.6*b*). It pushes the charges toward the edges of the wire, but in opposite directions on opposite sides of the loop. Since the charges can't leave the wire, they push it along with them and the entire loop experiences a torque. That torque is directed opposite the loop's angular velocity—it acts to slow the loop's rotation through the field. This is Lenz's law again: since the loop's induced current is caused by its rotation in a magnetic field, the effect of that induction is a torque that opposes the loop's rotation.

Evidently, a loop of wire has difficulty rotating in the presence of a stationary magnetic field. Remarkably, it has equal difficulty *not* rotating in the presence of a rotating magnetic field! That's how an induction motor spins its rotor: it uses electromagnets to make a magnetic field that rotates in space and its looplike rotor spins along with that field. While a single-loop rotor is sufficient, more effective rotors contain many conducting loops. The classic rotor is the "squirrel cage," so named because of its resemblance to an animal exercise wheel (Fig. 11.3.8).

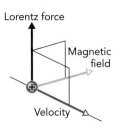

Fig. 11.3.7 A positive charge moving through a magnetic field experiences a force that's perpendicular to both its velocity and the magnetic field. A negatively charged particle would experience a force in the opposite direction.

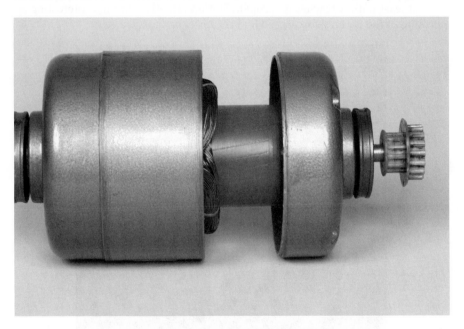

Fig. 11.3.8 This disassembled induction motor surrounds its red squirrel cage rotor with stationary electromagnetic coils. The coils' magnetic field rotates around the rotor and the rotor spins with it. To increase the motor's performance, the rotor has iron laminations within its conducting loops.

Although the rotor tries to follow the rotating magnetic field perfectly, it doesn't quite keep up. If it did, no current would flow through it and it would experience no electromagnetic torque at all. Even the friction in its bearings would slow the rotor down. Instead, the rotor always turns somewhat slower than the magnetic field. And the more torque the rotor exerts on the machinery it's turning, the more electromagnetic torque it needs and the slower it must spin in order to obtain that torque. When supplying its maximum rated mechanical power, the rotor of a typical induction motor turns a few percent slower than the motor's rotating field.

Unfortunately, creating a true rotating magnetic field requires complicated electromagnets and currents. Though routine in industrial settings, that level of sophistication is difficult to achieve at home. That's why most household induction motors make do with *alternating* rather than *rotating* magnetic fields. Amazingly enough, when the rotor is spinning at about the right rate, it responds to the alternating magnetic field as through it were a rotating one.

This effect resembles one you may have noticed while walking past a string of flickering holiday lights: if you travel at about the right speed, the pattern of light you see can appear to move with you even though the motionless bulbs are merely blinking on and off. Similarly, when the rotor spins inside a ring of AC electromagnets at about the right rate, the pattern of magnetism can appear to spin with it even though the motionless electromagnets are merely flipping their poles with each reversal of the AC power.

To be practical, an induction motor must add at least some rotating character to this alternating magnetic field. The rotating portion of the field gets the rotor started and up to speed, and the alternating portion takes over from there. In an induction motor that doesn't have to supply much starting torque, just a hint of rotating character is all it takes to coax the rotor gradually up to speed. That hint of rotation can be created with simple electromagnetic devices. But an induction motor that must exert a large torque while starting needs a true rotating magnetic field, at least until it reaches its proper rotation rate. (For another interesting application of the Lorentz force, see Fig. 11.3.9.)

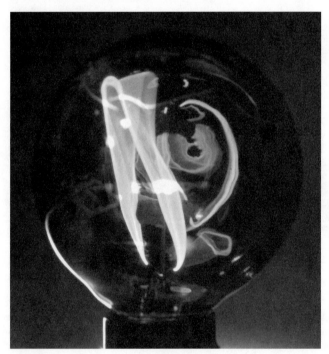

Fig. 11.3.9 This BalaFire incandescent lightbulb has a fine carbon filament that loops around a permanent bar magnet—visible just below the middle of this photograph. As alternating current flows through the filament, that current experiences the Lorentz force. Since the current is confined inside the filament, the Lorentz force affects the entire filament and the filament accelerates. Overall, the filament dances back and forth in a seemingly random fashion and resembles a candle flame.

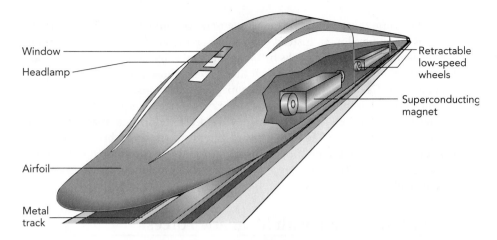

Window

Headlamp

Retractable
low-speed
wheels

Superconducting
magnet

Airfoil

Metal
track

SECTION 11.4 Magnetically Levitated Trains

While jet airplanes still carry most travelers between distant cities, modern trains have eroded a jet's advantage for intermediate distances. Bullet trains play important transportation roles in a number of countries, notably Japan (Shinkansen), France (TGV), Germany (ICE), Italy (TAV), South Korea (KTX), and Spain (AVE). These high-speed trains presently move their passengers at speeds of up to 300 km/h (187 mph) and they may reach 350 km/h (220 mph) soon. However, mechanical interactions between a bullet train and its rails ultimately limit its maximum speed. The next generation of ultra-high-speed trains will probably fly above the rails on cushions of magnetic force.

The Need for Magnetic Levitation

Trains operate more or less as they have since the early nineteenth century. Cars still roll on wheels, pulled forward by friction between locomotives and the rails beneath them. But while this scheme works well at speeds below about 160 km/h (100 mph), two serious problems develop at higher speeds.

First, a high-speed train has trouble following the rails because it has so little time to move up, down, left, or right. Sudden shifts in direction require rapid accelerations and involve large forces between the rails and the wheels. In general, these accelerations and forces are proportional to the square of the train's speed. For example, doubling the speed of a train quadruples the forces involved in riding over a bump in the rails. That's because the train has only half as much time to climb up the bump, necessitating twice the upward speed, and it has only half the time to acquire that doubled upward speed, necessitating four times the upward acceleration. The train and rails must therefore push on one another four times as hard.

To minimize the accelerations and forces at high speeds, the rails must be straight and exceedingly smooth, with only the most gradual curves. But even carefully constructed rails take a beating from the wheels and produce noise, vibration, and mechanical wear. Both the rails and the wheels must be replaced frequently so that, above about 350 km/h (220 mph), maintenance becomes prohibitively expensive.

Second, air resistance becomes so strong at high speeds that traction between the locomotive's wheels and the rails isn't enough to keep the train moving forward. The train can't maintain its cruising speed with friction alone and it can't stop quickly enough using friction, either. Just for safety reasons, conventional rail travel shouldn't exceed about 350 km/h (220 mph).

Fig. 11.4.1 These two disk magnets repel because their north poles are facing one another. The stick is needed, however, to prevent the top disk from falling off the magnetic cushion that supports it.

Fig. 11.4.2 This spinning magnetic top hovers above a repelling magnet, hidden in the wooden base. The top's dynamic stability involves gyroscopic effects that would be impractical for a maglev train.

To overcome these limitations, a whole new technology for ultra-high-speed trains is being developed. Instead of running on metal wheels, these new trains are suspended above their tracks by magnetic forces. Because these magnetically levitated or maglev trains don't touch their tracks, those tracks don't have to be very smooth. The trains ride forward on magnetic cushions.

Conventional propulsion won't work for a maglev train because it doesn't have any wheels or friction. Instead, a maglev train uses additional magnets on the train and track to propel the train forward and to stop it during braking. In effect, maglev trains unroll the motors (and generators) that we examined in the previous section and use them for linear rather than rotational motion. Motors of that sort are called linear motors.

Supporting a Train with Magnetic Forces

Instead of rolling forward on wheels, a maglev train rides above its track on a magnetic suspension. Magnets in the train and track exert forces on one another and support the train without any direct contact. Though simple in concept, this idea is fraught with complications. To start uncovering those complications, let's think about the simplest approach to magnetic suspension: strategically placed permanent magnets.

If you recall from the beginning of this chapter, magnetic poles always appear in equal but opposite pairs. There just don't seem to be any magnetic monopoles in our universe—no pure north poles or pure south poles. So the best you can do with permanent magnets is magnetic dipoles: a magnet with a north pole at one end and a south pole at the other end.

The forces between dipole magnets can be surprisingly strong. If you cover the top of a track and the bottom of a train with dipole magnets, turned so that their like poles face each other, the train and track will repel one another fiercely. This repulsion becomes stronger as the two approach one another and will exceed the train's weight at some distance. When it does, the train stops accelerating downward and can remain suspended over the track on a magnetic cushion. This is magnetic levitation.

Unfortunately, there are at least two problems with this simple levitation scheme. First, a track covered with magnets would attract magnetic junk—old steel cans, nuts and bolt, scraps of iron—making it a nightmare to keep clean. But much more importantly, a magnet suspended above another magnet is statically unstable (Fig. 11.4.1). The train will tend to slip sideways until it descends to the ground, just as a marble sitting on top of a dome will tend to roll sideways until it descends to the ground. The train may hover briefly on its magnetic cushion but it won't stay there for long without help. While there are a few exotic schemes in which levitating permanent magnets achieve dynamic stability, stability through motion (Fig. 11.4.2), those schemes are of no use to maglev trains.

Stability and Feedback

A train supported by permanent magnets can be at equilibrium—experiencing zero net force—but that equilibrium will always be unstable. An object in stable equilibrium experiences a restoring force after being disturbed and accelerates back toward that equilibrium (Fig. 11.4.3a). But as I explained in Section 4.1 on bicycles, an object in unstable equilibrium experience forces in the wrong direction after being disturbed and accelerates away from that equilibrium (Fig. 11.4.3b).

An observation known as Earnshaw's theorem states that no arrangement of electric charges can be in a stable equilibrium as the result of electrostatic forces alone. Similarly, no arrangement of magnetic poles can be in a stable equilibrium

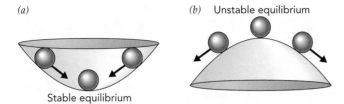

(a)

(b) Unstable equilibrium

Stable equilibrium

Fig. 11.4.3 (*a*) A ball disturbed from a stable equilibrium position experiences a restoring force that pushes it back toward equilibrium. (*b*) A ball disturbed from an unstable equilibrium doesn't return.

as the result of magnetostatic forces alone. No matter how you arrange permanent magnets on a train and a track, the train will be in an unstable equilibrium.

Just as people who try to invent perpetual motion machines are doomed to failure by the first and second laws of thermodynamics, people who try to invent stable magnetic levitation schemes that involve only permanent magnets are doomed to failure by Earnshaw's theorem. Although it might seem that some clever arrangement of permanent magnets could make the train stable, there's absolutely no way around Earnshaw's theorem.

In the simple case shown in Fig. 11.4.4, repulsion between the magnets can keep the train at a stable height, but the train's sideways motion is unstable. If the train isn't perfectly centered above the track, the repulsive forces will push it toward the side and it will fall off its magnetic cushion.

The only way to stabilize the train and keep it centered above the track is to use magnets that can be adjusted—turned on and off—so as to push the train back toward center if it starts to fall. Electromagnets can satisfy this need because their magnetic fields are created by electric currents and they can be turned on or off simply by changing those currents. To make electromagnets even stronger, their wires are usually wound around iron or other magnetic materials that will magnetize and thereby reinforce their coils' magnetic fields.

To keep the train stable, a control system must monitor the train's position and adjust its electromagnets accordingly. This technique of using information about the current situation to control how the situation changes is called feedback. Feedback is used frequently in engineering to stabilize systems that are inherently unstable and to modify the ways in which systems respond to external stimuli (Fig. 11.4.5). Feedback can effectively create restoring forces where none exist naturally.

Some of the maglev train systems currently under development or in trial use employ feedback-stabilized magnetic levitation (Fig. 11.4.6). Most support the

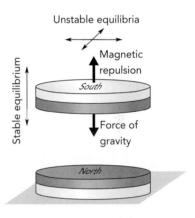

Fig. 11.4.4 If the repulsion between permanent magnets is used to suspend a train above a track, the train's height will remain stable but its horizontal position will not. If the train isn't perfectly centered, the repulsive forces will push the train sideways until it falls.

Fig. 11.4.5 This ball and magnet are suspended in midair by the electromagnet above them. A control system uses light to measure the position of the ball. It then adjusts the strength of the electromagnet so as to keep the ball floating at a constant height.

Fig. 11.4.6 This maglev train is supported by attractive forces between magnets located on the rail and magnets located in the train arms that reach beneath the rail. Feedback provides stability to this system.

train's weight with attractive forces between opposite poles and carefully control the electromagnets to maintain a constant separation between those attractive poles. The obstacles to success, however, are formidable and a number of feedback-stabilization maglev programs have failed over the years. Feedback stabilization requires very sophisticated control systems that must overcome both theoretical and technical challenges. Nonetheless, there has been much progress in the science and engineering of these train levitation techniques and there is a good possibility that feedback-stabilized maglev trains will be cruising the ground at aircraft speeds in the next decade or two.

Alternating Current Levitation

There's another magnetic technique that can provide restoring forces in all directions and can levitate a train without any control system or feedback. Instead of putting permanent magnets or electromagnets on both the train and track, some trains use electromagnetic induction to make conducting surfaces magnetic temporarily. A train employing this concept exposes its conducting track to a changing magnetic field and induces an electric current in the track. Since that induced current is itself magnetic, the train and track push on one another with magnetostatic forces. Properly done, those magnetostatic forces can support the train in a stable fashion.

The simplest approach to electromagnetic-induction magnetic levitation is to put an electromagnet in the train and connect it to a battery (Fig. 11.4.7a). Since the electromagnet will act as an inductor, the current flowing through it will rise gradually and its magnetic field will increase steadily. The nearby conducting track will experiencing both that increasing magnetic field and the resulting electric field (recall that changing magnetic fields produce electric fields). That electric field will push current through the conducting track and the track will become magnetic.

In accordance with Lenz's law, the track's magnetism will oppose the train's increasing magnetic field. The train and track will repel one another and the train will begin to float above the track. If the conducting track is curved into a slight trough and the train's electromagnet matches it, the levitation is actually stable. This maglev train is in a true stable equilibrium and won't fall off its magnetic cushion!

Unfortunately, this levitation effect has a problem. The track's electrical conductivity isn't perfect, so it always needs an electric field to keep its current flowing and remain magnetic. Since the train's changing magnetic field is responsible for that electric field, the train's magnetic field can't stop changing or the train will fall!

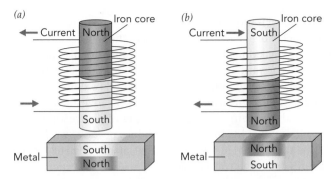

Fig. 11.4.7 (*a*) When an electromagnet turns on with its south pole down, its increasing magnetic field induces a current in the metal below it. That metal becomes magnetic with its south pole up and repels the electromagnet. (*b*) If the current in the electromagnet is reversed, all the poles reverse and the metal still repels the electromagnet.

Our simple train's magnetic field has been increasing steadily because the battery has been gradually sending more current through the inductor/electromagnet. But this situation can't go on. Eventually the electromagnet will be carrying more current than it can handle and it will burn up. Fortunately, there is a solution: reverse the current flow periodically (Fig. 11.4.7*b*). According to Lenz's law, the track's induced magnetism will oppose whatever change produces it, so connecting the train's electromagnet to a source of alternating current will keep the train and track repelling one another indefinitely.

Scientists and engineers have used alternating current magnetic levitation to suspend trains above trough-shaped tracks. In some cases, they have put the electromagnets on the train and in others they have put the electromagnets on the track. But while both of these schemes work well, currents in the metal surfaces and the electromagnets encounter electrical resistance effects that convert electric energy into thermal energy. Because that electric energy must be replaced continuously, these maglev trains need too much power to be cost effective.

Electrodynamic Levitation

A more practical induction-based levitation scheme is electrodynamic levitation, in which a train containing permanent magnets moves rapidly across a conducting track. As the magnets move, the magnetic field in the track changes and an electric field appears inside the track. This electric field propels an electric current through the track and the track becomes magnetic. The track's magnetic poles repel those of the train so the train rises on a cushion of magnetic force (Fig. 11.4.8).

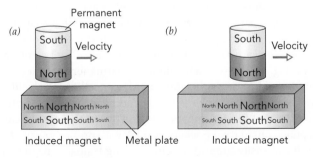

Fig. 11.4.8 When a permanent magnet moves rapidly across a metal surface, it induces currents in that surface. The metal becomes magnetic and repels the moving magnet. The magnetized region of the metal surface moves with the permanent magnet.

Electrodynamic levitation works best when the train moves very fast and its magnets pass quickly over each region of track. That's because the magnetic poles that appear in the track don't just support the train; they also push it backward as it approaches and forward as it departs. The forward push is particularly important, because without it the train would gradually slow down.

When it induces current in the track, the train invests energy in the track's magnetic field and the train experiences that energy investment as a backward force—the track's developing magnetic field does negative work on the approaching train. As the train subsequently leaves that magnetic patch of track, the train withdraws energy from the track's magnetic field and it experiences that energy return as a forward force—the track's diminishing magnetic field does positive work on the departing train.

The strength of the forward push depends on the train's speed. If the train moves quickly, the track's magnetic poles will push the train forward when it leaves almost as strongly as they pushed it backward when it arrived. But if the train moves too slowly, the currents in the track will have plenty of time to decay due to electrical resistance and the track will lose much of its magnetism while the train is still nearby. Since the train will receive only a weak forward push as it leaves, it will slow down and some of its energy will be left behind as thermal energy in the track.

Even at high speeds, the train experiences a net backward magnetic force. This force, called magnetic drag, is the sum of all the forward and backward magnetic forces on the train. Because the poles in the track ahead of the train's magnet are new and fresh, they're relatively strong. The poles in the track behind the train's magnet are weaker because they were created earlier and the currents that produce them have lost some energy. With a stronger repulsion from the poles ahead of it than from those behind it, the train experiences a backward force.

This magnetic drag force is most severe at about 30 km/h (19 mph), the minimum speed at which the train can levitate, and diminishes with increasing speed. The faster the train moves, the less time there is for the poles in the track to lose energy and the more the forward and backward magnetic forces on the train balance one another. Above about 300 km/h (190 mph), magnetic drag is insignificant when compared to ordinary air resistance.

Of course, electrodynamic levitation can't support the train when it's stationary, so the train needs retractable "landing gear" to support it while it's starting or stopping. The train leaves the station on wheels, picks up speed, and then "lifts off" onto the magnetic cushion. It retracts its wheels during high-speed movement and lowers them again before "landing" near the next station.

With suitable permanent magnets on board, an electrodynamically levitated train can hover more than 15 cm above an aluminum track. It easily clears small obstacles and provides an extremely smooth ride. Small imperfections in the railway can't be felt at all.

Recent developments in permanent magnets have made them considerably cheaper and stronger, so electrodynamic maglev trains that suspend themselves without using any electromagnets on either train or track are possible. If the train can keep moving forward fast enough, it will levitate. The invention of Halbach magnet arrays (Fig. 11.4.9)—arrangements of permanent magnets that project a strong magnetic field to one side while nearly cancelling that field on the other side—also makes it easier to project the suspending magnetic field only toward the track. That's important for passengers who don't want their credit cards erased or steel objects pulled out of their pockets. A track with properly shaped loops of wire embedded in it can easily support a train carrying Halbach magnet arrays.

The alternative to permanent magnets is superconducting electromagnets—electromagnets built from wires that have perfect electrical conductivities when cooled sufficiently close to absolute zero. Since a superconductor has zero electrical resis-

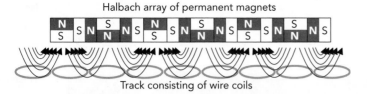

Halbach array of permanent magnets

Track consisting of wire coils

Fig. 11.4.9 In the Inductrak concept, a Halbach array of permanent magnets on the bottom of a train projects a strong, alternating magnetic field into the coils of a specially designed track. As the train moves across the track, the magnets induce currents in the track coils and the train and track repel one another strongly. The train is thus suspended above the track.

tance, a current introduced into it at the factory or maintenance facility will continue to flow through it indefinitely. The superconductor then behaves as a light, super-strong permanent magnet. A maglev train using superconducting magnets only needs to keep its magnets cold to keep them magnetic, so it can levitate easily without requiring much electric power.

Propulsion and Linear Motors

Regardless of how it's suspended, a maglev train still needs something to propel it forward. While a propeller or jet engine would do, most maglev trains use another novel magnetic concept: a linear motor. A linear motor is fairly complicated in practice, but simple in concept. Like an unrolled rotational motor, it uses attractive or repulsive forces between electromagnets on the track and train to push the train forward. These magnets are turned on and off at appropriate times so that the force on the train is always forward (Fig. 11.4.10).

In addition to pushing the train forward, the linear motor serves as a braking device. Changing the operating sequence of the electromagnets can reverse the direction of the horizontal force on the train and convert the linear motor into a linear generator. The linear motor can stop the train much more effectively than sliding friction can and it can generate electrical power at the same time. The linear motor converts electric energy into kinetic energy as the train accelerates forward and converts kinetic energy into electric energy as the train decelerates to a stop.

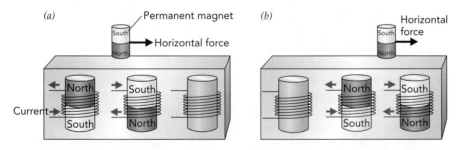

Fig. 11.4.10 A linear motor works by turning electromagnets on and off at the proper times. The moving permanent magnet always experiences a horizontal force toward the right. In (*a*), the left pair of electromagnets is active and in (*b*), the right pair is active. Reversing the current through its coil reverses the poles of the middle magnet.

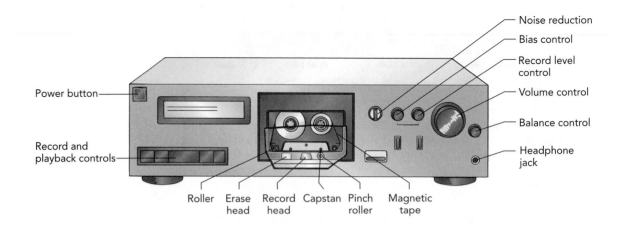

Power button

Record and playback controls

Noise reduction

Bias control

Record level control

Volume control

Balance control

Headphone jack

Roller Erase Record Capstan Pinch Magnetic
        head   head   roller tape

Section 11.5 **Magnetic Recording**

Magnetic recording has had a long and varied life. It was initially invented as a means for recording audio information, first on steel wires and later on coated plastic tapes. As television developed, magnetic recording adapted to storing video information on coated plastic tapes. And when computers needed a way to store vast amounts of digital information, magnetic recording was there, too, first with its coated plastic tapes and ultimately with coated disks. But despite all these changes in how magnetic recording is used, its basic concepts haven't changed all that much since the early days of wire recorders and reel-to-reel tapes.

Magnetic Audio Recording: Representing Sound

I'll start with magnetic audio recording. But before looking at how sound information can be recorded to and retrieved from magnetic tape, I should explain briefly what sound information is and how it can be represented at all. I'll resume that explanation when we come to audio players in the next chapter.

Recall that sound consists of compressions and rarefactions of the air. These disturbances from air's stable equilibrium density move outward from their source at about 331 m/s. In a region of compression, the air's density and pressure are somewhat higher than average and in a region of rarefaction, its density and pressure are somewhat lower than average. When these density fluctuations pass by your ear, you hear them as sound.

If a device measures these density fluctuations carefully enough and often enough, and records those measurements faithfully in some archival form, it will have collected enough information to reproduce the original sound. If the device then uses that sound information to recreate the density fluctuations and you listen to those fluctuations, you'll hear pretty much what you heard originally.

A magnetic audio recorder stores and retrieves sound information. But gathering that sound information requires microphones and playing it back requires speakers. A microphone is a pressure sensing device: it produces an electric current that's proportional to how much the local air pressure differs from its average value. When the air pressure is higher than average, current flows in one direction through the microphone's electrical circuit and when the air pressure is lower than average, current flows in the other direction. This electric current is representing the sound and it conveys that sound information to the magnetic audio recorder. The recorder's job is to represent that sound on a magnetic tape.

During playback, the recorder retrieves the sound information from the magnetic tape and prepares an electric current to represent that sound. This current is amplified and then flows through a speaker. A speaker is a pressure changing device: it produces an air pressure change that's proportional to how much current is flowing through it. The pressure and density changes produced by the speaker travel outward through the air and when they pass your ear, you hear them as a reproduction of the original sound.

Magnetic Recording Tape

Magnetic tape is a thin strip of plastic, coated on one side with a film of tiny permanent magnets. Each of these permanent magnets has a north magnetic pole at one end and a south magnetic pole at the other. During recording, a strong magnetic field can swap the poles of these particles, altering the tape in a way that can be detected during playback. This controlled modification of the tape, together with its detection at playback, is the basis for magnetic recording. (Early recorders didn't use tape; see ❶.)

As I noted while discussing household magnets on p. 356, permanent magnets are made from hard magnetic materials—materials that can be magnetized by external magnetic fields and that retain their magnetizations even after the external magnetic fields are gone. It makes sense to use permanent magnets to store information: if you magnetize them, they'll stay magnetized and their magnetizations can represent information.

Hard magnetic materials aren't all alike, however, and the permanent magnets used on plastic magnetic recording tapes differ from those used in the plastic sheet refrigerator magnets in at least two ways: in their types of magnetic order and in their sizes. I'll start by examining the magnetic order.

As I noted early in this chapter, while most atoms are magnetic, most solids are not. That's because the interactions between atoms as they bind together to form solids tend to cancel out all the atomic magnetism. Only a few solids avoid that total cancellation and retain their atomic magnetism. These magnetic materials are the basis for magnetic recording.

There are three major types of magnetic solids, ferromagnetic, antiferromagnetic, and ferrimagnetic, distinguished by the ordering of their individual magnetic atoms (Fig. 11.5.1). Iron, cobalt, nickel, and chromium dioxide are ferromagnetic solids, meaning that each atom's magnetic dipole is aligned with those of its neighbors. Chromium is an antiferromagnetic solid, in which the atoms' magnetic dipoles point in alternating or gradually varying directions and completely cancel one another. Ferrite (γ-iron oxide) and barium ferrite are ferrimagnetic solids, meaning that its atoms' magnetic dipoles alternate back and forth but, because they have different strengths, don't cancel one another completely.

Both ferromagnetic and ferrimagnetic materials can be used to make permanent magnets. Common plastic sheet magnets usually contain barium ferrites, which

❶ In 1898, while a research engineer for the Copenhagen Telephone Company, Danish engineer and inventor Valdemar Poulsen (1869–1942) developed a method for recording sound by magnetizing a steel wire. In his original recorder, the wire was stretched taut across his laboratory and the recording and playback devices were trolleys that traveled on that wire. The recording trolley would magnetize the wire to record sound and the playback trolley would measure the wire's magnetization to reproduce the sound. Poulsen would run along with the recording trolley and yell into the recording microphone. He would then listen to a reproduction of his voice as the playback trolley traveled along the wire.

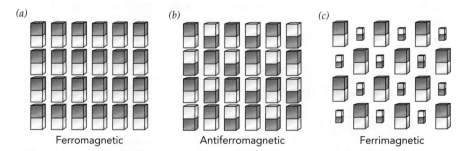

(a) Ferromagnetic   (b) Antiferromagnetic   (c) Ferrimagnetic

Fig. 11.5.1 (*a*) The atomic magnets in ferromagnetic solids point in the same direction. (*b*) In antiferromagnetic solids, they cancel one another completely. (*c*) In ferrimagnetic solids, there is partial cancellation.

are ferrimagnetic substances. Most plastic magnetic recording tapes, however, are coated with iron or other ferromagnetic materials. There isn't any profound reason for these choices; they're the consequences of technical and practical concerns.

Size, on the other hand, is a much more significant issue. The large barium ferrite particles contained in plastic sheet magnets have the magnetic domain structures that I discussed on p. 357. Magnetizing these barium ferrite particles involves enlarging all the domains pointing in the right direction and shrinking the ones pointing in the wrong direction. Since barium ferrite's domains don't change sizes easily, this material is moderately difficult to magnetize. Once magnetized, however, it stays magnetized indefinitely.

Magnetic recording media use a different type of hard magnetic material—tiny elongated particles of γ-iron oxide, chromium dioxide, cobalt-coated iron oxide, or pure iron. Each of these particles is so small that it contains only a single magnetic domain (Fig. 11.5.2). It has a north pole at one end and a south pole at the other, because that arrangement permits the opposite poles of adjacent magnetic atoms to be as close together as possible and minimizes the particle's total magnetic potential energy. With its magnetization pointing along its cigar shape, the particle is in a stable magnetic equilibrium: if something disturbs its magnetization direction, it will experience restoring effects that return its magnetization to that equilibrium direction!

Each of these particles is a tiny permanent magnet that can't be demagnetized. It's always magnetized in one direction or the other along its length. However the particle's magnetic poles can be interchanged by exposing it to a strong magnetic field. The particle itself won't flip, just its magnetic poles. This behavior makes such a particle ideal for magnetic recording. A magnetic recorder can choose a particle's magnetic orientation during recording and expect it to remain that way for an extremely long time.

Magnetic recording tape is produced by coating a Mylar film with a mixture of these magnetic particles, a binder, and a solvent. Immediately after coating, the wet film passes through a strong magnetic field, which rotates all of the tiny magnetic particles so that they're aligned along the direction in which the tape normally moves. The tape is then dried and pressed so that its magnetic coating is dense, smooth, and shiny. The final magnetic coating is about 5 microns thick on audiocassette tapes and videotapes and 10 microns thick on audio reel-to-reel tapes. Only the outer micron or so of these coatings is actually used for recording.

Magnetic particles are rated according to the magnetic fields needed to interchange their magnetic poles. Particles that are more resistant to interchange make better magnetic tapes because they are less affected by accidental magnetic fields. The conventional measure of this resistance to interchange is the oersted, with a higher number of oersteds meaning more resistance. Table 11.5.1 shows that iron particles make the best recording tapes. Although a large piece of iron is a soft magnetic material, with magnetic domains that resize easily, an elongated single-domain iron particle is an extremely hard magnetic material. Metal particle tapes are made from elongated iron particles.

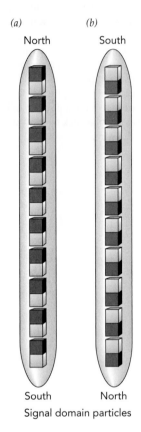

(a) North (b) South

South North

Signal domain particles

Fig. 11.5.2 A tiny, elongated magnetic particle has only a single magnetic domain, which aligns with the long axis of the particle. This magnetic alignment is hard to change, so that only a strong magnetic field can convert (a) into (b) or (b) into (a) without actually rotating the particle itself.

Table 11.5.1 Particles commonly used in magnetic tapes.

Type of Magnetic Particle	Resistance to Interchange of Poles
γ-iron oxide (γ-Fe_2O_3)	300 oersteds
chromium dioxide (CrO_2)	450 oersteds
cobalt-modified γ-iron oxide	600 oersteds
iron	1500 oersteds

Recording Sound Information onto Magnetic Tape

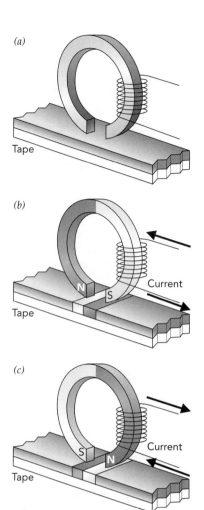

We're now ready to look at how a magnetic audio recorder records sound information on a magnetic tape. This is done by a miniature electromagnet in the recording head, a tiny ring of soft magnetic material with a coil of wire wrapped around it (Figs. 11.5.3 and 11.5.4). When a current flows through the coil, the coil's magnetic field temporarily magnetizes the ring.

The ring is actually incomplete; it has a tiny gap just at the point where it touches the magnetic tape (Fig. 11.5.3a). When current flows one way through the coil, the gap has a north pole on its left end and a south pole on its right end (Fig. 11.5.3b). When the current reverses, the poles also reverse (Fig 11.5.3c). During recording, sound is represented as a fluctuating current in the coil and therefore a fluctuating magnetic field in the ring's gap.

Fig. 11.5.4 This stereo recording head has two soft iron rings, the shiny rectangles above and below the horizontal dividing line. The vertical gaps in these rings are so thin that they can only be seen with a microscope. The head records and plays back two separate channels.

The coating on the magnetic tape must be demagnetized by an erase head prior to its encounter with the magnetic recording head. Although the individual particles in the coating are always magnetic, the erasing process makes sure that roughly half of their north poles point forward and half point backward (Fig. 11.5.5a). To achieve this demagnetization, the erase head exposes the tape to a rapidly reversing magnetic field that leaves the magnetic particles randomly oriented. The erase head is essentially a recording head with an extra wide gap and a large, high-frequency alternating current in its coil. As the magnetic coating passes through the rapidly reversing magnetic field in the gap, the magnetic alignments of its particles flip back and forth many times. About half end up oriented in one direction and half in the other.

While the erase head's job is to demagnetize the tape, the recording head's job is magnetize it: to reorient those tiny magnets so that their poles all point in the same direction (Fig. 11.5.5b). The particles don't actually flip over; their poles simply interchange. As the magnetic tape moves steadily past the recording head, the gap in the recording head magnetizes regions of the tape's magnetic coating in one direction or the other (Fig. 11.5.6). The depth to which the coating is magnetized

Fig. 11.5.3 (a) A tape head is a ring of soft magnetic material with a small gap in it and a coil of wire wound around it. The ring is temporarily magnetized whenever a current flows through the coil (b,c) and it permanently magnetizes any hard magnetic material near its gap.

(a)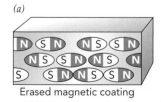

Erased magnetic coating

(b)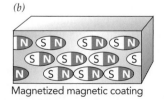

Magnetized magnetic coating

Fig. 11.5.5 In an erased magnetic coating (a), roughly half the particles have north poles on the left and half on the right. In a magnetized portion of coating (b), all of the particles have their north poles aligned on the same side.

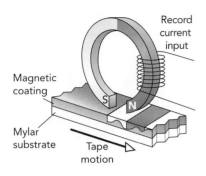

Fig. 11.5.6 The recording head magnetizes regions of the tape's magnetic coating as the tape moves past the head. The depth of this magnetization is proportional to the amount of current in the coil.

depends on the strength of the magnetic field in the gap, which is proportional to the current in the coil. Since that current represents sound and air pressure, the magnetization depth is greatest during loud parts and least during quiet parts. The direction of magnetization, north poles forward or backward, depends on the direction of current flow through the coil, which depends on whether the recorder is trying to represent a compression or a rarefaction of the air.

Audio recording has an important complication: the recording head has trouble flipping the magnetic alignments of particles during quiet portions of the sound. The particles are in stable magnetic equilibriums and they won't respond to magnetic fields that are too weak. Magnetizing the tape during loud passages isn't a problem, but during the quiet ones, the magnetic particles may not magnetize at all.

To solve this problem, the tape recorder sends an additional high-frequency alternating current through the coil in the recording head. This bias current flips the particles' magnetic alignments back and forth during the recording process. While the bias current doesn't give the tape any particular overall magnetization, it allows the other current in the coil, the current representing sound, to magnetize parts of the coating even during quiet passages. Tape recorders adjust the bias current according to the magnetic hardness of the particles in the coating, either type I (normal bias), type II (high bias), or type IV (metal bias).

Playback, Noise Reduction, and Stereo

The tape recorder recreates sound by passing the magnetic tape across a playback head. The playback head is similar to the recording head and the two are often the same device. Fancy tape recorders have a separate playback head located downstream from the recording head, so that you can listen to the tape as you are recording it. Since the roles of the recording and playback heads are different, separate heads that are specialized to the two tasks outperform a single head that must perform both duties.

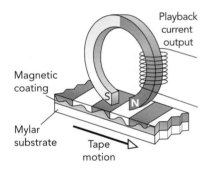

Fig. 11.5.7 As the magnetized tape passes across the gap in the playback head's ring, the ring is magnetized and the magnetic flux passing through the coil changes. Current representing the recorded sound is induced in the coil.

As the magnetized tape passes across the gap of the playback head, it temporarily magnetizes the ring (Fig. 11.5.7). The deeper the magnetization of the coating, the more the ring is magnetized and the more magnetic flux lines pass through the coil of wire wrapped around it. As the tape moves past the head, the changing number of flux lines induces current in the coil. The induced current is directly related to the original current that flowed through the recording head when the sound was recorded. This current is amplified and sent to a speaker, where it reproduces the compressions and rarefactions of the air that occurred in the original sound.

Unfortunately, a tape recorder doesn't reproduce sound perfectly. It introduces some noise into the sound because of imperfections in the recording and playback processes. This noise is due primarily to:

(1) imperfect erasure of the tape.
(2) nonuniformities in the distribution of magnetic particles in the coating.
(3) imperfections in the bias current aiding low-level recording.
(4) surface irregularities in the coating.

Noise also appears when highly magnetized portions of a tape imprint their magnetizations onto adjacent layers of tape when the tape is wound tightly on a spool. Noise from all of these sources is particularly disturbing during quiet passages and is mostly high-pitched hissing sounds.

To improve their fidelity, many tape recorders use noise reduction techniques such as Dolby A, B, and C. These techniques boost the volumes of high frequency sounds during recording and then reduce their volumes during playback. The desired sound is unchanged overall but any high-frequency noise that appeared between recording and playback is decreased in volume.

Many tape recorders work with more than one sound channel at a time. Stereo recorders write and read two or more audio tracks at once to represent sounds heard by microphones at different locations. These machines have at least two closely spaced recording heads and an equal number of playback heads. Each head writes or reads its own narrow track of magnetization on the tape.

Magnetic Audio and Video Recorders

Tape transport mechanisms have evolved during the past fifty years so that most magnetic tapes are now contained in plastic cassettes. A long, continuous strip of tape is transferred from one spool to another, passing through the transport mechanism and across the erase, recording, and playback heads.

The one transport component worth noting in an audio tape recorder is the capstan. The capstan is a cylindrical rod that rotates at a very steady rate and controls the tape's speed through the recorder. The tape is pressed against the capstan by the pinch roller and is drawn steadily forward as the capstan turns. The capstan has a flywheel attached to it to give it a large rotational mass so that it always turns at exactly the same rate. The resulting steady motion of the tape is important to maintaining pitch fidelity between the recorded sound and the reproduced sound. On a rare occasion when the capstan fails to engage during playback and the tape moves too quickly past the playback head, a recording of Louis Armstrong may suddenly sound like Alvin and the Chipmunks.

A magnetic video recorder is much more complicated than a magnetic audio recorder because storing image information as well as sound takes lots of room on a magnetic tape. Since the smallest region of magnetization that a recording head can create is about 1 micron long, a video recorder uses about 2 m of track length on the tape every second. Instead of moving a long tape quickly past stationary heads, a video recorder sweeps the heads quickly across a wide tape. The heads are contained in a spinning drum and the tape is pressed against that drum at an angle. As the tape is pulled slowly across the drum, the spinning heads write or read a series of stripes across the tape's width. Because these stripes appear on the tape at an angle, the recorder stores information across the entire surface of the tape.

Magnetic Data Storage

Recording digital information magnetically is simply more of the same. The digital information appears as a stream of binary bits, 0s and 1s, and recording them magnetically is mostly a matter of magnetizing single domain particles either one way or the other. The differences between magnetic audio or video recording and magnetic data storage are primarily technical details.

The dominant form of magnetic data storage is the hard drive—a rigid, magnetically coated disk that spins rapidly past magnetic recording and playback heads (Fig. 11.5.8). Since these heads can move toward or away from the disk's center as the disk spins, they can interact with any portion of the disk's surface. The hard drive stores information by magnetizing tiny patches of the disk's surface and it retrieves that information by measuring the magnetization of those same patches. Since the hard drive has no erase head with which to prepare the disk's coating, its recording head must be able to magnetize the coating all the way through to represent either a 0 or a 1. For that reason, the magnetic coatings used on hard drive disks are less than 1 μm thick.

But locating the bits on the disk's surface isn't easy and the hard drive always needs some indication of where one bit ends and the next bit begins. When the disk's magnetization suddenly changes directions, the drive knows that a new bit is starting. But if there are too many bits all magnetized in the same direction, the drive will have trouble distinguishing one bit from the next. To help it find the bits,

Fig. 11.5.8 This hard drive has four disks and eight magnetic recording surfaces. Each surface has a thin magnetic coating and its own read and playback head. The heads can swivel from the inner portion of the drive to the outer portion in a fraction of a second.

the drive manipulates the digital information it's recording in a way that helps it locate the bits. In effect, the hard drive occasionally adds an extra reversal of the magnetization direction. When the hard drive later retrieves bits from the disk's surface, it recognizes and ignores these extra reversals.

As a disk's surface turns past the recording head, that head magnetizes only a thin circular track on the disk's surface. By shifting the head toward or away from the disk's center, the hard drive can access thousands of different tracks. The hard drive uses an electromagnetic drive system to push its recording head quickly from one position to another. The head's tiny mass and the strong forces between the electromagnet and fixed permanent magnet allow the head to accelerate rapidly. It takes only a few thousandths of a second to leap from one track to another.

Three things determine the maximum number of bits that a computer can record on each square centimeter of disk or tape surface:

1. The size of the recording and playback heads.
2. The drive's ability to locate bits on the surface.
3. The size of the magnetic recording particles.

Each year, recording and playback heads shrink further so that they're able to record bits on smaller patches of surface. Detecting these tiny bits is difficult, so a new playback head technology has developed that uses materials that change their electric resistances when exposed to magnetic fields. These magnetoresistive heads have allowed bits to become significantly smaller than they used to be.

Because smaller bits are harder to find, the controls used to position recording and playback heads are getting more and more sophisticated. Electronic and optical feedback techniques are now often used to find and follow the narrow magnetic recording tracks on a disk.

The size of recording particles is now a serious problem. These particles can't be made smaller without becoming susceptible to the disordering effects of thermal energy. Although the elongated single-domain particles are in stable magnetic equilibriums, their restoring effects are proportional to their volumes and they become less stable as they become smaller. Random thermal agitation can flip the magnetizations of small magnetic particles and the smaller the particles, the more likely they are to flip.

Since hard drives must retain information reliably for at least 10 years, they can't use magnetic particles that experience lots of thermal flipping. At present, each bit of information is recorded as the magnetizations of several dozen tiny magnetic particles. That redundancy offers some immunity to thermal demagnetization. And orienting the magnetic particles perpendicular to the disk's surface, rather than parallel to it, allows those particles to squeeze more tightly together to save space. But scientists and engineers are rapidly reaching the limit of how tightly they can store information magnetically on a surface.

CHAPTER 12

ELECTRONIC THINGS

Electric currents and magnetized metals can do more than simply power light-bulbs or attach notes to refrigerators. They can also represent many different things, from sounds to video images to computerized information. Moreover, they can be used to manipulate that information in remarkable ways and at astonishing speeds. In this chapter, we'll see how electricity and magnetism have given rise to the field of electronics.

Electronic devices are tools that use electric currents to perform sophisticated tasks. They first appeared in the early twentieth century with the development of vacuum tubes and have been maturing ever since. Advances in electronics have frequently followed advances in quantum and solid-state physics, so it was hardly surprising that research into the physics of semiconductors should lead to interesting electronics. But the invention of the semiconductor transistor shortly after World War II started a revolution in electronics so profound that no one could have anticipated it. Advanced jointly by physicists and engineers, electronic devices have gradually become so inexpensive and so effective that they now permeate every facet of modern society.

While this brief survey can't go beyond the most basic issues in electronics, it should provide a good foundation on which to build. Once you understand how a few important electronic components work, it's not so hard to see how to combine them in ways that no one else has ever imagined. Like building blocks that control the movements of charged particles, electronic components can be put together to do almost anything.

Chapter Itinerary

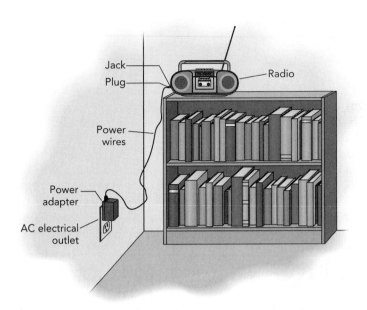

Jack

Plug

Radio

Power wires

Power adapter

AC electrical outlet

Section 12.1 **Power Adapters**

Virtually every electronic device you buy comes with its own power adapter, a small black cube or brick that either plugs directly into an electrical outlet or has a cord that does. These adapters obtain power from the century-old electrical grid and prepare it for this week's gadgets. Power adapters range widely in the voltages and currents they supply, the degree to which their voltages are smoothed and regulated, and even the connectors through which they deliver their power. The result is dozens of adapters that are rarely interchangeable and, consistent with Murphy's law, the adapter you need is always the one that fell behind the bookshelf.

But despite their differences, these power adapters all perform essentially the same conceptual task: they take relatively high voltage alternating current and use its power to prepare relatively low voltage direct current. They do this conversion efficiently, inexpensively, and with a high degree of reliability. In this section, we'll see how simple power adapters perform that conversion.

Producing DC Power from AC Power

While electrical devices such as toasters and incandescent lightbulbs can operate from either AC or DC power, most electronic devices require DC power. There are two reasons for their pickiness. First, they contain sophisticated electronic components that are sensitive to the direction of current flow and won't operate properly if that current is reversed. Second, electronic devices typically need a continuous supply of power and can't tolerate the brief moments of powerlessness that occur when alternating current is reversing.

Addicted as they are to direct current, electronic devices operate beautifully on batteries. Of course, those batteries must be installed in the right direction or they'll act to push current the wrong way. But batteries and electronics are practically made for each other.

Unfortunately, batteries eventually run out of stored energy and must be replaced or recharged. And some electronic devices are so power-hungry that operating them on batteries would cost a fortune. Unless you're an heiress or the nephew of a battery manufacturer, you often need a cheaper, more practical source of DC power. You need a power adapter.

A power adapter's task is simple: it uses AC power from an electrical outlet to provide DC power to an electronic device. More specifically, it delivers current to the electronic device through a positive power wire and receives that current back through a negative power wire. And the adapter maintains a specified average voltage rise from the negative wire to the positive wire, as though the adapter were a battery.

The steadiness of an adapter's voltage rise depends on its sophistication. Some adapters let their voltage rises fluctuate with the AC cycle while others use complicated electronics to regulate their voltage rises precisely. Intermediate between those two extremes are adapters that use simple electronics to smooth out most of the fluctuations due to the AC cycle, but make no further attempt to regulate their voltages. In this section, we'll examine these intermediate power adapters.

To be specific, we'll examine the components in a 9-V DC power adapter (Fig. 12.1.1) that operates from 120-V AC power and supplies power to an ordinary radio. Apart from wires, this adapter contains only a few components: a transformer, four diodes, and a capacitor. To understand this power adapter, we need only examine those components and how they interact. And since we studied transformers in Section 11.2, we already have a head start.

When you plug the transformer into an electrical outlet, its primary coil forms a circuit with the power company and an alternating current flows through it. The primary coil has 13⅓ times as many turns as the secondary coil, so the secondary coil provides an induced emf of 9 V AC.

The adapter then converts 9 V AC into 9 V DC; it uses 9-V alternating current from the transformer to provide 9-V direct current to the radio. That simple sounding conversion will keep us busy for the remainder of this section. In large part, that's because it involves two devices that we haven't encountered before: diodes and capacitors.

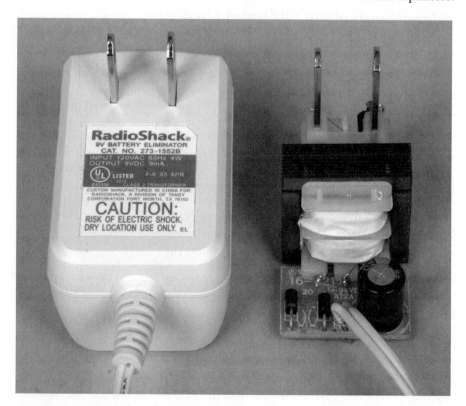

Fig. 12.1.1 These two power adapters use 120-V AC power from an electrical outlet to produce 9-V DC power for a radio. The adapter on the right has been removed from its plastic shell to reveal a transformer (top), two black diodes (lower left) and a cylindrical capacitor (lower right).

Diodes are one-way conductors of electric current. Unlike a wire, which carries current equally well in either direction (Fig. 12.1.2a), a diode allows current to flow through it only in one direction (Fig. 12.1.2b). It acts as a conductor when current tries to flow the allowed direction and as an insulator when current tries to flow the forbidden direction. This switching behavior is what the power adapter uses to obtain direct current from alternating current.

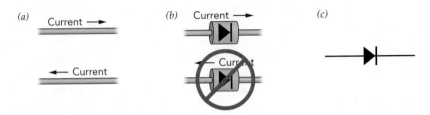

Fig. 12.1.2 (*a*) A wire can carry current in either direction. (*b*) A diode can carry current only in the direction shown symbolically by its arrow. (*c*) In a schematic diagram of an electronic device, the diode is represented by an arrow and bar.

Capacitors are devices that store separated electric charges. By accumulating positive charge on one side and negative charge on the other side, a capacitor stores both charge and energy. That storage helps the adapter steady its voltage rise and endure the moments when the reversing alternating current provides no power to the adapter.

A Few Words about Quantum Physics

Since a diode is sensitive to the direction of current flow, it can't be symmetric. It must have two distinct and different ends. To understand those two ends and how they differ, we need to examine the materials from which they're both made. We need to examine semiconductors.

Semiconductors are materials with properties intermediate between those of electrical conductors and those of electrical insulators. While charge is *always* mobile in conductors and *almost never* mobile in insulators, charge is *sometimes* mobile in semiconductors. In general, a semiconductor acts like an insulator when it is cold, pure, and in the dark, and it acts like a conductor when it is hot, impure, or exposed to light. Semiconductors are so important to diodes and to most of modern electronics that we'll spend the next several pages learning how charge moves through them. And we'll begin by discussing quantum physics, which has immense influence over the tiny particles from which semiconductors are built.

As quantum physics gradually revealed itself to the scientists of the early twentieth century, they found the experience both exhilarating and disorienting. Prior to that era, the physical world seemed to divide neatly into particles and waves: scientists viewed an electron only as a particle and light only as a wave. However, one of the most basic observations of quantum physics, and the one most relevant to our present topic, is that everything has both particle and wave characteristics. Put simply, everything begins and ends as a particle, but travels as a wave.

For an electron, the quantum surprise is that it travels as a wave. For light, the quantum surprise is that it is emitted and absorbed as a particle. Called the wave-particle duality, this observation that everything in nature has both particle and wave characteristics has left few areas of physics unaffected. But while quantum physics is now a basic and essential part of nearly all modern physics research, its effects are subtle and often nonintuitive. They are most apparent in the microscopic world and are visible to us only indirectly. No wonder they seem so strange.

We'll encounter quantum physics and its effects several times in the next few chapters. It figures prominently in the electronic properties of semiconductors, in the light emitted by atoms and lasers, and in the radioactive decays that release nuclear energy. Our examination of semiconductors will acquaint us with the wave nature of electrons and show how the wave phenomena that we studied in Chapter 9 apply to quantum physics. Our exploration of light from atoms and lasers will acquaint us with the particle nature of light and how the collision effects we explored in Chapters 1–3 apply to quantum physics. And in our examination of radioactivity, we'll uncover particle and wave effects that we would not even have anticipated without quantum physics. With each encounter, we'll take a small bite of the quantum apple—looking at how quantum effects manifest themselves in our everyday world.

Electrons in Solids

We learned in Chapter 10 that metals conduct electricity because they contain mobile electrons and that insulators don't conduct electricity because none of their electrons are mobile. Now it's time to see what controls electron mobility. As you might guess, the explanation lies in quantum physics.

In a nonquantum world, an electron in a solid would travel only as a particle and it would be able to move at any speed along any path. But ours is a quantum world and the electron does not travel as a particle at all; it travels as a wave. And like the waves on a violin string, drumhead, or basin of water, the electron waves in a solid have limited possibilities.

In Chapter 9, we observed that the most basic mechanical waves on a limited object are all standing waves—waves that effectively oscillate in place. This rule also applies to quantum waves. The electrons in a solid are best understood as standing waves in that solid. Each electron wave extends across part or all of the solid and has such wave characteristics as wavelength and frequency. Unlike a vibrating string or drumhead, the electron is a three-dimensional wave and its oscillation is internal. But it's still a wave.

This wave character has profound effects on the electronic structure of solids. Most significantly, it limits what electrons can do in those solids. We saw in Chapter 9 that a violin string's one-dimensional standing waves consist only of its fundamental mode (Fig. 9.2.3) and its harmonic modes (Fig 9.2.4) and that a drumhead's two-dimensional standing waves consist only of its fundamental mode and overtones (Fig. 9.2.7). Similarly, a solid's three-dimensional electron standing waves consist only of a fundamental mode and overtones. And while there are a great many overtone modes available, their possibilities are nonetheless limited.

The electron standing waves in solids are often called levels—a recognition that each standing wave has an amount or "level" of energy associated with it. The electron standing waves in atoms, another group of limited systems in which electrons exist as standing waves, are called orbitals—a nod to the orbiting nature of an atom's electrons. We'll see when we examine discharge lamps in Section 14.2 that each atom's limited orbital choices determine the colors of light it can emit or absorb. And we'll see in the present section that a solid's limited level choices determine its electrical conductivity.

Another remarkable observation of quantum physics is that every indistinguishable electron must have its own level or orbital, its own unique quantum wave. This law is called the Pauli exclusion principle, after its discoverer, Wolfgang Pauli ❶. The principle applies to a whole class of subatomic particles, the Fermi particles, that includes all of the basic constituents of matter: electrons, protons, and neutrons. For reasons that lie deep in an area of physics known as quantum field theory, two indistinguishable Fermi particles can never be in the same quantum wave.

❶ Austrian physicist Wolfgang Pauli (1900–1958) rose to fame at 21 by writing an article on relativity that impressed even Einstein. He went on to discover the exclusion principle, a fundamental part of quantum theory. He was renowned for his intensely critical attitude toward new ideas, considering them all "rubbish" until convinced otherwise. Pauli was also quite interested in psychology, corresponding with Carl Gustav Jung and even writing articles on the subject.

The Pauli Exclusion Principle

No two indistinguishable Fermi particles ever occupy the same quantum wave.

However, a peculiar property of electrons allows two of them to share an orbital or level. Electrons have two possible internal states, usually called spin-up and spin-down. Because a spin-up electron is distinguishable from a spin-down electron, one spin-up electron and one spin-down electron can share a single orbital or level. However, two electrons is the absolute maximum allowed by quantum physics and the Pauli exclusion principle.

Despite being a wave, an electron in a level has a specific total energy—the sum of its kinetic and potential energies. That energy, which depends on the shape and structure of the electron wave, also determines the wave's oscillation frequency. According to quantum physics, the electron's total energy and the oscillation frequency of its wave are exactly proportional to one another; a low-energy electron oscillates slowly while a high-energy electron oscillates quickly. We'll return to this observation in Chapter 14, but for now we'll note that each level, each quantum standing wave in the solid, has a specific frequency and energy. The solid's fundamental standing wave has the lowest frequency and energy, while the overtone waves have progressively higher frequencies and energies.

Physicists have come to view these standing waves as abstract placeholders, independent of the electrons that may or may not exhibit them at a given moment. The levels are then analogous to seats in a theater, each of which may or may not be occupied right now. Instead of saying that the *electron* is experiencing a particular standing wave or level, we say that the *level* is occupied by an electron. In this reversed perspective, the level plays the more important role. For the rest of this section, we'll view a solid in this manner; we'll think in terms of the levels it has available and whether or not those levels are occupied by electrons.

A solid contains an enormous number of electrons and there are always plenty of levels around to accommodate those electrons. But which levels do they occupy?

At sufficiently low temperature, electrons occupy those levels which have the least energy. For thermodynamic reasons, the electrons settle into the lowest energy levels available, two electrons per level. By the time all the electrons have been accommodated, they fill the levels up to a certain maximum energy. Halfway between the highest filled level and lowest unfilled level is the Fermi level—a hypothetical level that defines the top of this Fermi sea of electrons. The energy an electron would have in that hypothetical level is called the Fermi energy.

Our theater analogy can again provide insight into this level-filling process. It is analogous to what happens at a popular show: the seats fill from the orchestra level on up—everyone wants to sit in the lowest (and closest) seat. When the show starts, people have filled all the seats up to a certain highest seat. Halfway between that last filled seat and the next unfilled seat is the hypothetical Fermi seat.

If we represent the levels graphically by boxes and arrange them vertically according to energy (Fig. 12.1.3), then levels (boxes) below the Fermi level contain two electrons each, while those above the Fermi level are empty. Although thermal energy complicates this picture somewhat by shifting electrons about near the Fermi level, we can ignore that detail near room temperature or below.

Since levels are standing waves, they don't have sharply defined locations in space. But we can safely imagine that each level places its electrons near a particular location in the solid, as shown in Fig. 12.1.3. While this picture is somewhat oversimplified, it's accurate enough to illustrate much of the physics of charge motion in materials.

Of course, electrons aren't the only charged particles in a solid. The atoms also have positively charged nuclei. But those nuclei are essentially immobile and rarely

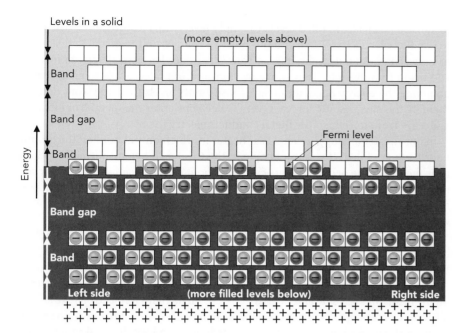

Fig. 12.1.3 The levels in a solid are grouped together in bands and filled from the lowest energy level up to the Fermi level. Each level can hold at most two electrons: one spin-up electron (light) and one spin-down electron (dark). The positive charges shown at the bottom are the nuclei of atoms that make up this electrically neutral solid.

participate in the flow of electricity. Instead, they form a uniform background of positive charge, shown schematically as pluses (+) at the bottom of Fig. 12.1.3, so that the object is roughly neutral throughout.

Metals, Insulators, and Semiconductors

The levels in a solid occur in groups called bands. Each band corresponds to standing waves with a particular type of structure. Since the levels in a band involve similar waves, they also involve similar energies. Between these bands of levels there are sometimes band gaps—ranges of energy in which no levels exist. The solid does not and cannot contain electrons with energies that lie within a band gap.

Bands and band gaps are what distinguish metals, insulators, and semiconductors. When the Fermi level is located in a band gap, it can prevent the electrons in a solid from responding to outside forces. To see how that happens, let's examine first a metal and then an insulator.

In a metal, the Fermi level lies in the middle of a band (Fig. 12.1.4). Because the band's empty levels are just above its filled levels, very little energy is needed

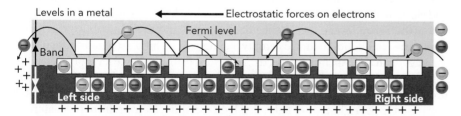

Fig. 12.1.4 In a metal, the Fermi level lies in the middle of a band. When you put electrons to the metal's right and positive charges to the metal's left, electrons shift leftward through the metal with the help of the empty levels. Since there is a net flow of charges through the metal, it is conducting electricity.

to shift electrons from filled levels to empty levels. This feature allows the metal to conduct electricity. When you put positive charges on the metal's left side and negative charges on its right, its electrons experience leftward electrostatic forces and begin to move left. They move by shifting from filled levels to empty levels, obtaining the energy needed to reach those empty levels from the work done on them by the electrostatic forces. Overall, electrons enter the metal from its right and leave from its left, so the metal conducts electricity!

In our theater analogy, a metal is a theater in which only about half the ground-floor seats are filled. If you ask people in the theater to begin shifting left, those near the top of the occupied seats can shift about easily. Each finds an empty seat nearby on the left and moves over. New people are then able to enter the theater from the right while others leave the theater from the left. This "metal" theater would be "conducting" people.

Unlike the situation in a metal, an insulator's Fermi level lies in the middle of band gap, between the top of one band and the bottom of another band (Fig. 12.1.5). With no easily accessible empty levels available, a great deal of energy is required to shift electrons from filled levels to empty levels. When you put positive charges on the insulator's left side and negative charges on its right, its electrons experience leftward electrostatic forces but are unable to move. To shift into one of the empty levels in the upper band, an electron in the lower band would need more energy than it can get from the electrostatic forces. Since no net charge flows through the insulator, it doesn't conduct electricity!

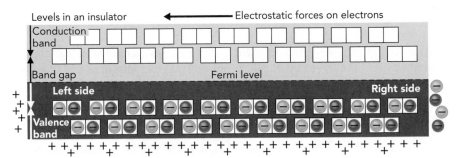

Fig. 12.1.5 In an insulator, the Fermi level lies in the middle of a band gap. When you put charges near the insulator, electrons in the filled band can't shift to produce a net flow of charge through the insulator. The insulator can't conduct electricity.

In our theater analogy, an insulator is a theater in which all the ground-floor seats are full and in which the balcony seats are empty. When you ask people in this theater to begin shifting left, they can't do it. All of the ground-floor seats to the left are filled and they can't reach the balcony to make use of its empty seats. This "insulator" theater would be unable to conduct people.

In a metal, the band of levels containing the Fermi level is only partially filled, and electrons can easily shift from filled to unfilled levels. In an insulator, the band below the Fermi level—the valence band—is full and the band above the Fermi level—the conduction band—is empty, making such shifts extremely difficult.

But even in an insulator, an electron can shift from a valence level (a level in the valence band) to a conduction level (a level in the conduction band) if something provides the necessary energy. One such energy source is light. When an insulator is exposed to the right type of light, that light can shift electrons from the material's valence band to its conduction band (Fig. 12.1.6).

Once electrons appear in the normally empty conduction band and empty levels appear in the normally full valence band, electrons can respond to electrostatic forces. They can shift from filled levels to nearby empty levels and thus travel

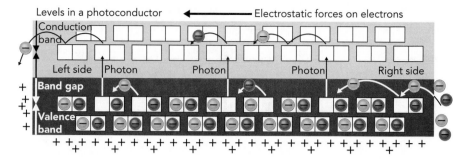

Fig. 12.1.6 When light strikes an insulator, the energy in its photons shifts some electrons from filled valence levels to empty conduction levels. Such shifts make it possible for electrons to move in response to electrostatic forces, so the insulator becomes an electrical conductor, a photoconductor.

through the material. Electrons can then enter the material from one side and leave from the other, so the material conducts electricity. And because light has made this insulator a conductor, we call the material a photoconductor.

Turning again to our analogy, light's role in the insulator theater is performed by a playful gorilla that walks about the ground floor, tossing patrons into the balcony. With some of the ground-floor seats suddenly empty and some of the balcony seats suddenly occupied by dazed theatergoers, the crowd can now respond to your request to move left. The gorilla has made the insulator theater a conductor of people—what you might call a "gorillaconductor."

Not all light causes photoconductivity in an insulator. That's because light is emitted and absorbed in energy packets or quanta called photons. As with an electron, a photon's energy is proportional to its frequency; the higher the frequency of light, the more energy each of its photons contains. To shift an electron across the large band gap in a typical insulator, high-energy, high-frequency light is needed; the insulator must be exposed to violet or even ultraviolet light.

But nature also provides materials with smaller band gaps that can be crossed with the help of low-energy, low-frequency red or even infrared light. These materials are called semiconductors because their properties lie somewhere between those of conductors and insulators. Semiconductors have small band gaps, making it relatively easy for light, heat, or other types of energy to shift electrons between valence and conduction levels. In our analogy, a semiconductor theater is an insulator theater with a low balcony, so that even a baby gorilla can toss people into it.

For half a century, scientists and engineers have worked with semiconductors to produce an astonishing array of electronic devices. By carefully tailoring the shapes and chemical compositions of semiconducting materials such as silicon, germanium, and gallium-arsenide, they have created virtuoso instruments for electron waves in solids that are every bit as remarkable as the instruments for musical waves found in great orchestras. And of all these electronic instruments, the simplest is the semiconductor diode.

Diodes

A diode is a one-way device for current; it allows current to flow through it in one direction but not in the other. Although diodes have taken many forms over the years, the diodes in power adapters and virtually all modern electronic devices are built from semiconductors.

A semiconductor diode is made by joining together two different semiconductors. These two semiconductors have been modified so that they don't have perfectly filled valence levels and perfectly empty conduction levels. Instead, they're doped

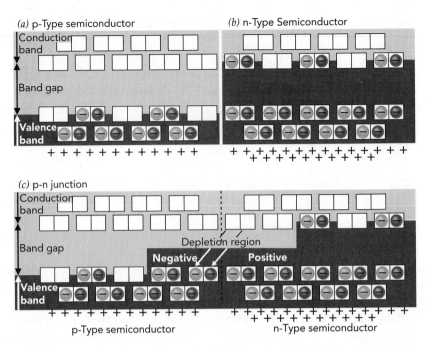

Fig. 12.1.7 (*a*) A p-type semiconductor has missing electrons (and missing positive atomic nuclei) and can conduct electricity through its partly filled valence band. (*b*) An n-type semiconductor has extra electrons (and extra positive atomic nuclei) and can conduct electricity through its partly filled conduction band. (*c*) When p- and n-type semiconductors touch, conduction level electrons shift from the n-type semiconductor to the p-type semiconductor, creating a thin, electrically polarized depletion region at the p-n junction. That p-n junction behaves as a semiconductor diode.

with atomic impurities that either create a few empty valence levels (p-type semiconductor, Fig. 12.1.7*a*) or place a few electrons in the conduction levels (n-type semiconductor, Fig. 12.1.7*b*). These empty valence levels or conduction level electrons allow p-type and n-type semiconductors to conduct electricity. The doping atoms bring with them just the right amount of positive charge in their nuclei to keep both p-type and n-type semiconductors electrically neutral.

But when a piece of p-type semiconductor touches a piece of n-type semiconductor, something remarkable happens: a p-n junction forms at the place where the two meet (Fig. 12.1.7*c*). To reduce their potential energies, higher energy conduction level electrons from the n-type semiconductor flow across the p-n junction and fill in empty lower energy valence levels in the p-type semiconductor. This electron flow creates separated charge. The n-type semiconductor acquires a positive net charge because it now has fewer electrons than positive charges. The p-type semiconductor acquires a negative net charge because it now has more electrons than positive charges. Electrostatic forces from this separated charge oppose the further flow of electrons across the junction and gradually bring that flow to a halt. Everything is then in equilibrium.

Near the p-n junction, there is now a depletion region—an area in which electron flow has emptied all of the conduction levels and filled all the valence levels. With no conduction level electrons or empty valence levels left, the depletion region can't conduct electricity and charge can't move across the p-n junction. The depletion region is an insulator, and the two pieces of semiconductor have become a diode.

In our theater analogy, the p-n junction is analogous to a theater with two halves. In the left or "p-type" half, the balcony is empty, and even the ground floor has some empty seats. In the right or "n-type" half, the ground floor is filled and there are even a few people in the balcony. Since these two halves touch, people in the right balcony notice the empty seats in the left ground floor, and a few of them near the center of the theater clamber down from the right balcony to the left ground floor to take advantage of the better seats. Near the center of the theater, the ground floor is now filled and the balcony is empty, forming a depletion region in which no one can move left or right. The theater can't conduct people!

Let's now look at what happens when we attach wires to each semiconductor half and try using a battery to push electrons across the p-n junction. If we push electrons leftward, adding them to the n-type side and removing them from the p-type side, the depletion region becomes thinner and eventually vanishes (Fig. 12.1.8a). We're adding electrons to the n-type conduction levels and pushing them toward the p-n junction. We're also removing electrons from the p-type valence levels and pulling them away from the p-n junction.

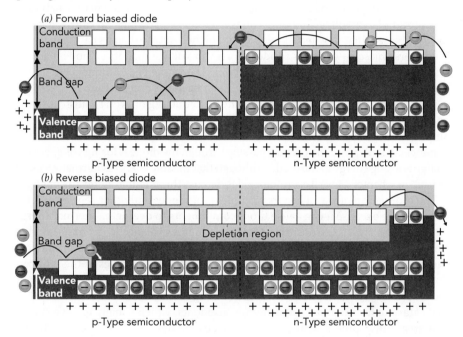

Fig. 12.1.8 (a) When you add electrons to the n-type side of a p-n junction and remove them from the p-type side, the depletion region vanishes and current can flow across the junction. (b) When you add electrons to the p-type side and remove them from the n-type side, the depletion region thickens and no current can flow across the junction.

The extra electrons on the n-type side and the missing electrons on the p-type side create a voltage difference between the diode's two halves. When the voltage of the p-type side reaches about 0.6 V above the voltage of the n-type side, the depletion region in a silicon diode disappears. Conduction level electrons in the n-type material then flow leftward across the junction and drop into empty valence levels in the p-type material. The p-n junction is conducting electric current.

In the theater analogy, we're adding people on the right to the n-type balcony and removing them on the left from the p-type ground floor. The new people in the n-type balcony can move about the empty seats and migrate toward the center of the theater. Similarly, the empty seats in the p-type ground floor allow people to shift about so that empty seats become available near the center of the theater. At that point, people in the n-type balcony can cross over to the p-type balcony and then climb down to the ground floor. There is a net leftward flow of people through the theater; the theater is conducting people from right to left.

But what happens when we try to send electrons backward through the diode, pushing them into the p-type side and removing them from the n-type side

(Fig. 12.1.8*b*)? In that case, the depletion region becomes thicker as we fill in empty valence levels in the p-type side and remove conduction level electrons from the n-type side. The widening depletion region prevents charge from moving and no current flows across the p-n junction. It remains an insulator.

In the theater analogy, we're removing people on the right from the n-type balcony and adding them on the left to the p-type ground floor. Soon the n-type balcony is virtually empty and the p-type ground floor is essentially full. The entire theater is now a depletion region and behaves like the insulator theater. No one can move and the theater can't conduct people.

Since it allows current to flow in one direction but not the other, the p-n junction is a diode. For historical reasons, the diode's p-type side is called the anode and its n-type side is called the cathode. Current, which is the flow of positive charge, can only pass through a diode from its anode to its cathode. Since current naturally flows from higher voltage to lower voltage, a diode carries current only when it is forward biased, that is, when its anode has a higher voltage than its cathode. When it is reverse biased—when its anode has a lower voltage than its cathode—no current flows through the diode. This ability to control the direction of current flow is fundamental to converting alternating current into direct current.

Even when a diode is forward biased, the depletion region won't vanish until the voltage of the anode is significantly higher than the voltage of the cathode. For example, an ordinary silicon diode must have a voltage drop of about 0.6 V in order to conduct current. Since the current passing through the diode is losing voltage, each charge is losing energy in the diode. Although that missing energy is normally wasted as heat, in Chapter 14 we'll examine diodes that use this missing energy to emit light.

The production of waste heat is a problem for diodes and other semiconductor components because it can cause those components to overheat. Thermal energy, like light, can shift electrons from a semiconductor's valence levels to its conduction levels, so that it conducts a small amount of current. Though insignificant near room temperature, this thermally induced conductivity increases as the semiconductor gets hotter. Above a certain temperature, currents flowing due to thermally induced conductivity lead to even more heating and eventually to a runaway thermal catastrophe. To avoid such thermal accidents, which usually involve smoke and great unhappiness, semiconductor electronic devices must not operate too hot and are often cooled by fans.

Capacitors

The other new electronic component in the power adapter is its capacitor, a device that stores separated electric charge (Fig. 12.1.9). A capacitor consists of two conducting plates separated by a thin insulating layer (Fig. 12.1.10). When one plate is positively charged and the other is negatively charged, the opposite charges attract

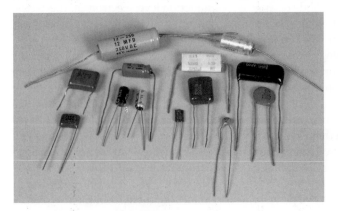

Fig. 12.1.9 Capacitors store separated electric charge. Each of these capacitors contains two conducting surfaces separated by a thin insulating layer.

one another. This attraction allows the plates to store large quantities of separated charge, while leaving the capacitor as a whole electrically neutral.

You can charge a capacitor's plates by transferring charge from its negative plate to its positive one. The work you do during this transfer is stored in the capacitor as electrostatic potential energy and is released when you let the separated charge get back together. A charged capacitor acts like a battery when connected to a circuit because it pushes charge through the circuit from its positive plate and collects that charge from the circuit with its negative plate.

Since (positive) charge has more electrostatic energy on the positive plate than on the negative plate, the voltage of the positive plate is higher than the voltage of the negative plate. The voltage difference between the plates is proportional to the separated charge on them; the more separated charge the capacitor is holding, the larger the voltage difference between its plates.

This voltage difference also depends on the structure of the capacitor. Enlarging the plates allows the like charges on each plate to spread out, so that they repel one another less strongly. Thinning the insulating layer between the plates allows the opposite charges on the two plates to move closer together, so that they attract one another more strongly. Both of these changes lower the separated charge's electrostatic potential energy and, consequently, the voltage difference between the plates. (For more on the structure of capacitors, see ❷.)

Because these changes allow the capacitor to store separated charge more easily, they increase its capacitance, that is, the separated charge it holds divided by the voltage difference between its plates. The SI unit of capacitance is the coulomb-per-volt, also called the farad (abbreviated F). A capacitor with a farad of capacitance stores an incredible amount of separated charge, even at a low voltage difference, while a capacitor with a billionth of a farad of capacitance is much more typical. The Greek letter µ in front of the F means millionths (µF or microfarads), the letter n in front of the F means billionths (nF or nanofarads), and the letter p in front of the F means trillionths (pF or picofarads). A capacitor's capacitance is marked on its wrapper, often in an abbreviated form.

The Complete Power Adapter

Our power adapter uses a transformer, four diodes, and a capacitor to provide low-voltage direct current to the radio. The connections between these components are shown photographically in Fig. 12.1.11a and schematically in Fig. 12.1.11b. Engineers and scientists often use such schematic diagrams to represent complicated electronic devices, assigning a specific symbol to each electronic component and denoting the electrical connections between those components with lines.

Let's examine Fig. 12.1.11 from left to right. The leftmost component is the step-down transformer. With its primary coil carrying alternating current from a 120-V AC outlet, its secondary coil develops an induced emf of 9 V AC. The secondary coil acts as a source of 9-V AC power.

If you were to connect a 9-V light bulb to the two ends of this secondary coil, the coil's alternating induced emf would push an alternating current through the bulb's filament and the bulb would light up. But if you were to connect the 9-V radio to this secondary coil, the coil's alternating emf would try to push an alternating current through that radio and it wouldn't work. The transformer alone can't supply the steady direct current the radio requires.

The diodes in the middle of Fig. 12.1.11 solve the alternation problem by steering current so that it always flows through the positive power wire to the radio and returns from the radio through the negative power wire. Although the transformer's secondary coil continues to experience an alternating induced emf and carries an alternating current, diodes D1 and D2 guide current from the secondary coil to

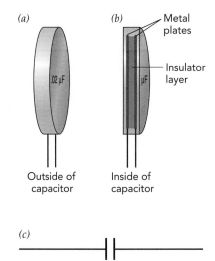

Fig. 12.1.10 (*a*) A capacitor is usually a disk or cylinder with two protruding wires. Its capacitance is printed on its surface. Inside (*b*), the wires are connected to two conducting plates that are separated by a thin insulating layer. (*c*) In a schematic diagram of an electronic device, the capacitor is represented by two parallel lines.

❷ While the simplest capacitors are just two flat plates separated by an insulator, other capacitors are made when a "sandwich" of metal and insulating sheets is rolled up like a jellyroll. Even a fairly small cylindrical capacitor formed in this manner may contain a square meter of sandwich inside. Some capacitors are created by a chemical process that forms enormous conducting surfaces separated by a fantastically thin chemical insulator. The two "plates" in this type of capacitor are not the same, since one of them is actually a conducting chemical called an electrolyte. Because of these plates differences, such electrolytic capacitors can only accommodate separated charge in one direction. One plate must always hold the positive charge while the other holds the negative charge.

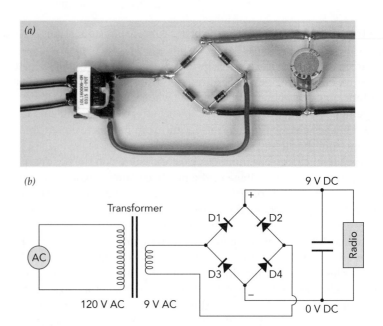

Fig. 12.1.11 (*a*) A particular power adapter consists of a transformer, four diodes, and a capacitor. (*b*) The same components shown schematically.

the radio and diodes D3 and D4 guide the returning current back to the secondary coil. Although the transformer itself is supplying AC power, the diodes allow the power adapter to supply DC power.

However, the diodes alone can't supply *steady* DC power. As the secondary coil's induced emf alternates, the voltage rise from the power adapter's negative power wire to its positive wire fluctuates up and down. That voltage rise peaks at about 12 V when the induced emf reaches its most positive or most negative value, but drops to zero while the induced emf changes sign. The transformer and diodes alone supply *pulsed* 9-V DC power—a form of direct current that offers the same average power as 9 V DC, but with severe voltage fluctuations. Since the radio will turn off between pulses, pulsed DC power won't do.

The capacitor solves the pulsing problem by storing charge and energy whenever the pulsed DC power is near its peak and releasing that charge and energy when the pulsed DC power is near its minimum. The capacitor acts like a rechargeable battery, charging whenever power is plentiful and discharging whenever its not.

When the secondary coil's induced emf is strong, one of the diodes D1 or D2 is forward biased and guides current from the secondary coil to both the radio and the capacitor's positive terminal. At the same time, one of the diodes D3 or D4 is forward biased and guides current from both the radio and the capacitor's negative terminal back to the secondary coil. The radio receives the DC power it needs and the capacitor gradually accumulates more separated charge. The voltage difference between the capacitor's terminals increases and it "charges."

When the secondary coil's induced emf is weak, all four diodes are reverse biased and current stops flowing through the coil. The capacitor takes over the job of supplying power to the radio. Current flows from the capacitor's positive terminal through the positive power wire to the radio and it returns from the radio through the negative power wire to the capacitor's negative terminal. The radio still receives the DC power it needs, but now the capacitor gradually loses its separated charge. The voltage difference between its terminals decreases and it "discharges."

By charging and discharging in this manner, the capacitor "filters" the pulsed DC provided by the transformer and diodes. The capacitor smooths away both the voltage peaks and the voltage zeros. Although the voltage of this filtered DC power fluctuates slightly as the capacitor charges and discharges, the radio can operate from it. Like many electronic devices that accept power adapters, the radio has its own system for accommodating minor voltage fluctuations so that they don't affect its performance. As long as the radio receives a continuous current with approximately the correct voltage, it will work as though on batteries.

Switching Power Adapters

Although the simple power adapters I have just described are inexpensive and reliable, they have several important limitations: they provide somewhat uncertain, non-constant voltages to the devices they power, they can't tolerate large changes in power line voltages, and they use relatively large, heavy transformers. These shortcomings are all absent in a more sophisticated device known as a switching power adapter.

A switching power adapter doesn't connect the power outlet directly to its transformer. Instead, it sends current from the outlet through a set of diodes to a capacitor. That capacitor accumulates separated charge and electrostatic potential energy, and the voltage difference between its terminals rises to a value that depends on the peak voltage difference of the power outlet. That value ranges from about 170 V for a 120-VAC outlet in the U.S. to about 330 V for a 230-VAC outlet in Europe.

The switching power adapter then employs electronic switches to convert the DC power provided by that charged capacitor into high-frequency AC power for its transformer. The use of high frequency AC (typically about 40,000 Hz) offers two tremendous advantages. First, it allows the power adapter to use a surprisingly small step-down transformer. Second, it allows the power adapter to regulate its output voltage simply by turning its synthesized AC power on or off occasionally.

A switching power adapter can use a tiny transformer because its magnetic field needn't store much energy. It only has to store enough energy to last for half a cycle of its AC power or about $1/80,000^{th}$ of a second. In contrast, a transformer that operates on 60 Hz or 50 Hz AC must store enough energy to last for $1/120^{th}$ or $1/100^{th}$ of a second, respectively. Since most of that energy is stored in an iron core, increasing the AC frequency by almost a factor of 1000 allows the transformer's iron core to be lighter by almost a factor of 1000. In general, the higher the frequency at which the switching power adapter can operate, the smaller its transformer can be and 40,000 Hz is a convenient, sensible choice. Switching adapters avoid frequencies below 20,000 Hz anyway because their transformers inevitably vibrate slightly and people can hear vibrations at frequencies below 20,000 Hz.

Current from the transformer's secondary coil passes through diodes to a capacitor, which then supplies DC power to whatever is connected to the adapter. The electronic switching system that synthesizes the adapter's high-frequency AC monitors the voltage difference between that capacitor's terminals and adjusts the switching process to keep this voltage difference at the correct value. If the voltage difference is too large, the switching system turns off the synthesized AC briefly so that the capacitor's store of separated charge decreases and the voltage difference decreases. If the voltage difference is too small, it turns on the synthesized AC to recharge the capacitor.

The overall electronic structure of a switching power adapter is relatively complicated. It has two sophisticated electronic sections: one that operates at power-line voltages upstream of its transformer and one that operates at low voltages downstream of that transformer. Moreover, those two electronic sections must communicate as the adapter monitors and controls its output voltage. Despite that complexity, these switching power adapters have become reasonably inexpensive and are widely used. Usually brick-shaped objects with interchangeable power cords, you can recognize them by their ability to operate over a wide range of outlet voltages and frequencies. If a particular adapter says it will accept 100–240 VAC and 50–60 Hz, you can be certain that it's a switching power adapter.

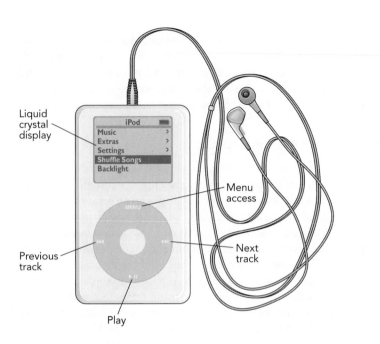

Liquid crystal display

Menu access

Previous track

Next track

Play

Section 12.2 Audio Players

Audio players have revolutionized portable music systems. Everywhere you look, people are sporting earpieces and listening to their favorite tunes with these little electronic marvels. Part computer and part stereo system, an audio player is a spectacular synthesis of some of the highest forms of modern electronic technology. Because they contain such a broad range of electronic components, audio players offer an excellent introduction to much of modern electronics.

To understand how audio players work, we'll need to look at how sound can be represented electronically and at how those electronic representations can be stored, retrieved, and ultimately used to recreate the sound itself. That exploration will take us all the way from the digital world of computers to the analog world of stereo amplifiers and headphones.

It will also expose us to that workhorse of modern electronics: the transistor. While early audio electronic devices were built with vacuum tubes, those relatively bulky components wasted power and aged quickly. Transistors have made audio electronics much more practical. They have also made computers so small and inexpensive that every audio player can have its own computer.

Representing Sound: Analog and Digital

An audio player doesn't store sound any more than a DVD stores flickering light. Instead, the audio player stores a representation of sound that it can use to recreate that sound on demand. Tucked into the player's memory is enough information to reproduce some number of songs almost perfectly. But between the microphones that originally collected that information and the headphones that finally reconstruct the sound itself are a number of fascinating electronic processes. In this section, we'll explore the journey that this sound information takes between recording and playback. In doing so, we'll encounter much of the electronic basis not only for audio electronics, but for digital computers as well.

Our first task in trying to follow sound information as it moves through an audio player is to understand the two different techniques the player uses to represent sound: analog and digital. Let's begin that task by thinking about the recording process and about what information that process gathers from the original sound. In Chapter 9, we learned that sound in air is a density wave and that you hear a sound wave through its associated air pressure fluctuations. Audio recording mimics your hearing by measuring the air pressure fluctuations over time and saving those measurements in some useful format. Audio reproduction uses those measurements to reproduce a precise facsimile of the original air pressure fluctuations and thereby recreates the sound of the artists themselves.

Recording starts with microphones, which are basically sensitive electromechanical air pressure sensors. Playback ends with headphones or earpieces, which are electromechanical devices for influencing air density and pressure. Between the microphones and the headphones are some serious electronics.

The microphones and headphones connect the world of true sound with the world of electronic audio representations. What leaves the microphone is an electronic representation of the sound, not the sound itself. As the air pressure at the microphone fluctuates up and down about atmospheric pressure, the microphone produces a current that fluctuates back and forth about zero. This fluctuating current is an analog representation of the sound, meaning that it is a continuously variable physical quantity (current) that represents another continuously variable physical quantity (air pressure). Like any analog representation, the microphone is drawing an *analogy* between two continuously variable physical quantities: current is serving as an *analog* for air pressure. The headphones draw the reverse analogy.

But while parts of the audio player continue to use analog representations for the sound, other parts use a different representation: digital. In digital representation, a continuously variable physical quantity is represented by a set of physical quantities—a set of digits—each of which can have only a limited number of discrete values. In a digital representation of sound, air pressure changes are measured numerically in some units and those numerical values are then represented by sets of digits. Digital representations are widely used in computers, which represent everything imaginable as numbers and then represent those numbers as sets of digits.

For example, suppose the present air pressure at a microphone is 124 units above atmospheric pressure. We first represent the air pressure increase by the number *124* and then represent the number *124* by a set of digits. For example, *124* could be decomposed into the decimal digits 1, 2, 4 and then these decimal digits could be represented as three separate physical quantities, such as charges, currents, or voltages. Because each charge, current, or voltage only has to represent the integers from 0 to 9, its value doesn't have to be very accurate. If a voltage of 4 V is supposed to represent the digit 4, then 3.9 V or 4.1 V will still be understood to mean 4.

In everyday life, we usually break numbers into ones, tens, hundreds, thousands, and so on—the powers of 10—because we think and work in decimal. But we could also break numbers into ones, twos, fours, eights, sixteens, and so on. Instead of using the powers of 10, as in decimal, we would be using the powers of 2. This system for representing numbers with the powers of 2 is called binary.

In decimal, *124* is written as 124, meaning that *124* contains 1 hundred (10^2), 2 tens (10^1), and 4 ones (10^0). When these pieces are added together, $100 + 20 + 4$, you obtain *124*. In binary, *124* is written as 1111100, meaning that *124* contains 1 sixty-four (2^6), 1 thirty-two (2^5), 1 sixteen (2^4), 1 eight (2^3), 1 four (2^2), 0 twos (2^1), and 0 ones (2^0). When these pieces are added together, $64 + 32 + 16 + 8 + 4$, they again total *124*. This apparently complicated way to represent even a fairly small number is actually quite useful. The number has been broken into pieces that have only two possible values. There either is a thirty-two in the number being represented or there isn't. The only two symbols you need when representing a number in binary are 0 and 1.

Because *124* is 1111100 in binary, you could represent *124* by the charge, current, or voltage of seven separate objects. The first five objects would contain 1's while the last two would contain 0's. For example, if the seven objects were capacitors, the first five might hold separated charge while the last two might hold no charge. A device that needed to know what number the capacitors represented would measure their charges. Finding separated charge in the first five capacitors (11111) and no separated charge in the last two (00), it would determine that the capacitors represent 1111100 binary or *124*.

Binary is useful because fast electronic devices that operate only between two extreme values are relatively easy to build: the current is on or off, the voltage is positive or negative, the charge is present or absent. It's much harder to build fast devices that deliver the specific currents, charges, or voltages needed for decimal or even analog representation. Analog representation is also susceptible to electronic imperfections and noise because an analog device that tries to represent *124* with a voltage of 124 V might accidentally produce 123 V or 125 V. Imagine a bank computer that couldn't tell $124 from $123 or $125! Although representing *124* in binary takes at least seven separate quantities, there is no confusion about which number is being represented.

Resistors

After that introduction to information representations, we're ready to look at how an audio player actually works with its information. We've already hinted at the fact that this information travels about the player as currents, voltages, and a few other physical quantities. But what tools does the player use to work with those physical quantities?

The player's tools are principally electronic ones and, happily, many are familiar to us. We have already encountered capacitors, inductors, transformers, and diodes, all of which are common in audio players and most other electronic devices. However, there are at least two other electronic components that no audio player or high-tech electronic device can be without: resistors and transistors.

Resistors complete a trio of electronic components that began with inductors and capacitors. As we saw in Chapter 11, an inductor uses its induced emf to produce a voltage rise that's proportional to how quickly current is changing with time. And in the previous section, we learned that a capacitor uses its ability to store charge to produce a voltage rise that's proportional to the sum of current over time. What we're missing is a device that uses its electrical resistance to produce a voltage rise that's simply proportional to current. That missing device is called a resistor.

A resistor is just two wires connected by an ohmic device, an imperfect conductor of electricity (Figs. 12.2.1 and 12.2.2). Since current flows through an ohmic device only if there's an electric field pushing it forward, a current-carrying resistor must have a voltage drop across it. In accordance with Ohm's law (see Section 10.3), that voltage drop is proportional to the current in the resistor.

But the voltage drop through a resistor is also proportional to the resistor's electrical resistance—to how imperfect a conductor it is. The larger its resistance, the less current flows through the resistor at a particular voltage drop. Some are relatively good conductors ("low" resistance) while others are relatively poor conductors ("high" resistance).

As we saw in Section 10.3, a resistor's electrical resistance is defined as the voltage drop through it divided by the current flowing through it and is measured in ohms (abbreviated Ω). A resistor with a few ohms of resistance is nearly a good conductor while a resistor with a few million ohms of resistance is nearly an insulator. A k in front of the Ω means thousands (kΩ or kiloohms), and an M in front of the Ω means millions (MΩ or megaohms). Thus a 100-kΩ resistor has a resistance of 100,000 Ω and a 10-MΩ resistor has a resistance of 10,000,000 Ω. A resistor's resistance is marked on its cylindrical body, often in the form of brightly colored stripes. Ten

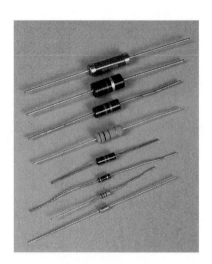

Fig. 12.2.1 Resistors are ohmic devices that impede current flow and produce a drop in voltage. They convert some of that current's electric power into thermal power. The larger resistors can handle more thermal power without overheating.

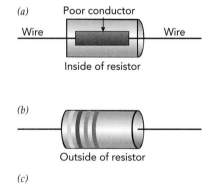

(a) Poor conductor

Wire — Wire

Inside of resistor

(b)

Outside of resistor

(c)

Symbol of resistor

Fig. 12.2.2 (a) A resistor is two wires with an imperfect conductor of electricity between them. (b) It's usually encased in a cylindrical shell, with colored stripes to indicate its resistance. (c) In a schematic diagram of an electronic device, the resistor is represented by a zigzag line.

different colors represent the digits 0 through 9, as well as various powers of ten, so that a resistor with stripes brown (1), black (0), and red (×100) is a 1000-Ω resistor.

Transistors

Invented in 1948 by three American physicists, William Shockley (1910–1989), John Bardeen (1908–1991), and Walter Brattain (1902–1987), transistors (Fig. 12.2.3) are key elements in nearly all modern electronic equipment. Like diodes, transistors are built from doped semiconductors—semiconductors such as silicon to which chemical impurities have been added. But unlike diodes, which simply prevent current from flowing backward through a single circuit, transistors allow the current in one circuit to control the current in another circuit.

While there are many types of transistors, the simplest and most important type is the field-effect transistor. Actually, even here there are several varieties, so we'll focus on the one that's most widely used in audio players, video equipment, and computers: the n-channel metal-oxide-semiconductor field-effect transistor or n-channel MOSFET. Despite its complicated name, the n-channel MOSFET is a relatively simple device, consisting principally of three semiconductor layers and a nearby metal surface (Fig. 12.2.4). The three layers are called the drain, the channel, and the source, while the metal surface is called the gate.

The drain and the source consist of strongly doped n-type semiconductor (many conduction level electrons) and the channel between them consists of lightly doped p-type semiconductor (a few empty valence levels). When those three layers touch, they form two back-to-back p-n junctions and conduction level electrons from the drain and source flow into the channel to fill its few empty valence levels (Fig. 12.2.5a). The completed transistor is thus left with a vast depletion region—a region devoid of empty valence levels or occupied conduction levels—extending all the way from its drain to its source. With nothing to convey charge through its channel, the transistor can't conduct current between its drain and source.

However, if more electrons could be coaxed into the channel somehow, those electrons would have to go into the channel's conduction levels and the channel would then behave like an n-type semiconductor. With an n-type channel sandwiched between an n-type drain and an n-type source, the p-n junctions would vanish and so would the depletion region. The three layers would become, in effect, a single piece of n-type semiconductor and the transistor would then be able to conduct current between its drain and source.

Drawing extra electrons into the channel from outside the transistor is the task of the metal-like gate. Separated from the channel by an incredibly thin insulating layer, the gate controls the channel's ability to carry current. When a tiny positive charge is placed on the gate through a wire, it attracts electrons into the channel's conduction levels and the transistor begins to conduct current (Fig. 12.2.5b). The more positive charge there is on the gate, the more electrons enter the channel and the more current can flow through the transistor. In effect, the transistor behaves like an adjustable resistor with a resistance that decreases as the positive charge on its gate increases.

We can now understand the n-channel MOSFET's name. "n-Channel" refers to the channel's n-type behavior when its gate is positively charged and the transistor can carry current. Although the channel is chemically p-type, it becomes electrically n-type when extra electrons are drawn into it and it acquires a negative net charge. "Metal-oxide-semiconductor" indicates that the metal gate is separated from the semiconductor channel by a thin insulating layer of oxide. This insulator is agonizingly easy to puncture, which is why so many electronic devices can be damaged by static electricity. "Field-effect transistor" indicates that the electric field from charge on the gate is what draws electrons into the channel and controls the current flow through the transistor.

Fig. 12.2.3 These MOSFETs make it possible for small electric charges to control large electric currents. The larger transistors can handle more electric power and thermal power without overheating.

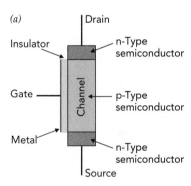

Inside of n-channel MOSFET

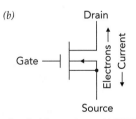

Symbol for n-channel MOSFET

Fig. 12.2.4 (a) In an n-channel MOSFET, the channel is normally a depletion region that cannot carry current between the source and drain. But when positive charge on the gate attracts electrons into the channel, the channel becomes an n-type semiconductor and allows current to flow. (b) The symbol representing an n-channel MOSFET in a schematic diagram.

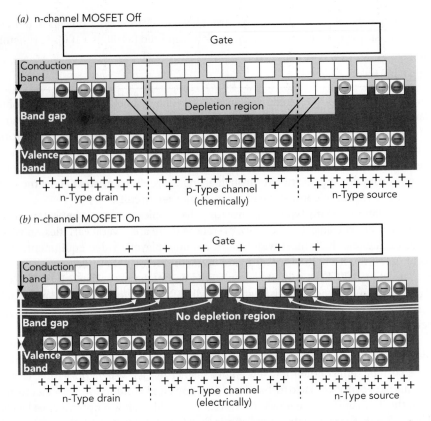

(a) n-channel MOSFET Off

(b) n-channel MOSFET On

Fig. 12.2.5 *(a)* When an n-channel MOSFET is formed, conduction level electrons from the n-type source and drain fill the empty valence levels of the p-type channel and form a vast, insulating depletion region. *(b)* But when positive charge is placed on the nearby gate, it attracts electrons from outside the MOSFET into the channel's conduction levels and turns the entire structure into a conducting n-type semiconductor.

Storing Digital Sound Information

An audio player is half computer and half stereo system. It stores and manipulates sound information in digital form like a computer but then amplifies that information for the headphones in analog form like a stereo system. We'll mirror that sequence when examining the player's electronics: we'll start with its digital memory and processing systems and finish with its audio amplifier.

Inside the digital portion of the audio player, air pressure measurements and other numbers are represented in binary form. How large or precise those numbers are determines how many binary digits are needed to represent them. Each binary digit is called a bit, and using more bits allows you to represent larger or more precise numbers. In general, the more detailed the information, the more bits are needed to represent it.

Eight bits can be used to represent any number from *0* (which is 00000000) to *255* (which is 11111111). Since there are fewer than 256 of many common objects, these objects can be identified by groups of eight bits. For example, the symbols used in ordinary text have been assigned numbers between *0* and *255*, with *65* denoting the letter "A." Since the eight bits 01000001 represent *65*, they also specify an A. Groups of eight bits are so common and useful that they are called bytes.

While it's possible to store sound information using one byte per air pressure measurement, a byte usually doesn't provide enough precision for quality sound reproduction. More often, digital audio is saved using two bytes per pressure measurement. These pressure measurements are made tens of thousands of times per second, usually from several microphones simultaneously in order to provide stereo or surround sound. Even when sophisticated data compression techniques are used to eliminate redundant or inconsequential information, it still takes a great many bits to represent an album. So an audio player needs a lot of memory.

There are several ways in which the audio player, like any computer, stores a bit. In its main working memory (often called random access memory or RAM), each bit is a tiny capacitor that uses the presence or absence of separated electric charge to denote a 1 or a 0. The player stores a bit by producing or removing separated charge and recalls the bit by checking for that charge.

Each capacitor is built right at the end of its own n-channel MOSFET. That MOSFET controls the flow of charge to or from the capacitor. To store or recall a bit, the audio player places positive charge on the gate of the MOSFET so that the MOSFET becomes electrically conducting. The memory system can then transfer charge to or from the bit's capacitor.

Storing the bit is relatively easy; the player simply sends the appropriate charge through the MOSFET to the capacitor. But recalling the bit is harder because the charge on the capacitor is extremely small. Sensitive amplifiers in the memory system detect any charge flowing through the MOSFET from the capacitor and report what they find to the audio player. Since this reading process removes charge from the capacitor, the memory system must immediately store the bit again.

Unfortunately, these tiny capacitors can't hold separated charge forever because it leaks out to their surroundings. Memory that uses charged capacitors to store bits is called dynamic memory and must be refreshed (read and restored) hundreds of times each second to ensure that a 1 doesn't accidentally switch to a 0 or vice versa.

Dynamic memory is also volatile—its contents are lost when the audio player turns off. To conserve its batteries, the player keeps its music information in non-volatile memory—memory which doesn't need power to retain its information. New possibilities for nonvolatile memory appear almost every year, but at present the three leading forms are flash, magnetic disk, and optical disk memories. We'll concentrate on flash and magnetic disk memory here and save optical memory for Section 15.2.

Flash memory resembles dynamic memory in that each bit is stored as the presence or absence of charge associated with a MOSFET. But in flash memory, that charge resides on the MOSFET's floating gate—a second, unattached gate located in the insulating layer between the channel and the normal gate. Since this floating gate is surrounded by insulator, it can keep its charge for decades. And as long as that charge is present, it will determine the MOSFET's conductivity and whether the bit is a 0 or a 1.

Reading bits from flash memory is easy, but storing them is a challenge. The same isolation that traps charge on the floating gate for years makes that charge difficult to change. To add or remove electrons from the floating gate, the memory system applies relatively large voltages to the MOSFET's source, drain, and normal gate and the resulting strong electric fields permit electrons to cross through the insulation separating the channel from the floating gate.

To add electrons to the floating gate, the electric fields are arranged so they accelerate channel electrons to such high speeds that those electrons simply burrow right through the insulating layer to the floating gate. To remove electrons from the floating gate, the electric fields are arranged so that the floating gate's electron standing waves are distorted into the insulator. When these distorted waves reach far enough into the insulator, electrons begin to leak through it into the channel via a process known as quantum tunneling. We'll return to explore quantum tunneling in Chapter 16.

Flash memory is fast to read but relatively slower to write. Moreover, the electron burrowing process causes cumulative damage to the insulating layer and limits the number of times flash memory can be written. An audio player uses a mixture of dynamic memory and flash memory: it does its computational work in dynamic memory but retains its long-term information in flash memory.

However, there's another memory concept that remains more cost-effective than flash memory for storing vast amounts of information: magnetic disk memory. Audio players that store tens of thousands of songs usually contain magnetic disks. Also called hard disks, these devices use the magnetic effects that we discussed in Chapter 11 to store music or other information.

Just as the magnetic strip on a credit card (Fig. 11.1.6) can store information in the locations of its magnetic poles, the surface of a magnetic disk can store information in the orientations of its magnetic poles. Actual hard disks are smooth aluminum platters which have been coated with hi-tech hard magnetic materials. Using microscopic electromagnets to write magnetic poles and sophisticated semiconductor magnetic sensors to read them, modern hard disks can pack almost a quarter-billion bits into a square-millimeter of surface (about 16 gigabytes per square-inch). Simply locating those microscopic bits on platters that rotate over 100 times per second is an electromechanical tour de force, yet these disks do it routinely even while you are jogging with your audio player.

The Audio Player's Computer

We've seen how sound information can be represented and stored as bits, so now let's look at how an audio player's computer works with those bits. This digital processing is done by electronic devices that take groups of bits as their inputs and produce new groups of bits as their outputs. Since their output bits are related to their input bits by the rules of logic, these electronic devices are called logic elements.

The simplest logic element is the inverter, which has only one input bit and one output bit. Its output is the inverse of its input (Fig. 12.2.6). If an inverter's input bit is a 1, then its output bit is a 0 and vice versa. Inverters are used to reverse an action—turning a light on rather than off or starting a song rather than stopping it. Inverters are also used as parts of more complicated logic elements.

But inverters aren't just abstract logic elements; they're real electronic devices. They act on electrical inputs and create electrical outputs. In an audio player's computer, inverters and other logic elements represent input and output bits with electric charge. Positive charge represents a 1 and negative charge represents a 0. Thus when positive charge arrives at the input of an inverter, the inverter releases negative charge from its output.

Inverters and other logic elements are usually constructed from both n-channel and p-channel MOSFETs. We've already seen that n-channel MOSFETs conduct current only when their gates are positively charged. p-Channel MOSFETs are just the reverse, conducting current only when their gates are negatively charged. The drain and source of a p-channel MOSFET are made from p-type semiconductor while the channel is made from n-type semiconductor. Since n-channel and p-channel MOSFETs are exact complements to one another, logic elements built from them are called complementary MOSFET or CMOS elements. An audio player's computer is built almost entirely from CMOS elements.

A CMOS inverter consists of one n-channel MOSFET and one p-channel MOSFET (Fig. 12.2.7). The n-channel MOSFET is connected to the negative terminal of the computer's power supply and controls the flow of negative charge to the inverter's output. The p-channel MOSFET is connected to the power supply's positive terminal and controls the flow of positive charge to the output. When negative charge arrives at the inverter's input and moves onto the gates of the MOSFETs, only the p-channel MOSFET conducts current and the output becomes positively charged. When positive charge arrives at the input, only the n-channel MOSFET conducts current and the output becomes negatively charged.

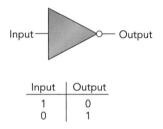

Input	Output
1	0
0	1

Fig. 12.2.6 An inverter, shown here symbolically, produces one output bit that is the inverse of its one input bit.

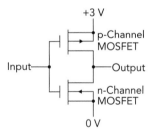

Fig. 12.2.7 When negative charge arrives at the input of a CMOS inverter, its p-channel MOSFET (top) permits positive charge to flow to the output. When positive charge arrives at the input, the n-channel MOSFET (bottom) sends negative charge to the output wire.

But a computer needs logic elements that are more complicated than inverters. One such element is the Not-AND or NAND gate. This logical element has two input bits and one output bit, and its output bit is 1 unless both input bits are 1's (Fig. 12.2.8). It's called a Not-AND gate because it's the inverse of an AND gate. An AND gate produces a 0 output unless both input bits are 1's. Simple memoryless logic elements are often called gates.

A CMOS NAND gate uses two n-channel MOSFETs and two p-channel MOSFETs (Fig. 12.2.9). The two n-channel MOSFETs are arranged in series—one after the next—so that current passing through one must also pass through the other. If either transistor has negative charge on its gate, no current can flow through the series. Components arranged in series all carry the same current, but they may experience different voltage drops.

The two p-channel MOSFETs are arranged in parallel—one beside the other—so that current can flow through either one of them to the output. If either transistor has negative charge on its gate, current can flow from one side of the pair to the other. Components arranged in parallel share the current they receive through one wire and deliver it together to the second wire. But while parallel components may share the current unevenly among themselves, they all experience the same voltage drop.

Parallel and Series Wiring

Connection in Series: Components that are wired in series, one after the other so that they form a chain, all carry the same current but can have different voltage drops. The total voltage drop from the start of the chain to its end is the sum of the individual voltage drops through the components.

Connection in Parallel: Components that are wired in parallel, one beside the other so that they all connect the same pair of wires, have the same voltage drop but can carry different currents. The total current flowing between those wires is the sum of the individual currents through the components.

If negative charge arrives at either input of the CMOS NAND gate, the series of n-channel MOSFETs will be nonconducting and one of the p-channel MOSFETs will deliver positive charge to the output. But if positive charge arrives at both inputs, both p-channel MOSFETs will be nonconducting and the series of n-channel MOSFETs will deliver negative charge to the output. Thus the CMOS NAND gate has the correct logic behavior.

These two logic elements, inverters and NAND gates, can be combined to produce any conceivable logic element. For example, they can be used to build an adder, a device that sums the numbers represented by two groups of input bits and produces a group of output bits representing that sum. These adders can themselves be used to build multipliers, and multipliers can be built into still more complicated devices. In this fashion, the simplest logic elements can be used to construct an entire computer.

Actually, a computer isn't built exclusively from NAND gates and inverters. To improve its speed and reduce its size, it uses a few other basic logic elements as well. Like the CMOS NAND gate and inverter, these elements are constructed directly from n-channel and p-channel MOSFETs.

All of these logic elements are wired together in an intricate pattern to create a complete computer. In an audio player, this computer retrieves and organizes music information and prepares it for the playback electronics, which are not digital. The computer's last act is to deliver the digital music information, the air pressure measurements, to a digital-to-analog converter or DAC. This electronic device is the interface between the two representations of information, digital and analog. The music information leaves the DAC as a voltage that's proportional to air

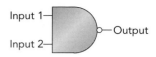

Input 1	Input 2	Output
1	1	0
1	0	1
0	1	1
0	0	1

Fig. 12.2.8 The output bit of a Not-AND or NAND gate, shown here symbolically, is a 1 unless both input bits are 1's.

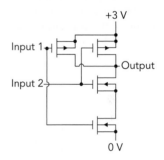

Fig. 12.2.9 A CMOS NAND gate has two input bits. When negative charge arrives through either input, the chain of n-channel MOSFETs (bottom) stops conducting current and one of the two p-channel MOSFETs (top) permits positive charge to reach the output. Only if both inputs are positively charged will negative charge reach the output.

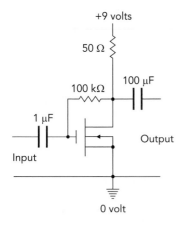

Fig. 12.2.10 A simple audio amplifier can be built with one n-channel MOSFET, two resistors, and two capacitors. A 9-V battery provides power to the device.

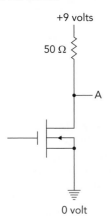

Fig. 12.2.11 The voltage at A depends on the resistance of the MOSFET. The lined triangle at the bottom signifies connection to ground (often the earth itself).

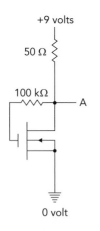

Fig. 12.2.12 The 100-kΩ resistor transfers positive charge to the gate until the voltage at A drops to 5 V.

pressure. This voltage is the input for the audio player's main analog component: its audio amplifier. Actually, the player has two complete analog audio systems so that it can produce stereo sound. But since those systems are identical, we'll focus on only one of them.

The Audio Player's Audio Amplifier

The fluctuating voltage provided by the audio player's DAC is often called an audio signal because it represents audio information. Many analog or digital representations of information are called signals, including video signals, data signals, and even turn signals. But while the player's audio signal contains all the information needed to reproduce the original sound, in convenient analog format, it doesn't have the power that the headphones need to produce that sound at a reasonable volume. Something must first enlarge the audio signal; it needs to be amplified.

Devices that enlarge various characteristics of signals are called amplifiers. An audio amplifier is an amplifier that's designed to boost signals in the frequency range that we hear or feel (20 Hz to 20,000 Hz). It has two separate circuits—an input circuit and an output circuit—and it uses the small current passing through its input circuit to control a much larger current passing through its output circuit. In this manner, the amplifier provides more power to its output circuit than it receives from its input circuit.

Figure 12.2.10 shows the schematic diagram for a simple audio amplifier, built from the components we've just studied. This amplifier has only five components: an n-channel MOSFET, two resistors, and two capacitors. It draws power from a 9-V battery (or an equivalent power adapter) and amplifies a tiny alternating current in its input circuit into a large alternating current in its output circuit.

To understand how this amplifier works, let's first remove everything but the MOSFET and the 50-Ω resistor (Fig. 12.2.11). These two components are in series, so any current that passes through one must also pass through the other. When the MOSFET doesn't conduct current, no current flows through the 50-Ω resistor and it experiences no voltage drop. So the voltage at A is 9 V. But if the transistor does conduct current, a voltage drop will appear through the 50-Ω resistor and the voltage at A will decrease.

The transistor will conduct current only if positive charge is put on its gate. That can be done by connecting the gate to A with a 100-kΩ resistor (Fig. 12.2.12). Since A is at 9 V, it is positively charged and pushes charge toward anything at lower voltage. Current flows slowly through the resistor from A to the gate. But as positive charge accumulates on the gate, the transistor begins to conduct current and the voltage at A drops. When the voltage at A reaches the voltage on the gate, current stops flowing through the resistor.

The amplifier is then in a stable equilibrium; A has a voltage of approximately 5 V and the transistor's gate has a modest amount of charge on it. The 100-kΩ resistor provides the transistor with feedback, that is, it provides the transistor with information about the present situation at A that the transistor can use to correct or improve that situation. Although the feedback is slowed by the resistor's large electrical resistance, it perpetually acts to return the voltage at A to its equilibrium value. If the transistor conducts too little current, charge flows onto its gate and makes it conduct more. If the transistor conducts too much current, charge flows off its gate and makes it conduct less.

The amplifier is now exquisitely sensitive to small changes in the charge on the transistor's gate. If you add just a tiny bit more positive charge to the gate, down goes the voltage at A. If you remove just a tiny bit of positive charge from the gate, up goes the voltage at A. Although current in the feedback resistor tries to undo these changes, it acts too slowly to oppose short-timescale variations. The amplifier's input signal successfully adds or subtracts positive charge from the gate and the amplifier's output signal emerges from A.

The amplifier has two input wires. An analog audio signal's current flows into the amplifier through one wire and returns through the other. But the audio signal is not connected directly to the gate. Instead, it's connected to the gate through a capacitor (Fig. 12.2.13). In addition to storing charge and energy, a capacitor can transfer current between two wires that have different voltages. Such voltage flexibility is important in battery-powered audio amplifiers that must do their amplifying exclusively with positive voltages. Aided by input and output capacitors, our audio amplifier can have an average operating voltage of about +5 V while having average input and output voltages of 0 V.

To see how a capacitor passes along current, let's watch input current flow rightward into our amplifier's input capacitor. As that current's positive charge accumulates on the capacitor's left plate, it attracts negative charge onto the capacitor's right plate and away from the gate. The capacitor remains electrically neutral but the gate becomes more positively charged. Overall, the capacitor has conveyed input current to the gate, even though no charge has actually passed through its insulating layer and its two plates remain at different voltages.

With the help of the input capacitor, the amplifier's fluctuating input current produces a fluctuating charge on the transistor's gate and the voltage at A fluctuates as a result. Even a tiny fluctuating current on the input wires creates a large fluctuating voltage at A.

This fluctuating voltage is responsible for sending a fluctuating current through the headphones. Although headphones are not truly ohmic devices, they respond to fluctuating voltages by carrying fluctuating currents. By applying a fluctuating voltage drop across the headphone's two wires, the amplifier can cause it to carry a fluctuating current and produce corresponding pressure fluctuations and sound.

But the voltage at A averages about 5 V, while the headphones expect an average voltage drop of 0 V. To convey the fluctuating voltages and currents from A to the headphones, while eliminating their large voltage difference, our amplifier connects them via an output capacitor (Fig. 12.2.14). As before, the fluctuations in current and voltage on the output capacitor's left plate are mirrored by current and voltage fluctuations on its right plate. Even though the amplifier operates at a high average voltage, the output signal for the headphones has an average voltage of 0 V.

Tiny fluctuating currents in our amplifier's input circuit produce large fluctuating currents in its output circuit. This amplifier works remarkably well, given its simplicity. If you connect a microphone to the input wires and headphones to the output wires, the headphones will do a surprisingly good job of reproducing the sound in the microphone.

However, our simple amplifier isn't perfect. It distorts the sound somewhat and it doesn't handle all frequencies or amplitudes of sound equally. It also wastes a large amount of electric power heating the 50-Ω resistor. The amplifiers in audio players carefully correct for these problems. Most use feedback to make sure that their output signals are essentially perfect replicas of their input signals, only larger. They sense their own shortcomings and correct for them.

But perfect replication of the input signal isn't always desirable. Sometimes you want to boost the volume for part of the sound. The treble and bass controls on an audio player allow you to selectively change the volumes for the high- and low-frequency portions of the sound, respectively.

An amplifier is typically rated according to the peak power it can supply to the headphones (or speakers) and its average power should never reach that value. But the amplifier can reach its peak power during a passage that's not particularly loud. That's because sound waves often interfere with one another (see Section 9.3 to review wave interference) and when their crests and troughs coincide at the microphone,

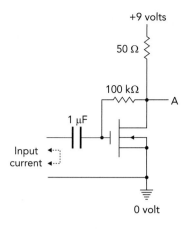

Fig. 12.2.13 Because current flowing back and forth through the two input wires affects the charge on the transistor's gate, it also affects the voltage at A.

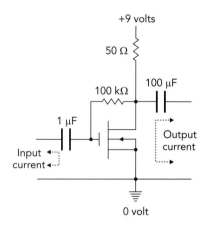

Fig. 12.2.14 The amplifier causes currents to flow back and forth through the two output wires. The alternating current in the output wires is a good replica of the alternating current in the input wires, only larger.

constructive interference can briefly produce enormous pressure fluctuations. To reproduce those overlapping waves properly, the audio amplifier must be able to provide several times its average power, though only for a moment. If the amplifier can't deliver that much power, the audio signal it sends to the headphones or speakers will be distorted and the sound will be unpleasant. That's why audiophiles often use powerful amplifiers even when they're playing quiet music.

As for the headphones themselves, they generally use electromagnetic effects to move a surface back and forth in sync with the amplifier's current fluctuations. In most cases, the amplifier's current is sent through a coil of wire that's immersed in a strong magnetic field and attached to a moveable surface. This current experiences a Lorentz force due to the magnetic field and that force drives the current, coil, and surface back and forth as the current fluctuates. The moving surface alternately compresses and rarefies the air, thereby reproducing the original sound.

The Limits to a Computer's Speed

Computers perform calculations extremely quickly. The simple NAND gate we examined above acts in as little as 120 ps (120 picoseconds or 120×10^{-12} s). That's the time it takes for the MOSFETs to start or stop conducting current and for the charge on the output wire to reach its proper value. Since scientists and engineers are continually improving the performances of MOSFETs, CMOS logic elements get faster every year.

But computer speed also depends on the time it takes for signals to travel between logic elements. While the signals from one element travel to other elements at almost the speed of light, that transfer speed limits how fast the computer can work. Each element must wait for its signals to arrive, and packing the elements closely together shortens the wait.

However, charged particles themselves travel much slower than light, so how can the signals move at nearly light speed? We can understand this by imagining a garden hose at the end of a closed water faucet. If you suddenly open the faucet, fresh water will begin to enter the hose. If the hose is already filled with water, water will begin to pour out of the hose long before fresh water from the faucet moves very far into the hose. In fact, water will begin to pour out of the hose almost as soon as you open the faucet. That's because the increase in water pressure that occurs when you open the faucet travels through the hose at the speed of sound.

Similarly, when charge enters a wire, its electric field pushes like charge ahead of it. Something resembling a pressure rise rushes through the wire at almost the speed of light, and like charge begins to flow out of the far end of the wire long before the original charge arrives there. Since light takes about 3 ns (3 nanoseconds or 3×10^{-12} s) to travel a meter or about 1 ns to travel a foot, putting the entire computer on a single, coin-sized integrated circuit is a good idea. The logic elements are then so close together that they can exchange signals extremely quickly.

But there's another problem that slows a computer. Whenever a logic element changes its output from a 1 to a 0, or a 0 to a 1, it must change the charge on its output wire. If that wire is long and wide, it will have a fair amount of the old charge on it. Removing this old charge and replacing it with the new charge takes time because only so much current will flow through the tiny MOSFETs in the logic element.

Since the wires connecting logic elements have much more surface area than the gates of the MOSFETs they deliver charge to, the wires end up with most of the charge. The wires act as capacitors, storing and releasing charge. The best way to speed up a computer is to make these wires as short and thin as possible. The wires in current computers are already 90 nm (90 nanometers) wide and getting thinner all the time. Smaller is faster. They're also becoming better conductors; copper wires now often replace aluminum ones.

Moving charge onto and off the wires and gates isn't just a matter of speed, it's also a matter of energy. Each time positive charge flows onto a wire and is then replaced by negative charge, there's a net transfer of positive charge from the power supply's positive terminal to its negative terminal. The computer thus consumes a little electric energy and converts it into thermal energy. If the computer can't get rid of this thermal energy quickly enough, it will overheat.

The electric power consumed by the computer depends on the current flowing through it and on the voltage drop across the logic elements. One way to reduce this power is to run the computer more slowly, changing the charge on its wires less frequently and transferring that charge more slowly from one terminal of the power supply to the other. But this power can also be reduced by shrinking the wire size and using lower voltages to push charge onto and off those wires. Some computers presently operate with voltage drops of less than 1 V.

◢◢◢ Synchronizing a Computer

Because each of the computer's logic elements takes a small but finite time to act, the computer needs something to keep faster logic elements from getting ahead of slower ones. For example, if the computer is trying to add two numbers that are themselves the results of other computations, its adder must be sure that those earlier computations are finished before producing a sum. To avoid any uncertainty, the computer needs a timekeeper to tell each logical element when it can be sure that all previous computations are finished.

This synchronizing device is the computer's clock. Just as an orchestra's conductor keeps its musicians working together, the steady electric beat of the computer's clock keeps all of its logic elements in step. At one beat of the clock, new computations begin. At the next beat, all those computations are certain to be completed and a second group of computations begins. At the third beat of the clock, this second group of computations is completed and a third group of computations begins.

The computer's clock must beat slowly enough that all of the logic elements can complete their tasks between beats. Since a few tasks are far more complicated than others, they may be permitted two or more beats to complete their work. Arithmetic division is an example of such a complex task. Rather than slowing the clock down so that division can finish in one beat, the computer allows the logic element handling division to work for several beats. Whenever the division element is activated, the computer's other logic elements generally wait for it to complete its work.

A computer that is specified as 3 GHz has a clock that beats 3 billion times each second. This means that there are 333 ps between beats and that most of the logic elements in the computer can complete their tasks in that time. The clocks in modern computers beat several billion times per second and as computers become smaller, these clock rates will continue to increase.

◢◢◢ Vacuum Tubes

Although vacuum tubes are seldom used in modern household electronics, they are still common in radio transmitters because they can handle enormous amounts of power. The basic vacuum tube has three parts: a filament, a grid, and a plate (Fig. 12.2.15). These parts are enclosed in a glass or ceramic shell that contains a vacuum—nothing but empty space. The filament is heated red hot by a current flowing through it and some of its electrons obtain so much thermal energy that they escape from the material. The electrons jump directly into the vacuum in a process called thermionic emission.

If the plate is positively charged, it attracts these free electrons and they travel through the vacuum toward the plate. A current therefore flows between the fila-

(a)

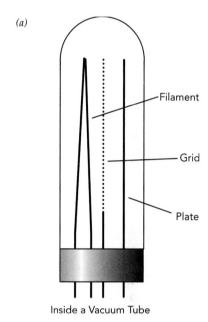

Inside a Vacuum Tube

(b)

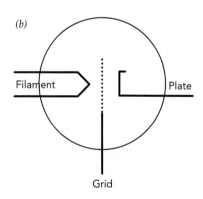

Symbol for Vacuum Tube

Fig. 12.2.15 The hot filament of a vacuum tube emits electrons, which are attracted toward the positively charged plate. But if the grid is negatively charged, it repels the electrons and no current flows.

ment and plate. Since current describes the flow of positive charge, the current officially flows from the plate to the filament. In fact, the vacuum tube acts like a diode because it conducts current in only one direction.

There are vacuum tube diodes that work in exactly this manner. But the vacuum tube in Fig. 10.1.18 also has a grid between the filament and the plate. This grid is a wire mesh that controls the passage of electrons from the filament to the plate. If the mesh has no charge on it, most of the electrons will fly past the grid and head for the positively charged plate. However, if the grid is negatively charged, it will repel the electrons and prevent them from reaching the plate. It only takes a small amount of negative charge on the grid to stop a large flow of current from the plate to the filament.

In this manner, a vacuum tube allows a small electric signal on its grid to control a large current between its plate and its filament. It amplifies the small signal on its grid to produce a similar but much larger signal in the circuit connected to its plate. Vacuum tubes are fine amplifiers but they require a hot filament that eventually wears out. When this filament fails, the entire tube must be replaced. Making a vacuum tube is complicated and expensive and having to replacing it is a nuisance. Moreover, the filament requires electric power all the time and the entire tube becomes hot. With all of these drawbacks, it's no wonder that transistors have replaced vacuum tubes in almost every application.

♒ Audio Speakers

An electromagnetic speaker uses an electric current representing sound to actually reproduce that sound. It includes a permanent magnet that's fixed in place and a coil of wire that's mobile (Fig. 12.2.16). The permanent magnet has a cylindrical channel cut into it and is magnetized with opposite magnetic poles on the inside and outside of that channel. The coil fits into the channel and as current flows through it, the Lorentz force pushes it into or out of the channel, depending on the current's direction.

The speaker coil is attached to the speaker cone, and the two move together. The cone is loosely supported at its periphery by the speaker's frame. The coil and cone have little mass, so they accelerate easily. Their main resistance to motion is the air itself. As the cone moves out it compresses the air in front of it, and as it moves in it rarefies the air in front of it. By moving in and out rapidly, it produces sound.

Unfortunately, speakers aren't perfect. A small speaker responds quickly to the electric current, even at high frequencies, but it has difficulty moving enough air to produce loud, low-pitched sounds. A large speaker can move plenty of air, but it's too massive to respond quickly enough to produce high frequencies. Headphones don't need much volume and use a single small speaker for each ear. Room sound systems, however, often employ multiple speakers: small tweeter speakers to produce high pitches, large woofer speakers to produce low pitches, and specialized subwoofers for pitches so low that you feel them more than you hear them.

Some small speakers aren't electromagnetic at all; they're piezoelectric. Like the quartz crystals used in watches (p. 300), piezoelectric materials respond mechanically to electric influences. By adding or removing charge from the surfaces of a piezoelectric disk or sheet, an audio player can cause it to move mechanically and produce sound. Piezoelectric speakers are commonly used as tweeters in sound systems and in countless electronic gadgets, including watches, cell phones, and palm-sized organizers.

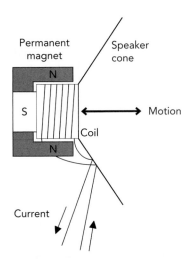

Fig. 12.2.16 A speaker allows a fluctuating electric current to produce sound by causing a paper or plastic cone to move. The cone is attached to a coil of wire that's inserted into a cylindrical channel in a permanent magnet. Current passing through that coil experiences the Lorentz force, so the coil is pushed in or out of the channel, depending on the direction of the current.

Chapter 13

Things That Use Electromagnetic Waves

Electric and magnetic fields are so intimately related that each can create the other even in empty space. In fact, the two fields can form electromagnetic waves, in which they recreate one another endlessly and head off across space at an enormous speed. These electromagnetic waves are all around us and are the basis for much of our communications technology, for radiative heat transfer, and for our ability to see the universe in which we live.

In this chapter, I'll explain the concepts behind electromagnetic waves and I'll concentrate on the portion of the electromagnetic spectrum that includes radio waves and microwaves. Our awareness of radio waves and microwaves is less than two hundred years old, yet we now take them completely for granted. We're inundated with wireless networking, satellite television and radio, and global positioning information, and some of us go through life with cell phones glued to our ears. Even when we use electrical cables to handle some of our communications needs, those cables are still carrying radio waves and microwaves in their enclosed environments. As the result of all this electromagnetic "communication," we now live in a world in which we're constantly in contact with everyone. Everyone, that is, except the person sitting across the table from us.

The spectrum of radio waves and microwaves is limited, so society tries to allocate it carefully. We now reuse frequencies and channels wherever possible, letting distance, directionality, and cabling keep the signals separate. But the ever increasing demands for connectivity are compelling us to squeeze more things into the radio wave and microwave spectrum every year. There's no end in sight.

Chapter Itinerary

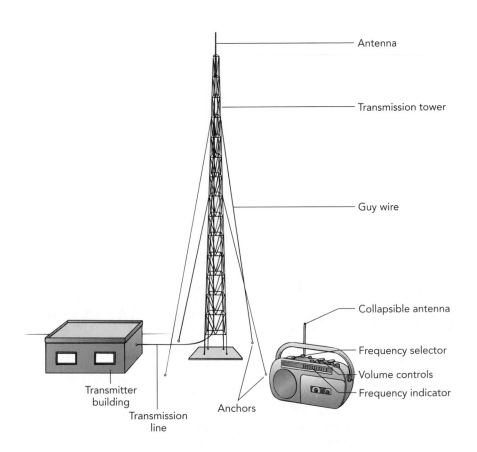

Antenna

Transmission tower

Guy wire

Collapsible antenna

Frequency selector

Volume controls

Frequency indicator

Transmitter
building

Transmission
line

Anchors

Section 13.1 **Radio**

We've seen that electric currents can represent sound and carry speech and music anywhere wires will reach. But how can we send sound to someone who is moving? We need a way to represent sound that doesn't involve wires. We need radio. This section describes how radio works. We'll look at how radio waves are transmitted and how they're received. We'll also examine the common ways in which sound is represented by radio waves so that it can travel through space to a receiver far away.

A Prelude to Radio Waves

Before we can examine radio and radio waves, let's take a moment to finish the introduction to electrodynamics that we began in Chapters 10 and 11. Although we've already learned most of the fundamental relationships between electricity and magnetism, the remaining one is about to become important. To refresh your memory, we have observed so far that electric fields can be produced by electric charges and by changing magnetic fields, and that magnetic fields can be produced by moving electric charges (Table 13.1.1).

In 1865, Scottish physicist James Clerk Maxwell (1831–1879) discovered a second source of magnetic fields: *changing electric fields*. That effect is subtle and scientists overlooked it for most of the nineteenth century. It wasn't until Maxwell was trying to formulate a complete electromagnetic theory that he uncovered this additional connection between electricity and magnetism. This final relationship completed the set shown in Table 13.1.1. Together, these relationships allowed Maxwell to understand one of the most remarkable phenomena in nature: electromagnetic waves!

3rd Connection between Electricity and Magnetism
Electric fields that change with time produce magnetic fields.

Table 13.1.1 Sources of Electric and Magnetic Fields

Sources of Electric Fields	Sources of Magnetic Fields
Electric charge	Moving electric charge
Changing magnetic fields	Changing electric fields

Since electric fields can create magnetic fields and magnetic fields contain energy, it's clear that electric fields must contain energy, too. The amount of energy in a uniform electric field is the square of the field strength times the volume of the field divided by 8π times the Coulomb constant. We can write this relationship for energy in an electric field as:

$$\text{energy} = \frac{\text{electric field}^2 \cdot \text{volume}}{8\pi \cdot \text{Coulomb constant}}. \tag{13.1.1}$$

With these observations, we have finished the prelude and are ready to see how radio works.

Antennas and Tank Circuits

A radio transmitter communicates with a receiver via radio waves. These waves are produced by electric charge as it moves up and down the transmitter's antenna and are detected when they push electric charge up and down the receiver's antenna. But what exactly are radio waves and how does charge on the antenna produce them?

We've already seen that electric charge produces electric fields and that moving charge produces magnetic fields. However, something new happens when charge *accelerates*. Accelerating charge produces a mixture of changing electric and magnetic fields that can reproduce one another endlessly and travel long distances through empty space. These interwoven electric and magnetic fields are known generally as electromagnetic waves. In the case of radio, the electromagnetic waves have low frequencies and long wavelengths and are known as radio waves.

But before we look at the structure of a radio wave and at how it travels through space, let's start with a much simpler situation. We'll look at how two nearby metal antennas affect one another. Figure 13.1.1 shows a radio transmitter and a radio receiver, side by side. Because of their proximity, electric charge on the transmitter's antenna is sure to affect charge on the receiver's antenna.

To communicate with the nearby receiver, the transmitter sends charge up and down its antenna. This charge's electric field surrounds the transmitting antenna and extends all the way to the receiving antenna, where it pushes charge down and up. Unfortunately, the resulting charge motion in the receiving antenna is weak and the receiver may have difficulty distinguishing it from random thermal motion or from motion caused by other electric fields in the environment. Therefore the transmitter adopts a clever strategy—it moves charge up and down its antenna rhythmically at a particular frequency. Since the resulting motion on the receiving antenna is rhythmic at that same frequency, it's much easier for the receiver to distinguish from unrelated motion.

Using this rhythmic motion has another advantage: it allows the transmitter and receiver to each use a tank circuit—a resonant electronic circuit consisting only of a capacitor and an inductor (Fig. 13.1.2). Charge "sloshes" back and forth through the tank circuit at a particular frequency, much as a child swings back and forth

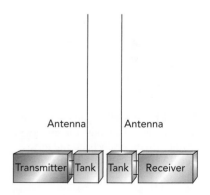

Fig. 13.1.1 Electric charge rushing on and off the transmitting antenna causes a similar motion of electric charge in the receiving antenna.

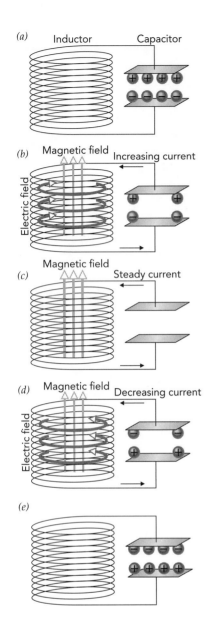

(a) Inductor   Capacitor

(b) Magnetic field   Increasing current

Electric field

(c) Magnetic field   Steady current

(d) Magnetic field   Decreasing current

Electric field

(e)

Fig. 13.1.2 A tank circuit consists of a capacitor and an inductor. Energy sloshes rhythmically back and forth between the two components.

on a swing. And just as you can get a child swinging strongly at the playground by giving her a gentle push every swing, so the transmitter can make charge slosh strongly through its tank circuit by giving that charge a gentle push every cycle. By helping the transmitter move larger amounts of charge up and down the antenna, the tank circuit dramatically strengthens the transmission.

A second tank circuit attached to the receiving antenna helps the receiver detect this transmission. Gentle, rhythmic pushes by fields from the transmitting antenna cause more and more charge to move through the receiving antenna and its attached tank circuit. While the motion of charge on this antenna alone may be difficult to detect, the much larger charge sloshing in the tank circuit is unmistakable.

We can understand how a tank circuit works by looking at how charge moves between its capacitor and its inductor. Let's imagine that the tank circuit starts out with separated charge on the plates of its capacitor (Fig. 13.1.2a). Since the inductor conducts electricity, current begins to flow from the positively charged plate, through the inductor, to the negatively charged plate. But the current through the inductor must rise slowly and, as it does, it creates a magnetic field in the inductor (Fig. 13.1.2b).

Soon the capacitor's separated charge is gone and all of the tank circuit's energy is stored in the inductor's magnetic field (Fig. 13.1.2c). But the current keeps flowing, driven forward by the inductor's opposition to current changes. The inductor uses the energy in its magnetic field to keep the current flowing, and separated charge reappears in the capacitor (Fig. 13.1.2d). Eventually, the inductor's magnetic field decreases to zero and everything is back to its original state—almost. While all of the tank circuit's energy has returned to the capacitor, the separated charge in that capacitor is now upside-down (Fig. 13.1.2e).

This whole process now repeats in reverse. The current flows backward through the inductor, magnetizing it upside down, and the tank circuit soon returns to its original state. This cycle repeats over and over again, with charge sloshing from one side of the capacitor to the other and back again.

A tank circuit is an electronic harmonic oscillator, equivalent to the mechanical harmonic oscillators we examined in Chapter 9. Like all harmonic oscillators, its period (the time per cycle) doesn't depend on the amplitude of its oscillation. Thus no matter how much charge is sloshing in the tank circuit, the time it takes that charge to flow over and back is always the same.

The tank circuit's period depends only on its capacitor and its inductor. The larger the capacitor's capacitance, the more separated charge it can hold with a given amount of energy and the longer it takes that charge to move through the circuit as current. The larger the inductor's inductance—its opposition to current changes—the longer that current takes to start and stop. A tank circuit with a large capacitor and a large inductor may have a period of a thousandth of a second or more, while one with a small capacitor and a small inductor may have a period of a billionth of a second or less.

Inductance is defined as the voltage drop across the inductor divided by the rate at which current through the inductor changes with time. This division gives inductance the units of voltage divided by current per time. The SI unit of inductance is the volt-second-per-ampere, also called the henry (abbreviated H). While large electromagnets have inductances of hundreds of henries, a 1-μH (0.000001 H) inductor is more common in radio.

Its resonant behavior makes the tank circuit useful in radio. That's because small, rhythmic pushes on the current in a tank circuit can lead to enormous charge oscillations in that circuit. In radio, these rhythmic pushes begin when the transmitter sends an alternating current through a coil of wire. Fields from this coil push current back and forth through the nearby transmitting tank circuit, causing enormous amounts of charge to slosh back and forth in it and travel up and down the transmitting antenna. That charge's electric field then pushes rhythmically on charge in the receiving antenna, causing substantial amounts of charge to travel down and up it and slosh back and forth in the receiving tank circuit (Fig 13.1.3). The receiver can easily detect this sloshing charge.

Energy flows from the transmitter to the receiver via resonant energy transfer—from the transmitter, to the transmitting tank circuit and antenna, to the receiving antenna and tank circuit, and finally to the receiver. This sequence of transfers can work efficiently only if all the parts have resonances at the same frequency. Tuning a radio receiver to a particular station is largely a matter of adjusting its capacitor and inductor so that its tank circuit has the right resonant frequency.

Radio Waves

When the two antennas are close together, charge in the transmitting antenna simply exerts electrostatic force on charge in the receiving antenna. But when the antennas are far apart, the interactions between them are more complicated. Charge in the transmitting antenna must then emit a radio wave in order to push on charge in the receiving antenna. Like a water wave, a radio wave is a disturbance that carries energy from one place to another. But unlike a water wave, which must travel in a fluid, a radio wave can travel through otherwise empty space, from one side of the universe to the other.

Like all electromagnetic waves, a radio wave consists only of a changing electric field and a changing magnetic field. These fields recreate one another over and over again as the wave travels through empty space at the speed of light—exactly 299,792,458 m/s (approximately 186,282 miles-per-second).

The radio wave is created when electric charge in the antenna accelerates. While stationary charge or a steady current produces constant electric or magnetic fields, accelerating charge produces fields that change with time. As charge flows up and down the antenna, its electric field points alternately up and down and its magnetic field points alternately left and right. These changing fields then recreate one another again and again, and sail off through space as an electromagnetic wave.

The wave emitted by a vertical transmitting antenna has a vertical polarization, that is, its electric field points alternately up and down (Fig. 13.1.4). We identify those "ups" as crests and the distance between adjacent crests is its wavelength. For radio waves, that wavelength is usually 1 m (3.3 ft) or more. The wave's magnetic field is perpendicular to its electric field and thus points alternately left and right.

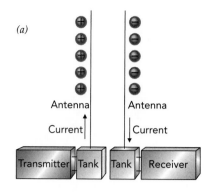

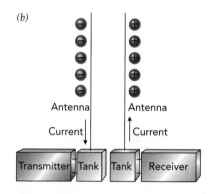

Fig. 13.1.3 (*a*) As current flows up the transmitting antenna, it causes current to flow down the receiving antenna and (*b*) vice versa.

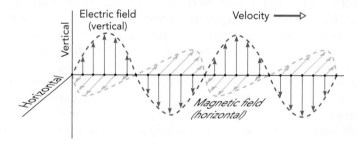

Fig. 13.1.4 In this vertically polarized electromagnetic wave, the electric field along the wave's level path points alternately up and down while the magnetic field along that path points alternately right and left. Each vertical arrow indicates the electric field at the point marked by the dot, while each horizontal arrow indicates the magnetic field at that point. This wave is heading to the right at the speed of light.

Had the transmitting antenna been tipped on its side, the wave's electric field would have pointed alternately left and right and the wave would have had a hori-

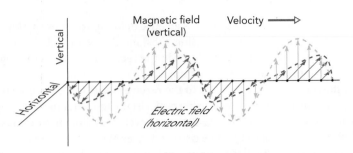

Fig. 13.1.5 In this horizontally polarized wave, the electric field along the wave's level path points alternately left and right while the magnetic field points alternately up and down. Each vertical arrow indicates the magnetic field at the point marked by the dot, while each horizontal arrow indicates the electric field at that point. This wave is heading to the right at the speed of light.

zontal polarization (Fig. 13.1.5). The wave's magnetic field would then point alternately up and down. Whatever the polarization, the electric and magnetic fields move forward together as a traveling wave, so the pattern of fields moves smoothly through space at the speed of light.

Common Misconceptions: Electromagnetic Waves and Undulations
Misconception: Since the fields of an electromagnetic wave appear wavy (Figs. 13.1.4 and 13.1.5), the light wave itself undulates; it actually undulates up and down or back and forth as it heads rightward!
Resolution: The arrows drawn to represent the fields in an electromagnetic wave are associated with points marked along the straight axis of the wave. Each wave in Figs. 13.1.4 and 13.1.5 is heading directly rightward along the axis line and the arrows indicate field values at points along that line.

If you stood in one place and watched this wave pass, you'd notice its electric field fluctuating up and down at the same frequency as the charge that created it. When the wave passes a distant receiving antenna, it pushes charge up and down that antenna at this frequency. If the receiving tank circuit is resonant at this frequency, the amount of charge sloshing in it should become large enough for the receiver to detect.

A radio station can optimize its transmission by using a transmitting antenna of the proper length. When that length is exactly a quarter of the wavelength of the radio wave it's transmitting, charge sloshes vigorously up and down the antenna in a natural resonance. Surprisingly enough, the antenna is another electronic harmonic oscillator, with a period that depends only on its length. (In fact, the antenna is the top half of a tank circuit, with its tip acting as one plate of a capacitor and its length acting as the top half of an inductor. Objects at the base of the antenna complete the tank circuit.) When the transmitting tank circuit and antenna have resonances at the same frequency, there's a resonant energy transfer from one to the other. As you might expect, these resonant effects help to produce a powerful radio wave.

The transmitting antenna sends the strongest portion of its radio wave out perpendicular to its length. That's not unexpected because the motion of charge on the antenna is most obvious when viewed from a line perpendicular to its length. Thus a vertical antenna sends most of its wave out horizontally, where people are likely to receive it. No wave emerges from the ends of an antenna.

Both electric and magnetic fields contain energy, so as the electromagnetic wave travels through space, it carries energy away from the transmitter. When a radio station advertises that it "transmits 50,000 watts of music," it's claiming that its antenna emits 50,000 J of energy per second or 50,000 W of power in its electromagnetic wave. The receiving antenna must absorb enough of this power to detect the wave. But the farther the wave gets from the transmitting antenna, the more spread out and weak it becomes. Trees and mountains also absorb or reflect some of the wave and hinder reception.

For the best reception, a listener should be located where the radio wave is strong and where there's an unobstructed path from the transmitting antenna to the receiving antenna. The receiving antenna should be a quarter-wavelength long and it should be oriented along the radio wave's polarization—vertical for a vertically polarized radio wave or horizontal for a horizontally polarized radio wave. Aligning the receiving antenna with the wave's polarization makes certain that the wave's electric field pushes charge *along* the antenna, not *across* it.

To ensure good reception regardless of receiving antenna orientation, many radio stations transmit a complicated circularly polarized wave that combines both vertical and horizontal polarizations. To form this wave, they need several quarter-wavelength antennas. For wavelengths under a few meters, these antennas can all be attached inexpensively to a single mast. That's why commercial FM and TV broadcasts, which use short-wavelength radio waves, are usually transmitted with circular polarization. However, commercial AM broadcasts, which use long-wavelength radio waves, are transmitted only with vertical polarization.

Because commercial FM radio waves usually include both polarizations, FM receiving antennas can be vertical or horizontal. Portable FM receivers often use vertical telescoping antennas while home receivers frequently use horizontal wire antennas. All of these antennas are roughly a quarter-wavelength long.

A quarter-wavelength antenna for commercial AM radio would have to be about 100 m (330 ft) long, so straight AM antennas (such as those on cars) are much shorter than optimal. That's why many AM antennas are designed to respond to the radio wave's horizontal magnetic field rather than to its vertical electric field. These magnetic antennas are horizontal coils of wire that experience induced currents when exposed to fluctuating magnetic fields.

Representing Sound: AM and FM Radio

A radio transmitter does more than simply emit a radio wave. It uses that radio wave to *represent sound*. Because sound waves are fluctuations in air density and radio waves are fluctuations in electric and magnetic fields, a radio wave can't literally "carry" a sound wave. However, a radio wave can carry sound information and instruct the receiver how to reproduce the sound.

To convey sound information, the radio station alters its radio wave to represent compressions and rarefactions of the air. The receiver then recreates those compressions and rarefactions. There are two common techniques by which a radio wave can represent those density fluctuations. One is called amplitude modulation and involves changing the overall strength of the radio wave. The other is called frequency modulation and involves small changes in the frequency of the radio wave.

In the amplitude modulation (AM) technique, air density is represented by the strength of the transmitted wave (Fig. 13.1.6). To represent a compression of the air, the transmitter is turned up so that more charge moves up and down the transmitting antenna. To represent a rarefaction, the transmitter is turned down so that less charge moves up and down the antenna.

The frequency at which charge moves up and down the antenna remains steady, so only the amplitude of the radio wave changes. The receiver measures the strength of the radio wave and uses this measurement to recreate the sound. When it detects a strong radio wave, it pushes its speaker toward the listener and compresses the air. When it detects a weak radio wave, it pulls its speaker away from the listener and rarefies the air.

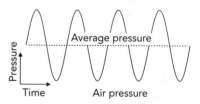

Fig. 13.1.6 When sound is transmitted using amplitude modulation, air pressure is represented by the strength of the radio wave. A compression is represented by strengthening the radio wave and a rarefaction is represented by weakening it.

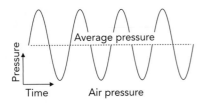

Fig. 13.1.7 When sound is transmitted by frequency modulation, air pressure is represented by changing the frequency of the radio transmitter slightly. A compression is represented by increasing that frequency and a rarefaction by decreasing it.

❶ Stereo FM broadcasts transmit the two separate audio channels, A and B, in clever way, one that allows a stereo receiver to play these channels separately while allowing a monaural receiver to play their combined sound through its single speaker. First, the stereo FM station adds the two signals (A+B) and uses that sum signal to frequency modulate its carrier wave. A monaural radio receiving the broadcast will detect only that sum signal and use it to produce sound from its speaker. At the same time, the stereo station subtracts the two signals (A–B) and shifts this difference signal up in frequency by 19 kHz. The station also frequency modulates its carrier with this shifted difference signal and with a small amount of the 19 kHz pilot tone involved in the up-shifting. A stereo radio receiving the broadcast will detect the sum signal, the up-shifted difference signal, and the pilot tone, and it will use the pilot tone to down-shift the difference signal back to its original frequencies. The stereo receiver will then add the sum signal and difference signal to obtain the original A signal and send it to one speaker, and it will subtract the difference signal from the sum signal to obtain the original B signal and send it to the other speaker.

In the frequency modulation (FM) technique, air density is represented by the frequency of the transmitted wave (Fig. 13.1.7). To represent a compression of the air, the transmitter's frequency is increased slightly so that charge moves up and down the transmitting antenna a little *more* often than normal. To represent a rarefaction, the transmitter's frequency is decreased slightly so that the charge moves up and down a little *less* often than normal. These changes in frequency are extremely small—so small that charge continues to slosh strongly in all the resonant components and reception is unaffected. The receiver measures the radio wave's frequency and uses this measurement to recreate the sound. When it detects an increased frequency it compresses the air, and when it detects a decreased frequency it rarefies the air.

Although the AM and FM techniques for representing sound can be used with a radio wave at any frequency, the most common commercial bands in the United States are the AM band between 550 kHz and 1600 kHz (550,000 Hz and 1,600,000 Hz) and the FM band between 88 MHz and 108 MHz (88,000,000 Hz and 108,000,000 Hz). Elsewhere in the spectrum of radio frequencies are many other commercial, military, and public transmissions, including TV, short wave, amateur radio, telephone, police, and aircraft bands. These other transmissions use AM, FM, and a few other techniques to represent sound and information with radio waves.

Bandwidth and Cable

An audio signal on a wire has a range of frequencies present in it, from zero frequency up to the highest pitch sound it is representing. Similarly, an audio signal traveling via a radio wave has a range of radio frequencies present in it, stretching from somewhat below the official frequency of the radio wave, the carrier frequency, to somewhat above that frequency. The wider the audio frequency range of the sound, the more sound information must be sent each second and the wider the range of radio frequencies needed to represent that sound. The range of frequencies needed to transmit such a stream of information is known as the transmission's bandwidth.

By international agreement, an AM radio station may use 10 kHz of bandwidth, 5 kHz above and below its carrier frequency. To stay within that bandwidth, the audio signal can't contain frequencies above 5 kHz. While this restricted frequency range is bad for music, it allows competing stations to function with carrier frequencies only 10 kHz apart, so that 106 different stations can operate between 550 kHz and 1600 kHz.

An FM radio station may use 200 kHz of bandwidth, 100 kHz on each side of its carrier frequency. This luxurious allocation permits FM radio to represent a very broad range of audio frequencies, in stereo ❶, which is why an FM radio station can do a much better job of sending music to your radio than an AM station can.

Because high frequency radio waves travel in straight lines between antennas, it's hard to receive a commercial FM station from more than about 100 km (60 miles) away. Even when the transmitting antenna sits on top of a tall tower, the earth's curvature and surface terrain severely limit the range of FM reception ❷.

But low frequency radio waves, such as those used by commercial AM stations, are reflected by charged particles in the earth's outer atmosphere so that portions of the radio wave that would otherwise be lost to space bounce back toward the ground. This returning power allows you to receive AM stations over a considerable distance, even when you have no direct line-of-sight to the transmitter's antenna. At sundown, these atmospheric layers become so effective at reflecting AM radio that you can hear a transmission from thousands of kilometers away as clearly as if it were a home town station.

The spectrum of electromagnetic waves is a limited resource and if it could only be used once, it would quickly run out of bandwidth. Fortunately, distance and enclosures make it possible to reuse the spectrum many times. Cell phones that are far from one another can share the same carrier frequencies and bandwidth because their radio waves weaken with distance and essentially don't overlap. But even nearby radio transmissions can use the same carrier frequencies by enclosing their electromagnetic waves inside cables.

Cable radio, television, and data networks are similar to broadcast networks except that they send electromagnetic waves through cables rather than through empty space. A typical radio or television cable consists of an insulated metal wire inside a tube of metal foil or woven metal mesh. This wire-inside-a-tube arrangement is called coaxial cable because its two metal components share the same centerline or axis. In contrast, a typical computer-data cable consists of a number of insulated metal wires that are twisted into several pairs.

Electromagnetic waves can propagate easily through a coaxial or twisted-pair cable, following its twists and turns from the transmitter that produces the waves to the receiver that uses them. The fact that wires are assisting these waves in their travels makes them more complicated than waves in empty space. However, they still involve electric and magnetic fields and still propagate forward at nearly the speed of light.

Because the electromagnetic waves inside a cable don't interact with those outside it, the transmitter and receiver can use whatever parts of the spectrum they chose, without concerns about sharing. A typical coaxial cable can handle frequencies up to about 1000 MHz and typical twisted-pair cable can reach 350 MHz, so either one can carry a great deal of information each second.

However, coaxial cables must now compete with optical fiber cables that guide light from one place to another. We'll examine optical fibers in Section 15.2. Like radio waves, light is an electromagnetic wave and can be amplitude or frequency modulated to represent information. But light's frequency is extremely high; the frequencies of visible light range from 4.5×10^{14} Hz to 7.5×10^{14} Hz. If we were to allocate FM radio channels 200 kHz apart throughout the visible spectrum, there would be about 1.5 billion channels available!

❷ Like FM radio, the high frequency electromagnetic waves used by television broadcasts travel in straight lines and are blocked by the horizon less than 100 km from ground-based antennas. To reach broader audiences, engineers at Westinghouse experimented with aircraft-based antennas in a program called "Stratovision." Developed in the late 1940s, this program would have used 14 B-29 aircraft, carrying receivers and transmitters, to relay television broadcasts to more than 75% of the population of the United States, including rural viewers. Despite successful test flights, the program foundered for political reasons and was never implemented.

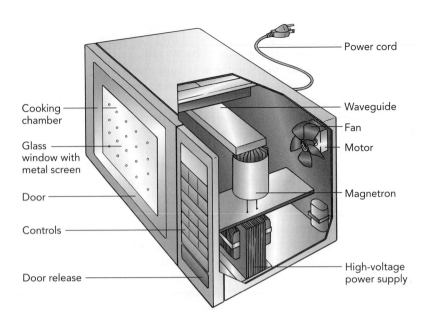

Power cord

Waveguide

Fan

Motor

Magnetron

High-voltage
power supply

Cooking
chamber

Glass
window with
metal screen

Door

Controls

Door release

Section 13.2 **Microwave Ovens**

In addition to carrying sounds from one place to another, electromagnetic waves can carry power. One interesting example of such power transfer is a microwave oven. It uses relatively high-frequency electromagnetic waves to transfer power directly to the water molecules in food, so that the food cooks from the inside out. This section discusses both how those waves are created and why they heat food.

Microwaves and Food

When studying incandescent lightbulbs in Section 7.3, we discussed the *wavelengths* of electromagnetic waves. While examining radio, we concentrated on the *frequencies* of electromagnetic waves. However, we know from Eq. 9.2.1 that wavelength and frequency of a wave aren't independent. A basic electromagnetic wave in empty space has both a wavelength and a frequency, and their product is the speed of light. That relationship can be written as:

$$\text{speed of light} = \text{wavelength} \cdot \text{frequency}. \tag{13.2.1}$$

Like Fig. 7.3.2, Fig. 13.2.1 shows the approximate wavelengths of many types of electromagnetic waves but it also shows their frequencies.

Radio broadcasts use the low-frequency, long-wavelength portion of the electromagnetic spectrum. Commercial AM radio is at frequencies of 550 kHz to 1600 kHz (wavelengths of 545 m to 187 m) and commercial FM radio is at frequencies of 88 MHz to 108 MHz (wavelengths of 3.4 m to 2.8 m). Because these waves have wavelengths longer than 1 m, they are called radio waves. But the electromagnetic waves used in microwave ovens have wavelengths shorter than 1 m and are called microwaves. Microwaves extend from wavelengths of 1 m (3.3 ft) down to 1 mm (0.04 inches).

To explain how a microwave oven heats food (❶), let's begin by looking at water molecules. Water molecules are electrically polarized—that is, they have positively charged ends and negatively charged ends. This polarization comes about because of quantum physics and the tendency of oxygen atoms to pull electrons away from hydrogen atoms. The water molecule is bent, with its two hydrogen atoms sticking up from its oxygen atom like Mickey Mouse's ears. When the oxygen atom

❶ Though he was orphaned as child and never completed grade school, American Percy Lebaron Spencer (1894–1970) had a brilliant career as a scientist and microwave engineer. In 1945, while visiting a magnetron testing laboratory, he leaned over an operating magnetron and the candy bar in his shirt pocket melted. Immediately recognizing what had happened, he soon had popcorn popping about the lab and even cooked an egg until it exploded. Cooking has never been the same since.

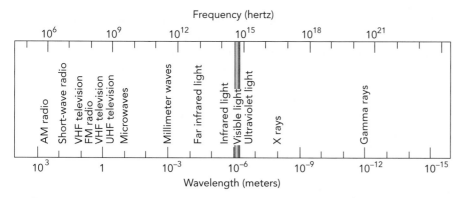

Fig. 13.2.1 The electromagnetic spectrum. Microwaves have wavelengths between about 1 m and 1 mm, corresponding to frequencies from 300 MHz up to 300 GHz.

pulls the electrons partly away from the hydrogen atoms, its side of the molecule becomes negatively charged while the hydrogen atoms' side becomes positively charged. Water is thus a polar molecule.

In ice, these polar water molecules are arranged in an orderly fashion with fixed positions and orientations. But in liquid water, the molecules are more randomly oriented (Fig. 13.2.2). Their arrangements are constrained only by their tendency to bind together, positive end to negative end, to form a dense network of coupled molecules. This binding between the positively charge hydrogen atom on one water molecule and the negatively charged oxygen atom on another molecule is known as a hydrogen bond.

If you place liquid water in a strong electric field, its water molecules will tend to rotate into alignment with the field. That's because a misaligned molecule has extra electrostatic potential energy and accelerates in the direction that reduces its potential energy as quickly as possible. In this case, the water molecule will experience a torque and will undergo an angular acceleration that makes it rotate into alignment. As it rotates, the molecule will bump into other molecules and convert some of its electrostatic potential energy into thermal energy.

A similar effect occurs at a crowded party when everyone is suddenly told to face the front of the room. People brush against one another as they turn and sliding friction converts some of their energy into thermal energy. If the people were told to turn back and forth repeatedly, they would become quite warm. The same holds true for water. If the electric field reverses its direction many times, the water molecules will turn back and forth and become hotter and hotter.

A microwave's fluctuating electric field is well suited to heating water. A microwave oven uses 2.45-GHz (2.45-gigahertz or 2,450,000,000-Hz) microwaves to twist food's water molecules back and forth, billions of times per second. As the water molecules turn, they bump into one another and heat up. The water absorbs the microwaves and converts their energy into thermal energy. This particular microwave frequency was chosen not because of any resonant effect but because it was not in use for communications and because it cooks food uniformly. If the frequency were higher, the microwaves would be absorbed too strongly by food and wouldn't penetrate deep into large items. If the frequency were lower, the microwaves would pass through foods too easily and wouldn't cook efficiently.

This twisting effect explains why only foods or objects containing water or other polar molecules cook well in a microwave oven. Ceramic plates, glass cups, and plastic containers are water-free and usually remain cool. Even ice has trouble absorbing microwave power because its crystal structure constrains the water molecules so they can't turn easily.

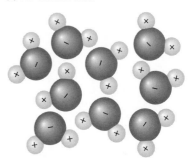

(a) No electric field

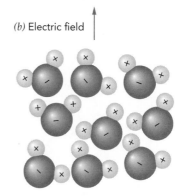

(b) Electric field

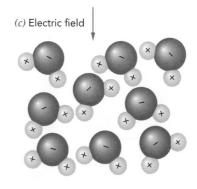

(c) Electric field

Fig. 13.2.2 (a) The water molecules in liquid water are randomly oriented when there's no electric field. (b,c) But an electric field tends to orient them with their positive ends in the direction of the field.

But while ice melts slowly in a microwave oven, the liquid water it produces heats quickly. This peculiar heating behavior explains why it's so easy to burn yourself on frozen food heated in a microwave oven. Portions of the food that defrost first absorb most of the microwave power and overheat while the rest of the food remains frozen solid. You never know whether your next bite will break your teeth or sear the roof of your mouth. To address this problem, many microwave ovens have defrost cycles in which microwave heating is interrupted periodically to let heat flow naturally through the food to melt the ice. Once the frozen parts have melted, all of the food can absorb microwaves.

Metal in a Microwave Oven

Contrary to popular lore, metal objects and microwave ovens aren't always incompatible. In fact, the walls of the oven's cooking chamber are metal, yet they cause no trouble when exposed to microwaves during cooking. Like most metal surfaces, the walls reflect microwaves. They do this by acting as both receiving and transmitting antennas. Electric fields in the microwaves cause mobile charges in the metal surfaces to accelerate and absorb the original microwaves. But as these charges accelerate, they emit new microwaves. The emitted microwaves have the same frequencies as the original ones, but they travel in new directions. The original microwaves have been reflected by the surface.

The cooking chamber walls reflect the oven's microwaves and keep them bouncing around inside. Even the metal grid covering the window reflects microwaves. That's because charge has enough time during a microwave cycle to flow around each hole in the grid and compensate for the hole's presence. As long as the wavelength of an electromagnetic wave is much larger than the holes in a metal grid, the wave reflects perfectly from that grid. In fact, if there's nothing inside the oven to absorb the microwaves, they'll bounce around inside it until they return to their source, a vacuum tube called a magnetron (Fig. 13.2.3), and eventually cause it to overheat.

While metal surfaces help confine the microwaves inside the oven, cooking your food and not you, extra metal inside the microwave can cause trouble. If you wrap food in aluminum foil, the foil will reflect the microwaves and the food won't cook. However, food placed in a shallow metal dish cooks reasonably well because microwaves enter the open top, pass through the food, reflect, and pass through the food again.

Sometimes metal's mobile charges do more than just reflect microwaves. If enough charge is pushed onto the sharp point of a metal twist-tie or scrap of aluminum foil, some of it will jump right into the air as a spark. This spark can start a fire, particularly when the twist-tie is attached to something flammable, like a plastic or paper bag. As a rule of thumb, never put a sharp metal object in the microwave oven.

Some metal objects heat up in a microwave oven. When microwaves push charge back and forth in a metal, the metal experiences an alternating current. If the metal has a substantial electrical resistance, this alternating current will experience a voltage drop and heat up the metal. While thick oven walls and cookware have low resistances and remain cool, thin metal strips quickly overheat. Metallic decorations on porcelain dinnerware are particularly susceptible to damage in a microwave oven, so that warming up coffee in Grandma's gold-rimmed teacup is sure disaster. When putting metal into a microwave oven, make sure that it is thick enough to conduct electricity well and that it has no sharp points.

Resistive heating in conducting objects can actually be useful at times. Since microwave ovens cook food inside and out at the same time, the food's surface never gets particularly hot and the food doesn't brown or become crisp. To improve their textures and appearances, some foods come with special wrappers that conduct just

Fig. 13.2.3 This oven's magnetron microwave source is located in the middle of the picture, just to the left of its cooling fan. Microwaves travel to the cooking chamber through the rectangular metal duct on top of the oven. The high-voltage transformer at the bottom right provides power to the magnetron.

enough current to become very hot in a microwave oven. These wrappers provide the high surface temperatures needed to brown the foods.

Another peculiar feature of microwave ovens is that they don't always cook evenly. That's because the amplitude of the microwave electric field isn't uniform throughout the oven. As the microwaves bounce around the cooking chamber, they pass through the same spot from several different directions at once. When they do, they exhibit interference effects (see Section 9.3). At one location, the individual electric fields may point in the same direction and experience constructive interference, so that food there heats up quickly. But at another location, those fields may point in opposite directions and experience destructive interference, so that food there doesn't cook well at all.

If nothing is moving in the microwave oven, the pattern of microwaves inside it doesn't move either. There are then regions in which the electric field has very large amplitudes and regions in which the amplitudes are very small. The larger the amplitude of the electric field, the faster it cooks food.

To heat food uniformly in such a microwave oven, you must move the food around as it's cooking. Many ovens have turntables inside that move the food automatically. Another solution to this problem is to stir the microwaves around the oven with a rotating metal paddle. The pattern of microwaves inside the chamber changes as the paddle turns and the food cooks more evenly. Still other microwave ovens use two separate microwave frequencies to cook the food. Because these two frequencies cook independently, it's unlikely that a portion of the food will be missed by both waves at once.

Creating Microwaves with a Magnetron

Clearly, changing electric fields cook the food as microwaves bounce around the inside of an oven. But how are these microwaves created? From the previous section

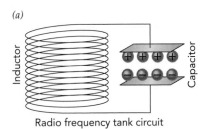

(a)

Inductor

Capacitor

Radio frequency tank circuit

(b)

Inductor

Capacitor

Microwave frequency tank circuit

Fig. 13.2.4 (*a*) At radio frequencies, a tank circuit's inductor is a coil of wire and its capacitor is a pair of separated plates. (*b*) At microwave frequencies, a tank circuit's inductor is merely the curve of a C-shaped strip and its capacitor is the tips of that strip.

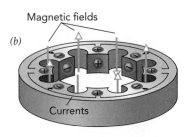

(a)

Magnetic fields

(b)

Currents

(c)

Magnetron resonators

Fig. 13.2.5 A typical magnetron has eight C-shaped resonant cavities arranged in a ring. (*a*) Separated charge on the tips of the cavities (*b*) flows as currents through the ring and (*c*) becomes reversed. As the currents flow, magnetic fields appear in the eight cavities, alternating up and down.

on radio, you might guess that the oven creates an alternating current at 2.45 GHz and that this current causes charge to slosh in a tank circuit and move up and down an antenna. That's pretty much what actually happens inside a magnetron tube.

A magnetron is a special vacuum tube—a hollow chamber from which all the air has been removed. Composed primarily of metal and ceramic parts, the magnetron uses beams of electrons to make charge slosh in a number of microwave tank circuits. These tank circuits have resonant frequencies of 2.45 GHz, the operating frequency of the oven. With the help of a tiny antenna, the magnetron emits the microwaves that cook the food.

The microwave tank circuits are arranged in a ring around the magnetron's evacuated chamber. For one of these tank circuits to oscillate naturally at 2.45 GHz, its capacitor must have an extremely small capacitance and its inductor must have an extremely small inductance. These requirements can be met by a C-shaped strip of metal (Fig. 13.2.4). Its curve is the inductor and its tips are the capacitor.

Electric charge sloshes back and forth on the C-shaped strip just as it does in a conventional tank circuit (Fig. 13.1.2). Known as a resonant cavity, this strip is another electronic harmonic oscillator and therefore has a period that doesn't depend on the amount of charge that's sloshing.

The magnetron of a microwave oven typically contains eight of these resonant cavities, each carefully adjusted in size and shape so that its natural resonance occurs exactly at 2.45 GHz. Because these cavities are arranged in a ring and each one shares its tips with those of its neighboring cavities, they tend to oscillate alternately (Fig. 13.2.5). At the start of an oscillatory cycle, half the metal tips are positively charged and half are negatively charged (Fig. 13.2.5a). Currents begin to flow through the ring and produce magnetic fields in the resonant cavities (Fig. 13.2.5b). These magnetic fields propel the currents around the ring even after the charge separations have vanished. Soon the charge separations reappear but with the positive and negative tips interchanged (Fig. 13.2.5c).

These currents oscillate back and forth around the cavities at 2.45 GHz, filling the magnetron with alternating electric and magnetic fields. But as the energy in these fields is extracted to cook the food or is lost to the imperfect conductivities of the cavities themselves, something must continuously replenish it. That replacement power is supplied to the cavities by four streams of energetic electrons.

At the center of the magnetron tube, surrounded only by empty space, is an electrically heated cathode that tends to emit electrons (Fig. 13.2.6a). A high-voltage power supply pumps negative charge onto this cathode so that a strong electric field points toward it from the positively charged cavity tips. If there were no other fields present in the magnetron, negatively charged electrons would emerge from the hot cathode and accelerate toward the positively charged tips as four beams of electrons (Fig. 13.2.6b).

However, the magnetron also includes a large permanent magnet. Why else would it be called a magnetron? This magnet creates a strong, steady magnetic field that points upward along the axis of the magnetron, parallel to the cathode itself (Fig. 13.2.6c). If there were no other fields inside the magnetron, electrons would experience only Lorentz forces perpendicular to their velocities and would circle around the magnetic flux lines in counterclockwise loops—a behavior known as cyclotron motion. The circling electrons would remain near the cathode and would never go near the cavities.

But in a real magnetron, the electric field of Fig. 13.2.6b and the magnetic field of Fig. 13.2.6c are present simultaneously. Because both of these fields exert forces on moving electrons, the paths the electrons follow are extremely complicated (Fig. 13.2.6d). The outward directed and circulating motions merge together into four electron beams which arc outward and rotate counterclockwise, like the spokes of a spinning bicycle wheel. An electron beam reaches each cavity not at its positively

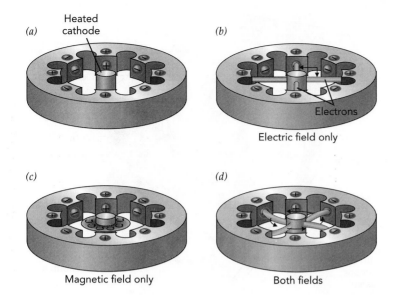

Fig. 13.2.6 (*a*) Electrons are emitted by the hot cathode in the center of a magnetron's ring of resonant cavities. (*b*) Electric fields alone would accelerate the electrons toward the positively charged cavity tips. (*c*) A magnetic field alone (pointing upward) would make the electrons orbit the cathode in counterclockwise loops. (*d*) Together, these fields create spokelike electron beams that circle the cathode counterclockwise and always strike negatively charged tips of the cavities.

charge tip, as it would without the magnetic field, but at its negatively charged tip. The electron beams actually add to the charge separations in the cavities!

The electron beams sweep around the cathode in perfect synchronization with the oscillating charge on the cavities. The beams sweep from one tip to the next in the same amount of time it takes for the charge separation on the tips to reverse. As a result, the beams always arrive on the negatively charged tip. By adding to the charge separations, the electron beams provide power to the oscillations in the cavities, keeping them going and allowing them to transfer power to the food. The electron beams actually initiate the oscillation in the cavities by adding energy to tiny random oscillations that are always present in electric systems.

But how does the oscillating charge inside the magnetron create microwaves inside the oven's cooking chamber? There are many ways to extract microwaves from the ring of cavities. One extraction method is to insert a single-turn wire coil into one of the magnetron's cavities. As the magnetic field in that cavity changes, it induces a 2.45-GHz alternating current in the coil. One end of this coil is attached to the ring but the other end passes out of the magnetron through an insulated, air-tight hole in the ring and connects to a quarter-wavelength antenna. This 3-cm (1.2-in) antenna emits microwaves into a metal pipe attached to the cooking chamber. These microwaves reflect their way through the pipe and into the cooking chamber, where they cook the food.

There are two types of high-voltage power supplies used in modern microwave ovens. The older style resembles a simple power adapter (pp. 411-412); it connects the primary coil of its heavy 60-Hz step-up transformer directly to the electrical outlet and uses diodes and a capacitor to produce high-voltage 60-Hz pulsed DC. The newer "inverter" style resembles a switching power adapter (p. 413); it synthesizes its own high-frequency AC and employs a small high-frequency step-up transformer, diodes, and a capacitor to produce high-voltage high-frequency pulsed DC. An oven's magnetron produce a brief but intense microwave pulse with each pulse of the high-voltage pulsed DC, and these microwave pulses cook the food.

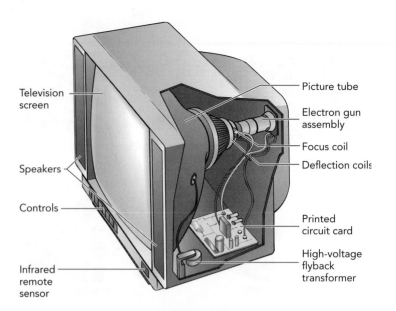

Television screen

Speakers

Controls

Infrared remote sensor

Picture tube

Electron gun assembly

Focus coil

Deflection coils

Printed circuit card

High-voltage flyback transformer

SECTION 13.3 **Television**

Once upon a time, entertainment meant getting out of the house for an evening at the movies, a trip to the ballpark, or a visit to the concert hall. But eventually entertainment followed us home and settled itself in our living rooms and dens. Community life hasn't been the same since. And while we may lament the isolation and loss of shared experience that accompany this boom in home entertainment, it's hard not to admire the science and technology behind it.

The problem with trying to describe the physics behind television is that television technology is changing so quickly that it's hard to keep up. Standards evolve, schedules change, and new innovations explode onto the market almost overnight. This discussion is at best a passing snapshot, one that will probably seem like ancient history a decade from now. While I've decided to include television here, in this chapter on things that use electromagnetic waves, it could appear almost anywhere in the latter half of the book.

Starting at the Beginning: Analog Television

One of the most impressive aspects of the ongoing digital television revolution is that it's taking place on top of the complicated, well-established infrastructure of analog television. The two systems, while not directly compatible with one another, will have coexisted long enough for an almost painless migration to digital. Moreover, this shift is being accomplished with few serious long-term compromises to the digital standard. Although analog television sets can't directly decode the digital transmissions and display only snow when you tune to a digital channel, those sets are surviving the transition, too. Set-top converter boxes are available that receive the digital transmissions and deliver them to television sets in analog format. While this hybrid approach doesn't provide the full resolution of digital video, the picture is still considerably sharper than with ordinary analog reception.

This sort of "open-heart surgery" on the television infrastructure happened once before when color television joined black-and-white back in the mid-1950's. In fact, examining that earlier revolution and the basics of analog television will make it easier to understand the transition to digital.

I'll start this story about television by explaining how a conventional analog television set works, one based on a device known as a picture tube. If you have

WARNING:
Televisions contain dangerously high voltages, even after they have been unplugged. Since a television uses beams of high-energy electrons to create the images that you see, it needs high voltages. To avoid any risk of shock, you should never open a television while it's operating. But even turning the unit off doesn't necessarily make it safe. Because the television uses capacitors to store separated electric charge in its high voltage power supply, high voltages can persist inside it for minutes or more after you turn it off and unplug it. Don't open a television until you're sure that it has no more stored electric energy in its capacitors.

a big, heavy, and deep television set with a thick glass screen, it's probably uses a picture tube. After introducing analog television as displayed by picture tubes, I could head off in two different directions: toward more modern video displays (e.g., LCD, LED, and plasma) or toward digital television. I've decided to head off toward digital television. Whatever I write is going to be out of date before the ink is dry, so I'll pick the path of slowest obsolescence.

Creating a Television Picture

A conventional analog television builds its picture out of tiny illuminated dots, arranged in a rectangular array on the screen. The number of dots in this array depends on the television standard, which varies with country. For the following discussion, I'll consider only the NTSC (National Television Systems Committee) analog color television standard used in the United States, which specifies an array 525 dots high by about 440 dots wide. Other analog television standards differ somewhat in the details but not in the concepts.

While a conventional NTSC analog television lights up its dots one by one, it finishes with all of them so quickly that it appears as though they're all illuminated at once. Your eye responds slowly to changes in light and you see the whole pattern of dots on the screen as a single picture, a television picture. To create this picture, the television starts at the upper left hand corner of the screen and scans through the dots horizontally from left to right. Every $1/15,750^{th}$ of a second, it starts a new horizontal line. Since there are 525 horizontal lines, the television completes the entire picture every $1/30^{th}$ of a second.

Actually, if the television worked its way from the top of the screen to the bottom in $1/30^{th}$ of a second, our eyes would sense a slight flicker. To reduce this flicker, the television builds the image in two passes from top to bottom: it illuminates the odd numbered lines during the first pass and the even numbered lines during the second pass. That way, the television scans down the screen once every $1/60^{th}$ of a second and there is essentially no flicker at all.

Black-and-White Picture Tubes

These dots of light are produced on the inside surface of a glass picture tube when electrons collide with a phosphor—a chemical that emits light when energy is transferred to it (Fig. 13.3.1). These electrons are emitted by a heated surface in the neck of the picture tube and accelerate toward positive charge on the phosphor-coated screen. When the electrons strike the phosphor, they transfer energy to it and it glows brightly.

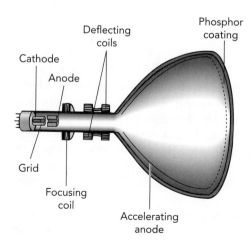

Fig. 13.3.1 The main components of a black-and-white picture tube. Electrons travel from left to right and illuminate the screen.

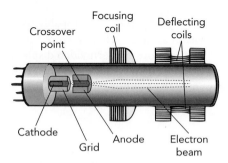

Fig. 13.3.2 The neck of a picture tube, showing how the electron beam comes to a focus once inside the anode and again at the screen.

Fig. 13.3.3 Two of the three electron sources in a color picture tube. Each source has its own cathode, grid, and anode and is responsible for illuminating one of the colored phosphors.

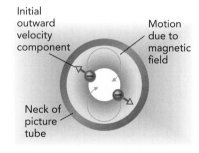

Fig. 13.3.4 As diverging electrons move down the neck of the picture tube toward the screen, the focusing magnetic field makes them accelerate to their right. They travel in a spiral and return together just as they hit the screen.

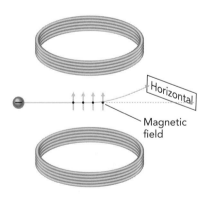

Fig. 13.3.5 The horizontal deflecting coils are located above and below the neck of the picture tube. The vertical magnetic field they produce deflects the electron beam either left or right.

But the television exerts several forces on the electrons as they fly through the empty space inside the picture tube. These forces focus the electrons into a narrow beam and also steer that beam to various points on the phosphor screen. This focusing and deflecting are done by electrical components in the neck of the picture tube, a region that's shown in detail in Figs. 13.3.2 and 13.3.3. For the moment, let's consider only a black-and-white picture tube.

The electrons emerge from a hot cathode, a device that uses thermal energy to emit electrons directly into the vacuum. This cathode is heated by a nearby filament and takes a few seconds to warm up when you turn on the television. If the filament breaks or burns out, the picture tube is ruined.

Surrounding the cathode is a hollow grid that's negatively charged. Since this grid repels electrons, most of the electrons leaving the cathode return to its surface. However, the grid has a small hole through which some of the electrons can escape. Once they escape through the grid, these electrons are attracted by a positively charged anode in front of them and accelerate forward.

The shape of the electric field between the cylindrical cathode, grid, and hollow anode has an interesting effect on the electrons: it focuses them to an extremely narrow spot inside the anode. Regardless of which way electrons were heading when they left the cathode's surface, they all accelerate toward the same point inside the anode, the crossover point.

But the electrons don't stop at the crossover point. Instead, they continue on through the anode and sail off toward the screen. After they have passed through the crossover point, the electrons are spreading away from one another. They must be brought back together again so that they all strike the screen at exactly the same spot. This second focusing action is done by Lorentz forces exerted on the electrons by a magnetic field.

A wire coil that circles the neck of the picture tube creates this focusing magnetic field—a field that points directly toward the screen. Since the electrons' velocities also point toward the screen, the Lorentz forces on them are small. Recall that the Lorentz force is strongest when a charge moves perpendicular to a magnetic field and is zero when the charge moves parallel to the magnetic field. When we look down the neck of the tube at the screen (Fig. 13.3.4), however, we see that these electrons have small outward components of velocity that take them away from the center of the tube and the crossover point they just left behind. These outward components of velocity are perpendicular to the magnetic field, so the electrons do experience Lorentz forces.

Because the Lorentz force pushes each electron toward its right as it flies through the magnetic field, each electron accelerates toward its right and travels in a spiral.

Viewed down the neck of the tube, this motion appears circular; it's the cyclotron motion we encountered on p. 440. Remarkably, the time each electron takes to complete one full circle of cyclotron motion doesn't depend on the diameter of that circle. Because all of the electrons circle once while traveling from the cross-over point to the screen, they all hit the screen at the same spot!

The television also uses magnetic fields and Lorentz forces to steer the electron beam to different parts of the screen. A pair of coils above and below the picture tube's neck produces a vertical magnetic field that deflects the electron beam horizontally (Fig. 13.3.5). By adjusting the currents in these horizontal deflecting coils, the television can control the horizontal position of the beam spot on the screen. A second pair of coils mounted to the left and right of the tube's neck produces a horizontal magnetic field that deflects the electron beam vertically (Fig. 13.3.6). The currents in these vertical deflecting coils determine the vertical position of the beam spot on the screen.

When the electron beam strikes the phosphor coating on the inside of the screen, it transfers energy to that phosphor, which then emits white light. Creating a bright image takes lots of energy, so the electron beam is accelerated on its way to the screen. A high voltage power supply (+15,000 V to +25,000 V) pumps positive charge onto the inside of the screen and the surrounding accelerating anode, and this charge attracts the electrons. By the time they hit, the electrons have enough kinetic energy to make the phosphor glow bright white.

But a television image isn't all white. To produce a gray or black spot, the television reduces the current of electrons in the beam. It controls this current by adjusting the charge on the picture tube's grid. The more negative charges on the grid, the harder it is for electrons to pass from the cathode to the anode and the fewer of them strike the phosphor. The television carefully adjusts this grid charge as it sweeps the electron beam back and forth across the screen and thus creates a complete television picture, one dot at a time.

Color Picture Tubes

Color picture tubes work much like black and white tubes, except that they have three electron beams and phosphors that emit red, green, and blue light. As we'll discuss in the next chapter, mixtures of red, green, and blue light can make you perceive any color. A color television appears full color by carefully mixing these three colored lights.

The inside surface of a color television screen is coated with thousands of tiny phosphor dots (Fig. 13.3.7a). Some of these dots emit red light, some green light, and some blue light. The television directs electrons at these dots through holes in a metal mask. Three separate electron beams, coming from three slightly different angles, pass through the holes and strike the phosphors. Since each beam can only strike one color of phosphor dots (Fig. 13.3.7b–d), each beam controls the brightness of one of the three colors.

While some picture tubes use phosphor stripes rather than dots, and a grille rather than a mask, the basic idea is the same. Only one electron beam can hit a particular phosphor stripe. These masks and grilles are carefully aligned inside picture tubes and because they must stay aligned even when heated by the electron beams, they're made from special thermally compensated metals—metals with small coefficients of volume expansion. Unfortunately, those special metals are ferromagnetic and easily magnetized, which is why you shouldn't hold a strong magnet near a color picture tube. The resulting magnetization would then deflect the electron beams and produce distorted images and colors. Fortunately, many modern televisions and computer monitors have built in degaussing systems that demagnetize their picture tubes when you turn them on.

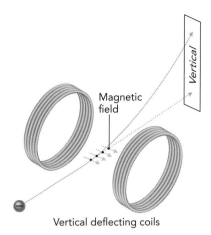

Fig. 13.3.6 The vertical deflecting coils are located to the left and right of the picture tube's neck. The horizontal magnetic field they produce deflects the electron beam either up or down.

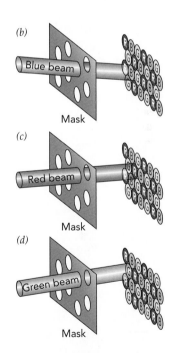

Fig. 13.3.7 (a) The pattern of red, green, and blue phosphor dots coating the inside of a picture tube. The holes in the mask allow (b) the "blue" electron beam to illuminate only blue dots, (c) the "red" beam to illuminate only red dots, and (d) the "green" beam to illuminate only green dots.

Analog Video Signals

At this point, I'll turn my attention to the analog video signal—the stream of information that tells an analog television which dots to illuminate and how brightly to make them glow. As I mentioned before, the NTSC standard in the U.S. specifies 60 partial images or "frames" per second, appearing in "interleaved" pairs having a combined total of 525 horizontal lines. The first frame in each pair contains only the even lines of the complete image, while the second frame contains only the odd lines. Since the first 20 lines of each frame aren't visible, occurring while the electron beam is returning to the top of the screen, the first partial image contains lines 41, 43,... , 525 while the second partial image contains lines 42, 44,... , 524. The number of dots per line is limited by the bandwidth of the video signal but rarely exceeds 440. Thus NTSC video forms complete images (both frames) of roughly 213,400 dots (485 visible lines with 440 dots/line) at a net rate of 30 images/second.

The NTSC monochrome video signal contains both brightness and synchronization information, a combination known as composite video. In a video cable such as you'd use to connect a DVD player's composite video output to an analog television set's composite video input, the video signal is represented as a fluctuating voltage on the cable's central wire. That voltage varies between –0.3 V and +0.7 V. Voltages ranging from 0.0 V to +0.7 V indicate brightness levels ranging from black to white, respectively. To synchronize the horizontal and vertical sweeping processes, however, the signal makes brief excursions to –0.3 V. One of these "blacker-than-black" sync pulses precedes each horizontal sweep and a string of such pulses precedes each vertical sweep.

But television developed in a broadcast environment that put strict constraints on bandwidth. When portions of the precious electromagnetic spectrum were set aside for black-and-white television in 1941, the FCC gave each channel only 6 MHz to carry both video and audio signals. The transmission scheme adopted by the NTSC involves two separate radio-frequency carriers, one for video and one for audio. Video information occupies the lower 5.25 MHz of the channel bandwidth while audio occupies the upper 0.75 MHz. In other words, the TV channel's video information was required to fit in a 5.25 MHz range of radio wave frequencies and its audio information had to fit in an adjacent 0.75 MHz range.

The phrase "was required to fit" implies some amount of difficulty, as though a 5.25 MHz portion of the radio spectrum can't easily "hold" a black-and-white video signal. Sure enough, it's a challenge. But for you to understand why it's a challenge, I must show that representing a real video signal requires a range of radio wave frequencies and that the range of frequencies gets broader as the video signal changes more rapidly with time.

According the NTSC scheme, the black-and-white video signal is transmitted using amplitude modulation of a radio wave. Just as the audio signal in AM radio modulates the amount of charge the radio station sends up and down its antenna at the carrier frequency, so the video signal in NTSC analog television modulates the amount of charge the television station sends up and down its antenna at the video carrier frequency.

When the video signal is constant, perhaps because someone turned off the video camera, the amount of charge moving up and down the antenna is constant and the antenna emits a steady radio wave—the carrier wave. Figure 13.3.8a shows the charge on the antenna as a function of time as it emits this steady wave.

With the video camera turned on, the video signal begins to vary with time. That signal modulates the amount of charge moving up and down the antenna and therefore the amplitude of the radio wave that the antenna emits. In general, the results of that modulation are complicated, so let's first examine a particularly simple

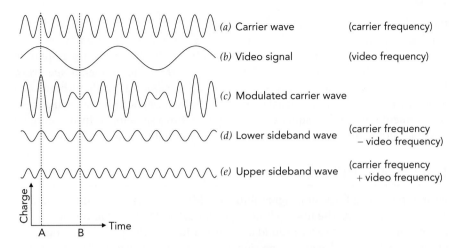

Fig. 13.3.8 A television station modulates its carrier wave (*a*) with its video signal (*b*) to obtain a modulated carrier wave (*c*). In the special case shown here, the video signal (*b*) is a sinusoidal function of time and therefore the modulated carrier wave (*c*) is the sum of three simple waves: the carrier wave (*a*), a lower sideband wave (*d*), and an upper sideband wave (*e*). If you add waves (*a*), (*d*), and (*e*) at a particular time, such as A or B, they'll sum to wave (*c*) at that time. Shown here are the amounts of charge on the transmitting antenna as it emits these various waves.

example. If the video signal varies sinusoidally as a function of time (Fig. 13.3.8*b*), with a frequency I'll call the video signal frequency, the resulting modulated carrier wave (Fig. 13.3.8*c*) resembles the carrier wave itself, except that its amplitude now has the sinusoidal shape of the video signal.

That choice of video signal is special because a sinusoidally modulated carrier wave can also be viewed as the sum of three steady waves: the carrier wave and two sideband waves: a lower sideband wave (Fig. 13.3.8*d*) and an upper sideband wave (Fig. 13.3.8*e*). The frequency of the lower sideband wave is the carrier wave frequency minus the video signal frequency, and the frequency of the upper sideband wave is the carrier wave frequency plus the video signal frequency. You can see that the carrier and two sideband waves sum to the modulated carrier wave in Fig. 13.3.8 by picking a specific time—such as A or B—and adding the antenna charges associated with those three waves; they sum to the antenna charge associated with the modulated carrier wave! In effect, the modulated antenna charge can be thought of as three separate portions of charge that move up and down the antenna at three slightly different frequencies.

Although a real video signal varies in an extremely complicated manner, it can always be thought of as the sum of many different sinusoidal pieces with frequencies ranging from approximately 0 Hz up to some maximum frequency. When this complicated video signal modulates the carrier wave, each of its sinusoidal pieces generates its own pair of sideband waves, one below the carrier frequency and one above it. The modulated carrier wave therefore includes frequencies ranging from the carrier frequency minus the highest frequency piece in the video signal to the carrier frequency plus the highest frequency piece in the video signal.

We can now begin to see why fitting a video signal into a 5.25 MHz portion of the radio frequency spectrum might be difficult. If a video signal contains a sinusoidal piece with a frequency of 2.625 MHz, the modulated carrier wave will have sidebands extending from 2.625 MHz below the carrier to 2.625 MHz above the carrier, thereby spanning the entire 5.25 MHz allocation. If the video signal has any sinusoidal pieces with frequencies *higher* than 2.625 MHz, it could cause trouble. Does it?

Unfortunately, the answer is yes. According to the NTSC scheme, the analog video signal must specify all 213,400 dots on the screen 30 times a second and it therefore contains sinusoidal pieces at frequencies as high as 5.5 MHz. With sideband waves extending 5.5 MHz below and 5.5 MHz above the carrier wave frequency, a black-and-white NTSC video signal transmitted by ordinary amplitude modulation will consume an 11.0 MHz portion of the radio frequency spectrum. But the official allocation is only 6 MHz, *including sound*. And if that weren't bad enough, a *color* NTSC video signal has to fit in that same allocation! Satisfying these requirements sounds like squeezing an elephant into a shoebox. It turns out to be possible, although it does reduce the quality of the picture ever so slightly.

The Elephant Fits in the Shoebox

Fitting the analog television signal into its 6-MHz allocation requires several ingenious techniques. The first technique makes use of the fact that the upper and lower sideband waves carry redundant information. Since a particular sinusoidal piece in the black-and-white video signal creates both a lower *and* an upper sideband wave, only one of these sideband waves is really needed. Moreover, transmitting both sidebands is a waste of radio transmitter power. So the television station uses electronic filters that remove most of the lower sideband waves from its radio transmission (Fig. 13.3.9a). When this filtering is done, the lower sideband waves extend no more than 1.25 MHz below the carrier and the carrier is moved down in frequency until it is only 1.25 MHz above the bottom of the channel. Television receivers are able to compensate for the missing lower sideband waves. Because its lower sidebands are so heavily filtered, this transmission technique is known as vestigial sideband (VSB) modulation, rather than amplitude modulation (AM).

To narrow the bandwidth of the television signal still further, the television station also filters out the highest frequency upper sideband waves. Instead of extending 5.5 MHz above the carrier frequency, these upper sideband waves extend no more than 4.0 MHz. This filtering reduces the picture sharpness slightly.

The television signal now occupies 5.25 MHz of bandwidth about the carrier wave frequency: 1.25 MHz below it and 4 MHz above it. The television station adds sound to its transmission by frequency modulating a separate carrier wave, just like in normal FM radio. This new audio carrier wave has a frequency that is 4.5 MHz above that of the video carrier wave.

With the inclusion of sound, the 6 MHz of allowed bandwidth is essentially full. Broadcast television channels are assigned to frequencies 6 MHz apart, although they are broken up into groups (VHF Low, VHF High, and UHF) in order to leave room in the terrestrial broadcast spectrum for FM radio, cellular telephones, and other communication activities. In cable television, there are no other services competing for bandwidth inside the cable and the whole radio frequency spectrum is available. Cable channels are therefore assigned 6 MHz apart over the cable's entire usable range of radio frequencies.

But what about color? When the NTSC issued a color video standard in 1953, it decided to overlay that new standard on top of the existing black-and-white standard in a way that would allow existing black-and-white sets to receive the color transmissions. The color standard used the same 6-MHz allocation and inserted color information in a way that black-and-white sets would simply ignore it. This was an absolutely amazing engineering feat.

The NTSC color video signal contains two separate signals: the original black-and-white video signal, now known as luminance, and a new color signal called chrominance. Luminance represents the screen's brightness patterns while chrominance represents the screen's color patterns. To keep these two signals separate, the chrominance signal is shifted up in frequency by 3.58 MHz—the chrominance subcarrier frequency—so that it shares frequencies with the highest upper sidebands of the luminance signal. For reasons relating to the time structures ❶ of

❶ When figuring out where to put the chrominance signal, the NTSC took advantage of a comb of holes in the luminance signal's frequency spectrum. Since adjacent horizontal lines are usually quite similar, many features of the luminance signal recur at the horizontal sweep rate. As a result of this repetition, most of the frequencies in the luminance signal appear at or very near the harmonics of the horizontal sweep rate. The frequencies between those harmonics are essentially absent because they represent the unlikely cases where adjacent horizontal lines contain brightness features that are exactly opposite to one another. Into this comb of quiet regions in the luminance signal, the NTSC inserted the chrominance signal. They placed a chrominance subcarrier half way between the 227th and 228th harmonic of the horizontal sweep rate and modulated this carrier with the two additional video signals needed to create full color images. The luminance and chrominance spectra interdigitate perfectly and share the channel's frequency space with almost no noticeable effects on one another.

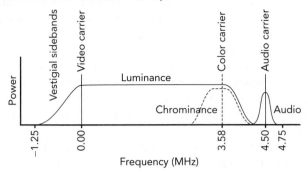

(a) NTSC Color Standard - Concept

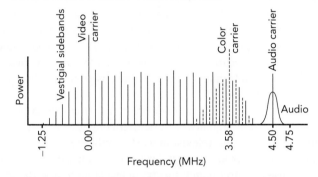

(b) NTSC Color Standard - Actual (simplified)

Fig. 13.3.9 NTSC analog television fills the 6-MHz channel with luminance, chrominance, and audio signals. *(a)* According to the concept, the lower 1.25 MHz of the channel is occupied by the useless vestigial sidebands, and the luminance (brightness) and chrominance (color) signals are overlapped. *(b)* In actuality, the luminance signal is strong only at harmonics of the horizontal sweep frequency and it forms a comb of powerful radio wave frequencies. The chrominance signal also forms a comb of powerful frequencies, which is interdigitated into the luminance comb.

the luminance and chrominance signals, these two electromagnetic waves coexist beautifully and a black-and-white television doesn't even notice that the chrominance signal is there (Fig. 13.3.9*b*).

The chrominance signal controls the three electron beams in a color analog television. It contains two streams of color instructions, which, together with the brightness instructions in the luminance signal, determine the full coloration of the screen. Since your eyes are much more aware of brightness details than of color details, the chrominance signal is simplified so that it occupies only a modest bandwidth in the television transmission.

The audio carrier, located 4.5 MHz above the video carrier, is frequency modulated by the audio signal and transmitted with both its sidebands intact. The techniques used are identical to those of stereo FM radio, although its stereo pilot tone is 15.734 kHz (15.750 kHz in the days before color television), rather than the 19 kHz of normal FM broadcasts. That 15.734 kHz frequency is the same frequency used for the horizontal sweep ❷.

Why Shift to Digital Television?

Yet despite its elegance and maturity, analog television is too limited to survive in this digital age. Low resolution, poor power efficiency in transmission, and a sensitivity to noise and interference all contribute to its demise. Digital television has none of these shortcomings and offers some amazing bonuses as well.

❷ When it added color in 1953, the NTSC made a slight shift in its frequency specifications. To avoid rhythmic beating problems between the audio and chrominance carriers, the horizontal sweep rate was decreased 0.1% from 15,750 to 15,734.264 Hz. This tiny shift made the 4.5 MHz frequency difference between the audio and video carriers equal to the 286th harmonic of the horizontal sweep rate and put the 3.579545 MHz chrominance carrier exactly halfway between the 227th and 228th harmonic of the horizontal sweep rate. With all these important frequencies—horizontal sweep, video carrier, audio carrier, and chrominance subcarrier—synchronized in this fashion, there are no beating problems. A side effect of this change was a reduction in the number of fields per second from 60 to 59.94. Fortunately, existing black-and-white sets tolerated those changes without a hitch and for nearly 50 years, NTSC analog color television has dominated the airwaves in the United States.

Analog television's resolution problem stems from its absence of memory: it squanders bandwidth transmitting redundant information. Since every dot is drawn the moment its information arrives at the television, there's no opportunity to save bandwidth. In contrast, digital television uses sophisticated compression techniques to find and exploit temporal and spatial patterns in the images. Redundant information is squeezed out of the information stream. As the result of this compression and other digital communication techniques, the same information bandwidth that supports low resolution analog television can support much higher resolution digital television.

Analog television's power inefficiency stems from the nature of its analog signals and the modulation techniques used to broadcast those signals. A large fraction of the transmitted power appears in its carrier waves and in the useless partial lower sideband waves of the video signal. Since television receivers obtain no information from the carrier or the lower sidebands, producing those waves is simply a waste of power. Moreover, analog television's frequency spectrum contains many peaks and valleys; some portions of that spectrum are badly congested while other portions are woefully underutilized. Its uneven spectrum makes analog television more susceptible to radio frequency noise in the environment than it would be if its spectrum were smoother.

In contrast, the spectrum of digital television is essentially flat—it uses all the available frequencies about equally, with neither congestion nor underutilization anywhere. In addition, digital television wastes only the most minimal amount of its transmitted power producing carrier waves and useless sidebands. Overall, digital television uses the frequency spectrum available to it with almost perfect efficiency and it transmits its signal with almost no wasted power.

Analog television's sensitivity to noise and interference is typical of all analog systems: they're unable to distinguish noise in the physical quantity representing information from that information itself. Electromagnetic noise in the video signal appears immediately on the screen as visual noise—snow, tick marks, and patterns are common. Likewise, delayed reflections of the signal itself produce ghost images.

Digital television is much less sensitive to noise because imperfections in the signals arriving at the television can be completely and reliably removed before they have any effect on the displayed images. By using discrete values for the physical quantities conveying information, the digital approach prevents low-level noise on those physical quantities from affecting the information. With the help of error correction and digital filtering, digital television is able to display perfect images even when the electromagnetic signals reaching the television have a fair amount of noise in them.

Digital Television Technology

While analog television deals with images dot by dot in real time, digital television treats the images as just another form of data to be compressed and sent through a high-speed communications channel. Compressed sound information uses the same channel. Since synchronization is achieved by decoding data at the proper rate, data transfers need not occur exactly in real time. This flexibility makes it possible to average out the load on the channel so that it's used with maximum efficiency. Since everything you see or hear passes through this channel, that's where I'll focus my attention.

In the United States, the specifications for digital television are set by the Advanced Television Systems Committee (ATSC), based on the work of a group known as "The Grand Alliance." In the ATSC standard, the same 6-MHz channels that have been carrying analog television for decades are now carrying digital television. Each digital channel carries 10.762 megasymbols/second, where a "symbol" is a digit in the numerical base used by the channel.

In binary (base-2) there are only two different symbols, 0 and 1, but digital transactions don't have to be in binary. For example, if a digital transaction were to use decimal (base-10), there would be 10 different symbols: 0, 1, 2, 3, 4, 5, 6, 7, 8, and 9. The ATSC decided that cable television will use hexadecimal (base-16) symbols (0, 1, 2, 3, 4, 5, 6, 7, 8, 9, A, B, C, D, E, and F) and that terrestrial broadcast television will use octal (base-8) symbols (0, 1, 2, 3, 4, 5, 6, and 7). Each of these representations is still digital, but now each symbol is carrying more than one bit worth of information. Base-8 symbols each carry three bits worth of information, and base-16 symbols each carry four bits worth of information.

During transmission via radio waves, the symbols must be represented electromagnetically. In digital television, the symbols are encoded as different discrete amplitudes of a carrier wave. Since noise makes it hard to distinguishing two closely spaced amplitudes, the size of the symbol group is limited by the signal to noise ratio at the receiver. If there is little noise, the symbol group can be large because even small differences in carrier amplitude are easily distinguished. But if there is much noise, a smaller symbol group must be chosen.

With its goal of less than one unrecoverable data packet error per second, ATSC found that it could use 16 different discrete amplitudes for cable transmissions, but not quite 8 different discrete amplitudes for terrestrial broadcast. Although the ATSC standard does use base-8 symbols (i.e., 8 different amplitudes) for terrestrial broadcasting, it adds a layer of encoding that reduces the information available per symbol to only two bits, as though it were using base-4 symbols. Although that encoding seems to throw away information, it makes the whole channel less vulnerable to radio frequency noise.

Digital broadcasts begin by digitizing and compressing the video and audio signals. Apart from analog-to-digital conversion issues, these steps are mostly mathematical and I won't dwell on them. The video compression involves the Moving Picture Experts Group MPEG-2 standard and the sound compression uses the Dolby AC-3 standard. While MPEG-2 compression loses some image information, compression by a factor of 10 is typically imperceptible and by a factor of 60 is usually quite acceptable. Following compression, the picture and sound are just packets of data that must be transferred to the viewer's television.

Before transmission, these data packets are processed and enlarged slightly to include error correction information. That additional information allows a receiver to detect and repair most transmission errors so that, even if the receiver misreads some of the symbols as they arrive, it can correct nearly all of those mistakes. After incorporating this error correcting information, the stream of symbols heads to the transmission system for broadcast.

The broadcast system is responsible for transmitting the stream of symbols. The technique used to squeeze them into a 6-MHz radio frequency channel is again VSB modulation. However, the filters that remove frequencies falling outside the 6-MHz channel are far more sophisticated in digital television than they are in analog television. In analog television, filter limitations required the carrier to be placed 1.5 MHz above the bottom of the channel and all the frequencies between the bottom and the carrier are wasted. In digital television, an advanced filter allows the carrier to be place only 0.31 MHz above the bottom of the channel (Fig. 13.3.10a). This filter therefore cuts off only the bottom 0.31 MHz of the channel. A similar filter cuts off only the top 0.31 MHz of the channel. Since the middle 5.38 MHz of the channel remains available for information, a remarkable 89.7% of the 6-MHz allocation is available for the transmission. Overall, this digital transmission scheme approaches the theoretical limit known as the Shannon limit in its use of bandwidth and power (Fig. 13.3.10b). It's nearly perfect.

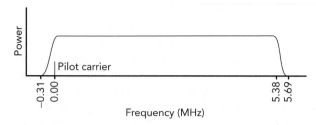

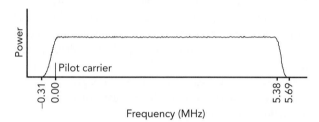

Fig. 13.3.10 ATSC digital television fills the 6-MHz channel with a digital datastream containing both video and sound information. (*a*) Sophisticated filters cut off the radio wave frequencies 0.31 MHz from each edge of the channel and the uniform spectrum of frequencies in the middle 5.38 MHz is utilized fully for information transfer. (*b*) The actual transmission is almost as perfect as the concept.

A receiver detects the radio frequency electromagnetic waves that fall within the channel and extracts the stream of symbols. After correcting any errors in that stream, it passes the symbols on to the presentation system. The presentation process is pretty straightforward—the MPEG-2 and AC-3 decoders are just special-purpose computers and the screen is simply a computer monitor. Because digital television decouples transmission from presentation, the image resolutions and display rates can vary according to need and several different video programs can often share the same 6-MHz channel.

The ATSC standards include image resolutions up to 1080 lines with 1920 dots/line, for a total of 2,073,600 dots/image—about ten times the resolution of NTSC analog television—and screen aspect ratios of both 4:3 (conventional) and 16:9 (wide-screen). The image rates include not only the nominal 30 Hz and 60 Hz standards of NTSC, but also the 24 Hz standard of motion picture film so movies and the many television shows recorded on film can be displayed at their best. Finally, the sound system is a huge improvement over stereo, delivering up to 5.1 audio channels of audio: left, right, center, left rear, right rear, and a special low-frequency effects channel (the extra 1/10th of a channel). So even if there's nothing worth watching, what is there will sure look and sound good.

CHAPTER 14

THINGS THAT INVOLVE LIGHT

While radio waves and microwaves are useful for communications and energy transfer, there's another portion of the electromagnetic spectrum that we find far more important: light. Light consists of very-high-frequency, very-short-wavelength electromagnetic waves. Light's frequencies are so high that normal antennas can't handle it. Instead, it's absorbed and emitted by the individual charged particles in atoms, molecules, and materials. Because of its special relationship with the charged particles in matter, light is important to physics, chemistry, and materials science. Moreover, it's one of the principal ways by which we interact with the world around us.

People have been making light since they first learned to handle fire, but recently they've discovered many new ways to produce it. Campfires, candles, oil lamps, and gaslights have given way to less romantic but more energy efficient and eye popping sources of light. We're surrounded by fluorescent and mercury vapor lamps, light-emitting diodes, and lasers. We now produce light for illumination, communication, entertainment, decoration, and we even use it as a tool to perform tasks such as metalwork, surgery, and hair removal. Light has become so popular that this season, "light is the new black."

In this chapter, I'll explore a number of different light sources and explain how each of them produces light. I'll also discuss three different types of light: thermal light, atomic resonance light, and coherent light. Hopefully, as you read through these topics, you'll find that you won't need to go to Las Vegas to shed some exciting new light on the world around you.

Chapter Itinerary

Blue sky   Red setting sun

Soap bubbles

Pool of water

Reflection of sky

Section 14.1 **Sunlight**

For thousands of years, people have marked the passage of time by the rising and setting of the sun over the horizon. The sun first appears as a red disk in the east every morning, rises white in the blue sky, and then sets once again as a red disk in the west. The sunlight that we see takes about 8 minutes to travel the 150,000,000 km (93,000,000 miles) from the sun to our eyes and provides most of the energy and heat that make life on earth possible. While the light in sunlight is really just another electromagnetic wave, and could be considered part of the previous chapter, it's so important to everyday life that it deserves special attention. And so we'll begin by looking at how sunlight interacts with our world.

Sunlight and Electromagnetic Waves

Electromagnetic waves can have any wavelength, from thousands of kilometers to a fraction of the width of an atomic nucleus. The radio waves and microwaves that we examined in the previous chapter have wavelengths longer than 1 mm. In this chapter, we'll turn our attention to shorter wavelength radiation. In particular, we'll study electromagnetic waves with wavelengths between 400 nm and 750 nm (recall that 1 nm or 1 nanometer is 10^{-9} m). These are the electromagnetic waves that we perceive as visible light and the principal components of sunlight.

Because the electromagnetic waves in sunlight have such short wavelengths, their frequencies lie between 10^{14} Hz and 10^{15} Hz (Fig. 14.1.1). As one of these waves of sunlight passes by, its electric field fluctuates back and forth almost 1,000,000,000,000,000 times each second. Since producing microwaves, which have much longer wavelengths and much lower frequencies, already requires specialized components and tiny antennas, what can possibly emit or absorb light waves? The answer is the individual charged particles in atoms, molecules, and materials. These tiny particles can move extremely rapidly, often vibrating about at frequencies of 10^{14} Hz, 10^{15} Hz, or even more. As these charged particles accelerate back and forth, they emit light waves. Similarly, passing light waves cause individual charged particles in atoms, molecules, and materials to accelerate back and forth, thereby absorbing the light waves as well.

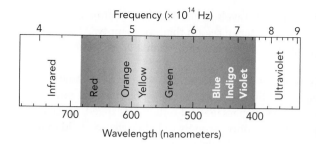

Fig. 14.1.1 The visible portion of the spectrum of sunlight. Each wavelength of visible light has a particular frequency and is associated with a particular color. At the ends of the visible spectrum are invisible infrared and ultraviolet lights.

Sunlight originates at the outer surface of the sun, in a region called the photosphere. There, atoms and other tiny charged systems (mostly atomic ions and electrons) jostle about at 5800 °C. Since these charged particles accelerate as they bounce around, they emit electromagnetic waves.

Because the sun's surface emits light through the random, thermal motions of its charged particles, the distribution of wavelengths it emits is determined only by its temperature. It emits a blackbody spectrum, like the incandescent lightbulbs we discussed in Section 7.3. Because the photosphere's temperature is 5800 °C, the jostling motions are extremely rapid and most of the sunlight falls in the visible portion of the electromagnetic spectrum (Fig. 14.1.2).

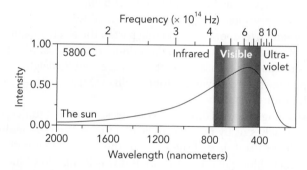

Fig. 14.1.2 Sunlight comes from the sun's photosphere, where the temperature is 5800 °C. This light has a blackbody distribution of wavelengths, with much of its intensity concentrated in the visible portion of the overall electromagnetic spectrum.

However, not all sunlight is visible. On the long-wavelength, low-frequency side of visible light is infrared light. We can't see infrared light with our eyes but we feel it when we stand in front of a hot object. In sunlight, infrared light is produced by charges that are accelerating back and forth more slowly than average.

On the short-wavelength, high-frequency side of visible light is ultraviolet light. We can't see ultraviolet light either, but we're aware of its presence because it induces chemical damage in molecules. It causes sunburns and encourages skin to tan. In sunlight, ultraviolet light is produced by charges that are accelerating back and forth more rapidly than average.

Sunlight's Passage to the Earth

Sunlight travels from the sun to the earth at the speed of light. But what sets the speed of light? Actually, as we learned in Section 4.2, it's one of the fundamental constants of nature, with a defined value of 299,792,458 m/s in empty space. While one could argue that the speed of light is set by the relationships between the electric and magnetic fields, that observation simply passes the buck. If you were then to ask what sets the relationships between the electric and magnetic fields, the answer would be the speed of light.

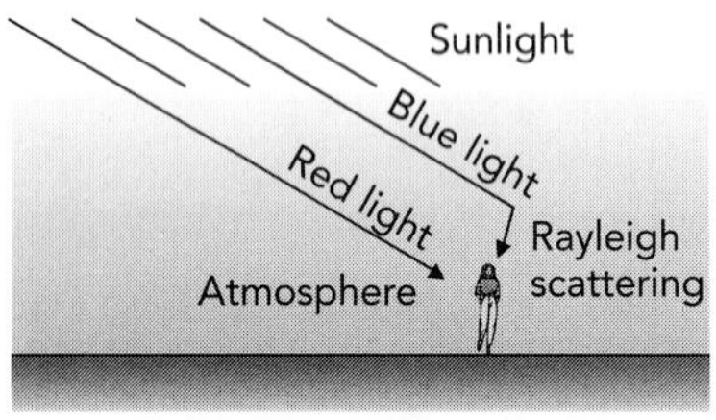

Fig. 14.1.3 As sunlight passes through the atmosphere, some of its blue light undergoes Rayleigh scattering from particles in the air. We see this redirected blue light as the diffuse blue sky. The remaining light reaches our eyes directly from the sun and tends to be reddish, particularly at sunrise and sunset.

Rather than justifying why sunlight travels as fast as it does in empty space, let's look at what happens to it when it enters a region that's not empty. After all, sunlight eventually reaches the earth's atmosphere and, when it does, several interesting things happen.

First, the sunlight slows down as its electric and magnetic fields begin to interact with the electric charges in the atmosphere. Light polarizes the molecules it encounters, a process that delays its passage and reduces its speed. Since most transparent materials respond much more strongly to light's electric field than to its magnetic field, we'll concentrate on only electric effects.

The factor by which light slows down in a material is known as the material's index of refraction. Light travels particularly slowly through materials that are easy to polarize, and some of them have indices of refraction of 2 or even 3. Because air near sea level is only slightly polarizable, however, its index of refraction is just 1.0003. While this reduction in light's speed is too small to notice, we do notice the polarized air particles that cause it. These polarized air particles are what makes the sky blue.

The particles in air consist of individual atoms and molecules, small collections of atoms and molecules, water droplets, and dust. As a wave of sunlight passes through one of these particles, the particle becomes polarized. Its electric charges accelerate back and forth as the sunlight's electric field pushes them around and they reemit a new electromagnetic wave of their own.

This new wave draws its energy from the original wave. In effect, the particle acts as a tiny antenna, temporarily receiving part of the electromagnetic wave and immediately retransmitting it in a new direction. This process, whereby a tiny particle redirects the path of a passing light wave, is called Rayleigh scattering, after the English physicist Lord Rayleigh (John William Strutt, 1842–1919) who first understood it in some detail.

While most sunlight travels directly to our eyes, some of it undergoes Rayleigh scattering and reaches us by more complicated paths. We see the direct light as coming from the brilliant disk of the sun, but the scattered light gives the entire sky a fairly uniform blue glow (Fig. 14.1.3). But why is this glow blue?

The sky's blue color comes about because the tiny air particles that Rayleigh scatter sunlight are too small to make good antennas for that light. We observed in Section 13.1 that an antenna works best when it is one-quarter as long as the wavelength of its electromagnetic wave. The air particles make particularly bad antennas for long-wavelength red light, so that very little red sunlight undergoes Rayleigh scattering on its way through the atmosphere. But the air particles are not such bad antennas for short-wavelength blue light. Some of the blue sunlight does Rayleigh scatter and reaches our eyes from all directions. We see this Rayleigh scattered light as the blue glow of the sky.

Rayleigh scattering not only makes the sky blue; it also makes the sunrises and sunsets red. As the sun rises or sets, its light must travel long distances through the earth's atmosphere in order to reach your eyes. Its path is so long that most of the blue light Rayleigh scatters away miles to your east or west and all you see is the remaining red light. Sometimes the whole local sky appears reddish because there simply isn't any blue light left to scatter toward you. Sunrises and sunsets are particularly colorful when extra dust or ash is present in the atmosphere to enhance the Rayleigh scattering. Air pollution, forest fires, and volcanic eruptions tend to create unusually red sunrises and sunsets.

In contrast, clouds and fog appear white because they're composed of relatively large water droplets. These droplets are larger than the wavelengths of visible light and scatter all of sunlight's wavelengths equally well. Although this scattering is often so effective that you can't see the sun's disk through a cloud, it doesn't give the cloud any color. The cloud simply looks white.

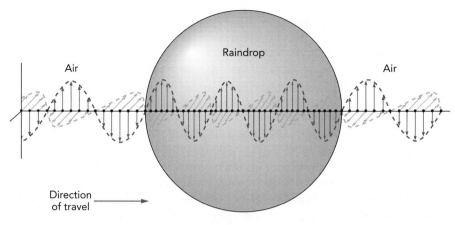

Fig. 14.1.4 As an electromagnetic wave enters a material, its speed decreases and the waves bunch up together. Its wavelength decreases.

Rainbows

Sometimes water droplets do separate the colors of sunlight. When sunlight shines on clear, round raindrops as they fall during a storm, these raindrops can create a rainbow. To understand how clear spheres of water can bend sunlight's path and separate it according to wavelength, we must understand three important optical effects: refraction, reflection, and dispersion. We encountered those same wave phenomena while studying water surface waves in Section 9.3, but now they appear in a new context: light waves!

Let's begin by looking at what happens when a wave of sunlight passes directly through a raindrop. Because water is more polarizable than air, the wave slows down inside the raindrop and its cycles bunch together (Fig. 14.1.4). While this bunching effect reduces the light's wavelength inside the drop, the light's frequency remains unchanged. The cycles don't disappear as they go through the raindrop, they just move more slowly.

If a narrow wave of sunlight is aimed directly through the center of the raindrop, it will follow a straight path and emerge essentially unaffected from the other side (Fig. 14.1.5a). But if that wave strikes the raindrop near the top, it will bend as it enters the water (Fig. 14.1.5b). This occurs because the lower edge of the wave will reach the water first and slow down; the upper edge will then overtake it and the wave will bend downward. The wave will head more directly into the water.

As the wave in Fig. 14.1.5b leaves the raindrop, its upper edge emerges first and speeds up while the lower edge lags behind. The wave bends downward even further and heads less directly into the air and away from the water.

This bending of sunlight at the boundaries between materials is refraction. It occurs whenever sunlight changes speeds as it passes through a boundary at an angle. If sunlight slows down at a boundary, it bends to head more directly into the new material. If sunlight speeds up at a boundary, it bends to head less directly into the new material. The amount of the bend increases as the speed change increases.

However, part of the sunlight striking a boundary doesn't pass through the boundary at all. Instead, it reflects. In Section 9.3, we attributed wave reflection specifically to changes in wave speed at a boundary. However, the more general cause of wave reflections is an impedance mismatch—an abrupt change in how the wave moves through its environment. In general, impedance is the measure of a system's opposition to the passage of a current or a wave. For an electromagnetic system, impedance measures how much voltage or electric field is needed to produce a particular current or magnetic field. In other words, electrical impedance measures how hard it is for electric activity to produce magnetic activity. Imped-

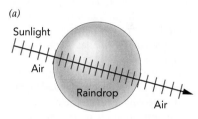

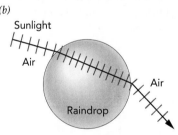

Fig. 14.1.5 A side view of two narrow waves of sunlight entering and leaving raindrops. The lines drawn across each light wave represent upward electric field maxima and are bunched together as light slows down in water.

❶ When electromagnetic waves travel through wires, they reflect from impedance mismatches. These mismatches occur whenever the relationship between electric and magnetic fields changes and must carefully be avoided in television wiring. If you don't provide an impedance matching device when connecting an antenna wire (300 V) to a video cable input (75 V), you'll have reflections in the wires and ghost images on the screen.

❷ Sand appears white because it redirects sunlight in all directions. One explanation for this effect is that the sand grains act as tiny antennas that respond to and reemit light's electromagnetic waves. A second explanation is that the sand grains present the sunlight with thousands of air–sand boundaries from which to reflect. However, both explanations are descriptions of exactly the same physics—the charged particles in the sand grains are electrically polarized by the waves passing through them. These waves are randomly redirected without being absorbed, and they give the sand its white appearance.

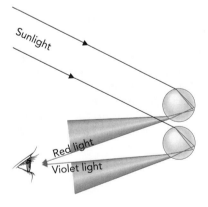

Fig. 14.1.6 As sunlight passes through spherical raindrops, its colors separate. Violet light bends more at each air/water boundary than does red light, and the two emerge from raindrops heading in different directions. You see red light coming toward you from the upper raindrops and violet light from the lower raindrops.

ance effects are common in nature (see **❶**) and also apply to mechanical waves and currents. When sound and water waves encounter impedance mismatches, they also partly reflect.

The impedance of empty space is high because an electric field there has nothing to aid it in producing a magnetic field. But inside most materials, the electric field has help. The electric field polarizes the material, which then helps to create the magnetic field. Because of this assistance, the impedance of most materials is much less than that of empty space. Since air is almost empty space, the boundary between air and water is an impedance mismatch for light.

Passing through an impedance mismatch upsets the balance between a light wave's electric and magnetic fields. To compensate for this imbalance, part of the incoming wave experiences reflection off the boundary. Thus some sunlight reflects each time it enters or leaves a water droplet. The fraction of light that reflects depends on the severity of the impedance mismatch but is typically 4% between air and most transparent materials, including water (for reflection from sand, see **❷**). In contrast, metals polarize so easily that their impedances are essentially zero and they reflect light almost perfectly.

There is one more important point about sunlight's passage through water: red light travels about 1% faster through water than violet light does. That's because higher-frequency violet light polarizes the water molecules a little more easily than lower-frequency red light, and that increased polarization slows down the violet light. This frequency dependence of light's speed in a material is dispersion. Dispersion affects refraction. The more light slows as it enters a raindrop, the more it bends at the boundary. Since violet light slows more than red light, violet light also bends more and the different colors of sunlight follow somewhat different paths through the raindrop.

A rainbow is created when raindrops separate sunlight according to color. To see the rainbow, you stand with the sun at your back and look up at the sky. When sunlight hits the raindrops, they redirect some of that light back toward you. Since each raindrop redirects light only in a narrow range of angles, you can't see light from every raindrop. Only the raindrops in a narrow arc of the sky redirect visible light toward you. This arc appears brightly colored because raindrops at the inner edge of the arc send violet light toward you while raindrops at the outer edge of the arc send red light toward you. In between, you see all the colors of the rainbow.

Figure 14.1.6 shows how a raindrop redirects different colors of light in different directions. While there are many possible paths light can take through the raindrop, this path is the one that produces rainbows. Sunlight enters near the top of the raindrop and bends inward. Violet light bends more than red light, so the sunlight begins to separate according to color. Some sunlight is also reflected from the raindrop but doesn't contribute to rainbows.

When the light inside the raindrop strikes the back surface, most of it leaves the drop and is lost. A small fraction of the light reflects from that surface, however, and continues to travel through the raindrop. When this light reaches the raindrop's front surface, most of it leaves the drop. Violet light bends more strongly than red light as they reenter the air, so the different colors of light leave the drop heading in different directions. Since violet light is redirected more upward than red light, you see violet light coming toward you from the lower raindrops. Red light is redirected more downward so you see it coming toward you from the upper raindrops. Thus the upper arc of the rainbow is red while the lower arc is violet.

Soap Bubbles

Soap bubbles also separate sunlight into its various colors (Fig. 14.1.7), but they use another wave phenomenon: interference. We encountered interference of me-

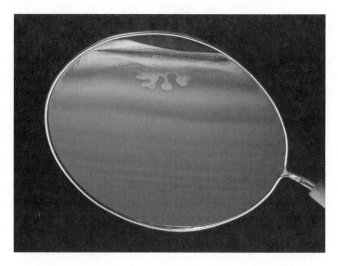

Fig. 14.1.7 Light reflected by the front and back surfaces of this soap film interferes with itself and gives the film its colorful appearance. Since the colors are determined by film thickness and since the film's thickness increases in the downward direction, the film displays horizontal bands of color.

chanical waves in Section 9.3 when we studied surf beat at the seashore (p. 322) and interference of electromagnetic waves in Section 13.2 when we considered the unevenness of microwave cooking (p. 439). Now we'll look at interference in another type of electromagnetic wave: light.

Light's interference effects stem from the summing or superposition of its electromagnetic waves. When several light waves overlap at a particular location, their electric fields sum together and so do their magnetic fields. If their individual fields all point in the same direction, the waves experience constructive interference—they sum together in a mutually assisting way and the light intensity at that location is enhanced (Fig. 14.1.8*a*). But when their individual fields point in opposing directions, the waves experience destructive interference—they sum together in a canceling way and the light intensity at that location is reduced (Fig. 14.1.8*b*).

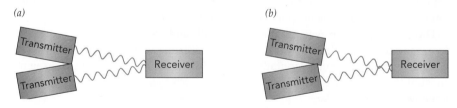

Fig. 14.1.8 (*a*) When waves from two separate paths arrive in phase at the receiver, constructive interference produces a strong effect on the receiver. (*b*) When waves arrive out of phase, destructive interference produces a weak effect on the receiver.

Both forms of interference occur when sunlight reflects from the outer skin of a soap bubble. As each wave of sunlight hits that thin film of soapy water, the film's front surface reflects about 4% of the wave and the film's back surface reflects another 4% (Fig. 14.1.9). Since both reflections travel in the same direction, the reflected light that you see reaches your eyes via two different paths, one from each reflection. If these two waves arrive in phase, that is, with their electric fields synchronized and assisting one another—you see the particularly bright reflection of constructive interference. If the two waves arrive out of phase—with their electric fields canceling one another—you see the particularly dim reflection of destructive interference.

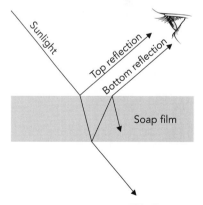

Fig. 14.1.9 Sunlight reflects from both the front and back surfaces of a soap film. The back surface reflection is delayed relative to the front surface reflection because it must pass twice through the soap film itself. If the two reflected waves arrive in phase, you see a bright reflection. If they are out of phase, you see a dim reflection.

Whether you see constructive or destructive interference depends on the wavelength of the sunlight. The back surface reflection has to travel twice through the soap film, so it's delayed relative to the front surface reflection. If the delay is just long enough for the wave to complete an integral number of cycles, then the two reflected waves are in phase with one another as they head toward your eye and you see a bright reflection. If the delay allows the back surface reflection to complete an extra half cycle, then the two reflected waves are out of phase with one another and you see a dim reflection.

Sunlight contains many different wavelengths of light, and these wavelengths behave differently during the reflection process. You see a colored reflection, consisting mainly of those wavelengths of light that experience constructive interference. Because the delay experienced by the back surface reflection depends on the thickness of the soap film, you can actually determine the film's thickness by studying its colors.

Sunlight and Polarizing Sunglasses

All sunglasses absorb some of the sunlight passing through them, but the best ones absorb horizontally polarized light much more strongly than vertically polarized light. These polarizing sunglasses dramatically reduce glare by eliminating most of the light reflected from horizontal surfaces.

When light strikes a transparent surface at right angles, about 4% of that light is reflected, regardless of its polarization. But when light strikes a horizontal surface at a shallow angle, horizontally polarized light reflects much more strongly than vertically polarized light. That's because horizontally polarized light's horizontal electric field pushes electric charges back and forth along the surface. Charges shift relatively easily in that direction and the surface becomes even easier to polarize as the angle becomes shallower. As a result, the surface reflects more horizontally polarized light at shallow angles than it does at steeper angles.

But vertically polarized light's vertical electric field pushes electric charges up and down vertically. At shallow angles, this field acts to lift the charges in and out of a horizontal surface. Since the charges can't leave the surface, the surface becomes harder to polarize as the angle becomes shallower. A horizontal surface doesn't reflect much vertically polarized light, and there is even a special angle, Brewster's angle, at which no vertically polarized light reflects at all.

Direct sunlight is an even mixture of vertically and horizontally polarized waves. Since horizontally polarized waves reflect most strongly from horizontal surfaces, sunglasses that absorb horizontally polarized light will prevent you from seeing most of this reflected light (Fig. 14.1.10). That's why polarizing sunglasses are so effective at reducing glare.

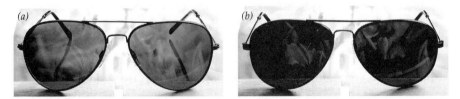

Fig. 14.1.10 (*a*) Ordinary sunglasses simply darken the scene. (*b*) Polarizing sunglasses block horizontally polarized light, the main component of glare.

Rayleigh-scattered light from the blue sky, however, tends to be mostly horizontally polarized light, which is absorbed by polarizing sunglasses. That's why the blue sky looks relatively dark when viewed through polarizing sunglasses.

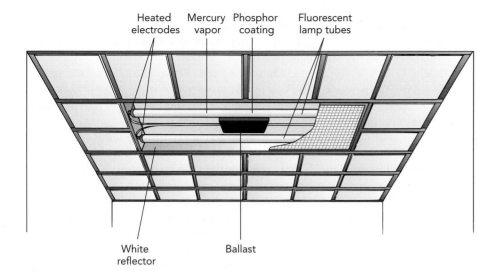

Heated electrodes   Mercury vapor   Phosphor coating   Fluorescent lamp tubes

White reflector   Ballast

SECTION 14.2 Discharge Lamps

Energy efficiency is crucial to modern lighting. While incandescent lamps provide pleasant, warm illumination, most of the power they consume is wasted as invisible infrared light. Fluorescent and other gas discharge lamps produce far more visible light with the same amount of electric power and now dominate office, industrial, and street lighting. In this section we'll explore several types of discharge lamps—fluorescent, mercury vapor, sodium vapor, and metal–halide lamps—that all share a common theme: current passing through a gas.

How We See Light and Color

Before examining discharge lamps, let's look at how our eyes recognize color. While it might seem that they actually measure the wavelengths of light, that's not the case. Instead, our retinas contain three groups of light-sensing cone cells that respond to three different ranges of wavelengths. One group of cone cells responds to light near 600 nm and lets us see red, another responds to light near 550 nm and lets us see green, and a third responds to light near 450 nm and lets us see blue (Fig. 14.2.1). These cone cells are most abundant at the center of our vision.

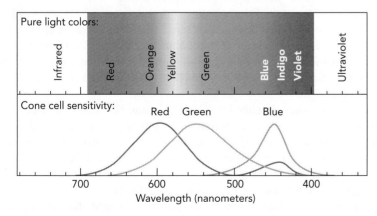

Fig. 14.2.1 The red-sensitive cells in our retina detect light near 600 nm, the green-sensitive cells near 550 nm, and the blue-sensitive cells near 450 nm. The red-sensitive cells also respond near 440 nm, so that we see violet.

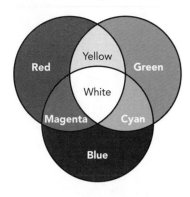

(a) Additive colors (light)

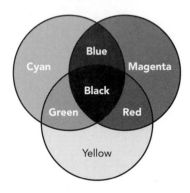

(b) Subtractive colors (pigment)

Fig. 14.2.2 (*a*) The primary colors of light or additive colors, red, green, and blue, can be combined to form any colors of light. (*b*) The primary colors of pigment or subtractive colors, cyan, magenta, and yellow, subtract red, green, and blue, respectively, from reflected or transmitted light and can be combined to form any colors of pigment.

While our retinas also contain rod cells, which are more light-sensitive than cone cells, rod cells can't distinguish color. They are most abundant in our peripheral vision and provide us with night vision.

Having only three types of color-sensing cells doesn't limit us to seeing just three colors. We perceive other colors whenever two or more types of cone cells are stimulated at once. Each type of cell reports the amount of light it detects, and our brains interpret the overall response as a particular color.

While light of a certain wavelength will stimulate all three types of cone cells simultaneously, the cells don't respond equally. If the wavelength is 680 nm, the red cone cells will respond much more strongly than the green or blue cells. Because of this strong red response, we see the light as red.

Other wavelengths of light stimulate the three types of cells somewhat more evenly. Light with a wavelength of 580 nm is in between red and green light. Both the red-sensitive and the green-sensitive cone cells respond about equally to this light and we see it as being yellow.

But we also see yellow when looking at an equal mixture of 640-nm light (red) and 525-nm light (green). The 640-nm light stimulates the red-sensitive cone cells and the 525-nm light stimulates the green-sensitive ones. Even though there is no 580-nm light entering our eyes, we see the same yellow color as before.

In fact, mixtures of red, green, and blue light can make us see virtually any color. For that reason, these three are called the primary colors of light or the primary additive colors (Fig. 14.2.2*a*). Color televisions and computer screens use tiny sources of red, green, and blue light to produce their full-color images.

While the idea of mixing primary colors also applies to paints, inks, and pigments, the palette is different. The primary colors of pigment or the primary subtractive colors are cyan, magenta, and yellow (Fig. 14.2.2*b*). When you apply one of these primary pigments to a white surface, it absorbs or subtracts one of the primary colors of light from the surface's reflection. Cyan subtracts the reflection of red, magenta subtracts the reflection of green, and yellow subtracts the reflection of blue. Color printers, photographs, magazines, and books use tiny patches of cyan, magenta, and yellow pigments to produce their full-color images.

More Light, Less Heat: Gas Discharges

When all three of our color-sensing cells respond about equally, we see white light. That's because our vision evolved under a single incandescent light source: the sun. Sunlight stimulates the red-, green-, and blue-sensitive cells in our eyes about evenly, so any other source of "white light" must do the same.

While an incandescent lightbulb makes a good attempt at producing white light, it suffers from two serious drawbacks. First, because its filament can't reach the temperature of the sun's surface (5800 °C), its blackbody spectrum contains too little blue light and appears redder than sunlight. Second, because most of its electromagnetic radiation is invisible infrared light, it doesn't use energy efficiently to produce visible light.

Fortunately, modern science has given us some alternative sources of light, sources that don't use heat and thermal radiation to produce their light. Among these sources are gas discharge lamps, which emit light when electric currents pass through their gases. While some discharge lamps are colored, others do excellent jobs of producing white light. And many of them are far more energy efficient at producing visible light than are incandescent lightbulbs.

To get an understanding of gas discharges and the lights they emit, let's start with one of the simplest examples: a neon sign tube. Though neon's rich red glow isn't suitable for lighting and isn't particularly energy efficient, it's great for signs and also relatively easy to understand.

A neon sign tube is a sealed glass tube that contains pure neon gas at a density less than 1% of that of the atmosphere outside the tube. It has metal electrodes at each end so that electric current can enter the gas through one electrode and leave through the other. Of course, gases are normally electrical insulators and neon is no exception. To transform the tube's neon into a conductor, a large voltage difference is applied between its two electrodes. As we saw in Section 10.2, a gas breaks down when exposed to large voltage gradients; its few naturally occurring ions initiate a cascade of ionizing collisions that quickly fill the gas with charged particles and render it conducting.

While ordinary air breaks down at a voltage gradient of about 30,000 V/cm, low-density neon breaks down at a much lower voltage gradient. That's because in a low-density gas, charged particles can travel farther and accumulate more kinetic energy before each collision. When about 10,000 volts is applied between the electrodes of a neon sign tube, the gas breaks down and begins to emit its familiar red glow.

The lamp is then experiencing a discharge, that is, current is flowing through the neon gas. This current consists mostly of electrons, flowing from the tube's negatively charged electrode to its positively charged electrode (Fig. 14.2.3). While these electrons collide frequently with neon atoms, they have so little mass that they usually just bounce off the atoms without losing much energy. Like a Ping-Pong ball rebounding from an elephant, the electron does most of the bouncing and then continues on its way.

Every so often, however, an electron will collide with a neon atom and something different will happen; the neon atom will rearrange internally and absorb part of the electron's kinetic energy. The electron will rebound with less energy than it had before and the neon atom will go on to emit light, probably red light. To understand why that light is probably red, however, we'll have to look at quantum physics and the structure of the neon atom.

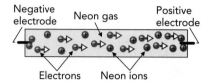

Fig. 14.2.3 A neon sign tube sends electrons through low-pressure neon gas. These electrons collide with neon atoms, transferring energy to them and causing them to emit primarily red light. Positively charged neon ions, created by particularly energetic collisions, keep the electrons from repelling one another to the walls of the tube.

Atoms, Light, and Quantum Physics

According to the wave-particle duality we encountered in Section 12.1, electrons have both particle and wave characteristics, and it is as waves that an atom's electrons are best understood. Like all objects in our universe, electrons travel as *waves* when they move from place to place, and it's only when you go looking for them that you find them as *particles* at particular locations (see ❶).

In an atom, each electron exists in a standing wave known as an orbital, with a specific energy and quantum frequency. Although these orbitals are three-dimensional standing waves, they resemble the two-dimensional standing waves of the drumhead shown in Fig. 9.2.7 on p. 309. And like the drumhead's vibrational modes, the orbitals are distinguished from one another by their patterns of nodes—surfaces along which the electron wave has no amplitude.

Like levels in a solid, orbitals in an atom can be viewed as placeholders that may or may not be occupied by electrons. From that perspective, an atom's orbitals fill with electrons from lowest energy on up. In accordance with the Pauli exclusion principle, each orbital accommodates two electrons, one spin-up and one spin-down, until all of the atom's electrons have been placed. The chemical nature of a particular atom, and its location in the periodic table of the elements (Fig. 14.2.4), is determined by how many electrons it has and how those electrons fill the available orbitals. Atoms are electrically neutral, so the number of negatively charged electrons in an atom is the same as the number of positively charged protons in its nucleus, its atomic number.

Just as there are patterns to levels, giving rise to the bands and band gaps that distinguish different solids, so there are patterns to orbitals, giving rise to the

❶ In 1927, American physicists Clinton Joseph Davisson (1881–1958) and Lester H. Germer (1896–1972) showed that electrons travel as waves by observing interference effects when electrons reflected from different atomic layers in a crystal of nickel metal. When the various electron waves arrived at a detector in phase, the detector found many electrons. When the waves arrived out of phase, the detector found few electrons. Their work was aided by a fortuitous accident in which air entered the experiment's glass vacuum tube. While carefully eliminating oxygen from their nickel sample after that leak, they managed to perfect its crystalline structure, making it possible to observe the interferences.

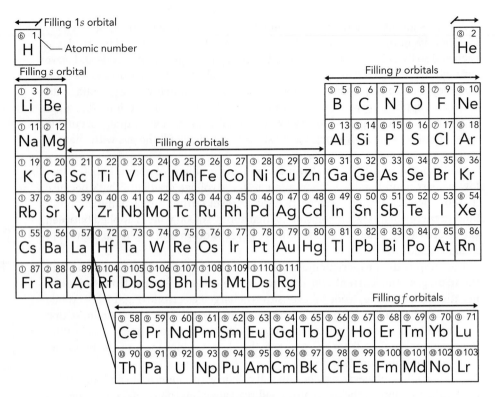

① Alkali Metals - Highly reactive, tend to donate lone *s* electron so as to empty outer shell
② Alkaline Earth Metals - Moderately reactive, tend to donate *s* electrons
③ Transition Metals - Common metals with similar properties, differ only in numbers of *d* electrons
④ Poor metals - Additional metals with differing properties
⑤ Metaloids - Intermediate between metals and non-metals
⑥ Non-metals - Semiconductors and insulators
⑦ Halogens - Highly reactive, tend to steal one *p* electron so as to complete outer shell
⑧ Noble gases - Non-reactive gases with completed outer shells
⑨ Lanthanides - Moderately reactive metals, differ only in numbers of *4f* electrons
⑩ Actinides - Moderately reactive metals, differ only in numbers of *5f* electrons

Fig. 14.2.4 The periodic table of the elements is organized by the filling of electron orbitals. Major shells of orbitals fill with electrons from left to right. The number of electrons in a neutral atom is equal to its atomic number.

shells and energy gaps that distinguish different atoms. Shells are groups of orbitals that have similar energies and tend to fill with electrons at about the same time. Although the orbital patterns and shells are complicated by the fact that charged electrons influence one another and distort one another's standing waves, many atomic properties are determined simply by which orbitals are occupied by an atom's electrons.

The atomic orbitals are identified primarily by an integer and a letter, both of which relate to their node patterns. The integer (1, 2, 3, …) is one more than the number of node surfaces in the orbital (0, 1, 2, …) while the letter (*s, p, d, f, g, h,* …) indicates how many of those node surfaces pass through the atom's center (0, 1, 2, 3, 4, 5, …). Although *s* orbitals appear one at a time, *p* orbitals appear in groups of three, *d* orbitals in groups of five, *f* orbitals in groups of seven, etc.

A neon atom has 10 electrons, so it takes 5 orbitals to accommodate them. The first two electrons go into the 1*s* orbital, the fundamental mode with zero nodes. Filling the 1*s* orbital completes the first major shell. The next two electrons go into the 2*s* orbital, which has one node surface that doesn't pass through the atom's center. Finally, neon's last six electrons go into the three 2*p* orbitals, each of which has one node surface passing through the atom's center. Since filling the 2s and 2p orbitals completes the second major shell, the neon atom (Ne in Fig. 14.2.4) is chemically inert. That is why neon exists as a gas of individual atoms!

An arrangement of occupied orbitals is called a state and the state we have just described is neon's ground state—the state with the lowest possible total energy. The neon atom has other states available, but they all involve additional energy. That's where the discharge enters the picture. When a charged particle collides with a ground state neon atom, there's a chance that the collision will knock an electron out of its usual orbital into one of the empty orbitals. The neon atom will then be in an excited state—it will have extra energy.

Suppose, for example, that a collision has just shifted one of the neon atom's electrons from a 2p orbital to a 3p orbital. The atom will quickly undergo a series of state-to-state transitions that eventually return it to its ground state. With each transition, the atom will drop to a lower energy state and emit a photon. The photon carries away the energy released in transitioning from the higher energy state to the lower energy state. In all likelihood, one of those transitions will shift an electron from a 3p orbital to the 3s orbital and produce a photon of red light, the red of a neon sign!

As long as the electron remains in the 3p orbital—a particular standing wave—it can't emit an electromagnetic wave. That standing wave has a quantum oscillation; however, the oscillation is internal to the electron and neither the electron nor its charge has any overall motion. Since charge must accelerate to emit electromagnetic waves (see Chapter 13), without such overall motion there can be no emission of electromagnetic waves.

But when the 3p electron begins its transition to the empty 3s orbital, its quantum wave begins to change with time. The wave moves rhythmically back and forth during the transition and its charge accelerates with it. The atom begins to emit an electromagnetic wave. By the time the transition is complete, the atom has emitted a single *particle* of light, a photon of red light. Like an electron, light travels as a *wave* but behaves as a particle when you try to locate it. While it's being emitted or absorbed by an atom, light exhibits its particle nature.

We've seen that hot objects emit light, a process called incandescence, but here the neon atom emits light without heat, a process called luminescence. That luminescence is the result of a radiative transition—a transition between states in which a photon of light is emitted or absorbed. In this case, the radiative transition emits a photon that carries away the energy released when the excited neon atom's 3p electron shifts into the empty 3s orbital. The difference in energy between those two states determines the photon energy, which in turn determines the frequency and color of the photon's light wave.

According to quantum physics, a photon's frequency is proportional to its energy. Specifically, the photon's energy is equal to the frequency of its electromagnetic wave times a fundamental constant of nature known as the Planck constant. This relationship between energy and frequency can be written as:

$$\text{energy} = \text{Planck constant} \cdot \text{frequency}. \tag{14.2.1}$$

First used by German physicist Max Planck (1858–1947) in 1900 to explain the light spectrum of a hot object, the Planck constant has a measured value of 6.626×10^{-34} J·s. And while we have just encountered the Planck constant and Eq. 14.2.1 in the context of light waves, it actually applies to all quantum waves. For example, the quantum frequency of an electron is related to its energy by Eq. 14.2.1.

The Planck constant is so tiny that even a photon of 10^{15}-Hz ultraviolet light has an energy of only 6.626×10^{-19} J. A typical beam of light thus contains so many photons that you can't see that they're arriving as particles. However, the energy in a single ultraviolet photon is substantial on a molecular scale. It can damage a molecule in your skin, contributing to a sunburn and inducing your skin to tan as a defensive response. X-rays have even higher frequencies, and their energetic photons can cause more severe molecular damage.

A neon atom can only emit photons corresponding to energy differences between two of its states, a constraint that severely limits its light spectrum. Neglecting nuclear issues that we'll discuss in Chapter 16, all neon atoms are identical and emit the same characteristic spectrum of light. The visible part of that spectrum is dominated by the warm red glow of photons emitted when electrons shift from $3p$ to $3s$ orbitals, which is why a neon sign glows red.

Since atoms of different elements have different numbers of electrons and different states, each emits its own unique spectrum of light after being excited. Copper atoms emit a blue-green spectrum, strontium atoms a deep red, and sodium a bright yellow-orange. Chemists, astronomers, and manufacturers rely on those emission spectra for information and applications. And, as we shall soon see, so do scientists and engineers working on illumination.

Fluorescent Lamps

If neon tubes appeal to you for illumination, you probably march to your own drummer. Most people opt for a somewhat better simulation of sunlight in their discharge lamps. As an energy-efficient source of artificial sunlight, it's hard to beat fluorescent lamps.

At the heart of a fluorescent lamp is a narrow glass tube filled with argon, neon, and/or krypton gases at about 0.3% of atmospheric density and pressure. The tube also contains a few drops of liquid mercury metal, some of which evaporates to form mercury vapor. About one in every thousand gas atoms inside the tube is a mercury atom, and it's these mercury atoms that are responsible for the light.

Like a neon sign tube, a fluorescent lamp uses a discharge in its gas to produce light. While fluorescent lamps occasionally use high voltages to initiate their discharges, most operate their discharges at household voltages and must rely on alternative techniques to render their gases conducting. These low-voltage lamps normally heat their electrodes so that thermal energy ejects electrons from their surfaces into the gas. But regardless of how the discharge is started, the result is current flowing through the gas and the emission of light.

But the fluorescent lamp has a problem. While its mercury atoms emit most of the light in its discharge, that light is almost entirely ultraviolet. The final radiative transition that returns each mercury atom to its ground state ($6p \rightarrow 6s$) releases a large amount of energy and produces a photon with a wavelength of 254 nm. This light can't go through the glass walls of the tube, and you couldn't see it if it did. So the fluorescent lamp converts it into visible light with the help of phosphor powder on the inside of the glass tube.

Phosphors are solids that luminesce—they emit light—when something transfers energy to them. Their behavior is similar to that of an atom: an energy transfer shifts the phosphor from its ground state to an excited state and it then undergoes a series of transitions that return it to its ground state. Some of those transitions are radiative ones and emit light.

In a fluorescent lamp, the phosphor is excited by ultraviolet light. This excitation or energy transfer is actually a radiative transition, but one in which the photon is absorbed by the phosphor; one of its electrons makes a transition from a lower energy level to a higher energy level. During this absorption, the light's electric field pushes the electron's wave back and forth rhythmically until it shifts to the new level. The photon disappears and the phosphor receives its energy.

Once the phosphor is in an excited state, its electrons begin to make transitions back to their ground state levels. Most of those transitions radiate visible light, the light that you see when you look at the lamp. However, some of the transitions radiate invisible infrared light or cause useless vibrations in the phosphor itself. Despite this wasted energy, phosphors are relatively efficient at turning ultraviolet light into visible light, a process called fluorescence.

The phosphors in a fluorescent lamp are carefully selected and blended to fluoresce over a broad range of visible wavelengths. While this light doesn't have the same spectrum as sunlight, it appears white because it stimulates the red-, green-, and blue-sensitive cells in our eyes about equally. The blending of phosphors is necessary because, like atoms, each phosphor fluoresces with a characteristic spectrum of light that's determined by the energy differences between its levels. Several different phosphors are needed to create the right balance between red, green, and blue lights. While some advertising and novelty lamps use brightly colored, unblended phosphors, fluorescent lighting tubes come in six standard color blends: cool white, deluxe cool white, warm white, deluxe warm white, white, and daylight.

When they were first introduced, fluorescent tubes used the daylight phosphor. This phosphor emits too much blue light and makes everything look cold and medicinal. Phosphors have improved over the years so that the most common phosphors today, cool white and warm white, look much more pleasant. Light from the "cool" phosphors resembles daylight, while that from the "warm" phosphors resembles incandescent lighting. Household fluorescent lamps often use deluxe cool white and deluxe warm white, which are even more pleasant versions of white but which are slightly less energy efficient.

A Few Practical Issues

A fluorescent tube needs a substantial electric field inside it to keep its electrons moving forward. Since this electric field is proportional to the voltage drop through the tube, longer tubes need higher voltages. Power line voltages (110 V to 240 V) are appropriate for tubes up to 3 m (10 ft) in length, but the longer and often colored fluorescent tubes used in artwork or advertising require much higher voltages.

To keep electrons from pushing one another into its walls, a fluorescent tube must also contain positively charged mercury ions. The discharge naturally produces these ions during particularly energetic collisions. The result, a gaslike mixture of positively charged ions and negatively charged electrons, is called a plasma. A plasma is distinct from a gas because its charged particles exert forces on one another over considerable distances. All operating discharge lamps contain plasmas, including the neon sign that we discussed earlier.

Since a typical fluorescent lamp must heat its electrodes red-hot in order to form its plasma, it runs current through filaments at each end of its tubes (Fig. 14.2.5). Once the discharge is operating, the heating current can usually be turned off because the electrodes will be kept hot by electrons that hit them from the discharge. However, some fluorescent fixtures can be dimmed, in which case electron heating alone won't sustain the plasma. Dimmable fluorescent fixtures must continue to pass heating currents through their filament/electrodes.

Unfortunately, the filament/electrodes are fragile. They're damaged by a process called sputtering, in which positive mercury ions from the plasma collide with the filament/electrodes and chip away their tungsten atoms. Because sputtering is particularly severe during start-up, a typical filament/electrode breaks after a few thousand starts. That's why you shouldn't turn a fluorescent lamp on and off more than once every few minutes.

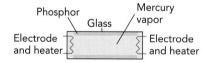

Fig. 14.2.5 In a hot-electrode fluorescent tube, the electrodes are actually filaments and are heated by running currents through them.

Mercury, Metal–Halide, and Sodium Lamps

While a low-pressure mercury discharge emits mostly ultraviolet light, a high-pressure mercury discharge emits more visible light than ultraviolet light. This change occurs because ultraviolet light becomes trapped in densely packed mercury atoms and only visible light is able to escape from the discharge.

Known as radiation trapping, this effect occurs because mercury atoms absorb 254-nm photons just as well as they emit them. The same radiative transition that causes a mercury atom to emit a 254-nm photon ($6p \rightarrow 6s$) can also run backward to absorb that photon ($6s \rightarrow 6p$). In a dense gas of mercury, whenever one mercury atom emits a 254-nm photon, another mercury atom snaps that photon up. So while the discharge keeps pouring energy into the mercury atoms, they can't get rid of it as 254-nm photons. Instead, they emit most of their energy through radiative transitions between other excited states. Since this light is much less likely to be captured by other mercury atoms, it emerges from the lamp as bluish visible light.

When you first turn on a high-pressure mercury lamp, most of the mercury is liquid and the pressure is low. The lamp starts like a small fluorescent tube without phosphor, so you see very little visible light. But the tube is designed to heat up during operation so that the liquid mercury evaporates to form a dense gas. As the gas pressure rises, the tube's color changes until it emits brilliant, blue-white light.

To make a lamp that's a little less bluish, some high-pressure mercury lamps contain additional metal atoms. These atoms are introduced into the lamps as metal–iodide compounds, making them metal–halide lamps. Sodium, thallium, indium, and scandium iodides all contribute their own emission spectra to the outgoing light and help to strengthen the red end of its spectrum. They give metal–halide lamps a warmer color than pure mercury lamps.

Pure sodium lamps resemble mercury lamps, except that they use sodium atoms. Sodium is a solid at room temperature, so both low- and high-pressure sodium lamps must heat up before they begin to operate properly. A low-pressure sodium

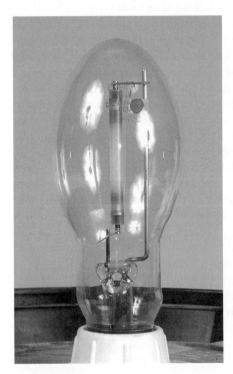

Fig. 14.2.6 The active component of a high-pressure sodium vapor lamp is a small translucent tube. As the lamp warms up, sodium metal in the tube evaporates to form a brilliant yellow discharge. The dense vapor of sodium atoms in the tube traps the 590-nm light so that the lamp emits a richer spectrum of wavelengths and a less monochromatic glow than a low-pressure sodium lamp.

lamp is extremely energy efficient because its 590-nm light comes directly from a sodium atom's strongest radiative transition. That transition is from sodium's lowest excited state to its ground state ($3p \rightarrow 3s$). Many highways are illuminated by the yellow-orange glow of low-pressure sodium lamps.

But this monochromatic illumination is unpleasant and permits no color vision at all. While it may be acceptable on a highway, you wouldn't want it near your home. That's why people buy high-pressure sodium lamps for home use (Fig. 14.2.6).

Remarkably enough, the 590-nm emission itself smears out at high pressure to cover a wide range of wavelengths, from yellow-green to orange-red. This spreading occurs because of the many collisions suffered by the densely packed sodium atoms as they try to emit 590-nm light. These collisions distort the atomic orbitals so that the photons emerge with somewhat shifted energies. Overall, a high-pressure sodium lamp emits remarkably little light exactly at 590 nm because the ground-state sodium atoms trap that light; they run the ($3p \rightarrow 3s$) transition backward and absorb the photons ($3p \rightarrow 3s$). This trapping is so effective that there is actually a hole in the lamp's spectrum right at 590 nm.

High-pressure discharge lamps suffer from a problem not found in low-pressure lamps: they're difficult to start when hot. It's much harder to initiate a discharge in a high-pressure gas than in a low-pressure gas, so they all start at low pressure and then evolve to high pressure. If the discharge in a high-pressure mercury, sodium vapor, or metal–halide lamp is interrupted and loses its plasma, the lamp must cool down before it can be restarted.

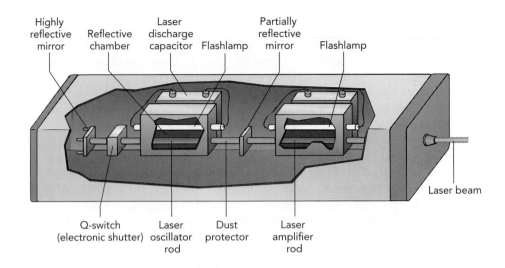

Highly reflective mirror · Reflective chamber · Laser discharge capacitor · Flashlamp · Partially reflective mirror · Flashlamp

Laser beam

Q-switch (electronic shutter) · Laser oscillator rod · Dust protector · Laser amplifier rod

Section 14.3 **Lasers and LEDs**

❶ While it might seem that a photon should have an exact wavelength and frequency, and travel in only one direction, that's not the case. Photons travel as electromagnetic waves and spread in more than a single direction. And because each photon has a beginning and an end, its wave contains more than a single wavelength or a single frequency. Thus while lasers can produce some of the most perfect electromagnetic waves imaginable, those waves still spread outward slightly and still have a range of wavelengths and frequencies.

Few devices have inspired our imaginations more than lasers. Since their invention in the late 1950s, lasers have found countless uses, from cutting metal and clearing human arteries to surveying land and playing CDs and DVDs. But lasers are more than just novel applications of old ideas. Instead, they bring together quantum and optical physics to produce a new type of light. This light is radically different from that produced by incandescent and fluorescent lamps, and its properties make it particularly useful for many applications. In this section we'll examine the nature of this new light and the ways in which lasers produce it.

Lasers and Laser Light

To understand lasers, you must understand how laser light differs from the normal light emitted by hot objects or by individual atoms in an electrical discharge. Each particle of normal light, each photon, is emitted willy-nilly without any relationship to the other light particles being emitted nearby. Because of this light's independent and unpredictable character, it's called spontaneous light and its creation is called spontaneous emission of radiation.

But theoretical work by Albert Einstein and others in the 1920s and 1930s predicted the existence of a second type of light, stimulated light, that can be created when an excited atom or atomlike system duplicates a passing photon. While this stimulated emission of radiation can occur only when the excited atom is capable of emitting the duplicate photon spontaneously, the copy that it produces is so perfect that the two photons are absolutely indistinguishable. Together, these two photons form a single electromagnetic wave.

To get a slightly better picture for how such stimulation occurs, think about an isolated atom in an excited state. That atom will eventually return to its ground state, but it must emit one or more photons for this to happen. That atom waits around in the excited state until it begins a spontaneous radiative transition. During the transition, one of the atom's electrons accelerates back and forth and the atom emits a photon.

But if a similar photon passes through the atom as it's waiting in the excited state, that photon's electric field can stimulate the radiative transition process through sympathetic vibration. The field pushes and pulls on the atom's electrons and makes them accelerate back and forth. While this effect is small, it may be enough to trigger the emission of light. If the atom does emit light, the photon it will produce will be a perfect copy of the stimulating photon.

When this stimulated emission process was first discovered, people immediately recognized that it made light amplification possible. If enough excited systems could be assembled together, a single passing photon could be duplicated exactly over and over again. Instead of a single particle of light, you would soon have thousands, or millions, or even trillions of identical light particles.

Implementing this idea had to wait until the late 1950s, however, when the technical details for how to actually achieve light amplification were worked out. In 1960, the first laser oscillators were constructed. These were devices that emitted intense beams of light, in which each particle of light was identical to every other particle of light. A single particle of light had been duplicated by the stimulation process into countless copies.

When individual excited atoms or atomlike systems emit light through spontaneous emission, the particles of light head off separately as many independent electromagnetic waves (Fig. 14.3.1a). Light consisting of many independent electromagnetic waves is called incoherent light.

But when that same collection of excited atoms or atomlike systems emit light by stimulated emission, all of the light particles are *absolutely* identical and form a single electromagnetic wave (Fig. 14.3.1b). Unlike electrons, which are Fermi particles, many identical photons can have the same quantum wave because photons are Bose particles and Bose particles don't obey the Pauli exclusion principle. Light consisting of many identical photons and a single electromagnetic wave is called coherent light. Because of its single wave nature, coherent light exhibits remarkable interference effects. These effects are easily seen in the coherent light emitted by lasers.

Light Amplifiers and Oscillation

Producing coherent light requires amplification. You must start with only one particle of light and duplicate it many times. The basic tool for this duplication of light is a laser amplifier (Fig. 14.3.2). When weak light enters an appropriate collection of excited atoms or atomlike systems—the laser medium—that light is amplified and becomes brighter. The new light has exactly the same characteristics as the original light, but it contains more photons.

When we think of lasers, however, we rarely imagine a device that duplicates photons from somewhere else. We usually picture one that creates light entirely on its own. To do that, the laser must produce the initial particle of light that it then duplicates to produce others. A laser oscillator is a device that uses the laser medium itself to provide the seed photon, which it then duplicates many times (Fig. 14.3.3). If a laser medium is enclosed in a pair of carefully designed mirrors, it's possible for the stimulation process to become self-initiating and self-sustaining. However, the mirrors must be curved properly and must have the correct reflectivities. One mirror must normally be extremely reflective, while the other must transmit a small fraction of the light that strikes its surface.

When the laser medium is placed between the two mirrors, there is a chance that a photon, emitted spontaneously by one of the excited systems, will bounce off a mirror and return toward the laser medium. As that returning photon passes through the laser medium, it's amplified. Because the photon was emitted by one of the excited systems, it has the right wavelength to be amplified by other excited systems. (For a discussion of a photon's properties, see ❶.)

By the time the original photon leaves the laser medium, it has already been duplicated many times. This group of identical photons then bounces off the second mirror and returns for another pass through the laser medium. It continues to bounce back and forth between the mirrors until the number of identical photons in the collection is astronomical.

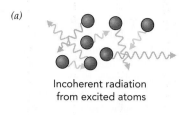

Incoherent radiation
from excited atoms

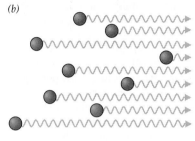

Coherent radiation
from excited atoms

Fig. 14.3.1 (*a*) Photons of incoherent light are created independently and have somewhat different wavelengths and directions of travel. (*b*) Photons of coherent light are produced by stimulated emission and are identical to one another in every way.

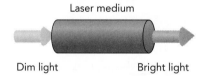

Laser medium

Dim light   Bright light

Fig. 14.3.2 A laser amplifier uses excited atoms or atomlike systems to increase the number of light particles leaving the laser medium. The incoming light is duplicated by stimulated emission.

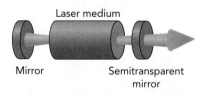

Laser medium

Mirror   Semitransparent mirror

Fig. 14.3.3 A laser oscillator is a laser amplifier enclosed in mirrors. Oscillation occurs when the laser medium spontaneously emits one photon in just the right direction. This photon bounces back and forth between the two mirrors and is duplicated many times. Some of the light is extracted from this laser by making one of the mirrors semitransparent.

❷ Semitransparent mirrors are often used for surveillance because they appear like normal mirrors in certain circumstances. As long as the subjects being watched are in bright illumination and their observers are in dim lighting, the subjects will see mostly their reflections in the semitransparent mirror and will have great difficulty seeing the observers. If the observers are brightly lit, however, the subjects will see them clearly. Television and movies frequently portray semitransparent mirrors as being truly one-way, which they are most definitely not. The observers must be careful not to illuminate themselves or they'll become visible to their subjects. The subjects can also make their observers visible by shining bright lights through the semitransparent mirror. If they put a flash camera right up against the mirror, they can take a picture right through it!

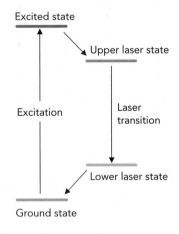

Fig. 14.3.4 An ideal laser system passes through four different states during the laser's operation.

Eventually there are so many identical photons that the laser medium is no longer able to amplify them. The laser medium has only so much stored energy and only so many excited systems in it. If the laser medium continues to receive additional energy, it may continue to amplify the light somewhat. But if it doesn't receive more energy, light amplification will eventually cease.

To let the light out of this laser oscillator, one of its mirrors is normally semitransparent, that is, some of the photons that strike the surface of the mirror travel through it rather than reflecting. The two-way glass used for surveillance is actually a semitransparent mirror ❷. This transmission creates a beam of outgoing light, a laser beam. The laser beam continues to emerge from the mirror as long as the amplification process can support it. Because this laser beam consists of duplicates of one original photon, it's coherent light. For technical reasons, many lasers duplicate more than one original photon simultaneously, so that their laser beams are a little less coherent than they might be. However, with suitable fine-tuning, one original photon can usually be made to dominate the laser beam.

When you focus a flashlight's beam with a lens, its independent photons won't end up exactly together at the focus of the lens. That's because the photons leave the flashlight heading in somewhat different directions and because their broad range of wavelengths leads to dispersion problems in the lens. But since virtually all of the photons in a laser beam are identical, they can all focus together to an extremely small spot. That's why a laser printer employs a laser; a laser beam can illuminate a very small spot on the photoconductor drum that is used in the xerographic process to produce a printed image.

How a Laser Medium Works

Obtaining the excited systems needed to amplify light is a critical issue for lasers. Ideally, a laser involves four different states of an atom or atomlike system: the ground state, an excited state, the upper laser state, and the lower laser state. The reason for having four separate states should become clear in a moment.

Let's consider an atom that acts as an ideal laser amplifier (Fig. 14.3.4). The atom starts in its ground state. A collision or the absorption of a photon shifts it to the excited state, giving it the energy it needs to amplify light. The atom then shifts to the upper laser state, either by emitting a photon or as the result of a collision. This preliminary shift is important because it prevents the excited atom from returning directly back to the ground state and avoiding the amplification process. Once it has shifted to the upper laser state, the atom is stuck there and will wait around long enough to amplify light.

When a suitable photon passes through the atom, that photon stimulates the emission of a duplicate photon and the atom undergoes a radiative transition to the lower laser state. So far, so good. However, if the atom remains in the lower laser state, it might absorb a photon of the laser light and return to the upper laser state. To prevent this sort of radiation trapping, the atom must quickly shift to the ground state, either by emitting a photon or as the result of another collision. The atom is then ready to begin the cycle all over again.

This four-state cycle, or something close to it, is found in nearly all lasers. The cycle helps the laser develop a population inversion between its upper and lower laser states—a situation in which there are more atoms in the upper laser state prepared to *emit* the laser light than there are atoms in the lower laser state prepared to *absorb* that light. Developing a population inversion is critical to laser amplification because without it, the laser medium is more absorbing than amplifying and there can't be a buildup of light intensity.

In each laser, something provides the energy needed to shift atoms or atomlike systems in the laser medium from their ground states to their excited states in order

Fig. 14.3.5 This flashlamp-pumped laser amplifier contains a purple neodymium:YAG rod. The rod is in the bottom half of the opened, gold-lined amplifier box and is protected by a glass tube. Light from a long flashlamp in the top half of the box excites the neodymium ions so that they can amplify infrared light passing horizontally through the rod.

to develop a population inversion. This transfer of energy into the laser medium to prepare it for amplifying light is called pumping. How a particular laser medium is pumped depends on the laser.

The most common pumping mechanisms are electronic and optical. In electronic pumping, currents of charged particles use their kinetic or electrostatic energies to excite the medium's atoms or atomlike systems from their ground states to their excited states. In optical pumping, intense light is shone on the laser medium, causing a similar excitation.

The most important examples of optical pumping are ion-doped solid-state lasers. These lasers are based on atomic ions embedded in transparent solids. Common ions are titanium (Ti), neodymium (Nd), and erbium (Er), and they are often embedded in sapphire, yttrium aluminum garnet (YAG), or glass. Ti:sapphire, Nd:YAG, and Er:glass lasers are important to modern research, technology, and optical communications systems. When these laser media are exposed to extremely bright light, their ions become excited, and they can act as laser oscillators or amplifiers (Figs. 14.3.5 and 14.3.6).

Variable Wavelength and High Intensity Lasers

Some laser media have many upper and lower laser states and can emit stimulated light over a range of wavelengths. A variable wavelength laser can be made by introducing a color filter into a laser based on such a medium. The color filter selects which wavelength of light can reflect back and forth between the mirrors, so that the laser emits only light of that wavelength (Fig. 14.3.7).

If the color filter is adjustable, so that you can choose which wavelength of light it allows through, then the wavelength of the laser's light will change as you adjust the color filter. These adjustable-wavelength lasers are frequently used in research and medicine, where specific wavelengths of light can be used to control chemical or physical changes in atoms, molecules, or materials.

A high intensity laser is one that temporarily prevents any seed photons from entering the laser medium. With no seed photons to amplify, the atoms or atomlike systems in the laser medium will remain in their upper laser states for a long time. This ability to store energy and wait makes it possible for a laser to emit extraordinarily intense bursts of light.

In addition to the two mirrors around the laser medium, these lasers include a shutter (Fig. 14.3.8). The shutter blocks light that is attempting to reflect from one

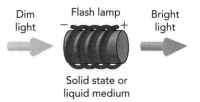

Fig. 14.3.6 In an optically pumped laser, intense light from a flashlamp, an arc lamp, or even another laser transfers energy to the laser medium. Atoms or atomlike systems inside the medium store this energy and use it to amplify light.

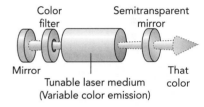

Fig. 14.3.7 Some laser media can amplify many different wavelengths of light. Inserting a color filter into the region between the mirrors makes the laser oscillator emit only a particular wavelength of light.

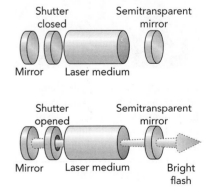

Fig. 12.3.8 Inserting a shutter into the region between the mirrors allows a laser oscillator to store energy in the laser medium. When the shutter is opened, a giant laser pulse is created as all of the stored energy in the medium is quickly converted into light.

of the two mirrors. Because light can't pass back and forth through the laser medium, the laser won't emit light while the shutter is closed and energy can be accumulated gradually by the laser medium. When the shutter opens, light starts to bounce back and forth between both mirrors and is amplified over and over again. The laser emits a brilliant burst of light. The shutter is usually an electrooptic device that uses the electric properties of matter to control the passage of light.

In some electrically controlled lasers, the pulse of light that's produced when the shutter opens is only a few billionths of a second long. But the light emitted during that brief period is brighter than all the lights in a medium-sized city. Even more sophisticated short-pulse lasers can create bursts of light that are only a few quadrillionths of a second long (10^{-15} s) and that are brighter than all of the other lights on earth.

Light-Emitting Diodes and Laser Diodes

The most common electronically pumped lasers are diode lasers. Found in pointers, barcode readers, printers, and CD and DVD players, these lasers are closely related to light-emitting diodes, which themselves resemble the ordinary diodes we discussed in Section 12.1. But whereas ordinary diodes are designed merely to control currents while wasting as little power as possible, laser and light-emitting diodes are optimized to produce light.

When a diode is forward biased and current flows from its anode to its cathode, conduction level electrons in the n-type cathode travel across the p-n junction to become conduction level electrons in the p-type anode. In effect, the anode is then in an excited state: it has conduction level electrons and empty valence levels.

What happens next depends on the characteristics of the diode. In a normal silicon diode, those conduction level electrons shift to empty valence levels without producing significant light. Silicon's band structure has characteristics that discourage light emission, so most of these electron transitions produce internal vibrations and heat the diode instead of producing light.

But in specialized diodes made from more exotic semiconductors, conduction level electrons in the p-type anode frequently undergo radiative transitions to empty valence levels and thereby emit light. Composed primarily from combinations of gallium, indium, aluminum, arsenic, phosphorus, and nitrogen, they are known as light-emitting diodes or LEDs. LEDs now come in just about any color of the rainbow, including infrared, red, orange, yellow, green, blue, violet, and ultraviolet (Fig. 14.3.9). Although white LEDs also exist, they're actually violet or ultraviolet LEDs with built-in phosphors that fluoresce white.

An LED's color is directly related to the energy released when an electron in its p-type anode shifts from a conduction level to a valence level. The most convenient unit in which to measure that energy is the electron volt (abbreviated eV)—the energy released when 1 elementary unit of electric charge experiences a 1 V decrease in voltage (1 eV is equal to 1.6021×10^{-19} J). In a typical red LED, an electron releases 1.9 eV as it shifts from a conduction level to a valence level and can produce a photon with an energy of 1.9 eV. Since energy and frequency are related by Eq. 14.2.1 and frequency and wavelength are related by Eq. 9.2.1, that 1.9-eV photon has a frequency of 4.6×10^{14} Hz and a wavelength of 650 nm.

To operate and produce these 1.9-eV photons, the red LED must be forward biased with a voltage drop of at least 1.9 V. The current-carrying diode uses that voltage drop to inject electrons into the anode's conduction band, where they have energies 1.9 eV above the valence band. Many of those electrons subsequently release their excess energies as 1.9-eV photons of light.

The shorter the wavelength of the light an LED emits, the more energy each electron must release as it shifts from a conduction level to a valence level and the

Fig. 14.3.9 These LEDs are connected in series, so that the same current flows sequentially through each of them. However, their different band gaps cause them to emit different colors of light. From left to right, they are red, orange, green, and blue.

larger the semiconductor's band gap must be. A violet LED that emits 400-nm light requires a band gap of about 3.1 eV to produce its 3.1-eV photons. That LED also needs a forward-bias voltage in excess of 3.1 V. The larger voltage drops required by LEDs near the violet end of the spectrum explain why those LEDs need higher voltage power sources.

Unfortunately, less than a quarter of the electrons sent across an LED's p-n junction succeed in lighting the room. Although a substantial fraction of those electrons emit photons, most of the photons are reabsorbed by the semiconductor before they leave the LED; the same radiative transitions that emit this light (conduction level → valence level) can also absorb it (valence level → conduction level). But despite those difficulties, modern LEDs can produce visible light with energy efficiencies comparable to those of fluorescent lamps. LED efficiencies continue to rise along with their operating lifetimes and it's only a matter of time before they become a primary form of illumination.

A laser diode is quite similar to an LED, except that a laser diode uses its radiative transitions to amplify light. Since that amplification can only occur when light emission exceeds light absorption, the laser diode must produce a population inversion between an upper laser state and a lower laser state.

The laser diode achieves such an inversion by concentrating current into a very narrow p-n junction made from heavily doped semiconductors. The intense current injects an enormous density of electrons into the anode's conduction band, where they quickly settle into the lowest energy conduction levels—the upper laser state. The heavy doping empties most of the anode's highest energy valence levels—the lower laser state. With many electrons in the upper laser state and few in the lower laser state, the diode has a population inversion and can amplify light.

Most laser diodes act as laser oscillators (Fig. 14.3.10), amplifying their own spontaneously emitted light until it forms an intense coherent beam. The ends of the anode itself are usually reflective enough to act as mirrors and form a complete laser oscillator. However, to concentrate the laser light in one direction and to control its beam characteristics, many laser diodes have complicated structures and coatings.

As in an LED, a laser diode's wavelength and color depend primarily on the band gap of its anode. Infrared and red laser diodes were developed first and quickly incorporated into a variety of household products. Developing laser diodes with larger band gaps and shorter wavelengths has been a long and arduous process. However, the spectrum of laser diode colors now extends into the ultraviolet and as these short-wavelength lasers become less expensive and more reliable, they begin to appear in everyday products.

Fig. 14.3.10 This tiny semiconductor chip is a diode laser that emits an intense beam of coherent light when a current flows through it.

Section 14.4 **Paint and Makeup**

Many of our perceptions of the world come to us through our eyes. When we look at objects, we see that they are shiny or dull, bright or dark, colored or gray. However, what we're often seeing is actually paint. Paints are used everywhere to change the appearances of things—in architecture, art, consumer products, packaging, communications, food, and even on people. Paints also protect objects and modify their surfaces. Like paint, makeup alters the appearances of things, but this time those things are people.

Formulating Paint

Paints are more than just colors we apply to surfaces. They are sophisticated coatings serving a variety of purposes. In addition to coloration, paints protect surfaces, change their shapes, and alter their physical properties. I'll begin by considering paints as coatings and turn later on to the issue of color.

Most paints include at least four groups of components: binders, volatile substances, pigments and dyes, and additives. Binders give paints their structures. They create the continuous coatings that stick to surfaces and remain in place indefinitely. In almost all cases, these coatings are polymers, the systems of giant string-like organic molecules that we'll discussed in Section 17.3. These molecules interlock with one another like noodles in a bowl of cold spaghetti to form rigid materials. Depending on the paint, these long molecules may already exist in the paint before application or they may form as the paint "dries."

Volatile substances are chemicals that keep the paints fluid enough to be applied easily and that subsequently evaporate during drying. In some cases, these volatile substances actually dissolve the binders and other chemicals so that the paint is nearly a uniform liquid. But other, more modern paints actually contain tiny solid binder particles that are suspended in the volatile substances. These particles become a continuous coating as the volatile substances evaporate.

Pigments are finely divided, insoluble powders that give a coating its color, opacity, and other optical properties. These powder particles are normally suspended in the paint when it's fluid and become trapped in the polymer molecules as the paint dries. There are some coatings that don't contain pigments, such as clear varnishes. There are also coatings in which the pigments do more than simply provide color, such as magnetic recording media. There are even pigments in which capsules break and release scents and essential oils when you scratch the paint.

Dyes are a supplement to pigments, adding color to a paint without introducing any opacity. Except in transparent stains, vanishes, and inks, dyes work in conjunction with pigments. Paints that use colored pigments and paints that mix white pigments with dyes are nearly indistinguishable and I'll discuss them interchangeably most of the time.

Additives help these three components work together to form a uniform coating. Some additives aid in the drying process by initiating the formation of polymer molecules out of much smaller molecules. Others help keep the binder or pigment particles from clumping together before the paint is used or help prevent freezing. Still others alter paint's viscosity to make it easy to apply.

Artist's Oil Paints

Three interesting examples of paint are artist's oil paints, oil-based house paints, and latex house paints. Artist's oil paints are principally pigment particles suspended in drying oil. The drying oil forms the binder and consists of moderately large

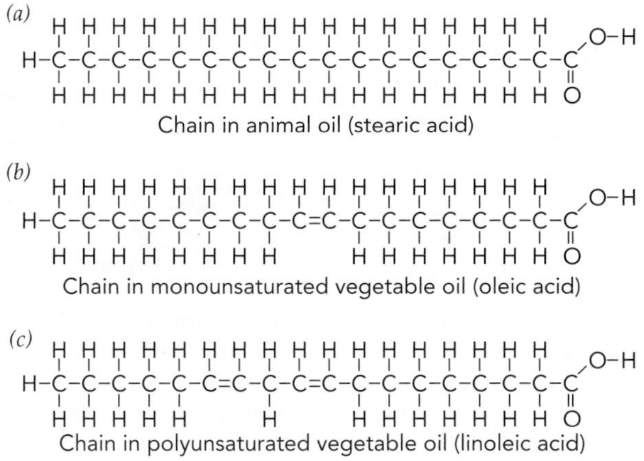

(a)
Chain in animal oil (stearic acid)

(b)
Chain in monounsaturated vegetable oil (oleic acid)

(c)
Chain in polyunsaturated vegetable oil (linoleic acid)

Fig. 14.4.1 Oils contain chains of carbon atoms (C) decorated with hydrogen atoms (H). (*a*) Saturated oils contain only single bonds between carbon atoms, while (*b*) monounsaturated oils have one double bond per chain. (*c*) Polyunsaturated oils have two or more double bonds per chain. These drawings are simplified because the chains aren't really straight or two-dimensional.

molecules that join together into long chains and tangled networks when exposed to oxygen for long periods of time.

Like all natural oil molecules, a drying oil contains chains of carbon atoms. These carbon atoms attach to one another by sharing their outermost or valence electrons. Such covalent bonds normally involve only a single valence electron from each carbon atom and are called single bonds. Since a carbon atom has four valence electrons, it can form single bonds with four atoms.

Many oil molecules, particularly those in animal oils, contain chains of carbon atoms connected exclusively by single bonds (Fig. 14.4.1*a*). Each carbon atom that's not at the end of a chain is attached to two other carbon atoms and two hydrogen atoms. These oil molecules are saturated because they contain as many hydrogen atoms as the carbon atoms can hold. Saturated oils are liquid only at relatively warm temperatures, so they're most common in warm-blooded animals and tropical plants. Saturated oils don't harden in air and aren't drying oils.

An unsaturated oil has at least one pair of carbon atoms that share two valence electrons from each atom. In that case, the carbon atoms are joined by a double bond. This double bond is slightly stronger than a single bond and it draws the two carbon atoms slightly closer together. But its most obvious effect is to decrease the number of hydrogen atoms in the molecule. With only two remaining valence electrons, each carbon atom involved in a double bond can only attach to two other atoms. Usually one is another carbon atom and the other is a hydrogen atom.

Since an oil molecule with double bonds has fewer hydrogen atoms than it would have if it had only single bonds, it's said to be unsaturated. An oil molecule that contains one double bond per chain of carbon atoms is monounsaturated while a molecule with more than one double bond per chain is polyunsaturated. Double bonds affect the shapes of oil molecules and make it more difficult from them to solidify. Unsaturated oils are liquid even at relatively low temperatures, so they're most common in plants from more temperate regions and in cold-blooded animals such as fish. In general, the colder a plant or cold-blooded animal's natural environment, the more double bonds are found in its oil molecules.

Double bonds are vulnerable to chemical attack and can link unsaturated oil molecules together. When a molecular fragment called a free radical attacks a double bond, the bond suddenly becomes a single bond. The free radical binds onto one of the carbon atoms and becomes its fourth partner. But the other carbon atom

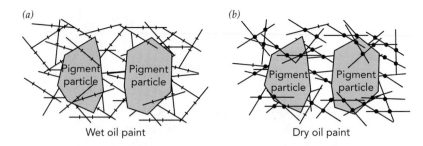

Fig. 14.4.2 (*a*) In wet oil paint, the oil molecules (represented by lines with tick marks at the double bonds) move as a liquid around the pigment particles. (*b*) In dry oil paint, the oil molecules have become attached to one another at the sites of the former double bonds. These drawings are highly simplified since oil molecules are more complicated than they appear here.

is left to seek a new fourth partner for itself. If it attaches to a carbon atom from another oil molecule, it permanently links the two oil molecules together. This cross-linking process, a form of polymerization, is what makes drying oils harden (Fig. 14.4.2).

Drying oils are polyunsaturated oils. Having several double bonds allows each chain to link with several other chains to form an intricate, interlocking network. The best drying oils have three or more double bonds per chain and these double bonds are well separated from one another. The classic drying oil is linseed oil, which has three evenly spaced double bonds in most of its chains. Linseed oil is obtained from flaxseed and was once a by-product of the linen-making process. Now that modern fabrics have replaced flax in linens, flax is grown primarily for the oil in its seeds.

The cross-linking process involves oxygen, which is essential to the formation of the free radicals. When a drying oil is exposed to air, it gradually hardens into a tough, clear, flexible plastic. This hardening converts chemical potential energy into thermal energy and the drying oil becomes warm. While not normally a problem, this warming can cause spontaneous combustion when paint-soaked rags are left to harden in a confined spaced.

While it normally takes weeks or months for pure drying oil to harden, adding various metal compounds to the paint can speed up the process. Certain metal atoms catalyze the formation of free radicals so that cross-linking occurs more frequently. Cobalt and manganese atoms help to harden the oil's surface while lead and zirconium help to harden the body of the oil. Because of lead's toxicity, it's no longer used as a drying agent.

Even with added drying agents, some time elapses before a drying oil begins to harden. Drying oils contain small amounts of antioxidants, which react with and eliminate free radicals. These antioxidants must be used up before the oil begins to harden. Similar antioxidants, such as vitamin E, slow such cross-linking processes in our own bodies and delay aging at a molecular level.

Artist's oil paints are made by mixing pigments with drying oils. The oils bind the pigment particles together and hold them onto the canvas. When first applied, the paint is a thick fluid that can be moved about easily. But the paint becomes thicker and thicker as the cross-linking progresses and it eventually solidifies completely.

Oil-Based House Paints

Oil-based house paints are somewhat different from artist's paints. Most oil-based paints contain polymer binders dissolved in organic solvents. These polymer molecules are formed in the factory but float about separately in the liquid solvent

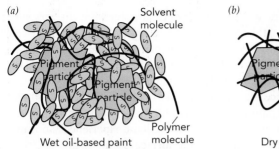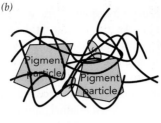

(a)
Solvent
molecule

(b)

Pigment
particle

Pigment
particle

Pigment
particle

Polymer
molecule

Wet oil-based paint

Dry oil-based paint

Fig. 14.4.3 (*a*) Wet oil-based paint contains pigment particles suspended in a solvent containing independent polymer molecules. The solvent molecules evaporate as the paint dries. (*b*) Once the paint is dry, the pigment particles and a few solvent molecules are trapped in a tangled network of polymer molecules. The polymer molecules have some mobility only above the polymer's glass transition temperature.

(Fig. 14.4.3). Since no one wants water-soluble house paints, paint polymers are insoluble in water and require appropriate solvent chemicals. While only enough solvent is added to make the paint spreadable, solvents usual exceed 25% of the paint's weight. When you apply the paint to a surface, the solvent molecules evaporate into the air and the polymer molecules are left behind as a thin plastic layer. Most acrylic paints are of this type.

For the polymer to dissolve in the solvent, its molecules must be attracted fairly strongly to those of the solvent. The polymer molecules must also be able to untangle from one another so that they can be carried away in the solvent. But while the polymer molecules of the binder dissolve completely in the solvent, the pigment particles in these paints do not. They remain solid and tend to settle to the bottom of the paint can as the paint sits on the shelf. That's why you must stir oil-based paints before you use them.

When you apply the paint, it must spread out on the surface rather than beading up into droplets (Fig. 14.4.4). The paint must wet the surface. Wetting occurs when the molecules in a liquid are attracted strongly enough to the surface molecules to bind to the surface molecules instead of one another. Not all liquids wet any particular surface. For example, water beads up on wax because water molecules aren't attracted strongly to wax molecules. Oil-based paints tend to wet most surfaces, although those surfaces should be clean and dry. Oil-based paints won't wet damp surfaces, just as water-based paints (the latex paints discussed below) won't wet oily surfaces. Sometimes primer coats are used to help the final paint wet and bond to the surface being painted.

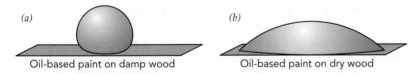

(a)

(b)

Oil-based paint on damp wood

Oil-based paint on dry wood

Fig. 14.4.4 (*a*) Oil-based paints bead up on damp surfaces but (*b*) wet most surfaces that are dry and clean.

Once you apply the paint, the solvent begins to evaporate from its surface. At first the solvent molecules travel easily through the open gaps between polymer molecules and the entire layer of paint dries together. Even the relatively large pigment particles drift freely about in the fluid. But as the solvent evaporates, the polymer molecules move closer together and the paint's viscosity increases. First the pigment particles stop moving, then the polymer molecules, and finally the solvent itself becomes immobile. At that point, the paint feels dry to the touch.

This drying sequence creates a thin layer of clear polymer at the very surface of the paint. As it evaporates, the solvent tends to carry polymer molecules past the larger pigment molecules to the surface of the paint. Even after the pigment molecules have become immobilized, the solvent and polymer still diffuse past them and create a micron-thick region of clear polymer at the surface.

This smooth, clear surface region is what gives many oil-based paints their glossy appearance. Most enamel paints achieve their high gloss by being oil-based. When light travels from air into the polymer layer, the light experiences an impedance mismatch and part of it reflects. Because the surface is smooth and flat, this reflection is specular (mirror-like) (Fig. 14.4.5). Light arriving at the surface from one direction reflects only in one direction.

An irregular surface with features that are comparable to a wavelength of light scatters light irregularly and has a dull reflection. Sometimes oil-based paints are made "semi-gloss" or even "flat" by including very fine particles of silicon dioxide

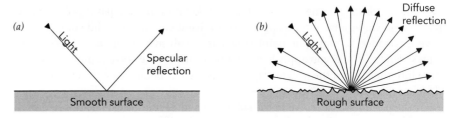

Fig. 14.4.5 (*a*) Surfaces that are smooth on the scale of a wavelength of light create a mirror-like or specular reflection. (*b*) Surfaces that are rough give a diffuse reflection and appear dull or "flat."

(quartz sand) in the mixture. These quartz particles are smaller than pigment particles and diffuse to the surface of the paint as it dries. They spoil the smoothness of the surface and weaken or eliminate the specular reflection. Similarly, when pigment particles begin to protrude through the polymer surface of weathered or worn paint, the appearance is flat.

Most oil-based paints never dry completely without baking. Although the solvent molecules diffuse rapidly to the surface and evaporate early on, they have more and more trouble leaving as the paint becomes more viscous. Eventually, the polymer tangle becomes so dense that even the tiny solvent molecules can't work their way through it. What appears to be dry paint actually contains a substantial amount of trapped solvent. The occasional solvent molecules that do escape give the paint its long-lasting odor.

The polymer tangle is a type of glass (a non-crystalline solid discussed in Section 17.2). To eliminate solvent molecules trapped in the glassy polymer, it must be warmed until its molecules begin to move past one another. Exceeding this glass transition temperature brings a degree of mobility to the molecules and allows the trapped solvent to escape. Paints and coatings that come into contact with foods must be baked after application to drive out the solvents.

Because of their rigid, impermeable natures, oil-based exterior house paints tend to bubble and crack. Water migrates through wood and gets under the layer of paint. This water is unable to escape through the paint and can create blisters when the paint is heated by sunlight. Changes in temperature and moisture content also cause the wood to expand and contract. Rigid oil-based paint experiences different relative expansions and contractions and it can buckle or crack.

But inside, oil-based paints are very durable. Their hardness and rigidity make them easy to clean. Dirt sits on the surface of these paints and can be removed with anything that doesn't damage the polymer itself. Professional painters, who don't mind cleaning their brushes and equipment with solvents, generally prefer oil-based interior paints. Do-it-yourselfers usually find it easier to use the latex paints discussed below.

Many oil-based house paints exhibit drying oil behavior, too. As the solvent evaporates and oxygen enters the paint, cross-linking occurs between the polymer chains. Most alkyd paints are of this type. The cross-linking process releases many tiny molecular fragments that evaporate as the paint dries and give the paint its characteristic odor.

These drying oil-like paints use drying agents to speed up the cross-linking between double bonds. Unfortunately, the drying agents of choice until the 1930's were lead compounds. Since these paints are too rigid and impermeable to tolerate changes in temperature and moisture content, they eventually flake off woodwork and present a terrible health hazard, particularly for young children. Modern drying oil-like paints use relatively non-toxic drying agents.

However, lead-free oil-based paints still present a health risk: solvent content. The volatile organic solvents they contain are toxic. Given that oil-based paints require that their water-insoluble polymers be fully dissolved, there seems to be no way of eliminating organic solvents from them. What is happening is that painters and coaters are gradually abandoning oil-based paints and coatings in favor of those that don't involve complete solution of the polymers. The most important paints that don't involve complete solution are latex paints.

Latex House Paints

Latex house paints aren't uniform liquids. Like oil-based paints, latex paints contain polymer molecules but these molecules aren't dissolved in a solvent. Instead, they are incorporated in tiny solid particles that float around in a liquid carrier (Fig. 14.4.6). The polymer molecules are actually insoluble in that carrier, which is usually water.

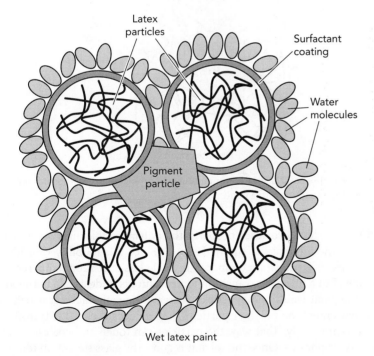

Wet latex paint

Fig. 14.4.6 Wet latex paint contains tiny spheres of polymer, coated with a surfactant, and suspended as an emulsion in water.

It may seem that such polymer particles could never get together to form a uniform coating, but they actually do. Once the paint is applied to a surface, the water evaporates and the polymer particles move closer together. They're pulled together by water as it leaves and their polymer chains begin to mingle together. By the time the water is gone, the molecules are tightly interwoven. They trap the paint's pigment particles and create a layer that's often more durable than that of an oil-based paint.

Latexes aren't simple to make. The polymer particles must remain suspended in the paint while it sits in the can and must not clump together before the paint begins to dry. Once the paint starts drying, the polymer particles should clump together. Obviously, this is a tricky business.

To keep the polymer particles suspended as an emulsion—one phase of material suspended as tiny particles in another phase of material—the particles are coated with special chemicals called surfactants. Surfactants are molecules such as soaps and detergents that naturally migrate to the interfaces between oils and water, and stabilize those interfaces (see Section 18.2 on Laundry). Surfactant molecules allow water to cling to the oil-like polymer particles. Each particle thus maintains at least a thin layer of water molecules around it, so that the polymer particles don't clump together while the paint is in the can.

But as water evaporates from drying latex paint, the remaining water pulls the polymer particles toward one another. The water molecules bind to one another and to the surfactant molecules with hydrogen bonds. The surfactant molecules cling to the polymer particles with van der Waals forces (see Section 18.1 on Oil Refineries). As water molecules leave the gaps between polymer particles, the remaining water molecules draw the polymer particles together. Eventually the polymer particles begin to touch and their polymer molecules become tangled together. The particles coalesce into a single material (Fig. 14.4.7).

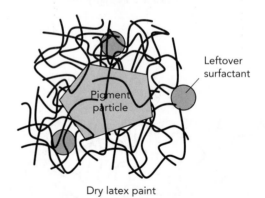

Leftover surfactant

Pigment particle

Dry latex paint

Fig. 14.4.7 As latex paint dries, the remaining water molecules pull the latex spheres together until they coalesce into a continuous film.

For the particles to coalesce properly, the polymer molecules must be able to move under the influence of thermal energy. Since this mobility vanishes below the polymer's glass transition temperature, that glass transition temperature must be very low. But a low glass transition temperature is undesirable in the long run because the paint would remain sticky. Instead, the manufacturer usually adds a solvent that enters the polymer particles and reduces their glass transition temperatures temporarily. That way, the particles can touch and coalesce only until the solvent evaporates. Once the solvent is gone, the glass transition temperature increases and the paint loses its stickiness. In latex house paint, this final drying process takes a few weeks.

But on a hot summer day, the sun can warm dry latex paint above its glass transition temperature. The paint's polymer molecules then begin to move around and the paint readjusts to its situation. This readjustment is a good thing because it allows the paint to relieve tension and avoid cracking or blistering. Water vapor can escape between the polymer chains and the paint "breathes." That's why latex exterior paints are more durable than oil-based exterior paints.

Unfortunately, this movement of the polymer chains allows latex paint to pick up dirt. Dirt particles that land on the paint's surface in hot weather can become caught in the polymer chains, spoiling the appearance of the paint. Also, the paint tends to stick to itself in hot weather, a behavior called blocking. While blocking is useful for polymer-based adhesives like contact cement, it's terrible for paint. If you touch two freshly painted latex surfaces together, even if they feel dry, they will often block together. All in all, it's best to leave latex paint alone after painting until all of the solvent has evaporated from its polymer particles and its glass transition temperature has become as high as possible.

Because latex particles are large, they don't diffuse to the surface of the paint as the water evaporates. As a result, latex paint doesn't form a clear, pigment-free layer at its surface. The presence of pigment particles all the way to its surface makes latex paint rough on the scale of a wavelength of light. That's why latex paints aren't as glossy as oil-based paints.

Because the principal liquid in latex paints is water, latex paints contain less volatile organic chemicals than oil-based paints. However, most latex paints contain some organic solvent to soften the polymer particles so that they can coalesce easily. Most also contain organic solvents that slow down drying. If the only liquid in latex paint were water, it would dry so quickly that you couldn't work with it. Each brush stroke would harden by the time you returned with the next stroke and you would have trouble blending the strokes into a smooth layer. To slow down the drying, most latex paints include large quantities of ethylene and propylene glycols. These two volatile organic chemicals are the main ingredients in automobile antifreezes. They resemble water as solvents but they evaporate much more slowly. They also protect the can of paint from freezing, an event that would squeeze the latex particles together and ruin the paint.

Paints also contain other environmentally questionable chemicals. Fungi and bacteria will attack almost anything and paint is no exception. To prevent such attacks, paints include fungicides and bactericides. One common and extremely effective fungicide and bactericide is phenylmercuriacetate, an organic mercury compound. Because mercury is toxic, it was banned from interior use in 1990. However, it's still used in exterior paints.

White and Metallic Pigments

Pigment particles give paint its opacity, color, or both. Since these two characteristics, opacity and color, are somewhat independent of one another, we'll examine them separately.

Even pure white paint has pigment in it but this pigment absorbs no light. Instead, its particles scatter light almost perfectly in random directions. White pigment particles are clear but have very high refractive indices. When these particles are embedded in a polymer layer, they produce countless impedance mismatches in that layer. As light tries to pass through the paint, part of it reflects at every boundary between polymer and pigment, and almost none of the light reaches the back of the layer. Because the pigment particles are rough and randomly oriented, they scatter the light everywhere and it leaves the paint's surface as a diffuse glow. The paint appears white.

The best white pigments are those that have the highest refractive indices and the least tendency to absorb visible light. The higher the refractive index, the more severe the impedance mismatches and the more light is reflected at each boundary between polymer and pigment. This high reflectivity gives paint its hiding power—its ability to prevent light from reaching the material beneath the paint and then returning to paint's surface. Paints with very high refractive index pigments are best at hiding the surfaces they cover.

Perfect clarity in these pigments is important because any absorption of light will give the paint a color. For example, granules of salt and sugar are almost perfectly clear, which is why they appear white as light reflects from them. But adding light-absorbing food dyes to sugar causes it to appear colored.

Tiny particles of calcium compounds (lime, chalk, and gypsum) were used to whiten walls for millennia. However, these compounds have refractive indices of between 1.5 and 1.7, so their hiding power is poor. Until the 1930's, the most common white pigment in paint was white lead (lead carbonate). This clear compound has a refractive index of 1.94, so its hiding power is modest but better than that of the calcium compounds. However, white lead is quite toxic and is a nightmare for people living in pre-1930's buildings. Fortunately, non-toxic white pigments with much better hiding powers have replaced it.

The most common white pigment in modern paints is titanium dioxide. This compound comes in two crystalline forms, rutile and anatase, each of which has a very high refractive index and near-perfect clarity. Rutile's refractive index is 2.76 while anatase's is 2.55. Its higher refractive index means that rutile particles are somewhat more reflective in paint than anatase particles. Because of its better reflectivity, rutile pigment has more hiding power than anatase pigment. However, rutile absorbs blue light slightly so that rutile-based white paint would have a yellow tint. For colored or off-white paints, rutile's slight coloration doesn't matter. However, pure white paints must use anatase.

Some paints use metal particles as pigments. These metallic pigments have enormous hiding powers because essentially zero light gets through a metal particle. But metals absorb a few percent of the light that hits them, converting light energy into thermal energy. Light entering a pile of round metal particles gets trapped bouncing between particles and is absorbed before it can work its way back out. Thus a pile of round metal particles appears black.

In contrast, metal flakes appear bright because they reflect light back out of the surface in a single reflection. Each flake acts as a tiny mirror and creates its own specular reflection. When metal flakes are incorporated in paint, they tend to lie roughly parallel to the surface and give the paint a shiny, metallic look. They don't create a single, mirror reflection because they're not perfectly aligned with one another. But they also don't scatter light uniformly in all directions like white pigment. Most metallic paints contain aluminum flakes or powders, although they may be colored to give the appearances of other metals.

Pearlescent and iridescent paints contain flakes of transparent materials that have been specially coated to enhance their reflectivities. Each surface of one of these coated transparent flakes partially reflects light and the interferences between those partial reflections give the flake a colored specular reflection. Because they transmit some of the light that strikes them, these flakes don't look metallic. They also lie roughly parallel to the surface in paint and give it an interesting appearance. Iridescent flakes are very popular in metallic car paints, giving them subtle colored highlights.

Colored Paints

Colored pigments and dyes give a paint its color by selectively absorbing some of the light striking the paint. Since the light reflected from the paint has a different spectrum of wavelengths than the light striking it, the reflected light looks colored.

For example, when you send white light at red paint, this paint absorbs light that would stimulate the green or blue sensors of your eyes. All that is left is light that stimulates your red sensors, so you see the paint as red. Green paint absorbs all but light that stimulates your green sensors and blue paint absorbs all but light that stimulates your blue sensors.

You see this reflected light because the paint's pigments reflect any unabsorbed light back from the paint. In some cases, the pigments are also the colored absorbers. In other cases, the pigments are white, typically titanium dioxide, and the colored absorbers are separate dye molecules—molecules that selectively absorb certain wavelengths of light and let other wavelengths pass unaltered.

Whether they use pigments alone or a combination of pigments and dyes, paints combine reflectivity and selective absorption to give a surface its colored appearance. By carefully adjusting which wavelengths of light are absorbed and the extent of that absorption, you can create a paint with any color you like.

Most colored paint pigments are based on specific molecules that absorb light in a particular range of wavelengths. Many metal compounds, including those of copper, chromium, iron, antimony, nickel, and lead absorb certain wavelengths of light and appear brightly colored. Lead and lead-chromium compounds are particularly beautiful pigments for paints, but they are just too toxic to use. Organic dyes are also extremely effective in paints, although they must be combined with white pigments, and have replaced metal compounds in many paints. (For history of organic dyes, see ❶.)

Of course, many colors of paint are created by mixing different pigments and dyes together in carefully controlled amounts. The rules governing this mixing of pigments and dyes are extended versions of the rules of color subtraction (p. 462). Normal subtractive color employs the three primary colors of pigment: cyan (which absorbs red), magenta (which absorbs green), and yellow (which absorbs blue). In principle, the proper mixture of those three pigments can create any color of paint you like. In practice, however, perfect cyan, magenta, and yellow pigments and dyes don't exist and near-perfect versions are expensive. Instead, paints work with a more varied set of basic pigments and dyes. Mixed according to various rules and recipes, these basic colors can produce nearly any color of paint.

Inks are similar to paints except that they contain only dyes. Because they contain no reflective pigments, inks tend to be transparent but colored and rely on the underlying paper to reflect light. Paper consists mainly of cellulose, a clear natural polymer. Because this cellulose is finely divided in paper, it reflects light at each surface and the paper appears white. Actually, paper's whiteness is often enhanced during manufacture by coating it with white pigments. Transparent stains are essentially inks that are applied to wood.

Normal color printing is done using black ink and three colored inks that are a close as is economically and technically feasible to the primary colors of pigment: cyan, magenta, and yellow. By varying the amounts of these four inks applied to each location on a printed page, the printer has almost complete control over the colors that you see when you look at it in white light. However, since each ink is applied separately to the page, the printer must be careful to overlap the inks perfectly. Careless color printing, such as often occurs in newspapers, creates blurry images when the yellow, cyan, magenta, and black ink images are shifted relative to one another.

High-quality, high-cost printing jobs are often done with more than four inks. Because the registration between the separate inks is never quite perfect, these fancy printing tasks are done with extra colors so that they can avoid having to mix colors directly on the page. Instead of trying to create a red letter "$\mathcal{H}$" by first printing a yellow "$\mathcal{H}$" and then a magenta "$\mathcal{H}$" directly on top of it, the printer will actually use a red ink to print the "$\mathcal{H}$" in a single step. The same printer may apply a

❶ As an 18-year-old student, English chemist William Henry Perkin (1838–1907) set about trying to synthesize the drug quinine. His technique was based on the simplistic understanding of organic chemistry at the time and couldn't possibly have worked. Instead, he discovered a dye he called aniline purple or mauve. Before his discovery, purple dyes were rare and costly. Purple clothing, once reserved for royalty, suddenly became available to the common person. Perkin established a company in Germany to manufacture this dye and thus founded the German dye industry.

clear varnish layer over every picture to give them a glossy appearance. This glossy look comes from the specular reflection at the varnish's smooth surface, caused by the impedance mismatch between the air and the varnish.

Problems with Paint

Unfortunately, the light that pigment molecules absorb often damages those molecules. Blue and ultraviolet light, which carry large amounts of energy in each photon are particularly damaging. Thus molecules that absorb blue light, such as those creating red or green colors, tend to be destroyed by long exposure to sunlight. You have probably seen pictures or photographs that have been bleached by the sun. The red color pigments are usually destroyed first and the pictures appear bluish.

Protecting paints and surfaces from light damage is actually very important. Sunlight can damage not only the pigment molecules, but also the binder that holds the paint together. Often the white or metallic pigments in paint protect it and the surface beneath from light damage by absorbing or reflecting ultraviolet light. Titanium dioxide is particularly good at absorbing ultraviolet light. Sunscreens perform a similar protective role, absorbing ultraviolet light and preventing most of it from damaging your skin.

However, many of the pigments and binders used in artwork are simply too fragile to tolerate long exposure to blue or ultraviolet light. These paintings are often kept in low light conditions or viewed only with long-wavelength, yellowish light. While yellow light is less damaging to the molecules than white light, it affects the appearance of the paints. Paint can't reflect light that isn't there. While blue paint is supposed to reflect blue light, it looks black when there is no blue light to be reflected. The blue paint absorbs the red and green lights in the yellow illumination and there is just nothing else left. Clearly the viewing light affects the perceived colors of paint.

This viewing illumination problem creates problems for color matching. Two paints may look like a perfect match in one type of illumination but may appear significantly different in another illumination. For example, under red illumination blue and black paint both appear black but under white illumination they look different. Similarly, two paints that look identical in incandescent lighting will look somewhat different in fluorescent lighting. This behavior is called metamerism and can only be avoided by using exactly the same pigments and dyes in the two paints being matched. If you can't match the paints chemically, you must do the color matching in the proper lighting. However, metamerism is occasionally helpful, such as in reading faded old letters. In the proper illumination, the contrast between the ink and the paper may be greatly enhanced.

Fluorescent and Phosphorescent Paints

In most cases, pigments and dyes selectively absorb certain wavelengths of light and convert the energy in that light into thermal energy. However, there are two important exceptions to this rule: fluorescent and phosphorescent paints. The dyes and pigments in these paints turn only part of the light energy into thermal energy and reemit the rest as a new color of light.

When this absorption and reemission process is rapid, it is fluorescence—the same process that allows a fluorescent lamp to convert ultraviolet light into visible light. In a fluorescent paint, an atom or atomlike system absorbs each photon, converts some of that photon's energy into thermal energy, and then emits a new, lower-energy photon. The light the fluorescent system emits is always longer in wavelength than the light it absorbed. For example, a red fluorescent dye or pigment might absorb blue light and emit red light as a result. If you expose the red fluorescent dye or pigment to blue light, it will glow red.

Fluorescent dyes and pigments are very popular in everything from clothes to marking pens. A fluorescent green shirt emits more green light than a white shirt does because the fluorescent green shirt actually converts the blue light that hits it into green light. Many white clothes or business cards appear whiter-than-white because they contain fluorescent pigments or dyes that convert ultraviolet light into white or blue-white light. They really do emit more visible light than a perfectly white, but non-fluorescent surface. Many laundry detergents include "brighteners" that are really fluorescent chemicals that cling to the clothes and turn ultraviolet light into visible light.

In phosphorescent paints, the reemission process occurs long after the absorption. The delay of this phosphorescence can be seconds, hours, or days and is a result of the atoms or atomlike systems getting stuck in excited states that can't undergo normal radiative transitions. Typically, the normal transition is forbidden because the electron trying to shift from the excited state to the ground state has the wrong spin and the Pauli exclusion principle prevents it from transitioning to the ground state unless it flips its spin at the same time. The exotic radiative process that ultimately returns this electron to the ground state is extremely slow. Paints containing phosphorescent dyes and pigments released light over a long period of time and are commonly known as glow-in-the-dark paints.

Makeup: Colors and Lighting

Just about everything I've said about paint also applies to makeup and hair color. Apart from its physiological purposes, such as moisturizing or soothing, makeup is primarily paint for the face or body. Similarly, hair color alters or replaces the natural coloration of hair and is basically dye or stain for hair.

Like a paint, a particular makeup product may include pigments, dyes, and binders. The pigments are microscopic particles, so they produce the many reflections that give a makeup hiding power. Each time light enters or exits a pigment particle, a fraction of that light reflects. If the pigment consists of colorless, transparent powder particles, the pigment will appear white. If the pigment particles are tiny metallic sheets, they'll give a metallic reflection. And if the pigment particles are tiny transparent sheets, they'll give a lustrous, pearlescent, and iridescent reflection. In each case, light won't penetrate far into the makeup, so the makeup will mask the wearer's natural skin tone.

Dyes selectively absorb certain wavelengths of light while allowing other wavelengths to continue through the makeup. Dyes alone have no hiding power, so a makeup that contains only dye molecules will permit the wearer's skin tone to show. A dye-based makeup will simply filter the light that reaches and reflects from the wearer's skin so that the skin appears darker or has a different color. A typical makeup product will combine pigments and dyes so that it has both hiding power and color.

Binders keep the pigments and dyes in place on the wearer's skin and contribute to the reflectivity of the pigments. Since a transparent powder appears white only when its particles are separated by a material having a different refractive index, the binder for a pigmented makeup typically has a lower refractive index than its pigments.

The binder also contributes to the surface character of the makeup. If the binder forms a smooth transparent coating on the wearer's outermost surface, it will appear shiny. Light will partially reflect from that glassy surface just as it does from a window pane. Although such a shiny appearance is perfect for lip gloss, it's not ideal for cheeks. To give skin a dull, matt appearance, the binder must allow pigment particles to cover its surface. One common way to create a low-sheen surface is to apply loose powder to a layer of binder already on the skin. That's why women dust powder onto shiny spots on their faces.

Because makeup only reflects or otherwise redirects the light that strikes it, makeup looks different under different lighting conditions. Some makeup mirrors allow you to choose between two separate light sources, incandescent and fluorescent, for exactly this reason. Since incandescent light emphasizes the red end of the spectrum and contains relatively little blue or violet light, almost any makeup illuminated by incandescent light will have a warm appearance: rich in reds, oranges, and yellows. A woman who is going to spend the evening bathed in candlelight or illuminated by incandescent lightbulbs should apply her makeup in similar lighting in order to obtain the effects she wants. If she applies her makeup in fluorescent lighting, the greens, blues, and violets that she sees while putting on her makeup will darken substantially when she is in incandescent light. She may be disappointed.

On the other hand, fluorescent light and sunlight include strong blue and violet components, so makeup illuminated by these sources tends to have a cold appearance: strong blues and violets. A woman who is planning to spend her time in fluorescent lighting or outdoors should apply her makeup in fluorescent lighting. If she applies her makeup in incandescent lighting, she may underestimate how much green, blue, and violet her face will reflect and overestimate its reflection of reds, oranges, and yellows. When she later enters fluorescent lighting or sunlight, she may find that her colors are out of balance.

Sunscreens, Fragrances, and Hair Color

Even colorless makeup products often contain dyes. These dyes absorb ultraviolet light and thereby introduce a measure of UV protection into the products. Since a photon of ultraviolet light carries enough energy to damage a typical biological molecule, letting these photons reach your skin is a bad idea. Over your lifetime, these ultraviolet photons cause cumulative damage to your skin cells and their genetic information. The single most important thing you can do to protect the health of your skin is to wear makeup, skin cream, or sunscreen that absorbs ultraviolet light year round.

Many makeup items and all perfumes and colognes contain scent molecules that diffuse through them to their surfaces and gradually evaporate into the air. The strengths of the chemical bonds between those scent molecules and the binders, together with the temperature of the wearer's skin, determine how quickly the scent molecules leave her skin for the air. Cleverly formulated scents will release their fragrances into the air gradually and in layers over the course of an evening.

Hair colors are dyes that reside in or on a wearer's hair. The dye molecules used in permanent hair colors are typically formed inside the hairs by chemical development processes that take place during coloring. Once these molecules are inside the hairs, they're extremely hard to remove and the hair color is permanent. The dye molecules used in temporary rinses usually stick to the outsides of the hairs and remain in place for a limited time.

Because hair colors don't include pigments, they have virtually no hiding power and must work with light reflected by the wearer's hair. A person with naturally light hair can apply a dark hair color relatively easily, but a person with naturally dark hair will probably have to bleach out that natural color in order to make the new one visible. I'll explain bleaching in Section 18.2 on laundry.

The exceptions are rinses that produce colored highlights. These rinses are dyes that lend color to the natural shine of even the darkest hairs. Like glossy black paint, a dark hair reflects some of the light that strikes it, particularly when that light arrives at a shallow angle. In an uncoated dark hair, that partially reflected light makes the hair look shiny. But if the hair is coated with colored dye molecules, its partially reflected light will be colored, too.

CHAPTER 15

OPTICAL THINGS

Many of the devices around us perform useful tasks by manipulating light: cameras record images of the objects in front of them, our eyes allow us to observe those objects directly, and eyeglasses and magnifying glasses help us to see details we'd miss with our eyes alone. Still other gadgets, including CD and DVD players, use light in ways that have nothing to do with vision. But all of these objects manipulate light using similar techniques—the techniques of optics.

People have been tinkering with optics since they first noticed that a clinging water drop offers a distorted view of a leaf or that a transparent quartz crystal can change the path of light passing through it. But many of the most memorable and important optical discoveries were made by the usual cast of characters: Galileo constructed one of the first successful telescopes, Newton experimented with prisms to separate white light into its colors, and Benjamin Franklin invented bifocal eyeglasses. The Greek scientist Archimedes also supposedly set fire to Roman ships with sunlight focused by curved mirrors.

Even if the Archimedes story is apocryphal, it remains something of a rite of passage for an optical experimenter to burn wood with sunlight and a magnifying glass. Naturally, I spent hours perfecting my solar woodburning skills and so did my son. I still recall my delight while reading Jules Verne's *The Mysterious Island* as a boy when the engineer relit the group's precious fire using a burning lens he had formed by filling the curved glasses of two watches with water.

While optical tools such as lenses and prisms have been around for hundreds of years, advances of modern technology have accelerated the development of optics. Just as the invention of transistors has sped the growth of the electronics industry, so the invention of lasers has sped the growth of the optics industry. The two fields aren't far apart in many ways, and there is hope that one day computers will be as much optical devices as they are electronic.

Chapter Itinerary

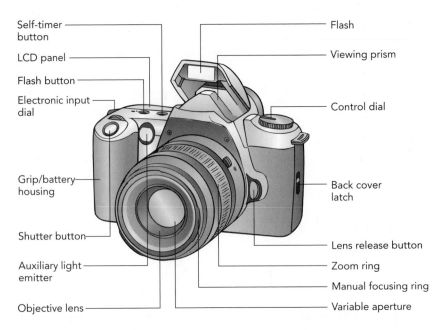

Self-timer button
LCD panel
Flash button
Electronic input dial
Grip/battery housing
Shutter button
Auxiliary light emitter
Objective lens

Flash
Viewing prism
Control dial
Back cover latch
Lens release button
Zoom ring
Manual focusing ring
Variable aperture

SECTION 15.1 **Cameras and Vision**

In the two centuries since their invention, cameras have become extremely easy to use. What started as a hobby for a few dedicated enthusiasts has evolved into an everyday activity. But despite all of the technological improvements, photography still employs many of the same principles it did in the 1800s. Cameras still use lenses to project images onto light-sensitive surfaces, and photographers still have to worry about getting the exposure right, focusing properly, and avoiding the blur of rapid motion. In this section, we'll explore some of the principles that make cameras work.

Lenses and Real Images

When you take a picture of the scene in front of you, the lens of your camera bends light from that scene into a real image on a light-sensitive surface. A real image is a pattern of light, projected in space or on a surface, that exactly reproduces the pattern of light in the original scene. Since the real image that's projected looks just like the scene you're photographing, recording the light in that image is equivalent to recording the appearance of the scene itself.

While that light-sensitive surface was once always photographic film, digital cameras are gradually replacing film with electronic image sensors. Fortunately, the two light-sensing surfaces are essentially interchangeable, so we can refer to them both as image sensors: one is electronic and one is photochemical.

Real images don't occur without help. When light from a candle falls directly on an image sensor, it produces only diffuse illumination (Figs. 15.1.1a). Similarly, you can't tell by looking at a sheet of paper what a candle looks like because the light that leaves the candle travels in all directions and is as likely to hit the top of the paper as it is to hit the bottom (Figs. 15.1.1b).

That's why a camera needs a lens, a transparent object that uses refraction to form images. The light passing through a lens bends twice, once as it enters the glass or plastic and again as it leaves. In a camera lens, this bending process brings much of the light from one point on the candle back together at one point on the sensor. As you can see in Fig. 15.1.2, the real image that forms is upside down and backward. This inversion of the real image relative to the object always happens when a single lens creates a real image.

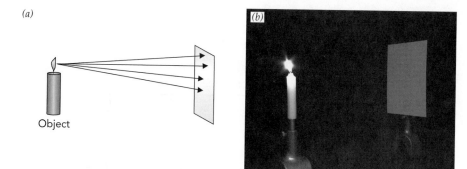

Fig. 15.1.1 (*a*) Without a lens, light from a candle uniformly illuminates an image sensor. (*b*) When candlelight falls directly on a sheet of paper, it produces no image.

The curved shape of the camera lens allows it to form a real image. Light passing through the upper half of the lens is bent downward while light passing through the lower half is bent upward. Because the camera lens bends light rays toward one another, it's a converging lens. You can see how it forms an image in Fig. 15.1.2*b* by following some of the rays of light leaving one point on the candle.

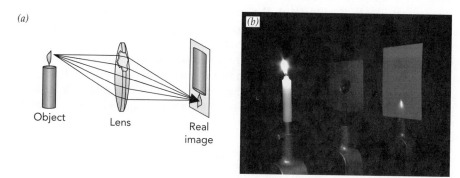

Fig. 15.1.2 (*b*) When a lens is introduced between the candle and the sensor, it brings light from each point on the candle back together on the sensor's surface, forming an up-side-down and backward real image of the candle. The distance between the lens and the sensor must be chosen correctly, or the image will be blurry. (*b*) A lens inserted between the candle and paper forms an inverted real image of the flame on the paper.

The upper ray from the candle flame travels horizontally toward the top of the lens. As it enters the lens and slows down, this ray of light bends downward. It bends downward again as it leaves the lens and travels downward toward the bottom of the image sensor.

The lower ray from the candle flame travels downward toward the bottom of the lens and bends upward as it enters the lens. It bends upward again as it leaves the lens and travels horizontally toward the bottom of the image sensor.

These two rays of light reach the image sensor at the same point. They are joined there by many other rays from the same part of the candle flame, so that a bright spot forms on the sensor. Overall, each part of the candle illuminates a particular spot on the image sensor, so the lens creates a complete image of the candle on the sensor.

However, the lens can bring the light back together to form a sharp image on the sensor only if the lens and sensor are separated by just the right distance (Fig. 15.1.3). If the sensor is too close to the lens, then the light doesn't have room to come together. If the sensor is too far from the lens, then the light begins to come apart

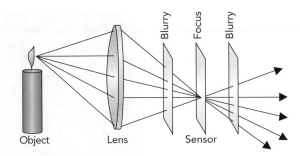

Fig. 15.1.3 The real image is in focus only when the image sensor is just the right distance from the lens. If the sensor is too near or too far from the lens, the image is blurry.

again before reaching the sensor. In either case, the image on the sensor is blurry. The candle's real image is only in focus at one distance from the lens.

If the candle moves toward or away from the camera lens, the distance between the lens and the image sensor must also change (Fig. 15.1.4). When the candle is far away, all of its light rays that pass through the lens arrive traveling almost parallel to one another and the inward bend caused by the lens makes those rays converge together quickly. The rays come into focus relatively near the lens and that's where the sensor must be (Fig. 15.1.4*a*). The candle's image on the sensor is much smaller than the candle itself because the light rays have only a short distance over which to move up or down after leaving the lens.

When the candle is nearby, its light rays that pass through the lens are diverging rapidly and the inward bend caused by the lens is just barely enough to make those rays converge at all. As a result, the rays come into focus relatively far from the lens (Fig. 15.1.4*b*). The candle's image on the sensor is quite large because the

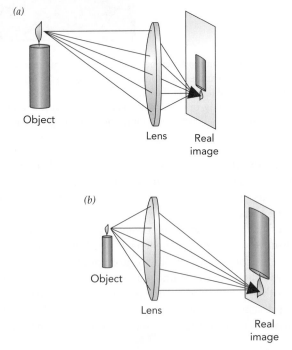

Fig. 15.1.4 (*a*) The light from a distant candle is traveling in almost the same direction and the lens focuses it easily. The real image forms close to the lens. (*b*) The light from a nearby candle diverges, and the lens has difficulty bending it back together. The real image forms far from the lens. If the candle is too close to the lens, no real image forms at all.

light rays have considerable distance over which to move up or down after leaving the lens.

Because distant and nearby objects form real images at different distances from the camera lens, they can't both be in focus on the same image sensor. When you take a picture of a person standing in front of a mountain, only one of them can be in sharp focus. However, if you're willing to compromise a little bit on sharpness, a lens can sometimes form acceptable images of both objects.

Focusing and Lens Diameter

A disposable camera is little more than a box with a lens. The lens projects a real image of the scene in front of it onto the camera's image sensor. Light in the real image exposes the image sensor, which records the image permanently. While it may also have a shutter that starts and stops the exposure, a flash to provide extra light, and a mechanism that prepares for the next photograph, there's little else to this simple camera.

However, there are limitations to the disposable camera design. One of the most severe limitations is that you can't focus it—the camera has a fixed distance between the lens and image sensor. Nonetheless, it manages to form relatively sharp real images on the sensor, even when there are objects at various distances from the camera. These simple cameras work because they use narrow (small-diameter) lenses. A narrow lens gathers less light than a wide lens but doesn't require focusing—adjusting the distance between the lens and the image sensor.

Because a wide (large-diameter) lens brings rays together from many different directions, you must focus it (Fig. 15.1.5a); if the image sensor is even slightly too near or too far from the lens, the recorded image will be blurry. But a narrow lens forms a reasonably clear image even without focusing. Any rays from one part of the scene that succeed in passing through the narrow lens must already be fairly close together. Their initial closeness means that these converging rays illuminate

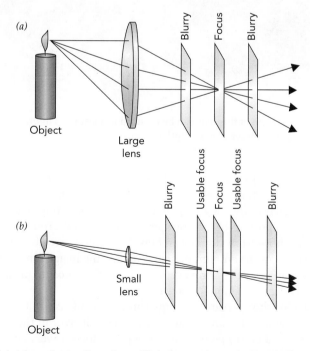

Fig. 15.1.5 (*a*) A large lens collects lots of light but its focus is critical. The image is blurry except at the actual focus. (*b*) A small lens collects less light but its image is relatively sharp anywhere near the focus.

only a small part of the image sensor even when the sensor isn't exactly the right distance from the lens (Fig. 15.1.5b). Since the image sensor can't record every minute detail anyway, the image that forms on it doesn't have to be in absolutely perfect focus. As a result, a camera with a narrow lens and no focus adjustment manages to take pretty good pictures.

Unfortunately, these simple cameras collect very little light and need extremely light-sensitive image sensors. These high-speed sensors can't record as sharp photographs as low-speed sensors. Furthermore, the pictures produced by simple cameras lack fine details; while everything is almost in focus, most things are a bit fuzzy if you look carefully or make enlargements.

More sophisticated cameras use wider lenses that gather more light and expose the image sensors much more rapidly. They also automatically adjust the distance between the lens and the sensor. They identify the object you are photographing and position the camera lens so that it projects a sharp image on the sensor. As the camera focuses, you can usually see components in the lens moving backward or forward to arrive at the correct distance from the sensor.

Even a camera with a wide lens can take advantage of the narrow lens trick for focusing. Its lens contains an internal diaphragm that reduces its aperture or effective diameter. The diaphragm is a ring of metal strips with a central opening. These strips can swing in or out, changing the diameter of the diaphragm's opening and with it the aperture of the lens (Fig. 15.1.6).

Fig. 15.1.6 The aperture of this lens can be reduced by closing its internal diaphragm, dimming its image but increasing its depth of focus.

When its lens aperture is narrow, the sophisticated camera imitates a simple camera. Nearly everything is essentially in focus simultaneously. In such a situation, the camera has a large depth of focus. Actually, the sophisticated camera can bring the most important object into perfect focus, so it produces pictures that are superior to those from simple cameras. However, narrowing the aperture of the lens also reduces the amount of light reaching the image sensor. The scene in front of the camera must either be very bright or the exposure must be relatively long. You don't get something for nothing.

While widening the aperture of a large lens makes full use of its light-gathering capacity, focusing then becomes crucial. Even a small error in the lens-to-sensor distance produces a blurry picture, so the depth of focus is very small. This trade-off, between light gathering and depth of focus, is a continual struggle for photographers. However, photographers sometimes take advantage of the small depth of focus in wide lenses to blur the background or foreground of a photograph deliberately. A camera's portrait setting adopts this strategy to produce sharp images of people against blurred backgrounds.

At other times photographers choose long exposures at narrow apertures in order to bring an entire scene into sharp focus. A camera's landscape setting takes this route, so that everything in the photograph shows full detail. And to capture fast motion while retaining a large depth of focus, photographers use a flash to brighten the scene and shorten the exposure. Unfortunately, a camera flash is ineffective at brightening a distant scene and it can produce unpleasant reflections from windows and eyes. A camera's sports setting emphasizes brief exposures to avoid speed blur, even though that may require a wide aperture and small depth of focus.

Focal Lengths and f-Numbers

Lenses are characterized by two quantities: focal length and f-number. The focal length of a lens is the distance between the lens and the real image it forms *of a very distant object*. For example, if a real image of the moon forms 100 mm (4 inches) behind a particular lens, then that lens has a focal length of 100 mm. The focal lengths of camera lenses range from less than 10 mm (0.4 inches) in many compact cameras to about 2 m (7 feet) in cameras used for nature photography.

When light from a scene passes through a short focal length lens, it comes to a focus near that lens and produces a relatively small image on the image sensor. Because a long focal length lens permits the light passing through it to spread out more before coming to a focus, it produces a larger real image on the sensor.

The "normal" lens for a particular camera has a focal length that allows all of the objects in your central field of vision to fit onto the image sensor. When you hold the finished photograph about 30 cm (1 foot) from your eyes, the objects in the picture appear about the same size they did when the photograph was taken. The focal length of a camera's normal lens is about 1.5 times the horizontal width of its image sensor.

Table 15.1.1 Several Cameras, the Widths of the Image Sensor They Use, and Their Normal Lenses

Type of Camera	Sensor Width	Normal Lens
Typical Digital Camera	8 mm	12 mm
35-mm Camera	36 mm	50 mm
2¼-inch Medium Format Camera	57 mm (2¼ inches)	80 mm
5-inch Portrait Camera	127 mm (5 inches)	180 mm

A wide-angle lens has a shorter focal length than the normal lens (Fig. 15.1.7). The image it projects onto the image sensor is smaller but brighter, and most of

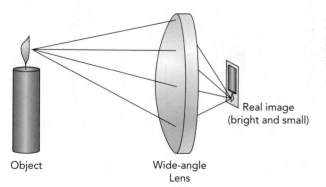

Object   Wide-angle Lens   Real image (bright and small)

Fig. 15.1.7 A wide-angle lens has a short focal length and forms a small, bright real image near the lens.

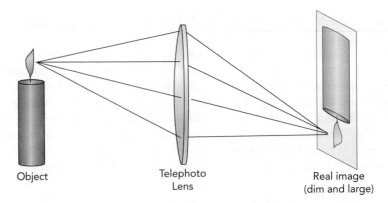

Object   Telephoto   Real image
                 Lens   (dim and large)

Fig. 15.1.8 A telephoto lens has a long focal length and forms a large, dim real image far from the lens.

the objects in your entire field of vision appear in the photograph. A telephoto lens has a longer focal length than the normal lens (Fig. 15.1.8). The image it projects onto the sensor is larger but dimmer, with only objects at the center of the scene appearing in the photograph.

In addition to indicating where the image of a distant object forms, the focal length of the camera lens relates the object distance to the image distance. The object distance is the distance between the lens and the object you're photographing. The image distance is the distance between the lens and the real image it forms (Fig. 15.1.9). The relationship is called the lens equation and can be written:

$$\frac{1}{\text{focal length}} = \frac{1}{\text{object distance}} + \frac{1}{\text{image distance}}. \qquad (15.1.1)$$

The Lens Equation

One divided by the focal length of a lens is equal to the sum of one divided by the object distance and one divided by the image distance.

According to the lens equation, the image distance for a distant object is equal to the focal length of the lens. That agrees with our earlier discussion of focal length. But when the object is nearby, the image distance becomes larger than the focal length. That's why a camera lens moves away from the image sensor as you focus

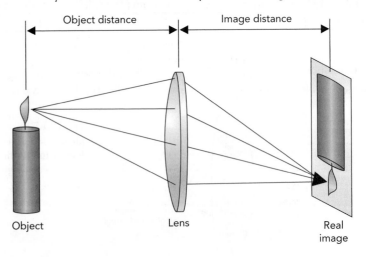

Object distance   Image distance

Object   Lens   Real image

Fig. 15.1.9 The relationship between the object distance, the image distance, and the focal length of the lens is given by the lens equation.

closer. And when the object distance becomes less than the focal length, the image distance becomes negative and no real image forms at all. That's why you can't focus on an object that's too close to the lens.

A lens's f-number characterizes the brightness of the real image it forms on the image sensor, with smaller f-numbers indicating brighter images. The f-number is calculated by dividing the lens's focal length by its diameter. Since long focal length lenses naturally produce larger and dimmer images on the image sensor, the f-number takes into account both the light-gathering capacity of the lens and its focal length. Increasing a lens's diameter increases its light-gathering capacity and decreases its f-number. Increasing a lens's focal length decreases the brightness of its real image and increases the f-number. Doing both at once, increasing the lens diameter and focal length equally, leaves the brightness and f-number unchanged.

Most sophisticated cameras use large-diameter lenses so that their f-numbers are generally less than 4. Since it's difficult to fabricate a lens that's larger in diameter than its focal length, the smallest practical f-number is about 1. And because long focal length lenses need large apertures to keep their f-numbers small, some telephoto lenses are huge.

The diaphragm inside a lens allows you to decrease the lens's aperture and thus increase its f-number. A factor of 2 increase in f-number corresponds to a factor of 2 decrease in the lens's effective diameter and a factor of 4 decrease in the lens's light-gathering area. Thus when you double the f-number of the lens, you must compensate by quadrupling the exposure time. Although closing the aperture increases the lens's depth of focus, it requires a longer exposure.

Improving the Quality of a Camera Lens

A high-quality camera lens isn't a single piece of glass or plastic. Instead, it's composed of many separate elements that function together as a single lens. This complexity improves the quality of the real image. To begin with, dispersion in a single element lens causes different colors of light to bend differently and focus at different distances behind that lens. Known as chromatic aberration, this problem can be fixed by using several lens elements made of different types of glass, with different amounts of dispersion. These elements compensate for one another so that the overall lens, known as an achromat, has very little dispersion and almost no color focusing problems.

After correcting for color and other technical image problems ❶, a sophisticated camera lens may contain more than ten individual elements. For the purposes of the lens equation, this complicated lens has an effective center from which to calculate object and image distances. But having so many separate elements creates reflection problems; each time light passes from air to glass or vice versa, some of it reflects. To avoid fogging the photographs with this bouncing stray light, the individual elements are antireflection coated with thin layers of transparent materials. The best coatings use interference effects to cancel out the reflected light waves and give the lens only a weak violet reflection.

Many modern cameras are equipped with zoom lenses. A zoom lens is a complicated lens that can change the size of the real image it projects onto the image sensor. By carefully moving its lens elements relative to one another, the zoom lens can adjust its effective focal length.

A common type of zoom lens contains three separate groups of lens elements and produces a sequence of three images (Fig. 15.1.10). The first lens group forms a first image of the scene in front of the camera. The second lens group forms a second image of that first image. And the third lens group projects a third, real image of the second image onto the image sensor. Zooming the lens, that is, changing its focal length, involves altering the spacings between the lens groups in order to

❶ Good lenses correct for a number of other image imperfections, always trying to project all of the light from one part of the scene onto one part of the image sensor. These image imperfections include spherical aberration, coma, and astigmatism. *Spherical aberration* occurs because lenses are normally ground with their two curved surfaces shaped like parts of large spheres. While spherical shapes are easy to form mechanically, they don't make ideal lenses. Aspheric lenses are usually better, but require sophisticated lens forming equipment. *Coma* occurs when trying to focus light from an object that's far from the center of the picture. Lenses have difficulty focusing light passing through them at shallow angles. *Astigmatism* occurs because the corners of the flat image sensor are actually farther from the lens than is the sensor's center.

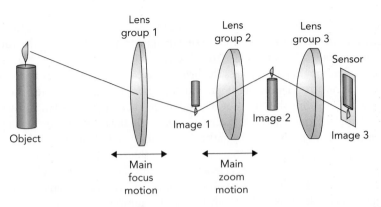

Fig. 15.1.10 A common type of zoom lens uses three lens groups to project a variable size real image on the image sensor. Zooming is done mostly by moving the second lens group to change its image and object distances. The first lens group is responsible for focusing and the third lens group projects the real image on the sensor.

vary the second lens group's object and image distances and thus the relative sizes of the first and second images.

As the zoom lens changes from short focal length to long focal length, the image it projects on the sensor becomes larger. This effect allows you to compose the picture so that the scene fills the photograph completely. A lens that can change its focal length while retaining the same f-number and still keep the real image in focus on the sensor is a truly remarkable achievement.

The Viewfinder and Virtual Images

SLR or single lens reflex cameras permit you to change their lenses so that you can choose a lens that's optimized for the task at hand. When you peer through the viewfinder of an SLR camera (Fig. 15.1.11), you're looking at the same real image that will be projected onto the image sensor during the exposure. The light you see travels through the camera's main lens, reflects off a mirror, and projects onto a translucent screen inside the top of the camera. You're simply inspecting this screen and the real image through a magnifying lens in the eyepiece. During the exposure, the mirror flips out of the way and the real image projects briefly onto the image sensor.

Since the screen and real image are only an inch or two from your eye, you can't focus on them without the help of the eyepiece lens. The eyepiece lens is converging, but in this case it doesn't form a real image. Instead, it forms a virtual image—an image located at a negative image distance, that is, on the wrong side of the lens!

The screen displaying the scene that you're photographing is so close to the eyepiece lens that the object distance is less than that lens's focal length. According to the lens equation, the image distance should be negative and it is; the image is located on the screen side of the eyepiece lens (Fig. 15.1.12). You can't put your fingers in the light and project this image on your skin because the image is virtual rather than real.

You can, however, see this image through the eyepiece. It's located farther away than the screen itself, so your eye can comfortably focus on it. And the image is magnified—the eyepiece lens is acting as a magnifying glass (Fig. 15.1.13). This lens provides magnification because when you look at the screen through it, the screen image covers a wider portion of your field of vision. This magnification increases as the eyepiece lens's focal length decreases. That's because a shorter focal length eyepiece lens must be quite close to the screen, where it can bend light rays coming from a smaller region so that they fill your field of vision. The eyepiece lens in

Fig. 15.1.11 The mirror in the center of this reflex camera directs light from the lens (removed for this photograph) onto the focusing screen above it. During the exposure, the mirror swings upward to permit light from the lens to strike the image sensor at the back of the camera.

a typical camera has been chosen so that the screen fills a comfortable portion of your visual field, allowing you to examine the virtual image in great detail and adjust the lens and camera settings until you have just the right picture in your view. Then all you have to do is take the photograph.

(a)

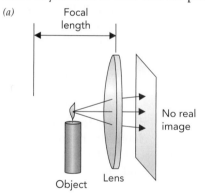

Fig. 15.1.12 (*a*) Light from an object very near a converging lens diverges after passing through the lens and no real image forms. (*b*) Your eye sees a virtual image that is large and far away.

(b)

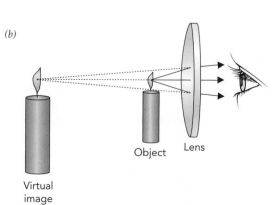

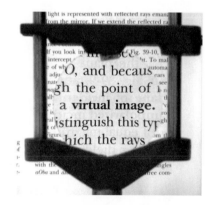

Fig. 15.1.13 This magnifying glass creates an enlarged virtual image located far behind the printed text. You can't touch the image or put your fingers in its light, but you can see it clearly with your eyes.

Cameras with fixed lenses often have two separate viewfinder systems. A typical digital camera has a small electronic viewfinder, which displays the real image being projected onto its image sensor. But many digital and all film cameras also have optical viewfinders. While optical viewfinders vary in style and sophistication, the best combine real and virtual images. In a real-image optical viewfinder, a system of lenses, mirrors, and/or prisms produces an erect real image of the scene and you then examine that real image through an eyepiece magnifying glass. The lenses projecting the real image zoom along with the camera's main lens so that what you see through the viewfinder is similar to what the camera's image sensor will record.

Image Sensors

Once the lens has projected its real image onto the image sensor, it's the image sensor's job to record that pattern of light. Interestingly enough, both film and electronic image sensors use semiconductors and both detect light when its photons shift electrons from valence levels to conduction levels. But how those two image sensors act on the electron transitions is quite different.

Photographic film detects light photochemically. Embedded in the film are tiny crystals of silver salts. Composed primarily of silver and halogen atoms, these semiconductor crystals are extremely sensitive to light. When a silver halide crystal absorbs a photon of visible light, it can undergo a radiative transition that shifts an electron from a valence level to a conduction level and eventually frees one silver atom from a silver halide molecule. After several nearby silver atoms have been freed by light, they can form a tiny particle of silver metal. When the film is developed, this silver particle transforms the entire silver halide crystal into metallic silver. The microscopically rough structure of that silver makes it appear black rather than shiny.

In black-and-white photography, the silver particles themselves form a negative image on the developed film. Wherever the film was struck by light, it acquires a dense, black pattern of silver particles. Wherever light was absent, the film becomes clear once the unexposed silver salts are washed away. Although the image on the developed film itself is negative—light is dark and dark is light—the process of preparing photographic prints reverses light and dark a second time so that the image on the prints is positive.

In color photography, the silver halide crystals are exposed to light through color filters and sensitizers, so that the film separately records its exposure to the three primary colors of light (see Section 14.2). During development, the silver itself is washed away, but a negative color image remains in the film. For example, wherever blue light struck the film, it acquires a yellow tint and therefore absorbs blue light. Again, the photographic printing process reverses the colors a second time so that the prints have positive images.

Electronic image sensors detect light with photodiodes—diodes that are optimized to detect light. A photodiode is approximately the reverse of a light-emitting diode (Section 14.3). Recall that a light-emitting diode *emits* light when conduction level electrons cross from the cathode to the anode and undergo radiative transitions to the valence levels. The LED is using an electric current to produce light. In contrast, a photodiode *absorbs* light when valence level electrons undergo radiative transitions to conduction levels and cross from the anode to the cathode. The photodiode is using light to produce an electric current.

An electronic image sensor uses a vast array of microscopic photodiodes to record the pattern of light in the real image. As it's exposed to light, each photodiode accumulates electrons on its cathode and the camera subsequently measures the accumulated charge on each of its photodiodes. To obtain color information, the image sensor's photodiodes are covered with a pattern of red, green, and blue filters so that each photodiode measures the intensity of only one primary color of light.

Eyes and Eyeglasses

Not all cameras involve modern technology. Most people are born with two of them: their eyes. Like the cameras we discussed above, each eye consists primarily of a converging lens and an image sensor (Fig. 15.1.14). In this case, the lens is a combination of the front surface of the eyeball, its cornea, and the internal lens just beneath the cornea. The image sensor is the retina, a vast pattern of light-sensitive cells and nerves at the back of the eyeball.

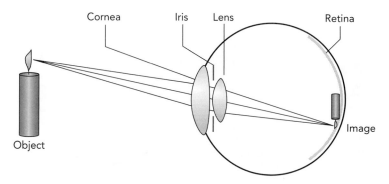

Fig. 15.1.14 An eye is a camera, with its cornea and lens forming a real image on the retina. The eye focuses by changing the curvature of its lens. The iris not only changes the brightness of the image, it also changes the eye's f-number and thus the depth of focus.

When you look at the scene in front of you, the cornea and lens of your eye project a real image of that scene onto your retina and your retina reports the resulting pattern of light to your brain. As usual, the real image is inverted and reversed left to right, but your brain compensates for that effect.

Since your eyeball can't alter the distance between the lens and the image sensor, it focuses the real image by adjusting the focal length of the lens. When you look at nearer objects, the lens in your eye becomes more highly curved and its focal length decreases. The light rays from that nearer object thus converge more sharply and form a real image on your retina. When you view a more distant object, the lens becomes less curved and its focal length increases.

Like a sophisticated camera, your eye has an iris within its lens system. When you view a bright scene, that iris shrinks to limit the amount of light striking your retina. As a side effect, your depth of focus increases and everything appears sharper. It's easier to focus when you read or work in a well-lighted environment.

But not all eyes are perfect and many need help forming sharp real images on their retina. Although modern laser surgical techniques can reshape corneas to improve image sharpness, the classic approach to better vision is to wear eyeglasses or contact lenses. An eye's lens system already consists of two components, the cornea and the lens, so adding a third component, eyeglasses, is no big deal.

A person who is farsighted can't see nearby objects sharply because her lens system has too long a focal length (Fig. 15.1.15a). While it can project real images of distant objects on her retina, nearby objects focus too far away from the front of her eye and the light strikes her retina before it forms a real image.

To compensate for farsightedness, she wears eyeglasses with converging lenses (Fig. 15.1.15b). These lenses begin the task of bending light rays together even before they enter her eyes. Her own lens system completes the bending of these

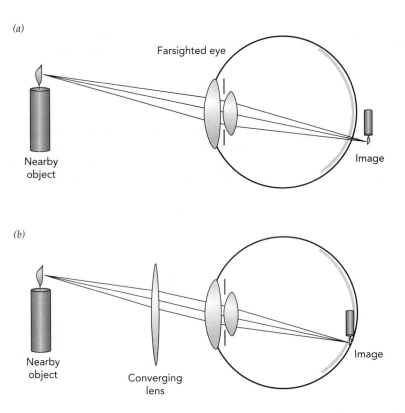

(a)

Farsighted eye

Nearby object

Image

(b)

Nearby object

Converging lens

Image

Fig. 15.1.15 (*a*) A farsighted eye bends light too weakly to focus on a nearby object. The real image forms beyond the retina. (*b*) A converging lens shifts the real image forward so that it focuses on the retina.

rays and the real images form closer to the front of her eyes. She is thus able to see distant objects clearly.

In contrast, a person who is nearsighted is unable to focus on distant objects because his lens system has too short a focal length (Fig. 15.1.16*a*). Real images of those distant objects form too close to the front of his eye and the light has already begun to spread apart by the time it reaches his retina.

To compensate for nearsightedness, he wears eyeglasses with diverging lenses (Fig. 15.1.16*b*). A diverging lens is one which bends light rays apart and therefore has a negative focal length. Typically thinner at its middle than at its edge, a diverging lens bends the nearly parallel rays of light from a distant object so that they diverge more rapidly. Those rays then appear to come from a much nearer object, actually a nearby virtual image, and his eyes are able to focus them properly on his retina.

⋔ Projectors

Projectors are nothing more than cameras in reverse. While a camera projects a real image of a distant scene onto a nearby film or image sensor, a projector projects a real image of a nearby film or image source onto a distant screen. But there are a few details that make a brief description of projectors worthwhile.

First, the real image that the projector forms on the screen must be bright enough to be visible to everyone in the theater. The larger this image is, the more spread out the illumination. The brightness of the image depends on its area and doubling the width of the image quadruples its area. To keep the image bright, the amount of light leaving the projector must increase as the square of the image width.

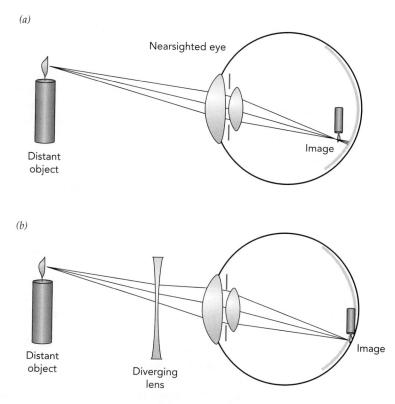

(a)

Nearsighted eye

Distant
object

Image

(b)

Distant
object

Diverging
lens

Image

Fig. 15.1.16 (*a*) A nearsighted eye bends light too strongly to focus on a distant object. The real image forms before the retina. (*b*) A diverging lens shifts the real image backward so that it focuses on the retina.

Even in a darkened theater, the light illuminating the piece of film or image source must be extremely bright. It also needs to be white so that it doesn't shift the colors that appear in the image. The only sources that provide light that is white enough and bright enough for this purpose are electric discharge lamps.

Until recently, the standard lamp used in movie projectors was a carbon arc lamp (Fig. 15.1.17). A carbon arc lamp uses an electric discharge between two carbon rods to produce a dazzling white light. The two rods are touched together briefly to start the discharge, which then continues across a small air-filled gap. The discharge is sustained as long as there are enough charged particles in the gap to carry an electric current from one rod to the other. Since the rods slowly burn up in the air, a motor gradually advances them toward one another.

More recently, carbon arc lamps have been replaced by xenon, krypton, and metal-halide arc lamps. These high-pressure electric discharge lamps create brilliant white light without having to replace burned carbon rods. However, these lamps are expensive and don't last long—material from a lamp's electrodes sputters onto its walls so that it darkens with use and must eventually be replaced.

In operation, light from the discharge lamp is collected with a parabolic mirror and directed through the film or image source. I'll discuss parabolic mirrors when examining telescopes later in this chapter. For now, all that matters is that the parabolic mirror collects all of the light originating in the small discharge and directs it in one direction. This process of directing light into an intense, parallel beam is called *collimation*.

Once the film or image source is brightly illuminated with white light, it serves as the object and the projection lens creates a real image of it on the screen. Because real images are always upside down and backward with respect to their objects, the image on the film or image source is upside down during projection.

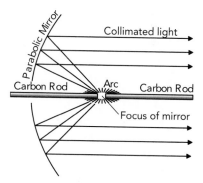

Fig. 15.1.17 Movie and video projectors use electric discharge lamps to illuminate the film or image source. Here, a discharge between two carbon rods produces an intense white light that is collimated by a parabolic mirror into a brilliant beam of white light.

For about a century, projectors worked with film. Still projectors worked with slides and movie projectors worked with spools of movie film. Film creates a bright object with which a projector can work by absorbing unwanted parts of the white light passing through it. The light leaving the film looks like the original scene and the projector casts an enlarged real image of that film image onto the screen. In a still projector, a single piece of film casts its real image on the screen for as long as you like. An overhead projector works in much the same way, although the film is a large piece of plastic and you can write on it.

Movie projection is a little more complicated than slide projection because it involves motion and sound. Instead of projecting a single image on the screen, the movie projector projects a sequence of images at a conventional rate of 24 images-per-second. This means that, 24 times a second, the light from the projector's discharge lamp is temporarily blocked, the current film frame is removed and replaced with a new film frame, and the light is restored. The image on the screen dims away to darkness and a new image appears in its place about a hundredth of a second later.

Persistence of vision prevents us from seeing the flicker between projected images, just as it keeps us from seeing how a television image is constructed. Our eyes simply can't tell the difference between a continuously moving object and a rapid sequence of still pictures of that object.

Persistence of vision was identified in 1824 by English physician Peter Mark Roget (1779–1869) and became the basis for a number of moving picture toys during the middle and late nineteenth century. But it wasn't until 1894 that Thomas Edison's Kinetoscope made it possible for a single person to observe a substantial (15 s) film of people and objects in motion.

In the Kinetoscope, the observer looked directly at a loop of celluloid photographic film that was passing over a blinking light source. The light turned on briefly as each film frame passed before the observer's eyes. The observer saw a rapid sequence of still images and perceived motion. By 1895, Edison and others had developed machines that could project real images onto screens. Edison's projector was called the Vitascope and was essentially a Kinetoscope with a projection lens. Within a decade of these inventions, the motion picture industry was in full swing.

Sound appeared in 1927 with the release of *The Jazz Singer*. While in the early years of sound movies the sound was recorded on a separate medium, it was later included as a separate region of the photographic film itself. The compressions and rarefactions of air that are sound were represented as clear and opaque regions on a sound track that appeared to one side of the photographic images. Light from an incandescent lamp was directed through this sound track onto a light sensor. The light sensor controlled the current flowing into an amplifier and speakers. Thus as the film and sound track moved past the light sensor, the current through the sensor fluctuated up and down and the speakers moved to reproduce the sound. Disney Studios called attention to the optical sound track in their 1940 film *Fantasia*, where the track appeared in the visual portion of the film as well as the audio portion.

Modern movies include a magnetic sound track, permanently bonded to the film itself. This magnetic sound track provides all of the quality and flexibility possible in modern magnetic tapes, including several separate sound channels and digital sound techniques. During filming, sound is recorded from many separate microphones and carefully mixed and blended to create the magnetic sound track.

Finally, color movies appeared in the 1930s. Like color photographs, color movies trick our eyes into seeing colors by stimulating our three types of color sensors, red, green, and blue, in the proper amounts. Our brains then perceive the original colors of light as they appeared in the original scene.

Early color movies didn't use the complete range of additive colors. Instead of using three primary colors of light, they use a more primitive grouping of two basic colors. In these movies, two separate film frames were exposed simultaneously, each recording a different range of wavelengths of light. For example, Technicolor recorded the red-orange colors on one strip of film and the green-blue colors on another strip of film.

These negative film images where printed onto new strips of film, forming positive images. The positive images were then tinted with colored dyes so that the film strip recording red-orange light became red-orange and the film strip recording green-blue light became green-blue. The two strips were then laminated together and the resulting filmstrip appeared to have nearly a full range of natural colors. Although three primary colors are needed to make us see a complete spectrum of visible colors, two color systems are quite effective at fooling our eyes into perceiving most colors.

But the two color systems were awkward, expensive, and couldn't reproduce every color. Eventually Technicolor was replaced by films such as Kodachrome, which record and reproduce all three primary colors of light simultaneously with a single strip of film. These and subsequent three color films consist of several thin light-sensitive layers on a clear plastic substrate, just like still camera film. To produce the actual movie projection film, the film negative recorded by the camera is printed onto another strip of color negative film, creating a positive copy. This copy appears red wherever red light was present in the scene, blue where blue light was present, and green where green light was present.

Film projection, however, is gradually being replaced by video projection, even in movie theaters. Instead of projecting the light emerging from a piece of film, a modern video projector projects the light emerging from an electronic image source. The first video projectors used picture tubes as their image sources and projected light produced by those picture tubes onto distant screens. Unfortunately, color picture tubes aren't bright enough, so these projectors had to employ black-and-white tubes. Each color video projector used three expensive, high-brightness black-and-white picture tubes, each with a different color filter, and three huge, carefully match projector lenses to form the overall color image on the screen. Overlapping the red, green, and blue images on the screen properly was a difficult task and color problems were common.

Fortunately, newer projectors use electronic imaging surfaces that act like electronic film. Each of these imaging surfaces has thousands or millions of tiny pixels and creates its source image by selectively transmitting or reflecting light from a collimated arc lamp. Because the imaging surface is small and extremely bright, a single, small projector lens is all that's required and many projectors have lenses with variable focal lengths (i.e., zoom lenses) so that the size of the projected image can be adjusted.

A modern video projector can use a single lens because all three primary colors of light originate from the same image source. In some cases the image source has color filters built into its pixels so that the source image is inherently colored all the time. In other cases, a filter wheel spins rapidly through the projector's optical path so that the projector works alternately with red, green, and blue light. As red light is passing through the projector, the electronic imaging surface forms the brightness pattern for red light. It does the same for green light and then blue light before starting over with red. This sequence repeats so rapidly that your eye doesn't see it.

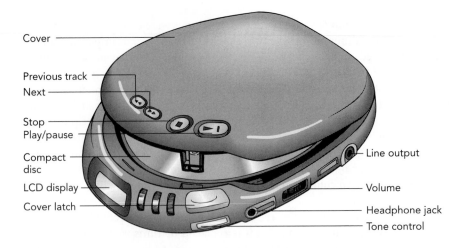

Cover

Previous track

Next

Stop

Play/pause

Compact disc

LCD display

Cover latch

Line output

Volume

Headphone jack

Tone control

Section 15.2 Optical Recording and Communication

Using light to convey information is as old as signal fires and as natural as sight itself. But advances in light sources, optical materials, and electronics have radically increased the possibilities for optical information systems. Optics and information go so well together that they're partly responsible for the current information revolution. The introduction of compact disc players in the early 1980s transformed the music industry virtually overnight, and optical fibers are knitting our world together at an astonishing pace. In this section, we'll look at how optical devices use light to manipulate information.

Digital Recording

Although analog techniques dominated audio recording for about a century, they were replace by digital techniques in less than a generation. It was mostly a matter of sound quality. Analog recording, in which a continuous physical quantity represents the density fluctuations that are sound in air, is susceptible to noise, wear, and a myriad of other imperfections. Phonograph records and analog tape recordings simply can't provide perfect sound.

In contrast, digital recordings can approach perfection. By representing sound in digital form, using many physical quantities with discrete values to represent density measurements numerically, digital recording can provide exact sound information that's free of noise and wear. As long as those density fluctuations are measured accurately, digital recording can recreate them with uncanny precision.

Like the audio players we examined in Section 12.2, audio CDs and DVDs represent sound's compressions and rarefactions of the air as numbers (Fig. 15.2.1). These numbers are essentially air density measurements taken over and over again during the recording process, and the player uses them to reproduce the recorded sound.

In the conventional CD format, density measurements are made 44,100 times each second for two independent audio channels and these measurements are recorded on the disc in binary form, using 16 bits for each measurement. Since the air density can go down as well as up, these bits represent the positive and negative integers from -32,768 to 32,767, which in turn represent how much the air density is above or below the average density. Density measurements with 16 bits of precision are sufficient to reproduce both loud and soft music with almost perfect fidelity.

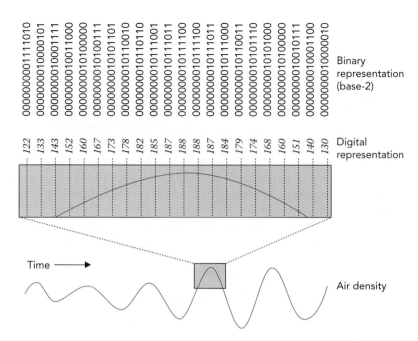

Fig. 15.2.1 Sound can be represented in digital form as a series of numbers. Each number in this digital representation corresponds to the air density at a particular moment in time, relative to the average density. Those numerical values can themselves be represented in binary form, using the digits 0 and 1.

Like CDs, DVDs record information digitally. But DVDs are newer technology and therefore more sophisticated. Audio DVDs can choose from several measurement rates, bits per measurement, and numbers of channels. A typical DVD might have five audio channels: left-front, center-front, right-front, left-rear, and right-rear. The three front channels might have 96,000 density measurements per second at 24 bits per measurement, and the two rear channels might have 48,000 measurements per second at 20 bits per measurement. While all of these samples, bits, and channels involve far more information than is stored on a CD, a DVD compresses that information before storing it. In contrast, a conventional CD's information is uncompressed although some more modern formats (e.g., mp3) do employ compression techniques.

In either case, air density measurements aren't simply recorded one after the next on a disc's surface. Instead, these numbers are extensively reorganized before they're stored. This reorganization allows the player to reproduce the sound perfectly even if the disc can't be read completely. As we'll see shortly, reading these discs is a technological tour de force and susceptible to various failures. To be sure that the sound (and video) can be reproduced completely and without interruption, the numbers are recorded in an encoded manner. They appear redundantly so that even if one copy of a number is illegible, there is still enough legible information along the same arc of the disc's spiral track to completely recreate that missing number. This duplication of information reduces the playing time of both CDs and DVDs but is essential for reliability.

Its encoding scheme leaves a CD or DVD almost completely immune to all but the most severe playback problems. In principle, you can damage or obscure a 2-mm-wide swath of the disc, from its center to its edge, and the player will still be able to reproduce the sound (and video) perfectly. But damage along an arc of the spiral track is far more threatening to the data. If the player can't read a long stretch of a single arc, it won't be able to recover the information. That's why you should always clean a CD or DVD from its center outward to its edge.

The Structure of CDs and DVDs

Standard CDs and DVDs are 120 mm (4.72 in) in diameter and 1.2 mm (0.05 in) thick. One side of a CD is clear and smooth but the other side contains a sandwich of layers: a thin film of aluminum, a protective lacquer, and a printed label (Fig. 15.2.2b). In contrast, a DVD is laminated from two 0.6-mm-thick clear plastic discs, with one, two, or four reflective layers of aluminum, gold, or silicon stacked up in between them (Fig. 15.2.2d). The more layers in the DVD, the more information it holds.

The reflective layers are the recording surfaces. These layers are so thin that they actually transmit a small amount of light. In the gold or silicon DVD layers, this semitransparency is essential because it allows the optical system that reads information to send light through the semitransparent layer to the aluminum layer beyond it. But even the aluminum layers transmit some light. While aluminum's electrons accelerate in response to the light's electric field and normally reflect that light completely, there aren't enough electrons in these 50 to 100-nm-thick layers to do the job and some light gets through.

The reflective layers aren't perfectly smooth. Instead, each has a narrow spiral track formed in its surface (Fig. 15.2.2b,d). This track is a series of microscopic pits, as little as 0.83 μm long on a CD or 0.40 μm long on a DVD. Adjacent arcs in the spiral track are only 1.6 μm apart on a CD and just 0.74 μm apart on a DVD (Fig. 15.2.2a,c). The lengths of the pits and the flat "lands" that separate them represent numbers. The player examines these pits and lands as the disc turns and converts their lengths into numbers, sound, and video.

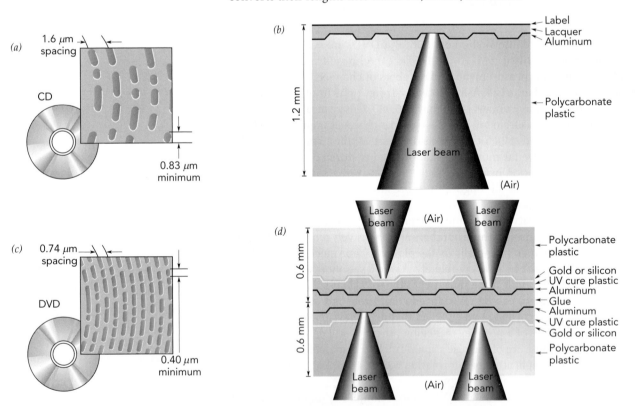

Fig. 15.2.2 (a,b) A CD contains a thin layer of aluminum on one side of a clear plastic disc. The aluminum layer has tiny pits that are detected by a 780-nm laser beam. (c,d) A DVD contains up to four aluminum, gold, or silicon layers sandwiched between two clear plastic discs. The gold or silicon layers are semireflective. Pits in those layers are detected by a 650-nm laser beam. The beam can focus either on the semireflective layer or, by passing through that layer, on the aluminum layer beyond it.

The pit lengths and the spacings between arcs weren't chosen arbitrarily. Since electromagnetic waves are unable to detect structures much smaller than their wavelengths (see Section 13.2), the laser beam's wavelength limits the smallest features on a disc. In a CD player, that beam's wavelength is 780 nm in air and 503 nm in polycarbonate plastic—short enough to detect the pits of a CD easily. In a DVD player, the laser beam's wavelength is between 635 and 650 nm in air and between 410 and 420 nm in plastic—just short enough to detect the pits in a DVD. The wavelength reduction inside the disc occurs because polycarbonate plastic has an index of refraction of 1.55, meaning that the light's speed in that plastic is reduced from its vacuum speed by a factor of 1.55. Its wavelength is reduced by the same factor.

The player detects a pit by bouncing light from the disc and determining how much of it reflects. As the focused laser beam passes over a pit, the reflection becomes dim, in part because the curved pit scatters light in all directions and in part because of interference effects. Light that's reflected back from a pit travels farther than light that's reflected from the flat region around it, so electric and magnetic fields in the two waves are shifted relative to one another. The pit depth was chosen so that the two reflected waves are approximately out of phase and they interfere destructively. Overall, the player's light sensors detect relatively little light when the laser beam is located over a pit.

A CD or DVD player uses a laser diode to produce its light. The 780-nm standard for CD players was adopted in 1980, when 780-nm infrared laser diodes were reliable but still fairly expensive. Technology has advanced since then, however, and the 635- to 650-nm standard for DVD players reflects the development of inexpensive red laser diodes. New standards follow technology, so optical recording systems based on blue, violet, and ultraviolet laser diodes are beginning to appear.

The Optical System of a CD or DVD Player

A CD or DVD player's optical system measures the lengths of the tiny pits as they move by on a spinning disc. That reading process requires incredible precision. Not only must the player focus its spot of laser light exactly on the reflective layer, but it must also follow the spiral track as it moves by. The disc itself is neither perfectly flat nor perfectly round, so the player must continuously adjust its reading unit during playback. The optical system must keep its laser beam focused on the reflective layer (autofocusing) and must follow the track as it passes (autotracking). These two automatic processes are beautiful examples of the use of feedback.

The basic structure of a typical CD or DVD player is shown in Fig. 15.2.3. Light from a laser diode passes through several optical elements on its way to the disc's reflective layer. It comes to a tight focus on that layer, where it illuminates only a single track. Some light reflects from the layer and returns through the optical elements. Finally, the reflected light turns 90° at a special mirror called a polarization beam splitter and focuses on an array of light detectors. The player measures the electric currents flowing through the detectors and uses those measurements both to obtain data from the disc and to control the focusing and tracking systems.

Let's examine this optical system one element at a time. After leaving the laser diode, light passes through a polarization beam splitter. This device analyzes the light's polarization. As we saw in Section 14.1, different polarizations of light reflect differently when they strike a transparent surface at an angle. In this case, polarized light from the laser passes through the 45° surface, but light of the other polarization reflects. The beam splitter is specially coated to separate the two polarizations almost perfectly.

Light from the laser diode diverges rapidly as it passes through the beam splitter. It's not that the laser diode is broken or poorly designed; it's that a light wave emerging from a small opening naturally spreads outward, like ripples on a pond.

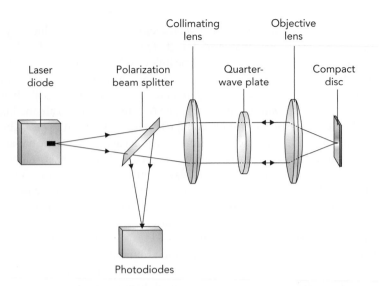

Fig. 15.2.3 In the optical system of a CD or DVD player, light from a laser diode passes through a polarization beam splitter, a collimating lens, a quarter-wave plate, and an objective lens before focusing on the reflective layer inside the disc. Reflected light turns 90° at the polarization beam splitter and focuses on an array of photodiodes.

This spreading is known as diffraction and occurs whenever a light wave is truncated by passage through an opening. The smaller the opening, the worse is the spreading. Because the emitting surface of the laser diode is essentially a very small opening, the laser beam spreads rapidly as it heads away from the diode. The player uses a converging lens located after the beam splitter to stop this spreading. At that point, the light beam is already wide enough that diffraction causes little additional spreading. The beam leaves the lens collimated, meaning that it maintains a nearly constant diameter after passing through the lens.

The laser light then passes through a quarter-wave plate. This remarkable device performs half the task of converting horizontally polarized light into vertically polarized light or vice versa. Horizontally and vertically polarized lights are said to be plane polarized because their electric fields always oscillate back and forth in one plane as they move through space. But the quarter-wave plate turns plane polarized light into circularly polarized light. We encountered circular polarization before in the radio transmissions from FM stations. In circularly polarized light, the electric field actually rotates about the direction in which the light is traveling.

Now the light passes through an objective lens that focuses it onto the reflective layer of the disc. On its way to the reflective layer, the light enters the plastic surface of the disc. At its entry point, the beam is still more than 0.5 mm in diameter, which explains why dust or fingerprints on the disc's surface don't cause much trouble. While contamination may block some of the laser light, most of it continues onward to the reflective layer.

The light comes to a tight focus just as it arrives at the reflective layer. Although it might seem that all of the light should converge together to a single point on that surface, it actually forms a spot roughly one wavelength in diameter. This spot size is limited by the wave nature of light. No matter how perfectly you try to focus light, you can't make a spot that's much smaller than the light's wavelength. Instead, the beam forms a narrow waist and then spreads apart (Fig. 15.2.4). This beam waist is about one wavelength of the light in diameter and a few wavelengths long, depending on the f-number of the converging lens. Since that waist is less than 2 microns long, the player's autofocusing system must keep the objective lens just the right distance from the reflective layer.

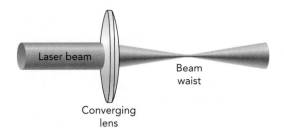

Fig. 15.2.4 When a converging lens focuses a laser beam, the light doesn't meet at a single point in space. Instead, it reaches a narrow waist with a diameter roughly equal to the wavelength of the light.

Laser beam

Beam waist

Converging lens

This fundamental limitation on how tightly a beam of light can be focused is another example of diffraction; the focusing lens truncates the light wave and inevitably introduces spreading. But even reaching this ideal focusing limit requires careful design and fabrication of the optical elements. While most other optical systems fall short of their ideal limits, a CD or DVD player's optical system does as well as can be done within the constraints of diffraction itself. Its optics are essentially perfect and are said to be diffraction limited.

The amount of light that reflects from the layer depends on whether or not the laser spot hits a pit. This reflected light follows the optical path in reverse. It's collimated by the objective lens and then passes through the quarter-wave plate again. The plate now finishes the job it started earlier; the light ends up plane polarized, but with the opposite polarization from what it had when it left the laser. Horizontally polarized light is now vertically polarized and vice versa.

The reflected light then passes through the collimating lens, which makes the light converge, and then strikes the polarization beam splitter. Because the light's polarization has changed, the beam splitter no longer allows the beam to pass directly through. Instead, it turns this reflected beam 90° and directs it toward the detector array. This clever redirection scheme is important for two reasons. First, it conserves laser light by allowing most of it to travel from the laser diode to the detector. Second, it prevents reflected light from returning to the laser diode, where that light would be amplified and cause the laser diode to misbehave.

The light comes to a focus on an array of photodiodes. This array allows the player to detect pits via the reflected light intensity and also to determine whether the objective lens is properly positioned relative to those pits. For generality, Fig. 15.2.3 omits optical elements involved in autofocusing and autotracking. However, because of those elements, the pattern of light hitting the detector array indicates which way the objective lens should move, if necessary. That lens is attached to coils of wire that are suspended near permanent magnets. By varying the currents flowing through those coils, the player uses Lorentz forces to move its objective lens about rapidly and keep it in the right place over the disc.

Optical Fibers

Optical playback of prerecorded discs is fine for music and movies, but it isn't of much use when you want up-to-the-minute information. The Internet and World Wide Web require communication links that operate at lightning speed. Yet even here, optics and light have an important role to play. The fastest way to send enormous amounts of information is to use optical fibers.

An optical fiber is a glass conduit that guides light from one place to another. Nearly every photon that enters the fiber at one end emerges from the other end moments later. In its simplest form, the fiber is made from two different glasses: a solid core of one glass surrounded by a cladding of the other glass. Both glasses are so incredibly transparent that light can travel through them for kilometers with little loss. For comparison, look through the edge of an ordinary piece of window glass and you'll see how dark the glass looks. It absorbs far too much light to be suitable for optical fibers. They're made of the purest glasses known.

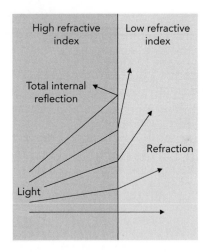

Fig. 15.2.5 When light traveling through one material enters a second material with a lower refractive index, it bends toward the boundary between those materials. If its approach angle is too shallow, the light will bend so much that it will simply reflect from the boundary. This effect is called total internal reflection.

If both glasses are almost perfectly transparent, what keeps the light from leaking out of the sides of the fiber? The answer is a phenomenon known as total internal reflection. As light tries to move from the inner glass core to the outer glass cladding, it's reflected perfectly and thus can't escape.

Total internal reflection is an extreme case of refraction. When light encounters the boundary between two materials with different indices of refraction, refraction causes that light to bend (Fig. 15.2.5). If the material it enters has a smaller index of refraction than the one it leaves, the light bends away from a line perpendicular to the boundary. The amount of this bend depends on the two indices of refraction and on the angle at which the light approaches the boundary. As long as the approach angle is steep enough, light will succeed in entering the second material. But if the approach angle is too shallow, the light won't enter the second material at all. Instead, it will reflect perfectly from the boundary. In fact, total internal reflection is far more efficient at reflecting light than a conventional metal mirror.

To keep the light inside the fiber's core, the core glass must have a higher index of refraction than the cladding glass. As light in the high-index core encounters the boundary with the low-index cladding, it experiences total internal reflection and bounces back into the core (Fig. 15.2.6a). As long as the fiber doesn't bend too sharply, the light bounces back and forth inside the core and can't escape. As a result, light entering the fiber core through one of its cut ends follows the fiber all the way to its other cut end.

But a large-diameter fiber (typically 50 μm or more) has a problem. Light rays bouncing through it at slightly different angles travel different distances during their passage through the fiber. Light heading almost straight down the center of the fiber rarely bounces and takes less time to complete its trip than light that bounces many times. Because this wide fiber has many bouncing paths or "modes" in which light can travel through it, a short pulse of light going through the fiber gets stretched out in time (Fig. 15.2.6a). This pulse broadening severely limits the rate at which information can be sent through a multimode fiber.

To reduce stretching problems, the core of a better performance optical fiber has a graded refractive index. The core glass is specially treated so that its index of refraction decreases smoothly away from its center toward the cladding. Instead of bouncing abruptly when it reaches the boundary between core and cladding, light in this graded index environment turns smoothly back toward the core (Fig. 15.2.6b). The path differences between the modes in a graded index multimode fiber aren't so different and a short pulse of light isn't stretched very much in a fiber of moderate length.

But in very long fibers, even small path differences add up and short pulses become blurred in multimode fibers. Therefore, the highest performance optical conduits are single-mode fibers. These fibers have very narrow graded index cores that permit light to travel only in one mode—effectively right down the center of the fiber (Fig. 15.2.6c). The core is typically only 9 μm in diameter. A pulse of light entering this narrow core broadens very little in time during its passage.

What little broadening occurs in a single-mode fiber isn't caused by the light taking different paths; it's caused by ordinary dispersion in the glass. To carry information, the light wave must change with time and thus must have a range of frequencies and wavelengths. The sidebands that we encountered with television transmission also appear in optical communication. As usual, the shorter wavelengths travel slower than the longer wavelengths and the pulses get stretched out in time. To minimize these dispersion effects, the highest speed optical fibers operate at wavelengths that minimize dispersion. They also operate at wavelengths that minimize loss of light through absorption in the glass. These two wavelengths coincide at 1550 nm in so-called dispersion shifted fibers, so this infrared wavelength is commonly used in long-haul optical communication.

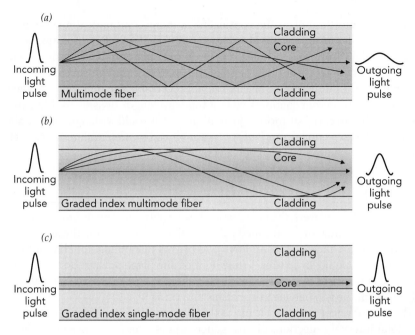

Fig. 15.2.6 Light trying to leave the high-index core of an optical fiber at a shallow angle undergoes total internal reflection at the boundary with the low-index cladding. (*a*) In an ordinary multimode fiber, a pulse of light can follow many paths through the core and becomes spread out in time. (*b*) A core with a graded refractive index exhibits less temporal spreading because the reflection process is more gradual. (*c*) The least spreading, however, occurs in a single-mode fiber. The small core diameter of this fiber provides light with only one mode of travel so that the only spreading occurs because of ordinary dispersion.

Optical Communication

A typical optical communication transmitter uses a 1550-nm laser diode to generate short pulses of light. These pulses carry information from the transmitter to a receiver somewhere far away. The transmitter produces its pulses by varying the current passing through the laser diode. Light emerging from the laser diode is focused into the exposed core of a single-mode optical fiber, and it follows the core all the way to the other end of the fiber. When the light emerges, it's gathered by a lens and focused onto the receiver's photodiode. Each pulse of light causes a pulse of current to flow through the photodiode, allowing the receiver to begin processing the information.

A laser diode and a single-mode optical fiber can send billions of bits of data per second for 50 km or 100 km without significant errors. At longer distances, the gradual absorption of light in the glass makes it difficult to receive the information reliably. The easiest solution to this problem is to receive the data before the light becomes too weak and then to retransmit it with another laser diode.

But instead of interrupting the optical transmission with a receiver and retransmitter, some long-haul communication systems employ erbium-doped fiber amplifiers (EDFAs). An EDFA is a piece of optical fiber that has about 0.01% erbium ions in its glass core. When the EDFA is exposed to 980-nm or 1480-nm light, it becomes a laser amplifier for 1550-nm light. As the weakened pulses of light from a long fiber pass through the fiber amplifier, the amplifier duplicates photons and brightens the pulses. These amplified pulses then continue through ordinary fiber before being amplified again. Undersea optical cables often splice fiber amplifiers into the fibers every 50 km or so. These amplifiers allow light to travel thousands of kilometers through a continuous path, from one side of an ocean to the other.

To get the most out of a single optical fiber, many communication systems use several laser diodes operating at somewhat different wavelength ranges around 1550 nm. Light from these diodes is merged together and focused into the fiber. When the light emerges at the far end of the fiber, its different wavelength ranges are split apart and directed onto individual receivers. The different wavelength ranges are like different channels, so that this wavelength-division multiplexing allows one fiber to carry far more information than it could with light from a single laser. Remarkably enough, an EDFA can amplify all of these different channels at once because erbium ions can copy a wide range of wavelengths.

Birefringent Polarization Beam Splitters

In the original CD players, the polarization beam splitter was made out of two pieces of a clear crystalline mineral called calcite (calcium carbonate). These two pieces are carefully cut and polished to form two prisms that almost touch one another in the completed beam splitter. While light of one plane polarization manages to pass from one prism to the other without difficulty, light of the other polarization reflects from the gap between the prisms and turns 90°.

The two physical effects that make this beam splitter work are birefringence and total internal reflection. Birefringence occurs when a material has different indices of refraction for the two plane polarizations of light. In a crystal such as calcite, the molecules are arranged in an orderly fashion, like stacks of apples in a grocery store. This orderly arrangement of molecules may respond differently to vertical and horizontal electric fields. If one of these fields produces more electric polarization in the crystal than the other, the crystal will respond differently to vertically and horizontally polarized lights and will have two different indices of refraction. Light of the plane polarization that affects the crystal most strongly will travel slower than light of the other polarization.

Calcite exhibits birefringence. The index of refraction for one plane polarization of light is 1.70. This light travels relatively slowly through the crystal and is said to be polarized along the crystal's slow axis (Fig. 13.2.7). The index of refraction for the other plane polarization is 1.52. This light travels relatively quickly and is said to be polarized along the crystal's fast axis. The calcite prisms of the polarization beam splitter are carefully cut so that horizontally and vertically polarized lights travel through them at different speeds.

The two polarizations of light also refract differently as they enter the air gap between the two prisms. Light polarized along the crystal's fast axis bends considerably as it leaves the first prism but still manages to cross the air gap and enter the second prism. However, light polarized along the crystal's slow axis bends so severely as it tries to leave the first prism that it never leaves at all. This light is perfectly reflected from the surface of the prism as the result of total internal reflection.

The polarization beam splitter is designed so that one plane polarization of light is able to negotiate the air gap between the calcite prisms while the other polarization of light experiences total internal reflection. Light polarized along the calcite's fast axis travels straight through the beam splitter with minimal disturbance. But light polarized along the calcite's slow axis is turned 90° and leaves the first calcite prism through a separate surface. In the compact disc player, light from the laser diode travels directly to the surface of the compact disc while light returning from the compact disc with the other polarization is turned toward the photodiodes. Although inexpensive, thin-film polarizing beam splitters are used in all modern optical disc systems, birefringent beam splitters are still used in high-powered laser systems and research laboratories.

Birefringence isn't unusual in crystals and it also appears in transparent glasses and plastics that are under stress. A piece of glass or plastic that is being bent, that

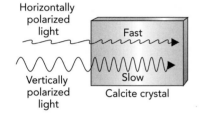

Fig. 15.2.7 The two plane polarizations of light travel at different speeds in calcite. The crystal shown above has been cut and oriented so that horizontally polarized light travels faster than vertically polar-ized light.

was cooled from its softened state carelessly, or that was deliberately tempered, may well exhibit birefringence. If it is birefringent, the material will refract or reflect different polarizations differently and may exhibit a mottled or striated appearance when examined with polarized light. Tempered rear car windows often produce blotchy reflections when viewed in the somewhat polarized light from the sky, particularly if you're wearing polarizing sunglasses. Glassblowers and engineers often used polarized light to look for birefringence in their materials in order to identify stressed regions.

♏ Phonograph Recordings

Long before CDs, DVDs, and audio players appeared on the scene, people enjoyed listening to phonograph records. For most of the twentieth century, phonograph records dominated the world of audio recording. They were a convenient and inexpensive way to retain and distribute sound. Phonograph records work on a principle pioneered by Thomas Edison: that hills and valleys on a surface can be used to represent compressions and rarefactions in the air. Although the materials used to make phonograph records changed over the years and the records went from monaural to stereo, the basic principles of phonograph recording remained the same throughout their reign.

A phonograph record has a narrow groove cut in its surface (Fig. 15.2.8). This groove forms a single continuous spiral from the outer edge of the record to its center. As the record rotates on a turntable, a carefully polished diamond stylus glides along the center of this groove. Because the groove's two surfaces, left and right, support the stylus, changes in the locations of those two surfaces affect the stylus's position in space.

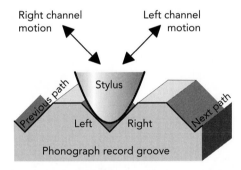

Fig. 15.2.8 A view down the groove of a phonograph record. As the record turns, the stylus glides along the spiral grove. Undulations in the groove's left and right surfaces move the stylus in two independent directions and are used to control the positions of the speaker cones and thus to recreate stereo sound.

The record represents sound as changes in the locations of the two sides of the groove. One side represents sound for the left stereo speaker and the other side represents sound for the right stereo speaker. Because the stylus has a tiny mass and the forces between it and the groove surfaces are large, the stylus easily follows the rises and falls in the two surfaces. The stylus is attached to a permanent magnet that moves inside two coils of wire. As the stylus and magnet move, the numbers of magnetic flux lines passing through the two coils change and currents are induced in those coils. These currents are amplified and used to power speakers. That's all there is to the playback process for a phonograph.

But the phonograph recording scheme has many shortcomings. Most importantly, anything that affects the exposed surface of the record diminishes the quality of the sound reproduction. Dust and fingerprints can accumulate in the grooves and alter the sound. You must clean records frequently to remove this debris, without adding new residues and without damaging the surface.

The records themselves are fragile and experience wear during the playback process. The sound quality of a record diminishes with every playing as sliding friction between the extremely hard stylus and the groove gradually wears away the record's surface. Scratches are also a concern because they cause abrupt accelerations of the stylus that produce loud noises during playback. Moreover, an unfortunately placed scratch can break down the barrier between adjacent turns of the spiral groove so that the record skips from one portion of the groove to another. The expression "like a broken record" refers to the nearly bygone era when a scratched record would play a passage over and over again because the stylus kept returning to a particular portion of the spiral groove.

But even when a record is in perfect shape, its sound reproduction is limited. Quiet sounds involve very small motions of the stylus while loud sounds involve very large motions. The groove must have both small and large undulations in its surfaces and there is simply a limit to how large or small those mechanical undulations can be. Furthermore, the phonograph record can't keep the left and right stereo channels separated perfectly. Undulations in the left surface of the groove cause the stylus to move back and forth at an angle that produces sound in the left speaker. Similar variations in the right surface of the curve produces sound in the right speaker. But the stylus's motion isn't perfect and two channels aren't really independent of one another. You don't notice this cross talk between the left and right stereo channels in most music, but it reduces the overall stereo effect.

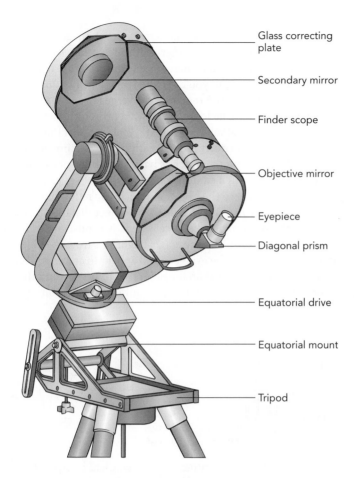

Glass correcting plate

Secondary mirror

Finder scope

Objective mirror

Eyepiece

Diagonal prism

Equatorial drive

Equatorial mount

Tripod

SECTION 15.3 Telescopes and Microscopes

Not everything that we wish to see is visible to the naked eye. We have trouble making out details in distant objects and can't resolve the tiny structures in nearby ones, either. We need help. To enlarge and brighten distant objects, we use telescopes, and to do the same with nearby objects, we use microscopes. Because these devices are almost identical in structure, though specialized to apparently different tasks, it's natural to describe them together.

Photographic and Visual Telescopes

I'll begin by looking at light telescopes, which form images of distant objects either on an image sensor or in front of your eyes. Most telescopes used by professional astronomers are photographic—essentially gigantic telephoto lenses that form real images of stars on pieces of film or electronic light sensors. However visual telescopes are still popular among amateur astronomers and for terrestrial work. Binoculars are essentially a pair of visual telescopes, carefully matched to one another and equipped with prism-based devices that produce upright images. A single visual telescope that produces an upright image is sometimes called a monocular.

A photographic telescope produces a real image in which light from one part of the sky is brought together on one part of the image sensor. But a visual telescope doesn't create a real image behind its final lens. Instead, it creates a virtual image in front of its final lens.

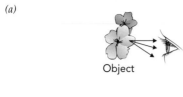

(a)

Object

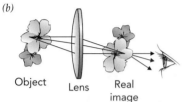

(b)

Object   Lens   Real image

Fig. 15.3.1 (*a*) The lens of your eye takes light diverging from an object and focuses it onto your retina. (*b*) When you look at a real image, you again see light diverging from a region of space and the lens of your eye focuses it onto your retina.

To understand a visual telescope, imagine removing the image sensor from a photographic telescope (or, equivalently, a camera with a telephoto lens). With nothing to stop it, light continues past the real image and forms a stream of diverging light. If you look into this stream of light, you'll see the real image floating there in space (Fig. 15.3.1). Because light from the real image diverges in the same way it would from a solid object, the real image looks like a solid object. In fact, it can be hard to tell whether you're looking at an object or at a real image of that object.

But in a visual telescope, you don't look directly at the real image; you look at it through an eyepiece or ocular lens, which is basically a high-performance magnifying glass. Like the eyepiece in a camera viewfinder, this eyepiece is effectively a converging lens that's so close to the object at which you're looking that a real image doesn't form *after* the lens (Fig. 15.3.2*a*). The object distance is less than the eyepiece's focal length, so the image distance is negative and a virtual image forms *before* the eyepiece (Fig. 15.3.2*b*). This means that light rays from the object continue to diverge after passing through the eyepiece and appear to come from an enlarged virtual image, located farther away than the object itself. While you can't touch this virtual image or put a piece of paper into it, it's quite visible to your eye.

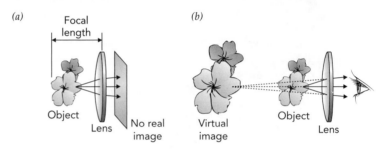

(a) Focal length   Object   Lens   No real image   (b) Virtual image   Object   Lens

Fig. 15.3.2 (*a*) Light from an object very near a converging lens diverges after passing through the lens. (*b*) Your eye sees a virtual image that is large and far away.

In a visual telescope, the object you look at through the eyepiece is actually the real image from the first part of the telescope. You use the eyepiece to enlarge the real image so that you can see its detail more clearly. You can focus the eyepiece, like any magnifying glass, by moving it back and forth. The farther the eyepiece is from the real image you're looking at, the less the light rays diverge after passing through the lens and the farther away the virtual image becomes. When the object distance is exactly equal to the eyepiece's focal length, the virtual image appears infinitely distant. You can then view the enlarged image with your eye relaxed, as though you were looking at something extremely far away.

The eyepiece provides magnification because when you look at an object through it, that object covers a wider portion of your field of vision. This magnification increases as the eyepiece's focal length decreases. That's because a short focal length eyepiece must be quite close to the object, where it can bend light rays coming from a small region so that they fill your field of vision. A long focal length eyepiece must be farther away from the object, where it can bend light rays coming from a large area to give you a broad but relatively low-magnification view of the object.

Refracting Telescopes

A common type of visual telescope uses two converging lenses: one to form a real image of the distant object and the second to magnify that real image so that you can see its fine detail (Fig. 15.3.3). The first lens is called the objective and the second lens is the eyepiece. Because this telescope uses refraction in its objective to collect light from the distant object, it's called a refracting telescope. More spe-

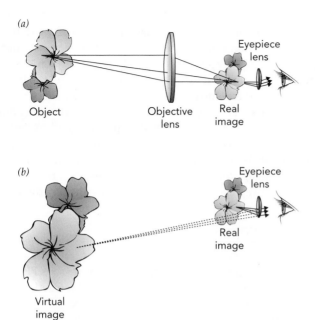

(a)

Eyepiece
lens

Object   Objective   Real
lens   image

(b)

Eyepiece
lens

Real
image

Virtual
image

Fig. 15.3.3 (*a*) A Keplerian telescope consists of two converging lenses. The objective lens forms a real image of the distant object. The eyepiece acts as a magnifying glass, allowing you to inspect that real image in great detail. (*b*) As you look through the eyepiece, you see a very large, far away, and inverted virtual image of the object.

cifically, it's called a Keplerian telescope, after the German astronomer Johannes Kepler (1571–1630). An alternative design that uses one converging lens and one diverging lens—a lens that bends light rays away from one another—is called a Galilean telescope, after the Italian scientist Galileo Galilei (1564–1642).

Because a Keplerian telescope's objective forms an inverted real image, the virtual image you see through the eyepiece is also inverted. This inversion is tolerable for astronomical use but is a problem for terrestrial use. It's hard to watch birds when they appear upside down and backwards. That's why binoculars, which are basically Keplerian telescopes, include erecting systems that make the final virtual image upright. These erecting systems fold the optical paths of the binoculars and give them their peculiar shape.

When you look at an object through a visual telescope, you see a magnified virtual image. The magnification of a Keplerian telescope depends on the focal lengths of the two lenses. As the focal length of the objective increases, its real image becomes larger and the telescope's magnification increases. As the focal length of the eyepiece becomes shorter, you look more closely at the real image so the telescope's magnification again increases. Overall, a Keplerian telescope's magnification is its objective's focal length divided by its eyepiece's focal length.

However, as we discussed in the section on cameras, simple lenses create poor quality images. Most telescope objectives and eyepieces are actually constructed from several optical elements. To avoid chromatic aberration (color focusing problems), these lenses are built from achromatic doublets, pairs of matched optical elements that are joined by clear cement. The two elements in a doublet are made from different types of glass so that their dispersions cancel and they act as a single lens that's almost free of chromatic aberration.

But even achromatic doublets don't create perfect images. That's why telescope eyepieces frequently contain four, five, or even six individual optical elements. These precision eyepieces produce images that are sharp from edge to edge. As in camera lenses, these individual elements are anti-reflection coated so that you don't see any reflections when you look into the telescope.

Telescope eyepieces come in a wide range of complexities and prices. An eyepiece usually has a black aperture that prevents you from seeing the distorted parts of the image outside the sharp central circle. In general, the more complex and expensive the eyepiece, the wider its usable field of view and the larger this open circle appears. The finest and most expensive eyepieces provide such broad, sharp views of the telescope's real image that you feel like you're looking directly at the scene in front of you—you can hardly see the aperture at all. Whether you're observing the stars or the landscape, the difference between the narrow porthole view offered by a simple, low-cost telescope or binoculars and the broad vista you see through a sophisticated, expensive telescope or binoculars is startling. Your first gaze through a good eyepiece on a good telescope can be breathtaking.

Reflecting Telescopes

But not all telescopes use lenses as their objectives. A reflecting telescope uses an objective mirror to create the initial real image. For reasons I'll discuss shortly, reflecting telescopes have almost completely replaced refracting telescopes for astronomical observation.

A flat mirror is already an interesting optical device. It forms a virtual image of any object (Fig. 15.3.4). Light diverging from the object bounces off the mirror and continues to diverge until it reaches your eye. This light appears to come from a virtual image located on the other side of the mirror, at just the same distance from the mirror as the object itself. This virtual image is reversed side-for-side from the object, so you see a "mirror image." When you view this mirror image, it looks the same size as the object itself.

If the mirror is curved rather than flat, it still creates virtual images. But the virtual images are no longer located exactly opposite the original objects and they don't appear the same size. A mirror that is bowed inward like the inside of a bowl is concave. When an object is located near a concave mirror, light diverging from the object bounces off the mirror and continues to diverge, but not as quickly (Fig. 15.3.5). It appears to come from a large virtual image that's located far behind the mirror (Fig. 15.3.6). This effect explains why makeup mirrors give you an enlarged view of your face. As long as your face is close to the bowl-shaped mirror, you see an enlarged virtual image of your face located far behind the mirror. (In contrast, a mirror that is bowed outward in the middle, a convex mirror, creates a reduced virtual image located just behind the mirror.)

However, when an object is located far from a concave mirror, light diverging from the object bounces off the mirror and converges to form a real image (Figs. 15.3.7

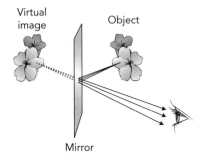

Fig. 15.3.4 A flat mirror forms a virtual image of any object in front of it. The virtual image is located exactly opposite the original object and is flipped side-for-side (or, more accurately, flipped front-for-back).

Fig. 15.3.6 A curved mirror produces an upright virtual image of a light bulb located just in front of it.

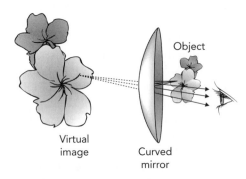

Fig. 15.3.5 A concave mirror forms a virtual image of any object in front of it and close to its surface. The virtual image appears larger than the original object and is more distant from the mirror than the object itself. Like the flat mirror, the virtual image is flipped side-for-side (or front-for-back).

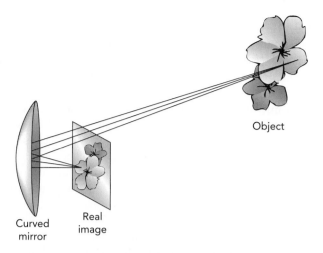

Object

Curved mirror

Real image

Fig. 15.3.7 When an object is far from a concave mirror, the mirror forms a real image of the object. In such a reflecting telescope, a photographic film or other light-sensitive surface is usually placed where the real image forms.

Fig. 15.3.8 A curved mirror produces an inverted real image of a light bulb located on a table far in front of it.

and 15.3.8). This real image is just like one formed by a lens except that it occurs *in front* of the mirror. You can project this real image onto a sheet of paper or onto an image sensor, which is how a photographic reflecting telescope works. As usual, the real image is inverted and flipped side-for-side. (For another use of concave mirrors and their real images, see ❶.)

A visual reflecting telescope simply adds an eyepiece, through which you can inspect the real image in great detail (Fig. 15.3.9). Through that eyepiece, you see an enlarged virtual image of the original object.

❶ Solar collectors are simply mirror telescopes pointed at the sun. Since sunlight comes from a great distance, its rays are essentially parallel and a parabolic mirror will focus them beautifully. The mirror forms a circular real image of the sun at its focus and that brilliant solar image can cook food, boil water, or process chemicals and materials.

(a)

Object

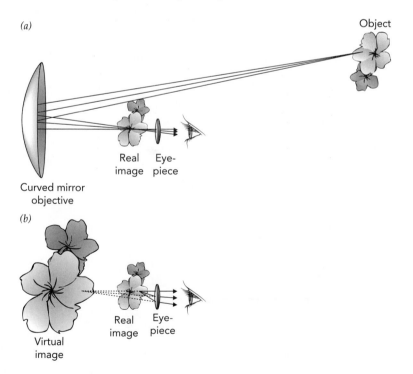

Curved mirror objective

Real image

Eye-piece

(b)

Virtual image

Real image

Eye-piece

Fig. 15.3.9 *(a)* A visual reflecting telescope consists of a curved mirror objective and a converging lens eyepiece. The concave mirror forms a real image of a distant object while the eyepiece allows you to inspect that real image in detail. *(b)* As you look through the eyepiece, you see an enlarged and inverted virtual image of the object out in the distance.

Like the objective lens of a refracting telescope, the mirror of a reflecting telescope must be carefully shaped to create a sharp real image. Since it uses reflection rather than refraction, it has no chromatic aberration and handles colors well. But a single mirror suffers from a variety of image imperfections, which, though usually negligible near the center of the real image, become progressively worse toward its periphery. The mirrors used for observing astronomical objects generally have parabolic shapes and create real images that are nearly perfect at their centers but relatively poor at their edges. Such telescopes normally only let you see the central portions of their real images, where those images are sharp and clear.

The visual reflecting telescope of Fig. 15.3.9 places your head in front of the curved mirror as you look into the eyepiece. While some giant telescopes allow you to sit in front of their mirrors to inspect their real images, most smaller telescopes try to get you out of the mirror's way. They redirect light from the objective mirror so that you aren't in front of it. The two most common ways of doing this are the Newtonian reflecting telescope (Fig. 15.3.10a) and the Cassegrainian reflecting telescope (Fig. 15.3.10b).

In a Newtonian telescope, a flat secondary mirror deflects light from the concave objective mirror by 90°, so that the real image forms at the side of the telescope. You can then look at this real image through an eyepiece without blocking light on its way to the primary mirror.

In a Cassegrainian telescope, a convex secondary mirror reflects light from the concave objective mirror back through a hole in the center of the primary mirror. Since the secondary mirror bends the light outward slightly, it delays the light's convergence so that a real image forms behind the telescope. You look at this real image through an eyepiece at the end of the telescope.

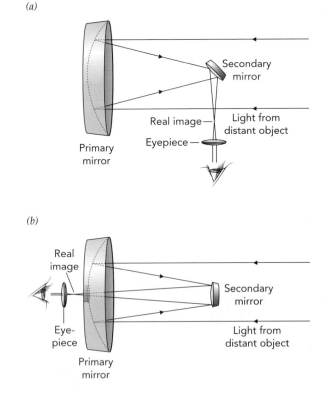

Fig. 15.3.10 (*a*) A Newtonian telescope bends light from the primary mirror by 90° so that you can view it from the side of the telescope. (*b*) A Cassegrainian telescope sends light from the primary mirror back through a hole in the center of the primary mirror so that you can view it from the end of the telescope.

In both the Newtonian and Cassegrainian telescope designs, something must support a small mirror in the center of the telescope's field of view. These rigid supporting structures or "spiders" often appear in the photographs of stars because of interference effects and give each star an artificial cross shape.

The Telescope's Aperture

The size of the telescope's lens or primary mirror determines two important features of the telescope: its ability to gather light and its ability to resolve two nearby objects from one another. The issue of light gathering is straightforward—the more light that the telescope can collect and focus into a real or virtual image, the brighter that image will be. Since objects in the night sky are dim and the telescope is spreading the light it collects over large images, the telescope should gather as much light as possible. High magnification is useless if all you see is enlarged darkness.

The other reason for building large mirrors is resolving power. Light travels as a wave. When a wave passes through a narrow opening, it tends to spread out as it travels into the region beyond the opening; it experiences diffraction. When the light waves coming from a distant star pass through the lens or mirror of a telescope, they diffract and begin to spread out somewhat. The larger the lens or mirror, the less diffraction occurs and the smaller the spread. This spreading may seem like a trivial detail. After all, the photographs that you take of a meadow or a schoolhouse appear extremely sharp and clear even if you use a small lens. But what if you were trying to count a person's eyelashes from a kilometer away? That's equivalent to what some telescopes try to do.

When the light from two nearby stars passes through a telescope, it diffracts. The spreading waves of light from one star may overlap with the spreading waves of light from the other star. In this case, the two stars will be unresolved. They will appear as a single bright spot on the image sensor or in the virtual image that you are viewing. To resolve the two stars, you need a lens or mirror that is large enough that light waves from the stars don't spread into one another. To reach the level of resolution needed by modern astronomers, telescopes have to be several meters in diameter.

Unfortunately, lenses more than about 0.5 m in diameter are extremely hard to fabricate because the glass becomes so thick and heavy that it deforms under its own weight. The glass also acquires stresses and strains as it cools from its molten state and these degrade its optical quality. To make matters worse, two separate elements are needed to make an achromatic doublet. While there are a few refracting telescopes with lenses more than 0.5 m in diameter, each of these lenses represents the lifetime's work of a master lens maker.

Because enormous lenses can't be made, astronomers have turned to mirrors to photograph or observe the dim, distant objects of our universe. Mirrors are easier to fabricate than lenses, although producing a mirror larger than about 3 m in diameter is still a heroic effort. Because mirrors don't have to be transparent, their internal structures can be optimized for strength and stability. Mirrors are often made of translucent glass-ceramics that tolerate temperature changes much better than clear glasses. In recent years, it's become possible to build composite mirrors, in which several mirror segments are designed and crafted by computer-controlled machines and positioned by computers and laser beams. These giant segmented mirrors can gather light from enormous areas.

Unfortunately, the earth's atmosphere limits the resolution of large mirror telescopes. As light from a star passes through the atmosphere, density variations in the air distort the wave just enough that it can't be brought to a single point in a real or virtual image. The most straightforward way to reach the full resolving potential of a giant mirror is to remove it from the atmosphere by putting it in space. The Hubble Space Telescope achieves this potential spectacularly.

However, adaptive optical techniques have made it possible for ground-based telescopes to achieve astounding resolution. These telescopes use computerized controls to alter the shapes of their flexible mirrors in order to compensate almost perfectly for atmospheric distortion. An adaptive telescope studies how the atmosphere distorts light from a real or laser-simulated star many times each second and uses that information to reshape its mirror so as to minimize atmospheric distortion. Although air still absorbs certain wavelengths of light and creates a background glow due to Rayleigh scattering, ground-based telescopes with adaptive optics are now quite competitive with space-based telescopes in their resolving powers.

Microscopes

Optical microscopes are almost identical to Keplerian telescopes, except for the focal lengths of their objective lenses and the locations of their objects. A Keplerian telescope uses a long focal length objective lens to project a large real image of a distant object while a microscope uses a short focal length objective lens to project a large real image of a nearby object (Fig. 15.3.11). The eyepieces in telescopes and microscopes are virtually identical and serve as high quality magnifying glasses through which to inspect the real images.

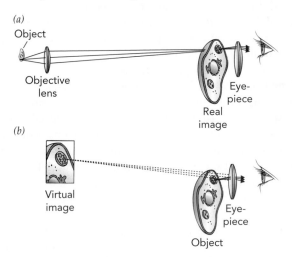

Fig. 15.3.11 (*a*) A visual microscope consists of a two converging lenses. The objective lens forms a real image of a tiny, nearby object. The eyepiece acts as a magnifying glass, allowing you to inspect that real image in great detail. (*b*) As you look through the eyepiece, you see a very large, far away, and inverted virtual image of the object.

Although microscopes are technically similar to telescopes, they have different constraints on their construction. While a telescope objective must be very large, a microscope objective must be very small. The microscope objective must move in close to the object, catch its rapidly diverging light rays, and make those rays converge together as a real image far from the lens. Because the objective's focal length must be even shorter than the object distance, the objective must be built from small, highly curved optical elements. Several elements are needed to form a high quality real image, one that will still appear sharp and crisp when you look at it through the magnifying eyepiece.

A microscope's magnification is difficult to calculate exactly, mostly because it depends on what you choose as the unmagnified view—an object held close to your eye looks bigger than one held farther away. Nonetheless, shortening the focal length of the objective lens always increases a microscope's magnification because the objective must then move closer to the object and will form a real image of a smaller region of the object's surface. Shortening the focal length of the eyepiece also increases the microscope's magnification because it then allows you to inspect the real image more closely.

Illumination is important for a microscope. In a telescope you must use the light that's available naturally, but in a microscope you can add whatever light you like. Most microscopes use a bright lamp to send light upward through the object or downward onto the object. What you observe in these two cases is different. If you send light through the object, you see how much light it absorbs in different places. If you bounce light off its surface, you see how much light it reflects in different places.

Many modern microscopes have two separate eyepieces so that you can use both eyes to look at the object. These microscopes have only one objective lens but light from this lens is separated into two parts with a partially reflecting mirror. Two identical real images form and each eyepiece magnifies its own real image. In fact, there are also microscopes that separate additional portions of the light for other viewers or for photography. In the latter case, some of the light from the objective lens is diverted and forms a real image directly on an image sensor.

Non-Light Telescopes and Microscopes

We observe objects in the universe almost exclusively through the electromagnetic radiation they send toward us. Electromagnetic waves travel easily through empty space and can cross the vast distances from one side of the universe to the other. But light isn't the only kind of electromagnetic wave. The earth is also exposed to radio waves, microwaves, X-rays, and gamma rays from sources elsewhere in the universe.

To study these electromagnetic waves, astronomers generally use reflecting telescopes. But the characteristics of those telescopes are somewhat different from light telescopes. A radio telescope uses a huge concave metal dish to form a real image of the electromagnetic waves from a distant object. In most cases, these waves are actually microwaves. Instead of using an optical image sensor or a magnifying glass to look at that real image, a radio telescope uses a small microwave receiver. The telescope studies the real image one spot at a time either by moving the microwave receiver around the image or by turning the dish and the receiver to point toward different regions of space.

Since diffraction effects are worse for longer-wavelength radio waves than they are for shorter-wavelength light, radio telescopes must use large diameter mirrors in order to produce detailed images of the sky. Fortunately, long wavelengths also permit radio telescope mirrors to be made from metal mesh rather than solid metal. That's because electromagnetic waves are essentially insensitive to holes that are substantially smaller than their wavelengths.

A home satellite dish antenna is just a small radio telescope. It forms a real image of the microwaves coming toward it from the sky and examines the center of that real image with a microwave receiver. When the antenna is pointed at a satellite, that satellite's waves come to a focus exactly on the receiver's antenna. And like its bigger relatives, a satellite dish can use metal mesh to reflect microwaves.

That electromagnetic waves are mostly unaffected by structures smaller than their wavelengths works against optical microscopes. Because a light microscope can't observe details much finer than the wavelength of visible light, you'll need another kind of microscope to see structures smaller than about 1 micron on a side.

Two important types of non-optical microscopes are electron microscopes and scanning probe microscopes. Electron microscopes use electrons to form images of small objects. Scanning probe microscopes use extremely sharp needles and tips to study the structures of surfaces, one point at a time.

The electrons in an electron microscope travel as waves and, like light, are unable to resolve structures substantially smaller than their wavelengths. However an electron's wavelength decreases as its kinetic energy increases so that the energetic electrons used in electron microscopes have wavelengths much less than atomic diameters. Some electron microscopes can resolve individual atoms.

Like optical microscopes, electron microscopes use lenses to form images with beams of electrons. However electron lenses aren't simply pieces of glass. Instead, they are electromagnetic devices that use electric and magnetic fields to bend and focus the moving electrons. In a transmission electron microscope, electrons pass through an object, through various electromagnetic lenses, and finally form a real image on a phosphor screen or image sensor. The focal lengths of the lenses are chosen so that this real image is thousands or even millions of times larger than the object.

A scanning electron microscope uses an electromagnetic lens to focus a beam of electrons to a tiny spot on the object and then detects reflected electrons. The microscope sweeps this spot back and forth across the object's surface and gradually creates an image of that surface.

A scanning electron microscope can also detect characteristic X-rays emitted by the atoms exposed to its electron beam. As I'll explain in the next chapter, atoms excited to high energies often fluoresce X-rays. Since atoms of each element emit their own unique spectrum of characteristic X-rays, the microscope can locate and identify elements on a surface with incredible precision. A scientist studying a tiny patch of surface can see not only its spatial structure, but also identify the elements present at any point on that surface.

Scanning probe microscopes originated in the early 1980s with the scanning tunneling microscope. The idea behind a scanning probe microscope is simple: move a sharp probe back and forth across a surface and use its interactions with that surface to study the surface. But few people imagined that a point could be made sharp enough to touch or interact with individual surface atoms.

Naturally, people were surprised when the German physicist Gerd Binnig (1947–) and the Swiss physicist Heinrich Rohrer (1933–) found that their metal probes could exchange electric currents with individual atoms. By scanning a probe across a surface, they could map out the locations of atoms and obtain images of the surface at the atomic scale. For their invention of the scanning tunneling microscope, this pair received the 1986 Nobel Prize in Physics.

Since then, a wide variety of scanning probe microscopes has been developed. In each case, an ultra-sharp probe or tip scans back and forth across a surface, interacting with a few atoms on the surface as it moves along. A computer measures these interactions and builds a map of the surface. These interactions include the exchange of electrons, electric and magnetic forces, and even friction.

CHAPTER 16

THINGS THAT USE
RECENT PHYSICS

In recent years, scientists have been looking deeper into the atom to see how it's made, farther into space to see how the universe works, and more carefully into matter and motion to see how complicated things can be understood in terms of simple laws. Among the most important tools that these scientists have to work with are quantum theory and the theory of relativity. This chapter will look at some of the ways in which modern physics affects our lives.

The applications of modern physics to everyday life have been a mixed blessing. On the one hand, physics has provided some truly marvelous tools for studying the human body and for working with it to treat disease and improve health. On the other hand, physics has given us some of the most horrifying weapons imaginable. If I present both aspects of modern physics' contributions to everyday life without commenting regularly on the positive and negative aspects of those contributions, it's not because I don't care; it's only that I'll be less effective as an explainer if I'm busy voicing my opinions. I'll give a brief rant here and then get down to business.

My concerns about science and technology are many and varied; we're heading off in so many different directions at once without pausing to consider the consequences. Some of our paths will lead us to wonderful destinations, improving our lives and futures enormously. Other paths will lead us only to trouble and strife, to the disintegration or destruction of relationships, communities, peoples, and ways of life. How much better the world would be if only we could figure out which paths were which at the outset and if only we would care enough to plan ahead or to be concerned. Most often, we only recognize the ethical and societal costs of our poor choices long after we've made them.

More perilous than ignoring the ethical and social consequences of science, however, is ignoring science itself. We live in a dangerous era in which science is regularly discounted or even dismissed as though it were opinion and in which pseudo-scientists casually make up data to support their vacuous theories because they find scientifically valid theories based on actual data inconvenient or threatening. For me one of the most satisfying aspects of physics is that it has real answers, real truths, that are proven again and again by careful measurements. Try as they might, non-scientists can't wave away those truths without waving away reality itself. Fortunately, a worldview grounded in science can be just as rich with art, history, humanity, and theology as one that excludes science. In every way, our's is a beautiful universe.

Chapter Itinerary

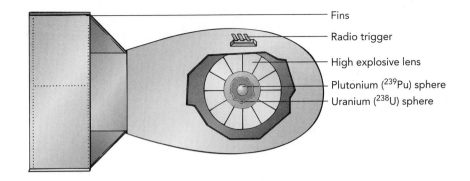

Fins
Radio trigger
High explosive lens
Plutonium (^{239}Pu) sphere
Uranium (^{238}U) sphere

SECTION 16.1 **Nuclear Weapons**

The atomic bomb is one of the most remarkable and infamous inventions of the twentieth century. It followed close on the heels of various developments in the understanding of nature, developments that in many ways made the invention of nuclear weapons inevitable. By the late 1930s, scientists had discovered most of the principles behind nuclear energy and were well aware of how those principles might be applied. The onset of World War II prompted concern that Germany would choose to follow the military path of nuclear energy. Propelled by fear, curiosity, and temptation, the scientists, engineers, and politicians of that time brought nuclear weapons into existence. The world has lived in the shadow of these terrible devices ever since.

Background

At the end of the nineteenth century, "classical physics" reigned supreme. Here classical physics means the rules of motion and gravitation identified by such people as Galileo, Newton, and Kepler, and the rules of electricity and magnetism developed by others including Ampère, Coulomb, Faraday, and Maxwell. It was generally felt that most of physics was well understood: physicists knew all of the laws governing the behavior of objects in our universe, and what was left was to apply those laws to more and more complicated examples. It was a time when physicists didn't know what they didn't know.

However, a few nagging problems remained—specific difficulties that couldn't be explained by the rules of classical physics. Among these were the spectrum of light emitted by a blackbody, the photoelectric effect in which electrons are ejected from metals by light, and the apparent absence of an ether or medium in which light traveled. At the beginning of the twentieth century, the whole of classical physics collapsed under the weight of these seemingly trivial difficulties, and a largely new understanding of the universe emerged. The major advances took 25 years, from 1901 to 1926, and the time since has largely been spent applying those new laws to more and more complicated examples.

The two main developments, both essential to the making of the atomic bomb, were the discoveries of quantum physics and relativity. Often these are called quantum theory and the theory of relativity. But while the word theory might imply that they're somehow on shaky ground, they're not theories in the sense of hypotheses waiting to be tested. In fact, they've been confirmed countless times since they were developed and have been shown to have enormous predictive power. Rather, they're theories in the sense of being carefully constructed and codified rules that model the behavior of the physical universe in which we live. Between them, these two theories made the discovery of nuclear forces and nuclear energy unavoidable. Finally, given peoples' love for gadgets and power, they also made nuclear weapons inevitable.

The Nucleus and Radioactive Decay

Though the name "atomic" bomb has stuck for more than half a century, the more correct name would be "nuclear" bomb. The items that are responsible for the energy released by nuclear weapons are not atoms, but tiny pieces of atoms—their nuclei (plural of nucleus). But before we can discuss nuclei, let's put them into context. Let's start by looking at atoms.

To get an idea of just how tiny atoms are, imagine magnifying a grain of table salt, 1 mm (0.04 inch) on a side, until it was the size of the state of Colorado. That grain would then appear as an orderly arrangement of spherical particles, each about the size of a grapefruit (Fig. 16.1.1). These spherical particles would be single atoms, and there would be about 7.2 million of them along each edge of the grain.

Like most solids, table salt is a crystal and its atoms are bound to one another by their outermost components: their electrons. Electrons dominate the chemistry of atoms and molecules. Sodium is a reactive metal because of its electrons, and chlorine is a reactive gas because of its electrons. When mixed, these two chemicals react violently to form table salt and release a considerable amount of light and heat. This, then, is a true "atomic bomb."

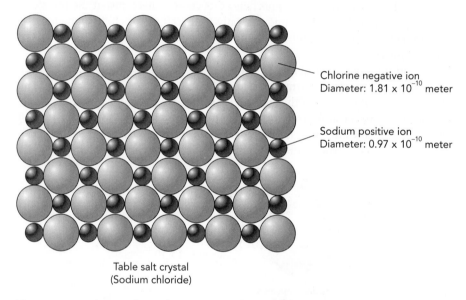

Chlorine negative ion
Diameter: 1.81×10^{-10} meter

Sodium positive ion
Diameter: 0.97×10^{-10} meter

Table salt crystal
(Sodium chloride)

Fig. 16.1.1 A salt crystal is an orderly array of positively charged sodium ions and negatively charged chloride ions. These ions are held together by the attractive forces between oppositely charged ions. The ions are so small that there are about 7.2 million ions on the edge of a 1-mm-wide salt crystal.

Obviously, something is missing here. If a crazy person could buy a kilogram or two of sodium and a tank of chlorine from a chemical company and destroy an entire city, rural living would be a whole lot more popular. Fortunately the energy released by chemical reactions is fairly limited. A kilogram of chemical explosives just can't do that much damage. But nuclear bombs tap an entirely different store of energy deep within the atoms.

While all of nuclear weaponry is often attributed to Einstein's famous equation, $E = mc^2$, that notion is vastly oversimplified. Nonetheless, this equation is quite significant. As we noted in Section 4.2, one of Einstein's discoveries at the beginning of the twentieth century was that matter and energy are in some respects equivalent. In certain circumstances mass can become energy or energy can become mass. This equivalence is part of the theory of relativity and has some interesting consequences. It implies that an object can reduce its mass by transferring energy to its surroundings. Thus, if you weigh an object before and after it undergoes some internal transformation, you can use any weight loss to determine how much energy was released from the object by that transformation.

Because of this equivalence, mass and changes in mass can be used to locate energy that's hidden within normal matter. This technique is important in nuclear physics, but it also applies to chemistry. When sodium and chlorine react to form table salt, their combined mass decreases by a tiny amount. What's missing is some chemical potential energy, which becomes light and heat and escapes from the mixture. In leaving, this chemical potential energy reduces the mass of the sodium and chlorine mixture by about 1 part in 10 billion. That tiny change in mass is too small to detect with present measuring devices, although scientists are working on techniques that will soon make it possible to measure mass changes due to chemical bonds.

Electrons are, however, by far the lightest part of an atom and thus have relatively little mass to release as energy. Most of an atom's mass is located in its nucleus. The nucleus is fantastically small—only a little more than 10^{-15} m in diameter. If you were to peer into one of the grapefruit-sized sodium ions of our giant salt crystal, you would see a tiny particle at its center. There, just at the threshold of visibility, would be the ion's nucleus, only 1 µm (0.00004 inch) in diameter. The remaining 99.9999999999999% of the ion is occupied only by its 10 electrons in their orbitals.

The sodium nucleus contains 11 protons and 12 neutrons (Fig. 16.1.2). Each of these nuclear particles or nucleons has about 2000 times as much mass as an electron, so that 99.975% of the sodium ion's mass is in this nucleus. Thus, while the electrons are certainly important to chemistry and matter as we know it, their contribution to the ion's mass is insignificant. The ion is mostly empty space, lightly filled with fluffy electrons and having a tiny nuclear lump at its center.

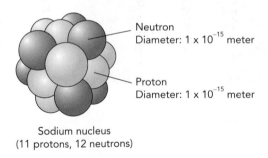

Neutron
Diameter: 1×10^{-15} meter

Proton
Diameter: 1×10^{-15} meter

Sodium nucleus
(11 protons, 12 neutrons)

Fig. 16.1.2 At the center of a sodium ion is a tiny nucleus containing about 99.975% of the ion's mass. It consists of 11 positively charged protons and 12 uncharged neutrons. The protons repel one another at any distance, but the protons and neutrons are bound together by the highly attractive nuclear force as long as they touch one another.

The nucleons that make up this nucleus experience two competing forces. The first of these forces is the familiar electrostatic repulsion between like electric charges. Because each proton in the nucleus has a single positive charge, they're constantly trying to push one another out of the nucleus. However, the second of these forces is attractive and holds the nucleus together. This new force is called the nuclear force, and at short distances it dominates the weaker electrostatic repulsion. However, the nuclear force only attracts the nucleons toward one another when they're touching. As soon as they're separated, they're on their own.

The competition between these two forces, the repulsive electrostatic force between like charges and the attractive nuclear force between nucleons, is analogous to what happens in a familiar toy (Figs. 16.1.3 and 16.1.4). This hopping toy has a

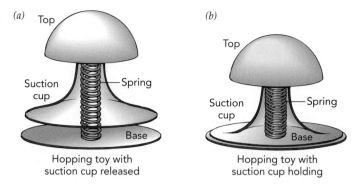

(a) Top — Suction cup — Spring — Base

Hopping toy with
suction cup released

(b) Top — Suction cup — Spring — Base

Hopping toy with
suction cup holding

Fig. 16.1.3 (*a*) A spring and a suction cup combine to form a toy that hops suddenly after a long wait. (*b*) After you press the two parts together, doing work on them and storing energy in the spring, the spring tries to separate the top from the base while the suction cup tries to hold the two parts together. The stored energy is released when the leaking suction cup eventually allows the spring to expand.

Fig. 16.1.4 This hopping toy stores energy as its spring is compressed and retains that energy while its suction cup grips its base. When the suction cup lets go, the toy leaps into the air.

suction cup attached to a spring, so that the spring tries to separate the toy's top from its base while the suction cup tries to keep the two parts together. When the two parts are well separated, only the spring exerts a force. But when the two parts touch, the suction cup begins to act and holds the two parts together.

What makes the hopping toy exciting is that its suction cup leaks. Eventually, the suction cup lets go and allows the spring to toss the toy into the air. But suppose that the suction cup didn't leak. Once pushed together, the pieces would never separate and the spring would retain its stored energy indefinitely. To get the suction cup to let go, you would have to pull it away from the base. Only then could the spring release its stored energy.

In effect, an energy barrier would be preventing the leak-free toy from hopping. Until you did a little work on it by pulling the suction cup off the base, it wouldn't be able to release its stored energy. Another example of a system that needs energy to release energy is a bottle of champagne, where you must push on the cork to help it out of the neck. After that initial investment of energy, a great deal of energy is released as gas blasts the cork across the room.

The nucleus is in a similar situation. The attractive nuclear force prevents the nucleus from coming apart, despite the enormous amount of electrostatic potential energy it contains. The nuclear force creates an energy barrier that prevents the nucleons from separating. Unless something adds energy to the nucleus to help the nucleons break free of the nuclear force, the nucleus will remain together forever. At least that's the prediction of classical physics.

Quantum physics, however, has an important influence on the behavior of the nucleus. One of the many peculiar effects of quantum physics is that you can never really tell exactly where an object is located, or at least not for long. That fuzziness is a manifestation of the Heisenberg uncertainty principle, which observes that some pairs of physical quantities, such as position and momentum or energy and time, are not entirely independent and cannot be determined simultaneously beyond a certain accuracy. This principle is a result of the partly wave and partly particle nature of objects in our universe. Since waves are normally broad things that occupy a region of space rather than a single point, objects in our universe normally don't have exact locations.

The smaller an object's mass, the fuzzier it is and the more uncertain its location. While the fuzzy nucleons in a nucleus will normally stay in contact with one another for an extremely long time, there's always a tiny chance that they'll find themselves temporarily separated by a distance that's beyond the reach of the nuclear force. The nucleons will then suddenly be free of one another, and electrostatic repulsion will push them apart in a process called radioactive decay. The quantum process that allows the nucleons to escape from the nuclear force without first obtaining the energy needed to surmount the energy barrier is called tunneling because the nucleons effectively tunnel through the barrier. We first encountered quantum tunneling in Section 12.2, when we saw that erasing flash memory requires that electrons tunnel through an insulating barrier.

Natural radioactive decay is a perfectly random process. Although half of a large population of identical radioactive nuclei will decay in a certain amount of time, you absolutely cannot predict in advance which of the original nuclei will have survived. Because of this randomness, radioactive decay is characterized simply by a half-life—the time required for half of the nuclei to decay. If you wait a second half-life, only a quarter of the original nuclei will remain (half of a half). After a third half-life, only an eighth will remain (half of a half of a half). And so on.

This halving of the population with each additional half-life is a type of exponential decay. In general, the fraction of nuclei remaining after a given amount of time is one half raised to the power of the time divided by the half-life, or:

$$\text{fraction remaining} = \left(\frac{1}{2}\right)^{\text{elapsed time}/\text{half-life}}. \tag{16.1.1}$$

While most radioactive nuclei have short half-lives and don't linger long in our environment, there are a few with half-lives of billions of years. It is those long-lived radioactive nuclei, particularly uranium and thorium nuclei, that have survived since the formation of the earth, remain abundant in nature, and gave rise to nuclear weapons.

Fission and Fusion

The more protons there are in a nucleus, the more they repel one another and the more likely they are to cause radioactive decay. Adding additional neutrons to the nucleus reduces this proton–proton repulsion by increasing the size of the nucleus without adding to its positive charge. However, adding too many neutrons also destabilizes the nucleus for reasons that we'll discuss in the next section. So constructing a stable nucleus is a delicate balancing act.

In nuclei with only a few protons, the attractive nuclear force wins big over the repulsive electrostatic force and the nucleons stick like crazy. These nuclei resemble hopping toys with weak springs and big suction cups: once brought together, the pieces never come apart. In fact, the average binding energy of the nucleons (the energy required to separate them from one another divided by the number of nucleons) would increase if these nuclei had even more protons and neutrons.

In nuclei with many protons, the electrostatic repulsion is so severe that the nuclear force can't hold the nucleons together for long. These nuclei decay rapidly. They resemble hopping toys with strong springs and small suction cups. The average binding energy of the nucleons would increase if these nuclei had fewer protons and neutrons.

In nuclei with roughly 26 protons, in between the two extremes we've just considered, the attractive nuclear force and repulsive electrostatic force are nicely balanced. These nuclei are extremely stable, and you can't increase the average binding energy of their nucleons by adding or subtracting nucleons. Smaller nuclei can release potential energy by growing to reach this intermediate size, while larger nuclei can release potential energy by shrinking toward the same goal.

For a small nucleus to grow, something must push more nucleons toward it. Electrostatic repulsion will initially oppose this growth, but once everything touches, the nuclear force will bind the particles together and release a large amount of potential energy. This coalescence process is called nuclear fusion.

For a large nucleus to shrink, something must separate its pieces beyond the reach of the nuclear force. Electrostatic repulsion will then push the fragments apart and release a large amount of potential energy. This fragmentation process is called nuclear fission.

The energies released when small nuclei undergo fusion or when large nuclei undergo fission are enormous compared to chemical energies. Uranium, a large nucleus, converts about 0.1% of its mass into energy when it breaks apart. Hydrogen, a tiny nucleus, converts about 0.3% of its mass into energy when it fuses with other hydrogen nuclei. Kilogram for kilogram, nuclear reactions release about 10 million times more energy than chemical reactions. Fortunately, they're much harder to start.

With that scientific background, let's follow a sequence of discoveries near the start of the twentieth century. Natural radioactive decay was discovered accidentally by French physicist Antoine-Henri Becquerel (1852–1908) in 1896. Intrigued by the recent discovery of X-rays, he began looking for materials that might emit X-rays after exposure to light. To his surprise, he found that uranium fogged photographic plates, even through an opaque shield and even without exposure to light. His discovery was soon confirmed and elaborated on by Polish-born French physicist Marie Curie (1867–1934) and French chemist Pierre Curie (1859–1906). This wife and husband team discovered several new radioactive elements, including polonium (named after Marie's homeland) and radium.

In 1911, British physicist Ernest Rutherford (1871–1937) discovered that atoms have nuclei. He subsequently found that nuclei sometimes shatter when struck by energetic helium nuclei. And in 1932, British physicist James Chadwick (1891–1974) discovered a fragment of the nucleus, the neutron, which has no electric charge and can thus approach a nucleus without any electrostatic repulsion. It was soon discovered that neutrons stick to the nuclei of many atoms.

But the crucial discovery that made the atomic bomb possible was neutron-induced fission of nuclei. In 1934, Italian physicist Enrico Fermi (1901–1954) and his colleagues were trying to solve a particular riddle about the nucleus, a radioactive decay process called beta decay. They were adding neutrons to the nuclei of every atom they could get hold of. When they added neutrons to uranium, with its huge nucleus, they observed the production of some very short-lived radioactive systems. They thought that they had formed ultraheavy nuclei and even went so far as to give these new elements tentative names.

Four years later, however, Austrian physicists Lise Meitner (1878–1968) and Otto Frisch (1904–1979) and German chemists Otto Hahn (1879–1968) and Fritz Strassmann (1902–1980) collectively showed ❶ that what Fermi's group had actually done was to fragment uranium into lighter nuclei (Fig. 16.1.5). Many of the fragments created by this induced fission were neutrons, which could themselves cause the destruction of other uranium nuclei.

❶ Austrian-born physicist Lise Meitner moved to Berlin in 1907 and soon began a 30-year collaboration with chemist Otto Hahn. In 1934, she convinced Hahn to join her in studying nuclear processes and they made great progress. Unfortunately, Meitner's Jewish ancestry made her a target of Nazi academics restriction and she fled to Sweden in 1938. Meitner continued to guide their collaboration through letters. Only months after she left, Hahn and his assistant Fritz Strassmann found that neutron irradiation of heavy elements was creating smaller rather than larger nuclei. Meitner and her nephew Otto Frisch soon developed a model of nuclear fission based on these measurements. Hahn, however, published the results without Meitner's name on the paper, ostensibly to avoid Nazi interference. As a result of this omission, the 1944 Nobel Prize in Chemistry was awarded to Hahn alone. Hahn went on to claim that Meitner was simply his assistant, rather than the leader of their joint effort. In recognition of this gross injustice, element 109 was named meitnerium in 1994.

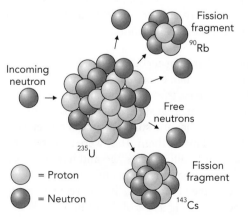

Fig. 16.1.5 When a neutron strikes a uranium nucleus, there's a good chance that the nucleus will fall apart into fragments. This process is called induced fission. Among the fragments of induced fission are other neutrons.

Chain Reactions and the Fission Bomb

Physicists quickly realized that a chain reaction was possible, a reaction in which the fission of one uranium nucleus would induce fission in two nearby uranium nuclei, which would in turn induce fission in four other uranium nuclei, and so on (Fig. 16.1.6). The result would be a catastrophic nuclear process in which many or even most of the nuclei in a piece of uranium would shatter and release fantastic amounts of energy.

In a sense, the remaining work toward both the atomic bomb and the hydrogen bomb was a matter of technical details. Only four conditions had to be satisfied in order for an atomic or fission bomb to be possible:

1. A source of neutrons had to exist in the bomb to trigger the explosion.
2. The nuclei making up the bomb had to be fissionable, that is, they had to fission when hit by a neutron.
3. Each induced fission had to produce more neutrons than it consumed.
4. The bomb had to use the released neutrons efficiently so that each fission induced an average of more than one subsequent fission.

Meeting the first condition was easy. Many radioactive elements emit neutrons. But meeting the second and third conditions was more difficult. Here is where uranium fit into the picture. It was known to be fissionable, and it was known to release more neutrons than it consumed.

But not all uranium nuclei are the same. While a uranium nucleus must contain 92 protons, so that it forms a neutral atom with 92 electrons and has all of the chemical characteristics of uranium (U in Fig. 14.2.4), the number of neutrons in that nucleus is somewhat flexible. Nuclei that differ only in the numbers of neutrons they contain are called isotopes, and natural uranium nuclei come in two isotopes: ^{235}U and ^{238}U, where the number specifies how many nucleons are in each nucleus. The ^{235}U nucleus contains 92 protons and 143 neutrons, for a total of 235 nucleons. In contrast, the ^{238}U nucleus contains 238 nucleons—92 protons and 146 neutrons.

It turns out that only ^{235}U is suitable for a bomb. It's marginally stable, with too many protons for the nuclear force to keep together, even with the diluting effects of 143 neutrons. Like many proton-rich nuclei, ^{235}U eventually undergoes alpha decay—it emits a helium nucleus (^{4}He) and thereby loses two protons and two neutrons. ^{235}U has a radioactive half-life of 710 million years. But when struck by a neutron, ^{235}U shatters immediately into fragments and this induced fission releases about 2.5 neutrons.

^{238}U is slightly more stable than the lighter isotope, and its half-life is 4.51 billion years. But this nucleus absorbs most neutrons without undergoing fission. Instead, it undergoes a series of complicated nuclear changes that eventually convert it into plutonium, an element not found in nature. It becomes ^{239}Pu, a nucleus with 94 protons and 145 neutrons. As we'll see later on, plutonium itself is useful for making nuclear weapons.

So ^{238}U actually slows a chain reaction rather than encouraging it. Since only ^{235}U can support a chain reaction, natural uranium had to be separated before it could be used in a bomb. But ^{235}U is quite rare. The earth's store of uranium nuclei was created long ago, in the explosion of a dying star. That supernova heated the nuclei of smaller atoms so hot that they collided together and stuck. Uranium nuclei were formed, with the supernova's energy trapped inside them. They were incorporated in the earth during its formation about 4 or 5 billion years ago and have been decaying ever since. The only uranium isotopes that remain in any quantity are ^{235}U and ^{238}U. Since ^{235}U is less stable, its percentage of the naturally occurring uranium nuclei has dwindled to only 0.72% . The remaining 99+% of the uranium is ^{238}U.

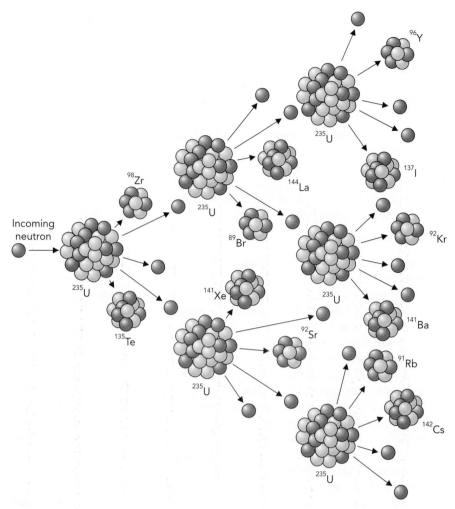

Incoming
neutron

^{98}Zr

^{235}U

^{235}U

^{89}Br

^{135}Te

^{141}Xe

^{235}U

^{235}U

^{92}Sr

^{96}Y

^{144}La

^{137}I

^{92}Kr

^{235}U

^{141}Ba

^{91}Rb

^{142}Cs

^{235}U

Fig. 16.1.6 A chain reaction occurs when the fragments of one fissioning nucleus induce fission in at least one additional nucleus, on average. Such a chain reaction is particularly easy in ^{235}U, the light isotope of uranium, where each fissioning nucleus releases an average of 2.5 neutrons.

Separating ^{235}U from ^{238}U is extremely difficult. Since atoms containing these two nuclei differ only in mass, not in chemistry, they can only be separated by methods that compare their masses. Because the mass difference is relatively small, heroic measures are needed to extract ^{235}U from natural uranium. During World War II and the Cold War era, the U.S. government developed enormous facilities for separating the two uranium isotopes. The need for such installations is one of the major obstacles to the proliferation of nuclear weapons.

The last condition for sustaining a chain reaction is that the bomb must use neutrons efficiently, so that each fission induces an average of more than one subsequent fission. That means that the bomb's contents can't absorb neutrons wastefully and that it can't let too many of them escape without causing fission. A lump of relatively pure ^{235}U wouldn't absorb neutrons wastefully, but it might allow too many of them to escape through its surface. For a chain reaction to occur, the lump must be large enough that each neutron has a good chance of hitting another nucleus before it leaves the lump. The lump also should have a minimal amount of surface. It should be a sphere.

But how large must that sphere be? Since atoms are mostly empty space, a neutron can travel several centimeters through a lump of uranium without hitting a nucleus. Thus a golf ball-sized sphere of uranium would allow too many neutrons to escape. For a bare sphere of ^{235}U, the critical mass required to initiate a chain reaction is about 52 kg (115 lbm), a ball about 17 cm (7 inches) in diameter. At that point, each fission will induce an average of one subsequent fission. But for an explosive chain reaction, in which each fission induces an average of much more than one fission, additional ^{235}U is needed: a supercritical mass. About 60 kg (132 lbm) will do it.

By 1945, the scientists and engineers of the Manhattan Project had found ways to meet these four conditions and were prepared to initiate an explosive chain reaction. They had accumulated enough ^{235}U to construct a supercritical mass. Carefully machined pieces of ^{235}U would be put into the bomb so that they would join together at the moment the bomb was to explode. When the critical mass was reached, a few initial neutrons would start the chain reaction. When the uranium became supercritical, catastrophic fission would quickly turn it into a tremendous fireball.

However, assembly was tricky. The supercritical mass had to be completely assembled before the chain reaction was too far along; otherwise the bomb would begin to overheat and explode before enough of its nuclei had time to fission. In pure ^{235}U, the time it takes for one fission to induce the next fission is only about 10 ns (10 nanoseconds or 10 billionths of a second). In a supercritical mass, each generation of fissions is much larger than the previous generation, so it takes only a few dozen generations to shatter a significant fraction of the uranium nuclei. The whole explosive chain reaction is over in less than a millionth of a second, with most of the energy released in the last few generations (about 30 ns).

To make sure that the assembly was complete before the bomb exploded, it had to be done extraordinarily quickly. In the ^{235}U bomb called "Little Boy" (Fig. 16.1.7a), exploded over Hiroshima on August 6, 1945 at 8:15AM and responsible for the deaths of about 200,000 Japanese citizens, the supercritical mass was assembled when a cannon fired a cylinder of ^{235}U through a hole in a sphere of ^{235}U (Fig. 16.1.8). When the cylinder was centered in the hole, it completed a 60-kg sphere of uranium, housed in a tungsten carbide and steel container. This container confined the uranium, holding it together with its inertia as the explosion began. An explosive chain reaction started immediately, and by the time the uranium blew itself apart, 1.3% of the ^{235}U nuclei had fissioned. The energy released in that event was equal to the explosion of about 15,000 tons of TNT.

But Little Boy was actually the second nuclear explosion. Its concept was so foolproof and its ^{235}U so precious that Little Boy was dropped without ever being tested. However, the Manhattan Project had also developed a plutonium-based bomb that involved a much more sophisticated concept. This bomb was much less certain to work, so it was tested once before it was used.

(a)

(b)

Fig. 16.1.7 (*a*) The Little Boy used a cannon to fire a cylinder of uranium into an incomplete sphere of uranium. (*b*) The Fat Man used high explosives to crush a sphere of plutonium to well above its normal density. Little Boy destroyed Hiroshima, while Fat Man destroyed Nagasaki.

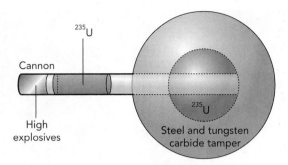

Fig. 16.1.8 The concept behind Little Boy was simple: a sphere of ^{235}U was divided into two parts so that neither was a critical mass on its own. One part was a hollow sphere, and the other part was a cylinder that would complete that sphere. When the bomb was detonated, a cannon fired the cylinder into the hollow sphere, creating a supercritical mass and initiating an explosive chain reaction.

"The Gadget," as the first atomic bomb was called, didn't use ^{235}U. Instead, it used plutonium that had been synthesized from ^{238}U in nuclear reactors. A nuclear reactor carries out a controlled chain reaction in uranium, and neutrons from this chain reaction can convert ^{238}U into ^{239}Pu.

Like ^{235}U, ^{239}Pu meets the conditions for a bomb. The ^{239}Pu nucleus is relatively unstable, with a half-life of only 24,400 years. It fissions easily when struck by a neutron and releases an average of 3 neutrons when it does. Thus ^{239}Pu can be used in a chain reaction. For a bare sphere of ^{239}Pu, the critical mass is about 10 kg (22 lbm)—a ball about 10 cm (4 inches) in diameter.

But ^{239}Pu has a problem. It's so radioactive and releases so many neutrons when it fissions that a chain reaction develops almost instantly. There is much less time to assemble a supercritical mass of plutonium than there is with uranium. The cannon assembly method won't work because the plutonium will overheat and blow itself apart before the cylinder can fully enter the sphere.

Thus a much more sophisticated assembly scheme was employed. At 5:29 AM on June 16, 1945, over 2000 kg (4400 lbm) of carefully designed high explosives crushed or imploded a sphere of plutonium (Fig. 16.1.9) inside The Gadget (Fig. 16.1.10) at Alamogordo, New Mexico. By itself, the 6.1-kg (13.4-lbm) sphere wasn't large enough to be a critical mass; it was a subcritical mass. But it was surrounded by a tamper of ^{238}U whose massive nuclei reflected many neutrons back into the plutonium like marbles bouncing off bowling balls. And the implosion process compressed the plutonium well beyond its normal density. With the plutonium nuclei packed more tightly together, they were more likely to be struck by neutrons and undergo fission.

Fig. 16.1.9 The concept used in The Gadget and Fat Man was relatively sophisticated. Carefully shaped high explosives crushed a baseball-sized sphere of ^{239}Pu inside a neutron-reflecting shell of ^{238}U. The plutonium was compressed to higher than normal density and quickly reached supercritical mass, initiating an explosive chain reaction.

Fig. 16.1.10 The first atomic bomb, nicknamed "The Gadget," was detonated on a tower at a remote desert site near Alamogordo, New Mexico, on June 16, 1945. Here employees of the top secret Manhattan Project are seen hoisting parts of the plutonium bomb onto the tower before the explosion.

The scheme worked. The chain reaction that followed caused 17% of the plutonium nuclei to fission and released energy equivalent to the explosion of about 22,000 tons of TNT. The tower and equipment at the Trinity test site disappeared into vapor, and the desert sands turned to glass for hundreds of meters in all directions. A nearly identical device named "Fat Man" (Fig. 16.1.7*b*) was dropped over Nagasaki, Japan on August 9, 1945, at 11:02AM, where it ultimately killed about 140,000 people.

In the years following the first fission bombs, development focused on how best to bring the fissionable material together. The longer a supercritical mass could be held together before it overheated and exploded, the larger the fraction of its nuclei that would fission and the greater the explosive yield. The crushing technique of The Gadget and Fat Man became the standard, and bombs grew smaller and more efficient at using their nuclear fuel. The implosion process reduced the amount of plutonium needed to reach a supercritical mass, so that very small fission bombs were possible. The smallest atomic bomb, the "Davy Crockett," weighed only about 220 N (50 lbf).

The Fusion or Hydrogen Bomb

Since fissionable material begins to explode as soon as it exceeds critical mass, this critical mass limits the size and potential explosive yield of a fission bomb. In search of a way around this limit, bomb scientists took a look back at small nuclei and figured out how to extract energy by sticking them together.

The fission bomb brought to the earth, for the first time, temperatures that had previously only been observed in stars. Stars obtain most of their energy by fusing hydrogen nuclei together to form helium nuclei, a process that releases a great deal of energy. Because hydrogen nuclei are protons and repel one another with tremendous forces, hydrogen doesn't normally undergo fusion here on earth. To

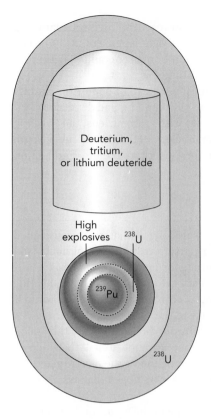

Fig. 16.1.11 A fusion bomb releases energy by fusing deuterium (^{2}H) and tritium (^{3}H) nuclei together to form helium (^{4}He) and neutrons. This fusion is initiated by heating the hydrogen to more than 100 million degrees Celsius with a fission bomb. High-energy neutrons released by the fusion process then induce fission in the ^{238}U tamper, releasing still more energy. Lithium (^{6}Li) produces tritium when exposed to neutrons.

Deuterium, tritium, or lithium deuteride

High explosives ^{238}U

^{239}Pu

^{238}U

cause fusion, something must bring those protons close enough for the nuclear force to stick them together. The only practical way we know of to bring the nuclei together is to heat them so hot that they crash into each other. That is what happens in a fusion bomb, also called a thermonuclear or hydrogen bomb.

In a fusion bomb, an exploding fission bomb heats a quantity of hydrogen to about 100 million degrees Celsius (Fig. 16.1.11). At that temperature, hydrogen nuclei begin to collide with one another. To ease the nuclear fusion processes, heavy isotopes of hydrogen are used: deuterium (^{2}H) and tritium (^{3}H). While the normal hydrogen nucleus (^{1}H) contains only a proton, the deuterium nucleus contains a proton and a neutron. The tritium nucleus contains a proton and two neutrons. When a deuterium nucleus collides with a tritium nucleus, they stick to form a helium nucleus (^{4}He) and a free neutron. Because this process converts about 0.3% of the original mass into energy, the helium nucleus and the neutron fly away from one another at enormous speeds.

Since hydrogen won't explode spontaneously, even in huge quantities, a hydrogen bomb can be extremely large. A fission bomb is used to set it off, but after that, the sky's the limit. Some enormous fusion bombs were constructed and tested during the early days of the Cold War. These bombs usually consisted of a fission starter and a hydrogen follower, all wrapped up in a tamper of ^{238}U. The tamper confined the hydrogen as the fusion began. Once fusion was underway, converting deuterium and tritium into helium and neutrons, the neutrons collided with ^{238}U nuclei in the tamper. These fusion neutrons were so energetic that they were able to induce fissions even in ^{238}U nuclei and release still more energy. Overall, this structure is sometimes called a fission–fusion–fission bomb.

A variation on this bomb is the so-called neutron or enhanced radiation bomb. That bomb has no ^{238}U tamper so that the energetic neutrons from the fusion process travel out of the explosion and irradiate everything in the vicinity. This bomb is lethal to humans, but because its blast is relatively weak, it is not particularly destructive to property.

Tritium is a radioactive isotope created in nuclear reactors. It has too many neutrons to be a stable nucleus and slowly decays into a light isotope of helium (^{3}He). Because tritium has a half-life of 12.3 years, fusion bombs containing tritium require periodic maintenance to replenish their tritium.

Many fusion bombs use solid lithium deuteride instead of deuterium and tritium gases. Lithium deuteride is a salt containing lithium (^{6}Li) and deuterium (^{2}H). When a neutron from the fission starter collides with a ^{6}Li nucleus, the two fragment into a helium nucleus (^{4}He) and a tritium nucleus (^{3}H). In the bomb, lithium deuteride is quickly converted into a mixture of deuterium, tritium, and helium, which then undergoes fusion.

Heat, Radiation, and Fallout

Once a nuclear weapon has exploded—after its fissionable material has fissioned and its fusible material has fused—what then? First, a vast number of nuclei and subatomic particles emerge from the explosion at enormous speeds, many at nearly the speed of light. These particles crash into nearby atoms and molecules, heating them to fantastic temperatures and producing a local fireball around the bomb itself. They also cause extensive radiation damage in the surrounding area.

Second, a flash of light emerges from the explosion, caused partly by the fission and fusion processes themselves and partly by the ultrahot fireball that follows. This light is not only visible light, but also every portion of the electromagnetic spectrum from infrared to visible to ultraviolet to X-rays to gamma rays. It burns things nearby, inside and out.

Third, the explosion creates a huge pressure surge in the air around the fireball. A shock wave propagates outward from the fireball at the speed of sound, knocking over everything in its path for a considerable distance. Fourth, the rarefied and superheated air rushes upward, lifted by buoyant forces, to create a towering mushroom cloud.

But the most insidious aftereffect of a nuclear explosion is fallout, the creation and release of radioactive nuclei. Fission converts uranium and plutonium nuclei into smaller nuclei. Each new nucleus has several dozen protons and its share of neutrons from the nucleus that fissioned. These new nuclei attract electrons and become seemingly normal atoms like iodine or cobalt. But while large nuclei such as uranium need extra neutrons to dilute their protons and reduce their electrostatic repulsions, intermediate and small nuclei like those of iodine and cobalt don't need as many neutrons. The new nuclei wind up with too many neutrons and are radioactive. They have half-lives that are anywhere from thousandths of a second to thousands of years.

Until they decay, the atoms that contain these nuclei are almost indistinguishable from normal atoms. They are radioactive isotopes of common atoms, and our bodies naively incorporate them into our tissues. There they sit, performing whatever chemical tasks our bodies require of them. But eventually these radioactive atoms fall apart and release nuclear energy. Because each radioactive decay that occurs near us or inside us releases perhaps a million times more energy than is present in a chemical bond, these decays cause chemical changes in our cells. They can kill cells or damage the cells' genetic information, potentially causing cancer.

This transmutation of elements—the restructuring of nuclei to transform atoms of one element into another—occurs in an uncontrolled fashion in nuclear weapons and produces a lethal mix of unstable isotopes. Those isotopes take years to decay out of the environment and all anyone can do is wait. Even nuclear weapons with poor explosive yields, so-called dirty bombs, can litter the surrounding landscape with radioactive debris. Nuclear reactors produce similar mixtures of radioactive isotopes in their fuel assembles and core structures, which is why disposing of spent nuclear fuel remains so problematic.

On the other hand, radioactive isotopes have been a boon to medicine and bio-chemistry, where many of them have found valuable and life-saving applications. Moreover, in controlled circumstances, elements can be transmuted systematically to generate primarily desirable isotopes rather than a random assortment. But such transmutation is difficult and expensive, and it involves nuclear rather than chemical processes. Although the alchemists' dream of transmuting lead into gold is finally possible, it's not a path to riches.

Common Misconceptions: Radiation and Radioactivity

Misconception: When a material such as food is exposed to microwave, radio wave, infrared, or ultraviolet radiation, it may become radioactive.

Resolution: To render a material radioactive, something must alter its nuclei so that they are no longer stable. Such an alteration requires vastly more energy than one of those low-energy photons can provide. The only forms of electro-magnetic radiation with enough energy per photon to affect nuclei are gamma rays and, occasionally, X-rays.

〰 Carbon Dating

Like all radioactive nuclei, carbon-14 or ^{14}C nuclei decay exponentially with time. Carbon-14 has a half-life of about 5,730 years. As a result, each time 5,730 years pass, about half the nuclei in a large collection of ^{14}C nuclei will decay away. In the next 5,730 years, about half of the remaining nuclei will decay, and so on.

In accordance with Eq. 16.1.1, the functional form for this decay in ^{14}C is given by the relationship:

$$\text{current population} = \text{original population} \cdot \left(\frac{1}{2}\right)^{\text{elapsed time}/5730\,\text{years}}.$$

At time zero, the fraction remaining is $(\frac{1}{2})^0$ or 1 and the current population is simply the original population. After 5,700 years, the fraction remaining is $(\frac{1}{2})^1$ or $\frac{1}{2}$ and the current population of ^{14}C nuclei is half its original value. After another 5,730 years (a total of 11,460 years), the fraction remaining is $(\frac{1}{2})^2$ or $\frac{1}{4}$ and the current popu-lation of ^{14}C nuclei is a quarter its original value. This trend continues indefinitely.

You don't have to wait for a full half-life to see a decrease in the population of ^{14}C nuclei. That decrease is a steady exponential function of time. If you know what the original population of ^{14}C nuclei was and how long it has been since that population was all present, you can predict about how many nuclei are still around. But more interestingly, if you know what the original population of ^{14}C nuclei was and what that population is now, you can determine roughly how long it has been since the original population was present. That is the basis for carbon-14 dating.

Carbon-14 or ^{14}C nuclei are formed at a steady rate by the interactions of cosmic rays with nitrogen atoms in our atmosphere. Because they are chemically indis-tinguishable from ordinary carbon nuclei, ^{14}C nuclei are taken up by plants in the process of photosynthesis. In that manner, ^{14}C nuclei enter the food chain. Any living organism that is sustained by photosynthesis or by eating other photosyn-thetic organisms will contain ^{14}C nuclei. More importantly, a certain fraction of its carbon nuclei will be ^{14}C nuclei.

While it's alive, the ratio of ^{14}C nuclei to ordinary carbon nuclei will be roughly constant. But as soon as the organism dies, that ratio will begin to decrease. Each time a ^{14}C nuclei decays, the ratio drops slightly. After 5,730 years, the ratio is half what it was at the organism's death. After 11,460 years, the ratio is down to a quar-ter its original value. By carefully measuring the ^{14}C to normal carbon ratio in the remains of an organism, it's possible to determine how long ago it died.

In this fashion, organic materials can be dated back about 50,000 years. Wood, clothing, leather, seeds, and other organic remains can be dated with considerable accuracy. The only uncertainties come from possible fluctuations in the rate at which ^{14}C is made in the atmosphere and slight preferences some plants have for the various isotopes of carbon. This last observation reflects the fact that while the various isotopes of carbon are "chemically indistinguishable," that indistinguishability isn't quite perfect. Some plants manage to concentrate one isotope relative to another by a tiny amount and this preferential behavior makes carbon-14 dating a little more complicated and inexact. Nonetheless, it is a very useful, accurate, and widely accepted method for determining the dates of organic objects.

When combined with tree-ring dating methods, carbon-14 dating becomes even more accurate. Living trees form annual growth rings that vary in thickness according to the weather. By matching the weather-induced sequences on modern trees with trees that died recently and at various times in the past, scientists have pieced together extensive sequences of tree rings and wood going back thousands of years. That dated wood allows them to measure the carbon-14 fraction in plant material from any year in the sequences. Those measurements of exact carbon-14 fractions in dated samples are used as references when trying to establish the age of an artifact of unknown vintage.

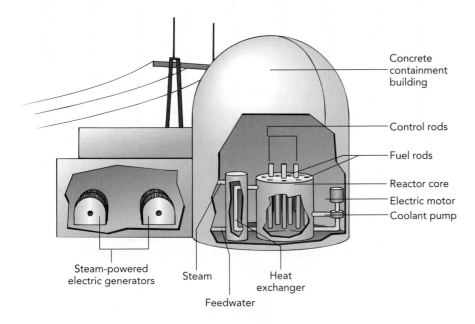

Concrete containment building

Control rods

Fuel rods

Reactor core
Electric motor
Coolant pump

Steam-powered electric generators

Steam

Heat exchanger

Feedwater

SECTION 16.2 Nuclear Reactors

Weapons weren't the only possibilities open to nuclear scientists and engineers at the end of the 1930s. While nuclear fission chain reactions and thermonuclear fusion were clearly ways to unleash phenomenal destructive energy, they could also provide virtually limitless sources of useful energy. By controlling the same nuclear reactions that occur in nuclear weapons, people have since managed to extract nuclear energy for constructive uses. In the half-century since their conception, nuclear fission reactors have developed into a fairly mature technology and have become one of our major sources of energy. Nuclear fusion power remains an elusive goal, but efforts continue to harness this form of nuclear energy as well.

Nuclear Fission Reactors

Assembling a critical mass of uranium doesn't always cause a nuclear explosion. In fact, it's rather hard to cause a big explosion. The designers of the atomic bomb had to assemble not just a critical mass but a supercritical mass and they had to do it in much less than a millionth of a second. That rapid assembly is not something that happens easily or by accident. It's much easier to reach a critical mass slowly, in which case the uranium will simply become very hot. It may ultimately explode from overheating, but it will not vaporize everything in sight.

This slow assembly of a critical mass is the basis for nuclear fission reactors. Their principal product is heat, which is often used to generate electricity. Fission reactors are much simpler to build and operate than fission bombs because they don't require such purified fissionable materials. In fact, with the help of some clever tricks, nuclear reactors can even be made to operate with natural uranium.

Let's begin by showing that a fission chain reaction doesn't always lead to an explosion. What's important is just how fast the fission rate increases. In an atomic bomb, it increases breathtakingly quickly. At detonation, the fissionable material is far above the critical mass so the average fission induces not just one, but perhaps two, subsequent fissions. With only about 10 ns (10 nanoseconds) between one fission and the two it induces, the fission rate may double every 10 ns. In less than a millionth of a second, most of the nuclei in the material undergo fission, releasing their energy before the material has time to blow apart.

But things aren't so dramatic right at critical mass, where the average fission induces just one subsequent fission. Since each generation of fissions simply reproduces itself, the fission rate remains essentially constant. Only spontaneous fissions cause it to rise at all. The fissionable material steadily releases thermal energy and that energy can be used to power an electric generator.

A nuclear reactor contains a core of fissionable material. Because of the way in which this core is assembled, it's very close to a critical mass. Several neutron-absorbing rods, called control rods, which are inserted into the reactor's core, determine whether it's above or below critical mass. Pulling the control rods out of the core increases the chance that each neutron will induce a fission and moves the core toward supercriticality. Dropping the control rods into the core increases the chance that each neutron will be absorbed before it can induce a fission and moves the core toward subcriticality.

A nuclear reactor uses feedback to maintain the fission rate at the desired level. If the fission rate becomes too low, the control system slowly pulls the control rods out of the core to increase the fission rate. If the fission rate becomes too high, the control system drops the control rods into the core to decrease the fission rate. It's like driving a car. If you're going too fast, you ease off the gas pedal. If you are going too slowly, you push down on the gas pedal.

The car driving analogy illustrates another important point about reactors. Both cars and reactors respond relatively slowly to adjustments of their controls. It would be hard to drive a car that stopped instantly when you lifted your foot off the gas pedal and that leaped to supersonic speed when you pushed your foot down. Similarly, it would be impossible to operate a reactor that immediately shut down when you dropped the control rods in and that instantly exploded when you pulled the control rods out.

But reactors, like cars, don't respond quickly to movements of the control rods. That's because the final release of neutrons following some fissions is slow. When a ^{235}U nucleus fissions, it promptly releases an average of 2.47 neutrons which induce other fissions within a thousandth of a second. But some of the fission fragments are unstable nuclei that decay and release neutrons long after the original fission. On average, each ^{235}U fission eventually produces 0.0064 of these delayed neutrons, which then go on to induce other fissions. It takes seconds or minutes for these delayed neutrons to appear and they slow the response of the reactor. The reactor's fission rate can't increase quickly because it takes a long time for the delayed neutrons to build up. The fission rate can't decrease quickly because it takes a long time for the delayed neutrons to go away.

To further ease the operation of modern nuclear reactors, they are designed to be stable and self-regulating. This self-regulation ensures that the core automatically becomes subcritical if it overheats. As we'll see later on, this self-regulation was absent in the design of Chernobyl Reactor Number 4.

Thermal Fission Reactors

The basic concept of a nuclear reactor is simple: assemble a critical mass of fissionable material and adjust its criticality to maintain a steady fission rate. But what should the fissionable material be? In a fission bomb, it must be relatively pure ^{235}U or ^{239}Pu. But in a fission reactor, it can be a mixture of ^{235}U and ^{238}U. It can even be natural uranium. The trick is to use thermal neutrons—slow moving neutrons that have only the kinetic energy associated with the local temperature.

In a fission bomb, ^{238}U is a serious problem because it captures the fast moving neutrons emitted by fissioning ^{235}U nuclei. Natural uranium can't sustain a chain reaction because its many ^{238}U nuclei gobble up most of the fast moving neutrons before they can induce fissions in the rare ^{235}U nuclei. The uranium must be highly enriched—most of its ^{238}U nuclei must be removed so that it contains far more than the natural abundance of ^{235}U.

But slow moving neutrons have a different experience as they travel through natural uranium. For complicated reasons, the ^{235}U nuclei seek out slow moving neutrons and capture them with remarkable efficiency. ^{235}U nuclei are so good at catching slow moving neutrons that they easily win out over the more abundant ^{238}U nuclei. Even in natural uranium, a slow moving neutron is more likely to be caught by a ^{235}U nucleus than it is by a ^{238}U nucleus. As a result, it's possible to sustain a nuclear fission chain reaction in natural uranium if all of the neutrons are slow moving.

Because ^{235}U nuclei emit fast moving neutrons when they fission, natural or slightly enriched uranium alone can't use the slow-moving-neutron effect to maintain a chain reaction. However, most nuclear reactors contain something else besides uranium, natural or otherwise. Along with uranium, they use a material called a moderator that slows the neutrons down so that ^{235}U nuclei can grab them. A fast moving neutron from a fissioning ^{235}U nucleus enters the moderator, rattles around for about a thousandth of a second, and emerges as a slow moving neutron, one with only thermal energy left. It then induces fission in another ^{235}U nucleus. Once the moderator is present, even natural uranium can sustain a chain reaction! Reactors that carry out their chain reactions with slow moving or thermal neutrons are called thermal fission reactors.

To be a good moderator, a material must simply remove energy and momentum from the neutrons without absorbing them. When a fission neutron leaves a good moderator, it has only thermal energy left. The best moderators are nuclei that rarely or never absorb neutrons and don't fall apart during collisions with them. Hydrogen (^{1}H), deuterium (^{2}H), helium (^{4}He), and carbon (^{12}C) are all good moderators. When a fast moving neutron hits the nucleus of one of these atoms, the collision resembles that between two billiard balls. Because the fast moving neutron transfers some of its energy and momentum to the nucleus, the neutron slows down while the nucleus speeds up.

Water, heavy water (water containing the heavy isotope of hydrogen: deuterium or ^{2}H), and graphite (carbon) are the best moderators for nuclear reactors. They slow neutrons down to thermal speeds without absorbing many of them. Of these moderators, heavy water is the best because it slows the neutrons quickly yet doesn't absorb them at all. However, heavy water is expensive because only 0.015% of hydrogen atoms are deuterium and separating that deuterium from ordinary hydrogen is difficult.

Graphite moderators were used in many early reactors because graphite is cheap and working with it is easy (see ❶). However, graphite is a less efficient moderator than heavy water, so graphite reactors had to be big. Furthermore, graphite can burn and was partly responsible for two of the world's three major reactor accidents. Normal or "light" water is cheap, safe, and an efficient moderator, but it absorbs enough neutrons that it can't be used with natural uranium. For use in a light water reactor, uranium must be enriched slightly, to about 2–3% ^{235}U.

The core of a typical thermal fission reactor consists of small uranium oxide (UO_2) fuel pellets, separated by layers of moderator (Fig. 16.2.1). A neutron released by a fissioning ^{235}U nucleus usually escapes from its fuel pellet, slows down in the moderator, and then induces fission in a ^{235}U nucleus in another fuel pellet. By absorbing some of these neutrons, the control rods determine whether the whole core is subcritical, critical, or supercritical. The ^{238}U nuclei are basically spectators in the reactor since most of the fissioning occurs in the ^{235}U nuclei.

In a practical thermal fission reactor, something must extract the heat released by nuclear fission. In many reactors, cooling water passes through the core at high speeds. Heat flows into this water and increases its temperature. In a boiling water

❶ The first nuclear reactor was CP–1 (Chicago Pile–1), a thermal fission reactor constructed in a squash court at the University of Chicago. Each of the graphite bricks used in this pile contained two large pellets of natural uranium. By December 2, 1942, the pile was complete and would reach critical mass once the control rods were removed. As Enrico Fermi, the project leader, directed the slow removal of the last control rod, the pile approached criticality and the neutron emissions began to mount. It was noon, so Fermi called a famous lunch break. When everyone returned, they picked up where they had left off. At 3:25PM, the pile reached critical mass and the neutron emissions increased exponentially. The reactor ran for 28 minutes before Fermi ordered the control rods to be dropped back in.

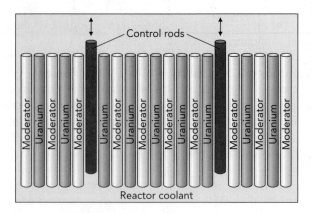

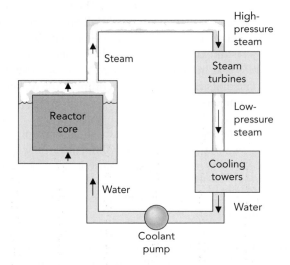

Fig. 16.2.1 The core of a thermal fission reactor consists of uranium pellets, interspersed with a moderator that slows the fission neutrons to thermal energies. Neutron-absorbing control rods are inserted into the core to control the fission rate. A cooling fluid such as water flows through the core to extract heat.

reactor, the water boils directly in the reactor core, creating high-pressure steam that drives the turbines of an electric generator (Fig. 16.2.2). In a pressurized water reactor, the water is under enormous pressure so it can't boil (Fig. 16.2.3). Instead, it's pumped to a heat exchanger outside the reactor. This heat exchanger transfers heat to water in another pipe, which boils to create the high-pressure steam that drives a generator (Fig. 16.2.4).

Fig. 16.2.2 In a boiling water reactor, cooling water boils inside the reactor core. It creates high-pressure steam that drives steam turbines and an electric generator. The spent steam condenses in a cooling tower and is then pumped back into the reactor.

When properly designed, a water-cooled thermal fission reactor is inherently stable. The cooling water is actually part of the moderator. If the reactor overheats and the water escapes, there will no longer be enough moderator around to slow the fission neutrons down. The fast moving neutrons will be absorbed by ^{238}U nuclei and the chain reaction will slow or stop.

Fast Fission Reactors

Thermal reactors require simple fuel and are relatively straightforward to construct. But they consume only the ^{235}U nuclei and leave the ^{238}U nuclei almost unaffected.

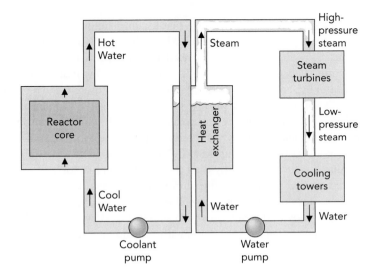

Fig. 16.2.3 In a pressurized water reactor, liquid water under great pressure extracts heat from the reactor core. A heat exchanger allows this cooling water to transfer heat to the water used to generate electricity. Water in the generating loop boils to form high-pressure steam, which then powers the steam turbines connected to the electric generators. The steam condenses back into liquid water and returns to the heat exchanger to obtain more heat.

Anticipating the day when ^{235}U will become scarce, several countries have built a different kind of reactor that contains no moderator. Such reactors carry out chain reactions with fast moving neutrons and are thus called fast fission reactors.

A fast fission reactor operates much like a controlled fission bomb and it requires highly enriched uranium fuel as well. While a thermal fission reactor can get by with natural uranium or uranium that has been enriched to about 2–3% ^{235}U, a fast fission reactor needs 25–50% ^{235}U fuel. That way, there are enough ^{235}U nuclei around to maintain the chain reaction.

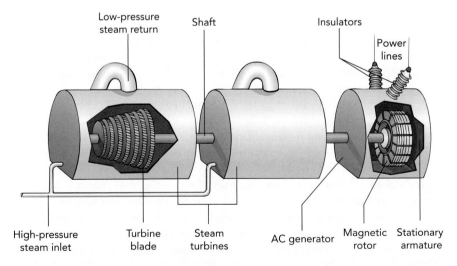

Fig. 16.2.4 High-pressure steam from a boiling water reactor core or the heat exchanger of a pressurized-water reactor enters a series of turbines. There it does work on the turbine blades, losing pressure and temperature in the process. It emerges from the turbines as low-pressure steam and goes to the cooling towers to condense back into liquid water for reuse. The rotary mechanical power provided by the turbines turns generators that produce AC electric power.

But there is a side effect to operating a fast fission reactor. Many of the fast fission neutrons are captured by ^{238}U nuclei, which then transform into ^{239}Pu nuclei. Thus the reactor produces plutonium as well as heat. For that reason, fast fission reactors are often called breeder reactors—they create new fissionable fuel. The ^{239}Pu can eventually replace the ^{235}U as the principal fuel used in the reactor.

A thermal fission reactor makes some plutonium, which is usually allowed to fission in place, but a fast fission reactor makes a lot of it. Because plutonium can be used to make nuclear weapons, fast fission reactors are controversial. However, because they convert otherwise useless ^{238}U into a fissionable material, they use natural uranium far more efficiently than thermal fission reactors.

One interesting complication of the unmoderated design is that fast fission reactors can't be water-cooled. If they were, the water would act as a moderator and slow the neutrons down. Instead, they are usually cooled by liquid sodium metal. The sodium nucleus ^{23}Na rarely interacts with fast moving neutrons so it doesn't slow them down.

Fission Reactor Safety and Accidents

One of the greatest concerns with nuclear fission reactors is the control of radioactive waste. Anything that comes in contact with the reactor core or the neutrons it emits becomes somewhat radioactive. The fuel pellets themselves are quickly contaminated with all sorts of fission fragments that include radioactive isotopes of many familiar elements. Some of these radioactive isotopes dissolve in water or are gases and all of them must be handled carefully.

The first line of defense against the escape of radioactivity is the large and sturdy containment vessel around the reactor. Because most of the radioactive materials remain in the reactor core itself or in the cooling fluid, they are trapped in the containment vessel. Whenever the nuclear fuel is removed for reprocessing, care is taken not to allow radioactive materials to escape.

The other great concern is the safe operation of the reactors themselves. Like any equipment, reactors experience failures of one type or another and a safe reactor must not respond catastrophically to such failures. Toward that end, reactors have emergency cooling systems, pressure relief valves, and many ways to shut down the reactor. For example, injecting a solution of sodium borate into the core cools it and stops any chain reactions. The boron nuclei in sodium borate absorb neutrons extremely well and are the main contents of most control rods. But the best way to keep reactors safe is to design them so that they naturally stop their chain reactions when they overheat.

There have been three major reactor accidents in the past half-century. The first of these accidents occurred in 1957 at Windscale Pile 1, one of Britain's two original plutonium production reactors. This thermal fission reactor was cooled by air rather than water and had a graphite moderator. During a routine shutdown, the reactor overheated. The graphite's crystalline structure had been modified by the reactor's intense radiation and had built up a large amount of chemical potential energy. When that energy was suddenly released during the shutdown, the graphite caught fire and distributed radioactive debris across the British countryside.

The second serious accident occurred at Three Mile Island in 1979. This pressurized water thermal fission reactor shut down appropriately when the pump that circulated water in the power-generating loop failed. Although this water loop wasn't directly connected to the reactor, it was important for removing heat from the reactor core. Even though the reactor was shut down with control rods and had no chain reaction in it, the radioactive nuclei created by recent fissions were still decaying and releasing energy. The core continued to release heat and it eventually boiled the water in the cooling loop. This water escaped from the loop through a pressure relief valve and the top of the reactor core became exposed. With nothing to cool it, the core became so hot that it suffered permanent damage. Some of the water from the cooling loop found its way into an unsealed room and the radioactive gases it contained were released into the atmosphere.

The third and most serious accident occurred at Chernobyl Reactor Number 4 on April 26, 1986. This water-cooled, graphite-moderated thermal fission reactor was a cross between a pressurized water reactor and a boiling water reactor. Cooling water flowed through the reactor at high pressure but didn't boil until it was ready to enter the steam generating turbines.

The accident began during a test of the emergency core cooling system. To begin the test, the operators tried to reduce the reactor's fission rate. However, the core had accumulated many neutron-absorbing fission fragments, which made it incapable of sustaining a chain reaction at a reduced fission rate. The chain reaction virtually stopped. To get the chain reaction running again, the operators had to withdraw a large number of control rods. These control rods were motor-driven, so it would take about 20 seconds to put them back in again.

The operators now initiated the test by shutting off the cooling water. That should have immediately shut down the reactor by inserting the control rods, but the operators had overridden the automatic controls because they didn't want to have to restart the reactor again. With nothing to cool it, the reactor core quickly overheated and the water inside it boiled. The water had been acting as a moderator along with the graphite. But the reactor was overmoderated, meaning it had more moderator than it needed. Getting rid of the water actually helped the chain reaction because the water had been absorbing some of the neutrons. The fission rate began to increase.

The operators realized they were in trouble and began to shut down the reactor manually. However, the control rods moved into the core too slowly to make a difference. As the water left the core, the core went "prompt critical." The chain reaction no longer had to wait for neutrons from the decaying fission fragments because prompt neutrons from the ^{235}U fissions were enough to sustain the chain reaction on their own. The reactor's fission rate skyrocketed, doubling many times each second. The fuel became white hot and melted its containers. Various chemical explosions blew open the containment vessel and the graphite moderator caught fire.

The fire burned for 10 days before firefighters and pilots encased the wreckage in concrete. Many of these heroic people suffered fatal exposures to radiation. The burning core releasing all of its gaseous radioactive isotopes and many others into the atmosphere, forcing the evacuation of more than 100,000 people.

Although not a true reactor accident, the September 30, 1999 disaster in Tokai-mura, Japan did involve a critical mass and a resulting chain reaction. At about 10:35, employees of the Conversion Test Facility of JCO Co., Ltd. Tokai Works were pouring a solution of uranyl (uranium) nitrate into a precipitation tank. Destined for an experimental fast fission reactor, this uranium had been enriched to about 18.8% ^{235}U. Although the equipment and facilities were designed to prevent the assembly of a critical mass, the workers decided to save time by circumventing the safeguards.

After pouring six or seven batches of uranyl nitrate solution into the stainless steel tank through a sampling hole, the solution suddenly reached critical mass. With about 16.6 kg of enriched uranium in the tank, a sudden burst of radiation was released. The water temperature leapt upward and its resulting expansion dropped the mixture below critical mass. But as heat flowed out of the mixture to the water-cooled jacket around the tank, the mixture again approached critical mass. An episodic chain reaction continued in the tank for about 20 hours, until draining the cooling jacket and its neutron-reflecting water finally put an end to the critical mass.

Nuclear Fusion Reactors

Nuclear fission reactors use a relatively rare fuel: uranium. While the earth's supply of uranium is vast, most of that uranium is distributed broadly throughout the earth's crust. There are only so many deposits of high-grade uranium ores that are easily turned into pure uranium or uranium compounds. Fission reactors also produce all sorts of radioactive fission fragments that must be disposed of safely. There is still no comprehensive plan for safe keeping of spent reactor fuels. These must be kept away from any contact with people or animals virtually forever. No one really knows how to store such dangerous materials for hundreds of thousands of years.

An alternative to nuclear fission is nuclear fusion. By joining hydrogen nuclei together, heavier nuclei can be constructed. The amount of energy released in such processes is enormous. However fusion is much harder to initiate than fission because it requires that at least two nuclei be brought extremely close together. These nuclei are both positively charged and they repel one another fiercely. To make them approach one another closely enough to stick, the nuclei must be heated to temperatures of more than 100 million degrees Celsius.

The sun combines four hydrogen nuclei (^{1}H) to form one nucleus of helium (^{4}He), a very complicated and difficult nuclear fusion reaction. For fusion to occur on earth, it must be done between the heavy isotopes of hydrogen: deuterium and tritium. These are the isotopes used in thermonuclear weapons. If a mixture of deuterium and tritium are mixed together and heated to about 100 million degrees, their nuclei will begin to fuse and release energy. The deuterium and tritium become helium and neutrons.

In contrast to fission reactions, there are no radioactive fragments produced. Tritium itself is radioactive, but can easily be reprocessed into fuel and retained within the reactor system. The dangerous neutrons can be caught in a blanket of lithium metal, which then breaks into helium and tritium. It's convenient that new tritium is created because tritium isn't naturally occurring and must be made by nuclear reactions. Thus, fusion holds up the promise of producing relatively little radioactive waste. If the neutrons that are released by fusion events are trapped by nuclei that don't become radioactive, then there will be no radioactive contamination of the fusion reactor either. This is easier said than done, but it's better than in a fission reactor.

Unfortunately, heating deuterium and tritium and holding them together long enough for fusion to occur isn't easy. There are two main techniques that are being tried: inertial confinement fusion and magnetic confinement fusion.

Inertial confinement fusion uses intense pulses of laser light to heat and compress a tiny sphere containing deuterium and tritium (Fig. 16.2.5). The pulses of light last only a few trillionths of a second, but in that brief moment they vaporize and superheat the surface of the sphere. The surface explodes outward, pushing off the inner portions of the sphere. The sphere's core experiences huge inward forces as a result and it implodes. As it's compressed, the temperature of this core rises to that needed to initiate fusion. In effect, it becomes a tiny thermonuclear bomb with the laser pulses providing the starting heat.

To date, inertial confinement fusion experiments have observed fusion in a small fraction of the deuterium and tritium nuclei. The technique is called inertial confinement fusion because there is nothing holding or confining the ball of fuel. The laser beams crush it while it's in free fall and its own inertia keeps it in place while fusion takes place. Unfortunately, the lasers and other technologies needed to carry out inertial confinement fusion are sufficiently complex and troublesome that it may never be viable as a source of energy. Nonetheless, these experiments provide important information on the behaviors of fusion materials at high temperatures and pressures.

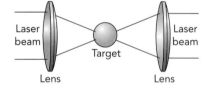

Fig. 16.2.5 In inertial confinement fusion experiments, several laser beams are focused onto a tiny sphere containing deuterium and tritium. These ultra-intense pulsed beams compress and heat the sphere so that fusion occurs.

The other technique being developed to control fusion is magnetic confinement. When you heat hydrogen atoms hot enough, they move so quickly and hit one another so hard that their electrons are knocked off. Instead of a gas of atoms you have a gas of free positively charged nuclei and negatively charged electrons, a plasma. A plasma differs from a normal gas because it's affected by magnetic fields.

As we saw in the section on television, moving charged particles tend to circle around magnetic field lines, the behavior called cyclotron motion (Fig. 16.2.6). If the magnetic field surrounding a charged particle is carefully shaped in just the right way, the charged particle will find itself trapped by the magnetic field. No matter what direction it heads, the charged particle will spiral around the magnetic field lines and will be unable to escape.

Magnetic confinement makes it possible to heat a plasma of deuterium and tritium to fantastic temperatures with electromagnetic waves. Since the heating is done relatively slowly, it's important to keep heat from leaving the plasma. Magnetic confinement prevents the plasma from touching the walls of the container, where it would quickly cool off.

One of the most promising magnetic confinement schemes is the tokamak. The main magnetic field of the tokamak runs around in a circle to form a magnetic doughnut or toroid, the geometrical name for a doughnut shaped object (Fig. 16.2.7). The magnetic field is formed inside a doughnut-shaped chamber by running an electric current through coils that are wrapped around the chamber. Plasma nuclei inside the chamber travel in spirals around the magnetic field lines and don't touch the chamber walls. They are confined inside the chamber and race around the doughnut endlessly. The nuclei can then be heated to the extremely high temperatures they need in order to collide and fuse.

Magnetic confinement fusion reactors have observed considerable amounts of fusion (Fig. 16.2.7). They can briefly achieve scientific break-even, the point at which fusion is releasing enough energy to keep the plasma hot all by itself. However, much more development is needed to meet and exceed practical break-even, where the entire machine produces more energy than it needs to operate.

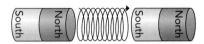

Fig. 16.2.6 When a charged particle moves in the magnetic field between two magnetic poles, it experiences the Lorentz force and undergoes cyclotron motion. It travels in a spiral path around the magnetic field lines connecting them. The particle is confined to a particular region of space.

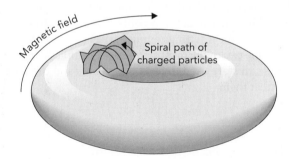

Tokamak fusion reactor

Fig. 16.2.7 A tokamak magnetic confinement fusion reactor consists of a doughnut shaped chamber with a similarly shaped magnetic field inside it. Plasma particles moving inside the tokamak's chamber travel in spirals and circle around the field lines inside the chamber. Because the plasma doesn't touch the walls of the tokamak, it retains heat well enough to reach temperatures at which fusion can occur.

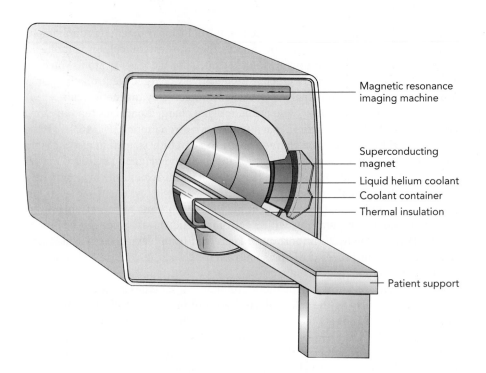

Magnetic resonance
imaging machine

Superconducting
magnet

Liquid helium coolant

Coolant container

Thermal insulation

Patient support

SECTION 16.3 Medical Imaging and Radiation

Some of the most important recent advances in health care have occurred at the border between medicine and physics. As scientists have refined their understanding of atomic and molecular structure and learned to control various forms of radiation, they have invented tools that are enormously valuable for diagnosing and treating illness and injury. The developments continue, with new applications of physics appearing in clinical settings almost every time you turn around. In this section, we'll examine two of the most significant examples of medical physics: the imaging techniques that are used to detect problems and the radiation therapies that are used to treat them.

X-rays

Since their discovery in 1895, X-rays have played an important role in medical treatment. Their usefulness was obvious from the very evening they were discovered. It was November 8 and the German physicist Wilhelm Conrad Roentgen (1845–1923) was experimenting with an electric discharge in a vacuum tube. He had covered the entire tube in black cardboard and was working in a darkened room. Some distance from the tube a phosphored screen began to glow. Some kind of radiation was being released by the tube, passing through the cardboard and the air, and causing the screen to fluoresce. Roentgen put various objects in the way of the radiation, but they didn't block the flow. Finally, he put his hand in front of the screen and saw a shadowed image of his bones. He had discovered X-rays and their most famous application at the same time.

The first clinical use of X-rays was on January 13, 1896, when two British doctors used them to find a needle in a woman's hand. In no time, X-ray systems became common in hospitals as a marvelous new technique for diagnosis. But this imaging capability was not without its side effects. Although the exposure itself was painless, overexposure to X-rays caused deep burns and wounds that took some time to appear. Evidently the X-rays were doing something much more subtle to the tissue than simply heating it.

X-rays are a form of electromagnetic radiation, as are radio waves, microwaves, and light. These different forms of electromagnetic radiation are distinguished from one another by their frequencies and wavelengths—while radio waves have low frequencies and long wavelengths, X-rays have extremely high frequencies and short wavelengths. But they're also distinguished by their photon energies. Because of its low frequency, a radio wave photon carries little energy. A medium-frequency photon of blue or ultraviolet light carries enough energy to rearrange one bond in a molecule. But a high-frequency X-ray photon carries so much energy that it can break many bonds and rip molecules apart.

In a microwave oven, the microwave photons work together to heat and cook food. The amount of energy in each microwave photon is unimportant because they don't act alone. But in radiation therapy, the X-ray photons are independent. Each one carries enough energy to damage any molecule that absorbs it. That's why X-ray burns involve little heat and appear long after the exposure—the molecular damage caused by X-rays takes time to kill cells.

Making X-rays

Medical X-ray sources work by crashing fast-moving electrons into heavy atoms. These collisions create X-rays via two different physical mechanisms: bremsstrahlung and X-ray fluorescence.

Bremsstrahlung occurs whenever a charged particle accelerates. This process is nothing really new, since we know that radio waves are emitted when a charged particle accelerates on an antenna. But in a radio antenna, the electrons accelerate slowly and emit low-energy photons. Bremsstrahlung usually refers to cases in which a charged particle accelerates extremely rapidly and emits a very high-energy photon. In X-ray tube bremsstrahlung, a fast-moving electron arcs around a massive nucleus and accelerates so abruptly that it emits an X-ray photon (Fig. 16.3.1). This photon carries away a substantial fraction of the electron's kinetic energy. The closer the electron comes to the nucleus, the more it accelerates and the more energy it gives to the X-ray photon. However, the electron is more likely to miss the nucleus by a large distance than to almost hit it, so bremsstrahlung is more likely to produce a lower energy X-ray photon than a higher energy one.

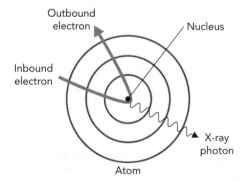

Fig. 16.3.1 When a fast-moving electron arcs around a massive nucleus, it accelerates rapidly. This sudden acceleration creates a bremsstrahlung X-ray photon, which carries off some of the electron's energy.

In X-ray fluorescence, the fast-moving electron collides with an inner electron in a heavy atom and knocks that electron completely out of the atom (Fig. 16.3.2). This collision leaves the atom as a positive ion, with a vacant orbital close to its nucleus. An electron in that ion then undergoes a radiative transition, shifting from an outer orbital to this empty inner one and releasing an enormous amount

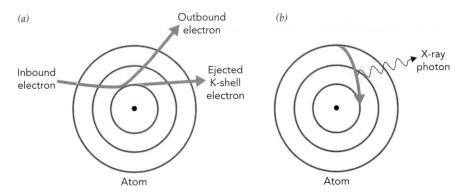

Fig. 16.3.2 (*a*) When a fast-moving electron collides with an electron in one of the inner orbitals of a heavy atom, it can knock that electron out of the atom. (*b*) An electron from one of the atom's outer orbitals soon drops into the empty orbital in a radiative transition that creates a characteristic X-ray.

of energy in the process. This energy emerges from the atom as an X-ray photon. Because this photon has an energy that's determined by the ion's orbital structure, it's called a characteristic X-ray.

To discuss the energies carried by X-ray photons, we'll use the energy unit we encountered in Section 14.2: the electron volt or eV. Photons of visible light carry energies of between 1.6 eV (red light) and 3.0 eV (violet light). Because the ultraviolet photons in sunlight have energies of up to 7 eV, they are able to break chemical bonds and cause sunburns. But X-ray photons have much larger energies than even ultraviolet photons.

In a typical medical X-ray tube, electrons are emitted by a hot cathode and accelerate through vacuum toward a positively charged metal anode (Fig. 16.3.3). The anode is a tungsten or molybdenum disk, spinning rapidly to keep it from melting. The energy of the electrons as they hit the anode is determined by the voltage difference across the tube. In a medical X-ray machine, that voltage difference is typically about 87,000 V, so each electron has about 87,000 eV of energy. Since an electron gives a good fraction of its energy to the X-ray photon it produces, the photons leaving the tube can carry up to 87,000 eV of energy. No wonder X-rays can damage tissue!

When the electrons collide with a target of heavy atoms, they emit both bremsstrahlung and characteristic X-rays (Fig. 16.3.4). The characteristic X-rays have specific energies so they appear as peaks in the overall X-ray spectrum. The bremsstrahlung X-rays have different energies but are most intense at lower energies.

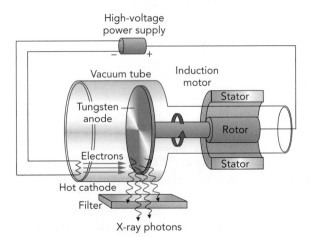

Fig. 16.3.3 In a medical X-ray machine, electrons from a hot filament accelerate toward a positively charged metal disk. They emit X-rays when they collide with the disk's atoms. A motor spins the disk to keep it from melting. The filter absorbs medically useless low-energy X-rays.

Because lower energy X-ray photons injure skin and aren't useful for imaging or radiation therapy, medical X-ray machines use absorbing materials, such as aluminum, to filter them out.

Using X-rays for Imaging

X-rays have two important uses in medicine: imaging and radiation therapy. In X-ray imaging, X-rays are sent through a patient's body to a sheet of film or an X-ray detector. While some of the X-rays manage to pass through tissue, most of them are blocked by bone. The patient's bones form shadow images on the film behind them. In X-ray radiation therapy, the X-rays are again sent through a patient's body, but now their lethal interaction with diseased tissue is what's important.

X-ray photons interact with tissue and bone through four major processes: elastic scattering, the photoelectric effect, Compton scattering, and electron–positron pair production. Elastic scattering is already familiar to us as the cause of the blue sky: an atom acts as an antenna for the passing electromagnetic wave, absorbing and reemitting it without keeping any of its energy (Fig. 16.3.5). Because this process has almost no effect on the atom, elastic scattering isn't important in radiation therapy. However, it's a nuisance in X-ray imaging because it produces a hazy background: some of the X-rays passing through a patient bounce around like pinballs and arrive at the film from odd angles. To eliminate these bouncing X-ray photons, X-ray machines use filters to block X-rays that don't approach the film from the direction of the X-ray source.

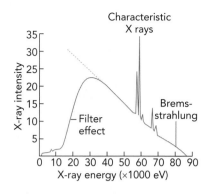

Fig. 16.3.4 When electrons with 87,000 eV of energy collide with tungsten metal, they emit X-rays via bremsstrahlung and X-ray fluorescence. While the bremsstrahlung X-rays have a broad range of energies, an absorbing filter blocks the low-energy ones. X-ray fluorescence produces characteristic X-rays with specific energies.

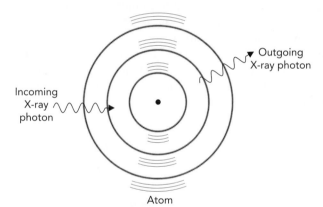

Fig. 16.3.5 When an X-ray photon scatters elastically from an atom, the whole atom acts as an antenna. The passing photon jiggles all of the charges in the atom, and these charges absorb the photon and reemit it in a new direction.

The photoelectric effect is what makes X-ray imaging possible. In this effect, a passing photon induces a radiative transition in an atom: one of the atom's electrons absorbs the photon and is tossed completely out of the atom (Fig. 16.3.6). If the atom were using the X-ray photon to shift an electron from one orbital to another, that photon would have to have just the right amount of energy. But because a free

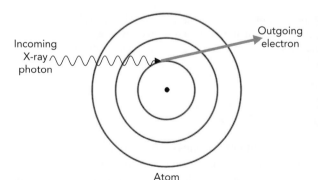

Fig. 16.3.6 In the photoelectric effect, an absorbed photon ejects an electron from an atom. Part of the photon's energy is used to remove the electron from the atom, and the rest becomes kinetic energy in the electron.

electron can have any amount of energy, the atom can absorb any X-ray photon that has enough energy to eject one of its electrons. Part of the photon's energy is used to remove the electron from the atom, and the rest is given to the emitted electron as kinetic energy.

However, the likelihood of such a photoemission event decreases as the ejected electron's energy increases. This decreasing likelihood makes it difficult for a small atom to absorb an X-ray photon. All of its electrons are relatively weakly bound and the X-ray photon would give the ejected electron a large kinetic energy. Rather than emit a high-energy electron, a small atom usually just ignores the passing X-ray photon.

In contrast, some of the electrons in a large atom are quite tightly bound and require most of the X-ray photon's energy to remove them. These electrons would depart with relatively little kinetic energy. Because the photoemission process is most likely when low-energy electrons are produced, a large atom is likely to absorb a passing X-ray. Thus, while the small atoms found in tissue (carbon, hydrogen, oxygen, and nitrogen) rarely absorb medical X-rays, the large atoms found in bone (calcium and phosphorus) absorb X-rays frequently. That's why bones cast clear shadows onto X-ray film. Tissue shadows are also visible, but they're less obvious.

Although one shadow image of a patient's insides may help to diagnose a broken bone, more subtle problems may not be visible in a single X-ray image. For a better picture of what's going on inside the patient, the radiologist needs to see shadows from several different angles. Better yet, the radiologist can turn to a computed tomography (CT) scanner. This computerized device automatically forms X-ray shadow images from hundreds of different angles and positions and produces a detailed three-dimensional X-ray map of the patient's body.

The CT scanner works one "slice" of the patient's body at a time. It sends X-rays through this narrow slice from every possible angle, including the two shown in Fig. 16.3.7, and determines where the bones and tissues are in that slice. The scanner then shifts the patient's body to work on the next slice.

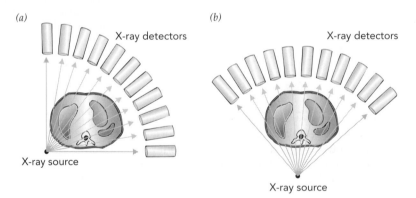

Fig. 16.3.7 A computed tomography or CT scan image is formed by analyzing X-ray shadow images taken from many different angles and positions. An X-ray source and an array of electronic X-ray detectors form a ring that rotates around the patient as the patient slowly moves through the ring.

Using X-rays for Therapy

Radiation therapy also uses X-rays, but not the ones used for medical imaging. Even though tissue absorbs fewer imaging photons than bone, most imaging photons are absorbed before they can pass through thick tissue. For example, only about

10% of the imaging photons make it through a patient's leg even when they miss the bone. That percentage is good enough for making an image, but it won't do for radiation therapy because most imaging X-rays would be absorbed long before they reached a deep-seated tumor. Instead of killing the tumor, intense exposure to these X-rays would kill tissue near the patient's skin.

To attack malignant tissue deep beneath the skin, radiation therapy uses extremely high-energy photons. At photon energies near 1,000,000 eV, the photoelectric effect becomes rare in tissue and bone, and the photons are much more likely to reach the tumor. Photons still deposit lethal energy in the tissue and tumor, but they do this through a new effect: Compton scattering.

Compton scattering occurs when an X-ray photon collides with a single electron so that the two particles bounce off one another (Fig. 16.3.8). The X-ray photon knocks the electron right out of the atom. This process is different from the photoelectric effect because Compton scattering doesn't involve the atom as a whole and the photon is scattered (bounced) rather than absorbed. The physics behind this effect resembles that of two billiard balls colliding, although it's complicated by the theory of relativity. The fact that it occurs at all is proof that a photon carries both energy and momentum and that these quantities are conserved when a particle of light collides with a particle of matter.

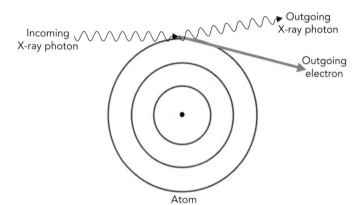

Fig. 16.3.8 In Compton scattering, an X-ray photon collides with a single electron and the two bounce off one another. The electron is knocked out of the atom.

Compton scattering is crucial to radiation therapy. When a patient is exposed to 1,000,000-eV photons, most of the photons pass right through them, but a small fraction undergo Compton scattering and leave some of their energy behind. This energy kills cells and can be used to destroy a tumor. By approaching a tumor from many different angles through the patient's body, the treatment can minimize the injury to healthy tissue around the tumor while giving the tumor itself a fatal dose of radiation.

But Compton scattering isn't the only effect that occurs when high-energy photons encounter matter. X-rays with slightly more than 1,022,000 eV can do something remarkable when they pass through an atom: they can cause electron–positron pair production. A positron is the antimatter equivalent of an electron. Our universe is symmetrical in many ways, and one of its nearly perfect symmetries is the existence of antimatter. Almost every particle in nature has an antiparticle with the same mass but opposite characteristics. A positron or antielectron has the same mass as an electron, but it's positively charged. There are also antiprotons and antineutrons.

Antimatter doesn't occur naturally on the earth, but it can be created in high-energy collisions. When an energetic photon collides with the electric field of an atom, the photon can become an electron and positron. In the previous section, we discussed matter becoming energy; pair production is an example of energy becoming matter. It takes about 511,000 eV of energy to form an electron or a positron, so the photon must have at least 1,022,000 eV to create one of each. Any extra energy goes into kinetic energy in the two particles.

The positron doesn't last long in a patient. It soon collides with an electron and the two annihilate one another—the electron and positron disappear and their mass becomes energy. They turn into photons with a total of at least 1,022,000 eV. So energy became matter briefly and then turned back into energy. This exotic process is present in high-energy radiation therapy and becomes quite significant at photon energies above about 10,000,000 eV. Not surprisingly, it also helps to kill tumors.

Gamma Rays

Producing very high-energy photons isn't quite as easy as producing those used in X-ray imaging. In principle, a power supply could create a huge voltage difference through an X-ray tube so that very high-energy electrons would crash into metal atoms and produce very high-energy photons. But million-volt power supplies are complicated and dangerous, so other schemes are used instead.

One of the easiest ways to obtain very high-energy photons is through the decay of radioactive isotopes. The isotope most commonly used in radiation therapy is cobalt 60 (^{60}Co). The nucleus of ^{60}Co has too many neutrons and, like many neutron-rich nuclei, ^{60}Co undergoes beta decay—one of its neutrons breaks up into a proton, an electron, and a neutrino (or more precisely, an antineutrino). Beginning with that beta decay, ^{60}Co undergoes a series of transformations that produce two high-energy photons: one with 1,170,000 eV and one with 1,330,000 eV. These photons penetrate tissue well and are quite effective at killing tumors.

Although the process by which ^{60}Co produces those two high-energy photons is complicated, beta decay itself shows that protons, electrons, and neutrons are not immutable and that there are other subatomic particles in our universe. Neutrons that are by themselves or in nuclei with too many neutrons are radioactive and experience beta decay. When that beta decay process occurs in a ^{60}Co nucleus, the negatively charged electron and neutral neutrino quickly escape from the nucleus but the newly formed proton remains. The nucleus thus becomes nickel 60 (^{60}Ni).

The neutrino is a subatomic particle with no charge and little mass. Neutrinos aren't found in normal atoms. Though important in nuclear and particle physics, neutrinos are difficult to observe directly because they travel near the speed of light and hardly ever collide with anything. Without charge, they don't participate in electromagnetic forces and, unlike the electrically neutral neutron, they don't experience the nuclear force. They experience only gravity and the weak force, the last of the four fundamental forces known to exist in our universe. (The other three fundamental forces are the gravitational force, the electromagnetic force, and the strong force—that's a more complete version of the nuclear force that we discussed in the previous section.) Because it's weak and occurs only between particles that are very close together, the weak force rarely makes itself apparent. One of the few occasions where it plays an important role is in beta decay.

With almost no way to push or pull on another particle, a neutrino can easily pass right through the entire earth. Neutrinos are detected occasionally, but only with the help of enormous detectors. That's why physicists first showed that neutrinos are emitted from decaying neutrons by measuring energy and momentum before and after the decay. The proton and electron produced by the decay don't have the same total energy and momentum as the neutron had before the decay. Something must have carried away the missing energy and momentum, and that something is the neutrino.

Once ^{60}Co has turned into ^{60}Ni, the decay isn't quite over. The ^{60}Ni nucleus that forms still has extra energy in it. Nuclei are complicated quantum physical systems just as atoms are, and they have excited states, too. The ^{60}Ni nucleus is in an excited state, and it must undergo two radiative transitions before it reaches the ground state. These radiative transitions produce very high-energy photons or gamma rays that are characteristic of the ^{60}Ni nucleus—one with 1,170,000 eV of energy and the other with 1,330,000 eV. These gamma rays are what make ^{60}Co radiation therapy possible.

Particle Accelerators

Electromagnetic radiation isn't the only form of radiation used to treat patients. Energetic particles such as electrons and protons are also used. Like tiny billiard balls, these fast-moving particles collide with the atoms inside tumors and knock them apart. As usual, this atomic and molecular damage tends to kill cells and destroy tumors.

However, obtaining extremely energetic subatomic particles isn't easy. High-voltage power supplies can be used to accelerate an electron or proton to about 500,000 eV, but that isn't enough. When a charged particle enters tissue, it experiences strong electric forces and is easily deflected from its path. To make sure that it travels straight and true, all the way to a tumor, the particle must have an enormous energy. Giving each charged particle the millions or even billions of electron volts it needs for radiation therapy takes a particle accelerator.

Particle accelerators use metal cavities that behave like the tank circuits and antennas we discussed in Section 13.1. Almost any metal structure can act simultaneously as a capacitor and an inductor and thus have a natural resonance for sloshing charge. In the resonant cavities of a particle accelerator, this sloshing charge creates huge electric fields that change with time. Those electric fields push charged particles through space until they reach incredible energies.

One important type of particle accelerator is the linear accelerator. In this device, the electric fields in a series of resonant cavities push charged particles forward in a straight line (Fig. 16.3.9). Each of these cavities has charge sloshing back and forth rhythmically on its wall. When a small packet of charged particles enters the first cavity through a hole, it's suddenly pushed forward by the strong electric field inside that cavity (Fig. 16.3.9a). The packet accelerates forward and leaves the first

(a)

Microwave cavities

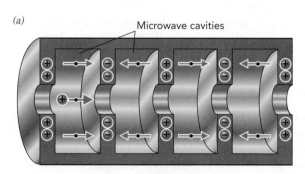

(b)

Electric fields

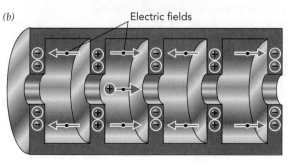

Fig. 16.3.9 In a linear accelerator, moving charged particles are pushed forward by electric fields that change with time. (a) While the moving positive charge is passing through the first of a series of microwave cavities, the field there pushes it forward. (b) By the time the moving charge has entered the second cavity, the fields have reversed and the field there once again pushes it forward through the device.

resonant cavity with more kinetic energy than it had when it arrived: the electric field in that cavity has done work on the packet.

If the fields in the cavities were constant, the electric field in the second cavity would slow the packet down. In Fig. 16.3.9*a*, you can see that the electric field in the second cavity points in the wrong direction. But by the time the packet reaches the second cavity, the charge sloshing in its walls has reversed and so has the electric field (Fig. 16.3.9*b*). The packet is again pushed forward, and it emerges from the second cavity with still more kinetic energy.

Each resonant cavity in this series adds energy to the packet, so that a long string of cavities can give each of the packet's charged particles millions or even billions of electron volts. This energy comes from microwave generators that cause charge to slosh in the accelerator's resonant cavities. The linear accelerator then only has to inject charged particles into the first cavity, using equipment resembling the insides of a television picture tube, and those charged particles will come flying out of the last cavity with incredible energies.

This acceleration technique, however, has a few complications. Most importantly, each cavity must reverse its electric field at just the right moment to keep the packet accelerating forward. For simplicity of operation, all the cavities have the same resonant frequency and reverse their electric fields simultaneously. Since the packet spends the same amount of time in each cavity and since it speeds up as it goes from one cavity to the next, each cavity must be longer than the previous one.

But as the packet approaches the speed of light, something strange happens. The packet's energy continues to increase as it goes through the cavities, but its speed stops increasing very much. This effect is a consequence of special relativity, the rules governing motion at speeds comparable to the speed of light. As we saw in Section 4.2, the simple relationship between kinetic energy and speed given in Eq. 2.2.1 isn't valid for objects moving at almost the speed of light; we must use Eq. 4.2.4 instead. As a further consequence of relativity, the packet can approach the speed of light but can't actually reach it. Though each charged particle's kinetic energy can become extraordinarily large, its speed is limited by the speed of light.

Because the packet's speed stops increasing significantly after it has gone through the first few cavities of the linear accelerator, the lengths of the remaining cavities can be constant. Only the first few cavities have to be specially designed to account for the packet's increasing speed inside them. The charged particles emerge from the accelerator traveling at almost the speed of light. They pass through a thin metal window that keeps air out of the accelerator and enter the patient's body. They have so much energy that they can penetrate deep into tissue before coming to a stop.

Magnetic Resonance Imaging

While X-rays do an excellent job of imaging bones, they aren't as good for imaging tissue. A better technique for studying tissue is magnetic resonance imaging or MRI. This technique locates hydrogen atoms by interacting with their magnetic nuclei. Since hydrogen atoms are common in both water and organic molecules, finding hydrogen atoms is a good way to study biological tissue.

The nucleus of an ordinary hydrogen atom, ^{1}H, is a proton. Protons, like electrons, have two possible internal quantum states, usually called spin-up and spin-down. Calling it spin is appropriate because spin-up and spin-down protons have equal but oppositely directed angular momentum. And when electric charge and rotation are both present, it's not surprising that magnetism is, too; electric currents are magnetic, after all. Sure enough, protons have magnetic dipoles—equal north and south poles at a distance from one another. A spin-up proton acts as though it has its north pole on top, while a spin-down proton acts as though it has its south pole on top.

When a proton is immersed in a magnetic field, it tends to align its magnetic dipole with that field. Doing so minimizes its magnetic potential energy. But while protons would align perfectly with the field at absolute zero, they are less successful near room temperature. Thermal energy agitates the protons so that, even in a strong, upward-pointing magnetic field, spin-up protons only slightly outnumber spin-down protons.

In that upward-pointing magnetic field, each proton has two possible quantum states: alignment with the field (spin-up) or anti-alignment (spin-down). Because alignment reduces the proton's magnetic potential energy, it's the ground state—the lower energy of the two possible states. The anti-aligned state is the excited state.

With its two possible states, ground and excited, a proton in a magnetic field can exhibit many of the behaviors we explored when looking at atoms in Section 14.2. Most importantly, the proton can experience radiative transitions between its two states. A ground state proton can *absorb* a photon while making a radiative transition to its excited state and an excited state proton can *emit* a photon while transiting to its ground state.

In Section 14.2, we saw that a given atom can absorb or emit only certain photons, photons carrying exactly the right amount of energy to shift the atom from one quantum state to another. For example, neon signs are red because neon atoms have states that are separated in energy by the energy of red photons. Similarly, a proton in a magnetic field can absorb or emit only certain photons, photons carrying exactly the right amount of energy to shift the proton from one quantum state to the other.

However, unlike a neon atom, which always interacts with red photons, a proton in a magnetic field interacts with photons that vary in "color" according to the strength of the magnetic field. That's because the energy separating the proton's two states is proportional to the magnetic field in which it resides. As a result, the photon energy needed to cause radiative transitions between the proton's two states is also proportional to the magnetic field. If the field changes, so does the photon energy.

When a patient enters the strong magnetic field of an MRI machine (Fig. 16.3.10), the protons in the patient's body respond to the field and a small excess of aligned protons develops. Only these excess aligned protons matter to the MRI machine because effects due to the remaining protons, which are equally aligned and anti-aligned, cancel completely. The excess aligned protons are in their ground state and they are what the MRI machine studies.

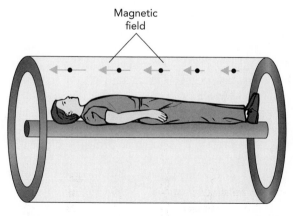

Magnetic field

Fig. 16.3.10 An MRI machine places the patient in a strong magnetic field. This field varies spatially, so that protons at different places in the patient's body experience different magnetic fields and absorb different radio wave photons.

The MRI machine interacts with these ground state protons using radio wave photons—photons with energies equal to the energy difference between their ground and excited states. The protons can absorb and subsequently emit those radio wave photons and they can also exhibit a variety of fascinating and useful quantum interference effects.

If the protons in the patient's body were all experiencing exactly the same magnetic field, they would all interact with the same radio wave photons. But the protons don't all experience the same field. The MRI machine introduces a slight spatial variation to its magnetic field. Because the magnetic field is different for different protons, only some of them can interact with radio wave photons of a particular energy. This selective interaction is how the MRI imager locates protons within a patient.

In its simplest form, an MRI machine applies a spatially varying magnetic field to the patient's body. It then sends various radio waves through the patient and looks for the radio wave photons to interact with protons. Since only a proton that is experiencing the right magnetic field can interact with a particular radio wave photon, the MRI machine can determine where each proton is by the photons with which it interacts. By changing the spatial variations in the magnetic field and adjusting the energies of the radio wave photons, the MRI machine gradually locates the protons in the patient's body. It builds a detailed three-dimensional map of the hydrogen atoms. A computer manages this map and can display cross-sectional images of the patient from any angle or position.

CHAPTER 17

THINGS THAT INVOLVE MATERIALS

Between natural resources such as water and air, and finished products such as bulbs and bicycles, lies a world of modified and synthesized materials. The study and development of these basic ingredients of technology have enormous influences on our daily environment and are part of a field called materials science.

Despite its optimistic title, this book can't do justice to everything that illustrates physics, let alone things that illustrate other fields of science. But the borders between those fields are fuzzy, and I cross their frontiers frequently in my research and my writing. That doesn't mean that I'm an expert in those other fields, but I admire the work of those who are and I love to connect those fields with physics. I also learn new things every time I take such a "field trip."

Materials science is rich with physics issues and many of its true experts began their trainings as physicists. Experimental physicists crave good tools while materials scientists thrill at making those tools. When we visit the hardware store, we're equally drawn to cutting implements that never dull or chip, glues that can stick to anything, glasses and plastics that won't break or scratch, and ropes that can support outrageous weights.

This chapter, like those preceding it, addresses the how and why questions of the physical world. This time, however, I'll explain both how various objects work and why the materials from which those objects are made behave as they do. In grade school, we learned to categorize substances simplistically into solids, liquids, or gases, and into mixtures or solutions. If that's still how you see the world of materials, I just want to say one word to you, just one word: plastics. In which category would you put vinyl upholstery? Or chewing gum? Or JELL-O®? The world of materials is evidently more complicated than it first appears.

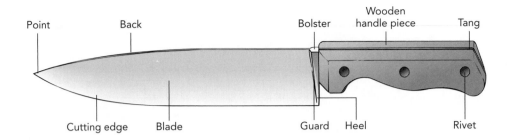

Point Back Wooden handle piece Tang Bolster

Cutting edge Blade Guard Heel Rivet

Section 17.1 Knives and Steel

A dull knife is useless in the kitchen—you might as well tear a tomato or a loaf of crusty bread apart by hand. But sharpening a knife isn't always easy because some knives take edges better than others. While there are many knives that simply won't stay sharp, a few wonderful knives never seem to dull. Though they are all made of steel, there is clearly something different about their blades.

Knives have always been one of the greatest tests of a steelmaker's skill, from the days when swordmaking was an art to the present era of science and technology. A knife's blade must be tough and flexible, while its cutting edge must be hard but not brittle. Giving steel these properties requires great control over its composition and processing. To understand a knife blade, you must understand its steel.

Stresses and Strains, Bends and Breaks

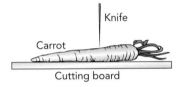

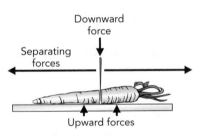

Fig. 17.1.1 When you push down on a knife to cut a carrot, the cutting board pushes up. Overall, the vertical forces on the carrot, including its weight, cancel completely. But the ramp-like surfaces of the knife blade create large horizontal forces on the two halves of the carrot and these forces divide the carrot.

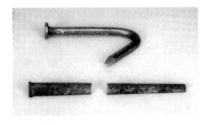

Fig. 17.1.2 Soft, low-carbon steel bends easily while hard, carbon-rich steel breaks instead of bending.

Knives are simple tools that we use to cut things into pieces. They are essentially wedges that use mechanical advantage to convert small forward forces into large separating forces. When you push down on the blade of a knife as you cut a carrot (Fig. 17.1.1), its cutting edge penetrates the carrot while the two inclined surfaces of the blade exert huge horizontal forces on the two halves of the carrot. One half moves to the left while the other half moves to the right and the carrot divides neatly in half.

But while the mechanical action of cutting is simple, the physical structure of the knife is not. The secret to its ability to cut through the carrot lies in the properties of its blade. This blade is almost certainly made of steel, so the story of how a knife works is really the story of how steel works.

Steel is not a specific material but rather a whole range of iron-based metals. These metals differ from one another in their specific chemical compositions and in how they have been processed. The variety of steels is so broad that it's difficult to encompass them all in a single definition. However steels are generally mixtures of iron and other elements that contain no more than 2.06% carbon by weight. A mixture with more carbon than this is usually called cast iron. It's an unfortunate historical accident that cast irons actually contain more carbon and less elemental iron than many steels.

Carbon content is important in distinguishing steel from cast iron because it affects the hardness, brittleness, and other characteristics of these metals (Fig. 17.1.2). Hardness is a measure of a material's resistance to penetration, deformation, abrasion, and wear. Brittleness is the tendency for a material to fracture while it's being deformed. A good knife blade should be hard but not brittle; distorting very little as you push it through the carrot but resisting fracture even if you use the knife to open a metal can.

Controlling the hardness, brittleness, and other characteristics of steel is a complicated task that involves all aspects of steel production. These characteristics can even vary within a single steel object—the cutting edge of a good knife blade is actually harder than the rest of the blade. To understand how these characteristics are controlled, we'll first look at how materials respond to outside forces and then examine the microscopic basis for steel's properties. What we learn in the process will apply not only to steel, but also to many other materials.

When you push gently on a solid object, that object distorts by an amount proportional to the force you exert on it. This relationship is simply Hooke's law, which we first encountered in Section 3.1 on spring scales. If the object is a steel block resting on the floor and you're stepping on it, then its tiny distortion is proportional to your weight. By measuring how much your weight distorts this block, you can learn something about the steel from which it's made.

But the distortion is also related to the block's dimensions. The broader its surface, the more your weight is spread out and the less the block distorts. Since we are trying to learn about the steel rather than the block, we divide your weight by the surface area of the block to obtain the stress on the steel (Fig. 17.1.3). Stress is the amount of force exerted on each unit of a block's surface area and is the measure of how much the steel is being squeezed.

However, the distortion also depends on the block's height. Each centimeter of metal shrinks a little bit so that a tall block shrinks more than a short block. Again, we are more interested in the steel than the block, so we divide the change in height by the original height of the block to obtain the strain in the steel (Fig. 17.1.3). Strain is the change in length per unit of a block's length and is a measure of how much the steel responds to being squeezed.

Hooke's law relates the stress on the steel to the strain that results:

$$\text{strain} = \frac{\text{stress}}{\text{Young's modulus}},\qquad (17.1.1)$$

where Young's modulus is a measure of how difficult it is to compress the steel. Strain is just a number and has no dimensions, while stress and Young's modulus are both pressures and are measured in pascals (Pa).

Young's modulus is related to the interatomic forces that hold a material together. The atoms in a solid exert attractive and repulsive forces on one another and these forces only balance when the atoms have the proper separations. If you push the atoms closer together or pull them farther apart, the forces no longer balance and they oppose your action. The stiffer the interatomic forces, the larger Young's modulus and the less the material shrinks as you squeeze it.

Because the interatomic forces in steel are primarily between iron atoms, it has roughly the same Young's modulus as iron: about 195 GPa (195 gigapascals or 195,000,000,000 Pa). This large value makes steel extremely difficult to compress. A 1-m cube of steel would lose somewhat less than a micron of height while supporting a city bus. A similar cube of lead would shrink 14 times more while a cube of tungsten would shrink only half as much.

Although we arrived at Eq. 17.1.1 while thinking about squeezing, this equation also applies to situations where the steel is exposed to gentle tension. In that case, the stress is negative (pulling apart rather than squeezing together) and the strain that results is also negative (stretching rather than compressing). Whether you squeeze or stretch steel, you are still measuring the stiffness of the interatomic forces and arrive at the same value for Young's modulus. Young's modulus is usually measured with tensile stress (stretching) rather then with compressive stress (squeezing) because compression can cause a thin piece of material to buckle while tension pulls it straight and true.

But compression and tension aren't the only stresses a steel block can experience. If you push the bottom of the block to the left and the top of the block to the right, the metal experiences shear stress (Fig. 17.1.4). This stress bends the block and causes shear strain in the metal. Shear strain is the angle of the bend caused

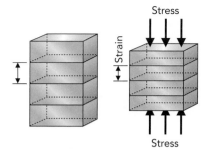

Fig. 17.1.3 Steel is compressed when it experiences an inward stress. Stress is equal to the inward force exerted on each unit of surface area. This stress causes a strain in the steel. This strain is equal to the fractional change in the steel's length—its change in length divided by its original length.

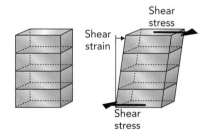

Fig. 17.1.4 A material bends when it experiences shear stress. Shear stress is equal to the sideways force exerted on each unit of surface area and causes a shear strain in the material. The shear strain is equal to angle of the bend.

by a shear stress. As long as the forces involved are relatively small, the shear strain is proportional to the shear stress,

$$\text{shear strain} = \frac{\text{shear stress}}{\text{shear modulus}}, \qquad (17.1.2)$$

where the shear modulus is a measure of how difficult it is to bend the metal.

Stress and strain help to characterize a particular steel. The crucial test for a knife is how it responds to stress. Because all steels have similar Young's and shear moduluses, it's hard to tell them apart with gentle stresses. The real differences between steels only appear when the stresses become large and Eqs. 17.1.1 and 17.1.2 stop being valid. That's when the steels begin to bend and break and when the good knives begin to distinguish themselves from the poor ones.

Plastic Deformation and Crystalline Materials

A spring has an elastic limit beyond which it stops obeying Hooke's law and bends permanently. A gentle force distorts the spring elastically and it bounces back while a strong force deforms it forever. The same holds true for a piece of steel. As long as the stress is small, steel undergoes elastic deformation—it distorts temporarily while stressed, according to Eqs. 17.1.1 and 17.1.2, and then returns to its original shape. However when the stress is large, steel undergoes plastic deformation—its atoms rearrange and it deforms permanently.

Steels differ considerably when it comes to plastic deformation. They all start off responding elastically to small stresses but gradually shift over to plastic deformation as the stresses increase. Just how much stress a particular steel can tolerate before it begins plastic deformation is its yield strength and is an important measure of that steel's load-bearing capacity. A weak steel yields easily to a modest stress while a strong steel responds elastically to all but enormous stresses. The cutting edge of a knife must tolerate severe stresses without yielding so that it doesn't become dull.

To see how plastic deformation occurs in a knife blade, we must examine the microscopic structure of steel. But since steel has a rather complicated structure, let's start with pure iron instead. Like most metals, iron is a crystalline solid. It may not have the pretty natural facets typical of some minerals but it does have an orderly arrangement of atoms. At room temperature, iron forms ferrite crystals. Ferrite is a ferromagnetic material in which the iron atoms are arranged in a body-centered cubic lattice (Fig. 17.1.5).

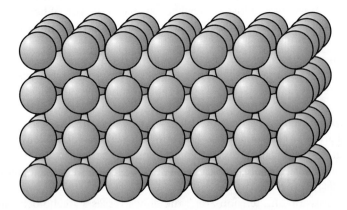

Fig. 17.1.5 In ferrite, iron atoms are arranged in a body-centered cubic lattice. This lattice is built out of eight-atom cubes (2 by 2 by 2 atoms) with an additional atom located in the center of each cube.

As you can see by looking at this lattice, it has smooth surfaces between sheets of atoms. When it experiences enough shear stress, this lattice can undergo a phenomenon called slip, in which sheets of atoms slide across one another. Slip is the most common mechanism for plastic deformation. When you squeeze, stretch, or bend a piece of pure iron so that it yields, you are probably causing slip. As the sheets slide across one another, the crystals change shape just like a stack of cards does when you push the top card to one side. The atoms lose track of their original positions and the iron doesn't return to its original shape when you stop pushing on it.

Slip occurs fairly easily in iron because the bonds that hold it together are relatively nondirectional. Its metallic bonds are formed by a general sharing of electrons between countless atoms. This sharing lowers both the kinetic and potential energies of the electrons and causes the atoms to cling to one another. Because these metallic bonds are rather insensitive to the relative positions of its atoms, iron crystals are quite susceptible to slip.

But there are other factors that affect when and where slip occurs. These factors appear because real metals are never perfect crystals. Nature tends to introduce randomness wherever possible so that even the purest crystals contain a few random mistakes. Typical iron crystals are filled with imperfections. One common defect in iron is a dislocation, a sheet of atoms that ends abruptly in the middle of the crystal (Fig. 17.1.6).

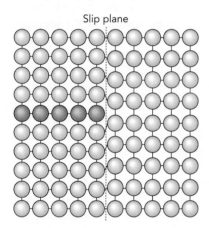

Slip plane

Fig. 17.1.6 Dislocations occur when a sheet of atoms ends abruptly, part way through a crystal. In this top view of an imperfect crystal, the darkened sheet of atoms on the left ends suddenly and the sheets on the right are stretched to compensate. Slip occurs relatively easily along the slip plane. Thermal energy allows dislocations to move about in many materials, further softening those materials.

In a perfect crystal, a whole sheet of atoms must slip at once. But a dislocation breaks up the uniformity of the crystal and allows the sheet to slip gradually, one row of atoms at a time. A crystal containing a dislocation has an unusually low yield strength along the slip plane, the sheet of atoms perpendicular to the end of the dislocation.

While dislocations weaken iron by easing slip, there are other imperfections that strengthen iron by preventing slip. One such imperfection is iron's polycrystalline structure—rather than being a single crystal, iron is normally composed of many individual crystallites that meet one another at random angles. These individual crystallites are called grains and the boundary layers of atoms between grains are called grain boundaries.

Grains and the grain boundaries strengthen iron. Within each grain, slip only occurs between particular sheets of atoms and along specific directions. Since the grains are randomly oriented, they aren't all able to slip along the same direction. They must coordinate their slips along many directions to allow the piece of iron to yield along one direction, a process that can only occur when the iron is experiencing enormous stress. As a result, iron with lots of tiny grains has a higher yield strength than iron with only a few large grains.

The sizes and shapes of the grains depend on how the iron has been handled and processed. Because the atoms in a grain boundary don't fit into the crystallites they connect, they have relatively high potential energies and create a surface tension around each grain. Surface tension appears whenever a surface has extra potential energy and it acts to make that surface as small as possible. Like a soap froth that tries to minimize the surface area of its soap bubbles, the iron tries to minimize the surface area of its grain boundaries. But the iron atoms can only rearrange while the iron is hot. Annealing iron—heating it to high temperature and then cooling it slowly—allows its larger grains to grow by consuming the smaller grains. Annealing is the principal method for softening iron, steel, and most other metals.

In contrast, deforming iron at low temperatures breaks up its grains and makes it harder. Like most metals, iron can be work-hardened by pounding, rolling, folding, and twisting. Wrought iron is a good example of work hardening, a technique that has been used for millennia to strengthen metals.

But annealing and work hardening have consequences beyond their effects on yield strength. Instead of yielding to stress, work hardened iron may fracture into pieces. This catastrophe is called brittle fracture because the metal doesn't yield before it breaks. Brittle fracture occurs in metals that are so hard that sheets of atoms separate completely rather than sliding across one another. Overworked iron may also suffer from metal fatigue, in which the metal tears as cracks work their way inward from surface defects.

Annealed iron yields before it breaks. In fact, it's ductile, meaning that it can be stretched quite a bit during plastic deformation. The main reason that the annealed iron breaks at all is that it gets thinner as it stretches and usually develops a narrow neck. Stretching work hardens the neck and the high stress there eventually pulls the atoms apart. Because the iron yields before it breaks, this type of breakage is called ductile fracture. The peak stress that the iron can withstand before such fracture occurs is its tensile strength.

Plastic deformation and ductility are actually quite useful in a knife. When the stress on the knife is too great for the blade to handle elastically, you would rather have it bend than break. A metal's ductility increases with temperature because thermal energy allows dislocations and other defects to move through crystals so that slip can occur between many different sheets of atoms. In a cold metal, those defects are immobile and slip is reduced. Because of its low ductility, cold iron is much more susceptible to brittle fracture than hot iron. (For some interesting examples of brittle fracture in cold metal, see ❶❷.)

Moreover, plastic deformation requires energy. When you push on an iron knife and it bends in the direction of your push, you are doing work on the blade and the blade absorbs energy. A metal's ability to absorb energy during deformation is called toughness. Toughness is extremely important in car bodies, swords, and armor, all of which dent during collisions to absorb energy. If they shattered instead of yielding, they wouldn't be nearly as useful. Glass windows and lenses are dangerous precisely because they don't exhibit plastic deformation. In contrast, plastic windows and lenses are much safer because they're able to dent and absorb energy without shattering.

Steel

Adding carbon to iron produces steel, an alloy or metallic mixture that is far tougher and harder than pure iron. But different amounts of carbon produce different types of steel. Moreover, the ways in which this material is processed mechanically and thermally can dramatically change its character.

The first small amount of carbon added to iron dissolves in the iron to form a solid solution. It may seem strange for a solid to have something dissolved in it, but

❶ When the Titanic struck an iceberg and sunk on April 14, 1912, it was thought that the iceberg had torn a long hole in its steel hull. However, recent explorations of the wreck have found that the real culprit was probably brittle fracture. Instead of denting when it collided with the ice at 41 km/h, the ship's cold steel hull appears to have shattered.

❷ When the Purity Distilling Company's 30-m tall molasses tank burst in Boston's North End on January 15, 1919, it released 2.2 million gallons of molasses in a wave that destroyed several city blocks and killed 21 people and dozens of horses. The noontime disaster was initially attributed to an explosion but an official investigation found that the tank's brittle steel had fractured in the 4 °C (40 °F) winter weather. Purity, a subsidiary of United States Industrial Alcohol Company, paid $1 million in damages. Ironically, the accident occurred one day before ratification of the 18th Amendment to the U.S. Constitution, prohibiting the sale of alcoholic beverages.

there is no rule that says a solvent must be liquid. Solution occurs whenever energy and entropy (randomness) make it favorable for one material to divide into atoms, molecules, or ions and become dispersed throughout a second material. Whether it's liquid or solid, iron can dissolve a small amount of carbon.

The iron remains in its ferrite form and the dissolved carbon atoms arrange themselves randomly in the interstitial spaces between iron atoms. Ferrite can dissolve up to 0.01% carbon by weight at room temperature and that carbon makes it somewhat harder than pure iron. The carbon atoms introduce local distortions in the ferrite crystals so that sheets of atoms can't slide as easily across one another. Since the dissolved carbon reduces slip, it increases the yield strength of the steel.

Only hydrogen, nitrogen, and carbon atoms are small enough to fit in between the iron atoms and cause such interstitial solution hardening of the steel. However, there are a number of other atoms that harden iron and steel by substituting for iron atoms in ferrite crystals. These atoms also distort the crystals, impeding slip and increasing the yield strength of the steel. Phosphorus, silicon, manganese, chromium, and nickel atoms are often added to steel to cause this substitutional solution hardening.

When the carbon fraction in ferrite exceeds 0.01%, it can't all remain dissolved in the ferrite at room temperature. A new material appears in the steel: iron carbide (Fe_3C). Iron carbide is an extremely hard and brittle crystalline material, a mineral called cementite. The presence of tiny cementite crystals dispersed throughout the ferrite impedes slip and increases the steel's yield strength. Strengthening of this sort is called dispersion hardening.

The arrangement of ferrite and cementite particles in the steel depends on the amount of carbon in the mixture and on the thermal and mechanical histories of the steel. When it's cooled slowly from high temperatures, steel contains pearlite, a material consisting of alternating layers of ferrite and cementite. Pearlite has a carbon content of about 0.8% by weight. If the steel has less than 0.8% carbon, it forms some extra ferrite and consists of pearlite interspersed with ferrite. If it has more than 0.8% carbon, it forms some extra cementite and consists of pearlite interspersed with cementite.

Good knives, however, use characteristics of steel that don't appear until steel is subject to more sophisticated heat treatments. The microscopic structure of steel changes remarkably as it's heated and cooled.

Above 723 °C, iron forms austenite crystals. Austenite is a nonmagnetic material in which the iron atoms are arranged in a face-centered cubic lattice (Fig. 17.1.7). Austenite can dissolve more carbon than ferrite; up to 0.8% by weight at 723 °C and up to 2.06% at 1148 °C. The latter figure sets the upper limit for what is considered steel. As you slowly heat steel above 723 °C, its ferrite, pearlite, and cementite all convert into austenite—a structural change called a solid-to-solid phase transition.

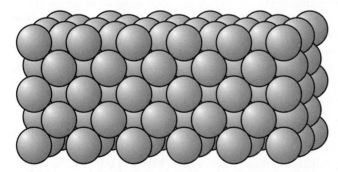

Fig. 17.1.7 Above 723 °C, the atoms in solid iron are arranged in a face-centered cubic lattice. This lattice is built out of eight-atom cubes (2 by 2 by 2 atoms) with an additional atom located in the center of every face of that cube.

When you cool the steel back down slowly, the reverse phase transition occurs and the austenite turns into ferrite, pearlite, and cementite.

But even more remarkable effects occur during rapid cooling. If the austenite is suddenly cooled to between 600 and 650 °C and kept there, it transforms into fine pearlite—pearlite with extremely thin layers. To form pearlite, carbon atoms must diffuse through the iron. At lower temperatures, they can't travel the long distances needed to form thickly layered coarse pearlite, so they form fine pearlite instead. Fine pearlite doesn't undergo slip as easily as coarse pearlite so it has a higher yield strength.

Austenite that is suddenly cooled to between 260 and 400 °C and kept there doesn't form pearlite at all. The carbon atoms diffuse such short distances that they form tiny nodules of cementite. These nodules are arranged between sheets of ferrite in a layered material called bainite. Bainitic steel has a higher yield strength than fine pearlite.

Austenite that is suddenly cooled below about 200 °C forms an entirely new material. At 200 °C, there is so little thermal energy around that carbon atoms don't diffuse through the iron at all, so cementite, pearlite, and bainite can't form. Instead, the austenite tries to turn into ferrite without first getting rid of all its dissolved carbon. What forms is martensite, a distorted ferrite that is stretched in one direction.

Since ferrite can't have more than about 0.01% carbon in it by weight, martensite is essentially ferrite with way too much dissolved carbon. A solution containing more dissolved material than is stable at the current temperature is said to be supersaturated. The dissolved material will eventually come out of solution but it may take quite a long time. Martensite is a supersaturated solution of carbon in ferrite that lasts almost forever at room temperature. Because martensite is very resistant to slip, steel containing it has an extremely high yield strength and hardness. However martensite also makes the steel brittle by preventing it from undergoing plastic deformation. Steel containing martensite isn't as tough as steel containing bainite or fine pearlite.

Sudden cooling or quenching of austenitic steel clearly produces a harder material than slow cooling. The faster you cool the austenite and the colder you take it in that first step, the harder the steel becomes. Unfortunately, red-hot carbon steel must be plunged into water to cool it quickly enough to form martensite. The steel contracts during this harsh treatment and traps stresses that weaken the metal. To relieve these internal stresses, quenched steel is often tempered by reheating it just enough to let some of the stresses resolve themselves. The hotter and longer this tempering process, the more stress is relieved but the softer the steel becomes.

Quenching carbon steel in water is dramatic, with steam billowing up from the hot metal, but only a thin surface layer cools quickly enough to harden properly. To ease the formation of martensite and reduce the cooling rate needed to harden steel, it's often alloyed with other elements. Such alloy steels harden easily and deeply when quenched unspectacularly in oils or air and are important in knives and other tools. Some alloy steels also undergo precipitation hardening, in which small crystals of various compounds precipitate out of solution in the steel as it cools and strengthen the metal. Titanium carbide, niobium carbide, and vanadium nitride frequently appear in precipitation-hardened steel.

Part of a knifemaker's skill comes in choosing just the right steel and just the right thermal processing to bring out that steel's best characteristics in the blade and its cutting edge. Knife blades require particularly careful heat treatment. A blade that retains too much stress will be hard but brittle while one that has been tempered at too high a temperature will dull easily.

In a fine knife, the cutting edge may be hardened more than the body of the blade by adjusting both the chemical composition of the edge and its heat treatment. Flames and laser beams are often used to reheat the edge and to add more carbon to the steel there. The result is a knife with the toughness of fine pearlite or bainite in its body and the hardness of carefully tempered martensite in its cutting edge. You can occasionally see evidence for this special treatment in the color and appearance of the blade. (Joining different materials to perform a task that neither could do alone is common in modern construction, see ❸.)

Stainless Steel Blades

One last feature of steel that's important in knife blades is corrosion resistance. Because carbon steel is susceptible to rust, knife blades are usually made of stainless steel. This corrosion-resistant material is formed by replacing at least 4% of steel's iron atoms with chromium atoms. Most stainless steels contain more than 11% chromium by weight and some contain nickel as well. Nickel enhances the steel's corrosion resistance and also makes it more ductile.

But not all forms of stainless steel are appropriate for knives. Alloying the steel with chromium and nickel affects more than just its chemical properties. The crystalline structures of these alloys depend on the precise mixtures of the various elements and on their thermal histories. Perhaps the most remarkable effect of adding chromium and nickel to the steel is that these elements can make the austenite crystal structure stable at room temperature rather than the ferrite crystal structure.

The most common stainless steel is called 18–8 stainless and contains roughly 18% chromium and 8% nickel. It consists exclusively of nonmagnetic austenite grains. Since 18–8 stainless is stable as austenite, even at room temperature, it doesn't form cementite or martensite and it thus can't be hardened by thermal processing. In fact, overheating it can spoil its corrosion resistance (see ❹). The only way to harden 18–8 stainless steel is to work harden it. Because it's difficult to make a good knife edge by work hardening, 18–8 stainless is a poor choice for knives. However 18–8 stainless is inexpensive, ductile, and easy to work with so it's often used in cafeteria grade cutlery. The next time a dull knife bends while you're cutting your food, you'll know that it's probably made from soft, austenitic stainless steel.

There are two ways to harden stainless steels: martensite formation and precipitation hardening. Martensitic stainless steels contain little or no nickel and moderate amounts of carbon. With too little nickel to stabilize austenite, martensitic stainless steels crystallize like normal steels and are magnetic. Quenching hot martensitic stainless steel produces tiny martensite crystals and hardens the metal.

Precipitation-hardening stainless steels contain alloying elements that form hard precipitate crystals within the steel during quenching. Titanium, niobium, aluminum, copper, and molybdenum dissolve in the steel when it's hot but form tiny precipitate crystals within the steel as it cools. These crystals make the stainless steel hard.

High-quality knives and cutlery are generally made of martensitic or precipitation hardening stainless steels. These metals are tough, keep their edges well, and are hard to bend. But knives and utensils made from these steels require careful heat treatment during manufacture, so they are considerably more expensive than common cafeteria dinnerware.

How Steel Is Made

Steel is produced on enormous scales using techniques that have become almost as high-tech as those involved in the semiconductor industry. The days of dirty, grimy steel mills are over because modern high-quality steels require exceptional chemical purity.

❸ A mixture of steel and concrete supports many buildings. These two materials enjoy a symbiotic relationship. Steel has enormous tensile strength but bends easily when compressed. Concrete can withstand immense compressive stress but fractures easily when stretched. However, they can be combined into a composite material that has the tensile strength of steel and the compressive strength of concrete. Pre-stressing the pair by stretching the steel while the concrete is drying further improves the performance of the composite.

❹ Austenitic stainless steels like 18–8 are susceptible to subtle damage when they're overheated. At temperatures between about 500 °C (932 °F) and 800 °C (1472 °F), the chromium and carbon in the steel can precipitate out at the grain boundaries as chromium carbide. This process depletes the grain edges of chromium and leaves them extremely vulnerable to corrosion. Corrosion at the grain edges cuts the grains right out of the metal so that the metal falls apart. That's why a badly overheated stainless steel pot is never the same again.

Steel is generally made in two steps: iron ore is converted into pig iron and pig iron is converted into steel. These two steps were once quite separate and followed by many further independent processing steps. However, modern steel mills convert iron ore into finished steel in an uninterrupted manner.

Iron ores are essentially iron oxides (Fe_2O_3, Fe_3O_4, FeO) so something must remove the oxygen atoms to convert iron ore into iron. That something is carbon. When iron oxide and carbon are heated together, the carbon reacts with the oxygen to form carbon monoxide and carbon dioxide. These two gases escape and leave behind pig iron, a mixture of iron and carbon.

The carbon needed for this process is obtained by heating coal to high temperatures. The coal cracks chemically and releases compounds such as acetylene, light oil, coal tar, and ammonia, all of which are collected for use outside the steel mill. What remains is nearly pure carbon, a material called coke.

But both the iron ore and the coke contain rocks and other undesirable materials that would contaminate the pig iron. To remove these contaminants, the iron ore and coke are mixed with lime (CaO) obtained from limestone ($CaCO_3$). The lime acts as a liquid flux at high temperature, floating on top of the molten pig iron and dissolving many of the undesirable materials.

The mixture of iron ore, coke, and lime is converted into pig iron in a blast furnace. The coke burns in a stream of air, providing the intense heat needed to initiate and sustain the chemical reactions between carbon and iron oxide. The lime flux carries away most of the unwanted materials as slag, which can then be used to make concrete.

Until about 1856, this was all the processing that iron received. In fact, most iron ore was converted to iron at relatively low temperatures so that it never melted at all. Carbon and carbon monoxide diffused through the iron ore and reacted with the oxygen atoms, thereby converting solid iron oxide into solid iron. This iron was then hammered into shape as wrought iron. It contained various amounts of carbon and lots of slag.

Good techniques for melting pig iron and removing carbon and slag from it to make steel appeared in the last century and a half. The Bessemer converter, the open-hearth process, and the basic oxygen process all remove excess carbon and other impurities by burning them out of the pig iron. The basic oxygen process now dominates the steel industry. It uses a water-cooled pipe or "lance" to inject pure oxygen gas into a vessel of liquid pig iron. The impurities burn away, leaving behind almost pure iron. Even oxygen and other gases are often removed from the iron by vacuum pumps. The conversion from pig iron to steel takes roughly 30 minutes and is monitored by disposable sensors that are plunged into the liquid metal to examine its composition.

To make the appropriate steel, alloying elements are added to the liquid iron and the resulting material descends out of the bottom of the converter through a series of containers. In a continuous casting machine, the steel is formed into a red-hot solid bar that descends from a mold. This bar is bent until it's horizontal and it then enters a series of rolling machines. These machines shape the steel into anything from beams to sheet metal. The whole procedure is so smoothly orchestrated that the steel making never stops. The mill just keeps on making more steel and it keeps flowing out to the finishing equipment. Even changes in the steel's composition are handled without interrupting the flow.

The finishing equipment rolls the steel into various shapes, gradually changing its dimensions with enormous pressures and high temperatures. While steel is occasionally worked with as a liquid, it's generally much easier to handle as a hot solid.

While the work hardened steel that emerges from these rolling machines is fine for construction, it must be softened before it can be molded into car body panels or cooking pots. Softening is done by annealing the steel in huge ovens. At the opposite extreme is spring steel, which is reheated and quench hardened so that it snaps back to its original shape after being deformed.

Many tools, including locks and wrenches, need a hard surface layer on a softer core. The hard surface makes the tool impenetrable while the softer core allows it to withstand energetic blows. This layered arrangement can be made by case hardening the tool—exposing its heated surface to extra carbon. The carbon atoms diffuse into the surface and increase its hardenability. Once it has been heat-treated, a case-hardened tool is extremely difficult to break.

SECTION 17.2 Windows and Glass

When you look through a clean glass window on a sunny day, you hardly know the glass is there. Light passes through each pane with so little loss and distortion that it's difficult to see the window at all. But you have no trouble feeling glass—it's a tough, rigid material that keeps out the wind and weather while admitting light. Because glass is also a good thermal insulator, it helps to keep buildings cool in the summer and hot in the winter. Produced from cheap and plentiful raw materials with minimal effort and equipment, glass is truly a remarkable substance.

What Is Glass?

Let's begin with what glass is not—it is not a crystal. The atoms in a crystal are organized in a regular, repetitive lattice so you need only locate a few atoms in order to predict where all of their neighbors are (Fig. 17.2.1*a*). The atoms are so neatly arranged that, except for occasional crystal defects, you can predict positions for thousands or even millions of atoms in every direction. This spatial regularity is called long-range order.

Glass is an amorphous solid, a material without long-range order (Fig. 17.2.1*b*). Locating a few glass atoms tells you next to nothing about where to find any other atoms. The atoms in glass are arranged in the random manner of a liquid because glass is essentially a frozen liquid. Its atoms are jumbled together in a sloppy fashion but they can't move about to form a more orderly arrangement. Glass arrives at this peculiar amorphous state when hot liquid glass is cooled too rapidly for it to crystallize.

If molten glass were an ordinary liquid, it would begin to solidify abruptly during cooling once it reached its freezing temperature. At that point, its atoms would begin to arrange themselves in crystals that would grow in size until there was no liquid left. That's what's normally meant by freezing.

However, some liquids are slow to crystallize when you cool them slightly below their freezing temperatures. While they may be cold enough to grow crystals, they must get those crystals started somehow. If crystallization doesn't start, a material's atoms and molecules will continue to move about and it will behave as a liquid. When that happens, the liquid is said to be supercooled.

Supercooling is common in liquids that have difficulties forming the initial seed crystals on which the rest of the liquid can crystallize. Because almost all of the atoms in a seed crystal are on its surface, it has a relatively large surface tension and surface energy. Below a certain critical size, a crystal is unstable and tends to fall apart rather than grow. However once the first seed crystals manage to form, a process called nucleation, the rest of the supercooled liquid may crystallize with startling rapidity. In Section 7.2, we saw how water can superheat when it fails to nucleate steam bubbles above its boiling temperature. Here the liquid can supercool when it fails to nucleate seed crystals below its freezing temperature.

Just below its freezing temperature, the atoms in glass don't bind to one another long enough to form complete seed crystals and nucleation takes almost forever. The glass is a supercooled liquid. At somewhat lower temperatures, seed crystals begin to nucleate, but glass's large viscosity (thickness) prevents these crystals from growing quickly. The glass remains a supercooled liquid for an unusually long time. At even lower temperatures, glass becomes so viscous that crystal growth stops altogether. The glass is then a stable supercooled liquid, one that will remain in that form indefinitely. At this temperature range, glass still pours fairly easily and can be stretched or molded into almost any shape, including windowpanes.

However, when you cool the glass still further, it becomes a *glass*. Here the word *glass* refers to a physical state of the material—a type of amorphous solid.

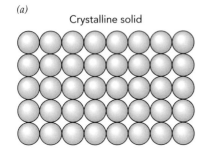

(a) Crystalline solid

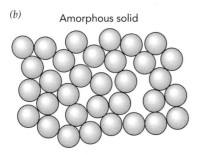

(b) Amorphous solid

Fig. 17.2.1 (*a*) A crystalline solid has long-range order, meaning that knowing where atoms are at one end of the crystal tells you where atoms are at the other end of the crystal. (*b*) An amorphous solid does not have long-range order.

To distinguish this use of the word *glass* from the common building material, it is italicized here and elsewhere in this section. Glass, the material, becomes a *glass*, the state, at the *glass* transition temperature (T_g). Below T_g, the atoms in the glass rarely move past one another; they continue to jiggle about with thermal energy but they don't travel about the material.

It's hard to tell by looking at the glass just when this *glass* transition takes place. It generally occurs when the glass's viscosity exceeds about 10^{12} Pa·s, where the Pa·s is the SI unit of viscosity. At that point, the great difficulty the fluid has flowing past itself reflects the microscopic difficulty the atoms have in rearranging. The atoms are practically frozen in place and the fluid flows so slowly that the glass is almost indistinguishable from a solid.

To see that glass behaves like a solid, let's look at how viscous fluids and solids respond to shear stress. If you exert shear stress on a viscous fluid, by pushing its top right and its bottom left, the top portion will flow right and the bottom portion will flow left (Fig. 17.2.2). But if you exert shear stress on a solid, it will experience shear strain. It won't flow anywhere. To get the top half of a steel bar to "flow" right while the bottom half "flows" left, you'd have to break it. In contrast, even a small shear stress will eventually cause a liquid to flow. Thus, there seems to be a fundamental difference between solids and liquids in their responses to shear stress: solids undergo shear strain while liquids flow.

But what happens when you suddenly expose an extraordinarily viscous liquid to shear stress? The liquid has so much trouble flowing that it undergoes shear strain instead. It bends elastically, just like a crystalline solid! If you release the stress quickly enough, the liquid will spring back almost to its original shape. However if you wait a while before releasing the stress, the liquid will have time to flow and the stress will go away on its own. You can see this effect by bending a stick of taffy. If you bend it briefly, the taffy is almost as springy as a solid. But if you bend it and wait, the taffy flows to relieve the shear stress.

The time a liquid takes to relieve shear stress by flowing increases with that liquid's viscosity. Thin liquids such as water relieve stresses quickly, but with a viscosity of 10^{12} Pa·s, it takes glass minutes or hours to relieve stresses and you would have to be very patient to detect anything nonsolid about that glass. It would feel just as hard and elastic as an ordinary crystalline solid unless you were willing to wait for hours to see it flow (Fig. 17.2.3).

There is one more change that occurs as the *glass* transition takes place: the glass stops shrinking like a liquid with decreasing temperature and starts shrinking like a solid. As I discussed on p. 239, liquids shrink rapidly as you cool them because their mobile atoms pack together more tightly as their thermal energies decrease. Solids shrink less rapidly as they're cooled because their atoms can't rearrange. Above the glass transition temperature, glass shrinks quickly as it's cooled, like a liquid. Below the glass transition temperature, glass shrinks slowly as it's cooled, like a solid.

What is in Glass?

With that introduction to the behavior of *glasses*, it's time to look at what's inside glass. Window glass is mostly silicon dioxide (SiO_2), the chemical found in quartz and quartz sand and commonly called silica. Silica is extremely common in nature, making up much of the earth's crust. It's hard and clear and resistant to chemical attack. It's also the best *glass*-forming material in existence.

To be a good *glass* former, a material must have trouble nucleating seed crystals and must prevent those seed crystals from growing quickly. Because silica is extraordinarily viscous at its freezing temperature, it's an excellent *glass* former and is easily converted into quartz *glass* or vitreous quartz.

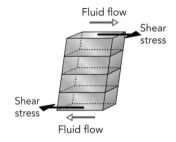

Fig. 17.2.2 Viscosity measures the difficulty a fluid has flowing past itself. The higher this fluid's viscosity, the more shear stress it must experience to keep its bottom layer flowing quickly to the left and its top layer flowing quickly to the right.

Fig. 17.2.3 (*a*) Below its *glass* transition temperature, this rod behaves as a solid and responds elastically to forces. (*b*) But above its *glass* transition temperature, the rod behaves as a viscous liquid and bends slowly under its own weight. The gas flame glows yellow near the hot soda-lime-silica glass because sodium atoms that are vaporized from the glass emit intense 590-nm light (see p. 469).

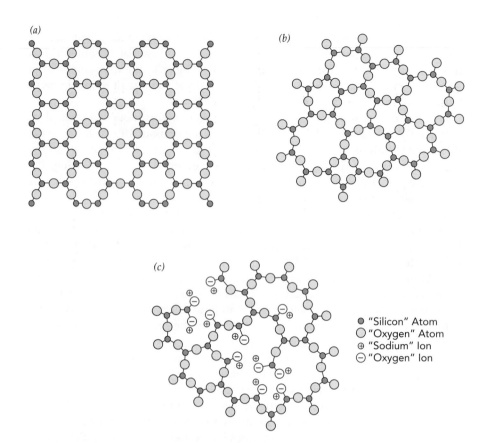

(a)

(b)

(c)

- ● "Silicon" Atom
- ○ "Oxygen" Atom
- ⊕ "Sodium" Ion
- ⊖ "Oxygen" Ion

Fig. 17.2.4 Two-dimensional equivalents of the structures of (*a*) crystalline silica, (*b*) liquid or glassy silica, and (*c*) glassy or vitreous silica containing some sodium oxide.

Silica is held together by covalent bonds, even as a liquid. A covalent bond forms between two atoms when they share a pair of electrons. This sharing reduces the electrostatic potential energy of the atoms by placing extra negative charge in between their positive nuclei. It also reduces the kinetic energies of the shared electrons by letting them spread out as standing waves onto both atoms. Their wavelengths increase as they spread and lengthening an electron's wavelength, like lengthening a photon's wavelength, reduces its energy.

Covalent bonds are highly directional. Each silicon atom in the silica forms covalent bonds with four oxygen atoms and orients those atoms at roughly the corners of a tetrahedron. Each oxygen atom in the silica forms covalent bonds with two silicon atoms and orients them at roughly two corners of a tetrahedron. The result is an intricate network of silicon and oxygen atoms in which each oxygen atom acts as a bridge between two adjacent silicon atoms. Because of its interlinking structure, silicon dioxide is often called a network former.

In crystalline silica, this networking process creates an orderly three-dimensional lattice that's hard to visualize. To help us discuss the basic character of this lattice, let's examine a two-dimensional lattice with similar features (Fig. 17.2.4*a*). The "silicon" atoms in this simplified lattice bind to only three "oxygen" atoms, at the three corners of an equilateral triangle. The "oxygen" atoms bind to two "silicon" atoms, at opposite ends. While this lattice is oversimplified, it illustrates how a quartz crystal is intricate and orderly. If you know the locations of two or three atoms in this lattice, you can predict where all of the other atoms will be found.

When you melt silica, its atoms remain covalently bound, but they change partners frequently. The orderly crystalline lattice vanishes and is replaced by a tangled network of interconnecting rings (Fig. 17.2.4*b*). While the silicon and oxygen atoms still have the right numbers of neighbors, they often form rings with the wrong numbers of atoms in them.

Once the crystalline order of silica has been destroyed, it's hard to recover. The tangled networks of the liquid are relatively stable and the atoms can't tell which rings have too many or too few members. Moreover, liquid silica is still held together by covalent bonds that constrain the motions of its atoms and make it an extremely viscous fluid. Silica is thus an ideal *glass* former and supercools all the way to a *glass* when you cool it rapidly.

Soda–Lime–Silica Glass

But pure silica has a serious drawback. Its covalent bonds hold it together so tightly that it doesn't melt until its temperature reaches approximately 1650 °C. That's far above the melting temperatures of most metals, including iron and steel. While pure quartz glass can be made, it's a specialty item and quite expensive. Instead, most common glass contains other chemicals that lower the mixture's melting temperature so that it's easier to work with.

The principal addition to window glass is sodium oxide or soda (Na_2O). Although soda contains oxygen atoms, it's held together by ionic rather than covalent bonds. Ionic bonds form when the atoms in a material become oppositely charged ions and attract one another. In this case, each oxygen atom in soda removes an electron from two nearby sodium atoms, producing a mixture of negatively charged oxygen ions and positively charged sodium ions.

When soda is added to silica and the two are heated, the mixture melts at a temperature that is below the melting temperature of either material. It's a eutectic, a mixture that melts more easily than the chemicals from which it's made. A mixture containing 25% soda by weight melts at a temperature of only 793 °C. Equally remarkable is the fact that a fine mixture of soda and silica powders can be melted together at temperatures near this value. The soda acts as a flux to melt the silica and the glass manufacturer never has to heat the mixture to silica's normal melting temperature.

The reason for the low melting temperature of this soda–silica glass is illustrated in Fig. 17.2.4c. The sodium atoms in the mixture donate electrons to the oxygen atoms, leaving the substance full of positively charged sodium ions and negatively charged oxygen ions. In this case, the oxygen atoms have just a single charge. But an oxygen atom with an extra electron only needs to share one electron to complete its electronic shell. As a result, it binds to only one silicon atom and doesn't form a bridge between a pair of silicon atoms.

Soda–silica glass is full of nonbridging oxygen atoms that end the network rather than extending it. Their presence weakens the glass and lowers its melting temperature. While this ease of melting is important for glass manufacturing, soda–silica glass is less robust in almost every way than pure silica glass. It's softer and weaker than silica glass so it scratches and breaks more easily. It has more internal friction so it wastes energy in bending and emits a dull tone when struck. It's also much less chemically resistant than quartz glass. In fact, it's water-soluble.

Sodium ions are so water-soluble that they allow water to enter the glass's molecular network and chop it to pieces. As a result, soda–silica glass can be dissolved in water and painted onto surfaces. It's called water glass and is used to seal the outsides of eggs so that they don't dry out. It's also included in laundry detergent, where it protects the washer from corrosion.

Soda–silica glass can be made much less water-soluble by adding calcium oxide or lime (CaO) to it. Calcium oxide is an ionic solid, but it's not soluble in water and makes the glass much more durable. Soda–lime–silica glass is almost insoluble in water and is the principal commercial glass. Windows, bottles, and jars are all made of soda–lime–silica glass.

In each of these glasses, silica is the *glass* or network former and soda and lime are modifiers. The *glass* former's job is to create the tangled network that gives the liquefied material its high viscosity and allows it to supercool all the way to the *glass* transition temperature. A modifier's job is to ease the path to the *glass* transition and to alter the properties of the glass that's produced. Most importantly, modifiers help to ensure that the *glass*-former doesn't crystallize during cooling, a serious problem called devitrification.

Other Glasses

There are many other compounds and mixtures that can form *glasses*. Most of these systems involve oxygen atoms, which are excellent bridging atoms and good at producing covalent networks. But there are some *glasses* that don't contain oxygen and there are even *glasses* that contain only metal atoms (see ❶). Because these exotic *glasses* are rarely used in windows, we'll focus instead on the more common oxygen-containing glasses (Table 17.2.1).

Silica isn't the only *glass*-forming chemical. Other materials that create tangled networks and form *glasses* during rapid cooling include phosphorus pentoxide (P_2O_5), germanium oxide (GeO_2), and boron oxide (B_2O_3). Of these, only boron oxide is commercially important and then only when mixed with silica.

The problem with pure boron oxide glass is that it's not very durable and dissolves easily in hot water. However, boron oxide–silica glasses are quite stable and play important roles in laboratories and kitchens under trade names such as Pyrex® and Kimax®. These borosilicate glasses tolerate temperature changes much better than soda–lime–silica glasses.

Soda and lime aren't the only modifiers used in commercial glasses. All of the alkali oxides and alkaline earth oxides act as modifiers when mixed with silica or other *glass*-formers. Changing the modifiers in soda–lime–silica glasses produces subtle alterations in the mechanical, chemical, and optical properties of the glass. The most common alternative modifiers are potassium oxide (K_2O), which is substituted for soda, and magnesium oxide or magnesia (MgO), which is substituted for lime. Barium oxide (BaO) and strontium oxide (SrO) are frequently substituted for lime in glass that must block X-rays, such as that on the front of a television picture tube.

There are also some chemical compounds that are intermediate between *glass*-formers and modifiers. While these compounds don't form *glasses* on their own, they participate in the tangled networks initiated by other compounds such as silica. The most important intermediate compounds are aluminum oxide or alumina (Al_2O_3) and lead oxide (PbO). Aluminosilicate glasses tolerate high temperatures better than soda–lime–silica glasses and are used in halogen lamps, furnaces, and fiberglass insulation. A small amount of alumina is included in most glasses to help them tolerate weather better.

Lead oxide is included in special optical glasses to increase their indices of refraction. It also makes them dense, hard, and X-ray absorbing. The lead crystal used in decorative glassware and windows isn't crystalline at all. Instead, it's a *glass* consisting mostly of soda, lead oxide, and silica. This soda–lead–oxide–silica glass refracts light strongly and disperses its colors. Because its networks are more complete than those in soda–lime–silica glass, lead crystal is stronger and it has less internal friction. Lead crystal rings beautifully when you strike it gently with a hard object.

One other important variation in window glass is color. The pure glasses that we've considered up until now are colorless because they have no way of absorbing photons of visible light. But impurities and imperfections can give these glasses colors. Many of the colored glasses that appear in stained-glass windows are that

❶ Amorphous or *glassy* metals are made by ultrafast cooling of liquid metals. In "splat" cooling, a thin stream of molten metal pours onto a spinning refrigerated wheel. The liquid cools and solidifies in millionths of a second to produce a paper-thin ribbon of metallic *glass*. However, even when cooled incredibly quickly, only special alloys form metallic *glasses*. Such *glasses* are much harder than normal metals because they have no crystalline structure and can't undergo slip.

way because they contain metal ions such iron, cobalt, or copper. The electrons in these ions have many empty orbitals to which they can move and they absorb photons of visible light. Just which colors these ions absorb depend on their structures and on how many electrons they're missing.

Because iron ions that are missing three electrons appear green, iron impurities in common window glass make it appear slightly green. Copper and cobalt ions that are missing two electrons appear blue (blue bottle glass). Manganese ions that are missing three electrons appear purple. Chromium ions that are missing three electrons appear green (green bottle glass). Vanadium ions that are missing four electrons appear red. And so it goes. There is also special ruby glass that contains tiny particles of gold metal and appears red because these metal particles absorb green and blue light.

Most well-mixed glasses are completely uniform throughout and have no internal variations in refractive index to scatter and reflect light. But there are also opal or milk glasses that are not uniform and scatter light like a cloud. These translucent glasses are used in privacy windows and are produced when particles of one type of glass precipitate out of solution in another type of glass during cooling. Materials in which crystalline particles precipitate out of solution in a glass during cooling are called glass-ceramics and are generally translucent rather than transparent.

Table 17.2.1 The Approximate Compositions and Uses of Various Common Forms of Glass.

Type of Glass	Composition (by weight)	Uses
Soda–Lime–Silica	73% silica, 14% soda, 9% lime, 3.7% magnesia, and 0.3% alumina	Glass windows, bottles, and jars
Borosilicate	81% silica, 12% boron oxide, 4% soda, and 3% alumina	Pyrex® cookware and laboratory glassware
Lead (Crystal)	57% silica, 31% lead oxide, and 12% potassium oxide.	Lead crystal tableware
Aluminosilicate	64.5% silica, 24.5% alumina, 10.5% magnesia, 0.5% soda.	Fiberglass insulation and halogen bulbs

Making Glass Windows

Making glass is relatively easy compared to making steel. In principle, all you have to do is mix the raw materials, melt them together into a liquid, and form the liquid into a finished product as it cools. As long as you cool it quickly through the temperature range at which seed crystals can nucleate and grow, it will turn into glass. Nonetheless, there are some interesting procedures involved in creating glass and its products, particularly windows.

The raw materials in soda–lime–silica glass are commonly occurring minerals. Silica (SiO_2) is obtained from quartz sand and sandstone. Soda ash (Na_2CO_3), which decomposes into soda (Na_2O) and carbon dioxide (CO_2) when heated, is a naturally occurring mineral called trona and can also be produced by reactions between limestone and salt. Limestone ($CaCO_3$), which decomposes into lime (CaO) and carbon dioxide (CO_2) when heated, is a ubiquitous sedimentary rock.

These three minerals are heated together in a large ceramic chamber to a temperature of roughly 1500 °C. At this temperature, the soda acts as a flux and helps the other two minerals to melt. Soon the entire mixture is liquid. Because it contains several different compounds, the liquid glass is carefully stirred to ensure that it's uniform throughout.

Gases released from the minerals, particularly carbon dioxide, water vapor, and trapped air, cause the liquid to froth. One of the hardest tasks in glassmaking is eliminating the bubbles. This process, called fining, involves waiting for the bubbles to float to the surface and adding small quantities of additional chemicals—often arsenic and antimony compounds—to help the bubbles escape.

Finally, the glass is cooled to a working temperature of about 800 °C, passing quickly through the temperature range in which it can crystallize. As it cools, the liquid's viscosity increases dramatically. In the melting region of the chamber, its viscosity is about 10 Pa·s or 1000 times that of water. At 800 °C, its viscosity has risen to 1000 Pa·s or 100,000 times that of water. It flows like a thick syrup and is ready for use in making bottles or windows.

Glass bottles are formed by injecting gobs of glass into molds and then blowing air into them to create their hollow interiors. The injection step creates the neck of the bottle, attached to a carefully shaped but uninflated glass bubble. The blowing step inflates that bubble to create the bottle's storage region. The glass is then cooled slowly through its glass transition temperature to produce a finished bottle.

Windowpanes are produced by pouring liquid glass onto a pool of molten tin (Fig. 17.2.5). This "float" method was developed in 1959 and revolutionized windowmaking. The biggest problem in windowmaking has always been producing perfectly flat surfaces on both sides. The top surface of liquid glass is naturally flat but the bottom surface conforms to whatever it's lying on. When the glass is poured onto a liquid metal, its bottom surface is supported by another perfectly flat surface. The finished plate of float glass is thus perfect on both sides.

That this process works is a wonderful confluence of behaviors. First, tin melts at the relatively low temperature of 232 °C but doesn't boil until 2260 °C. This range

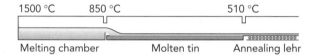

Fig. 17.2.5 Float glass is made by pouring liquid glass onto the surface of liquid tin. As the ribbon of liquid glass flows away from the melting chamber, it cools and hardens. Finally, it's annealed in a long tunnel-shaped lehr before being cut into sheets.

allows the liquid glass to harden while the tin remains liquid. Second, liquid tin is far denser than liquid glass so the glass floats easily on top, supported by the buoyant force. Third, tin and glass are immiscible—tin is held together by metallic bonds while glass is bound by covalent and ionic bonds. The two liquids don't bind strongly to one another so they keep to themselves. The glass remains chemically unaffected by the liquid tin below it.

As the hot glass flows out onto the tin, it naturally spreads until it's about 6-mm thick. A glassmaker can produce thicker windowpanes by corralling the spreading liquid so that it remains relatively thick. To produce thinner sheets, the glassmaker stretches the liquid to spread it more thinly. The float technique can produce extremely flat glass in thicknesses ranging from less than a millimeter to several centimeters.

The soda–lime–silica mixture used for float glass has a *glass* transition temperature of about 540 °C. When the glass leaves the liquid tin, it's already slightly below that temperature. However, its rapid change in temperature on the tin surface has caused it to shrink and this shrinkage creates stresses in the glass. If the glass were simply cooled to room temperature at this point, the stresses would remain trapped forever and would weaken the glass. Stresses also make the glass birefringent—it responds differently to different polarizations of light and can distort the light passing through it

To reduce the stresses trapped in the glass, it's carefully annealed in a long tunnel called a lehr. It's kept near the glass transition temperature for a long time so that the atoms in the glass can rearrange just enough to relieve the stress. When most of the stress has been eliminated, the finished glass is finally allowed to cool to room temperature. In the optical glass used in lenses and telescopes, the annealing is done much more slowly. In many cases, the glass is cooled through the *glass* transition temperature at a rate of less than 1 °C per hour. The glass disk of the 200-inch telescope mirror at Mt. Palomar was cooled from 500 °C to 300 °C at the phenomenally slow rate of only 1 °C per day.

Strengthening Windows

Once it leaves the annealing lehr, the float glass is ready for use. But first it must be cut to size. Cutting glass is a tricky business because it lacks a crystalline structure and its amorphous nature gives it some peculiar properties. To begin with, glass can't undergo any plastic deformation because it can't undergo slip. If you bend glass, it deforms elastically up until the moment it breaks via brittle fracture. When that break finally occurs, a crack propagates uncontrollably through the glass because the glass has no crystalline grains to stop it or redirect its path.

The crack always begins at the surface of the glass, where a defect weakens the molecular structure. Even the most perfect glass surface has a few defects in it and any one of these can initiate a crack when it's under tensile stress. (For an interesting way to minimize these defects, see ❷.) When you bend a piece of glass, you stretch one surface while compressing the other and create a tensile stress on the stretched surface. If this tensile stress is large enough, a defect will become a crack and will propagate through the glass so that it shatters.

Simply bending glass and hoping that it will break along a straight line isn't a practical scheme for cutting glass. Instead, you use a diamond scribe to scratch the glass along the intended break. The scratch introduces defects right where you want the cracks to occur. Wetting the scratch helps because water creates defects in glass and weakens it. If you then stress the glass carefully, either mechanically or by a sudden change in temperature, you can usually get it to crack along the scratch. Still, the glass often breaks in an undesired direction and you are left with useless fragments.

Another way to cut glass is with an abrasive saw. Here a rapidly turning disk containing extremely hard crystals chips out tiny fragments from the glass. The chipping exerts only a modest stress on the glass so that it doesn't crack. Although slow, this method is quite reliable. Similar abrasive techniques are used to cut decorative glass.

Once it's been cut to size, the window is finished. It's ready for installation in a house or building. But it's not ready for use in an oven or an automobile. This window would not be able to tolerate the sudden changes in temperature present in an oven and would pose a serious hazard during a collision in an automobile. So the glass must be heat stabilized and strengthened.

To make an oven window more tolerant of thermal shocks, its chemical composition must be changed. Because soda–lime–silica glass has a large coefficient of volume expansion, it expands considerably as its temperature rises and this expansion can produce huge stresses in the glass. If one half the oven window is suddenly heated, it will expand and try to stretch the other half. If the resulting tensile stress is large enough, that window will crack and break. However, the oven window is made out of a borosilicate glass such as Pyrex. Borosilicate glasses have coefficients of volume expansion that are about a third those of soda–lime–silica glasses. Thus, it takes a much larger thermal shock to break a Pyrex window or container. For that reason, most cookware and laboratory glassware is made out of borosilicate glass.

To strengthen a car window, its mechanical structure must be changed. Glass breaks when its surface begins to tear apart. If you modify the glass so that its

❷ Fiberglass is made by pulling hot aluminosilicate glass into thin strands. Hot glass is a simple liquid and will stretch without breaking to form fantastically thin fibers. These fibers are so narrow that they have almost no surface area and thus very few surface defects. As a result, glass fibers are difficult to break and are very strong. They are used as structural materials and as thermal insulation.

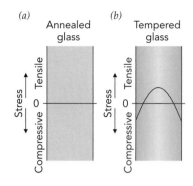

Fig. 17.2.6 (*a*) Properly annealed glass has no internal stresses. (*b*) Tempered glass is specially heat treated so that its surfaces are under enormous compressive stresses while its body is under substantial tensile stress. Tempered glass is very hard to break.

surface is normally under compressive stress, it will be much harder to get a tear started and the glass will be stronger.

The best way to put the glass's surface under compressive stress is to temper the glass. Tempering is done by heating the glass until it softens and then suddenly cooling its surfaces with air blasts. When the glass's surface cools through the *glass transition temperature*, it becomes solid-like and no longer shrinks rapidly with temperature. However the inside of the glass is still liquid-like and continues to shrink rapidly as it cools. As the glass inside shrinks, the surface layers are placed under enormous compressive stresses (Fig. 17.2.6).

This compressive stress makes it difficult to initiate a crack in the outer surface of tempered glass. You must first stretch that outer surface so much that the compressive stress disappears and becomes tensile stress. Breaking tempered glass takes about three times as much force as breaking ordinary glass.

But tempered glass has an interesting complication. When it breaks, it undergoes dicing fracture—the glass crumbles completely into tiny pieces less than a centimeter on a side. This catastrophic failure is what makes tempered glass safe in automobile windows. Tempered glass windows are hard to break but once they do break, they crumble into little cubes.

Dicing fracture occurs whenever the body of tempered glass loses the protection of the compressed surface layer. The body is under severe tensile stress so it will crack and tear whenever possible. As long as it's encased in a shell of compressed glass, the body won't crack. But any penetration into the body will cause the whole sheet to self-destruct. For that reason, tempered glass can't be cut in any way once it has been tempered. Car windows are cut to size before they're tempered because they would crumble if you tried to cut them.

Oven windows and refrigerator shelves are also tempered to give them additional resistance to thermal shocks. By tempering the glass, you make it harder for thermal shocks to put the glass surface under tension and cause it to break.

However, a car's front windshield is not tempered because it would crumble when struck by road debris. Instead, it's made by laminating a plastic sheet between two sheets of annealed glass. This three-layered safety glass sandwich can tolerate minor breaks without falling apart. The plastic keeps the glass together even if it does break and provides a barrier so that a crack in one sheet of glass can't propagate into the other sheet. Bulletproof glass is a natural extension of this idea, with many alternating layers of glass and plastic.

ᛗ Diamonds

If you wanted to sharpen a knife blade made of hard martensitic steel or scribe a piece of glass so that it will break along a straight line, what would you use? You would use something even harder than steel or glass, something like a diamond. Diamond is a crystalline form of pure carbon in which each carbon atom is bound to four other carbon atoms in a three-dimensional network that is extremely rigid and extremely hard. The other common crystalline form of diamond is graphite, in which each carbon atom is bound to two other carbon atoms in a two-dimensional sheet that is soft and flexible. How can two forms of the same element be so completely different?

Carbon is a unique element. It bonds with unparalleled flexibility to many other elements. The bonds it forms are strong covalent bonds, so that the molecules it creates are resistant to dissociation and have well-defined shapes. It is the backbone atom for the chemistry of life. Organic molecules are those that are built around carbon and their number greatly exceeds that built around any other element.

One of the flexible aspects of carbon is its ability to bond with two, three, or four neighboring atoms. In particular, it bonds well with three other carbon atoms to form a graphite sheet or with four other carbon atoms to form a diamond lattice. The change from two dimensional to three dimensional crystals is what makes graphite so different from diamond. In graphite, the individual sheets can slide across one another easily. As a result, graphite is a fine lubricant. Putting graphite between two surfaces allows those surfaces to slide relative to one another with reduced friction. Graphite-based lubricants are common and effective. The individual sheets of graphite crystal can bend easily so that there is no apparent hardness to graphite. Even though the bonds between atoms are strong, the overall structure is flexible and deformable.

Diamond, on the other hand, permits almost no movement. While it has atomic planes in several directions, slipping across those planes is nearly impossible. For one plane of atoms to slide across another, all of the covalent bonds connecting one plane with the other would have to break simultaneously and then reform after the plane has shifted. Unlike metals, covalent bonds don't permit such slippage. Instead, diamond cleaves along one of the atomic planes—the atoms separate from one another and never reattach.

Thus diamond resists elastic deformation strongly and shatters when the strain exceeds its elastic limit. It is the hardest material known, rivaled only by boron nitride. Boron nitride has a diamondlike character because it's composed of an equal mixture of the next lighter element, boron, and the next heavier element, nitrogen. The crystal structure of boron nitride is the same as that of diamond.

While graphite is common, diamond is rare. That's because graphite is the equilibrium phase of carbon at atmospheric pressure. To understand what that means, let me remind you about water near 0 °C (32 °F). Exactly at 0 °C liquid water and solid ice can coexist in phase equilibrium—water molecules leave the ice for the water and land on the ice from the water at equal rates, so both phases can be present at once. But if you increase the pressure, denser liquid water will become the equilibrium phase. The leaving rate will exceed the landing rate and the ice will gradually melt into liquid water. Lowering the pressure below atmospheric pressure will have the opposite effect; less dense ice will become the equilibrium phase, and the liquid water will gradually freeze into ice.

In the case of carbon, when graphite and diamond are in contact with one another, there is leaving and landing of carbon atoms. To say that these rates are slow at room temperature is a vast understatement, but the concept is correct. At atmospheric pressure, carbon atoms leave diamond for graphite more often than they leave graphite for diamond, so the diamond gradually transforms into graphite and less dense graphite is the equilibrium phase. Fortunately, for diamond lovers and jewelry store owners, this transformation takes so long at room temperature that it's not observable. If a diamond is heated to 1000 °C (in vacuum so that it doesn't burn), however, it will slowly turn into graphite. At 1700 °C, the conversion is quite fast.

At 50,000 atmospheres pressure, on the other hand, carbon atoms leave graphite for diamond more often than they leave diamond for graphite and the graphite gradually transforms into diamond. At these high pressures, denser diamond is the equilibrium phase! Sadly, this transformation is fantastically slow at room temperature and will only proceed at reasonable speeds at high temperatures.

The need for heat and pressure to form diamonds out of graphite accounts for their rarity. Natural diamonds are formed at great depths in the earth, where the pressures and temperatures are very high. As a matter of fact, there are a great many diamonds located a few dozen miles below the earth's surface. Unfortunately, we can't go down there to get them and must wait from them to come up to us. For them to reach the earth's surface, they must rise and cool quickly. Otherwise, they'll turn back into graphite on their way up. Natural diamonds are only found in kimberlite pipe and dikes. These formations are associated with volcanic activity where magma from deep in the earth rose swiftly to the surface and cooled quickly enough that the diamonds it carried upward didn't turn into graphite during the journey.

Synthetic diamonds are made by simulating the growing conditions deep in the earth. Graphite and diamond seed crystals are exposed to pressures of about 50,000 atmospheres at temperatures of about 1200 °C. This process is only barely practical for producing gem-sized diamonds, though innovations are gradually bringing synthetic gem diamonds onto the market. Diamond synthesis is quite practical, however, for producing small, useful stone for the cutting industry. Small synthetic diamonds are better than crushed natural diamonds for use on cutting devices because synthetic diamonds are compact crystals that have nice sharp crystal facets while crushed natural diamonds have strange, irregular shapes that don't cut well.

Another use of man-made diamonds is in surface coatings. One alternative to making a knife or a ball-bearing out of a single, hard material is to cover the surface of a softer material with an ultrahard coating. One such coating is a diamond film. Materials scientists are able to coat metals or ceramics with carbon films having hardnesses close to that of diamond. They have even succeeded in make these films at low pressures. A diamond film on a knife might keep it sharp indefinitely or allow a ball bearing to run for a hundred years without wearing out. Similar films could also provide non-stick, non-scratch surfaces for cookware.

Apart from rarity and hardness, one of the features of diamonds that make them particularly attractive for jewelry is their ability to bend light and separate white light into its colors. These properties follow from diamond's unusually high index of refraction and its strong dispersion. The speed of light in diamond is only about 40% of its vacuum speed and it slows violet light substantially more than red light.

Let me remind you about refraction and dispersion, topics I discussed in Chapter 9 with mechanical waves and Chapter 14 with light. When light enters a diamond perpendicular to the surface it travels straight ahead but slows down (Fig. 17.2.7a), and when it leaves the diamond perpendicular to the surface it travels straight ahead but speeds up (Fig. 17.2.7b). When light enters the diamond at an angle, however, it refracts toward the line perpendicular to the surface (Fig. 17.2.7c), and when it leaves the diamond it refracts away from the line perpendicular to the surface (Fig. 17.2.7d). Because of dispersion those bends toward or away from the perpendicular are stronger for violet light than they are for red light.

I should remind you about one last effect: total internal reflection. If light tries to leave the diamond at an angle too far from the line perpendicular to the surface, it fails to escape. Instead, the light reflects perfectly from the inside surface of the diamond (Fig. 17.2.7e). This total internal reflection is what makes the diamond so reflective. If you look inside a diamond, you'll see many apparently mirrored surfaces. They have no metal mirror coating at all; they reflect only through total internal reflection.

Because of diamond's enormous index of refraction, light trying to leave a diamond must approach its surface almost perpendicular in order to escape. In contrast, glass has a much lower index of refraction, so it allows a greater range of escape angles. Since glass costume jewelry can't use total internal reflection to obtain its sparkle, its back surfaces must be silvered. Even so, glass doesn't have enough dispersion to give costume jewelry much color.

A properly cut diamond (Fig. 17.2.8a) uses a combination of refraction at its front surfaces and total internal reflection at its back surfaces to return most of the white light entering its front facets as colored sparkles leaving its front facets. Sunlight separates into its various colors on entry into the front facets, reflects perfectly off the back facets, and emerges from the front facets in a spray of colored beams. Keeping the gem clean is important, especially the back facets. Grease and grime

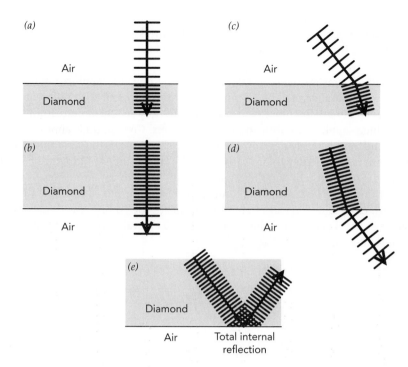

Fig. 17.2.7 (*a*) When light enters a diamond perpendicular to the surface, the light slows but continues straight. (*b*) When it leaves the diamond perpendicular to the surface, it speeds up but continues straight. (*c*) When light enters the diamond at an angle, it bends toward the line perpendicular to the surface. (*d*) When it leaves the diamond at an angle, it bends away from the line perpendicular to the surface. (*e*) When light tries to leave the diamond at an angle too far from the line perpendicular to the surface, the light instead reflects perfectly; it experiences total internal reflection.

on those back facets can spoil the total internal reflection and reduce the brilliance of the diamond's reflections.

A diamond that is cut either too deep (Fig. 17.2.8*b*) or too shallow (Fig. 17.2.8*c*) will fail to return some of the light striking its back facets and will appear less brilliant that a properly cut diamond. It won't have much sparkle to it. Since diamond cutters work with natural stones, they must plan carefully how to cut each stone to obtain the largest finished gem that has the proper geometry. Some stones are

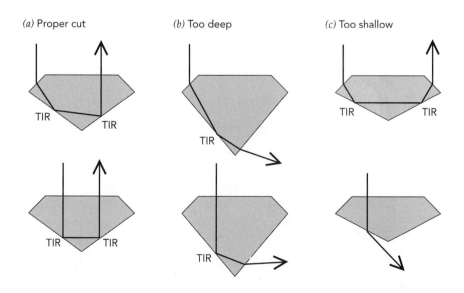

Fig. 17.2.8 (*a*) A properly cut diamond refracts light that arrives through its face or side facet and returns that light out the top of the diamond. It makes extensive use of total internal reflection (TIR) to keep light bouncing off of surfaces. When light enters or leaves the diamond at an angle, it refracts and dispersion causes its colors to follow separate paths. These effects, total internal reflection and dispersion, give diamonds their brilliance and color. A diamond that is (*b*) too deep or (*c*) too shallow doesn't return light well and has little brilliance. Much of the light reaching the back facets escapes from the diamond and fails to return to its front facets.

easy to cut into perfectly structured gems while others are poorly shaped and yield inferior gems. The cut of a diamond is as important to its appearance and value as is its clarity.

Most other gems are also crystalline minerals. Ruby and sapphire are both the mineral corundum, crystalline aluminum oxide (Al_2O_3), but with different impurities to give them their colors. The impurities in ruby are chromium atoms while those in blue sapphire are iron and titanium atoms. Corundum is almost as hard as diamond, and it rarely develop scratches or shows wear. But aluminum oxide crystals are relatively easy to grow, so synthetic rubies and sapphires are fairly inexpensive. Pure synthetic sapphire, colorless due to a lack of impurities, is useful in everything from laser windows to ultrahard ball bearings. Emerald is the mineral beryl, crystalline beryllium aluminum silicate ($Be_3Al_2SiO_6$), with chromium impurities. Beryl is less hard than corundum, so emeralds are somewhat susceptible to scratches.

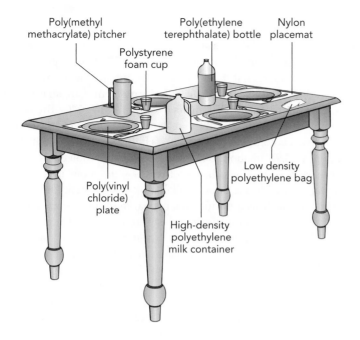

Poly(methyl methacrylate) pitcher

Polystyrene foam cup

Poly(ethylene terephthalate) bottle

Nylon placemat

Poly(vinyl chloride) plate

High-density polyethylene milk container

Low density polyethylene bag

❶ Polymers were discovered long before anyone understood what they were. It wasn't until 1920 that the German chemist Hermann Staudinger (1881–1965) made his macromolecular hypothesis, suggesting that polymers are actually giant molecules formed by the permanent attachment of countless smaller molecules. Through careful experiments, he proved his hypothesis to be correct and was award the 1953 Nobel Prize in Chemistry.

Section 17.3 **Plastics**

Until the middle of the nineteenth century, virtually all objects of everyday life were made from naturally occurring or naturally derived materials such as wood, glass, metal, paper, wool, and rubber. While this short list of available materials was sufficient for most purposes, there were situations that demanded something new. In 1863, a billiard ball manufacturing company, Phelan and Collander, offered a prize of $10,000 to anyone who could find a substitute for natural ivory in billiard balls. In response to this offer, American printer John W. Hyatt and his brother Isaiah figured out how to form billiard balls out of a recently discovered synthetic chemical called nitrocellulose. By 1871, Hyatt had established two companies to work with this new material, under the name celluloid, and the plastics industry was born.

The word plastic has a wide usage and a broad definition. Most often it refers to materials that, while solid in their final forms, take on liquid or shapeable states during earlier phases. Modern plastics can assume many shapes and forms and exhibit a rich variety of physical and chemical properties. Most importantly, the development of plastics has made it possible to design materials that exactly suit their uses.

Polymers

Plastics are based on polymers, enormous chain-like molecules containing thousands or even millions of atoms. Like all organic molecules, and the glasses of the previous section, the atoms in a polymer are held together by covalent bonds. But while propane, diesel fuel, and paraffin wax consist of chain-like molecules roughly 3, 16, and 30 carbon atoms long, respectively, the chain-like molecules of high-density polyethylene (HDPE) are between 1000 and 3000 carbon atoms long (Fig. 17.3.1). Once a molecule is more than about 1000 atoms long, it's considered a polymer. (For a history of the understanding of polymers, see ❶.)

Its simple structure makes HDPE a good starting point for a discussion of polymer structure. It's also a commercially important polymer because it's sturdy, nontoxic, and quite resistant to chemical attack. HDPE is used extensively in bottles and containers and is familiar as the cloudy white plastic of milk jugs.

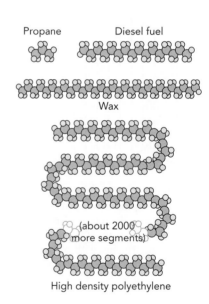

Propane

Diesel fuel

Wax

(about 2000 more segments)

High density polyethylene

Fig. 17.3.1 Paraffin chains are found in many materials. Each paraffin chain consists of a backbone of carbon atoms (the larger balls), decorated by pairs of hydrogen atoms (the smaller balls). A chain is terminated at each end by an extra hydrogen atom. Propane, diesel fuel, and wax contain relatively short paraffin chains while high-density polyethylene, the plastic used in milk containers, contains paraffin chains of between 1000 and 3000 carbon atoms.

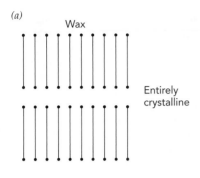

(a)

Wax

Entirely
crystalline

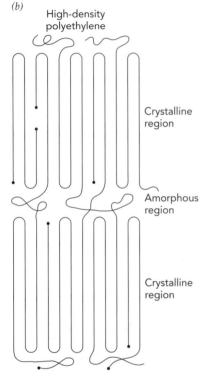

(b)

High-density
polyethylene

Crystalline
region

Amorphous
region

Crystalline
region

Fig. 17.3.2 (*a*) Wax is a crystalline material, held together only by van der Waals forces. (*b*) High-density polyethylene is about 80% crystalline and 20% amorphous, with entanglements helping to hold it together.

HDPE looks and feels like wax because the two are closely related chemically. In both materials, the molecules consist of hydrogen atoms attached to chains of carbon atoms. But the great lengths of these chains in HDPE distinguish it from wax in several important ways. First, HDPE has a substantially higher melting temperature than wax, which isn't surprising because its molecules are so much larger. In fact, what *is* surprising is that HDPE melts at only about 140 °C. The way in which it melts is complicated and is one of the reasons that polymers are so interesting.

Second, while wax is completely crystalline, HDPE is partly crystalline and partly amorphous (Fig. 17.3.2). In about 80% of the plastic, the molecules are neatly oriented like dry spaghetti in a box. In the other 20%, the molecules are coiled randomly and entangled with one another like cooked spaghetti. These amorphous regions are as unavoidable as the knots that appear in shoulder-length hair. HDPE's mixed structure gives it a non-uniform index of refraction—its denser crystalline regions slow light more than its amorphous regions. This non-uniformity gives solid HDPE a milky haziness that disappears when the plastic melts to become a uniform liquid.

Third, HDPE doesn't crumble like wax when you bend it. The amorphous regions of tangled polymer chains provide the plastic with considerable flexibility. When you stress HDPE, its coiled chains unwind. They can't cross one another where they're entangled but they still allow the plastic to stretch by as much as 50% in length. The material responds elastically to small stresses but undergoes plastic deformation when the stresses are severe. If you push gently on a milk container, it will bounce back, but if you crush it, it will stay crushed.

Finally, HDPE has a much greater tensile strength than wax. Wax molecules are relatively short chains, clinging to one another only with van der Waals forces. Van der Waals forces are weak, nondirectional intermolecular forces that are created by temporary fluctuations in the positions of electric charges in molecules. All molecules exert van der Waals forces on nearby molecules. Van der Waals forces are so weak that the wax molecules slide across one another easily and the whole crystal can be pulled apart. Although HDPE chains are also bound together only by van der Waals forces, their great lengths and frequent entanglements make the plastic hard to pull apart. You must actually break some of the molecules in order to divide the plastic at room temperature.

While low-density polyethylene (LDPE) is also built from long wax-like molecules, these molecules have short branches or side chains. Side chains get in the way of crystallization because they make it hard for the molecules to line up neatly. As a result, LDPE is about half amorphous—heaps of coils and tangles.

With its mechanical properties dominated by the amorphous regions, LDPE is limp and flexible at room temperature. However, it has considerable tensile strength because its entangled chains can't be separated easily. The major uses of LDPE are in trash bags, shopping bags, food and dry goods wrappers, and electric insulation. Both HDPE and LDPE are easily recycled. HDPE is ⚠ and LDPE is ⚠.

Polymers and Temperature

At room temperature, LDPE is neither a solid nor a liquid. It bends almost effortlessly so it isn't a normal solid. It doesn't flow so it isn't a normal liquid. Instead, LDPE lies in between solid and liquid in what is called the rubbery plateau regime. There are five temperature regimes for most polymers and the rubbery plateau is the middle one. At lower temperatures, there are the glassy regime and the glass–rubber transition regime and at higher temperatures, there are the rubbery flow regime and the liquid flow regime.

These five regimes exist because the long polymer chains can't crystallize completely at low temperatures and can't avoid entanglements at high temperatures.

With trouble forming either a normal solid or a normal liquid, a polymer's characteristics change rather gradually with temperature.

Within this evolving behavior, polymer scientists have identified the five separate regimes (Fig. 17.3.3). There are no sharp boundaries between these regimes, so a particular polymer moves smoothly from one to the next as it warms up. While the presence of crystalline regions in a polymer complicates this picture, those regions can be safely ignored in polymers that are substantially amorphous. We'll examine the five regimes in LDPE, which is amorphous enough that we won't have to worry about its crystalline regions.

At extremely low temperatures, LDPE is a hard, brittle solid. Because the atoms of LDPE can't move relative to one another when it's below its glass transition temperature (about −128 °C), it's a glass (Fig. 17.3.3a). Your shopping bag never gets cold enough to reach this glassy regime, but there are many common polymers that are glassy at room temperature. One of the most important of these is poly(methyl methacrylate) or Plexiglas®.

At about −80 °C, the atoms in LDPE have enough thermal energy to move relative to one another. However, the overall chains still can't move (Fig. 17.3.3b). LDPE is in the glass–rubber transition regime and has a leathery character—stiff but pliable. A familiar polymer that is in this regime at room temperature is poly(vinyl acetate), the principal polymer in most latex interior house paints.

Near room temperature, LDPE is in the rubbery plateau regime, with its atoms moving freely and even its chains moving slightly. Because the chains are still too entangled with one another to permit flow (Fig. 17.3.3c), the plastic has a rubbery feel—flexible and elastic. The chains pull taut when stretched and snap back to normal when they are let go. But if you pull too hard on the plastic, the chains break at the entanglements and the plastic stretches irreversibly or tears. These behaviors explain the familiar stretchiness of LDPE trash and shopping bags and the tearing that occurs when you overload them with sharp objects.

At about 100 °C, the chains in LDPE are mobile enough to disentangle themselves and the material is in the rubbery flow regime (Fig. 17.3.3d). The chains disentangle themselves through a process called reptation, in which thermal energy causes the chains to slide back and forth along their lengths until they leave entanglements behind them. This motion was first postulated by French physicist Pierre-Gilles de Gennes (1932–) and is named for its similarity to the motions of snakes (Figs. 17.3.4 and 17.3.5). A common polymer that is in this rubbery flow regime just above room temperature is chicle, the main constituent of chewing gum. Silly Putty®, a silicon polymer, is also in this regime.

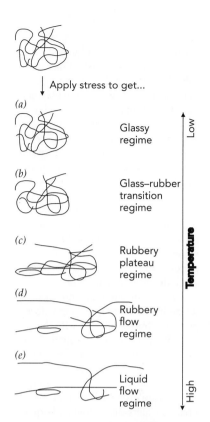

Fig. 17.3.3 As you heat a polymer, it goes through the five stages shown here from top to bottom. (a) When cold, its atoms don't shift during stress so it's glassy. (b) When cool, its atoms shift slowly but the chains remain fixed so it's between glassy and rubbery. (c) At intermediate temperatures, its atoms and chains move but entanglements prevent them from getting past one another so it's rubbery. (d) When warm, the chains begin to untangle so it exhibits rubbery flow. (e) When hot, the chains easily untangle and it exhibits liquid flow.

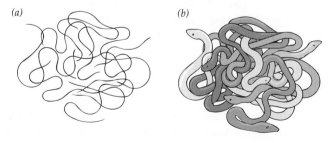

Fig. 17.3.4 (a) Tangled polymer chains can only untangle themselves by moving forward or backward along their lengths. (b) Because this motion is reminiscent of snakes, it's called reptation.

At higher temperatures, the chains in LDPE slide across one another freely and the polymer is in the liquid flow regime (Fig. 17.3.3e). Reptation occurs so rapidly in this regime that it's hard to distinguish a liquid polymer from any other viscous

❷ Many polymers are chemically attracted to water molecules, and water dissolves easily into them. Once dissolved in such a polymer, water acts as a plasticizer, lowering the polymer's glass transition temperature and decreasing its crystallinity. Among these hydrophilic (or water-loving) polymers is cellulose, the main constituent of wood, paper, and cotton. Because cellulose is built from sugar monomers, water dissolves in it extremely well. Although cellulose is glassy at room temperature, and it decomposes rather melts when you heat it, you can soften it considerably with water. Paper or cotton cloth that feels scratchy when dry softens when you wet it. That's why you should dampen cellulose-based cleaning clothes before polishing fragile surfaces with them. Most kitchen sponges are made from hydrophilic plastics such as cellulose and they exhibit this same softening effect when you wet them. Some sponges are even sold as thin compressed wafers that soften and expand to full size the first time they get wet. Water's ability to plasticize cotton makes water useful when ironing cotton clothing. Wetting cotton cloth, either by spraying water on it directly or by exposing it to steam, plasticizes its fibers and makes reshaping those fibers with the help of heat much more effective. Wool and hair are also hydrophilic polymer fibers that can be plasticized and softened by water.

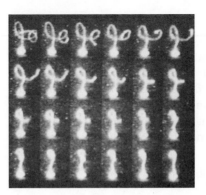

Fig. 17.3.5 While the theory of reptation correctly predicted many features of polymer behavior, reptation itself wasn't directly observed until 1993, when American physicist Steven Chu first watched it through a microscope. Chu and his colleagues coated a single polymer molecule with fluorescent dye and used a laser beam to drag that molecule through another polymer. The molecular chain snaked its way through entanglements, just as predicted by reptation theory.

liquid. One polymer that exhibits liquid flow at room temperatures is poly(dimethyl siloxane), a silicon-based compound. Because its long molecular chains slide across one another so easily, this liquid makes an excellent lubricant.

Even at high temperatures, polymers exhibit unusual flow characteristics as a consequence of reptation. Because the chains have to slither large fractions of their lengths to get past each entanglement, polymers with longer chains are more viscous than polymers with shorter chains. Reptation theory accurately predicts that a polymer's viscosity should be proportional to the third power of the average chain length.

It's sometimes possible to shift a polymer's behavior from one regime to another by adding chemicals to it. Plasticizers are chemicals that dissolve in a polymer and soften it. They lower its glass transition temperature (T_g) and usually decrease the sizes of any crystalline regions. Plasticizers added to poly(vinyl chloride) or "vinyl" convert it from a glassy solid at room temperature to the leathery material used in some upholstery. In hot weather, you can smell these plasticizers evaporating. When enough of them leave, the aging vinyl returns to its glassy state and becomes susceptible to cracking. (For another example of plasticizers at work, see ❷.)

Thermoplastics and Thermosets

Because the molecules in HDPE and LDPE can move independently at high temperatures, these polymers behave as liquids when heated. They are both thermoplastics, a class of plastics that can be reshaped at high temperatures. While some thermoplastics burn or char before they melt, even those materials can usually be dissolved in solvents, reshaped, and allowed to dry.

But not all polymers can be reshaped. Polymers known as thermosets have chemical cross-links between their chains that prevent them from reptating. A thermoset can't flow because the cross-links turn the entire plastic into one huge molecule. A thermoset won't melt when you heat it and it won't dissolve in solvents unless those solvents sever the cross-links.

Because they can't be reshaped, thermosets must be produced in their final forms. Many thermosets begin as thermoplastics that are then cross-linked, a process generally called vulcanization. In other contexts, this cross-linking is known as tanning (leather), curing (resins), and drying (oil paints). Many important polymers are cross-linked, as we'll soon see.

Most polymer molecules are built by tacking together much smaller molecules. These small molecules are referred to as monomers and the tacking together process is called polymerization. The final chains are then named after their monomers [e.g., polyethylene from an ethylene monomer, and poly(methyl methacrylate), from a methyl methacrylate monomer].

In many cases, a single monomer is used over and over again to form a homopolymer. Representing that monomer as the letter A, the finished homopolymer looks like AAAAAAAAAA…. However, there are also cases in which several different monomers are incorporated in the same chain. Such chains are called co-polymers. Representing two monomers by the letters A and B, the finished co-polymer might be ABABABABAB….

Distinguishing between homopolymers and co-polymers is particularly important when the various monomers in a co-polymer come together in random order. Then the finished co-polymer might be ABAABBBABAABBAAA…. The resulting molecular chains are different from one another and are unlikely to form crystals. Most of these statistical co-polymers are amorphous.

Even homopolymers can be disordered if their monomers randomly adopt different orientations during polymerization. If a monomer can enter the chain as either ↑ or ↓, then the homopolymer might be ↑↓↑↑↑↓↑↑↓↓↑…. These atactic homopolymers are amorphous.

Natural Polymers: Cellulose, Natural Rubber, and Hair

Several of the most important polymers occur naturally, most notably cellulose and natural rubber. Cellulose is the principal structural fiber in wood and plants and by far the most abundant polymer on earth. We use it as a building material, in making paper, clothing, rope, and chemicals, and as a fuel. Cellulose molecules are made of polymerized glucose sugar molecules (Fig. 17.3.6). These chains are unbranched and orderly so cellulose crystallizes easily. Because natural cellulose is about 70% crystalline and 30% amorphous, it appears translucent or white. But while it's built from sugar molecules, we can't digest it (see ❷).

❷ Most animals can't digest cellulose because they're unable to break its molecular chains into individual glucose molecules. Fortunately, bacteria and protozoans produce a catalytic enzyme that breaks up cellulose so that dead trees and plants decompose quickly. Cows and other ruminants carry bacteria and protozoans in their stomachs and let those tiny animals convert cellulose into small digestible sugar molecules.

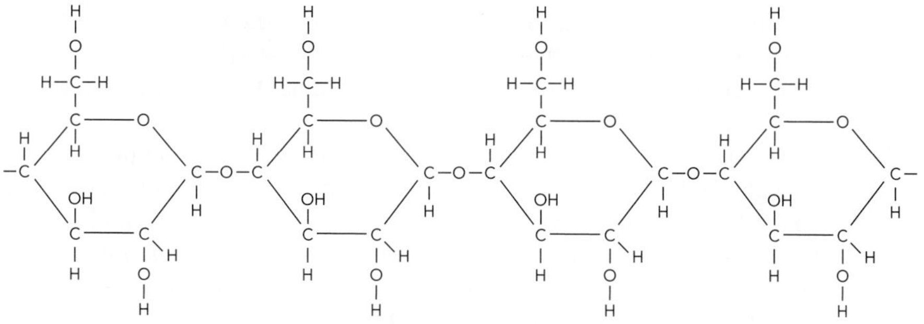

Cellulose

Fig. 17.3.6 Part of a cellulose molecule, showing five repetitions of the glucose monomeric units from which it's built. Each carbon atom (C) is bound to four other atoms, including oxygen atoms (O) and hydrogen atoms (H). The covalent bonds between the atoms are indicated by lines. Cellulose is a stiff, mostly crystalline solid at room temperature and is the main structural material in trees and plants.

The oxygen and hydrogen atoms in adjacent cellulose molecules form strong hydrogen bonds and bind the chains together so tightly that cellulose is unable to melt—it decomposes instead. That's why wood burns rather than turning into liquid in your fireplace. Cellulose's glass transition temperature is also quite high so wood is rigid. Cellulose is only flexible as extremely thin fibers, such as those in cotton. Cotton is nearly pure cellulose.

Since cellulose chains also form hydrogen bonds with water molecules, cellulose eagerly soaks up water. The water dissolves in cellulose, forming a solid solution. The fibers swell and soften as water enters them because water acts as a plasticizer. When the water dries up, the swollen fibers become hard and stiff again. These changes lead to the shrinkage that occurs when you launder cotton clothes in wa-

❸ Oxygen permeable contact lenses swell when wet, allowing oxygen molecules to dissolve in the plastic and diffuse to the surface of the eye. The plastic bottles used for carbonated beverages similarly dissolve water, oxygen, and carbon dioxide molecules. Because carbon dioxide and water slowly diffuse out of a bottle while oxygen diffuses in, the beverage it contains eventually goes flat.

❹ Poly(dimethyl siloxane) is a particularly important silicone polymer, with an extremely low glass transition temperature and thus a low viscosity near room temperature. Silicone rubber sealants are based on this polymer, but with an added twist—the ends of each polymer chain are treated so that they cross-link in the presence of moisture. After you apply the liquid sealant, it undergoes room temperature vulcanization (RTV) in air to form a soft silicone rubber. This process releases acetic acid, the main ingredient in vinegar, so you can smell that the rubber is vulcanizing.

❺ Genetic information is encoded in a naturally occurring co-polymer called deoxyribonucleic acid or DNA, where it's contained in the exact sequence of monomers. Medical, biological, and forensic scientists can now produce this co-polymer, using the polymerase chain reaction (PCR). This technique selectively polymerizes DNA monomers (nucleic acids) to copy portions of a particular DNA chain. PCR can locate particular patterns in the DNA and is coming into frequent use in trials as a genetic equivalent of fingerprinting.

ter. This shrinkage can be reduced by lubricating the fibers with fabric softeners. (For more on solid solutions in plastics, see ❸.)

Natural rubber is *cis*-polyisoprene (Fig. 17.3.7), a polymer that forms in the sap of several tropical trees. It consists of long chains of polymerized isoprene molecules. The prefix "*cis*-" indicates that the isoprene monomers alternate up and down, as shown in the figure. In the tree sap, *cis*-polyisoprene forms a latex of tiny polymer particles suspended in a watery liquid. As the sap dries, its remaining water pulls the polymer particles together and the rubber coagulates into an elastic solid.

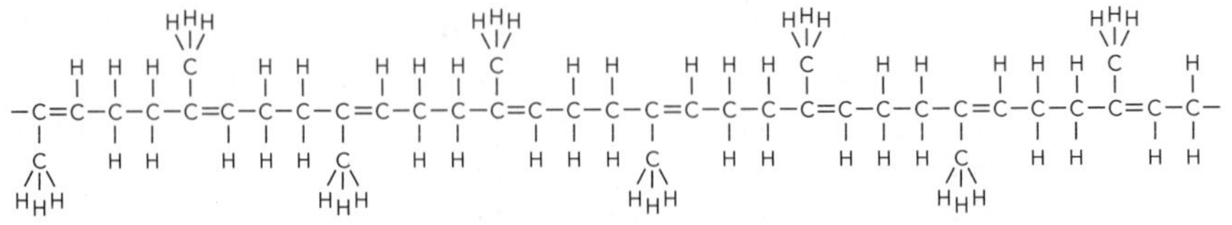

Natural rubber

Fig. 17.3.7 Part of a natural rubber molecule showing eight repetitions of the basic isoprene monomeric units from which it's built. The orientations of the isoprene units alternate up and down. The chains coil up randomly on their own but straighten when you stretch the rubber. The rubber reaches its elastic limit when the chains are straight.

The molecules in natural rubber stick together so weakly that its melting temperature is only 28 °C and its glass transition temperature is a frigid −70 °C. The molecular attraction is weak because *cis*-polyisoprene contains no oxygen atoms and thus can't form hydrogen bonds. However, even at low temperatures, rubber rarely crystallizes because its molecules naturally wind themselves up into random coils. These coils are what give rubber its wonderful elasticity. As you stretch a piece of rubber, its polymer molecules straighten out. When you let go, those molecules return to their randomly coiled shapes.

This coiling is an example of nature's tendency to maximize randomness and entropy. A straight rubber molecule is orderly. Once you stop pulling on it, thermal motions in its atoms give it a bend here and a twist there, and soon the molecule is wound into a jumble of random coils. While these thermal motions *could* straighten that molecule back out, the likelihood of that happening is fantastically small.

Natural rubber came into general use in 1820, when Englishman Thomas Hancock constructed the first rubber factory. Three years later, Scottish chemist Charles Macintosh began using rubber to waterproof fabrics. But natural rubber is sensitive to temperature and therefore hard to use. Cool weather puts rubber in the rubbery plateau regime, where it's a firm, elastic material. But hot weather takes it to the rubbery flow regime, where it's as gooey as glue.

The famous accident of the American inventor Charles Goodyear (1800–1860) changed all that. In 1839, after 10 years of trying to prevent rubber from softening at high temperatures, Goodyear accidentally dropped a mixture of rubber and sulfur onto a hot stove. The material that formed was exactly what he had been looking for. It was stiff but elastic and remained that way regardless of temperature. He called this process vulcanization, after the Roman god of fire.

What Goodyear had done was to form cross-links between individual molecular chains in the rubber (Fig. 17.3.8). He had turned natural rubber, which is a thermoplastic, into vulcanized rubber, a thermoset. Since the chains in the vulcanized rubber were interconnected, they couldn't flow and the rubber couldn't melt—it was a single giant molecule.

Vulcanized rubber exhibits the elastic behavior of the rubbery plateau regime over a broad range of temperatures. As long as there aren't too many cross-links between chains, the random coils can still wind and unwind, and the vulcanized

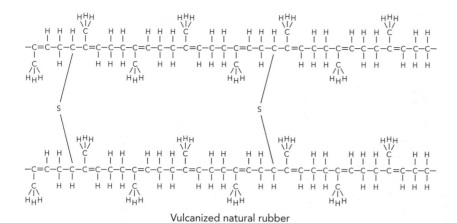

Vulcanized natural rubber

Fig. 17.3.8 Vulcanized rubber is formed by heating a mixture of natural rubber and sulfur. The sulfur links individual rubber chains to form a thermoset. Vulcanized rubber no longer melts the way natural rubber does. The more sulfur that's added to the natural rubber, the harder the vulcanized rubber becomes.

rubber still stretches when you pull on it. But the more sulfur you add to the mixture, the more cross-links are formed and the stiffer the rubber becomes. Sulfur content thus controls the stiffness of vulcanized rubber.

Because natural rubber is in limited supply, most modern rubber is synthesized from petroleum. Many of these synthetic rubbers or elastomers are chemically different from vulcanized natural rubber, however, they all have random coils in their polymer chains and cross-links between those chains. Their various molecular structures give them new and useful properties such as resistance to chemical attack and better stability at high temperatures. For a discussion of silicon rubber, see ❹, and for a note about another natural polymer, DNA, see ❺.

Hair is another natural polymer, one that is closely related to proteins molecules. Proteins themselves consist of limited chains of the 20 different amino acids, strung together in carefully chosen sequences to give these biologically important molecules their forms and functions. Hair is also made from chains of amino acids, but hair's chains are nearly endless helical ones that are known as α-keratin or α-helix. Fingernails are also amino acid chains, but in a folded sheet form known as β-keratin or β-sheet.

The most common amino acid in hair is the smallest one, glycine, but other amino acids are also present. The one amino acid worth mentioning is cysteine, which contains a sulfur atom and which tends to link to other cysteines on other α-keratin chains. The cysteine-cysteine linkages (or disulfide bonds) that develop within hair are what give hair its structure; they transform it from a thermoplastic material into a thermoset. The more disulfide bonding in a person's hair, the stiffer and more structured that person's hair is (Fig. 17.3.9). When you get a "permanent" at the hair salon, the chemicals used to reshape your hair first break apart the disulfide bonds to make the hair thermoplastic and then reestablish those bonds when the hair is in its new shape. The hair is once again a thermoset but in its new, permanent shape.

Synthetic Polymers: Celluloid, Plexiglas®, Nylon, and Teflon®

When Phelan and Collander offered $10,000 to anyone who could find a replacement for ivory in billiard balls, neither cellulose nor rubber was up to the task. Instead, the Hyatts used a chemically modified cellulose called nitrocellulose (Fig. 17.3.10) to produce the first practical synthetic plastic: celluloid.

Fig. 17.3.9 This woman's naturally curly hair obtains its structure from the disulfide bonds that cross-link the α-keratin chains in her hair. Those bonds interconnect all the chains in each strand of hair, making that strand a single, enormous molecule with the corkscrew shape that you see here. A person with naturally straight hair can curl her hair like this temporarily, using water and heat to soften it during the reshaping process. Water softens hair because it is chemically attracted to α-keratin and dissolves easily into hair. Once inside the hair, the water plasticizes it; the water acts as a molecular lubricant to render the wet hair soft and flexible. Heat softens hair by adding thermal energy so that segments of the hair molecules are able to move relative to one another. Thermal softening requires care because hair can scorch if it's overheated. Reshaping done only with water and heat fades with time, however, so to make the new curls permanent she would have to use chemicals. A permanent hair treatment breaks and reforms the disulfide bonds chemically, reshaping the strand's overall molecular structure.

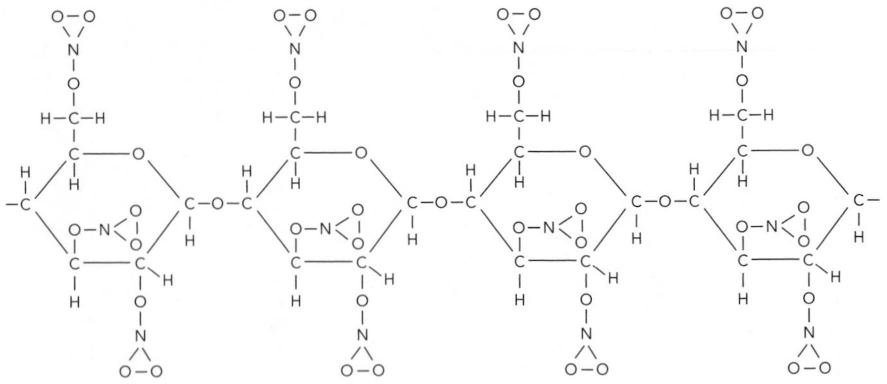

```
   O—O           O—O           O—O           O—O
    \ /           \ /           \ /           \ /
     N             N             N             N
     |             |             |             |
     O             O             O             O
     |             |             |             |
   H—C—H         H—C—H         H—C—H         H—C—H
     |             |             |             |
     C———O         C———O         C———O         C———O
   H/ \H         H/ \H         H/ \H         H/ \H
 —C      C—O—C      C—O—C      C—O—C      C—
   \    /H  \       /H  \       /H  \       /H  \
    O—N    O        O—N    O        O—N    O        O—N    O
    |  \ |          |  \ |          |  \ |          |  \ |
    C    C.H        C    C.H        C    C.H        C    C.H
    |    |          |    |          |    |          |    |
    H    O          H    O          H    O          H    O
         |               |               |               |
         N               N               N               N
        / \             / \             / \             / \
       O—O             O—O             O—O             O—O

                              Nitrocellulose or celluloid
```

Fig. 17.3.10 Nitrocellulose is formed in a chemical reaction between cellulose and an acid mixture. Unlike cellulose, nitrocellulose can be shaped and was the first synthetic plastic. However, nitrocellulose is both flammable and explosive.

In contrast to cellulose, which can't be reshaped, celluloid can be chemically softened or plasticized and then molded fairly easily. The objects that were produced with celluloid resemble those we use today, including transparent plastic films, combs, toys, and synthetic silk-like cloth. However, celluloid darkens with long exposure to light and is extremely flammable. Nitrocellulose itself is a high explosive and the main constituent of smokeless gunpowder. Movie projectors would occasionally catch fire when celluloid motion picture film overheated.

Instead of nitrating cellulose, chemists found that they could attach other useful chemical groups to cellulose to form useful and less dangerous plastics. Cellulose acetate was the first important alternative to nitrocellulose. Referred to as "acetate" or "triacetate," cellulose acetate continues to be used in many forms. Chemists also found that they could take cellulose through a series of chemical transformations that finally ended where it began, with regenerated cellulose. This chemical process allowed cellulose to be reshaped into fibers or films. Rayon and cellophane are both examples of regenerated cellulose.

Plexiglas® and Lucite® are made from the thermoplastic poly(methyl methacrylate) (Fig. 17.3.11). Like cellulose, poly(methyl methacrylate) contains oxygen atoms and forms hydrogen bonds between chains. These bonds are so strong that Plexiglas is a glassy solid below about 105 °C. It's hard, clear, and durable.

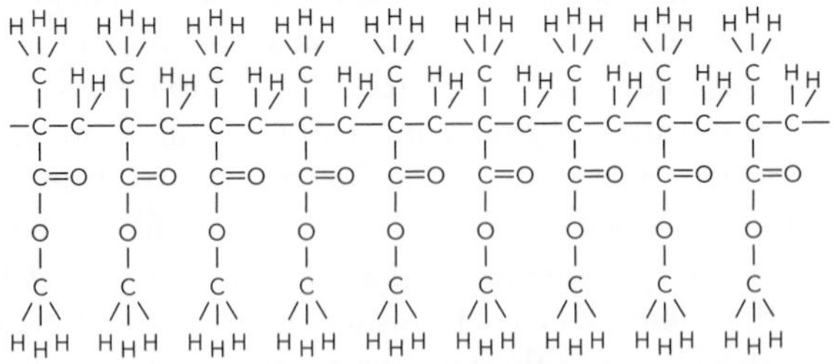

```
 HᴴH  HᴴH  HᴴH  HᴴH  HᴴH  HᴴH  HᴴH  HᴴH  HᴴH
  \|/   \|/   \|/   \|/   \|/   \|/   \|/   \|/   \|/
   C Hн C Hн C Hн C Hн C Hн C Hн C Hн C Hн C Hн
   | |/  | |/  | |/  | |/  | |/  | |/  | |/  | |/  | |/
 —C—C—C—C—C—C—C—C—C—C—C—C—C—C—C—C—C—C—
   |     |     |     |     |     |     |     |     |
  C=O   C=O   C=O   C=O   C=O   C=O   C=O   C=O   C=O
   |     |     |     |     |     |     |     |     |
   O     O     O     O     O     O     O     O     O
   |     |     |     |     |     |     |     |     |
   C     C     C     C     C     C     C     C     C
  /|\   /|\   /|\   /|\   /|\   /|\   /|\   /|\   /|\
 HнH  HнH  HнH  HнH  HнH  HнH  HнH  HнH  HнH

                         Plexiglass or Lucite
```

Fig. 17.3.11 Plexiglas is long chains of poly(methyl methacrylate). These chains bind relatively strongly to one another to form a glassy, transparent material at room temperature.

Instead of cracking like glass when you bend it too far, Plexiglas crazes. Voids appear on the stretched plastic surface but they're prevented from separating completely by taut polymer chains that span the open fissures. Because light's speed changes as it passes in and out of the voids, it reflects randomly from the crazed plastic and gives the plastic a hazy, whitish look. However, the Plexiglas doesn't break. This impact resistance makes Plexiglas a safer alternative to glass in many applications.

Plexiglas has a number of close relatives that are also useful. Poly(methyl acrylate) is more flexible and elastic than Plexiglas and is the principal polymer in acrylic paints. Poly(methyl cyanoacrylate) is a hard, glassy plastic that forms when methyl cyanoacrylate is exposed to moisture. Methyl cyanoacrylate is the active ingredient in "cyanoacrylate glues" such as Superglue® and Crazy Glue®.

Nylon was developed at Du Pont in 1931 by Wallace Carothers ❻ and Julian Hill. Their work was based on the chemical reaction that occurs when an organic acid group (of atoms) on one molecule encounters an organic base group on another molecule. This reaction releases a water molecule and leaves the two molecules bound permanently together. Carothers and Hill realized that hydrocarbon molecules with these reactive groups at both ends would tend to polymerize into long chains. To test their idea, they began producing double-ended organic acids and double-ended organic bases. Their work was a success. The acids and bases bound together to form chains of alternating monomers (Fig. 17.3.12). This strong, tough, elastic, and chemically inert co-polymer was named nylon.

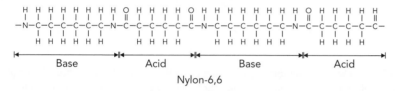

Nylon-6,6

Fig. 17.3.12 Nylon forms when a double-ended acid reacts with a double-ended base. The reaction permanently binds the acid to the base and they become alternating segments in long co-polymer chains. This particular nylon is called 6,6 because the double-ended base monomer contributes 6 carbon atoms to the co-polymer chain (the first 6) and so does the double-ended acid (the second 6).

Because the acid and base groups in the chains allow them to hydrogen-bond to one another, nylon's properties depend on the lengths of its monomer molecules. The shorter the monomer molecules, the more acid and base groups are incorporated into the nylon and the stiffer and harder the nylon becomes. Nylons are named according to the lengths of their base and acid monomers. Nylon–6,6 is made from a six carbon-atom-long base and a six carbon-atom-long acid and is stiffer than nylon–6,12, which uses a twelve carbon-atom-long acid.

Another plastic that's extremely resistant to chemical attack is polytetrafluoroethylene, known as Teflon® or PTFE (Fig. 17.3.13). Teflon's polymer chains resemble those in polyethylene except that its carbon chains are surrounded by fluorine atoms rather than hydrogen atoms. Because fluorine atoms are slightly larger than hydrogen atoms, they fully enclose the carbon backbones of the Teflon polymer chains. With its chains sheathed in fluorine atoms, Teflon is one of the most chemically inert materials in existence. It's also slippery and makes an excellent nonstick surface for cookware and laboratory equipment.

Polytetrafluoroethylene or Teflon

Fig. 17.3.13 Polytetrafluoroethylene or Teflon® is composed of enormous carbon chains, completely surrounded by fluorine atoms (F). The fluorine atoms make Teflon harder and even more resistant to chemical attack than its hydrogen-based relative, polyethylene.

❻ At 31, American chemist Wallace Hume Carothers (1896–1937) left an instructor's position at Harvard University to head an extraordinarily successful research group at Du Pont. Not only did his group invent nylon, they also prepared neoprene, one of the most important synthetic rubbers. His pioneering work even included polyesters, although his early fabrics melted during ironing. Despite his scientific successes, however, Carothers was plagued by terrible bouts of depression. Following the sudden death of his sister and three weeks after applying for a patent on nylon, he took his own life.

Unfortunately, Teflon pulls apart easily unless its molecules are at least 100,000 carbon atoms long. This length makes reptation so slow that Teflon is almost incapable of flowing. Even at its melting temperature of 330 °C, Teflon is so viscous that it appears solid. That's why Teflon objects are formed by filling a mold with powdered Teflon and heating it until it fuses together.

Liquid Crystal Polymers: Kevlar®

Not all polymers form tangles of random coils, even at high temperatures. Some polymers are stiff, rod-like molecules that line up with one another, even as they reptate forward and backward. As liquids, these polymers are liquid crystals and as solids, they are remarkably strong fibrous materials.

One of these liquid crystal polymers is poly(p-phenylene terephthalamide), also called PPD-T or Kevlar®. This co-polymer resembles nylon except that nylon's hydrocarbon chains are replaced by aromatic rings (Fig. 17.3.14). Because these rings don't bend, the co-polymer chains are rigid and straight.

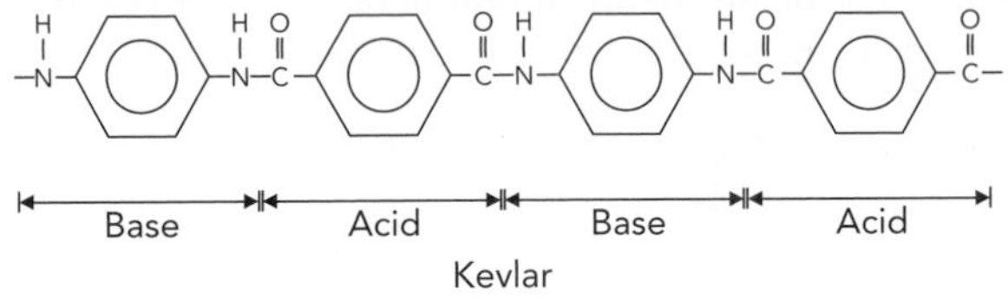

Fig. 17.3.14 Kevlar® resembles nylon, except that the flexible hydrocarbon chains of nylon are replaced by rigid aromatic rings. Each chain is stiff and straight, so the polymer forms a liquid crystal at high temperatures. When liquid Kevlar cools and solidifies, it becomes a fibrous material with extraordinary tensile strength.

At room temperature, the molecules of Kevlar form long, uniform crystals that give it an enormous tensile strength. Because the molecules are already straight, they don't have to uncoil or disentangle themselves when the Kevlar is subjected to tensile stress along the direction of the chains. Instead, they all oppose the stress together. To break a Kevlar fiber, you must break all of its molecules at once. Because Kevlar and other liquid crystal polymers stretch very little and are extremely hard to break, they are used in bulletproof vests, sails, parachutes, and ropes, and as reinforcement in ultrahigh-strength composites.

Even ordinary polymers can be strengthened by straightening their polymer chains, a step that often occurs when a fiber is spun. Spinning is done by squirting a liquid out of a fine nozzle and allowing it to solidify. Some polymers, such as nylon, are spun from hot liquids and solidify by cooling. Others, such as acrylics, are spun from solutions and solidify as the solvent dries. Still others, such as rayon, are spun from a solution that reacts with a chemical outside the spinneret to produce the final polymer fiber. In each case, the synthetic fiber is drawn out of the nozzle under tension. This tension stretches the fiber along its length and unwinds many of its polymer chains. When the fiber solidifies, it retains this modified structure and is stronger as a result.

Polymerization

Polymers form from monomers through the polymerization processes. Although these processes differ according to chemistry and situation, it's worth looking at an example.

Poly(methyl methacrylate) or Plexiglas is made by polymerizing methyl methacrylate monomer (Fig. 17.3.15). The methyl methacrylate molecule includes a

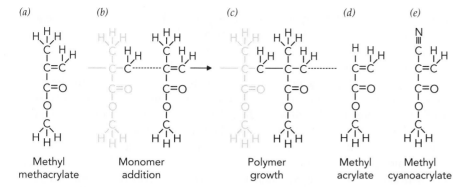

(a) Methyl methacrylate   (b) Monomer addition   (c) Polymer growth   (d) Methyl acrylate   (e) Methyl cyanoacrylate

Fig. 17.3.15 (*a*) Methyl methacrylate polymerizes in the presence of free radicals. (*b*) The free radical attacks the double bond between two carbon atoms and attaches itself to one of those carbon atoms. (*c*) The other carbon atom then becomes a free radical itself and begins searching for yet another monomer to attack. This process continues until a long chain is constructed. Alternative monomers include (*d*) methyl acrylate and (*e*) methyl cyanoacrylate.

double covalent bond between carbon atoms that's easily attacked by a free radical—an unpaired electron on a particular atom. When a chemical that produces free radicals is added to methyl methacrylate and the mixture is heated, a free radical attacks the double bond between two carbon atoms. The double bond becomes a single bond as the free radical binds to one member of the carbon atom pair. The other carbon atom becomes a free radical itself and begins to seek out something with which to bind. It soon attacks another methyl methacrylate molecule, which then attacks another, and so on. With each step, the molecule increases in length and eventually a polymer chain is produced.

Most glues are based on polymers and many of those glues harden via polymerization. While the simplest glues harden by evaporating their plasticizers, polymerization-hardening glues don't really "dry" at all. Instead, they form plastics for the first time through polymerization reactions. Cyanoacrylate glues, epoxies, ultraviolet-cured glues, and polyurethane foams are all monomers up until they begin to harden. At that point, chemical reactions assemble the small molecules into vast chains or networks and the glue hardens.

⚛ Gelatin

Gelatin is a highly processed animal product, containing billions of long, thin protein molecules. When it is mixed with water, it fills the water with these tiny fibers. As the mixture sits and cools, the fibers begin to cling together. They pile up like a "brush-heap" rather than clumping together by themselves. They then form a stiff network that extends over the entire volume of the mixture. The water sits in the voids created by the loosely piled fibers (Fig. 17.3.16). As long as there isn't too

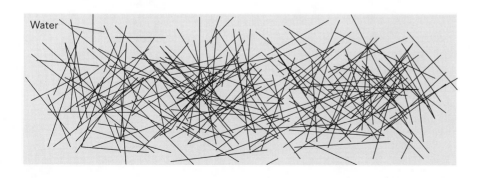

Water

Fig. 17.3.16 A gel is formed when gelatin is mixed with water. The gelatin molecules act like stiff fibers that pile up in a brush-heap. The water occupies the voids between fibers and is weakly held in place by the fibers.

much water, it becomes trapped in the brush-heap. The result is a soft but elastic "solid." Such a composite material, consisting of a solid network enclosing a liquid or a gas is called a gel.

Dessert gelatin such as JELL-O® is such a quasi-solid material. It holds its shape fairly well, even though it's mostly liquid water. If dessert gelatin sits too long, the water will begin to leak out of its voids and it will reduce in volume. That's why gelatin with a large fraction of water doesn't keep well. It tends to separate so that a layer of water appears above or around the gel fraction. Increasing the fraction of actual gelatin stiffens the material and prevents it from separating.

The same thickening processes occur when starches and other water-soluble polymers and fibers are added water. They also create networks of molecules that increase the viscosity of the liquid mixture. Thickeners are used frequently in cooking and in other household products where high viscosity is desirable. They have little effect on the chemistry or taste of the food or other product, but they change its texture. Next time you're drinking a super-thick fast-food milkshake, realize that what's making it so viscous isn't the richness of the ingredients, it's the chemical thickeners that have been added to make the drink more appealing.

CHAPTER 18

THINGS THAT INVOLVE CHEMICAL PHYSICS

Though chemistry is often seen as a field distinct from physics, the two sciences share much in common. The forces that bind atoms together as molecules can be explained in terms of physics, and the thermal, statistical, and even electrical behaviors of chemicals can all be examined from a physical perspective. The objects in this chapter can be viewed as chemical or physical or both.

I've always enjoyed chemistry. I started out in fifth grade with a classic Gilbert chemistry set, and I worked my way through at least several dozen of the 500 experiments that came with the set. But my father had an incredible handbook of chemistry that included an extensive section on dangerous chemicals, one that described in detail why each chemical was dangerous and what not to do with it. Naturally, when my father helped me place a big order to a real chemical company, it included just about every chemical in that section, except for those that were dangerous only as poisons or that were simply so unstable that no sane person would sell them, particularly to a sixth grader. Boy, did I have fun.

My clothes developed mysterious holes and our basement was the source of many strange odors and colorful gases, but I really learned to appreciate chemistry and chemical physics. As they say, what doesn't kill you makes you stronger and I came out stronger a whole lot of times. Somehow my parents never seemed to notice the orange clouds of nitrogen dioxide drifting out into the garden as I headed for the hills while holding my breath.

The topics in this chapter are a little more sedate than those, but much more practical and important. Laundry is one of my favorites, an amazing application of scientific issues to one of the most ordinary activities of life. Is it physics or chemistry? Who cares, it's simply a great topic with lots of science in it.

Chapter Itinerary

Section 18.1 **Oil Refineries**

Petroleum is one of our most versatile natural resources. In the past century and a half, petroleum has developed from a replacement for animal and vegetable oils in lighting and lubrication into one of the foundations of our economy. In addition to providing energy for transportation, petroleum is the source material for much of the chemical industry. Petroleum and petroleum products are so important to our society that they warrant a little more attention than the occasional trip to the gasoline pump.

What Is Petroleum?

To understand petroleum refining, you must first understand what petroleum is. It's a complicated mixture of chemicals, thought to have formed from the decay of ancient marine organisms. Most of the constituents of petroleum are hydrocarbon molecules—molecules composed exclusively of carbon and hydrogen atoms. However some of the organic molecules in petroleum also include oxygen, nitrogen, and sulfur atoms. Moreover, petroleum contains various metal salts as well.

Petroleum is found trapped in porous rocks beneath domes of impermeable rock. Because petroleum is less dense than water, it floats on water and becomes caught between the water below it and the impermeable surface above it. Sometimes the uppermost portion of the petroleum under a dome is natural gas. The water beneath the petroleum is saline (salt water) and probably came from the ancient sea in which the decaying organic matter was originally deposited.

With time and pressure, petroleum's chemical structure evolved into its present form and it migrated into the porous rocks in which it's now found. The mechanisms for its chemical evolution and its migration aren't well understood. But despite the uncertainties in its origins, petroleum exists and our society is now consuming it at a furious pace.

But what are the constituents of petroleum actually like? To answer that question, we must look at their molecular structures. The hydrocarbon molecules take principally four different forms: paraffins, olefins, cycloparaffins, and aromatics. These names, and a variety of equivalent names, describe the ways in which the carbon atoms bind to one another and to the hydrogen atoms.

In all four cases, the atoms are held together by covalent bonds. In a basic covalent bond, two adjacent atoms share a pair of electrons and become bound together at an equilibrium spacing. Since this pair consists of one spin-up electron and one spin-down electron, the Pauli exclusion principle allows the two distinguishable electrons to occupy the same standing wave. The electrons do exactly that, spreading out between the two atomic nuclei in a manner that reduces the atoms' total energy so that it takes work to separate them.

This energy reduction is partly due to electrostatic effects and partly due to quantum physics. Locating the electrons between the two nuclei creates stronger attractive forces than repulsive ones, reducing the overall electrostatic potential energy and helping to hold the two atoms together.

But the shared electrons also have a larger domain in which to reside; their standing waves extend over two atoms rather than one. Like photons, an electron's energy decreases as its wavelength increases and by spreading out between two atoms, an electron can increase its wavelength and reduce its kinetic energy. Overall, a covalent bond lowers both potential and kinetic energies, and makes it quite difficult to separate the atoms involved.

The most important atom in both petroleum and organic chemistry is carbon. A carbon atom has four valence electrons and needs four more to complete its electronic shell of eight electrons. This shell structure is a consequence of quantum physics and, when completed, forms a nearly uniform, spherical ball of electron standing waves. Completing the shell minimizes the energy of the atom and makes the molecule it resides in more chemically stable. To complete its electronic shell and achieve this stability, a carbon atom typically shares valence electrons with four adjacent atoms. It ends up with four pairs of shared electrons and a completed shell.

But those adjacent atoms can't be just anywhere. The sharing scheme only completes the carbon atom's electronic shell if the shared electron waves end up uniformly distributed around the atom. The simplest way to achieve this uniform distribution of electrons is to arrange the adjacent atoms on the four points of a tetrahedron (Fig. 18.1.1). A tetrahedron is an equilateral pyramid with a triangular base. This tetrahedral arrangement places the four atoms as far apart as possible and allows the electron waves to complete the electronic shell properly.

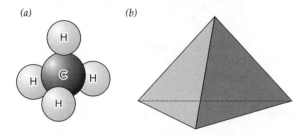

Fig. 18.1.1 The covalent bonds that hold organic molecules together are directional. When a carbon atom binds to four hydrogen atoms (*a*), the hydrogen atoms arrange themselves at the corners of a tetrahedron (*b*).

So a carbon atom's neighbors must be in the right places. In general, covalent bonds only work when the atoms involved are correctly oriented relative to one another. This directionality of covalent bonds gives organic molecules specific shapes and these shapes are important to the properties of petroleum. They are also critical to the functioning of biological systems, so that life couldn't exist without the directionality of covalent bonds. Since most atoms need four pairs of electrons to complete their electronic shells, tetrahedral arrangements of atoms are common in organic molecules.

To see how covalent bonds contribute to the characteristics of petroleum, let's look at the structures of the four different hydrocarbons listed above. The simplest case is the paraffins—chain-like molecules in which strings of carbon atoms are decorated with hydrogen atoms (Fig. 18.1.2). As you might expect, the four atoms surrounding each carbon atom in a paraffin molecule are located on the points of a tetrahedron. This arrangement gives the paraffins a zigzag structure. Some paraffin molecules have only a single chain (Fig. 18.1.2*c*) while others branch extensively (Fig. 18.1.2*d*). The branching is important for gasoline and diesel fuel.

Branching allows the carbon atoms in a paraffin molecule to arrange themselves in a variety of different ways. The 2,2,4-trimethylpentane molecule shown in Fig. 18.1.2*d* is just one of 18 ways in which 8 carbon atoms and 18 hydrogen atoms can join together to form a molecule. Two molecules that contain the same assortment of atoms but differ in the exact arrangements of those atoms are called isomers. Petroleum contains vast assortments of these different isomers.

While the paraffin molecules in Fig. 18.1.2 appear to be rigid, orderly structures, they actually have some freedom of motion left. The covalent bond between each pair of atoms allows those atoms to turn freely about the bond (Fig. 18.1.3). The atoms in a paraffin molecule can and do swivel about the bonds between them. As a result, paraffin molecules are quite floppy.

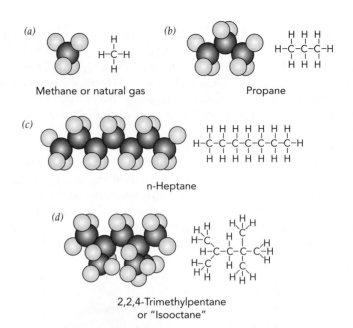

Methane or natural gas

Propane

n-Heptane

2,2,4-Trimethylpentane
or "Isooctane"

Fig. 18.1.2 Paraffins are chain-like hydrocarbons with single bonds between carbons. These molecules can be linear, as in (a) methane, (b) propane, and (c) heptane, or they can have branches as in (d) 2,2,4-trimethylpentane ("isooctane" or simply "octane").

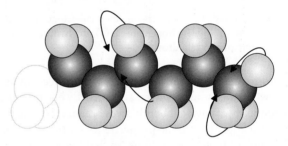

Fig. 18.1.3 Paraffin molecules aren't rigid because carbon atoms can rotate around the covalent bonds between them. The rightmost carbon atom in this paraffin molecule can swivel, just like the knob of a water faucet.

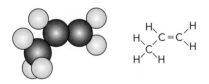

Propene

Fig. 18.1.4 Olefins have one or more double bonds between carbon atoms. Here the rightmost pair of carbon atoms in a propene molecule are connected by a double bond. One pair of valence electrons lies between the two carbon atoms, while the second pair of electrons orbits in front of and behind the line between the atoms.

Olefins are similar to paraffins except that they contain one or more double bonds between carbon atoms (Fig. 18.1.4). Instead of sharing one pair of valence electrons, the two carbon atoms in a double bond share two separate pairs of valence electrons. One pair of electrons occupies a standing wave concentrated directly between the atoms but the second pair occupies a standing wave located on either side of a line between the atoms.

The Pauli exclusion principle allows at most two electrons per standing wave, one spin-up and one spin-down. Since a covalent bond is a specific standing wave, it can hold at most one pair of electrons. With the first pair of electrons occupying a standing wave located between the two carbon atoms in the double bond of Fig. 18.1.4, the second pair of electrons occupies a standing wave located in front of and behind those two atoms. This broader arrangement of electron waves helps to complete the spherical electronic shells of both atoms, but it prevents the atoms from rotating about the double bond. The rightmost carbon atom in Fig. 18.1.4 can't swivel about the double bond. Because of their rigid double bonds, olefin molecules are stiffer than paraffin molecules.

The double bonds in olefin molecules make them susceptible to chemical attack. A double bond is particularly vulnerable to a free radical—an incomplete molecule containing an atom with an unpaired valence electron. This unpaired valence electron seeks out valence electrons on other atoms, attempting to form a new covalent bond to complete its electronic shell. While a free radical's best option for partnership is the unpaired electron on another free radical, it will sometimes attack an electron in an existing covalent bond, particularly a double covalent bond. This sort of attack changes the natures of the molecules involved.

A free radical attacks a double bond by grabbing one electron from the second pair, the pair that's not directly between the two atoms. The free radical forms a new covalent bond with one of the two carbon atoms. The former double bond becomes a single bond, leaving the second carbon atom with an unpaired electron. That second atom becomes a free radical itself.

Their double bonds make olefin molecules reactive and they tend to stick to one another permanently. Automobiles can tolerate olefin molecules in their gasolines as long as the olefins have only one double bond. But olefins with more than one double bond per molecule, polyolefins, can form gummy deposits in your car and are unsuitable for gasoline. While olefin molecules are rare in crude oil, they're created during the refining process. Part of the finishing work in a refinery is to remove polyolefin molecules to make the gasoline more stable against gum formation. Aircraft, which burn fuel in thin, high-altitude air, run into gum problems even with olefins containing only one double bond. Aviation fuels avoid olefins entirely.

In addition to the chain-like paraffin molecules, petroleum also contains ring-like molecules called cycloparaffins. Cycloparaffins occur because chain-like paraffins are floppy and can form loops and coils. The two ends of a typical chain-like paraffin molecule can touch one another and will bind together to form a ring if you remove two hydrogen atoms (Fig. 18.1.5). While the most commonly occurring rings contain five or six carbon atoms, rings with three, four, seven, or more carbon atoms are also found in petroleum. Molecules with more than one ring are also common.

The last important group of hydrocarbons found in petroleum are the aromatics. These molecules include a special type of six-carbon-atom ring—an aromatic

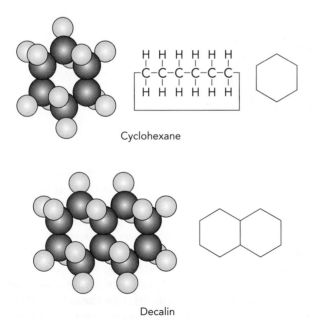

Cyclohexane

Decalin

Fig. 18.1.5 Cycloparaffins contain chains that close on themselves to form rings. Adjacent carbons bind to one another in a zigzag pattern so that these molecules aren't flat. The rings are often represented by polygons, in which each vertex corresponds to a carbon atom holding as many hydrogen atoms as it can.

ring. The simplest molecule containing an aromatic ring is the benzene molecule (Fig. 18.1.6). In this ring, two adjacent carbon atoms are held together by sharing one-and-a-half pairs of electrons. The first pair of electrons orbits between the two carbon atoms and forms a typical covalent bond. But the extra half pair of electrons is shared around the entire ring to form an extra half bond between each pair of carbon atoms. Each carbon atom contributes one electron to this special bonding arrangement, yielding six electrons overall. These six electron standing waves extend all the way around the ring, above and below the atoms themselves, and help to hold the six atoms together. With lots of room to move, these electrons have long wavelengths and low kinetic energies.

Aromatic rings are naturally flat. The tetrahedral structures that give paraffin and cycloparaffin molecules their zigzag shapes are absent in the aromatics. The carbon atoms in an aromatic ring still act to fill their electronic shells but the ring electrons occupy the tops and bottoms of those shells. As a result, the carbon atoms don't bond to atoms above or below the ring. Instead, each carbon atom bonds to three atoms at the points of an equilateral triangle and thus completes its electronic shell. Two of these atoms are other carbons in the ring. The third is typically a hydrogen atom. Because they are built out of equilateral triangles, the aromatic hydrocarbons are basically flat (Fig. 18.1.6).

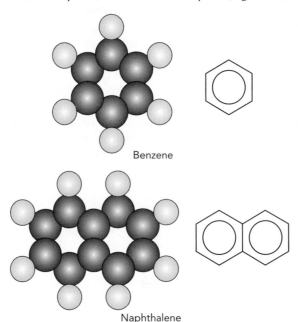

Benzene

Naphthalene

Fig. 18.1.6 Aromatic molecules include one or more special six-carbon-atom rings. The atoms in an aromatic ring are held together by one-and-a-half covalent bonds, with the half bond referring to six electrons that circulate about the ring above and below the lines between atoms. Aromatic rings are represented as hexagons with circles inside.

These four types of hydrocarbons account for most of the molecules in petroleum. However, petroleum also contains molecules that mix two or more of these types together. Such molecules include rings with side chains and aromatic rings attached to cycloparaffin rings. Systematic studies of crude oil have shown that it contains almost any hydrocarbon molecule you can imagine.

Some petroleum molecules also contain oxygen, sulfur, and/or nitrogen atoms. These three atoms, along with carbon and hydrogen, account for most of the organic chemicals in living organisms and presumably entered petroleum during its formation from decaying biological material. Like carbon and hydrogen, oxygen, sulfur, and nitrogen atoms form covalent bonds with their neighbors. However they are closer to completing their electronic shells and don't need to form as many covalent bonds as carbon atoms do.

Oxygen and sulfur atoms both have six valence electrons and need only two more to complete their electronic shells. These atoms normally form two covalent bonds with adjacent atoms, bringing in two additional shared electrons and completing a shell of eight electrons. Nitrogen atoms have five valence electrons and need three more to complete their electronic shells. They normally form three covalent bonds with adjacent atoms, bringing in three additional shared electrons and again completing a shell of eight.

Oxygen, sulfur, and nitrogen often substitute for carbon atoms in organic molecules but bind to fewer atoms (Fig. 18.1.7). While a carbon atom can bind to four hydrogen atoms to form methane, a nitrogen atom can bind to only three hydrogen atoms to form ammonia. Oxygen and sulfur can bind to only two hydrogen atoms, forming water and hydrogen sulfide (rotten egg gas), respectively.

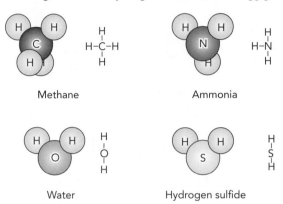

Methane

Ammonia

Water

Hydrogen sulfide

Fig. 18.1.7 Carbon, nitrogen, oxygen, and sulfur atoms can all form covalent bonds. A carbon atom can form four covalent bonds, a nitrogen atom three, and an oxygen or sulfur atom two. When these atoms bond to hydrogen atoms, they form four familiar molecules.

Substitutions of oxygen, sulfur, and nitrogen atoms in the molecules of crude oil are generally undesirable in finished petroleum products. Sulfur is particular bad because of its unpleasant smell and contribution to acid rain. Crude oil that contains substantial amounts of sulfur is called "sour crude" while that without much sulfur is referred to as "sweet crude." Oxygen, nitrogen, and sulfur atoms are often removed from petroleum molecules during the refining process by reacting them with hydrogen gas in a process called hydrotreating.

What Are Petroleum Products?

Unfortunately, crude oil isn't very useful in its raw form and must be processed extensively before it's marketable. This processing is the job of an oil refinery. But before we examine how oil refineries work, we must first consider the products they're trying to produce. Each petroleum product is an assortment of different molecules, selected and blended so that the finished mixture has the appropriate physical and chemical properties for the task it must perform. Here are some of the petroleum products made at refineries.

Let's start with gasoline for automobiles. To make gasoline, the refinery blends molecules that tend to be liquid at room temperature but gaseous at temperatures above about 200 °C, that burn easily and completely in the presence of sufficient air, and that are resistant to knocking. As we saw in Section 8.2, knocking is premature ignition that occurs when fuel and air are compressed in an automobile engine cylinder. Work done on the gaseous mixture of fuel and air during compression raises its temperature, so the mixture is in danger of igniting spontaneously before the spark plug fires. A properly formulated gasoline avoids this spontaneous ignition.

While gasoline should remain a liquid at room temperature to stay in the tank, it must become a gas in a hot engine to burn efficiently. Not every hydrocarbon molecule can meet these two requirements. Some hydrocarbon molecules are more volatile than others—converting easily into a gas. A hydrocarbon molecule's volatility is determined mostly by its size. Small hydrocarbon molecules evaporate more easily than large hydrocarbon molecules.

The size-dependence of volatility is related to the force holding the hydrocarbon molecules together as a liquid: the van der Waals force. This force is the result of tiny electric charge fluctuations that are present in all molecules. As electrons in two nearby molecules move about, they tend to arrange themselves so that the molecules attract one another (Fig. 18.1.8). At any given moment, the two molecules have small electrical dipoles that pull them toward one another. These dipoles come and go but they're still able to hold the molecules together.

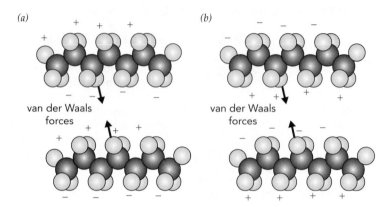

Fig. 18.1.8 Two nearby hydrocarbon molecules attract one another with van der Waals forces that are caused by tiny fluctuations in the distributions of electric charge. At one moment (*a*), the charges on the two molecules are arranged so that the molecules attract one another ever so slightly. At another moment (*b*), the charges have rearranged, but the molecules still experience a weak attraction.

The van der Waals forces between two molecules depend on their sizes and shapes. The larger the molecules are, the more electrons they contain and the more electrically polarizable they are. Large molecules experience stronger van der Waals forces than small molecules, which is why most small molecules are gases at room temperature while most large molecules are liquids or solids.

With gasoline, the van der Waals forces must be strong enough to keep it mostly liquid at room temperature, but weak enough to allow it to become mostly gaseous at about 200 °C. These requirements limit the sizes and shapes of the hydrocarbon molecules that gasoline can contain. The size of a hydrocarbon molecule is determined mostly by the number of carbon atoms it contains. For gasoline, the appropriate hydrocarbon molecules range from about 4 carbon atoms on the small end to about 12 carbon atoms on the large end.

Refineries adjust the precise balance of large and small molecules to give the gasoline just the right volatility over the normal range of operating temperatures. The large molecules bind together rather strongly and help to keep the gasoline liquid during storage. The small molecules are easily separated into a gas and quickly evaporate from an open container of gasoline. Butane molecules, which have only 4 carbon atoms, are included in the gasoline to make starting easy, even in a cold engine. This volatile chemical evaporates readily and is soon lost from stored gasoline. A car or lawnmower with an old tank of gas may not start because its gasoline no longer contains any butane.

Because gasoline's ideal volatility depends on the outdoor temperature, the oil refineries adjust their blends according to season and climate. In winter, they reduce the average molecule size so that the gasoline vaporizes more easily in cold weather. In summer, they increase the average molecule size so that the gasoline is less prone to unwanted boiling.

But volatility isn't the only criterion for the molecules in gasoline. The other critical issue for gasoline is its resistance to knocking. Unfortunately the unbranched paraffin molecules that are common in crude oil ignite much too easily to be the major components of gasoline. Instead, most gasoline molecules are highly branched paraffins, olefins, or aromatics that are difficult to ignite.

Resistance to knocking is normally characterized by a gasoline's octane number. The higher the octane number, the harder it is to make the gasoline knock. 2,2,4-Trimethylpentane (also called "isooctane" or simply "octane")—a highly branched paraffin molecule with 8 carbon atoms (Fig. 18.1.2d)—is particularly resistant to knocking and is the standard by which all other molecules are measured. Its octane number is defined as 100. n-Heptane, an unbranched paraffin molecule with 7 carbon atoms (Fig. 18.1.2c), knocks very easily and is the other standard. Its octane number is defined as 0.

These two hydrocarbons and their mixtures are used to assign octane numbers to gasolines. Each gasoline is compared to various mixtures of "octane" and n-heptane until a match is found. The percentage of "octane" in the matching mixture is then the octane number of the gasoline. For example, a gasoline that has the same knock resistance as a mixture of 90% "octane" and 10% n-heptane is given an octane number of 90. However, a gasoline's octane rating depends slightly on the conditions in which this comparison is made. The two standard conditions are *research*, corresponding to hard acceleration at low speeds, and *motor*, corresponding to zero acceleration at high speeds. Any gasoline has two different octane numbers, its research octane number (R) and its motor octane number (M). These two octane numbers are averaged, (R+M)/2, to give the octane number that appears on the pump.

In formulating a gasoline, the refinery blends various hydrocarbons to achieve an overall octane number of about 87 for regular or 93 for premium. Since octane number only measures resistance to knocking, two different gasolines with identical octane numbers may contain very different assortments of hydrocarbon molecules. Often anti-knock compounds are added to a gasoline to increase its octane number. These chemicals interfere with ignition. Tetraethyl lead was the anti-knock compound of choice until concerns about lead pollution sent it into disuse. Modern anti-knock additives include tert-butyl alcohol and methyl tert-butyl ether.

Kerosene is less volatile than gasoline and consists of molecules with between 10 and 15 carbon atoms. Since kerosene is often used inside houses in lamps and space heaters, it must burn easily and cleanly, without soot or noxious odors. It's normally made from unbranched paraffins and cycloparaffins. Olefins and aromatics are difficult to burn and tend to form soot. Aromatics also tend to have strong odors.

Diesel fuel, jet fuel, and heating oil are similar to kerosene, although they are even less volatile and contain hydrocarbons with between 12 and 20 carbon atoms. While heating oil can contain just about any hydrocarbon, diesel fuel and jet fuel have to be prepared with a little more care.

In a diesel engine, liquid fuel is injected into a cylinder filled with hot, high-pressure air (see Section 8.2). The fuel must ignite easily and spontaneously and burn completely in a short period of time. The same easy and rapid combustion is important in a jet engine. This requirement of easy ignition is just the opposite of that in a gasoline engine. The ideal diesel and jet fuel molecules are unbranched paraffin molecules such as n-cetane, which contains 16 carbon atoms. Diesel fuels are rated according to their cetane number; the extent to which the fuel resembles the easy-to-burn n-cetane and not the hard-to-burn heptamethylnonane, a highly branched paraffin molecule that also has 16 carbon atoms.

Lubricating oils and waxes are even less volatile than fuel oils and contain molecules of between 20 and 50 carbon atoms. Pure hydrocarbons with molecules this large are normally solids at room temperature. However, lubricating oils contain so many different molecules that they're unable to find the orderly arrangements needed to form crystals. The molecules don't fit together well enough to form a rigid structure and remain a thick, viscous liquid.

Only the longer unbranched paraffin molecules are able to join together to form crystalline solids. These solids are called paraffin waxes. With time, paraffin waxes settle out of lubricating oils and are usually removed. At lower temperatures, shorter unbranched paraffin molecules also settle out of lubricating oil. The semi-solid material that forms in cold lubricating oil is petrolatum or petroleum jelly.

The remaining fluid is lubricating oil. Inserted between two movable surfaces, lubricating oil prevents those surfaces from experiencing sliding friction and wear as they slide across one another. The oil molecules cling to the surfaces and to one another with van der Waals forces and keep the two surfaces from touching. While outside forces may try to push the two surfaces together, pressure in the oil pushes back and keeps the two surfaces apart.

Oil's slipperiness comes from the nature of the forces between molecules. A pair of oil molecules is drawn together by van der Waals forces and by whatever pressure is present in the oil. However, if the molecules approach one another too closely, they begin to repel. This repulsion appears when the electron waves of the two molecules begin to overlap. The Pauli exclusion principle doesn't allow identical electrons from both molecules to occupy the same wave so it keeps the two molecules separated at an equilibrium distance.

But these forces depend only on the distance separating the two molecules and don't prevent the two molecules from sliding across one another. In fact, the molecules in oil do slide across one another quite easily and it is this mobility that makes oil such a good lubricant (Fig. 18.1.9). The van der Waals forces are virtually unaffected by sideways motion in the oil molecules.

However, when two surfaces are pushed together by outside forces, they create pressure in the oil. In a completely sealed environment, this rise in oil pressure wouldn't matter. But most lubricated surfaces have openings to the outside, where the pressure is lower. Since fluids accelerate toward lower pressure, lubricating oil accelerates toward openings. The only thing preventing oil from squirting out from between two lubricated surfaces is the oil's viscosity—its difficulty flowing past itself. The more viscous the oil, the more it tends to remain between two surfaces to protect them from wear.

Using a lubricating oil with the right viscosity is important in many applications. If the oil isn't viscous enough, it will escape and leave the surfaces unprotected. If it's too viscous, energy will be wasted doing work against viscous forces, which turn that work into thermal energy.

An oil's viscosity depends strongly on the sizes of its molecules. The larger the molecules, the more viscous the oil. But molecular structure and temperature are also important. Some molecules, particularly cycloparaffins and aromatics, change their viscosities significantly as their temperatures change. Since most situations call for oils that don't change with temperature, most lubricating oils are composed primarily of branched paraffin molecules (Fig. 18.1.9).

Motor oils frequently contain additives to maintain their viscosities at higher temperatures. These additives are long molecules that roll up into compact balls at low temperatures but open up at high temperatures. In their open forms, these molecules thicken the oil and help it do its job. At very high temperatures, these additives and the oil itself fragment into smaller molecules and permanently lose much of their viscosity. That's why it mustn't be overheated.

The largest hydrocarbon molecules that leave an oil refinery are found in asphalt. Asphalt is what's left over when all of the other hydrocarbon molecules have been separated from crude oil. Asphalt molecules may have long paraffin chains or a

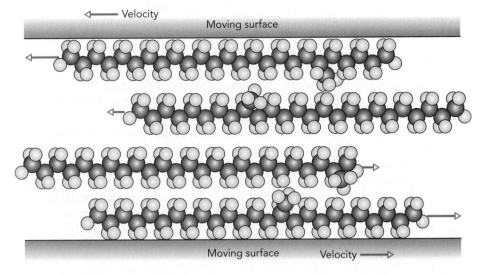

Fig. 18.1.9 Lubricating oil molecules cling to one another with nondirectional van der Waals forces and slide across one another easily. They prevent two surfaces from touching and dramatically reduce the amount of sliding friction between those surfaces.

number of interlocking rings, and frequently include atoms other than carbon and hydrogen. This crazy mixture of giant molecules is used mostly to pave roads. Asphalt molecules are large enough that the van der Waals forces between them prevent their motion at room temperatures. They form a stiff, structureless material that clings to surfaces and makes an excellent binder for the gravel in pavement.

The last major product of oil refineries is gases. These hydrocarbon molecules are so small that van der Waals forces can't keep them together at room temperature and atmospheric pressure, and they evaporate into gas. While many of these gaseous molecules are formed during the refining process, methane, with only 1 carbon atom (Fig. 18.1.2a), occurs naturally in underground reservoirs. Methane extracted from the ground is sent through pipelines and is sold as natural gas. It's a colorless, odorless, nontoxic gas that is significantly lighter than air. Breathing it is dangerous only because it contains no oxygen. Methane is very flammable, however, so a small amount of a sulfur-based odorant is added to it to draw attention to leaks.

Methane can only be liquefied by cooling it to very low temperature. This limitation makes natural gas difficult to store. However propane, with 3 carbon atoms (Fig. 18.1.2b), becomes liquid under pressure. Liquefied propane gas is stored in pressurized tanks and is used for heating and cooking. Liquefied petroleum gas (LP gas) contains both propane and butane.

Pressurizing the propane gas increases its density enough to sustain a liquid phase in a tank. Individual propane molecules continue to move back and forth between the liquid phase and the gaseous phase, but there's no net change in the amounts of liquid and gas. When you remove some of the propane gas from a tank to cook your food, some of the liquid evaporates to replace the missing gas molecules. This automatic replacement of the removed gas makes propane a very convenient fuel.

Oil Refining: Removing Water and Salts

After that long introduction, it's time to look at how an oil refinery works. The refinery must separate the various components of crude oil into specific petroleum products such as gasoline or lubricating oil. Unfortunately, the crude oil that arrives at the refinery rarely contains the right assortment of molecules for the products the refinery wants to produce. Thus, the refinery must usually modify the molecules it receives so that they fit its products. This purification and modification is an enormous undertaking and requires a large facility.

The refinery's first job is to remove water and salt from the crude oil. These contaminants are of no use to the refinery. Fortunately, water and hydrocarbons don't mix well because their molecules don't bind to one another strongly. The molecules in water cling to one another with hydrogen bonds, while the molecules in oil hold onto one another only with weaker van der Waals forces. When you put the two liquids together, the water molecules stay bound to water molecules and the oil molecules stay bound to oil molecules. They don't mix.

What ultimately makes oil and water so immiscible is the strength of the hydrogen bonds between water molecules. It takes far too much energy to separate water molecules for them to mix with the oil molecules. If you pour water and oil into a glass, the less dense oil floats on top of the water and a sharply defined interface forms between the oil and the water.

The water molecules at this interface are special. While the water molecules below them can form hydrogen bonds with neighbors in all directions, the water molecules at the interface have only oil molecules above them. These surface water molecules cling particularly tightly to one another and they create an inward tension along the water's surface. A surface tension of this type appears whenever one material ends and another begins. Surface tension is particularly strong in water because water molecules attract one another so strongly.

Surface tension always acts to minimize a liquid's surface area. The surface of the liquid behaves like an elastic membrane, stretching when you exert forces on it but always snapping back to a taut, smooth shape. Surface tension squeezes raindrops into tiny spheres and turns the surface of a calm lake into a trampoline for water bugs.

Surface tension minimizes the surface area between the water and the oil by making the interface flat and level. But if you cover the glass and shake it hard, the interface will stop being flat. Instead, the glass will become filled with droplets of oil in water and water in oil. You will have formed an emulsion, a situation in which droplets of one liquid are suspended in another.

Surface tension will quickly minimize the surface area of each droplet by making it spherical. But the emulsion will contain more surface area overall than it did before you shook the glass. Because it can further reduce the surface area by reducing the number of droplets, you will see the droplets touch and coalesce. Each time two droplets merge, their combined surface area goes down. Eventually all of the droplets will have joined together and the oil and water in the glass will have separated completely.

But the smallest droplets don't merge together easily. They experience large drag forces as they try to move through the surrounding liquid and travel extremely slowly. It takes a long time for them to find other droplets with which to coalesce. In thick, gooey crude oil, tiny water droplets form an emulsion that takes almost forever to settle. In fact, various chemical impurities in the petroleum actually surround the water droplets, so that they can't touch and coalesce. As a result, getting the water out of crude oil is quite difficult.

Oil refineries usually break the emulsions by heating the oil and passing it through settling tanks or filter columns. At elevated temperatures (90 to 150 °C), water's surface tension decreases and the water droplets are able to merge together more easily. In fact, the energetic water molecules bounce around so vigorously that they have trouble staying together at all. To keep molecules of water or oil from becoming gaseous, the hot crude oil must be kept under pressure. Heat also reduces the crude oil's viscosity and the water molecules are able to settle more easily.

As it settles, the water collects the salt molecules in the crude oil. Since salts are composed of electrically charged ions, they only dissolve in liquids that bind well with charged particles. Water molecules are polar and do a good job of dissolving most salts. Because hydrocarbon molecules are nonpolar—they have no electrically charged ends—they rarely dissolve salts. So the salts accumulate in the water as it settles to the bottom of a tank or filter column.

The smallest water droplets still have trouble settling out of the crude oil. Gravity and buoyant forces are sometimes just too weak to overcome drag forces. Many refineries use electrostatic precipitators to pull the water droplets through the oil. Since oil doesn't conduct electricity, it behaves like very thick air. Charged particles injected into the crude oil quickly attach themselves to water droplets and these electrically charged water droplets are pulled through the oil by electric fields.

Distilling the Crude Oil

Once water and salts have been removed from the crude oil, the refinery is ready to begin sorting its molecules. The principal sorting technique is distillation. Distillation is described in Section 8.4 on water purification, but here the job is somewhat different. In water purification, the goal is to separate a volatile chemical (water) from a nonvolatile chemical (salt) and the only molecule that becomes a gas at reasonable temperatures is water. But in petroleum distillation, almost all of the molecules can become gaseous in the right circumstances. So the refinery must adjust those circumstances in order to collect particular groups of molecules from the mixture.

The crude oil leaving the water separator is heated and then injected near the bottom of a tall distillation tower (Fig. 18.1.10). This tower contains a series of collecting trays, one above the other (Fig. 18.1.11). The temperature inside the tower is carefully controlled so that it's highest where the crude oil enters the tower and gradually decreases from the bottom to the top of tower. Thus, each collecting tray is a little cooler than the one beneath it.

As the hot oil enters the tower, all but the largest molecules evaporate and become gas. This gas gradually ascends the tower and its temperature decreases (Fig. 18.1.10). With each decrease in temperature, the molecules in the gas find it more difficult to stay apart. The larger molecules in the gas begin to stick to one another and form liquid in the tower's trays. Some of this liquid drips down from each tray to the tray below. Overall, gas moves up the tower from below and liquid drips down the tower from above.

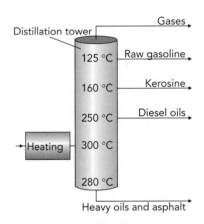

Fig. 18.1.10 The temperature in a distillation tower decreases from the crude oil inlet to the top of the tower. Liquids extracted from trays at various heights and temperatures contain different mixtures of molecules, and are appropriate for different petroleum products.

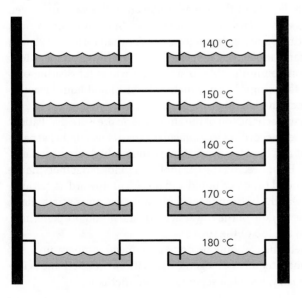

Fig. 18.1.11 Inside a distillation tower is a series of trays, each one cooler than the one below it. Gaseous oil molecules bubble up through each tray from below. As they do, the larger molecules condense into liquid. The liquid in each tray is different, with lower trays containing larger molecules than upper trays.

Each tray tends to accumulate those molecules that can be *either* gas or liquid at the tray's temperature. Any molecules that tend to be gaseous at that temperature will move up the tower to the trays above. Any molecules that tend to be liquid at that temperature will drip down the tower to the trays beneath. Thus, each tray concentrates a particular group of molecules.

However this concentrating process doesn't produce pure chemicals. The liquid in a particular tray still contains a number of different molecules. While one range of sizes is most likely to accumulate in that tray, it will also contain some smaller and larger molecules that manage to find their way into the liquid. In general, nature always tries to maximize the randomness of a liquid. The same statistical rules that govern the flow of heat and are responsible for the laws of thermodynamics also make it very difficult to purify chemicals completely.

Unlike oil and water, these hydrocarbon molecules mix easily with one another. They all stick together with van der Waals forces, regardless of how large their molecules are. Chemicals such as these that dissolve freely in one another are said to be miscible. While the smaller molecules will tend to evaporate from the liquid more easily than the larger molecules, they are all pretty much equal in the liquid itself.

Crude oil's first trip through a distillation tower separates it into several parts, including diesel oil, kerosene, and raw gasoline (Fig. 18.1.10). The diesel oil and kerosene are basically ready for consumer products, but the raw gasoline is not. It has a very low octane number and must be reformed and blended before it's ready for automobiles. Molecules that are too small to become liquid even at room temperature reach the top of the tower and are processed into propane and LP gases.

The largest molecules that enter the distillation tower rarely become gaseous below 300 °C and drip as a liquid to the bottom tray. It might seem reasonable to heat this residual liquid to a higher temperature to separate its molecules from one another. Unfortunately, temperatures above about 360 °C cause hydrocarbon molecules to decompose into fragments, a phenomenon called cracking. These fragments can then recombine to form gums that plug up the distillation equipment. To avoid cracking, the distillation columns must avoid excessive temperatures.

While the molecules in the residual liquid can still be separated by distillation, that distillation must be performed at very low pressures in a vacuum distillation tower (Fig. 18.1.12). The residual liquid from an atmospheric pressure tower is reheated to 350 °C and fed into a vacuum tower near its base. Gases move upward while liquid moves downward and each tray accumulates those molecules that can be either gaseous or liquid at its particular temperature.

Because the pressure and density of the gas are reduced in the vacuum tower, molecules don't have to be very volatile to become a gas. Since forming a thin, low-pressure gas of lubricating oil molecules is much easier than forming a dense, high-pressure gas of those same molecules, it occurs at a much lower temperature. Thus, the vacuum distillation column is able to separate various lubricating oils and waxes from molecules that simply aren't volatile. The residual liquid leaving the bottom of the vacuum column is asphalt.

The hydrocarbon gases that reach the top of the atmospheric pressure distillation column must still be separated according to molecular size. As usual, this separation involves distillation, but this time the distillation is done at high pressures and relatively low temperatures. By squeezing the gas molecules close together, the refinery encourages them to spend time as a liquid and they drift upward as gas and downward as liquid. Trays near the bottom of the high-pressure tower accumulate liquid butane, those near the middle of the tower accumulate liquid propane, and ethane and methane drift to the top of the tower.

Liquifying ethane and methane can only be done at reduced temperatures, so separating these two chemicals requires cryogenic distillation. Since ethane is important to the chemical industry, it's often separated from methane.

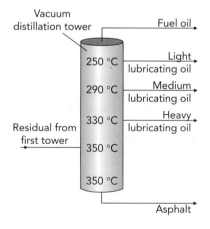

Fig. 18.1.12 In a vacuum distillation tower, the reduced pressure allows even relatively nonvolatile lubricating oils to become gaseous.

Thermal and Catalytic Cracking

Unfortunately, just sorting the molecules in crude oil isn't good enough for most refineries. The principal outputs of these refineries are transportation fuels and there is comparatively little market for the other molecules in crude oil. Since less than half of the molecules in crude oil are suitable for transportation fuels, the refinery has a problem. Moreover, it can't store the unmarketable molecules indefinitely. While the refinery burns some of the less useful molecules to provide its own power, it must sell everything else to make room for incoming crude oil. So large integrated refineries have facilities for converting the less useful molecules in crude oil into ones it can sell.

The original method for converting larger molecules into smaller molecules is thermal cracking. Above about 360 °C, hydrocarbon molecules decompose into fragments. At that temperature, the random thermal energy in a hydrocarbon molecule is occasionally large enough to break that molecule into two pieces. After a short time as a free radical, each fragment rearranges into something that's chemically stable. Most of the time the new molecules are smaller than the old molecules.

The higher the temperature, the more often such decompositions occur and the faster the petroleum cracks. While thermal cracking is a nuisance to be avoided in distillation, it's valuable when done in a controlled manner in a cracking tank. The big molecules that aren't suitable for gasoline generally decompose into smaller ones that are.

Moreover, thermal cracking produces many olefin molecules that have higher octane numbers than the usual contents of crude oil. These olefin molecules are made when the free radical fragments of original hydrocarbon molecules rearrange internally to form double bonds. If the last carbon atom in a chain has only three neighbors, it can complete its electronic shell by forming a double bond with the carbon atom next to it. This rearrangement causes the neighboring carbon to abandon a hydrogen atom, which immediately becomes part of a hydrogen molecule. So thermal cracking creates many smaller molecules, with double bonds at their ends, and hydrogen molecules.

But thermal cracking is difficult to control and also creates many large and useless molecules. As a rule, the higher the temperature in the cracking tank, the higher the octane of the gasoline it produces but the smaller the yield. To make premium gasoline by thermal cracking, the refinery might have to waste all but 20% of the hydrocarbons it feeds to the cracking tank. Because this waste is intolerable, thermal cracking has been replaced almost completely by fluid catalytic cracking and reforming.

In these processes, hot hydrocarbon molecules are brought into contact with silica-alumina catalysts. Like all catalysts, these materials facilitate chemical reactions by reducing the activation energies needed to complete them. When a hydrocarbon molecule attaches to the surface of the catalyst, the catalyst helps it rearrange (Fig. 18.1.13). The catalyst reduces the potential energies of the partially rearranged molecules so that less overall energy is needed to complete the rearrangement. Catalyzed rearrangements thus proceed at lower temperatures.

These catalysts also help to control the rearrangements. A particular catalyst will assist certain rearrangements more than others. Catalysts are particularly helpful in cracking larger molecules into smaller ones so that yields of gasoline molecules are much higher with catalysts than without.

Because all of the catalyst's work is done by its surface, most commercial catalysts are designed to have lots of surface area. The silica-alumina catalysts used in fluid catalytic cracking are actually small particles of porous materials. These particles are only about 50 microns in diameter and they swirl around with the fluid they are cracking.

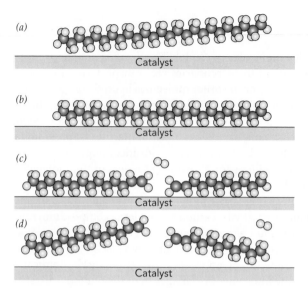

Fig. 18.1.13 A catalyst provides a special surface (a) to which a long unbranched paraffin molecule can attach (b). The catalyst helps the molecule crack into two parts (c)—the final pair of carbon atoms in each of the new chains is joined by a double bond and a hydrogen molecule is released. Once the rearrangement has occurred, the new molecules leave and the catalyst is left unchanged (d).

The reactions take only a few seconds to complete, after which the catalyst particles must be separated from the fluid. The mixture passes through a cyclone separator, where it moves very rapidly around in a circle. The acceleration causes the denser catalyst particles to migrate to the outside of the separator and the clear fluid can then be extracted from the middle of the device.

Unfortunately, the catalyst particles quickly accumulate a coating of very large molecules that don't react and can't be removed easily. Like most catalysts, they lose their catalytic activity when their surfaces become dirty. The only effective way to clean the surfaces of these particles is to burn the residue off them. That's just what the oil refinery does. This burning regenerates the catalyst particles and prepares them for their next trip through the fluid.

Improving the Quality of Petroleum Products

Catalysts also help individual hydrocarbon molecules to change their structures, processes called isomerization and reforming. Long unbranched paraffin molecules and cycloparaffins aren't useful in gasoline because they cause knocking. Isomerizing catalysts, usually platinum, help the unbranched molecules to rearrange into highly branched molecules (Fig. 18.1.14). Reforming catalysts, usually platinum and rhenium, assist in converting cycloparaffins into aromatics. In both cases, the octane numbers increase substantially. Much of the low octane, raw gasoline obtained from the first distillation tower is subsequently sent through catalytic isomerizing and reforming facilities to increase its octane number.

The goal of isomerization is to add more branches to a paraffin molecule by interchanging hydrogen atoms and carbon atoms. On the isomerizing catalyst's surface, one carbon atom and one hydrogen temporarily let go of the hydrocarbon molecule and exchange places. Several such interchanges turn the molecule into a highly branched paraffin with a high octane number.

Without a catalyst, this isomerizing process requires a great deal of energy. Two separate covalent bonds must break completely so that the pieces become free radicals. The carbon and hydrogen atoms must then exchange places and reattach to the main portion of the molecule. This complicated process is unlikely to happen, even at high temperatures.

The isomerizing catalyst facilitates the process by binding temporarily to the molecule and its fragments. The various pieces never become free radicals. Instead, they migrate along the surface of the catalyst and eventually reattach to one

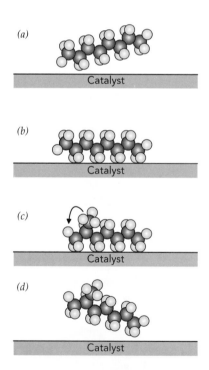

Fig. 18.1.14 An unbranched paraffin molecule (a) attaches to the surface of an isomerizing catalyst (b). This catalyst helps a carbon atom and a hydrogen atom exchange places (c) by stabilizing the pieces during the exchange. The branched paraffin molecule leaves and the catalyst is left unchanged (d).

another without ever being completely free. The catalyst even helps the fragments stay close enough together to exchange places. What would otherwise be an almost impossible event becomes rather likely.

A reforming catalyst helps cycloparaffin molecules get rid of hydrogen atoms and become aromatics (Fig. 18.1.15). Aromatics have higher octane numbers than cycloparaffins, so this reforming is important for gasoline. Although catalysts ease the removal of the hydrogen atoms as hydrogen molecules, the final product molecules have more chemical potential energy than the original molecules. Because this reaction converts a significant amount of thermal energy into chemical potential energy, heat must be added to keep it going.

In addition to isomerization and reforming, oil refineries also use catalysts to attach smaller molecules together to form larger molecules. Catalytic alkylation and polymerization are used to form gasoline molecules from smaller molecules that would otherwise be difficult to use. Both processes start with olefin molecules produced in thermal or catalytic cracking. The olefin molecules have reactive double bonds, and catalysts encourage them to stick to one another or to other molecules. These reactions produce highly branched, high-octane gasoline.

◈ Chemical Bond Types

The atoms in a molecule or material are held together by chemical bonds. To say there is a chemical bond between two atoms is to say that something must supply energy in order to completely separate those two atoms from one another. There are four major types of chemical bonds: covalent, ionic, metallic, and hydrogen. There is also a weaker attachment that occurs between any atoms or molecules and is associated with attractive forces known as van der Waals forces.

A covalent bond forms between two atoms when they share a pair of electrons. This sharing reduces the electrostatic potential energy of the atoms—the energy associated with charge forces—by placing extra negative charge in between the atoms' positive nuclei. It also reduces the kinetic energies of the shared electrons by letting them spread out as standing waves onto both atoms. Like all particles in our universe, electrons travel as quantum mechanical waves and the wavelengths of these waves increase as they spread out. Lengthening an electron's wavelength, like lengthening a photon's wavelength, reduces its kinetic energy.

Covalent bonds are highly directional. When atoms such as carbon, oxygen, hydrogen, and nitrogen stick to one another with covalent bonds, they find themselves held in nearly rigid structural order. The stiffness of the covalent bonds gives shape to most organic molecules and permits them to function in complicated ways as they create biological systems.

Ionic bonds form when atoms in a material become oppositely charged ions and attract one another. For example, when sodium and chlorine atoms approach one another, each chlorine atom removes an electron from a sodium atom, producing a mixture of negatively charged chlorine ions and positively charged sodium ions. These oppositely charged ions then order themselves neatly in rows and columns so that each sodium ion is as close as possible to several chlorine ions while each chlorine ion is as close as possible to several sodium ions. The ions are bound together in an ionic crystal, also called a salt, and energy is needed to separate the ions from one another. The ionic bonding that holds the crystal together is somewhat directional because it is strongest when like charges are far apart and opposite charges are close together.

Metallic bonds are formed by a general sharing of electrons between countless atoms. It only occurs in metallic system—materials in which the electrons can become delocalized so that they drift easily through the material. This sharing lowers both the kinetic and potential energies of the electrons and causes the atoms

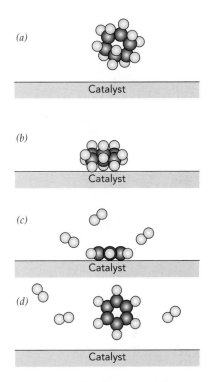

Fig. 18.1.15 A cycloparaffin (*a*) attached to the surface of a reforming catalyst (*b*). This catalyst helps in the removal of hydrogen molecules from the ring (*c*) and creates an aromatic molecule. This molecule leaves the surface and the catalyst is left unchanged (*d*).

to cling to one another. Because these metallic bonds are rather insensitive to the relative positions of its atoms, metallic bonds are non-directional and metal crystals are relatively easy to deform.

Hydrogen bonds occur when exposed hydrogen nuclei can get close to the extra electrons on certain types of atoms. When a hydrogen atom forms a covalent bond with a strongly electron-attracting atom such as oxygen, the electron-attracting atom pulls the hydrogen atom's electron toward it and leaves the hydrogen atom's positively charged nucleus, a proton, relatively exposed. If this exposed proton can get close to the pair of electrons in an oxygen atom's "lone pair orbital," that proton will find itself drawn into place. A hydrogen bond will form between the exposed hydrogen nucleus and the electrons on the back end of the oxygen atom and energy will be needed to pull the hydrogen and oxygen apart.

Finally, van der Waals forces appear between any two nearby atoms or molecules. These forces occur because the positive and negative charges in an atom or molecule can shift so that the particle becomes electrically polarized. When two particles are close enough, each one polarizes the other in a complicated, quantum mechanical way and they experience weak but important forces. These forces are strong enough to stick molecules together to form liquids when the temperature is low enough.

SECTION 18.2 Laundry

Whether you wash your clothes at home or send them to a laundry, you expect them to come out clean and crisp, looking and feeling as they did when they were new. The dirt, oils, perspiration, and stains should all have vanished, leaving the fabrics and their colors completely intact. Moreover, the clothes should retain their shapes, sizes, structures, and surface textures. With so many expectations, it's no wonder that the word "miracle" appears so frequently in advertisements for laundry detergents.

Achieving these many goals is something of a balancing act. The chemicals that contaminate clothes aren't always so different from those that give them their structures and colors. Trying to remove one chemical while leaving the other isn't easy and washday in the nineteenth century was hardly a treat for the garments being cleaned. However, in recent years laundering has developed from a simple art to an advanced technology. The miracles promised by the detergent commercials are almost reality. In this section, we'll examine physical and chemical mechanisms that make those miracles possible.

Soap

One of the most difficult problems in laundering clothes is how to remove all of the different soils in a single operation. Some soils consist of polar molecules, those that have electric charges or electrically charged regions, while others consist of nonpolar molecules, those that are effectively neutral throughout. These two types of soils are so different from one another that a liquid that dissolves one is unlikely to dissolve the other. To make things worse, there are also soils that don't dissolve well in anything, or at least not in anything that you could imagine putting on your clothes. Getting all of these soils out of the clothes without harming the clothes is what laundry is all about.

Polar soil molecules include salts from perspiration and ground dirt. These salts generally dissolve in water, where they become ions that are carried away in shells of water molecules. Because they are basically at home in water, these polar soils are described as hydrophilic (water loving). Because carbohydrates such as sugar have electrically charged regions and form hydrogen bonds with water, they dissolve easily in water and are thus also hydrophilic.

Nonpolar soil molecules include oils, fats, and waxes from skin, foods, and plants. These oily molecules tend to dissolve in nonpolar solvents such as gasoline or kerosene. Because they can't form hydrogen bonds with water molecules, they don't bind well with water and are essentially insoluble in it. These nonpolar soils are described as hydrophobic (water avoiding).

You could launder your clothes by first washing them in water to dissolve and remove hydrophilic soils and then laundering them in gasoline to dissolve and remove hydrophobic soils. But this would take a long time and would be very hard on the fabric. After the process was over, your clothes would have aged considerably, yet some of the soils would still remain. Cleaning clothes requires something more than water and gasoline. That's why we use soaps, detergents, bleaches, and brighteners in our laundry.

Soap is a peculiar type of salt. Like all salts, soap contains a mixture of positively and negatively charged ions (Fig. 18.2.1). There is nothing special about the positive ions, which are usually just sodium or potassium atoms that are missing an electron. What makes soap so unusual and so effective at cleaning is its negative ions. The negative ions in soap have the negative charge located at one end of a very long molecule. The other end of the molecule is an uncharged hydrocarbon chain such as those encountered in oil molecules.

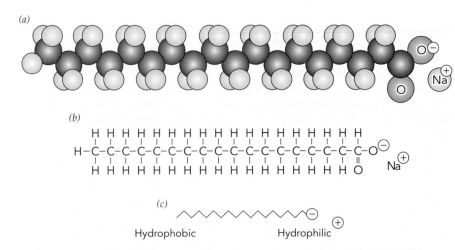

(a)

(b)

(c)

Hydrophobic   Hydrophilic

Fig. 18.2.1 Soap is a peculiar salt. Its negative ion consists of a long hydrophobic (water avoiding) hydrocarbon chain attached to a hydrophilic (water-loving) carboxylate group. A nearby positive ion, usually sodium (Na), balances the negative charge of the carboxylate group. A soap molecule can be represented as (a) balls, (b) letters, or (c) a zigzag hydrocarbon chain with charges attached to it.

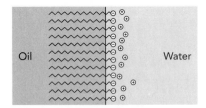

Fig. 18.2.2 Soap negative ions move spontaneously to interfaces between water and oil. Their hydrophobic ends project into the oil and their hydrophilic ends project into the water.

The negative soap ion is so long that its two ends operate independently. Its charged end is polar and hydrophilic. Water molecules cling to this end's electric charge and try to carry it into solution. But the soap ion's hydrocarbon end is nonpolar and hydrophobic. This end of the soap molecule is expelled from water but binds nicely to oil molecules.

The soap ion is a dual citizen. Its hydrophilic end is attracted to water while its hydrophobic oil end is attracted to oil. This split affinity causes soap ions to accumulate at interfaces between water and oil (Fig. 18.2.2). The negative ions spontaneously orient themselves at such an interface with their electrically charged ends in the water and their hydrocarbon ends in the oil. The positive ions hover around in the water near the interface to keep everything electrically neutral.

This tendency for soap ions to order themselves at interfaces is an example of self-organizing behavior. While mixtures of table salt and water are random and homogeneous, mixtures of soap and water are not. Even when there's no oil present, soap ions migrate to water's surface because individual soap ions don't mix freely with the water. Since water molecules don't bond well to the nonpolar hydrocarbon chains, the water molecules push the soap ions to the water's surface.

A tiny amount of soap added to a bowl of water soon creates an ultra-thin layer of soap ions on the surface of the water—a layer that's only a single molecule thick. The soap ions arrange themselves with their polar ends in the water and their nonpolar ends in the air. The uppermost water molecules in the bowl are then able to hydrogen bond to the soap ions above them and don't pull together as strongly as they would if they had only air above them. Thus, the water molecules contribute little to the liquid's surface tension.

The hydrocarbon chains of the soap ions now form the uppermost layer in the liquid. With nothing above them to stick to, these chains pull together and create surface tension. However they attract one another with van der Waals forces, not hydrogen bonds, and create a surface tension only about 30% that of water molecules. The soap's presence in the water significantly reduces its surface tension.

This reduced surface tension is soap's first contribution to the laundering process. Pure water keeps to itself, beading up on any surface that doesn't bind strongly to water molecules (Fig. 18.2.3*a*,*b*). Surface tension makes falling water droplets spherical and they remain almost spherical on oily, hydrophobic surfaces. But add-

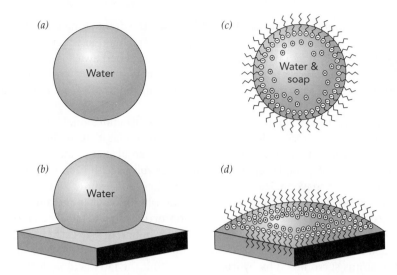

Fig. 18.2.3 (a) Surface tension in pure water causes its droplets to be spherical. (b) These water droplets remain almost spherical on many surfaces. (c) Soap ions coat the outside of a water droplet and dramatically reduce the surface tension. (d) The soapy droplet is able to spread out more easily and wets many surfaces completely.

ing just a tiny bit of soap to the water reduces each droplet's surface tension and allows it to wet the surface (Fig. 18.2.3c,d). A soapy droplet spreads outward because van der Waals forces attracting the droplet to the surface are strong enough to stretch it out into a flat puddle.

In effect, soapy water is "wetter" than pure water. Soapy water doesn't bead up on fabrics; it soaks right in. When you are cleaning clothes and want the water to wet every fiber in the fabric, you add soap to the water. Because soap helps water to wet surfaces, it's a wetting agent. It's also a surfactant or surface-active agent because of its tendency to modify the properties of surfaces or interfaces. There are other kinds of surfactants, but soaps and soap-like materials are the most important group.

However, not all soap molecules make it to the water's surface. If you put lots of soap in the water, or the surface is far away, the soap ions assemble themselves into spherical structures called micelles and remain inside the water (Fig. 18.2.4). In these micelles, all of the soap ions are oriented with their charged, polar ends pointing outward and their uncharged, nonpolar ends pointing inward. The water molecules stick to the micelles' polar outsides and carry the micelles about. As usual, the positive soap ions hover about nearby to keep everything electrically neutral.

These micelles are soap's second contribution to the laundering process. They tend to trap and collect oily soil molecules. The inside of a micelle is a nonpolar environment and ideal for oil molecules. When an oil molecule bumps into a micelle, the water pushes it into the center of the micelle and there it remains.

The micelles in soapy water move randomly, collecting any oil molecules they encounter in their travels. With a little thermal or mechanical agitation, micelles can even pluck oil molecules from the surfaces of fabrics. Naturally, this is helpful when you are doing laundry. Little by little, the oily soils in the clothes become trapped in micelles in the water.

Soap also helps to remove insoluble debris from clothes. Micelles form around dust particles and help the water to carry these particles away. Since most dust particles don't dissolve in any liquids, soap micelles are essential to their removal from clothing.

Since soap micelles are composed of negatively charged soap ions, they are negatively charged objects and tend to repel one another in the water. They remain

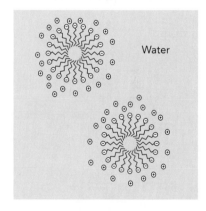

Fig. 18.2.4 In water, negative soap ions form spherical micelles. The hydrophobic chains form the centers of these micelles and tend to accumulate oily soil molecules.

(a)

```
   H  H  H  H  H
   |  |  |  |  |
 - C- C- C- C- C- C- O- H
   |  |  |  |  |  ‖
   H  H  H  H  H  O
```

Fatty acid

(b)

```
   H  H  H  H  H
   |  |  |  |  |
 - C- C- C- C- C- C- O⊖
   |  |  |  |  |  ‖
   H  H  H  H  H  O        H⊕
```

Fatty acid in water

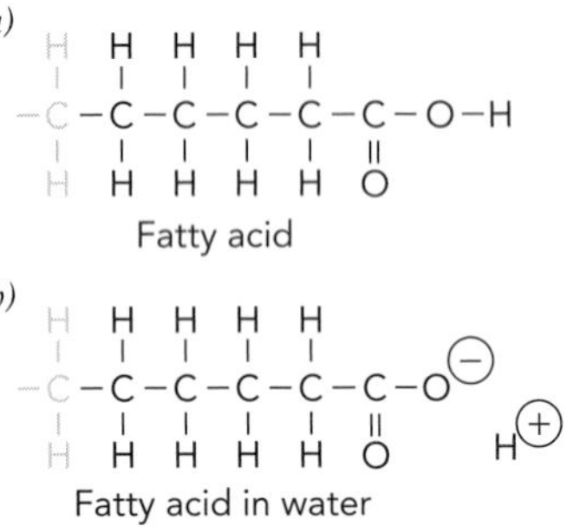

Fig. 18.2.5 (*a*) A fatty acid is a long hydrocarbon chain ending in a carboxylate group. (*b*) In water, the carboxylate group's hydrogen atom is carried away by water molecules as a positive ion (H⁺), leaving the rest of the molecule negatively charged.

separate and mobile and are easily washed down the drain along with their contents. In fact, most fabrics also become negatively charged in water. Their fibers include weakly attached hydrogen atoms that are carried away as positive ions by the water. The fibers are left with negative charges and they tend to repel the negatively charged soap micelles. This repulsion prevents soils from redepositing on the clothes.

This arrangement, soap micelles in water, is a stable emulsion. Unlike a simple mixture of oil and water, it doesn't separate when you let it sit. A surfactant that helps to form and stabilize emulsions is called an emulsifier. Emulsifiers are particularly important in food preparation, where egg yolks, lecithin, and various gums are used to make mayonnaise, chocolate, and other foods smooth and creamy.

Soap is clearly useful in laundering clothes, but where does it come from and what is its structure? Soap is made from naturally occurring oils and fats. Each molecule of oil or fat consists of three fatty acid molecules bound to a glycerin molecule. A fatty acid molecule resembles a paraffin or olefin molecule, with its long chain of carbon atoms surrounded by hydrogen atoms. But the fatty acid molecule has a special arrangement of carbon, oxygen, and hydrogen atoms, a carboxylate group, at one end that makes it an organic acid (Fig. 18.2.5).

An acid is a molecule that can easily lose a positively charged hydrogen ion (H⁺) when it is mixed with water. One of the hydrogen atoms at the special end of the fatty acid falls off easily because the adjacent oxygen atom has largely removed its electron. Oxygen and hydrogen form a covalent bond, with a pair of electrons between them, but the oxygen atom attracts the pair of electrons more strongly than the hydrogen atom does. As a result, the hydrogen atom's nucleus is relatively exposed and is easily carried away by passing water molecules. This loss leaves a negatively charged fatty acid ion.

In an oil or fat, these three fatty acid molecules are not ionized. Instead, they have reacted with a glycerin molecule like three large ships docking at a small port. The glycerin molecule has a chain of three carbon atoms and each of these carbon atoms plays host to one of the fatty acids (Fig. 18.2.6*a*). The resulting structure is called a triglyceride. Assembled in this manner, the triglyceride is nonpolar and virtually insoluble in water. It looks and feels like petroleum oil because both have the same long hydrocarbon chains. However, triglycerides are digestible while petroleum oils are not.

Triglycerides composed entirely of paraffin-like fatty acids are called saturated fats because they have only single covalent bonds and as many hydrogen atoms as possible. Such molecules experience strong van der Waals forces, forming fats that remain solid at relatively high temperatures. These fats are found in animals and tropical plants such as palms and coconuts.

Triglycerides containing olefin-like fatty acids are called unsaturated fats because they have double bonds that reduce their hydrogen atoms count. Double bonds stiffen the hydrocarbon chains and prevent them from bonding as strongly to one another. Oils containing these molecules melt at relatively low temperatures and are found in fish and temperate plants such as soybeans and corn. Unfortunately, people find the less healthy saturated fats more tasty and satisfying than the unsaturated fats. Converting unsaturated fats to saturated fats, a process called hydrogenation, is commonly used to stiffen oils for use in foods such as margarine and candy.

Soap enters into this picture when triglycerides react with sodium hydroxide (lye). Sodium hydroxide is a salt consisting of positive sodium ions (Na⁺) and negative hydroxyl ions (a hydrogen and an oxygen atom together, OH⁻) and it rapidly dissolves into independent ions when you put it in water. When you mix fat, water, and lye together, the hydroxyl ions from the lye attack the fat molecules and remove the fatty acids from the glycerin as negative ions (Fig. 18.2.6*b*). Soon the water is filled with glycerin molecules, negative fatty acid ions, and positive sodium ions. When

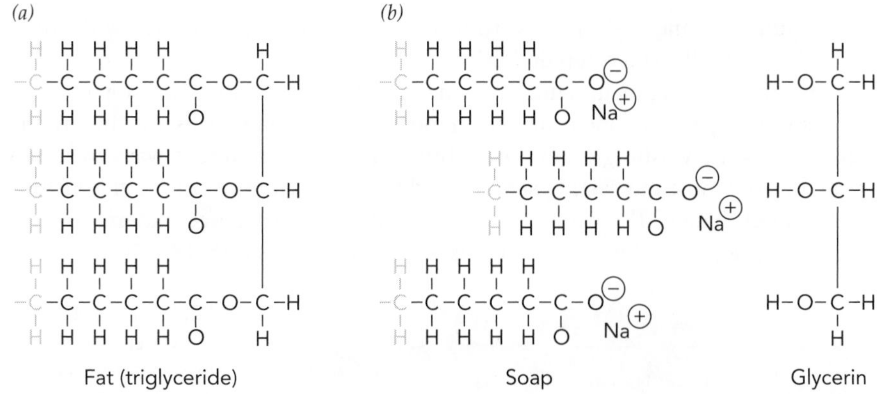

Fig. 18.2.6 (*a*) A fat molecule or triglyceride consists of three fatty acids bonded to a glycerin molecule. When the triglyceride reacts with sodium hydroxide (lye), the fatty acids break free of the glycerin and produce a mixture of soap and glycerin (*b*).

the reaction is complete and most of the water is removed, the result is soap. The glycerin may or may not be removed.

The hardness of the soap depends on the fats from which it was made. Saturated fats produce hard bar soaps while unsaturated fats produce soft liquid soaps. Soft hand soaps often include the glycerin. While most modern soaps are made with lye and are thus sodium salts, earlier soaps were made with potassium hydroxide obtained from wood ash and lime and were potassium salts.

Water Softening

Unfortunately, laundering clothes isn't quite this easy. While soap is wonderful at removing oils and fats from fabric, it has problems in hard water. Hard water is any water with more than about 120 mg of positively charged calcium and magnesium ions per liter. These two metal ions, and a few others, bind with the negative soap ions and form insoluble soap scums that deposit themselves on sinks, showers, bathtubs, washing machines, and clothing. If you try to launder clothes with soap in hard water, you are in for a messy surprise.

The problem occurs because calcium and magnesium ions behave differently from the ions of sodium and potassium normally found in soap. Sodium and potassium atoms each have one more electron than they need to complete an electronic shell—a quantum physical structure that is particularly stable. That extra electron is relatively easily removed, creating a positively charged ion that is easily drawn into solution in water. Water is so strongly attracted to sodium ions that almost every sodium salt in existence dissolves in water. Sodium's fantastic solubility explains why there is so much sodium in seawater. Potassium ions are almost as soluble as sodium ions. Salts consisting of sodium or potassium positive ions and soap negative ions are extremely soluble in water.

But calcium and magnesium atoms have two more electrons than they need to complete an electronic shell. They give up those two electrons somewhat reluctantly to form positively charged ions and aren't particularly soluble in water. While some calcium and magnesium salts are modestly soluble in water, calcium and magnesium soap salts are not.

When you put soap in hard water, the positive calcium and magnesium ions in the water combine with the negative soap ions and quickly form insoluble salts. These salts are pasty solids that cling to everything. These soap salts build up on sinks and often form bathtub rings. If you want to do laundry in a place where the water contains substantial amounts of dissolved minerals, you must either remove the calcium and magnesium ions from the water or replace the soap with something else. Actually, you often do both.

Removing the calcium and magnesium ions is the first option. This step is called water softening and is done routinely in most industrial laundries long before the water enters the washing equipment. There are several different ways to soften water, but the most interesting scheme and the one used most often in houses is called ion exchange. The water passes through an ion exchange material that replaces the calcium and magnesium ions with sodium ions (Fig. 18.2.7).

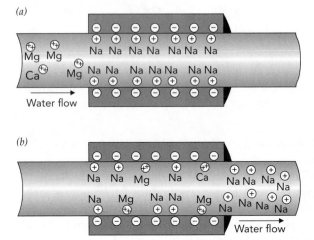

Fig. 18.2.7 (*a*) A fresh ion exchange water softener contains sodium ions (Na⁺), located near negatively charged sites in a resin or zeolite ceramic. (*b*) As water containing magnesium (Mg⁺⁺) and calcium (Ca⁺⁺) ions passes through the softener, the sodium ions are released and the magnesium and calcium ions remain behind.

The ion exchange material is a special ceramic (zeolite) or plastic resin with many negatively charged regions in its porous structure. To keep the material electrically neutral, a positive ion is located near each negative region. The negative regions are part of the material, so they can't go anywhere, but the positive ions are mobile.

When the ion exchange material is fresh, nearly all of the positive ions inside it are sodium ions. As hard water flows through the material, the sodium ions are gradually replaced by calcium and magnesium ions. Since the sodium ions are more soluble in water than the calcium and magnesium ions, the sodium ions tend to enter the water and the calcium and magnesium ions tend to leave it. Each calcium or magnesium ion that sticks to the ion exchange material releases two sodium ions, which leave the water softener for the water.

While the water leaving an ion exchange water softener still contains dissolved ions, they are sodium ions rather than calcium or magnesium ions. The sodium ions cause no trouble when laundering clothes or washing your skin, but people who are on low-sodium diets should avoid softened water. Moreover, you shouldn't use softened water in a steam iron.

When all of the sodium ions in the ion-exchange material have been replaced by calcium or magnesium ions, the water softener stops softening the water. To regenerate the ion-exchange material, you must flush it with very concentrated salt water. The many sodium ions in the salt water dislodge most of the calcium and magnesium ions and return the ion-exchange material to its original condition—it is once more full of sodium ions and ready to soften water.

Because many homes don't have water softeners, most laundry soaps soften the water themselves. They contain chemicals called builders that bind to the calcium and magnesium ions and keep them away from the soap ions. The most effective of these builders is sodium tripolyphosphate and, at one time, most household detergents contained large amounts of it. However, phosphates encourage the growth of algae, threatening the ecologies of rivers, lakes, and bays, and have been banned from detergents in many regions. Builders such as sodium carbonate, citric acid, and sodium citrate are now often used instead.

Another building technique used in some products is to incorporate small zeolite ceramic particles directly in the detergent. These builder particles exchange sodium ions for calcium and magnesium ions and soften the water directly in the washer. They fall to the bottom of the washer and are rinsed away

Detergents

But water softening is only half the solution. Because it's hard to eliminate all of the calcium and magnesium from water, the manufacturers also eliminate the soap from the laundry powder. That's right—most laundry detergents aren't soap at all. Instead, they are synthetic detergents that are structurally related to natural soap but aren't quite the same.

Actually, soap is a type of detergent. Detergents are a broad class of molecules that stabilize mixtures of oil and water. There are many other kinds of molecules that can perform this task and thus many types of detergents.

The most common laundry detergents are the linear alkylbenzenesulfonates (Fig. 18.2.8a). These petroleum products are sodium salts, just like most soaps, but the structures of the negative ions are different. Recall that a negative soap ion is a long hydrocarbon chain attached to a negatively charged carboxylate group. The detergent ion is also a long hydrocarbon chain, attached to an aromatic or benzene ring, attached to a negatively charged sulfonate group. The sulfonate group involves one sulfur atom and three oxygen atoms.

(a)

Linear alkylbenzenesulfonate detergent

(b)

Linear alcohol sulfate detergent

(c)

Linear alcohol ethoxysulfate detergent

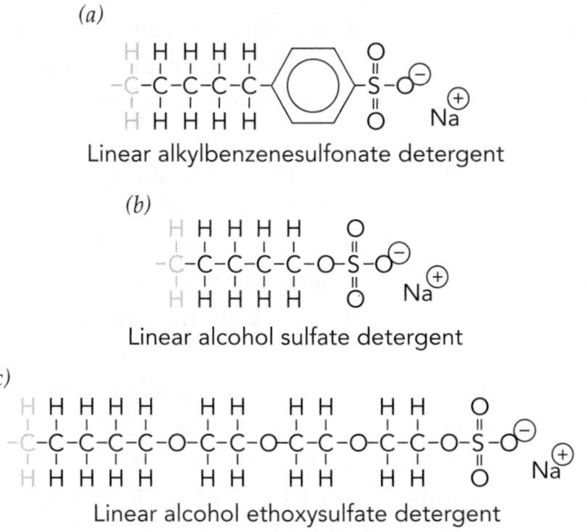

Fig. 18.2.8 (a) The most common laundry detergent has an aromatic or benzene ring connecting a long hydrophobic hydrocarbon chain to a hydrophilic sulfonate group. Common shampoo detergents connect the chain to the sulfonate with either (b) an oxygen atom or (c) a string of oxygen and carbon atoms.

The two important parts of the detergent molecule are the charged head and the long nonpolar tail. The charged end is a sulfonate group in which a sulfur atom attaches to four other atoms: a carbon atom and three oxygen atoms. This arrangement is roughly tetrahedral in shape. But sulfur normally attaches to only two atoms, so how is this arrangement possible?

First, the sulfur atom forms normal covalent bonds with the carbon atom and with one of the oxygen atoms. That oxygen atom has an extra electron, making it a negatively charged ion that can only form a single covalent bond.

Second, the sulfur atom allows each of the two other oxygen atoms to share a pair of its electrons. Because these shared electrons can spread out between both atoms, their wavelengths increase and their kinetic energies decrease. In this manner, the oxygen atoms become attached to the sulfur atom. The overall result is a negatively charged structure that's easily carried about by water molecules.

The detergent molecule's long nonpolar tail is essentially an unbranched paraffin chain, also referred to as a linear alkyl group. This chain provides the oily tail of the detergent molecule. While early alkylbenzenesulfonate detergents included branched paraffin chains, these proved to be less biodegradable than the linear versions. Bacteria can metabolize long linear chains because those chains are common in animals and plants, but branched chains are rare in nature and bacteria are unprepared for them. To keep detergent foam out of streams and lakes, manufacturers have learned to produce purely linear detergent molecules.

The last piece of the detergent molecule is the aromatic or benzene ring. This ring is a vestige of the manufacturing process. It's much easier to attach the linear alkyl tail and the sulfonate head separately to an aromatic ring than it is to attach them directly to one another. Unfortunately, the ring actually reduces the biodegradability of the molecule somewhat. There are other detergents, such as linear alcohol sulfates and linear alcohol ethoxysulfates, in which the aromatic ring is replaced by an oxygen atom (Fig. 18.2.8b) or a string of oxygen and carbon atoms (Fig. 18.2.8c). The most common linear alcohol sulfate is sodium lauryl sulfate, with 14 carbon atoms in its hydrophobic chain. Sodium laureth sulfate is a common linear alcohol ethoxysulfate, also with 14 carbon atoms in its chain. These detergents are derived from tropical oils and often used in shampoos and dishwashing liquids.

Since calcium and magnesium ions don't cause these sulfonate or sulfate detergents to form insoluble detergent scums, why do laundry detergents still worry about softening the water? Unfortunately, calcium and magnesium ions interfere with the micelles, making it difficult for them to extract soil from fabric and keep it suspended in water. Because each calcium or magnesium ion has twice the positive charge of a sodium or potassium ion, these highly charged ions approach the micelles closely and partially neutralize their surfaces. Since these neutralized micelles don't repel one another or the fabric well, they do a poor job of cleaning clothes. That's why laundries and laundry detergent still work to remove the calcium and magnesium ions.

Before leaving detergents, it's worth noting that not all detergents are negative ions (anions). It's also possible to construct detergent molecules that are positive ions (cations) and even ones that aren't ions at all. However, cationic detergents and surfactants aren't used in laundry detergents because they tend to stick to fabric—we'll discuss their use as fabric softeners later on. But nonionic detergents are often used to launder clothes.

The only requirement for a detergent is that its molecules stabilize a mixture of oil and water. Nonionic surfactant molecules don't have an electric charge, but they do have a hydrophilic end and a hydrophobic end (Fig. 18.2.9). Like most detergents, the hydrophobic end is just a long hydrocarbon chain. But the hydrophilic end is also a long chain, consisting of oxygen and carbon molecules attached one after the next and decorated with hydrogen atoms. The oxygen atoms in this special chain form hydrogen bonds with water molecules, giving that end of the molecule its hydrophilic character. These nonionic molecules form micelles and are very effective at removing grease.

Nonionic surfactants are unaffected by hard water and are actually better than anionic detergents at removing some soils—they are particularly good at removing skin oils from synthetic fabrics. However, they aren't salts and exist either as liquids or waxy solids. As a result, nonionic surfactants are difficult to formulate into powdered detergents but are common in liquid detergents.

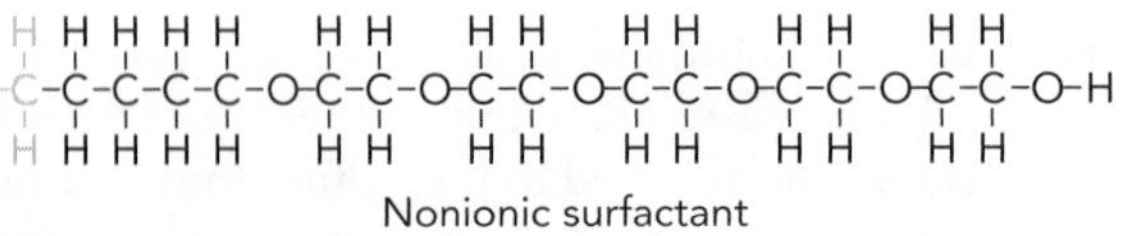

Nonionic surfactant

Fig. 18.2.9 Nonionic surfactants have a hydrophobic hydrocarbon chain (on the left) attached to a hydrophilic chain containing oxygen atoms (on the right).

Bleaches and Enzymes

Not all soils can be removed from fabrics with detergent and water. Molecules that form covalent bonds with the fabric create stains that can only be eliminated with bleaches or enzymes. Bleaches act to destroy the coloration of stain molecules or to cut them free from the fabric. Enzymes act to dice up large stain molecules into smaller fragments that can be washed away. With a little luck, these steps can be taken without destroying the fabric or its color.

Unlike soaps and detergents, bleaches react chemically with the soil molecules. They are particularly aggressive at converting double bonds to single bonds by attaching oxygen and chlorine molecules to the two atoms involved. Double bonds often give organic molecules their colors so this sort of rearrangement tends to make them colorless.

Just as atoms absorb and emit photons of light that are characteristic of their electronic states, so molecules absorb and emit photons that are characteristic of their electronic states. Each electron in a molecule is sensitive to passing electromagnetic radiation and responds to photons that can transfer it to some unoccupied state with more energy. If the electron finds such a photon, it may undergo a radiative transition to the excited electronic state and absorb the photon. The electron will eventually return to its original state, converting the extra energy into thermal energy, but the photon will be gone forever. If a particular molecule contains electrons that can absorb photons of visible light in this manner, it will appear colored.

In most single covalent bonds, the two electrons are bound so tightly between the two nuclei that any transition to a new electronic state requires more energy than a photon of visible light can provide. Only ultraviolet light can cause radiative transitions in these electrons. Molecules based entirely on single covalent bonds are normally unaffected by visible light and are thus colorless.

However, the outer electrons in a double covalent bond aren't so tightly bound and can be transferred to other electronic states relatively easily—a photon of visible light may well be able to cause the transfer. A double bond that absorbs blue photons from passing light appears yellow. One that absorbs red photons appears cyan. The usual rules of subtractive color apply.

Just how much energy it takes to cause this transfer depends on the chemical nature and environment of the double bond. Most important are the two atoms joined by the double bond. In addition to double bonds involving a pair of carbon atoms, there are also carbon-oxygen, carbon-nitrogen, nitrogen-oxygen, and nitrogen-nitrogen double bonds. These double bonds are often colored, particularly the latter two. Groups of atoms that give rise to color in molecules are called chromophores.

However the exact color of a double bond is determined by the detailed structure of the molecule around it. Since all of the electrons in that molecule affect one another through electrostatic forces and the Pauli exclusion principle, the whole molecular structure contributes to the color of the electrons in the double bond itself. A subtle change in a molecule's structure may change its color from red to orange. That's how organic dye manufacturers construct rich pallets of colors from a small number of different chromophores (Fig. 18.2.10).

Colored molecules are wonderful if you are an artist, but you don't want extraneous ones attached to your clothes. That's where bleach comes in. Bleach attacks double bonds, destroying the chromophores in the stain molecules. The molecules may remain on the fabric but they no longer absorb visible light.

The two major classes of bleaches are chlorine bleaches and oxygen bleaches. The chlorine bleaches tend to put chlorine and oxygen atoms on the two atoms involved in a double bond. The double bond vanishes as one of its atoms binds to a chlorine atom and its other atom binds to the oxygen atom of a hydroxyl group (OH).

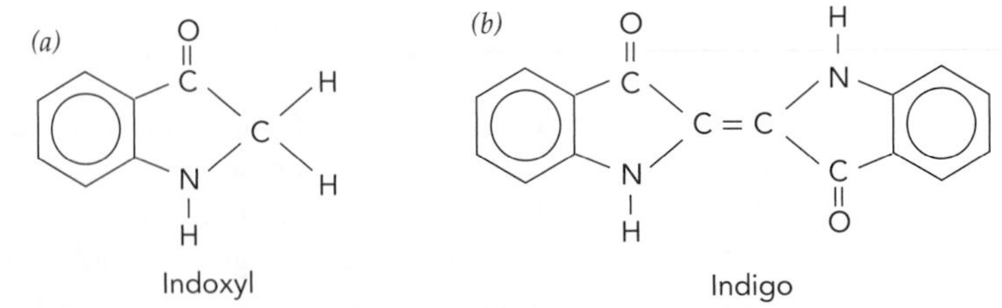

Fig. 18.2.10 (*a*) Indoxyl is a colorless, water-soluble chemical obtained from a fermented plant extract. When indoxyl is exposed to oxygen in the air, it reacts pairwise to form water-insoluble indigo (*b*), the blue dye used in blue jeans. The double-bonded carbon atoms at the center of this molecule are the chromophore and the rest of the molecule determines the precise color of the dye. Bleaches and ultraviolet light can destroy the double bond, giving blue jeans a faded look.

Unfortunately, chlorine bleaches are so effective at attacking chemicals that they damage the clothes, too. Sometimes they destroy the chromophores in dye molecules, turning colored fabric white. Other times they modify the dye molecules and change the fabric's color. But chlorine bleaches also damage natural fibers themselves, breaking up their molecules and weakening the fabric. While chlorine bleach may succeed in cutting stain molecules free from your clothing, it may also cut holes in the clothing itself.

Oxygen bleaches use hydrogen peroxide to attack double bonds. A hydrogen peroxide molecule is a water molecule with an extra oxygen atom inserted between the oxygen atom and one of the hydrogen atoms. This molecule decomposes in water and its fragments, either ions or free radicals, attack double bonds. Once again, they destroy chromophores and decolorize stains.

Hydrogen peroxide is less reactive than chlorine bleach and causes less damage to the fabric's polymer molecules, so the fabric lasts longer. It's also less damaging to commercial dye molecules than chlorine bleach. However, since hydrogen peroxide is often used to bleach hair, oxygen bleaches can obviously destroy the colors in some natural fibers.

Hydrogen peroxide itself is rather unstable and is often generated right in the washer by the decomposition of another molecule, sodium perborate. This decomposition occurs only above about 50 °C, so bleaching must be done in hot water. Some laundry detergents contain activators that help sodium perborate decompose in cooler water but it still works best in hot water.

Enzymes are biological catalysts. Your body uses a great many different enzymes to catalyze various chemical reactions that would otherwise rarely occur at body temperature or that might proceed along the wrong paths without help. Enzymes help to construct molecules, to take them apart, or to rearrange their components.

The enzymes used most often in detergents are those that degrade proteins. Protein molecules cling tightly to fabrics, are insoluble in water, and prevent detergents from penetrating to the fibers. They act as binders for other molecules, creating stains that are hard to remove. Familiar proteinaceous stains include blood, milk, eggs, and gravy.

The most effective way to remove these stains is by taking the protein molecules apart. This decomposition is done by proteolytic enzymes—enzymes that catalyze reactions between water and protein. In these reactions, protein molecules are broken up and water molecule fragments cap the severed ends. In time, proteolytic enzymes can dice up long protein molecules into their constituent parts: amino acids and short sequences of amino acids called peptides. The stain falls apart and is carried away by the water and detergents. Meat tenderizers operate in a similar fashion, using papain—a proteolytic enzyme extracted from unripe papaya—to degrade protein in meat before cooking.

However, proteolytic enzymes may have an effect on the people who use them. You certainly don't want the protein in your body decomposed while you do laundry or while you wear freshly laundered clothes. Although studies seem to indicate that the enzymes in household detergents pose no serious health threat, they're used sparingly in detergents to avoid any possible adverse effects.

Brighteners and Fabric Softeners

Not all laundry chemicals disappear down the drain when the wash is done. Brighteners and fabric softeners do their jobs by remaining on the clothes long after they leave the dryer. Brighteners affect the appearances of the clothes while fabric softeners affect their feels.

With age, white fabrics such as cotton tend to absorb more and more blue light and begin to look yellow (Fig. 18.2.11b). Cleaning and bleaching do little to reduce this effect. In fact, bleached fabric molecules tend to appear slightly yellow themselves. The old-fashioned solution to the yellowing problems was to add blueing to the wash. This blue dye absorbed red and green light, balancing the blue absorption of the fabric itself so that the fabric appeared colorless. To mask the yellowing of age, blueing darkened the whole fabric to a light gray—the amount of light reflected by the fabric was noticeably less than that striking it.

Instead of using blueing, virtually all modern laundry detergents add chemicals that optically brighten the fabric. These brighteners are actually fluorescent dyes, designed to absorb ultraviolet light and emit bluish-white visible light. Instead of absorbing red and green light to balance the white appearance of a fabric, the brighteners reintroduce the missing blue light (Fig. 18.2.11c). They work best in sunlight, which is rich in ultraviolet light. A brightener molecule absorbs a photon of ultraviolet light and re-emits it as a photon of blue light. The energy not reemerging from the molecule in the outgoing photon is converted through vibrations into internal energy in the clothes.

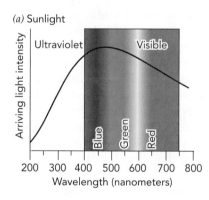

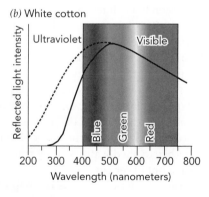

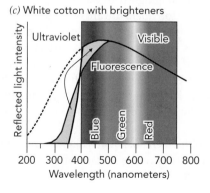

Fig. 18.2.11 When white cotton cloth is exposed to sunlight (*a*), it reflects a little less blue light than it should (*b*) and appears slightly yellow. This yellowness increases with age. But when fluorescent brighteners are added to the cotton (*c*), they convert ultraviolet light into blue light and make the cloth appear dazzlingly white.

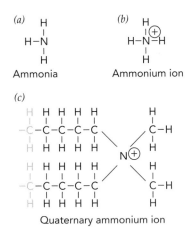

Fig. 18.2.12 (*a*) An ammonia molecule is three hydrogen atoms bound to a nitrogen atom. (*b*) An ammonium ion is four hydrogen atoms bound to a positively charged nitrogen ion. (*c*) The quaternary ammonium ion used in fabric softeners is formed by replacing the four hydrogen atoms with hydrocarbon chains.

❶ Softened fabrics were not always the most desirable. At one time, silk, cotton, and rayon were chemically treated to harden their fibers so that they would produce a rustling or swishing sound known as "scroop." This hardening increased the intra-fabric friction so that the garments would crackle as they rubbed across one another.

When clothes are washed in these fluorescent dyes, the dye molecules stick to the fabric to create brightened fabric. We can't see the ultraviolet light that the brightened fabric absorbs but we can see the bluish light that it emits. With the missing blue light restored by this fluorescence processes, the brightened fabric appears dazzlingly white. In fact, it may emit more blue light than it is exposed to, making it effectively "whiter than white." In a room illuminated only by ultraviolet light, the brighteners give clothes an eerie violet glow.

Fabric softeners are also chemicals that remain on fabrics after laundering. They are primarily cationic surfactants called quaternary ammonium compounds. These compounds are based on the positive ammonium ion, which is itself based on a positive nitrogen ion. A normal nitrogen atom has five valence electrons and must share three of these to reach the four pairs needed to complete an electronic shell. If it shares those electrons with three hydrogen atoms, it forms an ammonia molecule (Fig. 18.2.12*a*). But if it's missing an electron, the nitrogen atom must share four electrons to complete its shell and can actually bind to four hydrogen atoms. In that case, it forms a positive ammonium ion (Fig. 18.2.12*b*).

In quaternary ammonium compounds, a positive nitrogen ion forms covalent bonds with four other atoms. However, these atoms aren't necessarily hydrogen atoms. In fabric softeners, the central nitrogen ion binds to four hydrocarbon chains (Fig. 18.2.12*c*). Two of these chains are short, only one carbon atom long, while the other two chains may contain as many as 18 carbon atoms. These long chains are hydrophobic and have the same oily character as most lubricants. This oily character is what gives these compounds their fabric-softening ability.

When you apply the softener to wet fabric, its positively charged surfactant ions are drawn toward the negatively charged fibers and stick to them strongly. While anionic surfactants are repelled by wet fabric and help to clean it, cationic surfactants are attracted to wet fabric and help to soften it.

The surfactant molecules stick to the fabric with their long hydrophobic chains pointing outward. These molecules decorate every fiber in every thread of the clothing, giving them all an oily coating. The hydrocarbon chains lubricate the fabric so that each fiber slides easily within a thread and each thread slides easily within the fabric. This lubrication enhances the flexibility of the fabric and makes it feel softer and more flexible. (Softness wasn't always in vogue; see ❶.)

Fabric softeners also make fabric surfaces slightly hydrophobic, so that they dry more easily in the spin dry cycle of a washing machine. In this cycle, the clothes travel rapidly around in a circle, always accelerating toward the center of the circle and experiencing huge inward forces from the washer's metal drum. Water's inertia causes it to lag behind the accelerating clothes and it leaves the drum through perforations. By making the fabric slightly hydrophobic, the fabric softener helps the clothes to shed water as they spin, so that they don't have to spend as much time in a hot dryer later on.

Fabric softeners also raise the nap on cotton terry towels. Cotton fibers are normally hydrophilic and cling tightly to water droplets. As water droplets dry up, they shrink and pull the cotton fibers toward one another. By the time an untreated towel is dry, its fibers have been crushed together by these forces and it has little nap. But a towel that has been coated by quaternary ammonium compounds is hydrophobic enough that the water droplets can't pull its fibers together as they dry. The nap remains loose and thick, giving the towel a fluffy appearance and feel. Unfortunately, this same hydrophobic coating slightly reduces the towel's absorbency—a real problem for cotton diapers. To keep it under control, don't use too much fabric softener.

Despite their hydrophobic chains, quaternary ammonium compounds actually attract a few water molecules to the surface of the fabric. They are hygroscopic, meaning that they attract water molecules directly out of the air. Since water conducts electricity very weakly, fabric that has been treated with fabric softeners is very slightly conducting. This conductivity reduces the accumulation of static electricity on the fabric and eliminates static cling.

In a dryer, untreated clothes rub across one another and that contact transfers electric charge between the dissimilar fabrics. Large charge imbalances develop and the clothes leave the dryer clinging to one another with electrostatic forces. However, treated clothes allow the charge imbalances to dissipate through the moisture attracted by the fabric softener.

Quaternary ammonium compounds are also used in conditioners and shampoos to soften hair and reduce static electricity problems—they will coat and lubricate just about anything. They are actually bactericidal because they coat bacteria and smother them. These compounds also deactivate some of the enzymes in bacteria and upset their metabolisms. Some antiseptic throat lozenges and mouthwashes use quaternary ammonium compounds to kill germs.

Unfortunately, the positive charges of cationic quaternary ammonium compounds make them relatively incompatible with the negative charges of anionic detergents. When they're present together in the water, these two types of ions attract one another and may clump together. This clumping is avoided by keeping the two types of surfactants separate, which is why softeners are usually added during the rinse cycle, in the dryer, or in a separate conditioner when washing your hair. However, some detergent and shampoo formulators have successfully combined cationic softeners and anionic detergents.

Detergent Additives

Formulated detergents contain a number of important components that work together to clean clothes. We've already examined the anionic and nonionic detergents (surfactants), the builders (water softeners), the bleaches, and the brighteners. But there are also foam stabilizers, corrosion inhibitors, soil redeposition inhibitors, and processing agents.

Foam stabilizers are there to control bubble formation. These chemicals can either enhance or suppress foaming. Believe it or not, foam is unrelated to a detergent's ability to clean clothes. The same goes for shampoos and dishwashing detergents. However, the amount of foam a detergent produces may influence its use. If the detergent foams excessively, you may think the detergent is more powerful than it is and cut back on the amount you use. As a result, you may not use enough to clean your clothes properly. If the detergent doesn't foam much, you may think that it isn't working and buy another brand. So the detergent and shampoo manufacturers carefully control the foaminess of their products.

Air bubbles don't last long in pure water because water's surface tension causes them to tear. The final layers of water molecules on the bubble's outer and inner surfaces pull together so strongly that any tiny defect immediately initiates a rip that lets the air out of the bubble. By reducing water's surface tension, soaps and detergents remove its tendency to rip and stabilize air bubbles.

But how long each air bubble lasts depends on many features of the mixture and not on its ability to clean things. Some surfactant molecules make particularly stable and long-lasting bubbles while other molecules deliberately introduce defects that pop the bubbles. Methyl silicone polymers ("methicones") are particularly effective at weakening bubbles so that they tear and collapse. These polymers are common in antifoam additives and are even included in some antacid tablets.

Foam boosters are common in detergents and shampoos that are used by hand, where foam is regarded as a sign of effectiveness. Antifoaming agents are often used in washing and dishwashing machine detergents where you don't see the foam anyway and foam interferes with the machine's operation.

Corrosion inhibitors are important in detergent because the ions in detergent would otherwise quickly rust the steel in a washing machine. Rusting is an electrochemical reaction of the type explored in Section 18.3 on batteries. In normal rusting, the iron in steel is attacked by negatively charged hydroxyl ions. However, other negatively charged ions, including detergent ions, can also attack iron and rust it. So detergents include corrosion inhibitors. These compounds are usually sodium silicates—water-soluble glasses that are discussed in Section 17.2 on windows and glass. They form thin glassy coatings on the washer parts and inhibit rusting.

Soil redeposition inhibitors enhance the negative charge of wet fabric fibers. Some fabrics, particularly synthetic ones, don't acquire a strong negative charge in water. They need this electrostatic charge to keep the negatively charged detergent micelles from redepositing their soils on the fabric. So detergents include carboxymethyl cellulose, which attaches itself to the fibers and adds to their negative charge.

Finally, processing agents simply give the detergents the right structures in their boxes or bottles. Sodium sulfate helps to bulk up powdered detergent and make it pour easily. Sodium xylene sulfonate helps to keep all of the components of very concentrated liquid detergents in solution.

Dry Cleaning

Washing clothes in water isn't always a good idea. Fibers such as cotton, wool, silk, and rayon, are very hydrophilic and soak up water molecules like sponges. These fibers form hydrogen bonds with water molecules at various sites on their molecules and accumulate large quantities of water. This water takes up space and causes the fibers to swell. Cotton, wool, and silk fibers increase by about 1% in length and about 15% in thickness. Rayon expands even more, by 3% in length and about 25% in thickness. This swelling distorts the fabric and changes its structure. When the fabric eventually dries, it may have shrunk or wrinkled.

To avoid damage caused by this cycle of expansion and contraction, you can send your clothes to be dry-cleaned. Dry cleaning takes place in a nonpolar solvent. Since this solvent doesn't form hydrogen bonds, it's only weakly attracted to the fibers by van der Waals forces and doesn't cause them to swell. The clothes don't lose their shapes.

The solvents used in dry cleaning have evolved over the years since petroleum oils were first found to remove stains. Early dry cleaning was done with gasoline, resulting in many dramatic fires. In 1928, a less flammable solvent became available. The Stoddard solvent, named for the president of the National Institute of Dry Cleaning, W. J. Stoddard, is less volatile than gasoline because it contains larger hydrocarbon molecules. It's obtained by distilling crude oil and its vapor will not ignite in air at temperatures below 38 °C.

Nonetheless, Stoddard solvent is still dangerous during hot air drying so nonflammable nonpolar solvents have largely replaced it. The most common solvent in dry cleaning is now perchloroethylene. Its molecule consists of a pair of carbon molecules connected by a double bond and each attached to two chlorine atoms. The chlorine atoms bind so strongly to the carbon atoms that the molecule doesn't react with oxygen and forms a nonflammable liquid.

When you put clothes in either Stoddard solvent or perchloroethylene, the oily soils dissolve. These nonpolar solvents attract the oily molecules with van der Waals forces and carry them away. Chlorinated solvents clean better than hydrocarbons because they bind more strongly to oily soils. Chlorine atoms are more polarizable than hydrogen atoms and produce stronger van der Waals forces, which is why perchloroethylene doesn't boil until it is heated to 121 °C.

However, these nonpolar solvents are unable to dissolve salts and other polar soils. They are also poor at removing insoluble soils such as dust. To help in removing these other soils, dry cleaning solvents include detergents and a little water. The detergents form inside-out micelles in the nonpolar solvents, arranged with their nonpolar ends on the outside and their polar ends on the inside (Fig. 18.2.13). Each micelle surrounds a tiny droplet of water. Just as in water cleaning, detergents help to carry away substances that aren't soluble in the principal cleaning liquid.

The water in the dry cleaning mixture is carefully adjusted so that the clothes neither gain nor lose moisture during the cleaning process. In air, water molecules are continually leaving and returning to the clothing and an equilibrium is reached. At this equilibrium, the water molecules still move back and forth but the moisture in the clothing doesn't change significantly. The actual moisture level in the fabric then depends only on the relative humidity of the air, which is typically about 70% in a dry cleaning shop.

The same leaving and returning process takes place in the dry cleaning solvent. Water molecules move back and forth between the fabric and the solvent and establish an equilibrium. Like air, the dry cleaning solvent has a relative humidity and a dry cleaner tries to maintain this relative humidity at the same value as the air in the shop. That way, the fabrics don't accumulate too many water molecules and swell, nor do they lose too many water molecules and dry out. But the polar soils leave the fabrics, become trapped in the detergent micelles, and never return.

With the help of detergents, nonpolar dry cleaning solvents carry away nonpolar, polar, and insoluble soils from clothes without affecting the structure of the cloth. The dry cleaner then removes the solvent from the clothes by spinning them and drying them in hot air. Because solvents are expensive and environmentally damaging, dry cleaners collect the solvents for reuse. They do this by filtering and distilling the liquid solvents and by condensing the gaseous solvent molecules onto chilled surfaces. When this type of solvent recycling is done effectively, a dry cleaner can operate for a long time on the same supply of solvent.

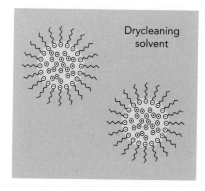

Fig. 18.2.13 Detergents form inside-out micelles in dry cleaning solvent. The polar hydrophilic ends of the molecules project inward, toward a tiny drop of water. The nonpolar hydrophobic ends project outward into the solvent. These micelles accumulate salts and other polar soil molecules.

SECTION 18.3 Batteries

Recent developments in electronics have revolutionized portable devices—you can now take radios, telephones, video recorders, and computers anywhere. These gadgets have created a need for portable electric power, most of which is supplied by batteries.

Batteries store energy in chemical form and deliver it when needed as electricity. Batteries are remarkable devices in their own right. They have no visible moving parts yet they manage to push electric currents through circuits. Batteries continue to improve from year to year, with designers always trying to increase energy capacity, reliability, and reusability while decreasing size, weight, and cost. Nonetheless, batteries remain the limiting factor in such technologies as electric vehicles and portable computers.

How a Battery Produces Electricity

A battery uses chemical potential energy to pump electrons from its positive terminal to its negative terminal. Since electrostatic forces push the electrons in the other direction, the battery must do work on the electrons as it moves them. Each time the battery transfers an electron, it uses up a small portion of its chemical potential energy. After transferring a certain number of electrons, the battery runs out of chemical potential energy and must be recharged or discarded.

But a battery sitting on the shelf stops transferring electrons long before it runs out of chemical potential energy. With each transfer, the negative terminal becomes more negatively charged and the positive terminal becomes more positively charged. The amount of separated charge on the terminals increases and so does the voltage rise across the battery—the battery must do more and more work to transfer each additional electron. Eventually, the electrostatic forces become so strong that the battery can't transfer any more electrons. The battery just sits on the shelf with negative charge on its negative terminal and positive charge on its positive terminal and it remains that way almost indefinitely.

However, when you install the battery in a flashlight and turn the flashlight on, an electric circuit connects the two terminals to one another (Fig. 18.3.1). Electrons flow from the negative terminal, through the lightbulb, to the positive terminal and the amount of separated charge on the battery's terminals decreases. The battery begins to pump electrons again. The battery pumps electrons onto the negative terminal and the flashlight returns those electrons to the positive terminal. The electrons flow around and around this circuit, receiving energy from the battery and delivering that energy to the lightbulb, until the battery's chemical potential energy is exhausted or you turn the flashlight off.

But how does a battery use its chemical potential energy to pump electrons from its positive terminal to its negative terminal? Many batteries are based on electron transfers from atoms of one element to those of another. These different atoms have different affinities for their outermost or valence electrons and many of the transfers result in releases of energy. When an atom that binds its valence electrons relatively strongly is missing some of them, it may extract electrons from another atom that binds them relatively weakly. Overall, the electrons move from one atom to the other and some potential energy is released. This process is the principal source of a battery's energy.

You can choose which atoms to use in a battery by examining their properties. These properties depend, in an orderly fashion, on the numbers of protons and electrons the atoms contain. One way to see this order is to arrange the atoms in the periodic table of elements (Fig. 18.3.2), as we did in Section 14.2. In this

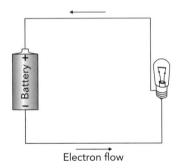

Fig. 18.3.1 A simple circuit in which a battery provides power for a lightbulb. The battery pumps electrons from its positive terminal to its negative terminal, giving each electron a certain amount of energy. These electrons flow through the lightbulb on their way back to the battery's positive terminal. The electrons give up their energies in the bulb, which then produces light.

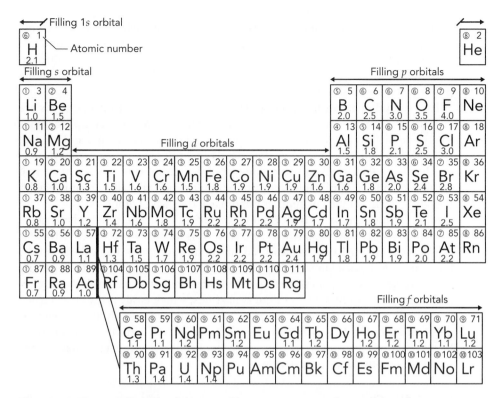

Fig. 18.3.2 The periodic table of elements. Elements are arranged according to their atomic numbers (indicated in the upper right corner). As the electron count increases, various electronic shells are filled and completed. The table is structured by these shell closings. A major closing occurs on the right-hand column, so that elements in the left-hand column are starting new shells. The number shown below each elemental symbol is that atom's Pauling electronegativity.

table, the atoms are arranged in horizontal rows according to their atomic numbers—the numbers of protons they contain. Since atoms are normally electrically neutral, their atomic numbers also indicate the numbers of electrons they contain. The atom with atomic number 1 is hydrogen (H), with atomic number 2 is helium (He), and so on.

As we saw in Section 14.2, the peculiar structure of the table comes from the way in which electrons fill the atomic orbitals. Because of the Pauli exclusion principle, all electrons of a particular spin—either spin-up or spin-down—must be in different orbitals. The electrons fill the orbitals from the lowest energy orbitals on up until the atom has the right number of electrons. The electrons in the last few orbitals filled determine most of the chemical properties of the atom, particularly the atom's behavior in a battery. This filling process is quite complicated, but there are a few simple observations we can make.

Some atoms have just enough electrons to completely fill a major electronic shell. These atoms are extremely stable, unwilling to give up any electrons and uninterested in any additional electrons. These atoms are the noble gases, helium (He), neon (Ne), argon (Ar), krypton (Kr), xenon (Xe), and radon (Rn), found in the rightmost vertical column of Fig. 18.3.2.

Some atoms have just one or two electrons more than are required to fill a major electronic shell and are relatively willing to give those electrons up. These atoms are the alkali metals, lithium (Li), sodium (Na), potassium (K), rubidium (Rb), cesium (Cs), and francium (Fr), and the alkaline earths, beryllium (Be), magnesium (Mg), calcium (Ca), strontium (Sr), barium (Ba), and radium (Ra), found in the leftmost and second to leftmost vertical columns of Fig. 18.3.2.

Still other atoms have almost enough electrons to complete a major electronic shell and are relatively aggressive at attracting more. These atoms are found just to the left of the noble gases on the right side of Fig. 18.3.2. They include nitrogen (N), oxygen (O), sulfur (S), and the halogens, fluorine (F), chlorine (Cl), bromine (B), iodine (I), and astatine (At).

The remaining atoms fall in between. While their major electronic shells are only partly complete, they tend to exchange electrons in order to complete minor electronic shells. The atoms in the long horizontal stretch between scandium (Sc) and zinc (Zn) are called the transition metals and differ from one another by how much of one minor shell they have completed. The atoms shown at the bottom of Fig 18.3.2, both the lanthanide series and actinide series, are called the rare earths; they differ by how much of another minor shell they have completed.

Many of these atoms are important for batteries. The main issue for batteries is just how strongly the atoms attract electrons. This tendency to attract electrons is called electronegativity and is measured in various ways. One scheme developed by the American chemist Linus Pauling (1901–1994) is called Pauling electronegativity. The more strongly an atom attracts electrons, the higher its Pauling electronegativity. Values range from 0.7 for cesium (Cs) atoms, which easily gives up electrons, to 4.0 for fluorine (F) atoms, which attracts electrons aggressively. Pauling electronegativities for the other atoms are shown in Fig. 18.3.2. Batteries generally work by transferring electrons from atoms with low Pauling electronegativities to ones with high Pauling electronegativities.

Daniell's Cell

Unfortunately, real batteries aren't simple. For one thing, the atoms in a battery aren't isolated objects—they are usually contained in solids and liquids, a situation that affects the transfers of charges between them. Furthermore, since a battery uses the energy released during electron transfers to pump electrons from its positive terminal to its negative terminal, the battery must carefully control electron transfers. Overall, a battery involves a number of interrelated processes. One way to see how all of these processes work together is to examine the operation of a *relatively* simple battery.

Daniell's cell (Fig. 18.3.3) is a simple battery that was invented in 1836 by English chemist John Frederick Daniell (1790–1845). Cells are the basic building blocks of batteries. Daniell's cell isn't a practical battery but it's one of the simplest cells to understand. It uses two metallic elements, copper and zinc. Because copper has a higher Pauling electronegativity than zinc, the cell releases energy by transferring electrons from zinc atoms to copper atoms.

Daniell's cell consists of two metal strips, each in its own container of salt solution. One of the metal strips is zinc and it sits in a water solution of zinc sulfate. This salt solution contains equal numbers of positively charged zinc ions (Zn^{++}) and negatively charged sulfate ions (SO_4^{--}), making it electrically neutral. The other metal strip is copper and it sits in a water solution of copper sulfate. This second salt solution contains equal numbers of copper ions (Cu^{++}) and sulfate ions (SO_4^{--}) and is also neutral.

When the two containers are isolated from one another, their contents don't appear to change. Their electrodes neither grow nor shrink and the solutions remain the same. But this constancy doesn't mean that nothing is happening in the containers. In fact, zinc and copper ions are constantly departing and arriving at the surfaces of the electrodes.

Zinc and copper ions are soluble because they polarize the water molecules, which then carry them around in solution. In contrast, zinc and copper atoms don't polarize the water molecules and are insoluble. Thus, for an atom to leave an electrode, it must give up two electrons to the electrode and enter the water as

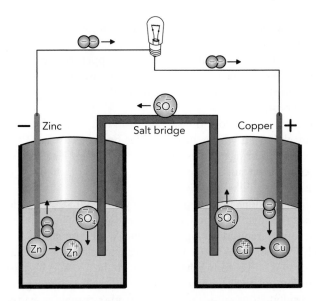

Fig. 18.3.3 Daniell's cell is powered by a transfer of electrons from zinc atoms to copper ions. A copper ion (Cu^{++}) picks up two electrons from the copper electrode and leaves the solution in the right container as a copper atom (Cu). To keep the solution electrically balanced, a sulfate ion (SO_4^{--}) flows over the salt bridge to the left container. There the sulfate ion allows a zinc atom (Zn) to give up two electrons to the zinc electrode and enter the solution as a zinc ion (Zn^{++}). The two electrons flow through wires and the lightbulb to the copper electrode.

an ion. Similarly, for an ion to return to an electrode, it must pick up two electrons from the electrode and leave the water as an atom.

In an isolated container, this exchange quickly reaches a chemical equilibrium, with equal numbers of ions leaving and returning to the electrode each second. If too many ions leave the electrode, the solution becomes positively charged and pushes ions back toward the electrode. If too few ions leave the electrode, the solution becomes negatively charged and pulls ions off the electrode. These balancing effects stabilize the chemical equilibrium. The containers remain electrically neutral.

But the two containers in Fig. 18.3.3 aren't isolated—a salt bridge connects them. This piece of damp cloth serves as a pathway for ions, allowing them to flow from one container to the other in response to electric fields. If the containers somehow develop a charge separation, ions will migrate over the bridge from one container to the other to balance the charge. Because it conducts an electric current through the movement of ions, the salt solution in the bridge is an electrolyte. So are the solutions in the two containers.

The electrodes, solutions, and salt bridge form an electrochemical cell, a Daniell's cell. This cell springs to life when wires and a lightbulb connect its zinc and copper electrodes. Suddenly an electric current begins to flow through the circuit that has formed. This circuit includes the cell, the wires, and the lightbulb. The cell pumps electrons from its copper electrode to its zinc electrode and these electrons return to the copper electrode by way of the lightbulb. Chemical potential energy in the cell becomes thermal energy and light in the lightbulb.

To see how the cell actually works, let's make our way around the circuit and watch the movements of atoms, ions, and electrons. We'll start at the copper electrode. Its surface is a busy place, with copper ions coming and going all the time. While the copper ions in the salt solution have relatively strong affinities for electrons, they are prevented from becoming atoms and returning to the electrode, on the average, because that would make the electrode positively charged and leave the salt solution negatively charged.

But suppose that some electrons suddenly arrive at the copper electrode through the wire attached to it. This delivery of negative charge upsets the chemical equilibrium in that container and allows some extra copper ions to return to the electrode. These ions pick up electrons and attach themselves to the electrode as atoms. The electrode now has more copper atoms than before and the salt solution has fewer copper ions.

The loss of copper ions from the salt solution leaves it with too many sulfate ions to be neutral. Here is where the salt bridge becomes useful. The extra sulfate ions migrate over this bridge and escape from the copper electrode's container. A proper balance between sulfate ions and copper ions is reached and the container returns to chemical equilibrium.

But the sulfate ions don't just disappear—they move into the zinc electrode's container and upset the chemical equilibrium there. The zinc electrode's surface is also busy with zinc ions coming and going all the time. While zinc atoms on the electrode have relatively weak affinities for two of their valence electrons, they are prevented from becoming ions and entering the solution, on the average, because that would make the solution positively charged and leave the electrode negatively charged.

When extra sulfate ions enter the zinc electrode's container via the salt bridge, they alter the balance of sulfate ions and zinc ions in the solution. Now extra zinc atoms can toss off their electrons and enter the solution as ions. The salt solution in the container quickly reaches the proper balance between sulfate ions and zinc ions. The electrode then has fewer zinc atoms than before and the salt solution has more zinc ions. But this change leaves the zinc electrode negatively charged. Electrons flow off the electrode through the wire attached to it and the container returns to chemical equilibrium.

The electrons that leave the zinc electrode pass through the lightbulb and soon arrive at the copper electrode. There they start the whole process over again. Thus our initial supposition of electrons suddenly arriving at the copper electrode is reasonable, because that's where the cell actually sends them. Overall, each pair of electrons that arrives at the copper electrode causes (1) one copper ion to leave the copper sulfate solution as an atom, (2) one sulfate ion to migrate over the salt bridge, (3) one zinc atom to enter the zinc sulfate solution as an ion, and (4) a pair of electrons to leave the zinc electrode.

It seems that this complicated process has simply absorbed a pair of electrons through one wire and released another pair through a second wire. But the pair of electrons that leaves the zinc electrode has more energy than the pair that arrived at the copper electrode. The copper ion pulled the two arriving electrons toward it, doing work on them. The zinc atom pushed the two leaving electrons away from it, doing work on them, too. All of this work is conveyed via electric fields to the two electrons as they leave the zinc electrode so that they head out with extra energy. They deliver this energy to the lightbulb, causing it to emit light, and then return to the copper electrode to participate in the next action.

You might have noticed that the electrons leaving the cell's zinc electrode don't seem to be the same ones that arrived at the cell's copper electrode. However, electrons are indistinguishable particles so it really isn't possible to tell which one is which. In effect, the cell pumps electrons from the copper electrode to the zinc electrode. Since the zinc electrode becomes negatively charged, it becomes the negative terminal of a battery containing Daniell's cell. The copper electrode becomes the positive terminal.

The cell can't keep pumping electrons forever. Pumping each pair of electrons decreases the number of copper ions in the copper sulfate solution by one and decreases the number of zinc atoms in the zinc electrode by one. When either the copper ions or zinc atoms run out, the cell stops pumping. The cell is then completely dead.

But even before the last ions or atoms are used up, the cell will be relatively ineffective. The problem is that the amount of energy the cell gives to each electron, the cell's voltage, depends on the cell's condition. Each cell has a natural voltage but its condition can push its actual voltage either above or below that natural voltage. A fresh cell tends to exceed the natural voltage while a heavily used cell tends to fall below it.

The natural voltage is determined by the total amount of chemical potential energy released in transferring each electron from the positive terminal to the negative terminal. In Daniell's cell, chemical potential energy is released when one copper ion picks up two electrons and leaves its solution and when one zinc atom loses two electrons and enters its solution. This total energy is shared by the two electrons that are pumped in the processes, giving an overall energy-per-charge—a voltage—of 1.10 V. Daniell's cell has a natural voltage of 1.10 V.

When Daniell's cell is fresh, with lots of copper ions in the copper sulfate solution and relatively few zinc ions in the zinc sulfate solution, its voltage is roughly 1.20 V. With so many copper ions available to leave the concentrated copper sulfate solution and so many zinc ions available to enter the dilute zinc sulfate solution, random thermal motions in the fresh cell assist its operation and contribute an extra 0.10 V to the cell's voltage. As the fresh cell pumps electrons, it creates disorder and entropy at a particularly rapid rate and thermodynamic effects assist this entropy production by increasing the cell's voltage. The extra energy comes from thermal energy.

But a well-used cell has already accumulated so much disorder and entropy that thermodynamic effects no longer help it function. They even begin to hinder its operation and convert chemical potential energy into thermal energy.

The voltage of Daniell's cell diminishes with use. The concentration of copper ions in the copper sulfate solution decreases and the copper electrode has more trouble converting these scarce ions into copper atoms. The concentration of zinc ions in the zinc sulfate solution increases and the zinc electrode has more trouble sending zinc atoms into the solution as ions. When the concentrations of copper ions and zinc ions in their respective solutions are equal, the voltage of Daniell's cell reaches its natural voltage of 1.10 V. As the cell's chemical potential energy becomes more severely depleted, its voltage diminishes to 1.00 V, 0.90 V, or even less and the cell eventually becomes useless.

Lead-Acid Batteries

Although Daniell's cell is an interesting and simple cell, it isn't a practical battery. It's unable to pump large numbers of electrons each second, or equivalently, to handle large electric currents. If you try to send too many electrons through any battery, much of the battery's chemical potential energy will be consumed in transporting the current through the battery itself and less will be given to the electrons as they leave.

In effect, each battery has an internal electrical resistance. If you try to send too much current through a battery, this internal resistance will convert the battery's own chemical potential energy into thermal energy. The battery will get warm and electrons will leave the negative electrode with relatively little energy. You may have noticed this effect when trying to operate a flashlight with old or damaged batteries. Their internal resistances are large, wasting most of their energy, and the lightbulb glows only dimly.

Starting an automobile requires an enormous electric current, far more than Daniell's cell can provide. That's why automobiles use lead-acid batteries. The chemical cells in these batteries resemble Daniell's cell in some respects, but they use different electrodes and only a single container of liquid (Fig. 18.3.4).

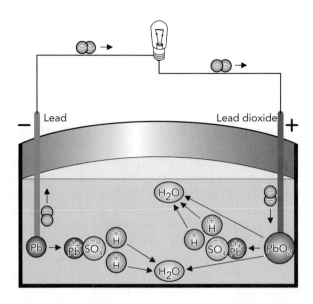

Fig. 18.3.4 A lead-acid cell is powered by a complicated electrochemical reaction between lead, lead dioxide, and sulfuric acid, which creates lead sulfate and water. A lead dioxide molecule (PbO_2) picks up two electrons from the lead dioxide electrode and reacts with four hydrogen ions (H^+) in the solution to form a lead ion (Pb^{++}) and two water molecules (H_2O). The number of positive charges in the solution decreases by 2, so a lead atom (Pb) can give up two electrons to the lead electrode to become a lead ion. The two electrons flow through the wires and lightbulb to the lead dioxide electrode.

The solution in a lead-acid cell contains sulfuric acid (H_2SO_4) in water. In water, sulfuric acid molecules dissociate into positively charged hydrogen ions (H^+) and negatively charged sulfate ions (SO_4^{--}). Since hydrogen ions carry only one positive charge and sulfate ions carry two negative charges, the electrically neutral solution contains twice as many hydrogen ions as sulfate ions.

The cell's negative electrode is a spongy form of lead metal, housed in a nonreactive lead-alloy lattice. The cell's positive electrode is lead dioxide, also housed in a protective lead-alloy lattice. The electrochemical process that pumps electrons from the lead dioxide electrode to the lead electrode is more complicated than in Daniell's cell, but it's still fairly simple to understand.

Let's start at the lead dioxide electrode. As before, we will assume that two electrons arrive at the lead dioxide electrode. They are picked up by a lead dioxide molecule, which reacts with four hydrogen ions and a sulfate ion to produce a lead sulfate molecule and two water molecules. The lead sulfate molecule consists of a positively charged lead ion and a negatively charged sulfate ion, but it's not soluble in water—the two ions bind together so strongly that water molecules can't separate them and carry them about in solution. The lead sulfate clings to the lead dioxide electrode as a solid.

Overall, the solution in the cell has lost four hydrogen ions and one sulfate ion. Because it has lost more positively charged ions than negatively charged ions, it can return to a proper balance by getting rid of one sulfate ion. So a lead atom on the lead electrode gives up two electrons and combines with a sulfate ion to form lead sulfate. This lead sulfate molecule is also insoluble and hangs onto the lead electrode. The solution is back to a proper balance, albeit with considerably fewer ions than before.

The two electrons that were released by the lead atom as it combined with a sulfate ion travel out through the wire, through the lightbulb, and return to the lead dioxide electrode where they initiate the process again. Overall, each pair of electrons that arrives at the lead dioxide electrode causes (1) one lead dioxide molecule to react with four hydrogen ions and a sulfate ion to form a solid lead sulfate molecule and two water molecules, (2) one lead atom to become a lead ion and react with a sulfate ion to form a solid lead sulfate molecule, and (3) a pair of electrons to leave the lead electrode.

The pair of electrons leaving the lead electrode have considerably more energy than the pair that arrived at the lead dioxide electrode. Lead sulfate is a very stable chemical, with little remaining chemical potential energy. Lots of energy is released in the electrochemical reactions that form lead sulfate and the cell gives this energy to the electrons it pumps. Its natural voltage is 2.04 V.

Like all batteries, the lead acid cell's voltage depends on its freshness. As the hydrogen and sulfate ions in its solution are used up, its voltage decreases. Because of the cell's structure, thermodynamic effects can't raise a fresh lead-acid cell's voltage above its natural voltage but they can decrease the voltage of a depleted cell. A fresh lead-acid cell has a voltage of about 2.03 V while a substantially discharged one has a voltage of about 2.00 V.

The solution in a lead-acid cell also becomes less dense as the cell's stored energy decreases. Sulfuric acid, the source of the hydrogen and sulfate ions, is more dense than water. You can use a density measuring tool called a hydrometer to measure the concentration of hydrogen and sulfate ions in the cell's solution and thus determine how fresh the cell is.

The lead-acid battery in an automobile has three other interesting features. First, it is a 12-V battery, not a 2-V one. Second, it seems to last for years before going dead. Third, it is able to deliver an enormous amount of current when you use it to start a car. What is going on?

A 12-V automobile battery actually contains 6 separate lead-acid cells connected in a series (Fig. 18.3.5). The electrons are pumped through each cell, one after the next, so that by the time they leave the sixth cell, they have considerably more energy than when they arrived at the first cell. Overall, the voltage rise from the battery's negative terminal to its positive terminal is 12 V. A few cars and motorcycles use 6-V batteries that contain only 3 cells rather than 6.

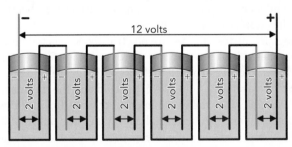

Fig. 18.3.5 A car battery produces a 12-V rise by sending current through a series of 6 separate lead-acid cells. Each cell contributes 2 V to the overall voltage.

The reason why an automobile battery lasts so long is that the car routinely recharges it. The car's electric system involves an electric generator that converts mechanical work from the engine into electric power. Some of this power is used to increase the battery's chemical potential energy. If everything is working well, an automobile's battery should be fully charged with energy almost all the time.

Recharging involves running the entire electrochemical process backward. Everything happens in reverse, driven by an outside source of electric power. Two electrons are pushed onto the lead electrode, where they bind with the lead ion in a lead sulfate molecule and convert it to a lead atom. The remaining sulfate ion then enters the solution in the cell. Two electrons are removed from the lead dioxide electrode, causing a lead sulfate molecule to react with water to produce a lead dioxide molecule, four hydrogen ions, and a sulfate ion. These ions also enter the solution in the cell.

As the recharging continues, the lead sulfate on the two electrodes is gradually converted into lead and lead dioxide. The concentration of hydrogen and sulfate ions in the solution slowly increases. Eventually the cell is back to its original condition, with concentrated sulfuric acid in the solution and very little lead sulfate on its electrodes.

Breaking up the stable lead sulfate molecules takes energy, energy supplied by the outside electric power source. Most of the energy supplied by that power source goes into returning the cell's chemicals to their original conditions. But a small amount of energy is lost to thermal energy. By acting to increase the randomness and entropy of the universe, thermodynamic effects ensure that some of the recharging energy is wasted. The faster you try to recharge the cell, the more severe the thermodynamic effects and the more energy is wasted. Overall, only about 75% to 85% of the electric power used to charge a lead-acid battery re-emerges from the battery during use.

What makes the lead-acid cell so easy to recharge is the solid nature of its electrodes—the lead atoms never leave. Because the electrodes retain their forms, the recharging process doesn't have to rebuild them. It just has to pull the sulfate ions out of the electrodes and put them back into the solution. In some other kinds of batteries, one of the electrodes actually disappears into solution as the battery discharges. Such batteries can't be recharged because the electrodes and their structures are gone and forgotten.

What allows a lead-acid battery to deliver the enormous currents needed to start a car is the vast amount of surface area in each cell's electrodes. The spongy lead and lead dioxide contained in the electrodes aren't smooth. They have lots of nooks and crannies so that they provide plenty of surface area on which electrochemical reactions can occur. Moreover, in a real car battery each cell actually contains a dozen or more electrode plates. These plates are arranged as a multilayered sandwich, with a negative (lead) plate next to a positive (lead dioxide) plate next to a negative plate and so on. A porous insulating material separates these individual plates so that they don't touch one another.

All of the negative plates in a cell are electrically connected so that they work together in parallel. All of the positive plates are, too. Overall, the cell has so much electrode surface area exposed to the sulfuric acid solution that it can pump enormous currents, even when the battery is cold and the thermal motions of its ions are diminished. Its internal resistance is extremely low and it wastes little of its chemical potential energy as thermal energy. A good automobile battery can handle more than 400 A of current while maintaining a 12-V voltage rise from its negative terminal to its positive terminal. The battery is then delivering 4800 W (12 V × 400 A) to the starter motor. No wonder the engine leaps into motion!

Of course, lead-acid batteries don't last forever. The charging and discharging cycles slowly damage the electrodes, which may eventually touch one another or fall to the bottom of the container. There is also the problem of electrolyzing water molecules in the solution, creating hydrogen and oxygen gases and depleting the liquid in the battery. This electrolysis usually occurs when charging the battery. Because of electrolysis, most lead-acid batteries need additional water periodically. "Maintenance-free" batteries include calcium in their electrodes. The calcium significantly reduces the electrolysis during charging and slows the loss of water.

Carbon-Zinc and Alkaline Batteries

Liquid-filled batteries, however, are impractical in many applications. Instead, most common household batteries are based on "dry cells." These cells may still contain water, but it's as a paste rather than a liquid. The most common dry cells, and the ones we'll examine, are carbon-zinc cells, alkaline cells, nickel-cadmium, and lithium cells.

There are actually two different types of carbon-zinc cells—one that contains ammonium chloride and one that contains zinc chloride. For many years, high-purity chemicals were expensive and commercial carbon-zinc cells employed the ammonium chloride system. This system is so tolerant of impurities that ore-grade chemicals right out of mines could be used to build batteries.

But these low-quality batteries didn't store much energy for their sizes and weights. They also tended to leak liquid electrolytes when their chemicals were used up, frequently damaging equipment. In recent years, purity control has improved so that most "heavy duty" batteries now use the zinc chloride system. This system packs more stored energy into the same size and weight as the ammonium chloride one and it's much less susceptible to leaking. However, higher purity chemicals are required to keep the electrochemical reactions proceeding properly. It's this zinc chloride system that we'll consider.

In the zinc chloride carbon-zinc cell (Fig. 18.3.6), a cylindrical carbon electrode is surrounded first by a layer of manganese dioxide (MnO_2) paste, then by a layer of zinc chloride ($ZnCl_2$) paste, and then by a cup-shaped zinc electrode. The pastes contain water, so the zinc chloride dissociates into zinc ions (Zn^{++}) and chlorine ions (Cl^-).

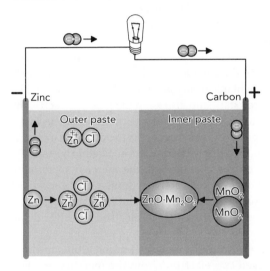

Fig. 18.3.6 In a zinc chloride carbon-zinc battery, two manganese dioxide molecules (MnO_2) pick up two electrons from the carbon electrode and react with a zinc ions (Zn^{++}) to form a molecule of zinc/manganese oxide ($ZnO{\cdot}Mn_2O_3$). The number of positive charges in the pastes decreases by 2, so a zinc atom (Zn) can give up two electrons to the zinc electrode to become a zinc ion (Zn^{++}). The two electrons flow through the wires and lightbulb to the carbon electrode.

The cell derives its energy from manganese dioxide's tendency to get rid of some of its oxygen atoms and in zinc's tendency to become ions. There are actually a number of different reactions that occur in the battery, but all of them pump electrons from the carbon electrode to the zinc electrode. We'll look at only one of the simpler reactions.

In this reaction, two manganese dioxide molecules (MnO_2) in the cell's inner paste pick up two electrons from the chemically inert carbon electrode. These molecules then react with a zinc ion (Zn^{++}) to form a complicated zinc/manganese oxide molecule ($ZnO{\cdot}Mn_2O_3$). With the loss of two positive charges from the pastes, it becomes possible for a zinc atom (Zn) to give up two electrons and become a zinc ion (Zn^{++}), dissolved in the pastes. The two electrons travel through the wires and lightbulb to the carbon electrode, where they cause the process to repeat.

There is also another important reaction, not shown in Fig 18.3.6, in which the zinc and zinc chloride react with water to produce solid zinc oxychloride [$ZnCl_2{\cdot}4Zn(OH)_2{\cdot}H_2O$]. This latter reaction is important because it uses up the water in the cell. When the cell finally runs out of chemical potential energy, it also runs out of water and is unlikely to leak in your flashlight.

The natural voltage of a carbon-zinc cell is about 1.5 V but it has a substantial internal resistance. It can't pump many electrons per second, particularly when it is cold, and thus is not very efficient in high-current applications. Furthermore, it's not rechargeable. The zinc electrode slowly goes into solution in the pastes and can't be rebuilt by running current backward through the cell. Still, carbon-zinc batteries have been around for a long time and continue to be the cheapest batteries for many situations. They're often included with consumer electronic devices that don't need much power, such as television remotes.

Alkaline batteries employ a somewhat more sophisticated electrochemical cell. Alkaline batteries look the same as carbon zinc batteries on the outside, but their electrochemistry is somewhat different (Fig. 18.3.7). To begin with, an alkaline battery is inside out, compared to a carbon-zinc battery. The alkaline battery's carbon and manganese dioxide positive electrode is on the outside and a powdered zinc electrode is at the center of the cell. The other main difference is that only one paste is used, consisting of potassium hydroxide (KOH) and water. The potassium hydroxide dissociates into potassium ions (K^+) and hydroxyl ions (OH^-). The large abundance of hydroxyl ions in the paste makes it alkaline.

The powdered zinc electrode at the center of the battery provides a large amount of surface area and increases the number of electrons this cell can pump each second. Since the alkaline cell has a low internal resistance, it's very efficient at converting its chemical potential energy into usable electric energy. Like the zinc chloride carbon-zinc battery, the alkaline battery uses up its water in a secondary reaction and rarely leaks as it ages.

Unfortunately, alkaline cells are very sensitive to contamination. The powdered zinc electrodes can corrode easily and the chemicals in the cell must be extremely pure. (For another example of electrochemical corrosion, see ❶.) One way to make the zinc electrode more corrosion-resistant and also keep the powder electrically connected is to alloy it with a few tenths of a percent of mercury metal. The resulting alloy powder works better than pure zinc but mercury is toxic and a dangerous pollutant. For environmental reasons, it is now rarely used in alkaline batteries.

There are two special variations on the alkaline cell: the mercury oxide/zinc cell and silver oxide/zinc cell. Both of these cells are very similar to the standard alkaline cell shown in Fig. 18.3.7 except that the carbon and manganese dioxide (MnO_2) electrode is replaced by a mercuric oxide (HgO) electrode or silver oxide (AgO) electrode.

Because oxygen has a very high electronegativity, the oxygen atoms in these oxide molecules have all but removed two electrons from the mercury or silver atom. However, mercury and silver also have relatively high electronegativities and seek out electrons. A mercuric oxide or silver oxide molecule can accept two electrons from the positive terminal of the battery and transfer its oxygen atom to a zinc ion (Zn^{++}) to form zinc oxide (ZnO) and a mercury (Hg) or silver atom (Ag). The missing zinc ion is replaced when a zinc atom gives up two electrons to the zinc electrode.

The electrochemical reactions in mercury and silver oxide batteries are simple and convert solids into solids. The concentrations of ions in the cells' electrolytes don't change so thermodynamic effects don't alter the batteries' voltages with use. A mercury battery's natural voltage is 1.35 V while that of a silver oxide battery is 1.50 V. In contrast to carbon-zinc and alkaline batteries, which lose their voltage with age, mercury and silver oxide batteries maintain their natural voltages until they are essentially exhausted. This steady voltage makes them well suited to use in watches, hearing aids, and cardiac pacemakers.

While all of these dry cells have voltages of between 1.35 and 1.50 V, many of the batteries that are based on them have higher voltages. These high-voltage commercial batteries simply contain several individual cells connected in series.

❶ Water drops rust iron through an electrochemical reaction with the air. Gaseous oxygen molecules at the edge of a water drop pick up electrons from the iron and react with water to form hydroxyl ions (OH^-). These ions in the water cause iron atoms near the center of the drop to give up two electrons and enter the drop as iron ions (Fe^{++}). Iron hydroxide molecules form and settle on the surface as rust. The electrons flow through the iron to the edge of the drop, where they cause the process to repeat.

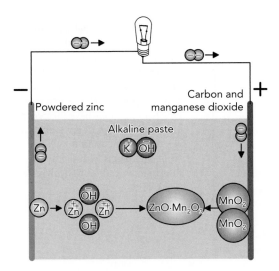

Fig. 18.3.7 The electrochemical reactions in an alkaline battery are very similar to those in a zinc chloride carbon-zinc battery, except that the positive ions in the electrolyte paste are potassium ions (K^+), accompanied by hydroxyl ions (OH^-). The powdered zinc electrode, normally located at the center of the battery, provides the large surface area needed for high-current applications.

The common 9-V batteries used in many electronic devices actually contain six tiny carbon-zinc or alkaline cells. Since these cells are small, they can't transfer as many electrons before running out of chemical potential energy. However, since all of the cells work together, they give each electron six times the energy of an individual cell. So these multicell batteries pump fewer electrons while giving each electron more energy.

Other Batteries

Two other important batteries are nickel-cadmium rechargeable batteries and lithium batteries. Nickel-cadmium batteries are based on the nickel atom's flexibility in forming bonds with other atoms. It can form stable molecules when sharing either two or three of its electrons with other atoms. This flexibility is also found in a number of other atoms, including the manganese atoms that appeared in the carbon-zinc and alkaline cells.

A nickel atom can share three of its valence electrons by joining with an oxygen atom (O) and a hydroxyl radical (OH). It then forms trivalent nickel hydroxide (NiOOH). Here the nickel atom shares two electrons with the oxygen atom and one electron with the hydroxyl radical (OH). A nickel atom can also share two of its valence electrons by joining with two hydroxyl radicals (OH). It then forms divalent nickel hydroxide [$Ni(OH)_2$]. The difference between these two nickel hydroxides is one hydrogen atom.

In a nickel-cadmium battery (Fig. 18.3.8), solid trivalent nickel hydroxide (NiOOH) picks up an electron from the battery's positive terminal and reacts with a water molecule (H_2O) to form solid divalent nickel hydroxide [$Ni(OH)_2$] and a hydroxyl ion (OH^-). When two of these reactions occur, they form two hydroxyl ions. The presence of two extra negative charges in the alkaline electrolyte causes a cadmium atom (Cd) to give up two electrons to the battery's negative terminal and form solid cadmium hydroxide [$Cd(OH)_2$]. As usual, the two electrons flow out through the wires and lightbulb, soon arriving at the positive nickel hydroxide electrode.

Oxygen atoms and hydroxyl radicals don't share electrons fairly with nickel or cadmium atoms. They pull the electrons so hard that they leave the nickel and cadmium atoms nearly ionic. But nickel has a higher electronegativity than cadmium. The nickel-cadmium battery's energy comes from a transfer of electrons from cadmium atoms to nickel atoms. As the battery pumps two electrons from its positive terminal to its negative terminal, a cadmium atom gives up two electrons and two nickel atoms get back one electron each. This electrochemical reaction releases enough chemical potential energy to give the cell a natural voltage of about 1.2 V.

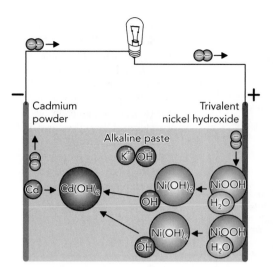

Fig. 18.3.8 In a nickel-cadmium battery, trivalent nickel hydroxide (NiOOH) reacts with water (H_2O) and an electron to form divalent nickel hydroxide [$Ni(OH)_2$] and a hydroxyl ion (OH^-). When two new hydroxyl ions have formed, a cadmium atom (Cd) gives up two electrons to the cadmium electrode and reacts with the hydroxyl ions to form cadmium hydroxide [$Cd(OH)_2$].

What makes nickel-cadmium batteries interesting is that they can be recharged many times. If you push electrons onto the cadmium electrode and pull them off the trivalent nickel hydroxide electrode, you can make the electrochemical reactions run backward. The electrodes both remain solid during charging and discharging. As you use the battery, its positive electrode goes from trivalent nickel hydroxide to divalent nickel hydroxide and its negative electrode goes from cadmium metal to cadmium hydroxide. As you charge the battery, the reverse occurs. Since the electrodes remain solid, they keep their shapes and the battery can be charged and discharged many times before failing.

To give nickel-cadmium batteries a low internal resistance and allow them to pump large currents, they are often built in a jelly-roll design. A long strip of insulating material is covered first with a thin layer of trivalent nickel hydroxide powder (the positive electrode), then with a thin porous separator soaked in a potassium hydroxide electrolyte, and then with a thin layer of cadmium metal powder (the negative electrode). This sandwich is rolled up around a cylindrical core to form a much larger cylinder. The cylinder is inserted into the battery housing, along with wires that attach the two electrodes to the battery's terminals and valves to prevent gases created by the cell during charging from bursting the sides of the battery.

However, nickel-cadmium batteries aren't perfect. They produce only about 1.2 V, rather than the 1.5 V produced by carbon-zinc and alkaline batteries and some equipment doesn't operate well on this reduced voltage. Moreover, nickel-cadmium batteries work best when their electrodes are finely divided powders. Unfortunately, these powders sometimes change their characteristics with use. If you only partially discharge a nickel-cadmium battery and then recharge it, you can cause some of the cadmium metal/cadmium hydroxide powder particles to grow in size. If you repeat this partial discharge cycle several times, some of the particles can become quite large.

These large particles are associated with a phenomenon called "memory," where the battery suddenly loses electric efficiency when it's discharged past the point where it has usually been recharged. At that point, all of the small cadmium particles have been used up and only the less efficient large particles are left. Memory can usually be remedied by fully discharging the battery before recharging it. Presumably this deep cycling breaks up the large particles and improves the battery's overall efficiency.

The other drawback to nickel-cadmium batteries is cadmium's toxicity. Nickel-cadmium batteries should be returned for recycling rather than being discarded. In recent years, manufacturers have begun using materials other than cadmium for the negative electrodes in their rechargeable nickel batteries. Those nickel-metal-hydride batteries have less memory problems and are more ecological.

Lithium batteries derive their energy from lithium metal's extremely low electronegativity. It gives up an electron so easily that it reacts violently with many chemicals. It reacts strongly with water so lithium batteries use various organic solvents and lithium salts rather than water-based electrolytes.

The structure of a lithium battery is similar to those we've already examined. The negative electrode is lithium metal and the positive electrode is one of a number of salts or plastics that undergo electrochemical reactions with lithium atoms. Electrons cause those salts or plastics to release negatively charged ions that then react with the lithium metal to form lithium salts. During the latter reactions, the lithium atoms give up electrons to become lithium ions. Overall, electrons are pumped from the positive terminal to the negative terminal.

Lithium's high reactivity is accompanied by a large chemical potential energy. Lithium batteries often have natural voltages of 3 V or more. Lithium is also one of the lightest elements, with a very light nucleus at the center of each atom, so lithium batteries pack a great deal of chemical potential energy into a very light package. They are thus ideal for watches and cameras, where reducing weight is important. (For an interesting alternative to normal batteries, see ❷.)

❷ Spacecraft often use fuel cells, a peculiar type of battery in which the active chemicals at the positive and negative electrodes are gases. The most basic fuel cell uses electrolysis in reverse. Oxygen gas arrives at the cell's positive terminal and hydrogen gas at the negative terminal. Electrochemical reactions, assisted by platinum and other catalysts, convert these gases into water. In the process, electrons are pumped from the cell's positive terminal to its negative terminal. Research is underway to develop fuel cells that are cheaper, more practical, and/or that use fuels other than hydrogen gas. There is considerable hope that such fuel cells will replace batteries and combustion engines in many applications.

Appendix A

Relevant Mathematics

Many of the quantities of physics are vectors, meaning that they have both magnitudes (amounts) and directions. Among these vector quantities are position, velocity, acceleration, force, torque, momentum, angular momentum, and electric and magnetic fields. Of these vector quantities, position is probably the easiest to visualize: you specify an object's position by giving its position vector—its distance and direction from a reference point. For example, you can specify the library's position with respect to your home by giving both its distance from your home (say 3.162 km or 1.965 miles) and its direction from your home (18.43° east of due north). That position vector is all the information someone would need to travel from your home to the library.

In illustrations, such as Fig. A.1, vector quantities are drawn as arrows. The length of each arrow indicates the vector's magnitude, while the direction of the arrow indicates which way the vector points. Suppose that you live in a city with major east–west and north–south streets spaced 1 km apart. Figure A.1 is four aerial views of your city. The vector **A** in Fig. A.1a shows the position of the library with respect to your home. It begins at your home and ends at the library, thus indicating both the magnitude and direction of the library's position.

Let's look at another position vector. The vector **B** in Fig. A.1b begins at the library and ends at your friend's home. This position vector shows the position of your friend's home with respect to the library and is 2.828 km long and points 45°east of south. If you happen to be at the library, you could use this vector to find your friend's home.

But how can you go from your home to your friend's home? To make this trip, you must add two vectors: you first follow vector **A** from your home to the library and then follow vector **B** from the library to your friend's home. This combined trip is shown as the upper path in Fig. A.1c. But you could also go directly from your home to your friend's home by following a new vector in Fig. A.1c—vector **C**. This vector from your home to your friend's home is the sum of vectors **A** and **B**, and is 3.162 km long and

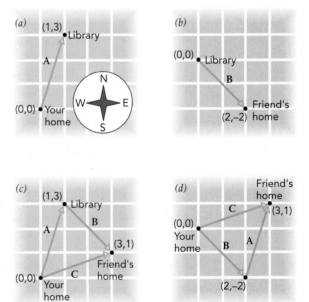

Fig. A.1 Four aerial views of your city, showing the major north–south and east–west streets that are spaced 1 km apart. (a) To go from your home to the library, you must travel 3.162 km at a direction 18.43° east of north, which is vector **A**. (b) To go from the library to your friend's home, you must travel 2.828 km at a direction 45° east of south—vector **B**. (c) You can go from your home to your friend's home either by going first to the library along vector **A** and then going to your friend's home along vector **B**, or you can travel directly along vector **C**, which is the sum of vectors **A** and **B**. (d) You can also reach your friend's home by going first along vector **B** and then along vector **A**. The sum of these two vectors is still vector **C**. However, you will not visit the library on that trip.

647

points 18.43° north of east. Using bold letters to indicate that **A**, **B**, and **C** are vectors, we can write **A**+**B**=**C**, meaning that vector **C** is the sum of vectors **A** and **B**.

Another interesting path from your home to your friend's home is shown in Fig. A.1*d*: you first travel along vector **B** and then along vector **A**. On this path, you would arrive at your friend's home without visiting the library. The first leg of this journey would take you into new territory but the second leg would leave you at your friend's home. The sum of vectors **B** and **A** is still vector **C**, or **B**+**A**=**C**. Thus vectors added in any order yield the same sum.

While you can estimate the sum of two vectors by drawing arrows on a sheet of paper, obtaining an accurate sum requires some thought. Adding their magnitudes is unlikely to give you the magnitude of the sum vector and adding their directions doesn't even make sense. To add two vectors, it helps to specify them in another form: as their *components* along two or three directions that are at right angles with respect to one another. In the present example of travel in a city, two right angle directions are all we need. If height were also important, we'd need three right angle directions.

For the two right angle directions we need in the city, let's choose east and north. We can then specify vector **A** as its component toward the east and its component toward the north. Its east component is 1 km and its north component is 3 km, so vector **A**, the position of the library with respect to your home, is 1 km to the east and 3 km to the north.

This new form for vector **A**, a pair of distances, is often more convenient than the old form, a distance and a direction. If you go to the library by walking 3.162 km in the direction 18.43° east of north, you'll have to pass through a lot of other buildings and backyards. Walking to the library by heading 1 km east and 3 km north allows you to travel on the sidewalks.

If we designate the position of your home as 0 km east and 0 km north, then the position of the library is 1 km east and 3 km north. These positions are labeled in Fig. A.1*a* as (0,0) and (1,3) respectively. The first number in the parentheses is the distance east, measured in kilometers, and the second number is the distance north, also in kilometers.

To go from the library to your friend's home, you must go 2 km east and 2 km south. That's the new form of vector **B**. Because a southward position has a negative component along the northward direction, the position of your friend's home with respect to the library is 2 km east and −2 km north. In Fig. A.1*b*, the library is at (0,0) and your friend's home is at (2,−2).

Now adding these vectors **A** and **B** is relatively easy. To go from your home to your friend's home by way of the library, you must move east first 1 km and then 2 km for a total of 3 km, and you must move north 3 km and then −2 km for a total of 1 km. Thus the position of your friend's home with respect to your home, vector **C**, is 3 km east and 1 km north. In Fig. A.1*c*, your home is at (0,0) and your friend's home is at (3,1).

Similarly, you could go from your home to your friend's home by following first vector **B** and then vector **A**, as shown in Fig. A.1*d*. This trip would take you through unknown regions, but you'd still arrive at the right place. You would move east first 2 km and then 1 km and you'd move north first −2 km and then 3 km. In the end, you'd be 3 km east and 1 km north of your home, the position of your friend's home. Thus the sum of vectors **B** and **A** is still vector **C**.

As you can see, vectors are useful for specifying physical quantities in our three-dimensional world. When you encounter vectors, remember that their directions are just as important as their magnitudes and that these directions must be taken into account when you add two vectors together.

Appendix B

Units, Conversion of Units

When you return from a camping trip and begin to describe it to your friends, there are a number of physical quantities that you may find useful in your conversation. Distance, weight, temperature, and time are as important in everyday life as they are in a laboratory. And when you explain to your friends how far you hiked, how much your backpack weighed, how cold the weather was, and how long the trip took, you'll have to relate those quantities to standard units or your friends won't appreciate just how difficult the trip was.

Most physical quantities aren't simple numbers like 7 or 2.9. Instead, they have units like length or time, and are specified in multiples of widely accepted standard units such as meters or seconds. When you say a door is 3.0 meters tall, you're comparing the door's height to the meter, a widely accepted standard unit of length. With this comparison, most people can determine just how tall the door is, even though they've never seen it.

But the meter isn't familiar to everyone; many people prefer to measure length in multiples of another standard unit: the foot. These people might be more comfortable hearing that the door is 9.8 feet tall. These quantities, 3.0 meters and 9.8 feet, are the same length.

Determining the door's height in feet doesn't require a new measurement because we can convert the one in meters to one that's in feet. To perform this conversion, we need to know how to express one particular length in both units. Any length will do. For example, Table B.1 states that 1 foot is the same length as 0.30480 meters. Because of that equality, we know that the following equation is true:

$$\frac{1 \text{ foot}}{0.30480 \text{ meters}} = 1.$$

We can multiply 3.0 meters, the height of the door, by this version of 1 and obtain the door's height in feet:

$$3.0 \text{ meters} \cdot \frac{1 \text{ foot}}{0.30480 \text{ meters}} = 9.8425 \text{ feet}.$$

Notice that the original units, meters, cancel and are replaced by the new units, feet. Since we only knew the door's height to 2 digits of precision in meters, we can't report the door's height to any more than 2 digits of precision in feet. So we round the result to 9.8 feet.

You can change the units of almost any physical quantity by multiplying that quantity by a version of 1. You should form this 1 by dividing new units by old units, where the number of new units in the numerator is equivalent to the number of old units in the denominator. You can obtain these pairs of equivalent quantities from Table B.1 below. When you do this multiplication, the old units will cancel and you'll be left with the physical quantity expressed in the new units.

One last note about units: when you use physical quantities in a calculation, make sure that you keep the units throughout the process. They're as important in the calculation as they are anywhere else. Some of the units may cancel, but in all likelihood the result of the calculation will have some units left and these units must be appropriate to the type of result you expect. If you're expecting a length, the units of your result should be meters or feet or another standard unit of length. If the units you obtain are seconds or kilograms, you've made a mistake in the calculation.

Table B.1 Conversion of Units*

1. Acceleration: (SI unit: 1 meter/second^2 or 1 m/s^2)

1 foot/second^2	= 0.30480 m/s^2

2. Angle: (SI unit: 1 radian)

1 degree (1°)	= 0.017453 radians

3. Area: (SI unit: 1 meter^2 or 1 m^2)

1 foot^2	= 0.092903 m^2
1 inch^2	= 6.4516 × 10^{-4} m^2

4. Density: (SI unit: 1 kilogram/meter^3 or 1 kg/m^3)

1 pound/foot^3	= 16.018 kg/m^3

5. Energy: (SI unit: 1 joule or 1 J)

1 Btu	= 1054.7 J
1 calorie, thermochemical	= 4.1840 J
1 electron-volt (1 eV)	= 1.6022 × 10^{-19} J
1 foot-pound	= 1.3558 J
1 kilocalorie (Food calorie)	= 4,186.8 J
1 kilowatt-hour	= 3,600,000 J

6. Force: (SI unit: 1 newton or 1 N)

1 pound	= 4.4482 N

7. Length: (SI unit: 1 meter or 1 m)

1 angstrom (1 Å)	= 10^{-10} m
1 centimeter (1 cm)	= 0.01 m
1 fermi (1 fm)	= 10^{-15} m
1 foot	= 0.30480 m
1 inch	= 0.02540 m
1 kilometer (1 km)	= 1,000 m
1 light year	= 9.4606 × 10^{15} m

1 micron (1 μm)	$= 10^{-6}$ m
1 mil	$= 2.5400 \times 10^{-5}$ m
1 mile	$= 1{,}609.3$ m
1 millimeter (1 mm)	$= 0.001$ m
1 nanometer (1 nm)	$= 10^{-9}$ m
1 picometer (1 pm)	$= 10^{-12}$ m

8. Mass: (SI unit: 1 kilogram or 1 kg)

1 gram (1 g)	$= 0.001$ kg
1 metric ton	$= 1000$ kg
1 pound-mass	$= 0.45359$ kg
1 slug	$= 14.594$ kg

9. Power: (SI unit: 1 watt or 1 W)

1 Btu/hour	$= 0.29307$ W
1 horsepower	$= 745.70$ W

10. Pressure: (SI unit: 1 pascal or 1 Pa)

1 atmosphere	$= 101{,}325$ Pa
1 millimeter of mercury (1 torr)	$= 133.32$ Pa
1 pound/inch^2 (1 psi)	$= 6{,}894.8$ Pa

11. Temperature: (SI units: Celsius or C; Kelvin or K)

Because temperature in the three common units, C, K, and F, aren't multiples of one another, you must use special formulas to convert between them:

Temperature in °C = 5/9·(Temperature in °F − 32)
Temperature in °C = Temperature in K − 273.15
Temperature in K = Temperature in °C + 273.15

12. Time: (SI unit: 1 second or 1 s)

1 day	$= 86{,}400$ s
1 femtosecond (1 fs)	$= 10^{-15}$ s
1 hour	$= 3{,}600$ s
1 microsecond (1 μs)	$= 10^{-6}$ s
1 millisecond (1 ms)	$= 0.001$ s
1 minute	$= 60$ s
1 nanosecond (1 ns)	$= 10^{-9}$ s
1 picosecond (1 ps)	$= 10^{-12}$ s

13. Torque: (SI unit: 1 newton-meter or 1 N·m)

1 inch-pound	$= 0.11298$ N·m
1 foot-pound	$= 1.3558$ N·m

14. Velocity: (SI unit: 1 meter/second or 1 m/s)

1 foot/second	$= 0.30480$ m/s
1 kilometer/hour (1 km/h)	$= 0.27778$ m/s

1 knot	= 0.51444 m/s
1 mile/hour (1 mph)	= 0.44704 m/s
1 mile/hour (1 mph)	= 1.6093 km/h

15. Volume: (SI unit: 1 meter^3 or 1 m^3)

1 cup	= 2.3659 × 10^{-4} m^3
1 fluid ounce	= 2.9574 × 10^{-5} m^3
1 foot^3	= 0.028317 m^3
1 gallon	= 0.0037854 m^3
1 liter (1 l)	= 0.001 m^3
1 milliliter (1 ml)	= 10^{-6} m^3
1 quart	= 0.00094635 m^3

*This table lists pairs of equivalent quantities, one in SI units and one in other units. Each pair can be used to convert measurements expressed as multiples of one unit into measurements expressed as multiples of the other unit. These pairs are grouped according to physical quantity and are given to a precision of 5 digits.

GLOSSARY

absolute temperature scale A scale for measuring temperature in which 0 K corresponds to absolute zero.

absolute zero The temperature at which all thermal energy has been removed from an object or system of objects. Because it's impossible to find and remove all the thermal energy from an object, absolute zero can be approached but is not actually attainable.

acceleration A vector quantity that measures how quickly an object's velocity is changing: the greater the acceleration, the more the object's velocity changes each second. It consists of both the amount of acceleration and the direction in which the object is accelerating. This direction is identical to the direction of the force causing the acceleration. The SI unit of acceleration is the meter-per-second^2.

acceleration due to gravity A physical constant that specifies how quickly a freely falling object accelerates and also relates an object's weight to its mass. At the earth's surface, the acceleration due to gravity is 9.8 m/s^2 (or 9.8 N/kg).

accelerometer A device that senses acceleration by measuring the position of a test mass suspended in stable equilibrium on springs or other elastic supports. As the overall device accelerates, the test mass shifts away from its equilibrium position because it requires a net force in order to accelerate with the device.

acid A chemical that when dissolved in water increases the density of hydrogen ions in solution and lowers the water's pH.

acidic Having a pH of less than 7.

activation energy The energy required to initiate a chemical reaction. This energy serves to break or weaken the bonds in the starting chemicals so that the reaction can proceed to form the reaction products.

adverse pressure gradient A region of fluid flow in which the fluid must flow toward higher pressure. The fluid's momentum and kinetic energy carry it through this situation, although the fluid does slow down.

aerodynamic forces The forces exerted on an object by the motion of the air surrounding it. The two types of aerodynamic forces are lift and drag.

aerodynamics The study of the dynamic (moving) interactions of air with objects.

airfoil An aerodynamically engineered surface, designed to obtain particular lift and drag forces from the air flowing around it.

airspeed An object's speed relative to the air through which it moves.

alpha decay A radioactive decay in which a helium nucleus (two protons and two neutrons) escapes from a larger proton-rich nucleus via quantum tunneling.

alternating current An electric current that periodically reverses its direction of flow. Abbreviated as AC.

amorphous solid A solid that is non-crystalline and has no long-range order.

ampere (A) The SI unit of electric current (synonymous with the coulomb-per-second). One ampere is defined as the passage of 6.25×10^{18} charged particles per second and is roughly the current flowing through a 100-W lightbulb operating on household electric power.

ampere-meter (A·m) The SI unit of magnetic pole.

amplifier A device that replicates an input signal as a larger output signal.

amplitude modulation A technique for representing sound or data by changing the amplitude (strength) of a wave.

amplitude The maximal displacement of an oscillator away from its equilibrium position.

analog representation The representation of numbers directly as continuous values of physical quantities such as voltage, charge, or pressure.

angle of attack The angle at which an airfoil is tilted relative to the airflow that is taking place around it.

angular acceleration A vector quantity that measures how quickly an object's angular velocity is changing: the greater the angular acceleration, the more the object's angular velocity changes each second. It consists of both the amount of angular acceleration and the direction about which the angular acceleration occurs. This direction is identical to the direction of the torque causing the angular acceleration. The SI unit of angular acceleration is the radian-per-second^2.

angular impulse The mechanical means for transferring angular momentum. One object gives an angular impulse to a second object by exerting a certain torque on the second object for a certain amount of time. In return, the second object gives an equal but oppositely directed angular impulse to the first object.

angular momentum A conserved vector quantity that measures an object's rotational motion. It is the product of that object's rotational mass times its angular velocity. The SI unit of angular momentum is the kilogram-meter^2-per-second.

angular position A quantity that describes an object's orientation relative to some reference orientation.

angular speed A measure of the angle through which an object rotates in a certain amount of time.

angular velocity A vector quantity that measures how rapidly an object's angular position is changing: the greater the angular velocity, the farther the object turns each second. It consists of both the object's angular speed and the direction about which the object is rotating. This direction points along the axis of rotation in the direction established by the right-hand rule. The SI unit of angular velocity is the radian-per-second.

anharmonic oscillator An oscillator in which the restoring force on an object is not proportional to its displacement from a stable equilibrium. The period of an anharmonic oscillator depends on the amplitude of its motion.

annealing Heating a material to relieve internal stresses and to allow recrystallization and grain growth, followed by gradual cooling to room temperature.

antiferromagnetic Composed of magnetic atoms that effectively alternate back and forth in magnetic orientation so that their magnetic fields cancel.

antimatter Matter resembling normal matter, but with many of its characteristics such as electric charge reversed.

aperture The diameter or effective diameter of a lens or opening.

apparent weight The sum of a person's weight plus their feeling of acceleration. All three quantities are vectors, so that apparent weight can be quite large if the weight and feeling point in the same direction or quite small if they point in opposite directions.

Archimedes' principle The observation that an object partially or wholly immersed in a fluid is acted on by an upward buoyant force equal to the weight of the fluid it's displacing.

atmospheric pressure The pressure of air in the earth's atmosphere. Atmospheric pressure reaches a maximum of about 100,000 Pa near sea level and diminishes with increasing altitude.

atom The smallest portion of a chemical element that retains the chemical properties of that element.

atomic number The number of protons present in an atomic nucleus and equal to the number of electrons in a neutral atom.

atomize Reduce a liquid to droplets that are so small that they are easily suspended in the air by drag forces.

axis of rotation The straight line in space about which an object or group of objects rotates. More specifically, the axis of rotation points in a particular direction along that line to reflect the sense of rotation according to the right-hand rule.

back emf The self-induced electromotive force that develops in an inductor when its current changes or in the coil of an electromechanical system such as a motor when its current causes magnets to move.

band A group of levels in a solid that involve similar standing waves and thus have similar energies.

band gap A range of energies over which a solid has no levels available.

bandwidth The range of frequencies involved in a group of electromagnetic waves.

base A chemical that when dissolved in water decreases the density of hydrogen ions in solution and increases the water's pH.

base of support A surface outlined on the ground by the points at which an object is supported.

basic Having a pH of more than 7.

Bernoulli effect The drop in pressure that occurs when the speed of an effectively incompressible fluid in steady-state flow increases as it moves along a streamline.

Bernoulli's equation An equation relating the total energy of an incompressible fluid in steady-state flow to the sum of its pressure potential energy, kinetic energy, and gravitational potential energy.

beta decay A radioactive decay in which the weak force allows a neutron in a neutron-rich nucleus to disintegrate into an electron, a proton, and an antineutrino. The proton remains in the nucleus, but the electron and antineutron escape.

binary The digital representation of numbers in terms of the powers of two. The number 6 is represented in binary as 110, meaning 1 four (2^2), 1 two (2^1), and 0 ones (2^0).

bit A single binary value, either a 0 or a 1.

black hole A region of space, normally spherical, within which the gravitational distortions of space and time are so severe that not even light can escape.

blackbody spectrum The distribution of thermal electromagnetic radiation emitted by a black object. This distribution is the amount of radiation emitted at each wavelength and depends only on the temperature of the black object.

blunt Not streamlined so that the fluid flowing around it stalls and experiences flow separation and pressure drag.

boiling Accelerated evaporation that occurs when stable gas-phase bubbles form and grow inside a material's liquid-phase.

boiling temperature The threshold temperature at which gas-phase bubbles first become stable within a material's liquid phase.

Boltzmann constant The constant of proportionality relating a gas's pressure to its particle density and temperature. It has a measured value of 1.381×10^{-23} Pa·m^3/(particle·K).

boundary layer A thin region of fluid near a surface that, because of viscous drag, is not moving at the full speed of the surrounding airflow.

bremsstrahlung The process in which a rapidly accelerating charge emits electromagnetic radiation, usually an X-ray or gamma-ray photon.

Brewster's angle The angle at which no vertically polarized light reflects from a transparent surface that is oriented horizontally. The precise angle depends on the surface's index of refraction.

brittle fracture Material fracture that occurs without plastic deformation when layers of atoms separate completely during stress.

brittleness A material's tendency toward brittle fracture during stress.

buoyant force The upward force exerted by a fluid on an object immersed in that fluid. The buoyant force is actually caused by pressure from the fluid. That pressure is highest below the object so the force exerted upward on the object's bottom is greater than the force exerted downward on the object's top.

byte Eight binary bits that collectively can represent a number from *0* to *255*. Bytes are often used to represent letters and other characters, where a convention associates each character with a specific number.

calibration The process of comparing a local reference object to a generally accepted standard.

capacitance The amount of separated charge on the plates of a capacitor divided by the voltage difference across those plates. The SI unit of capacitance is the farad.

capacitor An electronic component that stores separated electric charge on a pair of plates that are separated by an insulating layer.

carrier wave An electromagnetic wave with only a single frequency. This wave is modulated in order to represent an audio or video signal, or digital information.

catalyst A chemical surface that assists a chemical reaction by reducing the activation energy needed to initiate the reaction.

Celsius A temperature scale in which 0 °C is defined as the melting point of water and 100 °C is defined as the boiling point of water at sea level. Absolute zero is −273.15 °C.

center of gravity The unique point about which all of an object's weight is evenly distributed and therefore balanced. Because weight is proportional to mass, the center of mass is identical to the center of gravity for objects that are much smaller than the earth. For larger objects, the centers of mass and gravity differ slightly. An object suspended from its center of gravity will balance and will experience no net torque due to gravity. In many situations, you can accurately predict an object's behavior by assuming that all of the object's weight acts at its center of gravity.

center of mass The unique point about which all of the object's mass is balanced. The center of mass is the natural pivot point for a free object. In the absence of outside forces or torques, a rigid object's center of mass travels at constant velocity while the object rotates at constant angular velocity about this center of mass.

center of percussion The special spot on a bat or racket where a collision with a second object will not cause any acceleration of the bat's handle.

centripetal acceleration An acceleration that is always directed toward the center of a circular trajectory.

centripetal force A centrally directed force on an object. A centripetal force is not an independent force but rather the sum of other forces, such as gravity, acting on the object.

chain reaction A process in which one event triggers an average of one or more similar events so that the process becomes self-sustaining.

chaos Unpredictable behavior in which minute changes in a system's initial arrangement lead to very different final arrangements. These differences grow more dramatic with each passing second.

chaotic system A dynamic system that is exquisitely sensitive to initial conditions. Minute changes in how you set up a chaotic system can lead to wildly different final configurations.

characteristic X-ray An X-ray emitted by X-ray fluorescence from an atom. The energy of the characteristic X-ray is determined by the atom's orbital structure.

charges Objects, particularly small particles, that carry electric charge.

chemical bond An energy deficit that holds two or more atoms together to form a molecule and that must be repaid in order to separate the atoms. Chemical bonds form when chemical potential energy is released during a molecule's assembly.

chemical potential energy Energy stored in the chemical forces between atoms. Those chemical forces are electromagnetic in origin.

chemical reaction An encounter between two or more atoms and molecules that results in a rearrangement of the atoms to form different atoms and molecules.

circularly polarized A light wave in which the electric and magnetic fields rotate about the direction in which that wave is heading.

closed circuit A complete electric circuit through which electric current can flow continuously.

coaxial cable A two-conductor electric cable in which an insulated central conductor is surrounded by a cylindrical outer conductor. Electromagnetic waves can travel within the gap between the two conductors, and they propagate along the cable at almost the speed of light.

coefficient of restitution The measure of a ball's liveliness, determined by bouncing the ball from a rigid, immovable surface. It's the ratio of the ball's rebound speed to its collision speed.

coefficient of volume expansion The fractional change in an object's volume caused by a temperature increase of 1 °C.

coherent light Light consisting of identical photons that together form a single electromagnetic wave.

collision energy The amount of kinetic energy removed from two objects as they collide.

color temperature The temperature at which a black object will emit thermal electromagnetic radiation with this particular distribution of wavelengths.

components The portions of a vector quantity that lie along particular directions.

compressible A substance that changes density significantly as its pressure changes. A gas is compressible since its density is proportional to its pressure.

compressive stress A stress involving compression.

Compton scattering The process in which a photon bounces off a charged particle, usually an electron. The photon and charged particle exchange energy and momentum during the collision.

concave mirror A mirror that is bowed inward at the center, like the inside of a bowl.

condensation The phase transformation whereby a gas becomes a liquid.

conduction band The group of quantum levels in an insulator that lies above the Fermi level.

conduction level A quantum level in an insulator that requires more energy than the Fermi level and that is normally unoccupied by electrons.

conduction The transmission of heat through a material by a transfer of energy from one atom or molecule to the next. The atoms themselves don't move with the heat. In metals, mobile electrons also contribute to heat conduction.

conserved quantity A physical quantity, such as energy, that is neither created nor destroyed within an isolated system when that system undergoes changes. A conserved quantity may pass among the objects within an isolated system, but its total amount remains constant.

constructive interference Interference in which two or more waves arrive at a location in space and time in phase with one another and produce a particularly strong effect.

convection cell A loop of fluid flow that is propelled by convection. Fluid in a convection cell normally rises in a hotter region and descends in a colder region.

convection current A fluid flow propelled by convection.

convection The transmission of heat by the movement of a fluid. Convection normally entails the natural circulation of the fluid that accompanies differences in temperatures and densities.

converging lens A lens that bends the individual light rays passing through it toward one another so that they either converge more rapidly than before or at least diverge less rapidly from one another. A converging lens has a positive focal length and often produces real images.

convex mirror A mirror that is bowed outward at the center, like the outside of a bowl.

corona discharge A faintly glowing discharge that surrounds a small, highly charged object in the presence of a gas. In the discharge, electric charge is transferred from the object to the gas molecules.

coulomb (C) The SI unit of electric charge. One coulomb is about 1 million times the charge you acquire by rubbing your feet across a carpet in winter.

Coulomb constant The fundamental constant of nature that determines the electrostatic forces two charges exert on one another. Its measured value is 8.988×10^9 N·m^2/C^2.

Coulomb's law The magnitudes of the electrostatic forces between two objects are equal to the Coulomb constant times the product of their two electric charges divided by the square of the distance separating them. If the charges are like, then the forces are repulsive. If the charges are opposite, then the forces are attractive.

covalent bond A strong chemical bond between atoms in which the two atoms share one or more pairs of electrons in the region of space between them.

crest A peak positive excursion of an extended system that is experiencing a wave.

critical mass A portion of a fissionable material that is able to sustain a fission chain reaction. The amount of material required depends on its mass, shape, and density.

crystal An orderly arrangement of atoms or molecules that has both positional and orientational orders. The particles are organized in a simple, geometrical pattern that repeats endlessly in all directions.

crystalline Having its atoms arranged in an orderly pattern that extends for many atomic spacings in all directions.

current The amount of electric charge flowing past a point or through a surface per unit of time. The SI unit of current is the ampere.

cycle-per-second (1/s) The SI unit of frequency (synonymous with hertz).

cyclotron motion The circular or spiral motion of a charged particle in a magnetic field. The charged particle tends to loop around the magnetic flux lines.

decimal The digital representation of numbers in terms of the powers of ten. The number *124* is represented in decimal as 124, meaning 1 hundred (10^2), 2 tens (10^1), and 4 ones (10^0).

demagnetization The disappearance of magnetic polarization in a material.

density The mass of an object divided by its volume. The SI unit of density is the kilogram-per-meter^3.

depletion region The nonconducting region around a p-n junction in which all of the valence levels are filled with electrons and there are no conduction level electrons.

deposition The phase transformation whereby a gas becomes a solid.

destructive interference Interference in which two or more waves arrive at a location in space and time out of phase with one another and produce a particularly weak effect.

diffraction A wave phenomenon that limits the focusability of light and alters the way in which it travels after passing through an opening.

digital representation The representation of numbers by decomposition into digits that are then individually represented by discrete values of physical quantities such as voltage, charge, or pressure.

diode A semiconductor device that allows charge to flow through it only in one direction. Diodes are commonly formed by bringing n-type silicon into contact with p-type silicon.

direct current An electric current that always flows in one direction. Abbreviated as DC.

direction The line or course on which something is moving, is aimed to move, or along which something is pointing or facing.

discharge A flow of electric current through a gas.

dislocation An incomplete sheet of atoms that appears as a defect in a crystal.

dispersion The dependence of light's speed through a material on the frequency of that light.

dissolve Separate into individual atoms, molecules, or ions and disperse throughout a fluid.

distance The length between two positions in space. The SI unit of distance is the meter.

diverging lens A lens that bends the individual light rays passing through it away from one another so that they either converge less rapidly than before or don't converge at all. A diverging lens has a negative focal length and often produces virtual images.

domain wall A boundary surface between magnetic domains having different directions of magnetic orientation.

doped Modified by adding chemical impurities that change its physical properties.

Doppler effect A difference between the frequency at which a wave is sent and the frequency at which that wave is received caused by relative motion between the sender and receiver.

drag forces The frictionlike forces exerted by a fluid and a solid on one another as the solid moves through the fluid. These forces act to reduce the relative velocity between the two.

ductile Capable of undergoing extensive plastic deformation before fracturing.

ductile fracture Material fracture that occurs during plastic deformation when layers of atoms separate completely during stress.

dynamic stability An object's stability when it's in motion.

dynamic variations Changes in a physical quantity such as pressure that are caused by motion.

elastic collision A collision in which all of the kinetic energy present before the impact is again present as kinetic energy after the impact.

elastic deformation A stress-induced change in shape that doesn't involve permanent rearrangement of the atoms in a material. The material returns to its original shape when the stress is removed.

elastic limit The most extreme distortion of an object from which it can return to its original size and shape without permanent deformation.

elastic potential energy The energy stored by the forces within a distorted elastic object.

elastic scattering The process in which two particles bounce off one another without losing any of their kinetic energies.

electric charge An intrinsic property of matter that gives rise to electrostatic forces between charged particles. Electric charge is a conserved physical quantity. A specific charge can have a positive amount of electric charge (a positive charge) or a negative amount (a negative charge). The SI unit of electric charge is the coulomb.

electric charges Objects, particularly small particles, that carry electric charge.

electric circuit A complete loop of conductors, loads, and power sources through which an electric current can flow continuously.

electric current The movement or flow of electric charge.

electric field An attribute of each point in space that exerts forces on electrically charged particles. An electric field has a magnitude and direction proportional to the force it would exert on a unit of positive charge at that location. While electric fields are often created by nearby charges, they can also be created by other electromagnetic phenomena. The SI unit of electric field is the volt-per-meter or, equivalently, the newton-per-coulomb.

electric polarization A distribution of electric charge that is nonuniform so that the object has a region of positive charge and a region of negative charge.

electrical conductor A material that permits electric charges to flow through it.

electrical insulator A material that prevents any net movement of electric charge through it.

electrical resistance The measure of how much an object impedes the flow of electric current. The SI unit of electrical resistance is the ohm.

electromagnet A coil of wire, with or without an iron core, that becomes a magnet when an electric current flows through the coil.

electromagnetic waves Waves consisting of electric and magnetic fields that travel through empty space at the speed of light. These waves carry energy and momentum and are emitted and absorbed as particles called photons. Radio waves, microwaves, infrared, visible, and ultraviolet light, X-rays, and gamma rays are examples of electromagnetic waves.

electron volt (eV) A unit of energy equal to the energy obtained by an elementary charge (electron or proton) as it moves through a voltage difference of 1 V. One electron volt is equal to about 1.602×10^{-19} J.

electronic shell A quantum mechanical structure that exists within an atom wherein the electron standing waves have related structures and similar energies. Atoms that complete their electronic shells become very stable chemically.

electron–positron pair production The formation of an electron and a positron during an energetic collision.

electrons The tiny negatively charged particles that make up the outer portions of atoms and that are the main carriers of electricity and heat in metals.

electrostatic force The force experienced by a charged particle in the presence of other charged particles.

electrostatic potential energy Energy stored in the forces between electric charges.

elementary unit of electric charge The basic quantum of electric charge, equal to about 1.6×10^{-19} C.

emissivity A surface's capacity to emit or absorb thermal radiation, relative to that of a perfectly black object at the same temperature.

energy The capacity to do work. Each object has a precise quantity of energy, which determines exactly how much work that object could do in an ideal situation. The SI unit of energy is the joule.

English system of units An assortment of antiquated units that were used throughout the English colonies and remain in common use in the United States today. Units in this system include feet, ounces, and miles-per-hour.

entrainment The phenomenon in which a particle or portion of fluid is carried along in the flow of another fluid.

entropy The physical quantity measuring the amount of disorder in a system. The system's entropy would be zero at absolute zero.

equilibrium position The point at which an object experiences zero net force and doesn't accelerate.

equilibrium The state of an object in which zero net force (or zero net torque) acts on it. An object that is stationary or in uniform motion is in equilibrium.

escape velocity The speed a spacecraft needs in order to follow a parabolic orbital path and escape forever from a particular celestial object.

eutectic An alloy or solution of two or more materials that has a lower melting temperature than any of the individual materials.

evaporation The phase transformation whereby a liquid becomes a gas.

excited state A configuration of a system having excess energy; its electrons (or other particles) are in an arrangement of quantum waves (e.g., orbitals or levels) that has more than the least possible energy.

exhaust velocity The velocity of exhaust gas relative to the rocket engine from which it emerged.

explosive chain reaction A chain reaction in which each fission induces an average of much more than one subsequent fission and the fission rate skyrockets.

eyepiece The final lens of a telescope or microscope; the lens that prepares the light for your eye.

Fahrenheit A temperature scale in which 32 °F is defined as the melting point of water and 212 °F is defined as the boiling point of water at sea level. Absolute zero is −459.67 °F.

farad (F) The SI unit of electric capacitance. A 1-farad capacitor will have a voltage difference between its plates of 1 volt when storing 1 coulomb each of separated positive and negative charge.

feedback The process of using information about a system's current situation to control changes you are making in that system.

feeling of acceleration A person undergoing acceleration experiences a weight-like sensation in the direction exactly opposite the direction of acceleration. The amount of this feeling of acceleration is proportional to the amount of the acceleration.

Fermi energy The energy of an electron in the Fermi level.

Fermi level A hypothetical level located halfway between the highest occupied level and the lowest unoccupied level in a solid.

Fermi particles A class of fundamental particles that includes electrons, protons, and neutrons and that obeys the Pauli exclusion principle.

ferrimagnetic Composed of magnetic atoms that alternate back and forth in magnetic orientation but have different magnetic strengths so that their magnetic contributions don't cancel completely.

ferromagnetic Composed of magnetic atoms that all have the same magnetic orientation within a magnetic domain.

firm Having a large spring constant and thus experiencing large restoring forces in response to small distortions.

first law of thermodynamics The change in a stationary object's internal energy is equal to the heat transferred into that object minus the work that object does on its surroundings. This law restates the conservation of energy.

fissionable Able to undergo induced fission.

fluid A substance that has mass but no fixed shape. A fluid can flow to match its container. Gases and liquids are both fluids.

fluorescence An emission of light that immediately follows an absorption of light.

f-number The ratio of a lens's focal length to its effective aperture.

focal length The distance after a converging lens at which the real image of a distant object forms. The focal length of a diverging lens is negative and is the distance *before* the lens at which the virtual image of a distant object forms.

force An influence that if exerted on a free body results chiefly in an acceleration of the body and sometimes in deformation and other effects. A force is a vector quantity, consisting of both the amount of force and its direction. The SI unit of force is the newton.

forward biased A p-n junction in which the voltage of the p-type semiconductor has been raised relative to the voltage of the n-type semiconductor.

freezing The phase transformation whereby a liquid becomes a solid.

frequency modulation A technique for representing sound or data by changing the exact frequency of a wave.

frequency The number of cycles completed by an oscillating system in a certain amount of time. The SI unit of frequency is the hertz.

friction The force that resists relative motion between two surfaces in contact. Frictional forces are exerted parallel to the surfaces in the directions opposing their relative motion.

fundamental forces The four basic forces that act between objects in the universe—the gravitational force, the electromagnetic force, the strong force, and the weak force.

fundamental vibrational mode The slowest and often broadest vibration that an extended object can support.

gamma rays Extremely high-energy photons of electromagnetic radiation, often produced during radioactive decays.

gas or **gaseous** A form of matter consisting of tiny, individual particles (atoms or molecules) that travel around independently. A gas takes on the shape and volume of its container.

general theory of relativity The physical rules governing all motion, even motion involving speeds comparable to the speed of light and occurring in the presence of massive objects.

glass A non-crystalline material having no long-range order.

glass transition temperature The temperature below which the atoms in a liquid are no longer able to rearrange and the material responds elastically to shear stress.

glass–rubber transition regime A regime in which the atoms of an amorphous polymer can move freely enough to make it pliable and leathery.

glassy regime A regime in which the atoms of an amorphous polymer can't move relative to one another, making it rigid and hard.

grain A tiny crystallite that is part of a polycrystalline material.

grain boundary The thin layer of atoms between two separate crystals or grains of a polycrystalline material.

gravitational constant The fundamental constant of nature that determines the gravitational forces two masses exert on one another. Its value is 6.6720×10^{-11} N·m^2/kg^2.

gravitational mass The mass associated with the gravitational attraction between objects.

gravitational potential energy Potential energy stored in the gravitational forces between objects.

gravity The gravitational attraction of the mass of the earth, the moon, or a planet for bodies at or relatively near its surface. All objects exert gravitational forces on all other objects.

ground state The lowest energy configuration of a system; its electrons (or other particles) are in the arrangement of quantum waves (e.g., orbitals or levels) that has the least possible energy.

gyroscope A spinning disk which will continue to point in one direction in space as long as it is not subjected to any outside torques.

half-life The time needed for half the nuclei of a particular radioactive isotope to undergo radioactive decay.

hard magnetic material A material that is relatively difficult to magnetize and that retains its magnetization once the magnetizing field is removed. Hard magnetic materials are suitable for permanent magnets.

hardness A material's ability to resist penetration.

harmonic An integer multiple of the fundamental frequency of oscillation for a system. The second harmonic is twice the frequency of the fundamental, and the third harmonic is three times the frequency of the fundamental. In principle, harmonics can continue forever.

harmonic oscillator An oscillator in which the restoring force on an object is proportional to its displacement from a stable equilibrium. The period of a harmonic oscillator doesn't depend on the amplitude of its motion.

heat capacity The amount of heat that must be added to an object to cause its temperature to rise by 1 unit.

heat engine A device that converts thermal energy into work as heat flows from a hot object to a cold object.

heat exchanger A device that allows heat to flow naturally from a hotter material to a colder material without any actual exchange of those materials.

heat pump A device that pumps heat against its natural direction of flow, transferring it from a cold object to a hot object. To satisfy the second law of thermodynamics, a heat pump normally converts some ordered energy into thermal energy.

heat The energy that flows from one object to another as a result of a difference in temperature between those two objects.

Heisenberg uncertainty principle A quantum physical law that states that an object's position (a particle characteristic) and momentum (a wave characteristic) can't be sharply defined at the same time. This principle gives objects with small masses a fuzzy character.

henry (H) The SI unit of inductance. A 1-henry inductor will experience a 1-ampere change in the current flowing through it each second when subjected to a 1-volt voltage drop.

hertz (Hz) The SI unit of frequency (synonymous with cycles-per-second).

higher-order vibrational mode A vibrational mode that is more complicated than the fundamental mode and in which different parts of the extended system move in opposite directions.

Hooke's law The general law covering spring and elastic behavior. Hooke's law states that a spring exerts a restoring force that is proportional to the distance the spring is distorted from its equilibrium length.

horizontal polarization An electromagnetic wave in which the electric field always points left or right (horizontally). The magnetic field always points vertically.

hydrogen bond A moderately strong chemical bond in which an uncovered hydrogen nucleus is attracted to the electrons of a nearby oxygen atom.

ideal gas law The law relating the pressure, temperature, and particle density of an ideal gas. An ideal gas is one that is composed of perfectly independent particles. The particles don't stick and bounce perfectly from one another.

image distance The distance between the lens and the image that the lens creates. Real images form at positive image distances while virtual images form at negative image distances.

impedance A measure of a system's opposition to the passage of a current or a wave.

impedance mismatch An abrupt change in the opposition to a wave's passage, typically accompanied by reflections.

impulse The mechanical means for transferring momentum. One object gives an impulse to a second object by exerting a certain force on the second object for a certain amount of time. In return, the second object gives an equal but oppositely directed impulse to the first object.

in phase The relationship between two waves in which they complete the same portions of their oscillatory cycles at the same time and place.

incandescence The emission of thermal radiation from a hot object.

incoherent light Light consisting of individual photons, each its own independent electromagnetic wave.

incompressible A substance that doesn't change density significantly as its pressure changes. Liquids and solids are incompressible since their densities change very little as their pressures change dramatically.

index of refraction The factor by which the speed of light in a material is reduced from its speed in empty space, equal to the speed of light in empty space divided by light's speed in the material.

induced drag The drag force that occurs when a wing deflects the stream of air passing across it to obtain lift.

induced emf An overall voltage difference between the ends of a coil produced by a changing magnetic field in that coil and the resulting electric field.

induced fission A fission event that's caused by a collision, usually with a neutron.

inductance The voltage drop across an inductor divided by the rate at which current through that inductor is changing with time. The SI unit of inductance is the henry.

inductor An electronic component that stores magnetic energy in a coil of wire and opposes changes in current in that wire.

inelastic collision A collision in which some of the kinetic energy present before the impact is no longer present as kinetic energy after the impact.

inert gas A gas consisting of atoms that are chemically inactive and rarely bond permanently with other atoms or molecules. Inert gases include helium, neon, argon, krypton, and xenon.

inertia A property of matter by which it remains at rest or in uniform motion in the same straight line unless acted on by some outside force.

inertial frame of reference A frame of reference that is not accelerating and is thus either stationary or traveling at constant velocity. The laws of motion accurately describe any situation that is observed from an inertial frame of reference.

inertial mass The mass associated with an object's inertia, its resistance to acceleration.

inertial Moving because of inertia alone and therefore not accelerating.

infrared light Invisible light having wavelengths longer than about 750 nanometers.

insulator A solid in which the Fermi level falls within a band gap.

interference A wave phenomenon in which waves passing through the same location from different directions reinforce or oppose one another.

interference pattern A pattern of intensity variations in time and space that occurs when two or more waves are superposed and experience constructive and destructive interferences.

internal energy The sum of an object's thermal energy and any additional potential energy stored entirely within the object.

internal kinetic energy The portion of an object's kinetic energy that involves only the relative motion of particles within the object and that excludes the object's overall translation or rotation.

internal potential energy The portion of an object's potential energy that involves only forces between particles within the object and that excludes the object's interactions with its surroundings.

ion An atom or molecule with a net electric charge.

ionic bond A chemical bond in which ions are held together by their opposite electric charges.

isotopes Chemically indistinguishable atoms containing nuclei that differ only in their numbers of neutrons.

joule (J) The SI unit of energy and work (synonymous with newton-meter). Lifting 1 liter of water upward 10 centimeters near the earth's surface requires about 1 joule of work.

joule-per-kilogram-kelvin (J/kg·K) The SI unit of specific heat.

joule-per-second (J/s) The SI unit of power (synonymous with watt).

Kelvin The SI scale of absolute temperature, in which 0 K is defined as absolute zero. The spacing between units is the same as that used in the Celsius scale.

Kepler's first law All planets move in elliptical orbits, with the sun at one focus of the ellipse.

Kepler's second law A line stretching from the sun to a planet sweeps out equal areas in equal times.

Kepler's third law The square of a planet's orbital period is proportional to the cube of that planet's mean distance from the sun.

kilogram (kg) The SI unit of mass. (The standard kilogram is a platinum–iridium cylinder kept at the International Bureau of Weights and Measures near Paris.) A liter of water has a mass of about 1 kilogram.

kilogram-meter^2 (kg·m^2) The SI unit of rotational mass. One kilogram-meter^2 is roughly the rotational mass of your forearm as it pivots about your elbow.

kilogram-meter^2-per-second (kg·m^2/s) The SI unit of angular momentum. One kilogram-meter^2-per-second is about the angular momentum of a 7.3 kg (16 lbm) bowling ball spinning 34 times/second as it rolls down the lane.

kilogram-meter-per-second (kg·m/s) The SI unit of momentum. One kilogram-meter-per-second is about the momentum in a baseball traveling 25 km/h (16 mph).

kilogram-meter-per-second^2 (kg·m/s^2) The SI unit of force (synonymous with newton).

kilogram-per-meter^3 (kg/m^3) The SI unit of density. One kilogram-per-meter^3 is about the density of air at 2000 m (about 1 mile) above sea level.

kinetic energy The form of energy contained in an object's translational and rotational motion.

laminar flow Smooth, predictable fluid flow in which nearby portions of the fluid remain nearby as they travel along.

laser amplifier A device that amplifies weak incoming light to produce brighter outgoing light. The outgoing light is a brighter copy of the incoming light.

laser medium An assembly of excited atoms or other quantum systems that is capable of amplifying passing light through stimulated emission.

laser oscillator A laser amplifier that is surrounded by mirrors so that it can amplify one or more spontaneously emitted photons to form an intense beam of coherent light.

latent heat of evaporation The heat required to transform a unit mass of material from liquid to gas without changing its temperature.

latent heat of fusion Latent heat of melting.

latent heat of melting The heat required to transform a unit mass of material from solid to liquid without changing its temperature.

latent heat of vaporization Latent heat of evaporation.

lattice A regular geometrical pattern that repeats endlessly in all directions.

law of universal gravitation Every object in the universe attracts every other object in the universe with a force equal to the gravitational constant times the product of the two masses, divided by the square of the distance separating the two objects.

laws of thermodynamics The four laws that govern the movement of heat between objects.

lens A transparent optical device that uses refraction to bend light, often to form images.

lens equation The equation relating a lens's focal length to the object and image distances.

Lenz's law When a changing magnetic field induces a current in a conductor, the magnetic field from that current always opposes the change that induced it.

level An electron standing wave in a solid, one of the basic electron wave modes allowed in a solid by quantum physics.

lever arm The directed distance from the pivot or axis of rotation to the point at which the force is exerted.

lift forces Forces exerted by a fluid on a solid that are at right angles to the fluid flow around that solid.

light See visible, infrared, and ultraviolet light.

linear momentum A conserved vector quantity that measures an object's motion. It is the product of that object's mass times its velocity. The SI unit of linear momentum is the kilogram-meter-per-second.

liquid A form of matter consisting of particles (atoms or molecules) that are touching one another but that are free to move relative to one another. A liquid has a fixed volume but takes the shape of its container.

liquid crystal A liquid consisting of rod-like or disk-like molecules that exhibit orientational order.

liquid flow regime A regime in which the chains of an amorphous polymer reptate quickly enough to support liquid flow. The polymer is a viscous liquid.

longitudinal wave A wave in which the underlying oscillation is parallel to the wave itself.

long-range order Uniform spatial structure that extends for many nanometers in a solid.

Lorentz force The force experienced by a charged particle when it moves through a magnetic field.

lumen A common unit of total radiated light as perceived by a human eye.

luminescence The emission of light by any means other than thermal radiation.

magnetic dipole A pair of equal but opposite poles separated by a distance.

magnetic domains Regions of uniform alignment within a magnetic material.

magnetic field An attribute of each point in space that exerts forces on magnetic poles. A magnetic field has a magnitude and direction proportional to the force it would exert on a unit of north magnetic pole at that location. The SI unit of magnetic field is the tesla.

magnetic flux lines Abstract strands following along the local magnetic field direction and having a density proportional to that local field. Flux lines can only begin at north poles and end at south poles.

magnetic induction The process whereby a time-changing magnetic field initiates or influences an electric current.

magnetic monopole An isolated magnetic pole, either north or south. None has ever been observed.

magnetic polarization A distribution of magnetic poles that is nonuniform so that the object has a region of north pole and a region of south pole.

magnetic pole A property of nature that gives rise to magnetostatic forces between magnetic poles. A specific pole can have a positive amount of magnetic pole (a north pole) or a negative amount (a south pole). The SI unit of magnetic pole is the ampere-meter.

magnetic poles Objects that carry magnetic pole.

magnetization The development of magnetic polarization in a material.

magnetostatic force The force experienced by a magnetic pole in the presence of other magnetic poles.

magnitude The amount of some physical quantity.

Magnus force A lift force experienced by a spinning object as it moves through a fluid. The Magnus force points toward the side of the ball moving away from the onrushing airstream.

mass The property of a body that is a measure of its inertia or resistance to acceleration, that is commonly taken as a measure of the amount of material it contains, and that causes it to have weight in a gravitational field. The SI unit of mass is the kilogram.

mechanical advantage The process whereby a mechanical device redistributes the amounts of force and distance that go into performing a particular amount of mechanical work.

mechanical wave A natural and often rhythmic motion of an extended object about its stable equilibrium shape or situation.

melting temperature The temperature at which a material's solid and liquid phases can coexist in stable equilibrium.

melting The phase transformation whereby a solid becomes a liquid.

metal A solid in which the Fermi level falls within a band of levels.

metal fatigue Material fracture that occurs when manipulation allows surface blemishes to propagate into a metal as tears.

metallic bonds The form of bonding that occurs in metals. Electrons are shared generally among many atoms and this sharing binds the atoms together.

meter (m) The SI unit of length or distance. (One meter is formally defined as the distance light travels through empty space in 1/299,792,458th of a second.) One meter is about the length of a long stride or about 3.28 feet.

meter^2 (m^2) The SI unit of area. One square meter is about twice the area of an opened newspaper.

meter³ (m³) The SI unit of volume. One cubic meter is about the volume of a four-drawer file cabinet.

meter-per-second (m/s) The SI unit of velocity or speed. One meter-per-second is a typical walking pace or about 2.2 mph.

meter-per-second² (m/s²) The SI unit of acceleration. One meter-per-second² is about the acceleration of an elevator as it first begins to move upward.

microwaves Electromagnetic waves with wavelengths between about 1 meter and 1 millimeter.

mode A basic pattern of distortion or oscillation.

moderator A material with which fast neutrons can collide and give up energy without being absorbed. Neutrons leave a moderator with only thermal energies.

modulated carrier wave A wave that has been modulated so that it contains not only the carrier frequency, but also video, audio, or other information.

molecule A particle formed out of two or more atoms. A molecule is the smallest portion of a chemical compound that retains the chemical properties of that compound.

moment of inertia Rotational mass.

momentum Linear momentum.

monomer A small molecule from which a polymer forms.

natural resonance A mechanical process in which an isolated object's energy causes it to perform a certain motion over and over again. The rate at which this motion occurs is determined by the physical characteristics of the object.

net electric charge The sum of all charges on an object, both positive and negative. Positive charges increase the net charge while negative charges decrease it. Net charge can be negative.

net force The sum of all forces acting on an object, considering both the magnitude of each individual force and its direction. The magnitude of the net force is often less than the sum of the magnitudes of the individual forces, since they often oppose one another in direction.

net magnetic pole The sum of all poles on an object, both north and south. Since there are no isolated magnetic poles, an object's net magnetic pole is always zero.

net torque The sum of all torques acting on an object, considering both the magnitude of each individual torque and its direction. The magnitude of the net torque is less than the sum of the magnitudes of the individual torques, since they often oppose one another in direction.

neutral Having zero net electric charge.

neutrinos Chargeless and nearly massless particles created during radioactive decays and other nuclear events. They rarely interact with matter.

neutrons The electrically neutral subatomic particles that, together with protons, make up atomic nuclei.

newton (N) The SI unit of force (synonymous with the kilogram-meter-per-second^2). Eighteen U.S. quarters have a weight equal to about 1 newton. The common English unit of force, the pound, is about 4.45 newtons.

Newton's first law of motion An object that is free from all outside forces travels at a constant velocity, covering equal distances in equal times along a straight-line path.

Newton's first law of rotational motion An object that is not wobbling and is free from all outside torques rotates with constant angular velocity, spinning steadily about a fixed axis.

Newton's second law of motion An object's acceleration is equal to the force exerted on that object divided by the object's mass. This equality can be manipulated algebraically to state that the force on the object is equal to the product of the object's mass times its acceleration (Eq. 1.1.2).

Newton's second law of rotational motion An object's angular acceleration is equal to the torque exerted on that object divided by the object's rotational mass. This equality can be manipulated algebraically to state that the torque on the object is equal to the product of the object's rotational mass times its angular acceleration (Eq. 2.1.2). The law doesn't apply to wobbling objects.

Newton's third law of motion For every force that one object exerts on a second object, there is an equal but oppositely directed force that the second object exerts on the first object.

Newton's third law of rotational motion For every torque that one object exerts on a second object, there is an equal but oppositely directed torque that the second object exerts on the first object.

newton-meter (N·m) The SI unit of energy and work (synonymous with the joule). Also the SI unit of torque, exerted by a 1-newton force located 1 meter from the axis of rotation. One newton-meter is about the torque exerted on your shoulder by the weight of a baseball held in your outstretched arm.

newton-per-ampere-meter (N/A·m) The SI unit of magnetic field (synonymous with tesla).

newton-per-coulomb (N/C) The SI unit of electric field (synonymous with volt-per-meter).

newton-per-meter^2 The SI unit of pressure (synonymous with pascal).

normal Directed exactly away from (perpendicular to) a surface. A line that is normal to a surface meets that surface at a right angle.

normal force Support force.

n-type semiconductor A semiconductor such as silicon that contains impurity atoms such as phosphorus, arsenic, antimony, or bismuth that place electrons in the semiconductor's conduction level.

nuclear fission The shattering of a heavy nucleus into smaller fragments. During fission, the positively charged fragments repel one another and release energy.

nuclear force An attractive force that binds nucleons together once they touch one another.

nuclear fusion The merging of two small nuclei to form a larger nucleus. During fusion, the nuclear force binds the nucleons together and releases energy.

nucleation Forming an initial seed of one material phase in the midst of another material phase.

nucleon A general name given to the particles that make up atomic nuclei: protons and neutrons.

nucleus The positively charged central component of an atom, containing most of the atom's mass and about which the electrons are arranged. Plural is nuclei.

object distance The distance between the lens and the object that it is imaging.

objective The first lens or mirror of a telescope or microscope; the first significant optical component that light encounters after leaving the object.

ohm (Ω) The SI unit of electrical resistance. A 1-ohm resistor exhibits a voltage drop of 1 volt when 1 ampere of current flows through it.

Ohm's law The observation that the voltage drop across an ordinary electrical conductor is proportional to both the electric current passing through it and to its electrical resistance.

ohmic Exhibiting a voltage drop that's proportional to current, consistent with Ohm's law.

open circuit An incomplete electric circuit where a gap in the electrical conductors stops electric current from flowing.

orbit The path an object takes as it moves in the presence of a centripetal force.

orbital An electron standing wave in an atom, one of the basic electron wave modes allowed in an atom by quantum physics.

orbital period The time required to complete one full orbit.

ordered energy Energy that can easily be used to do work.

orientational order The characteristic of particles that all point in the same direction.

oscillation A repetitive and rhythmic movement or process that usually takes place about an equilibrium situation.

out of phase The relationship between two waves in which they complete opposite portions of their oscillatory cycles at the same time and place.

parallel (wiring arrangement) An arrangement in which the current reaching two or more electric devices divides into separate parts to flow through those devices and then joins back together as it leaves them. Current experiences the same change in voltage in each device.

particle density The number of particles in an object divided by its volume. The particle density of water is about 3.35×10^{28} molecules-per-meter^3. The particle density of air at sea level is about 2.687×10^{25} molecules-per-meter^3.

pascal (Pa) The SI unit of pressure (synonymous with newton-per-meter^2). Atmospheric pressure at sea level is about 100,000 pascals. A 1-millimeter-high water droplet exerts a pressure of about 10 pascals on your hand.

Pascal's principle A change in the pressure of an enclosed incompressible fluid is conveyed undiminished to every part of the fluid and to the surfaces of its container.

Pauli exclusion principle An observed property of nature that indistinguishable Fermi particles must each have their own unique quantum wave.

period The time required to complete one full cycle of a repetitive motion.

permanent magnet An object that can be magnetized and that retains that magnetization for a long time.

permeability of free space The defined constant that relates two poles and the magnetostatic forces they exert on one another. Its value is $4\pi \times 10^{-7}$ N/A^2.

pH A measure of the hydrogen ion concentration, equal to the negative of the log base-10 of the molar hydrogen ion concentration.

phase A form of matter, notably solid, liquid, gas, and plasma.

phase equilibrium A situation in which two material phases coexist stably, neither one growing at the expense of the other.

phase transition A transformation from one material phase to another.

phosphor A solid that luminesces (emits light) when energy is transferred to it by light or by a collision with a particle.

photoconductor A solid that is an electrical insulator in the dark but that becomes an electrical conductor when exposed to light of the correct wavelength.

photodiode A diode that permits current to flow backward across the p-n junction when exposed to light. Light provides the energy needed to move charges across the junction's depletion region in the wrong direction. The current flowing in the reverse direction through a photodiode is proportional to the light intensity.

photoelectric effect The process in which an atom absorbs a photon in a radiative transition that ejects one of the electrons out of the atom.

photon A particle or quantum of light, having energy and momentum but no mass.

pitch The frequency of a sound.

Planck constant The fundamental constant of quantum physics, equal to the energy of an object divided by the frequency of its quantum wave. It is about 6.626×10^{-34} J·s.

plane polarized A light wave in which the electric field (and the magnetic field) fluctuates back and forth in a plane as the light travels through space.

plasma A gaslike phase of matter consisting of electrically charged particles, such as ions and electrons. The strong electromagnetic interactions between its particles distinguish a plasma from a gas.

plastic deformation A stress-induced change in shape that involves permanent rearrangement of the atoms in a material. The material doesn't return to its original shape when the stress is removed.

plasticizer A chemical additive that lowers a polymer's glass transition temperature and reduces its crystallinity.

p-n junction The interface between an n-type semiconductor and a p-type semiconductor that gives the diode its unidirectional characteristic for electrons.

Poiseuille's law The volume of fluid flowing through a pipe each second is equal to $(\pi/128)$ times the pressure difference across that pipe times the pipe's diameter to the fourth power, divided by the pipe's length times the fluid's viscosity.

polar molecule A molecule that is electrically polarized, effectively having a positive end and a negative end.

poles Objects that carry a magnetic pole.

polycrystalline Composed of many individual crystallites.

polymer A giant molecule that forms when many monomer molecules bind together in a permanent chain. To be called a polymer, the chain must be about a thousand or more atoms in length.

polymerization The process in which individual monomer molecules are chemically bound together to form a polymer chain.

population inversion A nonequilibrium population of quantum systems in which more are in a higher energy state than in a lower energy state.

position A vector quantity that specifies the location of an object relative to some reference point. It consists of both the length and the direction from the reference point to the object.

positional order The characteristic of particles that are uniformly spaced and aligned so that they form a lattice.

positron The antimatter counterpart of the electron. The positron is positively charged.

potential energy The stored form of energy that can produce motion. Potential energy is stored in the forces between or within objects.

power The measure of how quickly work is done on an object. The SI unit of power is the watt.

precession The change in orientation of a spinning object's rotational axis that occurs when it's subject to an outside torque.

pressure drag The drag force that results from higher pressures at the front of an object than at its rear.

pressure gradient A distribution of pressures that varies continuously with position.

pressure potential energy The product of a fluid's volume times its pressure. However, this energy isn't really stored in the fluid. Instead, it's energy that's provided by a pump (or other source) when the fluid is delivered.

pressure The average amount of force a fluid exerts on a certain region of surface area. Pressure is reported as the amount of force divided by the surface area over which that force is exerted. The SI unit of pressure is the pascal.

primary colors of light The three colors of light (red, green, and blue) that are sensed by the three types of color-sensitive cone cells in our eyes. Mixtures of these three colors of light can make our eyes perceive any possible color.

primary colors of pigment The three colors of pigment (cyan, magenta, and yellow) that absorb the three primary colors of light (red, green, and blue, respectively). Mixtures of these three pigments can be applied to a white surface to make it reflect any possible mixture of the three primary colors of light and thus to make our eyes perceive any possible color.

principle of equivalence The principle that gravitational mass and inertial mass are truly identical and therefore that no experiment you can perform in a small region of space can distinguish between free fall and the absence of gravity

proton The positively charged subatomic particles found in atomic nuclei.

p-type semiconductor A semiconductor such as silicon that contains impurity atoms such as boron, aluminum, gallium, indium, or tellurium that remove electrons from the semiconductor's valence levels.

quanta The fundament, discrete units in which an item is emitted, absorbed, or otherwise observed, reflecting the particulate character of that item.

quantized Existing only in discrete units or quanta. Quantized physical quantities are only observed in integer multiples of the elementary quantum.

quenching Cooling rapidly from high temperature so that the material doesn't have time to adopt its normal low-temperature state.

radian The natural unit in which angles are measured. There are 2π radians in a full circle, so 1 radian is $180/\pi$ degrees or approximately $57.3°$.

radian-per-second (1/s) The SI unit of angular velocity or angular speed. An object turning at 1 radian-per-second completes a full revolution in just less than 6.3 seconds.

radian-per-second² (1/s²) The SI unit of angular acceleration.

radiation The transmission of heat through the passage of electromagnetic radiation between objects.

radiation trapping The phenomenon in which a particular wavelength of light has trouble propagating through a material that eagerly absorbs and emits it. The light passes from one atom or atomlike system to the next and makes little headway.

radiative transition The shift of an atom or atomlike system from one state to another through the emission or absorption of an electromagnetic wave.

radio waves Electromagnetic waves, usually with wavelengths longer than about 1 m.

radioactive decay The spontaneous decay of a nucleus into fragments.

ramp An inclined plane that allows work to be done over a longer distance, thereby requiring less force.

Rayleigh scattering The redirection of light due to its interaction with small particles of matter.

real image A pattern of light, projected in space, that exactly reproduces the pattern of light at the surface of the original object. A real image forms after the lens that creates it and can be projected onto a surface.

rebound energy The amount of kinetic energy returned to two objects as they push apart following a collision.

reflection The redirection of all or part of a wave so that it returns from a boundary between media.

refraction The bending of a wave's path that occurs when the wave crosses a boundary between media and experiences a change in speed.

relative humidity The actual humidity as a percentage of the humidity required to achieve phase equilibrium between liquid and gaseous water.

relative motion The movement of one object from the perspective of another object. Two objects that are moving relative to one another have different velocities.

relativistic energy An object's energy according to the relativistic laws of motion and including its rest energy.

relativistic laws of motion The laws of motion in the special theory of relativity. They correct deficiencies in the Newtonian laws of motion that appear primarily at speeds comparable to the speed of light.

relativistic momentum An object's momentum according to the relativistic laws of motion.

reptation A thermally induced slithering motion of polymer chains, backward and forward, that allows them to get past entanglements in order to flow. Reptation is so named because of its similarity to the motion of snakes.

resistor An electronic component that impedes the flow of electric current, converting some of its energy into heat.

resonant cavity A simple resonant circuit consisting of a carefully shaped conducting strip or shell and equivalent to a capacitor and an inductor. Energy flows back and forth between the cavity's electric and magnetic fields.

resonant energy transfer The gradual transfer of energy to or from a natural resonance caused by small forces timed to coincide with a particular part of each oscillatory cycle.

restoring force A force that acts to return an object to its equilibrium shape. A restoring force is directed toward the position the object occupies when it's in its equilibrium shape.

reverse biased A p-n junction in which the voltage of the p-type semiconductor has been lowered relative to the voltage of the n-type semiconductor.

Reynolds number A dimensionless number that characterizes fluid flow through a system. At low Reynolds numbers a fluid's viscosity dominates the flow, while at high Reynolds numbers a fluid's inertia dominates.

right-hand rule The convention whereby the specific direction of an object's angular velocity is established. According to this rule, if the fingers of your right hand are curled to point in the direction of the object's rotation, your thumb will point in the direction of the angular velocity.

root mean square (RMS) voltage A measure of AC voltage defined as the DC voltage that would cause the same average power consumption in an ohmic device.

rotational equilibrium The state of an object in which zero net torque acts on it. An object that has constant angular momentum is in rotational equilibrium.

rotational inertia A property of matter by which it remains at rest or in steady rotation about the same rotational axis unless acted on by some outside torque.

rotational mass The property of a body that is a measure of its rotational inertia. An object's rotational mass is determined by its mass and by how far that mass is from the axis of rotation. The SI unit of rotational mass is the kilogram-meter^2.

rotational motion Motion in which an object rotates about an axis. The orientation of an object undergoing only rotational motion will change, but its position will remain unchanged.

rubbery flow regime A regime in which the chains of an amorphous polymer can reptate just enough to flow very slowly past one another. The polymer is elastic and flows very slowly.

rubbery plateau regime A regime in which the individual atoms of an amorphous polymer can move freely but the chains can't flow past one another because of entanglements. The polymer is soft and elastic.

saturated In phase equilibrium with another material phase. The gaseous phase of a material is saturated when it is in phase equilibrium with that material's liquid and/or solid phase.

schematic diagram A symbolic picture of the conceptual structure of an electronic device.

second (s or sec) The SI unit of time. (One second is formally defined as the duration of 9,192,631,770 periods of the radiation corresponding to the transition between two hyperfine levels of the ground state of the cesium 133 atom.)

second law of thermodynamics The entropy of a thermally isolated system of objects never decreases. This law recognizes that creating disorder is easy; restoring order is hard.

Seebeck effect A charge transfer that occurs between the two ends of a metal when one end is hotter than the other. It's caused by the thermal motion of the conduction electrons, which tend to drift away from the hot end and accumulate at the cold end.

seed crystal An initial crystal on which a solidifying liquid can crystallize.

selective reflection Reflection only of light with a particular wavelength or color.

semiconductor An insulator with a small band gap, so that only a modest amount of energy is needed to shift an electron from an occupied valence level to an unoccupied conduction level.

series (wiring arrangement) An arrangement in which the current reaching two or more electric devices flows sequentially through one device after the next before leaving them. Current may experience different changes in voltage in the different devices.

Shannon limit The maximum amount of digital information that can be carried by a signal each second; equal to the bandwidth of that signal times $\log_2(1+\text{signal/noise})$.

shear strain The angle by which a material bends when experiencing a shear stress.

shear stress A stress in which the top of a material is pushed to the left while the bottom is pushed to the right.

shell A group of atomic orbitals having similar energies.

shock wave A narrow region of high pressure and temperature that forms when the speed of an object through a medium exceeds the speed at which sound, waves, or other vibrations travel in that medium.

short circuit A defect in a circuit that allows current to bypass the load it's supposed to operate.

SI units A system of units (Système Internationale d'Unités) that carefully defines related units according to powers of 10. SI units are now used almost exclusively throughout most of the world, with the notable exception of the United States.

sideband waves Waves at frequencies above and below the carrier wave that are produced when the carrier wave is modulated.

signal An electrical or optical representation of information.

simple harmonic motion The regular, repetitive motion of a harmonic oscillator. The period of simple harmonic motion doesn't depend on the amplitude of oscillation.

sliding friction The forces that resist relative motion as two touching surfaces slide across one another.

slip Plastic deformation of a crystal that occurs when one sheet of atoms slides across another sheet.

soft Having a small spring constant and thus experiencing small restoring forces in response to large distortions.

soft magnetic material A material that is relatively easy to magnetize and that loses its magnetization once the magnetizing field is removed. Soft magnetic materials are suitable for electromagnets.

solid A form of matter consisting of particles (atoms or molecules) that touch and that are not free to move relative to one another. A solid has a fixed volume and shape.

solvation shell A cloud of oriented liquid molecules that surrounds a molecule or ion and carries it around a liquid in solution.

sound In air, sound consists of density waves, patterns of compressions and rarefactions that travel outward from their source at the speed of sound.

special theory of relativity The physical rules governing all motion, even motion involving speeds comparable to the speed of light.

specific heat The amount of heat that must be added to a unit mass of a material to cause a unit rise in its temperature. The SI unit of specific heat is the joule-per-kilogram-kelvin.

speed A measure of the distance an object travels in a certain amount of time. The SI unit of speed is the meter-per-second.

speed of light The speed with which an electromagnetic wave travels through space. In empty space, a vacuum, the speed of light is exactly 299,792,458 m/s.

speed of sound The speed at which sound's compressions and rarefactions travel in a medium such as air or water.

spontaneous emission of radiation Light emission that occurs when an excited atom or atomlike system releases stored energy randomly through a radiative transition. The photon that results is independent and unique.

spring constant As a measure of the stiffness of an elastic object, the spring constant relates the object's distortion to the restoring force it exerts. The larger the spring constant, the stiffer the spring.

springlike force A force that is proportional to displacement, consistent with Hooke's law.

sputtering Ejection of atoms from a surface caused by the impact of energetic ions, atoms, or other tiny projectiles.

stable equilibrium A state of equilibrium to which an object will return if it's disturbed. At equilibrium, the object is free of net force or torque. If an object is moved away from that equilibrium state, however, the net force or torque that will then act on it will tend to return it to equilibrium.

stall When a fluid flow stops and spoils steady-state flow. In the aerodynamic flow around an airfoil, stalling refers to airflow separation triggered by a stall in the flow near the airfoil's surface.

standard units Agreed on amounts of various physical quantities, which define a system in which those quantities are subsequently measured.

standing wave A wave in which all the nodes and antinodes remain in place.

state A possible arrangement of electrons (or other particles) in a quantum system.

static friction The forces that resist relative motion as outside forces try to make two touching surfaces begin to slide across one another.

static stability An object's stability when it's not in motion.

static variations Changes in a physical quantity such as pressure that are not caused by motion.

steady-state flow A situation in a fluid where the characteristics of the fluid at any fixed point in space don't change with time.

Stefan–Boltzmann constant The constant of proportionality relating a surface's radiated power to its emissivity, temperature, and surface area. It has a measured value of 5.67×10^{-8} J/(s·m²·K⁴).

Stefan–Boltzmann law The equation relating a surface's radiated power to its emissivity, temperature, and surface area.

stiffness A measure of how rapidly a restoring force increases as the system exerting that force is distorted.

stimulated emission of radiation Light emission that occurs when an excited atom or atomlike system releases stored energy through a radiative transition by duplicating a photon passing through that system.

strain The stress-induced change in a material's length divided by its original length.

streamline The path followed by a particular portion of a flowing fluid.

streamlined Carefully tapered so that the fluid flowing around it doesn't stall and doesn't experience flow separation or pressure drag.

stress A compressive or tensile force exerted on a material divided by the area over which that force is exerted.

strong force The fundamental force that gives structure to nuclei and nucleons and is the basis for the nuclear force.

subatomic particles The fundamental building blocks of the universe, from among which atoms and matter are constructed.

subcritical mass A portion of fissionable material that is too small to sustain a chain reaction.

sublimation The process by which atoms or molecules go directly from a solid to a gas.

superconductor An electrical conductor that permits electrons to flow without losing any of their kinetic energy to thermal energy. Electrons will continue to flow in a superconductor indefinitely. Materials only become superconducting at extremely low temperatures.

supercooled Below the temperature at which a phase transition should have occurred. Supercooling results from a failure to nucleate the new phase and/or a loss of mobility in the material before the new phase could fully develop.

supercritical mass A portion of fissionable material that is well in excess of a critical mass so that it undergoes an explosive chain reaction.

superheated Above the temperature at which a phase transition should have occurred. Superheating results from a failure to nucleate the new phase.

superposition The overlapping of two or more waves so that their amplitudes add together and they form a combined wave.

supersaturated Containing more of something than phase equilibrium will allow. A phase separation should occur, but it may be delayed by a failure to nucleate the second phase and/or inadequate mobility within the existing supersaturated phase.

support force A force that is exerted when two objects come into contact. Each object exerts a force on the other object to keep the two from passing through one another. Support forces are always normal, or perpendicular, to the surfaces of objects.

surface area The extent of a two-dimensional surface bounded by a particular border. The SI unit of surface area is the meter^2.

surface tension An inward tension that appears whenever a surface has extra potential energy. It acts to make that surface as small as possible.

surface waves Disturbances in the stable equilibrium shape of a surface.

sympathetic vibration The transfer of energy between two natural resonances that share a common frequency of oscillation.

tank circuit A simple resonant circuit consisting of a capacitor and an inductor. Energy flows back and forth between these two devices repetitively.

temper (glass) To soften glass at high temperature and then cool its surface quickly so as to put that surface under compressive stress.

temper (metals) Bake a material at moderate temperatures to relieve trapped stresses.

temperature A measure of the average internal kinetic energy per particle in a material. In a gas, temperature measures the average kinetic energy of each atom or molecule.

tensile strength The maximum tensile stress that a material can withstand before fracturing.

tensile stress A stress involving tension.

tension Outward forces on an object that tend to stretch it.

terminal velocity The velocity at which an object moving through a fluid experiences enough drag force to balance the other forces on it and keep it from accelerating.

tesla (T) The SI unit of magnetic field (synonymous with newton-per-ampere-meter).

thermal conductivity The measure of a material's capacity to transport heat by conduction from its hotter end to its colder end.

thermal energy A disordered form of energy contained in the kinetic and potential energies of the individual atoms and molecules that make up a substance. Because of its random distribution, this disordered energy can't be converted directly into useful work. Other names for thermal energy include internal energy and heat.

thermal equilibrium A situation in which no heat flows in a system because all of the objects in the system are at the same temperature.

thermal motion The random motions of individual particles in a material due to the internal or thermal energy of that material.

thermal neutrons Neutrons that move with only thermal energies so that their kinetic energies are consistent with the local temperature.

thermionic emission The emission of electrons that occurs when thermal energy in a hot surface ejects those electrons from the surface.

thermistor A semiconductor temperature sensor that experiences a dramatic increase in electric conductivity as its temperature increases.

thermocouple A metallic temperature sensor in which two dissimilar metals joined at two junctions respond electrically to temperature differences between the junctions.

thermoplastic A polymer composed only of individual chains. In principle, a thermoplastic can melt and flow, although some thermoplastics decompose or burn before melting.

thermoset A polymer in which the individual chains have been cross-linked with covalent bonds so that it's one giant molecule. A thermoset can't melt or flow.

third law of thermodynamics As an object's temperature approaches absolute zero, its entropy approaches zero. This law points out that absolute zero is the unattainable state in which an object has no disorder.

thrust A forward, propulsive force.

tidal forces The differences between one celestial object's gravity at particular locations on the surface of a second object and the average of that gravity for the entire second object. Tidal forces tend to stretch the second object into an egg shape.

timbre The mixture of tones in an instrument's sound that are characteristic of that instrument.

torque An influence that if exerted on a free body results chiefly in an angular acceleration of the body. A torque is a vector quantity, consisting of both the amount of torque and its direction. The SI unit of torque is the newton-meter.

total internal reflection Complete reflection of a light wave that occurs when that wave tries unsuccessfully to leave a material with a large refractive index for a material with a small refractive index at too shallow an angle.

toughness A measure of how much mechanical energy a material can absorb without fracturing.

traction The largest frictional force that an object can obtain in its present situation.

trajectory The path taken by an object as it moves.

transformer A device that uses magnetic fields to transfer electric power from one circuit to another circuit. The two circuits are electrically isolated since no charges actually travel between the two circuits.

transistor An electronic component that allows a tiny amount of electric charge, either moving or stationary, to control the flow of a large electric current.

translational motion Motion in which an object moves as a whole along a straight or curved line.

transmutation of elements Changing the atoms of one element into another via nuclear processes that alter the numbers of protons in their nuclei.

transverse wave A wave in which the underlying oscillation is perpendicular to the wave itself.

traveling wave A wave that moves steadily through space in a particular direction.

trough A peak negative excursion of an extended system that is experiencing a wave.

tunneling Because of the Heisenberg uncertainty principle, small objects have somewhat ill-defined positions and occasionally move through energy barriers to places they can't reach classically. That quantum process is tunneling.

turbulence The unpredictable swirls and eddies of turbulent fluid flow.

turbulent flow Irregular, fluctuating, unpredictable fluid flow in which nearby portions of the fluid quickly become widely separated.

ultraviolet light Invisible light having wavelengths shorter than about 400 nanometers.

uniform circular motion Motion at a constant speed around a circular trajectory. An object undergoing uniform circular motion is accelerating toward the center of the circle.

unstable equilibrium An equilibrium situation to which the object will not return if it's disturbed. At equilibrium, the object is free of net force or torque. However, if the object is moved away from that equilibrium situation, the net force or torque that will then act on it will tend to accelerate it further away from the equilibrium situation.

vacuum tube A device in which the flow of electrons through empty space between a filament and a plate is controlled by a small amount of charge on one or more control grids.

valence band The group of quantum levels in an insulator that lies below the Fermi level.

valence electrons The outermost electrons of an atom. These atoms are principally responsible for the chemical properties of the atom.

valence level A quantum level in an insulator that requires less energy than the Fermi level and that is normally occupied by electrons.

van der Waals bond An extremely weak chemical bond in which nearby molecules are held together by quantum charge fluctuations.

van der Waals forces Weak, non-directional intermolecular forces that are created by temporary quantum fluctuations in the positions of electric charges in atoms or molecules.

vector quantity A quantity, characterizing some aspect of a physical system, that consists of both a magnitude and a direction in space.

velocity A vector quantity that measures how quickly an object's position is changing: the greater the velocity, the farther the object travels each second. It consists of both the object's speed and the direction in which the object is traveling. The SI unit of velocity is the meter-per-second.

Venturi effect The increase in speed and drop in pressure that occur when an incompressible fluid in steady-state flow passes through a narrow channel.

vertical polarization An electromagnetic wave in which the electric field always points up or down (vertically). The magnetic field always points horizontally.

vibration A spontaneous repetitive and rhythmic movement about an equilibrium position.

vibrational antinode A region of a vibrating object that is experiencing maximal motion.

vibrational node A region of a vibrating object that is not moving at all.

virtual image A pattern of light that appears to come from a particular region of space and reproduces the pattern of light at the surface of the original object. A virtual image forms before the lens that creates it and can't be projected onto a surface.

viscosity The measure of a fluid's resistance to relative motion within that fluid.

viscous drag A drag force that results from viscous forces on a moving surface immersed in a fluid.

viscous forces The forces exerted within a fluid that oppose relative motion. Layers of fluid that are moving across one another exert viscous forces on each other.

visible light Light having wavelengths between about 400 nanometers (violet) and 750 nanometers (red). This small portion of the electromagnetic spectrum is all that we are able to detect with our eyes.

volt (V) The SI unit of voltage (synonymous with joule-per-coulomb). The voltage on the positive terminal of a common battery is about 1.5 volts above that on its negative terminal.

voltage drop The amount of electrostatic potential energy that each coulomb of positive charge loses in passing through a device. It's equal to the voltage of the charges entering the device minus the voltage of the charges leaving that device.

voltage gradient A gradual slope in the voltage across a region of space. A voltage gradient is an electric field.

voltage rise The amount of electrostatic potential energy that each coulomb of positive charge receives in passing through a device. It's equal to the voltage of the charges leaving the device minus the voltage of the charges entering that device.

voltage The electrostatic potential energy of each unit of positive electric charge at a particular location. The SI unit of voltage is the volt.

volt-per-meter (V/m) The SI unit of electric field (synonymous with newton-per-coulomb).

volume The extent of a three-dimensional region of space bounded by a particular enclosure. The SI unit of volume is the meter^3.

vortex A whirling region of fluid that is moving in a circle above a central cavity.

vulcanization A process in which the individual chains of a thermoplastic are cross-linked to produce a thermoset.

wake deflection force A lift force experienced by a spinning ball when it deflects its turbulent wake to one side. The wake deflection force points toward the side of the ball moving away from the onrushing airstream.

wake The trail left behind by an object as it moves through a fluid.

water hammer The impact of a moving mass of water that is suddenly stopped.

watt (W) The SI unit of power, equal to the transfer of 1 joule-per-second. One watt is the power used by the bulb of a typical flashlight.

wave velocity The speed and direction of the moving crests of a wave.

wavelength A structural characteristic of a wave, corresponding to the distance separating adjacent crests or troughs.

wave-particle duality The observation that everything in nature has both particle and wave characteristics. An item is primarily particlelike when it is emitted, absorbed, or otherwise observed and primarily wavelike as it travels through time and space.

weak force The fundamental force that allows electrons and neutrinos to interact and that's responsible for beta decay.

weight (near the earth's surface) The downward force exerted on an object due to its gravitational interaction with the earth. An object's weight is equal to the product of that object's mass times the acceleration due to gravity. The direction of the weight is always toward the center of the earth.

work hardening Hardening that occurs when extensive manipulation reduces the size of grains in a polycrystalline material.

work The mechanical means of transferring energy. Work is defined as the force exerted on an object times the distance that object travels in the direction of the force. A large force exerted for a short distance or a small force exerted for a long distance can perform the same amount of work. The SI unit of work is the joule.

X-ray fluorescence The process in which an electron in one of the outer orbitals of an atom undergoes a radiative transition to an empty inner orbital, emitting an X-ray photon.

X-rays Very high-energy photons of electromagnetic radiation.

yield strength The stress at which a material first begins to undergo plastic deformation.

zeroth law of thermodynamics Two objects that are each in thermal equilibrium with a third object are also in thermal equilibrium with one another. This law is the basis for a meaningful system of temperatures.

Photo Credits

Chapter 1

Pages 3, 6, 15: Courtesy Lou Bloomfield. Page 7: Courtesy Jerry Ohlinger's Movie Material Store. Pages 25, 26, 27, 28: Courtesy Lou Bloomfield.

Chapter 2

Page 33: Courtesy Lou Bloomfield. Page 34: Courtesy NASA. Pages 39, 47, 48: Courtesy Lou Bloomfield. Page 57 (top): Courtesy US NHTSA. Page 57 (bottom), 58: Courtesy Lou Bloomfield.

Chapter 3

Pages 63, 65: Courtesy Lou Bloomfield. Page 66: Courtesy NASA. Pages 70, 71, 75: Courtesy Lou Bloomfield. Page 83: Courtesy NASA. Pages 84: Courtesy Karen Bloomfield.

Chapter 4

Pages 94, 95: Courtesy Lou Bloomfield. Page 98: Courtesy NASA. Page 100: Courtesy New York Public Library. Page 108: Courtesy NASA. Pages 199, 122: Courtesy Lou Bloomfield.

Chapter 5

Pages 131, 134 (top): Courtesy Lou Bloomfield. Page 134 (bottom): Courtesy Fredrik Fatemi. Page 135: Courtesy NASA. Page 141: Department of Water and Power of the City of Los Angeles.

Chapter 6

Pages 154, 155 (top): Courtesy Lou Bloomfield. Page 155 (bottom) Courtesy Peter Bradshaw, Stanford University. Page 163: Courtesy Lou Bloomfield. Page 164: Courtesy Rebus, Inc. Pages 171, 172: Courtesy Thomas Miller. Pages 173, 175, 176, 177, 182: Courtesy Lou Bloomfield.

Chapter 7

Page 198: Courtesy Travis Industries. Page 201: Courtesy Lou Bloomfield. Page 202: Courtesy Bryant Heating and Cooling. Pages 204, 209, 213, 214, 216, 217, 218, 219, 224, 225, 226, 231: Courtesy Lou Bloomfield. Page 232: Courtesy NASA. Pages 234, 235, 236, 242, 243, 246: Courtesy Lou Bloomfield.

Chapter 8

Page 255: Courtesy of Bryant Heating and Cooling. Page 257: Courtesy Lou Bloomfield. Page 265: Courtesy of BMW Corporation. Pages 272, 273: Courtesy Lou Bloomfield.

Chapter 9

Pages 297, 299, 301, 312, 313, 314, 320, 321: Courtesy Lou Bloomfield.

Chapter 10

Pages 330, 334, 338, 339, 342, 344, 347: Courtesy Lou Bloomfield.

Chapter 11

Pages 356, 357, 359, 360, 361, 364, 367, 373, 379, 381, 383, 384, 386, 387, 388, 395, 398: Courtesy Lou Bloomfield.

Chapter 12

Pages 401, 410, 412, 416, 417: Courtesy Lou Bloomfield.

Chapter 13

Pages 439, 444: Courtesy Lou Bloomfield.

Chapter 14

Pages 459, 460, 468, 473, 474: Courtesy Lou Bloomfield. Page 475: Courtesy SDL, Inc.

Chapter 15

Pages 491, 494, 499, 520, 521: Courtesy Lou Bloomfield.

Chapter 16

Page 531: Courtesy Lou Bloomfield. Page 636 (left): Courtesy U.S. Department of Energy. Page 636 (right): Courtesy Defense Nuclear Agency. Page 538: Courtesy: Los Alamos Scientific Laboratory, University of California.

Chapter 17

Pages 564, 575: Courtesy Lou Bloomfield. Page 590: Courtesy of Steven Chu. Page 593: Courtesy Lou Bloomfield.

Chapter 18

INDEX